SECOND EDITION

PRINCIPLES OF MUCOSAL IMMUNOLOGY

SECOND EDITION

PRINCIPLES OF MUCOSAL IMMUNOLOGY

Edited by

PHILLIP D. SMITH

Mary J. Bradford Professor in Gastroenterology, Professor of Medicine
and Microbiology, Department of Medicine (Gastroenterology & Hepatology),
University of Alabama at Birmingham, Birmingham, Alabama

RICHARD S. BLUMBERG

Jerry S. Trier Distinguished Chair in Gastroenterology
Professor of Medicine, Harvard Medical School, Boston, Massachusetts

THOMAS T. MACDONALD

Professor of Immunology and Immunobiology, Blizard Institute,
Barts and the London School of Medicine and Dentistry, London, United Kingdom

SOCIETY FOR MUCOSAL IMMUNOLOGY

CRC Press
Taylor & Francis Group
Boca Raton London New York

CRC Press is an imprint of the
Taylor & Francis Group, an **informa** business

A GARLAND SCIENCE BOOK

CRC Press
Taylor & Francis Group
6000 Broken Sound Parkway NW, Suite 300
Boca Raton, FL 33487-2742

© 2020 by Taylor & Francis Group, LLC
CRC Press is an imprint of Taylor & Francis Group, an Informa business

No claim to original U.S. Government works

Printed on acid-free paper

International Standard Book Number-13: 978-0-367-34894-6 (Hardback)
978-0-8153-4555-8 (Paperback)

Visit the Taylor & Francis Web site at
http://www.taylorandfrancis.com

and the CRC Press Web site at
http://www.crcpress.com

Visit the eResources at
https://www.crcpress.com/9780367348946

Dedicated to Anne Ferguson, Jiri Mestecky, Delphine M.V. Parrott, Warren Strober, Cornelis (Cox) P. Terhorst, and Sharon M. Wahl, who taught and inspired us, and so many others

Contents

Preface

The *Society for Mucosal Immunology* was established in 1987 to promote excellence in research and education in mucosal immunology and to foster communication among investigators studying unique aspects of the mucosal immune system. These goals emerged in response to the explosion in new information on the immunology at mucosal surfaces and the role of the mucosa in systemic diseases such as obesity and metabolic syndrome, autoimmune diseases such as diabetes and rheumatoid arthritis, local diseases such as Crohn's disease and ulcerative colitis, and the development of the systemic immune system.

To fulfill its educational mission, in 2007, the *Society for Mucosal Immunology* Board of Councilors, under the direction of *Society* presidents Richard Blumberg and then Thomas MacDonald, decided to undertake the development of a concise, concepts-oriented, and user-friendly textbook entitled *Principles of Mucosal Immunology*, the first edition of which was published in 2013. The book is intended for graduate students receiving advanced training in mucosal immunology, doctoral students, and post-doctoral fellows investigating topics in mucosal immunology, medical and dental students, and immunologists as well as scientists and clinicians in general who seek a broad perspective of the field. The book provides a framework for specialized courses in mucosal immunology.

To achieve the goals established by the *Society*, we obtained input from the Board of Councilors in the development of the book's content and sought the scientific expertise of internationally recognized leaders from throughout the world in the field of mucosal immunology, who contributed chapters related to their research and practice. This collective work, thus, addresses the principles and basic processes involved in mucosal immunology and mucosal diseases.

Principles of Mucosal Immunology, 2nd Edition, examines the commonalities of the mucosal immune system, focusing on those compartments for which the largest body of information is available, such as the small and large intestines, lung and upper airways, genital and urinary systems, and the ocular system. Some areas of the mucosal immune system, including those associated with the pancreaticobiliary tree, stomach, inner ear, and mammary glands, are discussed in relation to other topics. As our intention is to provide a strong basis for understanding general mucosal immunology, the book introduces readers to the topics in a way that provides a launching point for considering these and related areas in greater depth.

The book is organized into seven parts. Parts One through Three contain chapters on the development and structure of mucosal tissues, the cellular constituents of the mucosal immune system and their function in mucosal homeostasis, and their relationship with commensal microbes. This group of chapters emphasize the structural compartments and cellular components of the intestinal immune system, key for understanding the role of the commensal microbiota in mucosal homeostasis and disease processes. Parts Four and Five contain chapters that focus on the immune systems of the genitourinary, airway (nose, upper airway, oral cavity, and lung), and eyes. These chapters expand the review of the concepts of mucosal immune function in these important systems and organs. Parts Six and Seven chapters present the principles of mucosal immunity in the context of the major infectious and immune-mediated diseases and processes of mucosal surfaces, providing a paradigm for understanding immunopathology at these sites. In each of these sections, the textbook highlights major themes that are at the core of understanding the unique functions that characterize mucosal immunity such as the specialized anatomic organization and its relationship to the microbiota, secretory immunity, peripheral immune regulation, antigen handling, leukocyte homing, and mucosal cell signaling in immune and non-immune cells. This is especially relevant to the polarized epithelial cells which are a defining feature of all mucosal tissues.

This second edition includes updates of each chapter, new chapters on innate lymphoid cells, T regulatory cells and mucosal responses to helminths, and updated Further Reading sections. Key features of the book include the elegant and demonstrative illustrations. The illustrations and high-impact readings are valuable resources for instructors who use the book to guide their courses or training. Consequently, we believe this new edition is a unique asset for teaching mucosal immunology courses, a topic of increased popularity across academic campuses.

We and the *Society for Mucosal Immunology* are delighted to team up with Taylor & Francis—specifically the CRC Press/Garland Science imprint—the publisher of the acclaimed *Janeway's Immunobiology*. Accordingly, we envision that the book will serve as a companion to *Immunobiology* and hope that the opportunities that are derived from this partnership will benefit readers and instructors alike. We hope that *Principles of Mucosal Immunology*, 2nd Edition, helps expand your knowledge, as it has ours, of this fascinating field. We thank our contributors, as well as our students and colleagues, who share our passion for this topic and who inspire us on a daily basis with their enthusiasm and discoveries.

Finally, the Editors acknowledge with sadness the passing of four contributors to the first edition of this book—Per Brandtzaeg, Arlette Darfeuille-Michaud, Leo Lefrancois, and Lloyd F. Mayer—esteemed colleagues and dear friends.

Phillip D. Smith
Birmingham, Alabama

Richard S. Blumberg
Boston, Massachusetts

Thomas T. MacDonald
London, United Kingdom

Contributors

William Agace PhD
Mucosal Immunology Group
Department of Health Technology
Technical University of Denmark
Lyngby, Denmark

and

Immunology Section
Lund University
Lund, Sweden

Kenneth W. Beagley PhD
School of Biomedical Science
Queensland University of Technology
Brisbane, Australia

Diane Bimczok DVM, PhD
Department of Microbiology and Immunology
Montana State University
Bozeman, Montana

Stephan C. Bischoff MD
Department of Nutritional Medicine and Immunology
University of Hohenheim
Stuttgart, Germany

Richard S. Blumberg MD
The Jerry S. Trier Distinguished Chair in Gastroenterology
Professor of Medicine
Harvard Medical School
Boston, Massachusetts

Sarah Elizabeth Blutt PhD
Molecular Virology and Microbiology
Molecular and Cellular Biology
Baylor College of Medicine
Houston, Texas

Per Brandtzaeg PhD
Department of Pathology
Oslo University Hospital, Rikshospitalet
Oslo, Norway

Jonathan Braun MD, PhD
Inflammatory Bowel and Immunobiology
    Institute
Cedars Sinai Medical Center
Los Angeles, California

Elke Cario MD
Professor of Medicine
Division of Gastroenterology and Hepatology
University Hospital of Essen
Essen, Germany

Rachel R. Caspi PhD
Laboratory of Immunology
Immunoregulation Section
National Eye Institute
National Institutes of Health
Bethesda, Maryland

Nadine Cerf-Bensussan MD, PhD
INSERM U1163 and Institut Imagine
Laboratory of Intestinal Immunity
University of Paris
Paris, France

Hilde Cheroutre PhD
La Jolla Institute for Immunology
La Jolla, California

Mike Curtis BSc, PhD, FDSRCS
Faculty of Dentistry, Oral and Craniofacial
    Sciences
King's College London
London, United Kingdom

Satya Dandekar PhD
Department of Medical Microbiology
    and Immunology
School of Medicine
University of California
Davis, California

Antonio Di Sabatino MD

Department of Internal Medicine
San Matteo Hospital Foundation
University of Pavia
Pavia, Italy

Gerard Eberl PhD

Microenvironment and Immunity Unit
Institut Pasteur
Paris, France

Charles O. Elson MD

Basil I Hirschowitz Chair in Gastroenterology
Department of Medicine
University of Alabama at Birmingham
Birmingham, Alabama

Mary K. Estes PhD

Departments of Molecular Virology and Microbiology
    and Medicine
Baylor College of Medicine
Houston, Texas

Edda Fiebiger PhD

Department of Pediatrics
Boston Children's Hospital
Harvard Medical School
Boston, Massachusetts

Kohtaro Fujihashi DDS, PhD

International Research and Development Center for
    Mucosal Vaccines
Department of Medical Science
University of Tokyo
Tokyo, Japan

Glenn T. Furuta MD

Department of Pediatrics
Gastrointestinal Eosinophilic Diseases Program
Section of Pediatric Gastroenterology, Hepatology and
    Nutrition, Mucosal Inflammation Program
University of Colorado School of Medicine
Aurora, Colorado

Valérie Gaboriau-Routhiau PhD

INSERM U1163 and Institut Imagine
Laboratory of Intestinal Immunity
University of Paris
Paris, France

William Gause PhD

Center for Immunity and Inflammation
Rutgers New Jersey Medical School
Newark, New Jersey

Harry B. Greenberg MD

Joseph D. Grant Professor of Medicine and Microbiology and
    Immunology
Stanford University School of Medicine
Stanford, California

Richard Grencis PhD, FRSB, MAE

Lydia Becker Institute for Immunology and Inflammation
Wellcome Trust Centre for Cell Matrix Research
School of Biological Sciences
Faculty of Biology, Medicine and Health
University of Manchester
Manchester, United Kingdom

Jan Holmgren MD, PhD

Institute of Biomedicine
University of Gothenburg and Gothenburg University
    Vaccine Research Institute
Gothenburg, Sweden

Kenya Honda MD, PhD

Department of Microbiology and Immunology
School of Medicine
Keio University
Tokyo, Japan

and

Laboratory for Gut Homeostasis
RIKEN Center for Integrative Medical Sciences
Kanagawa, Japan

Shohei Hori PhD

Laboratory of Immunology and Microbiology
Graduate School of Pharmaceutical Sciences
University of Tokyo
Tokyo, Japan

Elaine Y. Hsiao PhD

Department of Integrative Biology and Physiology
University of California, Los Angeles
Los Angeles, California

Akiko Iwasaki PhD

Waldemar Von Zedtwitz Professor of Immunobiology and of
    Molecular, Cellular and Developmental Biology
Howard Hughes Medical Institute
Yale University
New Haven, Connecticut

Bana Jabri MD, PhD

Department of Medicine
University of Chicago
Chicago, Illinois

Edward N. Janoff MD

Mucosal and Vaccine Research Program Colorado
Infectious Diseases
University of Colorado Denver
Denver VA Medical Center
Aurora, Colorado

Bruce A. Julian MD

Department of Medicine (Nephrology)
University of Alabama at Birmingham
Birmingham, Alabama

Charlotte S. Kaetzel PhD

Department of Microbiology, Immunology and Molecular
    Genetics
University of Kentucky College of Medicine
Lexington, Kentucky

Arthur Kaser Dr.med.univ. PD

Department of Medicine
University of Cambridge
Cambridge, United Kingdom

Brian L. Kelsall MD

Laboratory of Molecular Immunology
National Institute of Allergy and Infectious Diseases
National Institutes of Health
Bethesda, Maryland

Hiroshi Kiyono DDS, PhD

CU-UCSD Center for Mucosal Immunology, Allergy and
    Vaccines (cMAV)
Division of Gastroenterology
Department of Medicine
School of Medicine
University of California San Diego
San Diego, California

and

Division of Mucosal Immunology
IMSUT Distinguished Professor Unit
Institute of Medical Science
University of Tokyo
Tokyo, Japan

and

Mucosal Immunology and Allergy Therapeutics
Institute for Global Prominent Research
Graduate School of Medicine
Chiba University
Chiba-ken, Japan

Bart Lambrecht PhD

VIB Inflammation Research Center
Ghent University
and
Department of Respiratory Medicine
University Hospital Ghent
Ghent, Belgium

Anthony St. Leger PhD

Ophthalmology and Immunology
University of Pittsburgh School of Medicine
Pittsburgh, Pennsylvania

Wayne Lencer MD

Longwood Professor of Pediatrics
Harvard Medical School
Division Chief, Pediatric GI
Boston Children's Hospital
Director, Harvard Digestive Disease Center
Boston, Massachusetts

Nils Lycke MD PhD

Department of Microbiology and Immunology
Institute of Biomedicine
University of Gothenburg
Gothenburg, Sweden

Thomas T. MacDonald PhD, FRCPath, FMedSci

Center for Immunobiology
Blizard Institute
Barts and the London School of Medicine and Dentistry
London, United Kingdom

David A. MacIntyre PhD

Institute of Reproductive and Developmental Biology
March of Dimes European Preterm Birth Research Centre
Imperial College London
London, United Kingdom

Michael A. McGuckin PhD

Faculty of Medicine, Dentistry and Health Sciences
The University of Melbourne
Victoria, Australia

Jiri Mestecky MD, PhD

Department of Microbiology
School of Medicine
University of Alabama at Birmingham
Birmingham, Alabama

Robert D. Miller PhD

Center for Evolutionary and Theoretical Immunology
Department of Biology
University of New Mexico
Albuquerque, New Mexico

Giovanni Monteleone MD, PhD

Department of Systems Medicine
University of Rome Tor Vergata
Rome, Italy

Anne Müller PhD

Institute of Molecular Cancer Research
University of Zurich
Zurich, Switzerland

Cathryn Nagler PhD

Biological Sciences Division
University of Chicago
Chicago, Illinois

Markus F. Neurath MD

Faculty of Medicine
Friedrich-Alexander Universitat
Erlangen, Germany

Marian Neutra PhD

Emeritus Professor
Harvard Medical School
and
Boston Childrens Hospital
Boston, Massachusetts

Luigi D. Notarangelo MD

Laboratory of Clinical Immunology and Microbiology
National Institute of Allergy and Infectious Diseases
National Institutes of Health
Bethesda, Maryland

Jan Novak PhD

Department of Microbiology
University of Alabama at Birmingham
Birmingham, Alabama

Hiroshi Ohno MD, PhD

Laboratory for Intestinal Ecosystem
RIKEN Center for Integrative Medical Sciences (IMS)
Yokohama, Japan

Jodie Ouahed MD, MMSc

Division of Gastroenterology, Nutrition, and Hepatology
Boston Children's Hospital
and
Department of Pediatrics
Harvard Medical School
Boston, Massachusetts

Andre J. Ouellette PhD

Department of Pathology and Laboratory Medicine
Keck School of Medicine
University of Southern California
Los Angeles, California

Oliver Pabst PhD

Institute of Molecular Medicine
RWTH Aachen University
Aachen, Germany

Reinhard Pabst MD

Institute of Immunomorphology/Centre Anatomy
Hannover Medical School
Hannover, Germany

Nicholas Powell MBChB, PhD

Faculty of Medicine
Imperial College London
and
Centre for Inflammation Biology and Cancer Immunology
King's College London
London, United Kingdom

Maria Rescigno PhD

Department of Biomedical Sciences
Humanitas University
Humanitas Clinical and Research Center – IRCCS
Milan, Italy

Irene Salinas PhD

Center of Evolutionary and Theoretical Immunology
Department of Biology
University of New Mexico
Albuquerque, New Mexico

Philippe J. Sansonetti MS, MD

Institut Pasteur
and
Collège de France
Paris, France

Pamela Schnupf PhD

INSERM U1163 and Institut Imagine
Laboratory of Intestinal Immunity
University of Paris
Paris, France

Britta Siegmund MD

Med. Klinik m.S. Gastroenterologie, Infektiologie,
    Rheumatologie
Charité – Universitätsmedizin Berlin
Berlin, Germany

Phillip D. Smith MD

Mary J. Bradford Professor in Gastroenterology
Department of Medicine (Gastroenterology
    & Hepatology)
University of Alabama at Birmingham
Birmingham, Alabama

Lesley E. Smythies PhD

Department of Medicine (Gastroenterology
    & Hepatology)
University of Alabama at Birmingham
Birmingham, Alabama

Scott Snapper MD, PhD

Egan Family Foundation Professor of Pediatrics
    in the Field of Translational Medicine
Brigham and Women's Hospital
and
Children's Hospital
Harvard Medical School
Boston, Massachusetts

Ludvig M. Sollid MD, PhD

Department of Immunology
Oslo University Hospital – Rikshospitalet
University of Oslo
Oslo, Norway

Jo Spencer PhD, FRCPath

School of Immunology and Microbial Sciences
King's College London
London, United Kingdom

Andrew Stagg PhD

Centre for Immunobiology
Blizard Institute
Barts and the London School of Medicine
    and Dentistry
Queen Mary University of London
London, United Kingdom

Jerrold R. Turner MD, PhD

Pathology and Medicine
Brigham and Women's Hospital
Harvard Medical School
Boston, Massachusetts

Dale T. Umetsu MD, PhD

The Prince Turki al Saud Professor of Pediatrics
Harvard Medical School
Boston Children's Hospital
Boston, Massachusetts

Thomas E. Van Dyke DDS, PhD

Vice President of Clinical and Translational
    Research
Department of Applied Oral Sciences
The Forsyth Institute
Harvard Dental School
Boston, Massachusetts

Valerie Verhasselt MD, PhD

University of Western Australia
School of Molecular Sciences
Crowley, Western Australia

Ifor R. Williams MD

American Association for the Advancement of Science
Washington, DC

Jenny M. Woof PhD

School of Life Sciences
University of Dundee
Dundee, United Kingdom

Gary D. Wu MD

Perelman School of Medicine
University of Pennsylvania
Philadelphia, Pennsylvania

DEVELOPMENT AND STRUCTURE OF MUCOSAL TISSUE

PART ONE

Overview of the mucosal immune system structure

REINHARD PABST AND PER BRANDTZAEG[*]

Pioneering experiments in the early 1960s showed that the large lymphocytes (lymphoblasts) that enter the bloodstream from the thoracic duct migrate into the intestinal lamina propria and undergo terminal differentiation into plasmablasts and plasma cells. Many of the circulating lymphoblasts express surface IgA (sIgA), and in the gut contain cytoplasmic IgA. These lymphoid cells were thought to be derived mainly from Peyer's patches (PPs), because transfer studies demonstrated that PPs and the draining mesenteric lymph nodes (MLNs), in contrast to peripheral lymph nodes and the spleen, were enriched precursor sources for IgA-producing plasma cells in the gut mucosa. It was also shown that plasma-cell differentiation occurs during dissemination of the mucosal B cells. Thus, the fraction of cells with cytoplasmic IgA increased from an initial 2% in PPs to 50% in MLNs and 75% in thoracic duct lymph, and finally 90% in the intestinal lamina propria.

These seminal studies led to the introduction of the term "IgA cell cycle," and subsequent research showed that B cells bearing other sIg classes than IgA, as well as T cells, when activated in PPs, also exhibit gut-seeking properties. It later became evident that different secretory effector sites can receive activated memory/effector B cells from a variety of mucosa-associated lymphoid tissue (MALT).

This work gave rise to the notion that the mucosal immune system is divided into distinct inductive sites and effector sites. The inductive sites are the organized MALT structures together with mucosa-draining lymph nodes, whereas the effector sites are the mucosal epithelia and the underlying lamina propria, which contains stromal cells and associated connective tissue stroma. The mucosae and related exocrine glands harbor by far the largest activated B-cell system of the body, and the major product—J-chain-containing dimeric IgA together with some pentameric IgM—is immediately ready for external transport by the polymeric immunoglobulin receptor (pIgR) across the secretory epithelium and into the mucus layers at mucosal surfaces, to provide antibody-mediated immunity (see Chapter 9).

IMMUNE INDUCTIVE LYMPHOID TISSUE

The MALT concept was introduced to emphasize that solitary organized mucosa-associated lymphoid follicles and larger follicle aggregates have common features and are the origin of T and B cells that traffic to secretory effector sites. This functional distinction is important because while the different tissues can be identified and discriminated by histology, single cell suspensions prepared from mucosal surfaces contain a mixture of cells from

---

[*] Per Brandtzaeg passed away in September 2016.

small MALT structures that cannot be dissected out, and connective tissue. This is particularly a problem in man, because solitary follicles cannot be seen in resected bowel.

1.1   Mucosa-associated lymphoid tissue is different from lamina propria or glandular stroma

MALT is subdivided according to anatomical regions (Table 1.1), and the cellular content of these lymphoid structures depends on whether the tissue is normal or chronically inflamed, superimposed upon striking age and species differences (Figure 1.1). The PPs in the distal small intestine of humans, rodents, and rabbits are classical MALT structures. PPs are inductive compartments generating conventional (B-2) sIgA-expressing memory/effector B cells and T cells, which, after a journey through lymph and peripheral blood, enter the gut mucosa. This "homing" appears to be antigen independent, but local antigens penetrating into the lamina propria contribute to local retention, proliferation, and differentiation of the extravasated lymphoid cells. Thus, although the lamina propria is considered an effector site, it is clearly important for the expansion and terminal differentiation of T and B cells. Some expansion of mucosal memory/effector T cells can also occur in the mucosal surface epithelium, and there is considerable "cross talk" between these two effector compartments.

The major component of human MALT comprises the gut-associated lymphoid tissue (GALT), including the PPs, the appendix, and numerous solitary follicles (see Figure 1.1) now termed "solitary isolated lymphoid tissue" (SILT) (Herbrand et al. 2008). Induction of mucosal immune responses can also take place in nasopharynx-associated lymphoid tissue (NALT), the tonsils (pharyngeal, palatine, lingual, pharyngeal), and bronchus-associated lymphoid tissue (BALT) as described later. Moreover, small MALT-like lymphoid aggregates are present in the conjunctiva and are associated with the larynx and various ducts such as those connecting the ocular and nasal compartments.

MALT resembles lymph nodes with B-cell follicles, interfollicular T-cell areas, and a variety of antigen-presenting cells (APCs) but lacks afferent lymphatics and a capsule. MALT therefore samples exogenous antigens directly from the mucosal surfaces through a follicle-associated epithelium (FAE), which contains "membrane" or "microfold" (M) cells (see Chapter 13). These specialized thin epithelial cells do not act as APCs but are effective

TABLE 1.1 DIFFERENT REGIONS OF MALT AND THEIR COMPONENTS

Region	Components
GALT (Gut-associated lymphoid tissue)	Peyer's patches (PPs) and isolated lymphoid follicles constitute the major part of GALT, but also the appendix is included.
NALT (Nasopharynx-associated lymphoid tissue) and tonsils	Rodents lack tonsils but do have paired NALT structures dorsally in the floor of the nasal cavity.
	In humans, tonsils consist of the lymphoid tissue of Waldeyer's pharyngeal ring, including the adenoids (the unpaired nasopharyngeal tonsil) and the paired palatine tonsils. Scattered isolated lymphoid follicles may also occur in nasal mucosa (NALT).
BALT (Bronchus-associated lymphoid tissue)	Not generally detectable in normal lungs of adult humans.

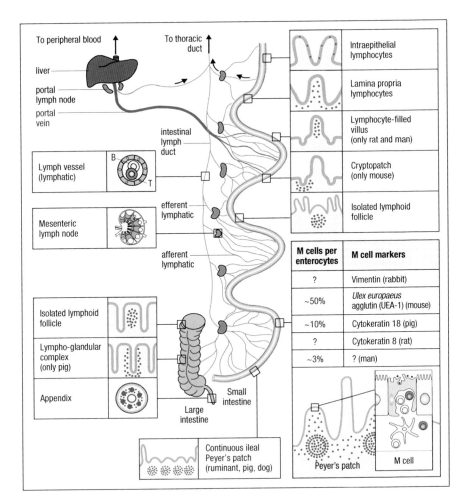

Figure 1.1 Distribution of lymphoid cells in various tissue compartments of the gut wall with species differences indicated. Lymphocytes leave the gut wall via draining lymphatics afferent to mesenteric lymph nodes, or via portal blood reaching the liver. Important regulation of immunity takes place in these organs, particularly induction of tolerance. The frequency of M cells in the follicle-associated epithelium of Peyer's patches is highly variable, and a reliable marker for this specialized cell has not been identified in humans. In contrast to the antigen-dependent activation of B cells that takes place in Peyer's patches of most species, the continuous ileal Peyer's patch present in ruminants, pigs, and dogs appears to be a primary lymphoid organ responsible for antigen-independent B-cell development, similar to the bursa of Fabricius in chicken (not shown). This Peyer's patch can be up to 2 meters long and constitute 80%–90% of the intestinal lymphoid tissue. (Adapted from Brandtzaeg, P., and Pabst, R. *Trends Immunol.* 2004, 25:570–577. With permission from Elsevier Ltd.)

in the uptake of microorganisms and other particulate antigens; they are also vulnerable "gaps" in the mucosal barrier. Studies in the mouse have, in addition, shown that dendritic cells (DCs) in the dome region of MALT send processes through the FAE to sample gut antigens.

The distinction between mucosal inductive and effector sites is not absolute, as the signals for extravasation and accumulation of naive versus memory/effector B and T cells are different. It is therefore confusing when MALT is used to refer to the mucosal effector compartments (e.g., the lamina propria and the surface epithelium of the gut and its diffusely distributed immune cells, Figure 1.2). This is in conflict with the classical definition of a lymphoid tissue, and so the Society for Mucosal Immunology and the International Union of Immunological Societies have agreed on the terminology listed in Table 1.1. Accordingly, the mucosa-draining local/regional lymph nodes should not be referred to as MALT because they do not sample antigens directly from mucosal surfaces. Thus, although being part of the intestinal immune system, MLNs should not be considered as part of GALT. Likewise, it is appropriate to refer to head-and-neck-draining cervical lymph nodes as part of the upper-airway immune system, but these are not MALT. Finally, because of the variety of homing molecules imprinted in the different MALT sites, the old term "a common mucosal immune system" should be avoided, although some integration of the migration of activated lymphoid cells to the various effector sites does occur. Therefore, we prefer the term "integrated mucosal immune system." An excellent example for the integration in the appearance of specific IgA antibodies in the secretions of the salivary and submandibular glands is after oral infection with an *Escherichia coli* strain in healthy volunteers (Aase et al. 2016).

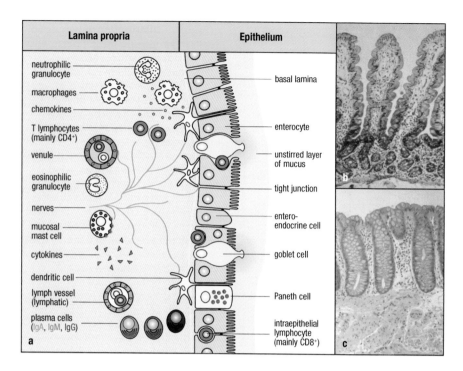

a

Lamina propria	Epithelium

- neutrophilic granulocyte
- macrophages
- chemokines
- T lymphocytes (mainly CD4⁺)
- venule
- eosinophilic granulocyte
- nerves
- mucosal mast cell
- cytokines
- dendritic cell
- lymph vessel (lymphatic)
- plasma cells (IgA, IgM, IgG)

- basal lamina
- enterocyte
- unstirred layer of mucus
- tight junction
- entero-endocrine cell
- goblet cell
- Paneth cell
- intraepithelial lymphocyte (mainly CD8⁺)

Figure 1.2 The cellular elements within the two effector compartments of the gut mucosa (lamina propria and epithelium). (a) The surface mucus, the monolayered epithelium with enterocytes and other cell types, the basal lamina (basement membrane), and the diffusely distributed cellular components, as well as mediators, within the lamina propria. IgA⁺ plasma cells (green) are normally dominating. The different cell types are in reality much closer to each other and can modulate the microenvironment by secreted chemokines (e.g., derived from the epithelium) and cytokines (e.g., derived from mast cells), as well as by the interaction with juxtaposed nerve fibers. Note that dendritic cells can penetrate the basal lamina and epithelial tight junctions to actively sample antigens from the mucosal surface, while maintaining barrier integrity by expressing tight junction proteins. (b) The histology (H&E staining) of normal human duodenal mucosa. (c) The histology (H&E staining) of normal human colonic mucosa. (Adapted from Brandtzaeg, P., and Pabst, R. *Trends Immunol.* 2004, 25:570–577. With permission from Elsevier Ltd.)

1.2 Mucosal immune system contains different types of gut-associated lymphoid tissue

In humans and rodents, PPs occur mainly in the ileum and less frequently in the jejunum. By the original definition coined by Cornes in the 1960s, a PP consists of at least five aggregated lymphoid follicles but can contain up to as many as 200 (**Figure 1.3**). Human PPs begin to form at 11 weeks of gestation, and discrete T- and B-cell areas occur at 19 weeks; but no germinal centers appear until shortly after birth, reflecting dependency on exogenous stimulation—a process that also induces follicular hyperplasia. Thus, macroscopically visible PPs increase from approximately 50 at the beginning of the last trimester to 100 at birth and 250 in the mid-teens, then diminish to some 100 at old age. Slightly over 50% of human PPs occur in the ileum.

In the mouse, there is a patch in the cecum. This seems to be responsible for forming precursors of IgA production in the colon, whereas small intestinal PP are more for exporting IgA precursors to the small intestine. The cecal patch of the mouse may be equivalent to the appendix in other species.

Human small intestinal mucosa may harbor at least 30,000 isolated lymphoid follicles, increasing in density distally. Thus, the normal small intestine reportedly contains on average only one follicle per 269 villi in the jejunum but one per 28 villi in the ileum. In the normal large bowel, the density of isolated lymphoid follicles also increases distally. Isolated lymphoid tissues

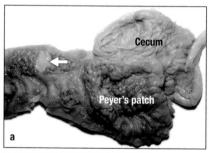

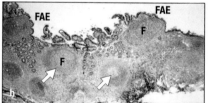

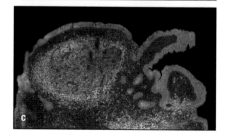

Figure 1.3 Peyer's patches in the terminal ileum of a young child. (a) A patch where a tissue specimen has been excised (arrow). (b) Histology of the excised specimen (H&E staining) displays several lymphoid follicles (F), and some are sectioned through the germinal center (arrows). The domes are covered with specialized follicle-associated epithelium (FAE) that lacks villi. Original ×25. (c) Two-color immunofluorescence staining for B cells (CD20, green) and IgD (red) in a cryosection from human Peyer's patch. Note that surface IgD (traditional marker of naive B cells) is mainly expressed on B cells in the mantle zones of the lymphoid follicle and to some extent on B cells scattered in the extrafollicular areas, whereas CD20 also decorates the germinal center. Epithelium is stained blue for cytokeratin. Original ×25.

(SILTs) have been characterized immunologically in mice and show features compatible with induction of B cells for intestinal IgA responses.

Interestingly, the organogenesis of murine isolated lymphoid follicles commences after birth, in contrast to the PPs. The isolated lymphoid tissue may actually play a special role in the induction of secretory IgA (SIgA) responses controlling the commensal microbiota, and their number and size depend on the intestinal bacterial level. These structures can function in a T-cell independent manner and cannot fully compensate for the absence of PPs in T-cell dependent antibody responses. Isolated lymphoid tissue may indeed reflect the bridge between the ancient innate and the modern adaptive immune systems. SILT is largely dependent on microbiota in the gut lumen. In the lamina propria, there is an enormous heterogeneity of macrophages with a great influence of the microbiota in the gut lumen. Bain et al. (2014) documented by experiments of parabiosis combined with modern molecular biological techniques that after weaning, monocytes from the blood constantly replenish macrophages in the mouse lamina propria. More recently, innate lymphoid cells (ILC) have been documented in the lamina propria and the gut epithelium subdivided into different subsets. These are difficult to localize because these cells are mainly characterized by regulatory effects, e.g., producing interferon-γ in response to IL-12, IL-15, etc. (Fuchs et al. 2013). Sometimes natural killer (NK) cells have also been included in this group of immune cells.

1.3 Organogenesis of murine Peyer's patches, isolated lymphoid follicles, and nasopharynx-associated lymphoid tissue is not synchronous

Experiments in knockout mice have largely helped in understanding how inductive lymphoid tissue develops, and there are remarkable differences among sites. The fetal organogenesis of PPs and MLNs depends on multiple interactions between hematopoietic lymphoid tissue inducer (LTi) cells and mesenchymal stromal cells. The development of LTi cells requires two transcription factors: the inhibitory basic helix-loop-helix transcription factor Id2 and the retinoic acid–related orphan nuclear receptor RORγt.

The cellular interactions involve lymphotoxins (LTs), which are members of the tumor necrosis factor (TNF) superfamily, as well as the two TNF receptors (TNF-Rs) I and II (55 and 75 kDa). A crucial differentiating event is membrane expression of $LT\alpha_1\beta_2$ on the $CD3^-CD4^+CD45^+IL-7R^+$ LTi cells after stimulation of the receptor activator of nuclear factor (NF) κB (RANK)–TNF receptor associated factor family 6 (TRAF6) pathway and—especially in PPs—alternative cytokine signaling through the interleukin-7 receptor (IL-7R). This event is followed by LTi cell interaction with the LTβ receptor (LTβR)$^{-positive}$ stromal cells, which are positive for vascular cell adhesion molecule (VCAM)-1 and are called lymphoid tissue organizer cells but display a different activation profile in the organogenesis of lymph nodes and PPs. Knockout mice deficient in LTα or LTβ virtually lack lymph nodes and have no detectable PPs. The organ development also involves several feedback loops with chemokines and adhesion molecules such as CXC chemokine ligand 13 (CXCL13)–CXCR5 and CXCR5-induced $\alpha_4\beta_{1integrin}$–VCAM-1 interactions.

Specifically, at 15 days of gestation, VCAM-1$^+$ stromal cells accumulate in fetal mouse small intestine. At 17.5 days, LTi cells are present, and mucosal addressin cell adhesion molecule-1 (MAdCAM-1) is expressed on vessels. T and B cells migrate into PPs from 18.5 days to form T- and B-cell zones. PPs and MLNs are, by unknown mechanisms, programmed to develop in the sterile environment of the fetus. Conversely, murine isolated lymphoid follicles develop only after birth in response to innate signaling from the gut microbiota, but they nevertheless require $LT\alpha_1\beta_2$–$LT\beta R$ interactions and TNF-RI function, as well as typical LTi cells. Also, CXCL13 has been shown to be important for isolated lymphoid follicle formation by promoting accumulation of LTi cells.

Interestingly, the kinetics of the appearance of lymphoid tissue in the mouse nasal mucosa are different, as development is completely postnatal. Unlike GALT structures, murine NALT organogenesis does not depend on the $LT\alpha_1\beta_2$–$LT\beta R$ signaling pathway, although NALT anlagen apparently requires the presence of $CD3^-CD4^+CD45^+$ cells and the transcription factor Id2. Nevertheless, NALT occurs in $ROR\gamma t$-deficient mice that lack the $CD3^-CD4^+IL$-$7R^+$ PP-type LTi cells. However, to become fully developed with follicular dendritic cells (FDCs) and optimal function, murine NALT as well as inducible murine BALT structures depend on the LT–$LT\beta R$ pathway.

Detailed ontological studies linked to function are only available for mouse and man but reveal huge differences. Mice possess unique structures, termed "cryptopatches," which are clusters of c-kit$^+$, lineage-negative common lymphoid precursor cells. Cryptopatches can, however, mature into isolated lymphoid follicles, where primary and secondary adaptive mucosal IgA responses can be generated. Cryptopatches and isolated lymphoid follicles are dependent on LTi cells and are absent in $ROR\gamma t$ null mice. Cryptopatches, but not isolated lymphoid follicles, are present in the fetus and in germ-free mice. Evidence indicates that isolated lymphoid follicle development requires signaling from the microbial flora via nucleotide-binding oligomerization domain 1 (NOD1) in epithelial cells to attract LTi cells that then form a lymphoid follicle; therefore, in terms of classical immunology, cryptopatches may be considered primary lymphoid organs and isolated lymphoid follicles secondary to lymphoid structures. Cryptopatches and isolated lymphoid follicles appear to form a continuum, but in the mouse they may be more like the rabbit, where GALT functions as a primary lymphoid organ before puberty and matures into a secondary organ later in life.

Whether the same sequence of events occurs in humans is difficult to determine. An LTi cell, which is $CD127^+$, $RORC^+$ and $CD4^-$, has been described in fetal human tissues. The lack of CD4 on the human LTi cell is somewhat surprising because many years ago human fetal gut was shown to contain clusters of $CD4^+$ $CD3^-$ cells, which were presumed to be the earliest PP anlagen. However, immunostaining for VCAM-1 is extremely strong in putative PPs beginning at 11 weeks of gestation, suggesting that even in man, VCAM-1 is involved in mucosal lymphoid tissue organogenesis. Beginning at 14–15 weeks of gestation, human small intestine contains loose accumulations of T cells and B cells and strongly major histocompatibility complex (MHC) class II$^+$ cells, but the mucosa lacks follicular structure. By 19 weeks, organized small PPs with a FAE, primary B-cell follicles, and T-cell zones can be seen. A population of $CD11c^+$ DCs clusters below the FAE, as shown in adult PPs. Although little information on the development between 20 weeks and birth is available, analysis of available specimens shows large well-organized Peyer's patches in 1- to 2-day-old tissues, containing what appear to be early secondary follicles with IgM$^+$ blasts in the developing germinal centers, presumably in response to intestinal colonization with microbes.

Cryptopatches are not present in humans and should not be confused with lymphocyte-filled villi (see **Figure 1.1**). Whether human isolated lymphoid follicles have properties similar to the murine counterparts is not known. However, irregular lymphoid aggregates, which may have germinal centers with follicular dendritic cells, are induced in chronic inflammatory bowel disease lesions, although the function of these isolated lymphoid-follicle-like structures remains elusive. It has been speculated that the isolated follicles in the colon also play a role in carcinogenesis in the colon.

1.4   Lymphoid tissue is present in nose, pharynx, and bronchi

GALT is the largest and best-defined part of MALT, but other potentially important inductive sites for mucosal B-cell responses are BALT, NALT, and tonsils in humans, particularly the unpaired nasopharyngeal tonsil (often referred to as adenoids) and the paired palatine tonsils. These structures make up most of Waldeyer's pharyngeal lymphoid ring and may play a more important role in mucosal immunity in human airways than BALT (Figure 1.4a) because the latter is rare in normal lungs of adults and detectable in only 40%–50% of small bronchial biopsy samples from adolescents and children but is regularly encountered in large autopsy specimens. Similarly, human nasal mucosa contains only occasional isolated lymphoid follicles, and they are difficult to detect in infants.

Rodents lack tonsils but possess two paired bell-shaped NALT structures laterally to the nasopharyngeal duct in the floor of the nasal cavity, dorsal to the cartilaginous soft palate. A regionalized protective IgA response can be induced by nasal vaccination in mice. Indeed, murine NALT can drive an IgA-specific enrichment of high-affinity memory B cells but also gives rise to a major germinal center population of IgG-producing cells—quite similar to that seen in human tonsils. In contrast to tonsils, however, the anlagen appear at the same fetal age as those of PPs, and the organogenesis of murine NALT begins after birth, similar to murine isolated lymphoid follicles. Murine BALT is also inducible, as is human BALT and nasal isolated lymphoid follicles. BALT has therefore also been classified as a tertiary lymphoid tissue and/or compared to tonsils in other species.

Like GALT in all mammalian species, rodent NALT has a smooth surface covered by a FAE. Human tonsils have deep and branched antigen-retaining crypts with a reticular epithelium containing M cells (Figure 1.4b and c). This difference probably explains why germinal centers develop shortly after birth in human NALT, similar to the heavily microbe-exposed GALT, whereas the absence of an abundant nasal microflora means that rodent NALT requires infection or a danger signal such as cholera toxin to drive germinal center formation. This is an intriguing species difference important in the induction of B-cell diversity and memory by vaccination via the nasal route.

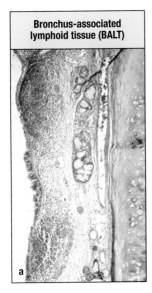

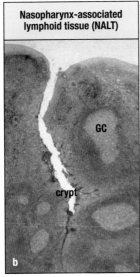

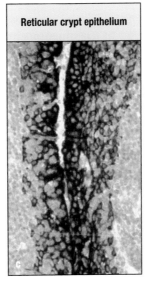

Figure 1.4 Human BALT and NALT. (**a**) Histology (H&E staining) of human bronchus-associated lymphoid tissue (BALT) from diseased lung of an adult; note submucosal glands. (**b**) Human nasopharynx-associated lymphoid tissue (NALT) represented by section through a crypt region of palatine tonsil; note abundant germinal centers (GC) in lymphoid follicles. (**c**) Parallel section immunostained for cytokeratin to demonstrate loose structure of reticular crypt epithelium.

1.5   Mucosal B cells may be derived from tissues other than MALT

In mice, activated T cells rapidly obtain gut-homing properties during antigen priming in MLNs. Most likely, regional mucosa-draining lymph nodes generally share immune inductive properties with related MALT structures from which they receive antigens via afferent lymph and antigen-transporting DCs. Numerous DCs are present at epithelial surfaces where they can pick up luminal antigens by penetrating tight junctions with their processes (see **Figure 1.2**). In the murine small intestine, a CX_3CR1^+ DC subset has been shown to actively sample luminal bacteria in this manner. Importantly, the human nasal mucosa is extremely rich in DCs, both within and beneath the epithelium, and a subepithelial band of putative APCs is present below the surface epithelium, as well as in the FAE of PPs in the human gut. Importantly, murine DCs that have sampled commensal bacteria in PPs do not normally migrate beyond the draining MLNs, thereby restricting the antibody response to the mucosal immune system of clean pathogen-free mice. This is clearly different in human beings where systemic antibodies to commensals are normally present in low titers.

The peritoneal cavity is recognized as yet another source of mucosal B cells in mice, perhaps providing up to 50% of the intestinal IgA^+ plasma cells. The precursors are self-renewing $sIgM^+$ B-1 ($CD5^+$) cells, which give rise to T-cell independent polyreactive ("natural") SIgA antibodies, particularly directed against polysaccharide antigens from commensal bacteria. However, the relationship of murine B-1 cells to the conventional bone marrow derived B-2 cells remains elusive. Notably, rather than being encoded only in the germline, B-1 cells from mice may show hypermutation of Ig heavy-chain V-region (V_H) genes as a sign of selection.

The location of the B-1 subset differentiation of IgA plasma cells is controversial, although the murine intestinal lamina propria has been suggested as a site for class switching. Notably, however, evidence that peritoneal B-1 cells contribute to mucosal IgA production in humans is not available. But considerable levels of cross-reactive SIgA antibodies directed against self as well as microbial antigens are present in human external secretions (see Chapter 9). The reason for this could be microbial polyclonal activation of B cells in GALT, independent of B-cell receptor (BCR)-mediated antigen recognition. In man, a subset of IgA producing plasma cells ($CD45^-$) persists for decades in the small intestine.

B-cell activation in MALT

MALTs are on the forefront of the recognition of, and responses to the many pathogens that colonize mucosal surfaces or use mucosal surfaces to enter the body. They also are probably important in controlling the intensity and texture of the response to harmless environmental antigens and foods. While there is a large literature on MALT in experimental animals, studies in humans are relatively uncommon.

1.6   Follicle-associated epithelium is an important site of antigen uptake in the gut

In the gut, M cells of FAE sample luminal antigens and, notably, enteropathogenic bacteria such as *Salmonella* spp. and *Yersinia* spp., viruses such as poliovirus and reovirus, and prions use M cells as portals of entry into the lymphoid tissue of GALT and may spread further throughout the body. Also, M cells of murine NALT have been shown to serve as a gateway for the entry of pathogens (group A streptococci), particularly relevant to airway infections.

M cell pockets clearly represent an intimate interface between the external environment and the mucosal immune system (**Figure 1.5**). In mice, the

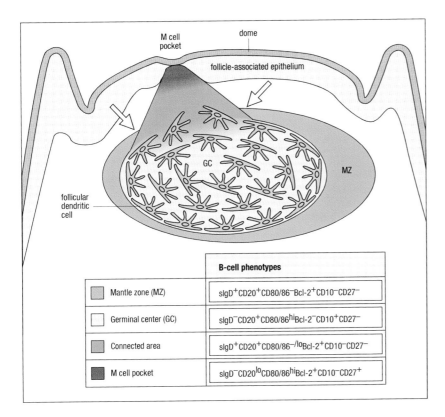

	B-cell phenotypes
Mantle zone (MZ)	sIgD$^+$CD20$^+$CD80/86$^-$Bcl-2$^+$CD10$^-$CD27$^-$
Germinal center (GC)	sIgD$^-$CD20$^+$CD80/86^{hi}Bcl-2$^-$CD10$^+$CD27$^-$
Connected area	sIgD$^+$CD20$^+$CD80/86$^{-/lo}$Bcl-2$^+$CD10$^-$CD27$^-$
M cell pocket	sIgD$^-$CD20^{lo}CD80/86^{hi}Bcl-2$^+$CD10$^-$CD27$^+$

Figure 1.5 Relationship between elements of secondary (activated) B-cell follicle and M cellpocket in human Peyer's patch. The germinal center (GC) is mostly surrounded by the mantle zone (MZ) and filled with surface (s) IgD$^-$CD20$^+$CD80/86^{hi}Bcl-2$^-$CD10$^+$CD27$^-$ B cells. The mantle zone consists of naive sIgD$^+$CD20$^+$CD80/86$^-$Bcl-2$^+$CD10$^-$CD27$^-$ B cells but is broken (open arrows) beneath the M cellpocket. This connected area is only seen in a restricted part of the follicle as defined by reduced immunostaining for sIgD toward the M cell. A follicular dendritic cell network defined as CD21$^+$CD20$^-$ phenotype, shows a topographically similar extension but does not reach inside the M cellpocket. This pocket contains both naive sIgD$^+$CD20$^+$CD80/86$^{-/lo}$Bcl-2$^+$CD10$^-$CD27$^-$ and memory (or recently stimulated) sIgD$^-$CD20^{lo}CD80/86^{hi}Bcl-2$^+$CD10$^-$CD27$^+$ B-cell phenotypes, the latter being predominant. (Adapted from Brandtzaeg, P. *Immunol. Invest.* 2010, 39:303–355. With permission from Taylor & Francis Group.)

FAE expresses chemokines (e.g., CCL9 and CCL20) that may attract the professional APCs that accumulate in the PP domes. Numerous putative APCs also occur immediately underneath the human FAE, while the M cellpockets are dominated by memory T and B lymphocytes in approximately equal proportions. Interestingly, activated B cells may induce the epithelial M cell phenotype from classical gut enterocytes. Furthermore, studies in germ-free and conventionalized rats have demonstrated that bacterial colonization drives the accumulation and differentiation of T and B cells in the M cell pockets, apparently with an initial involvement of antigen-transporting DCs, followed by germinal center formation.

M cell pockets may represent specialized germinal center extensions designed for rapid recall responses. The most likely cell type to mediate MHC class II interaction with cognate T cells in these microcompartments is the long-lived sIgD$^-$IgM$^+$Bcl-2$^+$CD27$^+$ memory B-cell (see Figure 1.5). In human tonsils, memory B cells colonize the antigen-transporting reticular crypt epithelium, and, by rapid upregulation of costimulatory B7 molecules, they acquire potent antigen-presenting properties. Likewise, memory B cells present in the M cell areas of human PPs relatively often express B7.2 (CD86) and sometimes B7.1 (CD80), perhaps a prerequisite for stimulation of productive immunity (see Figure 1.5). These memory cells are phenotypically similar to the so-called extrasplenic "marginal-zone B cells," which generally show a high frequency of BCR IgV$_H$ mutations, suggesting extensive antigen stimulation.

The FAE does not express pIgR, but an uncharacterized apical receptor for IgA has been identified on M cells. SIgA may exploit this receptor for targeting of antigens to DCs in PPs. Many commensal bacteria in human saliva and feces are coated with SIgA, which may facilitate their M cell–mediated uptake. Most of the bacteria that cross the epithelium are destroyed by subepithelial macrophages, but some are taken up by DCs in tiny amounts (0.001%), sufficient to induce a murine immune response restricted to GALT and MLNs. On the whole, however, the mucosal antibody response limits the penetration of bacteria into the body, with GALT "shielded" by the SIgA antibodies. Nevertheless, the positive-feedback mechanism implied by the apical M cell

IgA receptor may target relevant luminal antigens complexed with cognate maternal SIgA antibodies to the GALT of suckling newborns. Many reports show that breastfed babies over time develop enhanced secretory immunity.

Considerable effort has been directed to the characterization of M cell receptors that may be exploited for vaccine and adjuvant targeting to GALT. Murine M cells express toll-like receptors (TLRs), thus making TLR agonists potential mucosal vaccine adjuvants. For instance, outer membrane proteins of *Neisseria meningitidis* mixed with the TLR2 agonist PorB enhanced the microparticle transport capacity of M cells and induced migration of DCs into FAE. Also glycoprotein 2 expressed apically on murine M cells has been shown to serve as a transcytotic receptor for commensal and pathogenic bacteria carrying FimH, a component of type I pili.

1.7   Gut bacteria are important for molecular interactions needed for germinal center formation

Primary lymphoid follicles contain recirculating naive B lymphocytes (sIgD+IgM+) that pass into the network formed by antigen-capturing FDCs. The origin of FDCs remains obscure, but both their development and the clustering that allows follicle formation depend on LT signaling, and the B cells are one important source of LT. Among the actions of the soluble homotrimer LTα are augmentation of B-cell proliferation and expression of adhesion molecules; mice deficient in transmembrane LTβ have no detectable FDCs.

The germinal center reaction causes the transformation of primary follicles into secondary follicles. In humans, this process has been extensively studied in tonsils, but relevant mechanistic information relies on observations of lymph nodes and spleen from immunized animals. In general, germinal centers are of vital importance for T-cell dependent stimulation of conventional (B-2) memory/effector B cells, affinity maturation of BCR, and Ig-isotype switching (see Chapter 8). Naive B cells are first stimulated in the T-cell zone, just outside the primary follicle, by cognate interactions with activated CD4+ T cells, and exposed to processed antigen complexed with MHC class II molecules on interdigitating DCs. The B cells then reenter the follicle to become proliferating sIgD+IgM+CD38+ germinal center "founder cells" as observed in human tonsils.

The initially activated B cells produce unmutated IgM (and some IgG) antibody that binds circulating antigen with low affinity. B cells may then use their complement receptors to carry opsonized antigen or immune complexes into the follicles to become deposited on FDCs. Antigen is retained in this network for prolonged periods to maintain B-cell memory. A role for IgM in the induction of secondary immune responses with antibody affinity maturation has been strongly supported by observations in knockout mice lacking natural ("nonspecific") background IgM antibodies.

The complement receptors CR1/CR2 (CD35/CD21) are crucial in the germinal center reaction. CD21 is expressed abundantly on both B cells and FDCs and may thus function not only by localizing antigen to the FDCs but also by lowering the threshold of B-cell activation via recruitment of CD19 into the BCR. Activation of complement on FDCs bearing immune complexes is controlled by regulatory factors, but some release of inflammatory mediators may induce edema that facilitates dispersion of FDC-derived "immune complex-coated bodies," or iccosomes, thereby enhancing the BCR-mediated uptake of their contained antigens by the B cells.

Several other components of the innate immune system may be involved in the germinal center formation. Under certain experimental conditions, IgA differentiation driven by gut bacteria may even bypass the usual sIgM (or sIgD) BCR requirement. Nevertheless, there always appears to be a dependency on some follicle-like aggregates of B cells, which interestingly may lack antigen-retaining FDCs and germinal centers. Normally, the germinal-center reaction is driven by competition for a limited amount of antigen, but there may also be an innate drive as discussed earlier for isolated lymphoid follicles. The resulting B cells will then survive with a restricted repertoire and rather low BCR affinity.

In germ-free rabbit appendix, only certain commensal bacteria efficiently promote GALT development, which depends on stress responses in the same bacteria, suggesting a nonspecific impact on GALT. Other experiments have shown that GALT development can occur independently of BCR engagement, apparently because commensal bacteria promote the germinal center reaction in PPs and MLNs by interacting with innate immune receptors. This information helps explain the enormous IgA drive provided by the gut microbiota in the absence of high-affinity BCR development, thereby leading to the production of large amounts of IgA with restricted repertoire and capacity to bind with low affinity to redundant epitopes on commensal bacteria (see Chapter 8).

Thus, the gut microbiota is required and plays a critical role in the activation of GALT and normal intestinal plasma-cell development, although food proteins may also contribute to this development. Notably, commensal bacteria shape the BCR repertoire of the host. Perhaps the germinal center reaction originated evolutionarily in GALT to generate a protective antibody repertoire that was not antigen specific but instead cross-reactive. The indigenous microbiota might in this context act as polyclonal B-cell activators through several mechanisms, including TLR signaling. The induction of IgA by commensal bacteria seems to represent a primitive mechanism for limiting bacterial colonization and penetration through the epithelial barrier without eliminating them from the gut (Pabst et al. 2016). Superimposed on this "innate-like" defense, the B-2 system has the capacity to undergo germinal center-driven high-affinity BCR selection for particular antigens of pathogens to clear infections (see Chapter 8).

B-cell differentiation in MALT germinal centers

The most obvious morphologic features of MALT in tissue sections are the numerous large, activated germinal centers. This is especially the case in the ileum. In mice, the expansion of the germinal centers allows Peyer's patches to be identified from the serosal surface so they can be dissected out for functional studies. Typically, below each FAE there is a large germinal center.

1.8   IgA expression and class switching are dependent on activation-induced cytidine deaminase

As previously discussed, the IgA switch in murine B-1 cells may circumvent the need for T-cell help that provides CD40−CD154/CD40L interaction. Instead, the B-1 cells may rely on proliferation- and survival-inducing cytokines of the TNF family such as BAFF/BLyS (B-cell-activating factor of the TNF family/B-lymphocyte stimulator) and APRIL (a proliferation inducing ligand) secreted by activated DCs or macrophages, for instance, after exposure to lipopolysaccharide (LPS). Significantly reduced IgA antibody responses are present in APRIL-deficient mice, and it is possible that BAFF and APRIL also contribute to T-cell dependent IgA switching. In the human gut where class switch appears principally to be part of a germinal center reaction—but to some extent may also be T-cell independent—the role of BAFF and APRIL remains unclear (see Chapter 8).

Class switch in germinal centers is initiated after activation of the germinal-center founder cells, which change their BCR composition from sIgD$^+$IgM$^+$ to become sIgD$^-$IgM$^+$ memory/effector cells following affinity maturation. They may then switch to another class such as IgG or IgA by recombination of the intronic Cμ-switch (S) region with the S region of a downstream isotype, either directly or sequentially (**Figure 1.6**). During plasma-cell differentiation, the

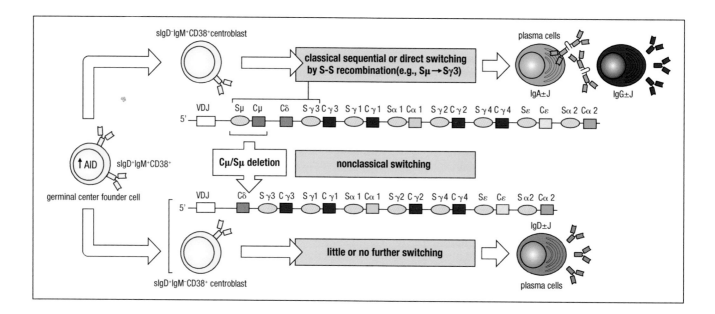

BCR is gradually lost with several other B-cell markers, particularly CD20 and then CD19 (in mice also B220).

The switch process requires a variety of transcription factors and enzymatic activity expressed by cell type–specific and more generalized DNA repair enzymes, particularly activation-induced cytidine deaminase (AID, see Figure 1.6). AID is present during class-switch recombination and may link class-switch recombination to somatic hypermutation of Ig V-gene segments, which takes place during the germinal center reaction (see Chapter 8). AID-deficient mice show dramatic hyperplasia of isolated lymphoid follicles in response to intestinal overgrowth of the indigenous microbiota, similar to the lymphoid hyperplasia that occurs in people with immunoglobulin deficiency who often are infected with the parasite *Giardia lamblia*. This could reflect an inadequate compensatory antibody repertoire in the gut due to lack of somatic hypermutation in the SIgM that partially replaces SIgA in these mice.

Specific alterations at any germline C_H locus prior to class-switch recombination include opening of chromatin structures—a key event in transcriptional activation. Thus, the switch to IgA1 and IgA2 in humans requires activation of the $I\alpha1$ and $I\alpha2$ promoters located upstream of the respective S regions of the C_H-gene segments of the two subclasses. DNA between the S sites then is looped out and excised, thereby deleting $C\mu$ and other intervening C_H genes. After direct switch to IgA1, the germline $I\alpha1$–$C\mu$ circular transcripts derived from the excised recombinant DNA are gradually lost through dilution in progeny during proliferation.

1.9 Different class-switching pathways operate in different mucosal sites

Human IgA responses are dominated by IgA1 both in tonsils and in the regional secretory effector sites, which suggests that mucosal B-cell differentiation in these sites mainly takes place from $sIgD^-IgM^+CD38^+$ centrocytes by sequential downstream C_H-gene switching (see Figure 1.6). Conversely, the relatively enhanced IgA2 expression in PPs and the distal human gut, including the MLNs (Figure 1.7), could reflect direct switch from $C\mu$ to $C\alpha2$. Murine PP B cells are able to switch directly from $C\mu$ to $C\alpha$, and in human B cells this direct pathway may preferentially lead to IgA2 production.

The germinal center reaction generates relatively more intrafollicular J chain–expressing IgA^+ plasmablasts in human PPs and appendix than in tonsils. Furthermore, immediately outside of GALT follicles, IgA^+ plasma

Figure 1.6 Class-switch recombination leading to differentiation of B cells in mucosa-associated lymphoid tissue. This process takes place after activation of germinal center founder cells, resulting in elevated levels of the enzyme activation-induced cytidine deaminase (AID). In the classical pathway, the centroblast progeny change their Ig heavy-chain constant-region (C_H) gene expression from $C\mu$ to one of the downstream C_H genes, either sequentially or directly, by recombination between switch (S) regions (comprising repetitive sequences of palindrome-rich motifs), followed by looping-out of the intervening DNA fragment containing $C\mu$ and other C_H genes from the chromosome. This pathway gives rise to plasma cells of various IgG and IgA isotypes (and IgE, not shown), with or without concurrent J-chain expression ($\pm$J). A minor fraction of tonsillar centroblasts undergo nonclassical switching by deleting a variable part of $S\mu$ and the complete $C\mu$ region. This pathway selectively gives rise to IgD-producing plasma cells and little or no further switching because the $C\delta$ lacks a separate S region. (Adapted from Brandtzaeg, P. *Immunol. Invest.* 2010, 39:303–355. With permission from Taylor & Francis Group.)

cells are equal to, or exceed numerically, their IgG⁺ counterparts, whereas in tonsils there is a more than twofold dominance of extrafollicular IgG-producing plasma cells. Therefore, the drive for IgA switch and J-chain expression is clearly much more pronounced in GALT than in tonsils (see Figure 1.7). The reason for this disparity remains elusive, as the tonsillar crypt epithelium can secrete BAFF and thymic stromal lymphopoietin—a cytokine that further promotes class-switch recombination and a broad reactivity of local B cells by activating BAFF-producing DCs. One possibility is that the continuous superimposition of novel stimuli from the gut microbiota in PPs enhances the development of early memory/effector B-cell clones with an increased potential for IgA and J-chain production.

A regionalized microbial impact on mucosal B-cell differentiation is further exemplified by the unique sIgD⁺IgM⁻CD38⁺ centroblasts identified in the dark zone of tonsillar germinal centers. This subset shows deletion of the Sμ and Cμ gene segments, therefore selectively giving rise to IgD⁺ plasma cells by "nonclassical" C_H-gene switching (see Figure 1.6). Molecular evidence has been provided for a preferential occurrence of B cells with Cμ deletion in the normal Waldeyer's ring (human NALT) and secretory effector sites of the upper aerodigestive tract, but homing of these cells to the small intestinal mucosa appears to be quite rare. Such compartmentalized B-cell responses explain the relatively high frequency of IgD⁺ plasma cells in the respiratory tract, and particularly that IgD⁺ plasma cells are often seen in IgA deficiency. Most strains of *Haemophilus influenzae* and *Moraxella catarrhalis*, which are frequent colonizers of the nasopharynx, express outer-membrane IgD-binding factors that can activate sIgD⁺ B cells by cross-linking sIgD/BCR. In this manner, it seems likely that sIgD⁺ tonsillar centrocytes are stimulated to proliferate and differentiate polyclonally, thereby driving V-gene hypermutation and Sμ/Cμ deletion. Conversely, LPS that is abundantly present in the distal gut may inhibit IgD expression, which—in addition to compartmentalized homing mechanisms—might contribute to the virtual lack of IgD-producing plasma cells in the gastrointestinal tract.

Microbial influence on B-cell differentiation is supported by the observation that Sμ/Cμ deletion is more frequently detected in diseased than in clinically normal tonsils and adenoids, and extrafollicular IgD⁺ plasma cells are relatively numerous in recurrent tonsillitis and adenoid hyperplasia. But there are large individual variations, and the mean proportion of IgD-producing plasma cells is well below 5% of all isotypes in the extrafollicular tonsillar compartment.

Interestingly, sIgD⁺IgM⁻ B cells generated by nonclassical switching appear to express V-gene repertoires that may allow considerable cross-reactivity, including autoimmunity, but the biological significance of this observation remains unknown. Although numerous antimicrobial and other IgD antibody activities have been detected in mouse and human serum, the protective or pathogenic role of circulating IgD has only recently been explored. Because IgD does not activate the classical complement pathway, it likely blocks other antibody defense functions within the mucosae and reduces the immune exclusion efficiency of SIgA and SIgM antibodies in the upper airways in the face of bacterial infections that drive local IgD production. The ability of IgD to bind to monocytes/macrophages and basophils and induce release of pro-inflammatory cytokines, including interleukin (IL)-1, IL-6, IL-8, and TNF-α, may add to the pathogenic potential of IgD antibodies. Of note, IgA-deficient individuals who exhibit substantial replacement with only IgM⁺ and IgG⁺ plasma cells in their nasal mucosa have fewer clinical problems in their airways than those with an abundance of local IgD⁺ plasma cells.

How mucosal immune cells home

Chapter 14 contains extensive information on lymphocyte homing to mucosal surfaces; however, there are some areas that are more appropriately covered

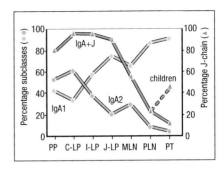

Figure 1.7 Subclass distribution and J-chain expression of IgA-producing plasma cells in various normal extrafollicular tissue compartments from adults. PP, Peyer's patch; C-LP, I-LP, and J-LP, distant colonic, ileal, and jejunal lamina propria, respectively; MLN, mesenteric lymph node; PLN, peripheral lymph node; and PT, palatine tonsil. For the latter organ, J-chain data from healthy children (who have had no inflammation in their tonsils) are also indicated (connected by broken line). Based on published immunohistochemical data from Brandtzaeg's laboratory. (Adapted from Brandtzaeg, P. *Immunol. Invest.* 2010, 39:303–355. With permission from Taylor & Francis Group.)

here. In general, immune cells exit from inductive sites through draining lymphatics. These vessels are believed to start blindly with a fenestrated endothelium, and the egress of lymphoid cells depends on a signal mediated by sphingosine-1-phosphate. In addition, lymphatic endothelium shares with high endothelial venules (HEVs) the expression of both CCL21/SLC and other adhesion molecules such as intercellular adhesion molecule (ICAM). For example, in the gut, activated T and B cells leave PPs, enter the afferent lymphatics, and drain into the MLN. From there the cells leave in efferent lymphatics, eventually draining into the thoracic duct that empties into the blood at the junction of the left subclavian vein and left jugular vein. In humans, approximately 4 L of lymph drains into the blood every day, much of it from the gut.

1.10 Naive and activated immune cells occupy different microenvironments in GALT

In human GALT, memory B (sIgD$^-$) and T (CD45RO$^+$) lymphocytes with a high level of $\alpha_4\beta_7$ integrin are often located near and within the microlymphatics, together with some CD19$^+$CD38^{hi}$\alpha_4\beta_7$^{hi} B-cell blasts. However, the lymph vessels contain mainly naive $\alpha_4\beta_7$^{lo} lymphocytes. Cytochemical and flow-cytometric analyses of human mesenteric lymph have provided similar marker profiles; also notably, the small fraction of identified B-cell blasts (2%–6%) reportedly contain cytoplasmic IgA, IgM, and IgG in the proportions 5:1:<0.5.

The $\alpha_4\beta_7$^{hi} subsets exiting through lymphatics in human GALT probably represent the first homing step to seed particularly the intestinal lamina propria with activated lymphoid cells. Relatively few memory cells express high levels of CD62L/L-selectin in intestinal and mesenteric lymph; those that do either reenter GALT or extravasate in MLNs, peripheral lymph nodes, or Waldeyer's ring together with naive cells by binding to peripheral lymph node addressin (PNAd) on HEVs.

The homing of activated lymphoid cells to the intestinal lamina propria clearly depends on the high surface level of $\alpha_4\beta_7$ in the absence of CD62L/L-selectin. The expression of $\alpha_4\beta_7$ allows binding of the circulating cells to unmodified MAdCAM-1 expressed apically on endothelial cells of the lamina propria microvasculature. The $\alpha_4\beta_7$^{hi} phenotype is predominantly induced on antigen-specific B cells appearing in peripheral blood after enteric (peroral) immunization in humans, whereas the counterparts elicited by systemic immunization express preferentially CD62L/L-selectin but relatively little $\alpha_4\beta_7$. Although interaction of MAdCAM-1 with CD62L has been explored mainly in mice, the virtual absence of CD62L-bearing cells in human intestinal lamina propria strongly suggests that this molecule does not bind to MAdCAM-1 outside of GALT. Many large B cells retain high levels of $\alpha_4\beta_7$ after migration into the intestinal lamina propria, despite abundant coexpression of CD38 and cytoplasmic IgA. Therefore, it is possible that $\alpha_4\beta_7$ may both mediate extravasation and, together with CD44, contribute to local retention of effector cells.

In addition, the migration of both T and B cells into, and/or retention within, the small intestinal lamina propria appears to be mediated by the thymus-expressed chemokine (CCL25/TECK). Notably, this chemokine that interacts with CCR9 is selectively produced by the crypt epithelium in this part of the normal gut in humans, as well as mice (**Figure 1.8**). Several mouse studies have suggested that small intestinal tropism directed by high $\alpha_4\beta_7$ and CCR9 expression combined with minimal CD62L is imprinted on T cells by DCs in PPs and MLNs.

In the healthy small intestine, $\alpha_4\beta_7$ and CCR9 homing molecules determine the extravasation of memory/effector B cells. Acquisition of such gut-homing properties apparently depends at least partly on retinoic

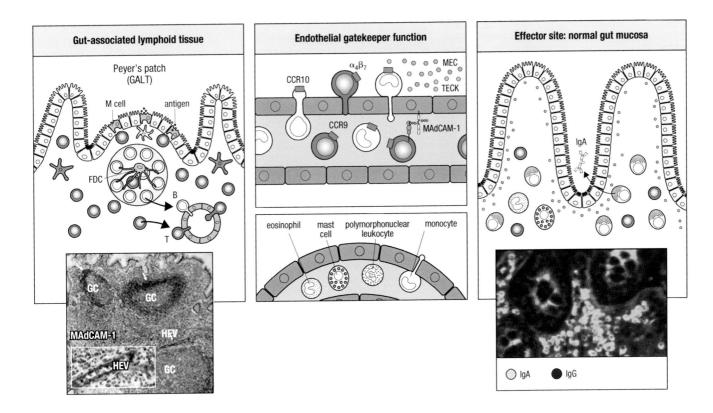

Figure 1.8 Homing mechanisms that attract gut-associated lymphoid tissue (GALT)-derived B and T memory/effector cells to the small intestinal lamina propria. A GALT structure with its epithelial M cells and antigen-presenting cells such as follicular dendritic cells (FDC), is depicted in the top left panel. Once activated, T and B cells in Peyer's patch migrate into draining lymph (solid arrows). Interactions between the multidomain unmodified (containing no L-selectin-binding O-linked carbohydrates) mucosal addressin cell adhesion molecule (MAdCAM)-1 on flat lamina propria venules preferentially targets mucosal $\alpha_4\beta_7$-bearing memory/effector B and T cells to the normal gut mucosa (middle panel). Selectively produced by the epithelium of the small intestine, the chemokine TECK (CCL25) attracts GALT-derived B and T cells that express CCR9 to this segment of the gut, whereas MEC (CCL28) is a more generalized chemokine interacting with CCR10 on mucosal B cells. The access to the mucosal effector site is normally very limited for circulating pro-inflammatory cells such as monocytes, polymorphonuclear leukocytes, mast cells, and eosinophils, whereas the favored GALT-derived cells promote mucosal immunity including polymeric Ig receptor (pIgR)-dependent secretory IgA (SIgA) generation (top right panel). The bottom left panel shows histology of a Peyer's patch with secondary lymphoid follicles containing germinal centers (GC); the insert shows that MAdCAM-1 (with L-selectin-binding capacity) is expressed on a high endothelial venule (HEV) to attract naive lymphocytes for priming. Right bottom panel shows paired immunofluorescence staining for IgA- and IgG-producing plasma cells in normal colonic human mucosa and crypts with selective transport of IgA to the lumen. (Adapted from Brandtzaeg, P. *Scand. J. Immunol.* 2009, 70:505–515. With permission from John Wiley & Sons.)

acid derived by oxidative conversion from vitamin A. Macrophages and DCs in GALT, MLNs, and intestinal mucosa express retinal dehydrogenase that drives this conversion. In the large intestine, $\alpha_4\beta_7$ appears, instead, to be assisted by the mucosae-associated chemokine (CCL28/MEC) as a decisive cue for attracting IgA$^+$ lymphoblasts with high levels of CCR10. Because the epithelial expression of CCL28 is higher in the colon than in the small intestine and appendix, this chemokine probably plays a compartmentalized role in intestinal B-cell homing. Also, interestingly, both CCL25 and CCL28 enhance integrin α_4-dependent adhesion of IgA$^+$ plasmablasts to MAdCAM-1.

The cues for gut homing of GALT- and MLN-derived B cells from the circulation may also contribute to lateral migration from GALT structures directly into the intestinal lamina propria. Vascular connections between PPs and surrounding lamina propria may be used for trafficking B cells, possibly explaining why the first IgA$^+$ plasma cells accumulate around PPs when germ-free mice are transferred to conventional conditions. Similar pathways between isolated lymphoid follicles and the surroundings are suggested by the observation that the number of intestinal IgA⁺ plasma cells is maintained by local proliferation when the B-cell traffic via the thoracic duct is diverted by lymph cannulation in rats. Similar observations of highly localized B-cell responses have been made in the multiple-intestinal-loop model in lambs. However, some of the plasmablasts that normally settle immediately adjacent to GALT follicles apparently belong to exhausted B-cell clones with decreased J chain–expressing potential and disproportionately increased class switch to IgG.

1.11 Homing molecules are important in homing of cells to extraintestinal sites

Dissemination of GALT-derived B cells to extraintestinal secretory effector sites is well documented, but only limited homing from NALT or BALT to the gut has been demonstrated by immunization or infection experiments in rodents and pigs. However, indirect evidence suggests that IgA⁺ plasmablasts disperse from Waldeyer's ring (human NALT) to regional secretory effector sites. Direct evidence for such regional homing has been obtained by immunization of murine NALT and rabbit palatine tonsils. A dichotomy between the upper and lower body regions (**Figure 1.9**) is further supported by the dispersion of human tonsillar sIgD⁺IgM⁻CD38⁺ plasmablasts. In the circulation, these cells are variably CCR10⁺ but show a homing-receptor profile that does not favor the small intestinal mucosa ($\alpha_4\beta_7^{int/lo}$CCR9^{lo}CCR7^{hi}CD62L^{hi}), and they are virtually excluded from this effector site. Conversely, such "marker cells" appear in lymph nodes and the bone marrow. The tissue distribution of these cells likely reflects the migratory properties of all plasma-cell precursors with a mucosal phenotype (J chain⁺) primed in Waldeyer's ring. In keeping with this notion, activated human tonsillar B cells transferred intraperitoneally to SCID (severe combined immunodeficiency) mice migrate to the lung but not to the gut mucosa.

Figure 1.9 Model for homing of activated B cells from mucosal inductive sites with secondary lymphoid follicles, to secretory effector sites in the integrated human mucosal immune system. The specialized follicle-associated epithelium contains M cells with antigen-transporting properties. Putative compartmentalization in the trafficking is indicated—the heavier arrows representing preferential B-cell migration pathways. In the small intestinal lamina propria, attraction and/or retention of CCR9-expressing cells is mediated by the chemokine CCL25/TECK (see **Figure 1.8**), while CCL28/MEC interacting with CCR10 is important in the large bowel. Other adhesion molecules, such as CD62L/L-selectin and $\alpha_4\beta_1$ that bind to endothelial PNAd and VCAM-1, respectively, may be employed mainly by B cells primed in nasopharynx-associated (NALT) and bronchus-associated (BALT) lymphoid tissue. Human NALT is composed of the various lymphoepithelial structures of Waldeyer's ring, including the nasopharyngeal tonsil (adenoids) and palatine tonsils. In this region, abundantly produced epithelial CCL28/MEC attracts activated B cells via CCR10. The female genital tract (cervix mucosa) may employ similar molecular homing mechanisms to the upper aerodigestive tract and the large bowel—therefore probably receiving activated B cells from inductive sites in both of these regions. Also, lactating mammary glands appear, by shared homing mechanisms, to receive primed cells from NALT as well as gut-associated lymphoid tissue (GALT)—and much more efficiently than the female genital tract. Homing molecules integrating airway immunity with systemic immunity are shown. (Adapted from Brandtzaeg, P. *Scand. J. Immunol.* 2009, 70:505–515. With permission from John Wiley & Sons.)

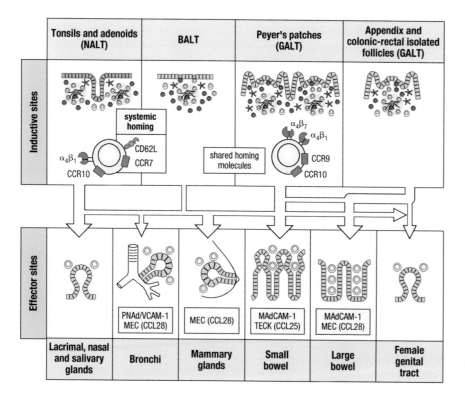

Earlier studies indicate that $\alpha_4\beta_7$ plays little or no role in homing of lymphoid cells to the airways in humans, mice, or sheep. In this connection, intranasal immunization induces an insufficient level of $\alpha_4\beta_7$ on antigen-specific human B cells to make them gut seeking, whereas antibody production is evoked in both adenoids and nasal mucosa. The circulating specific B cells show substantial coexpression of CD62L and $\alpha_4\beta_7$, in contrast to the high surface levels of $\alpha_4\beta_7$ induced by peroral (enteric) immunization. Mouse experiments have indeed suggested that CD62L is involved in B-cell homing to extraintestinal mucosal effector sites, and a similar role has been implied for CCR7 in a study of murine airways.

A nonintestinal homing receptor profile might explain migration of B cells from NALT to the urogenital tract. This is reflected by the titers of IgA and IgG antibodies in cervicovaginal secretions of mice, monkeys, and humans after intranasal immunization with a variety of antigens. The relatively consistent level of CD62L and CCR7 on NALT-derived B cells, allowing them to bind to PNAd, CCL21/SLC, and CCL19/ELC on HEVs in lymph nodes, could likewise explain the striking integration between mucosal immunity in the upper aerodigestive tract and systemic immunity. This is an attractive feature for protection against many pathogens and has been documented by immunization in rodents, humans, and nonhuman primates. Migration to secretory tissues beyond the gut might also involve $\alpha_4\beta_1$ (CD49d/CD29) expressed by NALT-derived B cells. The chief counter-receptor for this integrin is VCAM-1, which may be present on the microvascular endothelium in human bronchial, nasal, and uterine cervix mucosa.

In addition to directing activated mucosal B cells to the intestinal lamina propria, CCR10 appears to be a unifying chemokine receptor contributing to homing of plasmablasts to extraintestinal secretory effector sites. Several recent studies have shown CCR10 is expressed by IgA$^+$ plasmablasts (and less so by IgA$^+$ plasma cells) at every mucosal effector site in humans and mice. This expression pattern also characterizes IgD$^+$ plasmablasts/plasma cells in the upper airways, as well as IgM$^+$ and IgG$^+$ plasmablasts/plasma cells replacing the IgA$^+$ plasma cells in the gut of IgA-deficient subjects. As previously mentioned, CCL28/MEC, the CCR10 ligand, is constitutively produced by gut epithelium, especially in the large bowel, and is also expressed at relatively high levels by secretory epithelia in the upper aerodigestive tract and lactating mammary glands. Interestingly, CCL28 (but not CCL25) was shown to attract tonsillar IgA$^+$ plasmablasts *in vitro*. Therefore, graded tissue site-dependent CCR10–CCL28 interactions, together with insufficient levels of classical gut-homing molecules, most likely explain the observed dispersion dichotomy for memory/effector B cells derived from Waldeyer's ring. Because bone marrow stromal cells reportedly produce CCL28, interactions of this chemokine with CCR10$^+$ B cells may furthermore contribute to integration between mucosal and systemic immunity.

SUMMARY

The mucosal immune system can be divided into the inductive MALT structures and the effector sites with diffusely dispersed immune cells such as the lamina propria and epithelium of gut and airway mucosae as well as the stroma of exocrine glands. Abundant evidence supports the notion that human intestinal plasma cells are largely derived from B cells initially activated in GALT. However, the uptake, processing, and presentation of luminal antigens that accomplish priming and sustained expansion of mucosal B and T cells are incompletely understood. Also unclear is how the germinal center reaction in GALT so prominently, compared with other MALT structures, promotes class switch to IgA and expression of J chain, although the commensal microbiota appears to contribute to both diversification and memory. B- and T-cell migration from GALT to the intestinal lamina propria is guided by rather well-defined adhesion molecules and chemokines/chemokine receptors, but the signals directing homing to secretory effector sites beyond the gut require better definition. In this respect, the role of Waldeyer's ring (including adenoids and palatine tonsils) as regional human MALT must be better defined, although the balance of evidence suggests that it functions as NALT like the characteristic NALT structures in rodents. Thus, the remarkable compartmentalization of the mucosal immune system must be taken into account in the development of effective local vaccines to protect the small and large intestine, the airways, and the female genital tract.

Note added in press: The functional roles of tertiary lymphoid organs in the small intestine and colon have been summarized by Buettner and Lochner 2016. The differences between dendritic cells from the colon versus the ileum in human being was studied by Mann et al. 2016: CD103+, CD11b+ dendritic cells drive mucosal T helper 17 cell differentiation (Persson et al. 2013). The transcriptional and functional profiling is different in mouse and human intestinal differentiation (Watchmaker et al. 2014). The longevity of antibody secreting plasma cells in the human intestine is of great clinical relevance (Landsverk et al. 2017). Memory CD4+ CDR5+ T cells are already abundantly present in the gut wall of newborns, which is of great relevance in mother to child transmission of HIV-1 (Bunders et al. 2012). Resident memory CD8Tcells persist for years in the human small intestine, as documented in transplanted patients (Bartolomé-Casado et al. 2019).

The activation of M cells is an aspect in oral vaccination has been so far studied in mice (Kimura et al. 2019). The maturation steps from the stem cells in the bottom of the crypts to the mature M Cells have been summarized by Nakamura et al. 2018).

Macrophages in the human gut wall can be characterized by transcriptional and functional profiling into for different groups. Using material from transplanted patients, the turnover and localization in the duodenal walls was studied: The types were replaced within 3 weeks and the others showed a more slower turnover (Bujko A et al. 2017).

The innate lymphoid cells (ILS) have been studied recently in more detail (e.g. Penny et al. 2018). A major problem is that the localization within the lamina propria e.g. crypt or villous part has not been identified because these cells are characterized functionally but not by a cell surface marker to do immunohistology (Kortekaas Krohn et al. 2018). The composition of ILS subsets differs in their localization of different parts of the intestine and are significantly altered in HIV-infected patients (Krämer et al. 2017). Sometimes, natural killer cells are included in the ILS family. During the first years of life of human being, the number of NK cells decreased gradually. These cells were characterized by Eosomes, perforin, and granzyme B expression (Sagebiel et al. 2019).

FURTHER READING

Aase, A., Sommerfelt, H., Petersen, L.B. et al.: Salivary IgA from the sublingual compartment as a novel noninvasive proxy for intestinal immune induction. *Mucosal Immunol.* 2016, 9:884–893.

Bain, C.C., Bravo-Blas, A., Scott, C.L. et al.: Constant replenishment from circulating monocytes maintains the macrophage pool in the intestine of adult mice. *Nat. Immunol.* 2014, 15:929–937.

Bartolomé-Casado, R., Landsverk, O.J.B., Chauhan, S.K. et al.: Resident memory CD8 T cells persist for years in human small intestine. *J. Exp. Med.* 2019, 216: 2412–2426.

Brandtzaeg, P.: Function of mucosa-associated lymphoid tissue in antibody formation. *Immunol. Invest.* 2010, 39:303–355.

Brandtzaeg, P., Farstad, I.N., Johansen, F.-E. et al.: The B-cell system of human mucosae and exocrine glands. *Immunol. Rev.* 1999, 171:45–87.

Brandtzaeg, P., and Johansen, F.-E.: Mucosal B cells: Phenotypic characteristics, transcriptional regulation, and homing properties. *Immunol. Rev.* 2005, 206:32–63.

Brandtzaeg, P., and Pabst, R.: Let's go mucosal: Communication on slippery ground. *Trends Immunol.* 2004, 25:570–577.

Buettner, M., and Lochner, M.: Development and function of secondary and tertiary lymphoid organs in the small intestine and the colon. *Front Immunol.* 2016, 7:342.

Bujko, A., Atlasy, N., Landsverk, O.J.B. et al.: Transcriptional and functional profiling defines human small intestinal macrophage subsets. *J. Exp. Med.* 2018, 215:441–458.

Bunders, M.J., van der Loos, C.M., Klarenbeek, P.L. et al.: Memory CD4(+)CCR5(+) T cells are abundantly present in the gut of newborn infants to facilitate mother-to-child transmission of HIV-1. *Blood* 2012, 120:4383–4390.

Cella, M., Fuchs, A., Vermi, W. et al.: A human natural killer cell subset provides an innate source of IL-22 for mucosal immunity. *Nature* 2009, 457:722–725.

Fuchs, A., Vermi, W., Lee, J.S. et al.: Intraepithelial type 1 innate lymphoid cells are a unique subset of IL-12- and IL-15-responsive IFN-γ-producing cells. *Immunity* 2013, 38:769–781.

Heier, I., Malmström, K., Sajantila, A. et al.: Characterisation of bronchus-associated lymphoid tissue and antigen-presenting cells in central airway mucosa of children. *Thorax* 2011, 66:151–156.

Herbrand, H., Bernhardt, G., Förster, R. et al.: Dynamics and function of solitary intestinal lymphoid tissue. *Crit. Rev. Immunol.* 2008, 28:1–13.

Juelke, K., Romangnani, C.: Differentiation of human innate lymphoid cells (ILCs). *Curr. Opin. Immunol.* 2016, 38:75–85.

Kimura, S., Kobayahi, N., Nakamura, Y. et al.: Sox8 is essential for M cell maturation to accelerate IgA response at the early stage after weaning in mice. *J. Exp. Med.* 2019, 216:831–846.

Kiyono, H., and Fukuyama, S.: NALT- versus Peyer's-patch-mediated mucosal immunity. *Nat. Rev. Immunol.* 2004, 4:699–710.

Kortekaas Krohn, I., Shikhagaie, M.M., Golebski, K. et al.: Emerging roles of innate lymphoid cells in inflammatory diseases: Clinical implications. *Allergy* 2018, 73:837–850.

Krämer, B., Goeser, F., Lutz, P. et al.: Compartment-specific distribution of human intestinal innate lymphoid cells is altered in HIV patients under effective therapy. *PLoS Pathog.* 2017, 13:e1006373.

Kunkel, E.J., and Butcher, E.C.: Chemokines and the tissue-specific migration of lymphocytes. *Immunity* 2002, 16:1–4.

Landsverk, O.J., Snir, O., Casado, R.B. et. al: Antibody-secreting plasma cells persist for decades in human intestine. *J. Exp. Med.* 2017, 214:309–317.

Landsverk, O.J., Snir, O., Casado, R.B. et al.: Antibody-secreting plasma cells persist for decades in human intestine. *J. Exp. Med.* 2016, 214:309–317.

Liu, Y.J., and Arpin, C.: Germinal center development. *Immunol. Rev.* 1997, 156:111–126.

Lügering, A., Kucharzik, T.: Induction of intestinal lymphoid tissue: The role of cryptopatches. *Ann. N. Y. Acad. Sci.* 2006, 1072:210–217.

Macpherson, A.J., McCoy, K.D., Johansen, F.-E. et al.: The immune geography of IgA induction and function. *Mucosal Immunol.* 2008, 1:11–22.

Mann, E.R., Bernardo, D., English, N.R. et al.: Compartment-specific immunity in the human gut: Properties and functions of dendritic cells in the colon versus the ileum. *Gut* 2016, 65:256–270.

Masahata, K., Umemoto, E., Kayama, H. et al.: Generation of colonic IgA-secreting cells in the caecal patch. *Nat. Commun.* 2014, 5:3704.

Mowat, A.M., and Bain, C.C.: Mucosal macrophages in intestinal homeostasis and inflammation. *J. Innate Immun.* 2011, 3:550–564.

Nakamura, Y., Kimura, S., Hase, K.: M cell-dependent antigen uptake on follicle-associated epithelium for mucosal immune surveillance. *Inflamm. Regen.* 2018, 38:15.

Neutra, M.R., Mantis, N.J., and Kraehenbuhl, J.P.: Collaboration of epithelial cells with organized mucosal lymphoid tissues. *Nat. Immunol.* 2001, 2:1004–1009.

Newberry, R.D., and Lorenz, R.G.: Organizing a mucosal defense. *Immunol. Rev.* 2005, 206:6–21.

Pabst, O., Cerovic, V., Hornef, M.: Secretory IgA in the coordination of establishment and maintenance of the microbiota. *Trends Immunol.* 2016, 37:287–296.

Pabst, R., and Tschernig, T.: Bronchus-associated lymphoid tissue. An entry site for antigens for successful mucosal vaccination? *Amer. J. Respir. Cell Mol. Biol.* 2010, 43:137–140.

Pabst, R.: Mucosal vaccination by the intranasal route. Nose-associated lymphoid tissue (NALT)—Structure, function and species differences. *Vaccine* 2015, 33:4406–4413.

Penny, H.A., Hodge, S.H., Hepworth, M.R.: Orchestration of intestinal homeostasis and tolerance by group 3 innate lymphoid cells. *Semin Immunopathol.* 2018, 40:357–370.

Persson, E.K., Uronen-Hansson, H., Semmrich, M. et al.: IRF4 transcription-factor-dependent CD103(+) CD11b(+) dendritic cells drive mucosal T helper 17 cell differentiation. *Immunity.* 2013, 38:958–969.

Quiding-Järbrink, M., Nordström, I., Semmrich, M. et al.: Differential expression of tissue-specific adhesion molecules on human circulating antibody-forming cells after systemic, enteric, and nasal immunizations. A molecular basis for the compartmentalization of effector B cell responses. *J. Clin. Invest.* 1997, 99:1281–1286.

Sagebiel, A.F., Steinert, F., Lunemann, S.: Tissue-resident Eomes[+] NK cells are the major innate lymphoid cell population in human infant intestine. *Nat. Commun.* 2019 10:975.

Sipos, F., and Muzes, G.: Isolated lymphoid follicles in colon: Switch points between inflammation and colorectal cancer? *World J. Gastroenterol.* 2011, 17:1666–1673.

Tait Wojno, E.D., and Artis, D.: Emerging concepts and future challenges in innate lymphoid cell biology. *J. Exp. Med.* 2016, 213:2229–2248.

Watchmaker, P.B., Lahlki, K., Lee, M. et al.: Comparative transcriptional and functional profiling defines conserved programs of intestinal DC differentiation in humans and mice. *Nature Immunol.* 2014, 15:98–108.

Phylogeny of the mucosal immune system

2

ROBERT D. MILLER AND IRENE SALINAS

All animals, vertebrate or invertebrate, face the challenge of combating pathogens while maintaining a tolerant relationship with symbiotic microorganisms. Tolerance to symbiotes is not a static or inert interaction but, rather, requires continuous active regulation; the frontlines for these interactions are the mucosal surfaces. A fundamental question is whether or not specialized immune cells or organs have evolved in all animals to cope with the unique problems of mucosal defense, or whether specialized mucosal immunity is unique to vertebrates such as mammals. This question can be answered by investigating the phylogeny of the mucosal and systemic immune system, from invertebrates, to fish, amphibians, reptiles, and birds, to prototherians (egg-laying mammals), the metatherians (marsupials), and the eutherians (placental mammals). There is also an extremely practical reason for studying phylogeny, namely, that many species are important human foods and also suffer from diseases of mucosal surfaces, so that producing mucosal vaccines against fish pathogens is arguably as important to humankind as producing vaccines against the mucosal pathogens of mammals.

Most of the paradigms for immune defense and symbiosis were established from studies of humans and a few other mammalian species such as rodents. Although these studies have provided more than a century of discovery and progress, much has also been learned from nonmammalian vertebrates (such as amphibians and fish) as well as invertebrates (flies, snails, and worms), particularly over the past two decades. Although most metazoan species have a gut and some specialized organs for performing gas exchange, it does not seem that a specialized mucosal immune system is universal. Crustaceans and worms often have a higher percentage of immune cells, such as hemocytes, surrounding the gut, but these are not organized into discrete tissues as in mammals or other higher vertebrates.

Studies in invertebrates, however, have been extremely informative. It was through studies in the fruitfly *Drosophila melanogaster*, for example, that toll-like receptors and their human homologs were identified, among the more significant discoveries in immunology in the past decade. Similarly, studies of the sea urchin, *Strongylocentrotus purpuratus*, have revealed key steps in the evolution of the complement system, in particular the alternative pathway involving C3 and factor B.

With regard to the mucosal immune system, invertebrates continue to shed light on the conserved interactions between the microbiota and the intestinal immune system. The gut of most invertebrates is lined by the peritrophic matrix, which is composed by chitin fibrils and glycoproteins and physically excludes the gut epithelium from external abrasions and is analogous to the mammalian mucus layer. Both insects and prochordates like the sea squirt *Ciona intestinalis*, produce chitin binding proteins in their gut. *C. intestinalis*, in turn, produces variable chitin binding proteins (VCBPs). VCBPs contain an immunoglobulin-like domain, bind chitin fibrils, and facilitate phagocytosis of bacteria.

In *D. melanogaster,* antimicrobial peptide and Duox-dependent reactive oxygen species production are key innate immune mechanisms in the gut. The gut microbiota of mosquitoes regulates antimicrobial peptide and C-type lectin expression via the immune deficiency (Imd) pathway. Mosquito C-type lectins regulate antimicrobial peptide expression, coat bacterial surfaces, and contribute to the maintenance of the gut microbiota in a similar manner to vertebrate mucosal immunoglobulins.

PHYLOGENY OF RECEPTOR DIVERSITY

The unique aspect of the adaptive immune system is the generation of diverse receptors from germline DNA. Among the animal phyla, the chordates (Phylum Chordata) arguably exhibit the greatest diversity in immune systems.

Vertebrates are divided into two major living lineages: the jawed (gnathostomes) and the jawless (agnathans) (**Figure 2.1**). These two lineages have adaptive immune systems that followed very different evolutionary paths while maintaining some common features. Both lineages use lymphocytes as the major cell type mediating adaptive immunity. These lymphocytes express cell-surface, antigen-specific, and clonally unique receptors.

2.1 T-cell receptors are similar in all jawed vertebrates

In all jawed vertebrates, the T-cell antigen receptors (TCRs) are composed of a heterodimer of either paired αβ or γδ chains (Table 2.1). The genes encoding these four chains are highly conserved both in sequence and organization from sharks to mammals, leading to the conclusion that the TCR performs common functions across all the gnathostome lineages, principally as recognition and signaling receptors and the unique way in which they recognize antigens. Unlike antibodies that serve both as signaling receptors and are also secreted, the TCR serves solely to recognize antigenic epitopes when they meet a threshold of binding affinity. The requirement of recognizing antigenic peptides and glycolipids presented on molecules encoded by the major histocompatibility complex (MHC) has further constrained the evolution of the TCR, the αβ TCR in particular. This relationship between TCR and MHC is also ancient in the gnathostomes and conserved in all living species (see Table 2.1). Specific mucosal T-cell subsets and functional aspects of T cells in mucosal surfaces are poorly studied in nonmammalian vertebrates.

2.2 Immunoglobulins have evolved in different ways in different lineages

The evolution of different classes of immunoglobulins has occurred through both gene duplication and gene reorganization. In the bony fish and the tetrapods (amphibians, birds, reptiles, and mammals), the immunoglobulin (Ig) genes are organized in what is termed a "translocon" arrangement. The translocon arrangement is generally a single heavy-chain locus with clusters of separate variable (V), diversity (D), and joining (J) segments upstream of the exons encoding the constant domains. The significance of this translocon arrangement is that it facilitated the specialization of different antibody heavy-chain isotypes through gene duplication, ultimately leading to the evolution of different isotypes, such as IgG and the mucosa-associated antibodies IgT, IgX, and IgA (see Table 2.1). The evolution of class switching, however, did not drive isotype diversification, because multiple antibody classes are also found in cartilaginous fishes (sharks, rays, and skates) as well as bony fishes.

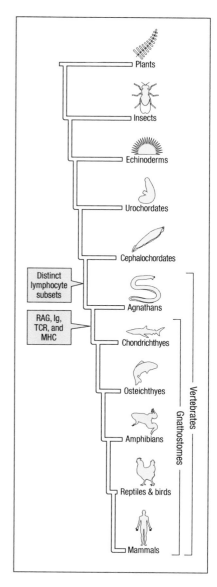

Figure 2.1 Evolutionary relationship of the animal lineages mentioned in this chapter. The branching pattern of the evolutionary "tree" shown represents the relative order of divergence of the different lineages. For example, the plants and animals diverged from a common ancestor before the separation of the animals into the protostomes (represented by the insects) and the deuterostomes (echinoderms and chordates). The evolution of different lymphocyte subsets seems to have occurred before the divergence of the vertebrates into the jawless (agnathans) and jawed (gnathostome) lineages. However, an adaptive immune system based on the recombination-activating gene (RAG) recombinase enzymes, including the immunoglobulins (Ig), T-cell receptors (TCR), and major histocompatibility complex (MHC), is unique and common to all the gnathostomes.

TABLE 2.1 DIVERSITY OF CELLS, RECEPTORS, AND IMMUNOLOGIC MACHINERY USED BY VERTEBRATES

Lineage	Age[a]	VLR	Ig types	Ig class switch	Specialized mucosal Ig	TCR	MHC	AID[b]	RAG[c]
Agnathan									
Lamprey, hagfish	~500	VLRA, VLRB	No	No	Undetermined	No	No	Yes	No
Gnathostomes									
Chondrichthyes (sharks, rays, skates)	~450	No	IgM, IgW, IgNAR	No	Undetermined	$\alpha\beta$, $\gamma\delta$	Class I, II	Yes	Yes
Osteichthyes (bony fish)	~400	No	IgM, IgD, IgT	No	IgT	$\alpha\beta$, $\gamma\delta$	Class I, II	Yes	Yes
Amphibians (frogs, toads, salamanders)	~360	No	IgM, IgD, IgY, IgX	Yes	IgX	$\alpha\beta$, $\gamma\delta$	Class I, II	Yes	Yes
Lepidosaurs (snakes, lizards)	~300	No	IgM, IgD, IgY, IgA	Yes	IgA	$\alpha\beta$, $\gamma\delta$	Class I, II	Yes	Yes
Archosaurs (birds, crocodilians)	~300	No	IgM, IgY, IgA	Yes	IgA	$\alpha\beta$, $\gamma\delta$	Class I, II	Yes	Yes
Mammals									
Monotremes (platypus, echidna)	~170	No	IgM, IgD, IgG, IgE, IgA	Yes	IgA	$\alpha\beta$, $\gamma\delta$, μ[d]	Class I, II	Yes	Yes
Marsupials (opossum, kangaroo)	~135	No	IgM, IgG, IgE, IgA	Yes	IgA	$\alpha\beta$, $\gamma\delta$, μ[d]	Class I, II	Yes	Yes
Eutherians (human, mouse, cow)	0	No	IgM, IgD, IgG, IgE, IgA	Yes	IgA	$\alpha\beta$, $\gamma\delta$	Class I, II	Yes	Yes

[a] Time in millions of years since last common ancestor with eutherian (placental) mammals such as humans and mice.
[b] Activation-induced cytidine deaminase or related cytidine deaminases used to diversify antigen receptors.
[c] Recombination-activating genes encoding the V(D)J recombinase.
[d] TCRμ is a unique TCR chain related to TCRδ.
Abbreviations: Ig, immunoglobulin; MHC, major histocompatibility complex; TCR, T-cell receptor; VLR, variable lymphocyte receptor.

However, class switching was a significant evolutionary innovation because it allowed B cells to change the functional attributes of the secreted antibody without changing antigen specificity.

2.3 T and B cells are present in lower vertebrates

In jawless vertebrates such as lampreys and hagfish, and the jawed vertebrates such as sharks, bony fish, birds, and mammals, there are discrete lymphocyte subsets. One subset does not secrete its antigen receptor but is capable of producing cytokines that regulate other cell types. In gnathostomes these are the T cells, whereas in agnathans they are the VLRA+ lymphocytes. Also present is a subset that is capable of producing its antigen receptor both as a cell-surface receptor and as a secreted molecule, often in a polymeric form (Figure 2.2). These are the B cells in gnathostomes and VLRB+ lymphocytes in agnathans. Therefore, the paradigm of the dichotomy between B and T cells in mice and humans probably arose in a common vertebrate ancestor more than 500 million years ago, before the evolution of the thymus, spleen, TCRs, Igs, or other structures with which most (conventional) immunologists are familiar. The strategy of having two

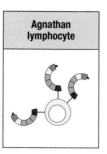

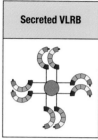

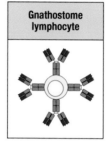

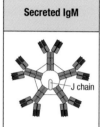

Figure 2.2 Comparison of the lymphocyte and secreted antibody forms from jawless and jawed vertebrates. Both jawless (agnathan) and jawed (gnathostome) vertebrates have lymphocytes that express a cell-surface receptor that is diversified through some process of somatic gene rearrangement. In the agnathans, these are the variable lymphocyte receptors (VLRs) containing leucine-rich repeat (LRR) domains; in the gnathostomes, they are the immunoglobulins and T-cell receptors containing Ig domains. Gene rearrangement in the VLRs occurs by gene conversion, resulting in variable numbers of LRR units being present, whereas in the gnathostomes it is mediated by the cutting and splicing of gene segments, the V(D)J recombination. Similarly to IgM, VLRB can be secreted in either tetrameric or pentameric forms.

distinct lymphocyte lineages, with one playing a regulatory role while the other produces antibodies, is an ancient defense strategy.

In the hagfish and lamprey, diversity occurs at a gene cluster encoding a class of receptors called the variable lymphocyte receptors (VLRs). The VLRs are the functional equivalent to the antibodies in jawed vertebrates. However, instead of being based on immunoglobulin protein domains, they are composed of leucine-rich repeat (LRR) domains similar to toll-like receptors. Both hagfish and lampreys have two sets of VLR genes, VLRA and VLRB, which define the distinct VLRA$^+$ and VLRB$^+$ lymphocyte lineages. VLRA$^+$ cells express a nonsecreted cell-surface receptor and produce cytokines and chemokines. VLRB$^+$ cells express a cell-surface receptor that, upon cell activation, is secreted either as tetramers or pentamers, much like IgM in gnathostomes, and functions as an antibody. VLRs differ from Igs and TCRs not only in the protein structure that is being used but also in the genetic mechanisms used to generate diversity. Ig and TCR diversity is generated via the assembly of an exon encoding the variable domain from the V, D, and J gene segments by means of the cutting and splicing of DNA using site-specific endonucleases and DNA repair mechanisms. In contrast, the VLR gene system in lampreys and hagfish generates diversity using a gene conversion mechanism that is dependent on DNA damage induced by cytidine deaminase. The cytidine deaminases expressed in agnathan lymphocytes are related to the activation-induced cytidine deaminases (AIDs) that participate in primary antibody diversification, affinity maturation, and isotype switching in birds, mammals, and other gnathostomes. Therefore, the involvement of such deaminases in antigen receptor genetics is also ancient in the vertebrate lineage (see Table 2.1). Through gene conversion, a variable number and combination of LRR domains are added to the expressed locus. As with V(D)J recombination, the VLR system is capable of generating many (in the order of 10^{14}) possible receptors.

Phylogeny of lymphoid tissue

It is generally agreed that primary lymphoid tissues appeared earlier in vertebrate evolution than the secondary lymphoid tissues. Many of the familiar tissues labeled as secondary lymphoid organs in mammals, such as lymph nodes, are only found in "higher" vertebrates such as birds and mammals. Defined mucosal lymphoid structures akin to Peyer's patches in eutherian mammals are found only in birds, monotremes (echidna and platypus), and marsupials (Figure 2.3). However, the evolutionary forerunners of these structures are already present in African lungfish.

2.4 Lymphoid tissues in agnathans seem to have evolved from mucosa-associated lymphoid tissues

Studies in jawless vertebrates support the evolution of early primary lymphoid tissues from mucosa-associated lymphoid tissues (MALTs). The

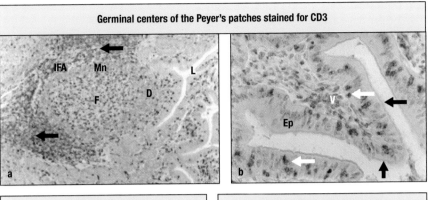

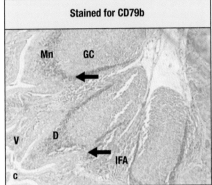

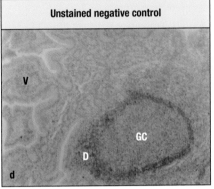

Figure 2.3 **Juvenile eastern gray kangaroo gut stained for T cells and B cells.** (**a**) The anti-CD3 stained cells were apparent in the germinal centers of the Peyer's patches and (arrows) in the interfollicular areas (IFA). A rare cell was stained in the mantle (Mn) of the follicles (F) and in the dome (D). The lumen (L) is also observed (×100). (**b**) Higher magnification (×400) of panel (a), showing clearly defined T cells (white arrows) throughout the villi (V); some were in close association with epithelial cells (Ep) (black arrows). (**c**) The germinal centers (GC), mantles (Mn), and dome regions (D) heavily stained by anti-CD79b (arrows). The interfollicular area (IFA) was not stained. There seems to be no staining of B cells in the villi (V) (×100). (**d**) Representative negative control slide with no staining observed. The dome (D) of the germinal center (GC) is easily seen, as are the villi (V) (×100). (Photographs courtesy of Julie Old.)

primary hematopoietic organ in lampreys is the typhlosole, an organ unique to the agnathans. The typhlosole develops as an invagination in the larval intestine and remains associated with the intestinal wall throughout the life of the organism. In addition, the lamprey pharyngeal gill epithelium contains accumulations of lymphocytes that express VLR genes. Furthermore, a thymus candidate has been proposed in lamprey and coined as "thymoid." The lamprey thymoid is located in the tips of the gill filaments and secondary lamellae within the gill basket. It is worth noting that this hypothesis is consistent with the development of the thymus in jawed vertebrates from the pharyngeal pouches. The discovery of an adaptive immune system in agnathans makes this question all the more interesting, because the generation of highly diverse VLR antigen receptors in lampreys and hagfish should create the need for some mechanism of self-tolerance, presumably via negative selection. The paradigm of the gnathostome thymus suggests the need for a specific organ in which cells undergo selection. Consistent with this idea, the pharyngeal/gill-associated lymphoid tissue (thymoid) in lampreys has the highest percentage of lymphocytes that are VLRA⁺, those lymphocytes that share the greatest number of characteristics with gnathostome T cells. Throughout the rest of the lamprey tissues, VLRB⁺ cells are far more prevalent than VLRA⁺ cells. Furthermore, there is recent evidence that the lamprey *Foxn4L* gene expressed in developing gill epithelium is the agnathan ortholog of *Foxn1*, which is expressed in the developing thymus of jawed vertebrates.

2.5 Gnathostome lymphoid tissues differ between lineages

All jawed vertebrates, from sharks to mammals, use the same basic components for their adaptive immune system, including immunoglobulin domain-based antibodies and T-cell receptors, and an antigen presentation system using the molecules of the MHC. Similarly, all jawed vertebrates have a thymus with clearly delineated cortical and medullary regions. The thymus is always associated with the pharyngeal regions or upper thoracic regions and develops from embryonic pharyngeal pouches. The conservation of the thymus is probably due to its critical role in the positive and negative selection of developing T cells. All jawed vertebrates, including cartilaginous fish, also have a spleen with clearly defined red and white pulp areas. However, fish do not have MALT, but the intestine does contain T cells and B cells in the epithelium and lamina propria.

In contrast to the conserved nature of the thymus, the primary lymphoid organs in gnathostomes vary between lineages. In cartilaginous fish, hematopoiesis takes place in the epigonal organ, whereas in bony fish it is in the head kidney. The role of the bone marrow as the primary hematopoietic tissue evolved in tetrapods and is found in amphibians, birds, and mammals. In well-known species such as mice and humans, B-cell development occurs primarily in the bone marrow. However, in a few mammals and all birds, the gut-associated lymphoid tissue (GALT) has an essential role in B-cell ontogeny. An extreme example of this is birds in which there is a unique avian GALT, the bursa of Fabricius. In fact, B cells acquired their name from the discovery that bursectomized chicks failed to develop antibody responses. Although, phylogenetically speaking, birds and crocodilians (crocodiles and alligators) are closely related, the latter do not have a bursa, and unfortunately little is known regarding B-cell development or the role of the GALT in this lineage.

In birds, B cells that have undergone V(D)J recombination in the bone marrow must migrate to the bursa to complete their development. In the bursa of Fabricius, avian B cells complete their development by introducing mutations into their recombined V genes using AID-mediated gene conversion. This is an essential step because of the lack of repertoire diversity generated by V(D)J recombination alone. The chicken IgH locus contains only a single functional V gene and a large array of nonfunctional V pseudogenes. Although the pseudogenes cannot be recombined and expressed, they can be used as a source of donor sequences for gene conversion (**Figure 2.4**).

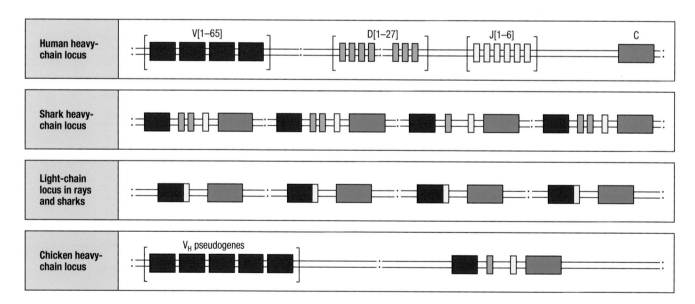

Figure 2.4 The organization of immunoglobulin genes in cartilaginous fish and other gnathostomes is different from that in mammals. The organization of the Ig heavy-chain genes in mammals, in which there are separated clusters of repeated V, D, and J gene segments, is similar to the way in which these genes are also organized in birds, reptiles, amphibians, and bony fish. This is referred to as the translocon organization. In sharks, the Ig loci consist of multiple clusters of a V gene, one or two D genes, a J gene, and the exons encoding the constant domains. These clusters can be in tandem arrays as shown or unlinked on separate chromosomes, depending on the species. In chickens, there is a single functional V gene that is rearranged in all B cells, limiting the amount of antibody diversity that V(D)J recombination alone can generate. To further diversify the antibody repertoire, B cells complete their development in the bursa of Fabricius, where the rearranged V is further mutated by gene conversion using an array of V pseudogenes upstream of the functional one as the donor sequences. (Adapted from Murphy, K.: *Janeway's Immunobiology*, 8th ed. New York, Garland Science, 2012.)

There is also enormous species variation in the functional role of GALT in mammals. Sheep and rabbits increase diversity by somatic hypermutation in the GALT (Peyer's patches and appendix in sheep; Peyer's patch, appendix, and sacculus rotundus in rabbits); GALT involutes in sheep at sexual maturity, but becomes secondary lymphoid tissue in the rabbit. Why do some species use a GALT site to complete B-cell development whereas others do not? Species that use GALT for B-cell development all seem to have another common characteristic in that their immunoglobulin heavy-chain and light-chain genes are unable to generate sufficient antibody diversity from V(D)J recombination alone. The migration of the B cells to a GALT site apparently facilitates further mutation of the expressed V genes, increasing the diversity of the primary antibody repertoire. The mutational processes seem to be dependent on both the presence of the GALT tissue and colonization of the gut flora. Germ-free rabbits, for example, do not generate as diverse an antibody repertoire as normal rabbits. In contrast, species such as humans and mice, which have diverse germline V gene repertoires, do not require such additional mutation to generate sufficient diversity and have not evolved the need for using GALT sites for primary B-cell development.

2.6   Origins of organized mucosa-associated lymphoid tissue lie in sarcopterygian fish

Organized mucosal lymphoid structures were traditionally thought to be an immune innovation of endotherms. Examples of these structures include Peyer's patches and tonsils, and they support germinal center (GC) formation and affinity maturation of antibodies during the course of an immune response. Moreover, these canonical organized MALT structures have distinct B- and T-cell zones. The evolutionary forerunners of organized MALT appear to be the lymphoid aggregates found in the gut and nasopharyngeal epithelium of sarcopterygian fish such as the African lungfish (*Protopterus dolloi*). This implies that the origins of organized MALT predate the emergence of tetrapods. Lungfish mucosal lymphoid aggregates are primarily composed of lymphocytes, but these are not segregated into B- and T-cell zones. No evidence of GC formation or AID expression was found in these structures, thus questioning whether affinity maturation of antibodies is the purpose of lungfish lymphoid aggregates. However, mucosal lymphoid aggregates expand in size in response to infection with both B- and T-cell numbers increasing. Finally, *de novo* formation of smaller, inducible lymphoid aggregates takes place in response to bacterial infection resembling the tertiary MALT of mammals.

Phylogeny of mucosal antibodies

Several different classes or isotypes of Igs have evolved in the jawed vertebrates (see Table 2.1). This has involved the evolution of novel heavy-chain and light-chain isotypes; however, it is the appearance of new classes of heavy chains that has had the greatest functional impact on immune responses.

2.7   Immunoglobulin A is the secretory antibody in amniotes

Immunoglobulin A, the antibody isotype primarily associated with mucosal secretions, was discovered in mammals in the 1960s. IgA is secreted in a polymeric form as a dimer associated with a conserved J chain and is transported across mucosal membranes by the polymeric Ig receptor (pIgR). Clear homologs of mammalian IgA have been identified in birds and reptiles but not in amphibians and fish (see Table 2.1). Similarly, birds have a pIgR homolog that, in chickens, is expressed in the liver, gut, and bursa of Fabricius. As in mammals, the pIgR interacts with the avian J chain and is cleaved during transcytosis across the epithelial cells. Therefore, IgA and its associated receptor appeared before the divergence of birds, reptiles, and mammals more than 300 million years ago.

2.8   Amphibians have a unique antibody, immunoglobulin X

Although a direct homolog of IgA has not been found in amphibians, that does not mean they lack specialized mucosal antibodies. Amphibians such as *Xenopus laevis* contain at least three Ig heavy-chain isotypes: IgM, IgY, and IgX (see Table 2.1). As in birds, IgY is functionally analogous to IgG in mammals. Amphibian IgX is believed to be the functional analog of the IgA found in birds, reptiles, and mammals. Like avian and mammalian IgA, frog IgX is mainly expressed at mucosal sites such as the intestine and is secreted in a polymerized form, in this case as a hexamer. However, IgX is not related to IgA and is more similar in sequence to IgM. Amphibian species such as *Xenopus* also express J chain and use it to form polymerized secretory IgM. They also have a pIgR that binds J chain suggesting the role of pIgR in IgX transport across epithelial surfaces. Curiously, however, *Xenopus* IgX is unable to interact with J chains owing to a stop codon that truncates the last constant domain. IgX therefore polymerizes in the absence of J chain. IgX responses in *Xenopus* larvae gut appear to be T-independent since they are not affected by thymectomy. Thymectomy also does not alter the composition of the *Xenopus* gut microbiome.

2.9   Teleosts have a unique secretory antibody, immunoglobulin T

Until recently, teleost fish were thought to have only IgM and IgD and to lack an antibody class specialized for mucosal sites. However, a new Ig heavy-chain isotype called IgT (T or tau [τ] for teleost, and also referred to as IgZ in the literature as a result of its previous discovery in the zebrafish, *Danio rerio*) that may have an important role in mucosal immunity was recently discovered in teleost fish. Serum IgT exists as a monomer, whereas IgT in mucosal secretions is a polymer, similar to IgA and IgX.

In teleost fish, both IgM- and IgT-producing B cells are phagocytic and bactericidal and are critical to defense against infections. Recently, both IgM and IgT cells have been found in the GALT, skin-associated lymphoid tissue (SALT), gill-associated lymphoid tissue (GIALT), and nasopharynx-associated lymphoid tissue (NALT) of trout, a common teleost model species of economic importance (Figure 2.5). However, during infection with gut parasites, IgT B cells undergo the greatest expansion, far outnumbering the IgM B cells. In gills, IgT has been shown to respond to both parasitic and bacterial infections and IgT B cells proliferate locally within the gills in response to infection. IgT is also more abundant in the mucosal secretions of the gut, skin, gill, and nose than in circulation and has been found to bind both mucosal pathogens and bacteria that comprise the normal trout flora. This is analogous to what has been shown for IgA in mammals and further demonstrates the ancient role that the mucosal immune system has played in maintaining homeostatic relationships with microbial flora. Teleost mucosal antibodies are also transported across epithelia via the pIgR transport system despite the absence of a covalent bond between this molecule and the antibodies.

Bony fish do not undergo class switching in the way that it is understood from tetrapods. As in mammals, IgD in teleosts is produced through alternative mRNA splicing with IgM transcripts rather than a true class switch involving DNA deletion. The exons encoding the IgT constant-region domains are unusual in that they are upstream of those encoding IgM/D and seem to

Uninfected control	Infected with *Ceratomyxa shasta*
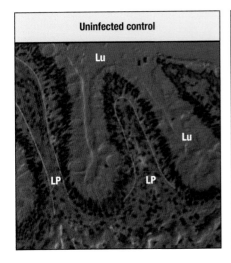	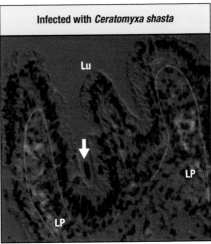

Figure 2.5 IgT-producing B cells in the gut of trout expand in numbers during parasite infection. Immunofluorescence staining of trout gut from uninfected fish (left) and fish infected with the parasite *Ceratomyxa shasta* (right). Cell nuclei are stained with DAPI (blue). IgT⁺ B cells (stained green) have expanded in number in the lamina propria (LP) in infected individuals relative to IgM⁺ B cells (stained red). The arrow points to a parasite located in the gut lumen (Lu). (Reprinted from Yong-An, Z. et al.: *Nature Immunol.* 2010, 11:827–835. With permission from Macmillan Publishers Ltd.)

be expressed with their own unique set of D and J gene segments that are also upstream of the D and J genes used by IgM/D. Because of the organization of the genes and the lack of the class-switch mechanisms, B cells expressing IgT are a separate subset from those expressing IgM/D. In other words, during ontogeny, B cells must make a decision to produce either only IgT or only IgM/D. What determines this cell fate is unknown, and it is possible that it depends on which D and J genes are productively recombined. If the D and J upstream of the IgM/D are used, the IgT constant regions are deleted. If the D and J upstream of IgT are used, the B-cell becomes an IgT-expressing cell.

It is worth emphasizing here that the mucosal antibodies found in bony fish, amphibians, and amniotes (birds, reptiles, and mammals) are not closely related. Analyses of their gene sequences suggest origins for IgT, IgX, and IgA from independent gene duplications. Therefore, the role of these antibodies at the mucosa is presumably the product of convergent evolution due to selective pressures for each species to have at least one class of antibody specialized for mucosal sites and secretions. This is an ancient evolutionary pressure dating back at least 400 million years.

SUMMARY

Not long ago, immunologists were convinced that the "higher" vertebrates such as birds and mammals had a monopoly on adaptive immunity. It is clear that many of the "lower" vertebrates not only are capable of developing diverse antigen receptors and specialized cell types but have also evolved specialized mucosal immune systems. Analyses of the immune systems of these lower vertebrates have supported both the ancient origins of the mucosal immune system as well as its role in the evolution of the primary lymphoid tissues known from mice and humans. The ancient origin of GALT in vertebrates probably illustrates how important a role that defense against pathogens at mucosal sites played in the evolution of the immune system.

FURTHER READING

Amemiya, C.T., Saha, N.R., and Zapata, A.: Evolution and development of immunological structures in the lamprey. *Curr. Opin. Immunol.* 2007, 19:535–541.

Flajnik, M.F.: Comparative analyses of immunoglobulin genes: Surprises and portents. *Nat. Rev. Immunol.* 2002, 2:688–698.

Flajnik, M.F., and Kasahara, M.: Origin and evolution of the adaptive immune system: Genetic events and selective pressures. *Nature Rev. Genet.* 2010, 11:47–59.

Flajnik, M.F.: Re-evaluation of the immunological Big Bang. *Curr. Biol.* 2014, 24:R1060–R1065.

Hofmann, J., Greter, M., Du Pasquier, L., and Becher, B.: B-cells need a proper house, whereas T-cells are happy in a cave: The dependence of lymphocytes on secondary lymphoid tissues during evolution. *Trends Immunol.* 2010, 31:144–153.

Pancer, Z., and Cooper, M.: The evolution of adaptive immunity. *Ann. Rev. Immunol.* 2006, 24:497–518.

Stavnezer, J., and Amemiya, C.T.: Evolution of isotype switching. *Sem. Immunol.* 2004, 16:257–275.

Tacchi, L., Larragoite, E.T., Muñoz, P., Amemiya, C.T., and Salinas I.: African lungfish reveal the evolutionary origins of organized mucosal lymphoid tissue in vertebrates. *Curr. Biol.* 2015, 25:2417–2424.

Woof, J.M., and Mestecky, J.: Mucosal immunoglobulins. *Immunol. Rev.* 2005, 206: 64–82.

Zhang, Y-A., Salinas, I., Li, J., Parra, D., Bjork, S., Xu, Z., LaPatra, S.E., Bartholomew, J., and Sunyer, J.O.: IgT, a primitive immunoglobulin class specialized in mucosal immunity. *Nat. Immunol.* 2010, 11:827–835.

Immunologic and functional differences among individual compartments of the mucosal immune system

3

HIROSHI KIYONO, KOHTARO FUJIHASHI, AND JIRI MESTECKY

The mucosal immune system is present in anatomically and physiologically diverse tissues, including the gastrointestinal tract, nasopharynx, oral cavity, lung, eye, and urogenital tract. Although these compartments share many common features, the mucosal immune systems of the tissues also display distinct characteristics, likely reflecting the anatomical, functional, and environmental requirements at diverse mucosal sites. This chapter discusses the common and unique features of the mucosal immune systems in these organ tissues, focusing on the distribution of mucosal antibodies (Figure 3.1) and immune inductive sites.

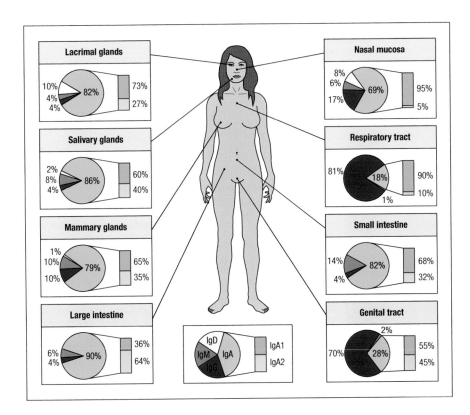

Figure 3.1 **Immunoglobulin (Ig) and IgA subclass antibody distribution in mucosal compartments.** IgA is the dominant Ig isotype in all mucosal secretions, except those of the lower respiratory and genital tracts, where IgG is the major Ig. In humans, two subclasses of IgA—IgA1 and IgA2—are present in different proportions in compartment secretions. IgA1 exceeds IgA2 in all secretions except those of the large intestine.

COMMON AND DISTINCT FEATURES OF MUCOSAL IMMUNE SYSTEM IN DIFFERENT TISSUES

3.1   Ocular immune system

Tears contain relatively high levels of SIgA, the dominant Ig in ocular fluid, with small amounts of monomeric IgA (mIgA) and IgG antibodies. Reflecting the dominance of IgA antibodies in tears, lacrimal glands contain high numbers of IgA-producing cells. The proportion of IgA1-producing cells (53%–81%) relative to IgA2-producing cells (19%–47%) in lacrimal grands corresponds to the proportion of IgA1 to IgA2 in tears. Approximately 10% of the Ig-secreting cells produce IgD of unknown functional importance. Importantly, lacrimal gland acini and ducts express the polymeric Ig receptor (pIgR), a key element in the formation of SIgA and its transport into tears.

The administration of antigen onto ocular mucosa can induce antigen-specific SIgA responses in the ocular and nasal cavities, as well as systemic IgG antibody responses. The tear-duct-associated lymphoid tissues (TALT) in the conjunctival sac are connected via the tear duct to the nasal cavity (Figure 3.2) and play an important role in the generation of these antigen-specific responses. Locally administered antigens enter the tear duct and then the nasal cavity with subsequent antigen sampling by M cells in the follicle-associated epithelium (FAE) covering the apical surface of TALT and nasopharynx-associated lymphoid tissue (NALT). Thus, the ocular mucosal immune system is involved in the induction of immune responses, similar to ingestion- and inhalation-induced responses.

3.2   In oral cavity, SIgA dominates, and sublingual tissue is a potential inductive site

Saliva consists of fluids derived from large salivary (parotid, sublingual, and submandibular) glands, small salivary (labial and buccal) glands, and crevicular fluid. The variable contribution of these tissues and crevicular fluid to the Ig pool in saliva depends on the periodontal and tooth health as well as composition of commensal microbiota of the oral cavity. SIgA is dominant in secretions of all salivary glands, with a composition of approximately 60% IgA1 and 40% IgA2. IgG and IgM are present in small quantities. In contrast, the crevicular fluid contains mainly plasma-derived proteins, and IgG is the dominant Ig isotype. In the oral cavity, mucosal and systemic Ig contributions

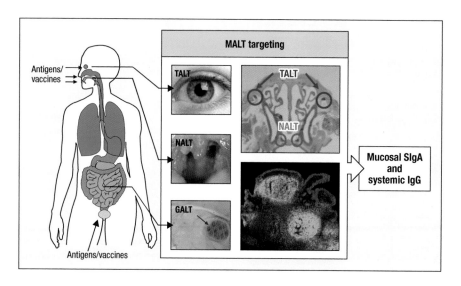

Figure 3.2 Lymphoid structures of mucosa-associated lymphoid tissue (MALT) involved in the induction of antigen-specific immune responses. Tear-duct-associated lymphoid tissue (TALT), nasopharynx-associated lymphoid tissue (NALT), and gut-associated lymphoid tissue (GALT), specifically Peyer's patches (red arrow), induce antigen- and vaccine-stimulated responses, leading to antigen-specific mucosal SIgA and systemic IgG production. (Immunofluorescence panel courtesy of T. Yamanaka and P. Brandtzaeg.)

depend on the state of oral health. In advanced periodontal disease, the proportion of plasma-derived IgG antibodies in the Ig pool in whole saliva increases substantially.

The local application of antigens to the buccal mucosa, labial mucosa, or gingiva stimulates very low antigen-specific immune responses. In contrast, the ingestion of antigens, intranasal immunization, or even rectal immunization induces antigen-specific SIgA-associated responses in saliva. Several studies have implicated the tonsils as a source of the B cells that populate salivary glands. Other studies also indicate that the sublingual application of antigen stimulates immune responses in both systemic and mucosal compartments, including the female genital tract (see Figure 3.2). Thus, among oral cavity tissues, the sublingual (SL) immune apparatus is a potential inductive site, reflected in the ability of vaccine antigen deposited at this site to elicit antigen-specific immune responses in both mucosal and systemic compartments. Oral deposition of vaccine antigen thus may lead to the stimulation of at least three inductive sites, including SL, adenoids/ tonsils, and gut-associated lymphoid tissue (GALT) for the induction of antigen-specific mucosal and systemic immune responses in humans.

3.3   Upper and lower respiratory tract display distinct immunologic features

In nasal secretions, which are a major component of the surface barrier in the upper respiratory tract, IgA constitutes approximately 70% of the Ig pool. Reflecting the dominance of IgA, nasal mucosa contains high numbers of IgA-producing cells, particularly of the IgA1 isotype. Approximately 20% of Ig in nasal secretions is represented by IgG, derived mostly from the circulation with limited local production. Thus, antigen-specific IgG antibodies appear in nasal secretions after systemic immunization. In sharp contrast to the predominance of IgA in nasal secretions, IgG is the major Ig isotype in bronchoalveolar fluid in the lower respiratory tract, contributing approximately 80% of the Igs to lung fluid. The IgG appears to be derived predominantly from the circulation. The dominance of IgA in the upper respiratory tract and IgG in the lower tract is consistent with the mucosal immune system playing the major role in protection of the upper respiratory tract and the systemic immune system playing the main role in the lower tract. For the control of aberrant immune responses associated with the airway tract, such as pollen allergy, SL immunization has been shown to be effective for the desensitization of pollen-induced allergic responses. Thus, the SL immune apparatus is an attractive site for the induction of active immunity and desensitization (or tolerance) against infectious and allergic diseases, respectively.

The respiratory tract is equipped with several organized lymphoid structures that may serve as inductive sites for the initiation of antigen-specific immune responses against inhaled antigens. Intranasal immunization results in the induction of antigen-specific humoral and cellular immune responses in the systemic compartment, as well as in mucosa-associated tissues and secretions at anatomically remote sites such as saliva, intestinal fluid, and female and male genital tract fluids. Lymphoid tissues in Waldeyer's ring in the nasopharynx appear to participate in local and distant immune responses in several species, including humans. In experimental animals, including mice, rats, and rabbits, NALT and bronchus-associated lymphoid tissue (BALT) serve as mucosal inductive sites. BALT was originally identified in rodents, but its presence is controversial in humans; in normal conditions the tissue is usually not present. However, inducible BALT (iBALT) was recently described as an ectopic lymphoid tissue that can develop throughout the lung in both humans and rodents following infection and/or inflammation. These tissues can take up inhaled antigens to induce local immunity and are involved in the maintenance of memory-type cells. These iBALTs have been shown to be involved in allergic inflammatory disorders, since memory-type pathogenic Th2 cells are induced and maintained in iBALTs. Although their anatomical location may differ in humans compared to rodents, such tissues (e.g., NALT) are clearly inductive sites, as they contain antigen-sampling M cells, DCs, Th cell subsets, IgA-committed cells, and cytotoxic T lymphocytes (CTLs) for the induction of antigen-specific humoral and/or cellular immunity.

Unlike Peyer's patch high endothelial venules (HEVs), which are present in T-cell zones, murine NALT HEVs are present in B-cell zones and express peripheral lymph node addressin (PNAd) either alone or associated with mucosal addressin cell adhesion molecule-1 (MAdCAM-1). Antibodies to L-selectin, but not MAdCAM-1, block the binding of naïve lymphocytes to NALT HEVs, suggesting a role for L-selectin and PNAd in the binding of naïve lymphocytes to these HEVs. Early induction of vascular cell adhesion protein-1 (VCAM-1), E-selectin, and P-selectin in the pulmonary vasculature was reported during pulmonary immune responses, with initial increased expression of P-selectin ligand by peripheral blood CD4$^+$ and CD8$^+$ T cells. The number of cells expressing P-selectin ligand subsequently declined in the blood as the cells accumulated in the bronchoalveolar lavage fluid. The very late antigen-4 (VLA-4) could be an important adhesion molecule involved in the migration of activated T cells into the lung, since migration of VLA-4$^+$ cells into bronchoalveolar fluids is impaired after treatment with anti-α_4 antibody. Antigen-specific L-selectin^{low} CTL effectors also have been shown to rapidly accumulate in the lung after adoptive transfer to naive mice with reduced pulmonary viral titers during early infection. Analysis of tissue-specific adhesion molecules showed that after systemic immunization, most effector B cells expressed L-selectin, and only a few cells expressed the gut homing molecule $\alpha_4\beta_7$, whereas after enteric (oral or rectal) immunization, the opposite occurred. Interestingly, effector B cells induced by nasal immunization displayed a more promiscuous pattern of adhesion molecules, with a large majority expressing both L-selectin and $\alpha_4\beta_7$. Thus, the upper respiratory immune system is an effective site for the delivery of vaccine antigen, such as through nasal and/or oral spray that targets the NALT.

3.4   Upper and lower intestinal tracts are major source of immunoglobulin and major site for induction of immune responses

The largest proportion of Ig-producing cells in the body (70%–80%), especially those secreting IgA, is present in the gastrointestinal tract mucosa. The majority of these cells are J-chain-positive and their product, pIgA, is transported into the lumen as SIgA by a pIgR-mediated mechanism through the intestinal epithelium. Approximately 3 g of SIgA (total production of IgA ∼5 g/day) enters the intestinal lumen each day. In some species, including rodents and rabbits, but not humans, the pIgR expressed on surfaces of hepatocytes is involved in the efficient transport of pIgA from the circulation into bile and thus contributes to intestinal SIgA. In humans, the dominant Ig isotype in bile is IgG. The transport of IgG antibodies by FcRn also contributes limited quantities to the intestinal secretion of IgG. IgA1-producing cells are dominant in human gastric and small intestinal mucosa. However, in contrast to the predominance of IgA1 in all other mucosal secretions, the proportion of IgA2-secreting cells in the large intestine exceeds that of IgA1. IgA2-producing cells are dominant in the large intestine, possibly related to the large numbers of colonizing gram-negative bacteria, which appear to stimulate mainly IgA2. Unlike IgA1, IgA2 is resistant to most bacterial IgA proteases, and the IgA2 subclass may be a critical component of antibody-mediated humoral immunity in the lower digestive tract.

Oral ingestion and rectal administration of antigen stimulate immune responses at the site of antigen uptake, as well as in extraintestinal mucosal sites, including the lacrimal, salivary, and lactating mammary glands, and the female genital tract (see **Figure 3.2**). Tissues that participate in immune response induction include Peyer's patches in the small intestine, scattered solitary lymphoid follicles, crypt patches, and accumulations of Peyer's patch–like structures in the colon and terminal large intestine called colonic patches and rectal tonsils, respectively. Among the GALT in both small and large intestines, Peyer's patches have been extensively characterized as an important inductive site for the initiation of antigen-specific humoral (e.g., SIgA) and cell-mediated (e.g., CTL) immunity. The dome-shaped Peyer's patches are covered by FAE-containing M cells that participate in sampling mucosally encountered antigens. Beneath the FAE, antigen-processing and -presenting DCs are located for the immediate capture, processing, and presentation of antigen taken up by M cells to initiate antigen-specific immune responses. The Peyer's patches also exhibit several B-cell follicles containing germinal centers where isotype switching from IgM to IgA occurs under the influence of follicular helper T (Tfh) cells, leading to the generation of IgA committed B (Bα) cells. The follicles are surrounded by the T-cell–enriched areas for the generation of Th1, Th2, Th17, Treg, and CTL responses. In addition to the induction of antigen-specific Th cells and Bα cells initiated by the M cell–DC antigen capture and presentation axis, Peyer's patch DCs educate the antigen-specific lymphocytes to acquire gut-imprinting molecules (e.g., CCR9 and $\alpha_4\beta_7$) via vitamin A, retinoic acids. Thus, Peyer's patches initiate antigen-specific immune responses to orally administered vaccines with the resultant B cells dispatched to distant effector sites, including the lamina propria of the intestine for the formation and production of SIgA antibodies. Importantly, oral and rectal immunization induce antigen-specific immune responses of variable strength at respective gut mucosal sites. For example, oral immunization elicits higher antigen-specific SIgA antibody responses in the proximal small intestine when compared with those in the distal colon, whereas rectal immunization is favor for the induction of specific immunity in the colorectal tissues, than that in the jejunum and duodenum.

3.5 Mammary glands are an important source of SIgA

Early milk, called colostrum, and milk collected at later stages of lactation contains high levels of SIgA and small amounts of mIgA, secretory IgM (SIgM), and IgG, which provide humoral protective immunity in newborn babies. The subclass distribution varies among donors, and on average, the proportion of IgA1 slightly exceeds that of IgA2 and thus resembles the distribution of IgA subclasses in the adult small intestine. Antigen-specific humoral immune responses can be induced in mammary glands by the oral administration of antigens; the effectiveness of intranasal, rectal, or sublingual immunization routes in the induction of SIgA antibodies in milk has not been evaluated in humans. The spectrum of SIgA antibodies in human milk reflects maternal exposure to orally and intestinally encountered antigens and thus provides an appropriate passive protection for breastfeeding infants. Further, human milk antibodies have been shown to influence the development of commensal microbiota in infants.

3.6 IgG is major Ig in genitourinary tract secretions

Similar to lower respiratory tract secretions (bronchoalveolar fluid), the dominant Ig isotype in secretions of the female and male genital tract and urine is IgG. Thus, IgA is the major Ig isotype in all mucosal secretions except the lower respiratory and female and male urogenital tracts. In females, the uterus is the most important source of Ig in cervicovaginal secretions, which explains the profound decrease in genital Igs after hysterectomy. IgG is derived locally from numerous antibody-forming cells in the uterine endocervix and from the circulation. Consistent with the systemic source of the IgG, systemic immunization induces IgG in genital fluids. In this regard, the mucosal immune system of the reproductive tract is distinct from that of the digestive tract. Genital tract IgA, represented by SIgA, pIgA without secretory component, and mIgA, also originates from both local and systemic sources. In the female genital tract secretions, IgA2 levels slightly exceed those of IgA1. The levels of Igs of all isotypes vary with the stage of the menstrual cycle, with the highest levels present during menstruation and shortly before ovulation and the lowest levels present during and a few days following ovulation. Consequently, for the induction of protective humoral immunity in female reproductive tissues, the immunologic effect of the menstrual cycle must be considered in vaccine development.

In the female genital tract, humoral immune responses to sexually transmitted pathogens or intravaginally inoculated antigens are generally low. However, the introduction of antigens in the upper female genital tract, as demonstrated in animal experiments, induces measurable antigen-specific Ig responses. Importantly, immunization through the intranasal, sublingual, rectal, or oral routes also stimulates IgA responses in female genital secretions. Among those different mucosal antigen-delivery routes, intranasal immunization has been shown to be a potent path for the generation of antigen-specific immunity in the reproductive tissue. An important difference between the genital tract and other mucosal tissues such as those of the gastrointestinal tract is the striking absence of M cells and inductive sites in the genital tract. Consequently, a considerable proportion of local IgG is derived from the circulation, and systemic immunization leads to the local appearance of antigen-specific IgG antibodies in the genital secretions. An FcγRn expressed on epithelial cells in the female genital tract mucosa is involved in the transport of this IgG.

In the male genital tract, IgG is the dominant isotype in semen. IgA is represented by approximately equal amounts of SIgA, pIgA devoid of secretory component, and mIgA. The ratio of IgA1 to IgA2 in semen resembles that in serum. Systemic and oral immunizations induce easily detectable IgG and IgA responses in semen; other immunization routes, including the intranasal and rectal routes, have not been explored.

Human urine contains low levels of IgG and even lower levels of IgA as SIgA, pIgA, or mIgA. The origin of urinary SIgA has not been firmly established. Ascendent bacterial infections induce urinary and systemic humoral responses, suggesting that urine may contain both local and systemic antibodies.

Antigen sampling in mucosal surfaces

One of the unique characteristics of the mucosal immune system is the presence of antigen-sampling mechanisms in certain mucosal tissues. The initiation of antigen-specific immune responses in inductive sites such as GALT (Peyer's patches), NALT, and TALT begins after the sampling of ingested, inhaled, or eye-dropped antigens by M cells located in the FAE covering those inductive tissues. In addition to the Peyer's patch M cells, similar antigen-sampling M cells, termed "villous M cells," have been identified in the villous epithelium of mice without underlying organized lymphoid structures, suggesting an alternative antigen-sampling pathway (**Figure 3.3**). These villous M cells generally share the morphologic, immunologic, and functional characteristics of Peyer's patch M cells. Whether villous M cells contribute to the induction of SIgA or systemic IgG is not known. Although clearly identified

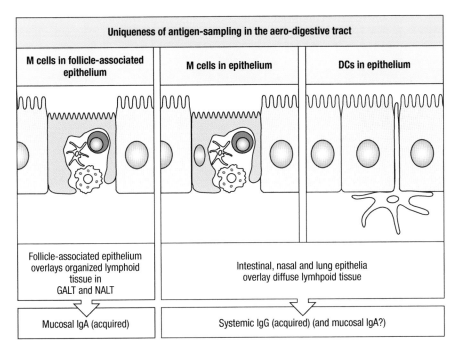

Uniqueness of antigen-sampling in the aero-digestive tract

M cells in follicle-associated epithelium	M cells in epithelium	DCs in epithelium
Follicle-associated epithelium overlays organized lymphoid tissue in GALT and NALT	Intestinal, nasal and lung epithelia overlay diffuse lymphoid tissue	
Mucosal IgA (acquired)	Systemic IgG (acquired) (and mucosal IgA?)	

Figure 3.3 **Antigen sampling in the digestive, nasal, and respiratory tracts.** Mucosa-associated lymphoid tissue (MALT)-dependent M cells are located in the follicle-associated epithelium and overlay organized lymphoid structures in gut-associated lymphoid tissue (GALT) and nasopharynx-associated lymphoid tissue (NALT). MALT-independent M cells (no underlying organized lymphoid structures) are present in villous epithelium, nasal cavity, and lung. Dendritic cells (DCs) are present in the intestinal and respiratory mucosa, where they extend dendrites between epithelial cells into the lumen to sample antigens. MALT-dependent antigen sampling is thought to induce antigen-specific local IgA antibody responses, whereas MALT-independent antigen sampling is thought to induce antigen-specific serum IgG responses.

in mice and nonhuman primates, villous M cells have not yet been detected in human small intestinal mucosa. However, recent studies suggest that these M cells can be investigated using primary human organoids generated by stem cell organogenesis.

In addition to M cells, intestinal DCs sample luminal antigen encountered at the mucosal surface. Studies performed in the mouse have shown that lamina propria DCs extend their dendrites between epithelial cell junctions into the lumen to take up antigenic material. These DCs are capable of initiating antigen-specific, systemic IgG responses, whereas antigen uptake and transport by M cells in Peyer's patches initiate local IgA production.

M cells also have been identified in the respiratory tract. In the upper respiratory tract, M cells present in the nasal mucosa (NALT M cells) are capable of taking up both particulate and soluble antigens from the nasal mucosa. In addition, similar to the intestinal villous M cells, respiratory M cells have been identified in the nasal mucosa away from NALT epithelium (e.g., nasal turbinate). M cells also are present in the airway mucosa from first bronchial bifurcation to the bronchioles, as shown in mice. These M cells serve as an alternate to alveolar macrophages as a route by which pathogens such as *Mycobacterium tuberculosis* gain entry to the mucosa. Analogous to intestinal intraepithelial DCs, some airway DCs extend their dendritic processes into the lumen of the respiratory tract for antigen sampling. Taken together, the nasal and digestive mucosal immune systems have an array of antigen-uptake and presenting systems that, depending on the tissue site and species, include Peyer's patch, NALT, and TALT M cells, as well as villous and respiratory M cells, and intraepithelial DCs (see Figure 3.3).

B-cell subsets in nasal and gut-associated tissues

At least two B-cell subsets, B-1 and B-2, are involved in the mucosal immune system. IgA-producing cells, which in mice originate from both B-1 and B-2 B cells, are present in the nasal and intestinal lamina propria. B-1 B cells in the murine peritoneal cavity and intestinal lamina propria develop from a common B-cell pool and may represent a lineage separate from that of conventional Peyer's patch B cells. Most Peyer's patch B cells belong to the B-2 cell family; similarly, NALT consists of B-2 B cells. Murine intestinal IgA antibodies with specificity for commensal bacteria are produced by B-1

B cells in a T-cell–independent manner. Based on the murine system, B-1 B cells appear to be an important source of IgA-producing cells in the intestinal mucosa, in addition to the GALT B-2 linage of IgA plasma cells. However, in humans, the frequency of B-1 B cells is very low in the intestinal mucosa. Whether human B-1 B cells contribute to IgA1 and/or IgA2 subclass antibody production remains unknown.

MALT organogenesis

MALT consists of the major organized lymphoid structures and inductive sites in the lacrimal, respiratory, and digestive tracts. MALT shares common immunologic characteristics and functions as the initiation site for antigen-specific humoral and cell-mediated immune responses. For the organogenesis of the three well-characterized MALTs, the cellular and molecular tissue genesis appears to be different (Figure 3.4), although they share some common characteristics of the organogenesis process including the necessity of CD3⁻CD4⁺CD45⁺ lymphoid tissue inducer (LTi) cells (or group 3 innate lymphoid cells: ILC3) for the initiation of MALT formation. For the development of LTi cells, transcriptional regulators of core binding factor (CBF) β2 are required, and the lack of CBFβ2 resulted in the deficiency of LTi cells leading to the absence of Peyer's patches, NALT, and TALT structure development.

Chronological examination of Peyer's patches, NALT, and TALT during development has revealed that organogenesis of most of the secondary lymphoid tissues, including Peyer's patches and peripheral lymph nodes, is initiated during the prenatal stage of development. In contrast, the genesis of NALT and TALT begins after birth. In addition to these differences, the requirement for essential organogenesis molecules also differs among the three MALTs. The critical molecules required for the organogenesis of Peyer's patches and other peripheral lymphoid tissues are dispensable for

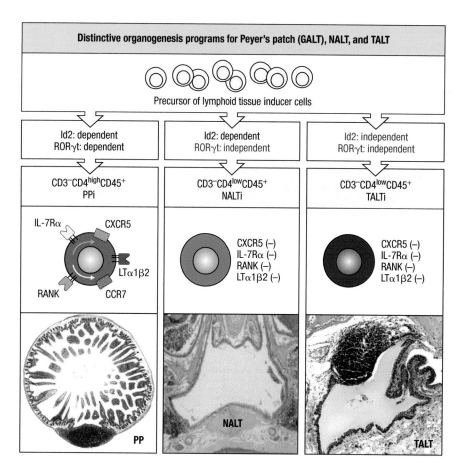

Figure 3.4 **Organogenesis programs for Peyer's patches (PP), nasopharynx-associated lymphoid tissue (NALT), and tear-duct-associated lymphoid tissue (TALT).** Despite their common features, PP, NALT, and TALT undergo separate organogenesis pathways regulated by the indicated transcriptional factors, integrins, and receptors. NALTi, NALT inducer; PPi, PP inducer; TALTi, TALT inducer.

the development of NALT. In mice, the transcription factor retinoic orphan receptor RORγt is required for the development of Peyer's patches but not for NALT or TALT. Similarly, neither the IL-7/IL-7Rα cytokine family nor the LTα₁β₂–LTβR signaling pathways, which are involved in the genesis of Peyer's patches and peripheral lymph nodes, are essential for NALT and TALT organogenesis. Interestingly, even Id2, another well-known transcriptional molecule for lymphoid tissue genesis, is essential for the development of Peyer's patches, peripheral lymph nodes, and NALT but dispensable for TALT genesis. Thus, at least three distinct tissue genesis programs operate in the development of MALTs associated with the intestine, respiratory tract, and tear duct: the organogenesis of Peyer's patches during gestation and that of NALT and TALT beginning during the postnatal stage (see Figure 3.4).

Influence of aging on mucosal immunity

In addition to the developmental differences in NALT, TALT, and GALT (Peyer's patches), it is not surprising that age-associated differences in these mucosal tissues also occur (Figure 3.5). In this connection, nasal, but not

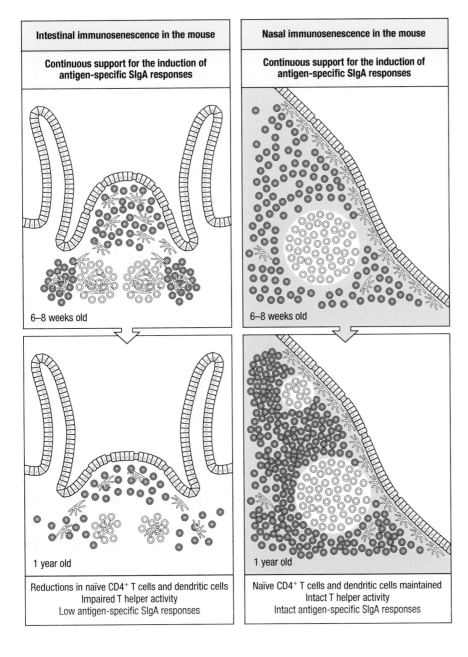

Figure 3.5 **Effect of aging on nasopharynx-associated lymphoid tissue (NALT) and gut-associated lymphoid tissue (GALT).** In the intestine, SIgA responses decrease during aging, at least in mice; this is reflected in the reduced size of Peyer's patches. Smaller numbers of naïve CD4+ T cells and follicular DCs also occur in the Peyer's patches in aging mice. In contrast, in the nasal cavity, specific SIgA (and systemic IgG) antibody responses in 1-year-old (aging) mice are similar to those in young adult (6- to 8-week-old) mice. The absolute numbers of naïve CD4+ T cells in the NALT are maintained in aging mice.

oral, immunization induced normal levels of antigen-specific immune responses in aging mice, supporting the notion that age influences GALT more than NALT. Reduced frequencies of CD4$^+$CD45RB$^+$ T cells are present in aging mice; however, the number of naïve CD4$^+$ T cells in the NALT of aging mice is higher than that in young adult mice. These findings suggest that a distinct immune aging process occurs in GALT compared to NALT, possibly contributing to differences in the induction of antigen-specific mucosal IgA and parenteral IgG responses in advanced age. Thus, nasopharynx administration may be a more effective route for the induction of antigen-specific mucosal and systemic immune responses in the elderly.

SUMMARY

The immune systems in the different mucosal tissues share a common mission: first-line defense against the entry of pathogens and undesired antigens and allergens. To accomplish this mission, mucosal surfaces have developed tissue-specific sites for the induction of antigen-specific SIgA antibody and/or cell-mediated immune responses. Locally produced and serum-derived IgG antibodies contribute immunologically to mucosal immune defense in certain compartments, including the oral cavity, lower respiratory, and genitourinary tissues. Distinct anatomic and immunologic characteristics in the various mucosal compartments influence the development and maintenance of the common and unique immunologic features of the mucosal immune system.

FURTHER READING

Brandtzaeg, P.: Regionalized immune function of tonsils and adenoids. *Immunol. Today* 1999, 20:383–384.

Eberl, G., Marmon, S., Sunshine, M.J., Rennert, P.D., Choi, Y., and Littman, D.R.: An essential function for the nuclear receptor RORγ(t) in the generation of fetal lymphoid tissue inducer cells. *Nat. Immunol.* 2004, 5:64–73.

Fujihashi, K., and Kiyono, H.: Mucosal immunosenescence: New developments and vaccines to control infectious diseases. *Trends Immunol.* 2009, 30:334–343.

Iwata, M., Hirakiyama, A., Eshima, Y., Kagechika, H., Kato, C., and Song, S.Y.: Retinoic acid imprints gut-homing specificity on T cells. *Immunity* 2004, 21:527–538.

Jang, M.H., Kweon, M.N., Iwatani, K. et al.: Intestinal villous M cells: An antigen entry site in the mucosal epithelium. *Proc. Natl. Acad. Sci. (USA)* 2004, 101:6110–6115.

Kiyono, H., and Fukuyama, S.: NALT- versus Peyer's-patch-mediated mucosal immunity. *Nat. Rev. Immunol.* 2004, 4:699–710.

Macpherson, A.J., Gatto, D., Sainsbury, E., Harriman, G.R., Hengartner, H., and Zinkernagel, R.M.: A primitive T cell-independent mechanism of intestinal mucosal IgA responses to commensal bacteria. *Science* 2000, 288:2222–2226.

Mestecky, J., and McGhee, J.R.: Immunoglobulin A (IgA): Molecular and cellular interactions involved in IgA biosynthesis and immune response. *Adv. Immunol.* 1987, 40:153–245.

Mestecky, J., Raska, M., Novak, J., Alexander, R.C., and Moldoveanu, Z.: Antibody-mediated protection and the mucosal immune system of the genital tract: Relevance to vaccine design. *J. Reprod. Immunol.* 2010, 85:81–85.

Nagatake, T., Fukuyama, S., Kim, D.Y. et al.: Id2-, RORγt-, and LTβR-independent initiation of lymphoid organogenesis in ocular immunity. *J. Exp. Med.* 2009, 206:2351–2364.

Nagatake, T., Fukuyama, S., Sato, S. et al.: Central role of core binding factor β2 in mucosa-associate lymphoid tissue organogenesis in mouse. *PLOS ONE* 2015, 10:e0127460.

Randall, T.D.: Bronchus-associated lymphoid tissue (BALT) structure and function. *Adv. Immunol.* 2010, 107:187–241.

Rescigno, M., Urbano, M., Valzasina, B., Francolini, M., Rotta, G., Bonasio, R., Granucci, F., Kraehenbuhl, J.P., and Ricciardi-Castagnoli, P.: Dendritic cells express tight junction proteins and penetrate gut epithelial monolayers to sample bacteria. *Nat. Immunol.* 2001, 2:361–367.

Shinoda, K., Hirahara, K., Iinuma, T. et al.: Thy1$^+$ IL-7$^+$ lymphatic endothelial cells in iBALT provide a survival niche for memory T-helper cells in allergic airway inflammation. *Proc. Natl. Acad. Sci. (USA)* 2016, 94:5267–5272.

Song, J.H., Nguyen, H.H., Cuburu, N., Horimoto, T., Ko, S.Y., Park, S.H., Czerkinsky, C., and Kweon, M.N.: Sublingual vaccination with influenza virus protects mice against lethal viral infection. *Proc. Natl. Acad. Sci. (USA)* 2008, 105:1644–1649.

van de Pavert, S.A., and Mebius, R.E.: New insights into the development of lymphoid tissues. *Nat. Rev. Immunol.* 2010, 10:664–674.

Yamamoto, S., Kiyono, H., Yamamoto, M., Imaoka, K., Fujihashi, K., Van Ginkel, F. W., Noda, M., Takeda, Y., and McGhee, J.R.: A nontoxic mutant of cholera toxin elicits Th2-type responses for enhanced mucosal immunity. *Proc. Natl. Acad. Sci. (USA)* 1997, 113:E2842–E2851.

Secreted effectors of the innate mucosal barrier

MICHAEL A. MCGUCKIN, ANDRE J. OUELLETTE, AND GARY D. WU

4

The mucosal surfaces, including the surface of the eye and the linings of the gastrointestinal, respiratory, reproductive, and urinary tracts, are the major interface between mammalian tissues and the potentially hostile external environment. Mucosal surfaces have evolved a well-regulated barrier to protect against chemical, physical, and microbial insults. In the intestinal tract, this barrier consists of epithelial cells, a secreted layer of mucus produced by goblet cells, and antimicrobial peptides released by the gastric and colonic epithelium and by small intestinal Paneth cells (Figure 4.1). In this chapter, we describe the components of this barrier and their regulation by innate and adaptive immunity, focusing mainly on the intestinal tract because it is the most thoroughly studied mucosal tissue and is continuously exposed to potential infection.

The small intestinal epithelium absorbs nutrients, regulates water and electrolytes, and helps prevent most luminal microbes from becoming significant populations. To facilitate nutrient absorption in the small intestine, mammals maximized absorptive membrane surface area by evolving villi and an apical brush border. However, the lumen from which nutrients are absorbed is also colonized by a resident microflora and transiently exposed to microbes and foreign antigens from the diet. Although the large absorptive area of the small intestine appears to favor mucosal colonization by microorganisms, few bacteria populate the small intestine relative to populations in the cecum and colon, suggesting that the small intestine has mechanisms that counter microbial colonization of the epithelium.

Many factors influence the growth of bacteria in the small intestine. For example, extensive energy is invested by villus enterocytes in the transcytosis of antigen-specific secretory IgA from the circulation to the small intestinal lumen. Other factors including gastric acidity, digestive enzymes, bile salts, peristalsis, the resident commensal microflora, exfoliation of enterocytes during epithelial renewal, and CD8$^+$ intraepithelial T lymphocytes contribute to mucosal health. In addition to adaptive immune responses and the potential bystander effects resulting from gastrointestinal physiologic processes, innate host defense mechanisms confer immediate biochemical protection against infection and colonization by pathogenic or opportunistic microorganisms. The secretory products of specific epithelial lineages, including mucins and antimicrobial peptides, are primary mediators of this biochemical barrier.

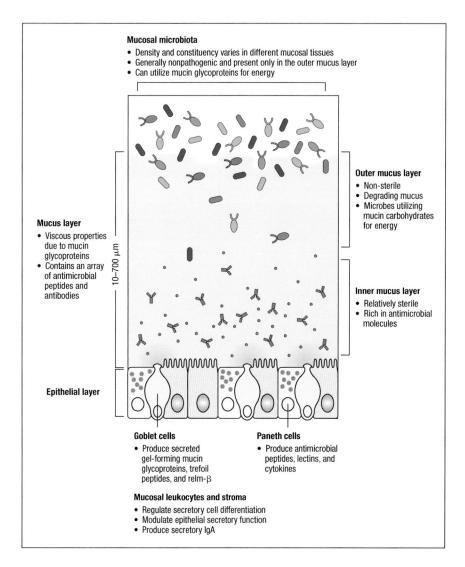

Mucosal microbiota
- Density and constituency varies in different mucosal tissues
- Generally nonpathogenic and present only in the outer mucus layer
- Can utilize mucin glycoproteins for energy

Outer mucus layer
- Non-sterile
- Degrading mucus
- Microbes utilizing mucin carbohydrates for energy

Mucus layer
- Viscous properties due to mucin glycoproteins
- Contains an array of antimicrobial peptides and antibodies

10–700 μm

Inner mucus layer
- Relatively sterile
- Rich in antimicrobial molecules

Epithelial layer

Goblet cells
- Produce secreted gel-forming mucin glycoproteins, trefoil peptides, and relm-β

Paneth cells
- Produce antimicrobial peptides, lectins, and cytokines

Mucosal leukocytes and stroma
- Regulate secretory cell differentiation
- Modulate epithelial secretory function
- Produce secretory IgA

Figure 4.1 The mucosal barrier. The secreted mucosal barrier showing the specialized secretory epithelial cells and the spatial relationships between the epithelial cells, the secreted mucus layer, and the microbes in the external environment.

INTESTINAL EPITHELIUM: STEM CELLS, SELF-RENEWAL, AND CELL LINEAGE ALLOCATION—NEW ADVANCES AND PRACTICAL APPLICATIONS

The intestinal epithelial barrier, along with specialized cells responsible for the secretion of mucus, antimicrobial peptides, and enteric hormones, is constantly renewed every 4–5 days. The primary function of the intestinal tract is the digestion and absorption of nutrients, electrolytes, and water. In the small intestine, the crypt compartment contains undifferentiated, proliferating progenitor cells, while the villi are populated by nonproliferating cells with specialized functions. Stem cells located in the base of the small intestinal crypt give rise to four differentiated cell types that populate the villus—absorptive enterocytes, enteroendocrine cells, tuft cells, and goblet cells—and a single differentiated cell type at the base of the crypt, Paneth cells. In specific regions of the intestine, namely in the terminal ileum, progenitor cells differentiate into specialized M cells that overlay Peyer's patches and play a role in internalizing luminal antigens for presentation to lymphocytes. Although villi are absent in the colon, similar compartmentalization exists with proliferating cells located in the lower two-thirds of the crypt and differentiated cells found higher in the crypt and at the epithelial surface.

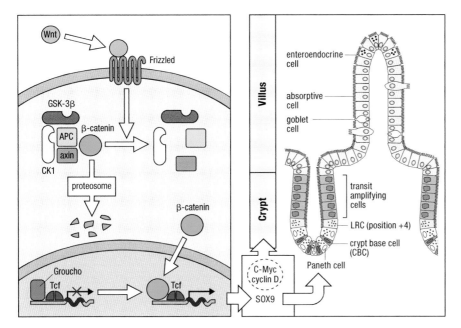

Figure 4.2 The role of the Wnt signaling pathway in the regulation of epithelial proliferation and differentiation in the small intestine. Components of the canonical Wnt signaling pathway are shown on the left. Activation of Wnt signaling by the binding of Wnt ligands to Frizzled receptors ultimately activates genes that drive cellular proliferation that is important for the maintenance of the intestinal stem cell and crypt progenitor compartment. Wnt signaling also activates the expression of SOX9, a transcription factor critical for the development of Paneth cells. CK1, casein kinase 1; GSK-3β, glycogen synthase kinase 3β; LRC, label-retaining cell. (Reprinted from Pitari, G. et al. *Clin. Pharmacol. Ther.* 2007, 82:441–447. With permission from Macmillan Publishers Ltd.)

Molecular mechanisms that regulate intestinal epithelial homeostasis include defining the stem cell niche, regulating epithelial cell proliferation in the intestinal crypt, and epithelial cell differentiation and lineage allocation (Figure 4.2).

All intestinal epithelial cell lineages arise from stem cell progenitors located near the base of the intestinal crypts in proximity to Paneth cells. These cells, termed "crypt-based columnar cells" (CBCs), have been proposed to be intestinal stem cells by Clevers' group in 2007. Lineage tracing of LGR5/GPR49 expressing cells has shown that CBCs are (1) multipotent for all terminally differentiated intestinal epithelial cell types, (2) long-lived, (3) self-renewing, and (4) radiation resistant, all features characteristic of stem cells. The notion that CBCs represent the true intestinal stem cells is contrary to the classic model of the stem cells being located at position +4, initially proposed by Cheng and Lelond in 1974, relative to the crypt bottom where Paneth cells occupy the first three positions. These cells were characterized as DNA label-retaining cells (+4LRCs). Although additional investigation will be required to fully understand the relationship between CBCs and +4LRCs, current evidence favors the notion that CBCs represent the intestinal stem cell compartment. Single Lgr5-positive CBCs are capable of building self-organizing crypt-villus organoids *in vitro* in the absence of a *n*-epithelial cellular niche. Nevertheless, there is some evidence that this niche is occupied by early differentiating secretory cells suggesting that there is some cell plasticity involved in crypt cell homeostasis leading to heterogeneity of cells with stem cell features.

4.1 Wnt signaling regulates intestinal epithelial cell positioning and differentiation

Intestinal stem cells give rise to rapidly proliferating "transit amplifying" cells that generate progeny that then migrate up the crypt compartment, producing the terminally differentiated cells that constitute the villus epithelium. The Wnt pathway is the primary driving force regulating the proliferation of the intestinal epithelium (see Figure 4.2). Indeed, *Lgr5* was initially identified as a potential intestinal stem cell marker because of its inclusion as a target of the Wnt signaling pathway. In the absence of Wnt activation, β-catenin is targeted to undergo proteasomal degradation by sequential phosphorylation at the

N-terminus through its interaction with a degradation complex consisting of axin, adenomatous polyposis coli, glycogen synthase kinase 3β, and casein kinase 1. The canonical Wnt pathway, activated by interactions between Frizzled and low-density lipoprotein receptor-related protein receptors with one of the many different Wnt ligands, results in disruption of the β-catenin destruction complex. The subsequently accumulating cytosolic β-catenin translocates to the nucleus and binds to TCF/LEF transcription factors. The newly formed active transcriptional complexes displace transcriptional repressors such as Groucho, enabling the transcription of Wnt target genes, many of which have roles in cellular proliferation, such as *c-myc* and *CCND1*.

The Wnt pathway regulates cellular differentiation of the intestinal epithelium in several ways. First, the β-catenin/TCF transcriptional complex activates genes expressing Ephrin-B ligands and their receptors EphB, which have a role in establishing migratory pathways as well as maintaining cellular boundaries. The gradient of EphB receptor expression within the crypt maintains the correct cellular architecture with CBCs and Paneth cells located at the base of the crypt compartment. Deletion of EphB3 results in aberrant Paneth cell localization. Second, signaling through the Wnt pathway is required for Paneth cell maturation and their expression of antimicrobial peptides. Furthermore, the Wnt-dependent *SOX9* gene is critical for Paneth cell development, because mice with a conditional deletion of *Sox9* lack Paneth cells completely. Finally, Wnt signaling may have an effect on cell fate determination by impeding terminal differentiation of secretory cell lineages. TCF4$^{-/-}$ mice have goblet cells and enterocytes but lack enteroendocrine cells, whereas transgenic overexpression of Dkk1, an inhibitor of Wnt signaling, ablates the secretory cell lineages while absorptive enterocytes remain normal.

4.2 Notch signaling determines cell lineage specification in the intestine

Notch signaling, which controls cell fate decisions in many different tissues, also determines secretory (goblet, enteroendocrine, tuft and Paneth cells) versus absorptive cell lineage development in the intestinal epithelium (**Figure 4.3**). Activation of one of four Notch receptors by any one of several ligands, from either the Delta or Jagged/Serrate families, results in the proteolytic cleavage and liberation of the Notch intracellular domain from

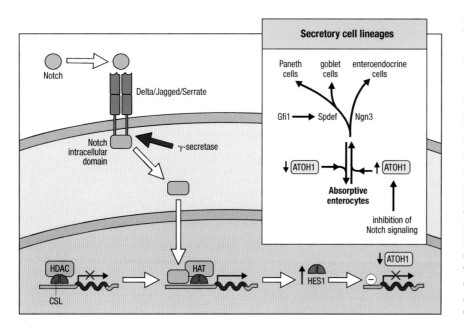

Figure 4.3 The Notch signaling pathway determines cell fate decisions in the intestinal epithelium. Activation of Delta, Jagged, or Serrate Notch receptors results in the release of the Notch intracellular domain by γ-secretase. Subsequent induction of HES1 expression, a basic helix-loop-helix (bHLH) transcriptional repressor, inhibits ATOH1, another bHLH factor, resulting in the suppression of secretory cell lineages and enhancement of absorptive enterocyte development. By contrast, inhibition of Notch signaling enhances the expression of ATOH1, resulting in the development of secretory cell lineages. Gfi1, a zinc finger transcriptional repressor, induces the expression of Spdef, an Ets-domain transcription factor that is important in the development of both goblet and Paneth cells. Similarly, inhibition of Notch leads to the induction of neurogenin 3, a bHLH transcription factor, promoting the development of enteroendocrine cells. HAT, histone acetyltransferase; HDAC, histone deacetylase; Ngn3, Neurogenin 3.

the plasma membrane by γ-secretase. The Notch intracellular domain subsequently translocates into the nucleus where it forms a transcriptional activation complex with RBP-jk (also known as CSL), displacing histone deacetylase corepressors and recruiting histone acetyltransferases, leading to transcriptional activation of Notch target genes such as hairy/enhancer of split (*HES*). HES1, a basic helix-loop-helix (bHLH) transcriptional repressor, inhibits the expression of another bHLH factor, ATOH1 (Math1 for mouse, Hath1 for human), suppressing secretory cell lineage differentiation in the intestinal epithelium and leading to disproportionate numbers of absorptive enterocytes. The intestinal epithelium of Math1 knockout mice is populated only by absorptive enterocytes. By contrast, the inhibition of Notch signaling by CSL gene knockout, or with γ-secretase inhibitors, induces Math1 expression, leading to the conversion of all epithelial cells into goblet cells. HES1 knockout mice, although embryonic lethal, show an increase in goblet, enteroendocrine, and Paneth cells, with decreased numbers of absorptive enterocytes. Likely downstream of Math1 is the zinc finger transcriptional repressor, Gfi1, which is involved in the generation of Paneth and goblet, but not enteroendocrine, cells. Interestingly, the loss of Kruppel-like factor 4 (Klf4), a zinc finger transcription factor that is repressed by Notch signaling, leads to reduced goblet cell numbers.

Spdef, an Ets-domain transcription factor and Notch pathway target gene downstream of Gfi1, plays an important role in the maturation of both goblet and Paneth cell lineages. Spdef is also the major driver of goblet cell phenotype in the respiratory tract. Spdef knockout mice show not only defects in the terminal differentiation of both goblet and Paneth cells but also alterations in the expression of a specific subset of Paneth cell genes including α-defensins, matrix metalloproteinase 7 (Mmp7), angiogenin-4 (Ang4), and kallikreins. In addition, Spdef induces the transcription of a suite of genes encoding proteins required for synthesis of the major components of the mucus barrier. Spdef regulates production of transcription factors such as Fox3A, mucin proteins, the endoplasmic reticulum (ER)–resident protein disulfide isomerase AGR2 required for correct mucin folding, multiple Golgi-resident glycosyltransferases required for complex mucin *O*-glycan assembly, proteins involved in mucin packaging and secretion, and several nonmucin proteins that are co-secreted with mucins.

Enteroendocrine cells account for approximately 1% of intestinal epithelial cells with about 15 different subtypes that express various enteric hormones. Downstream of the Notch–Math1 pathway is the bHLH transcription factor Neurogenin 3, which is required for development of the enteroendocrine cell lineage. Downstream of Neurogenin 3, additional bHLH and homeodomain transcription factors have been shown to be important for the development of specific enteroendocrine subtypes. Another relatively infrequent differentiated epithelial cell type is tuft cells that have a unique morphology with long and thick microvilli projecting into the lumen. These cells share morphologic and functional features with taste-receptor cells that express the cation channel Trpm5 and play a role in type 2 immune responses initiated by protozoa and helminth parasite infections, where they are the source of IL-25.

4.3 Intestinal epithelial cell differentiation and response to microbial short-chain fatty acids

Fermentation by the gut microbiota leads to the production of short-chain fatty acids (SCFAs) such as butyrate that can achieve millimolar concentrations in the colonic lumen, where it is the primary source of energy for the colonic epithelium. Butyrate is not only a substrate for energy production but also has epigenetic effects through its ability to inhibit histone deacetylases (HDAC) leading to histone hyperacetylation and alterations in gene expression. Fatty acid oxidation of butyrate by the differentiated superficial epithelium plays an important role in maintaining mucosal homeostasis via several mechanisms. First, mitochondrial β-oxidation of butyrate by the differentiated surface cells of the colonic epithelium prevents the inhibition of HDAC and impairment of cellular proliferation by stem cells in the colonic crypt. Second, aerobic glycolysis in neoplastic colonic cells via the "Warburg effect" reduces mitochondrial β-oxidation of butyrate making it available for HDAC inhibition leading to epigenetic effects and the inhibition of cellular proliferation. Finally, oxidative metabolism of butyrate by the colonic epithelium may help to maintain a steep oxygen gradient across the colonic mucosa, thereby helping to maintain the anerobic nature of the colonic lumen and preventing the expansion of aerotolerant Proteobacteria bacterial taxa, a signature of the dysbiotic microbiota in inflammatory bowel disease (IBD).

4.4 Technical applications of advances in intestinal epithelial stem cell biology: Development of enteroid technologies

A very practical outcome stimulated by advances in the understanding of the intestinal epithelial stem cell niche has been the development of technologies to develop three-dimensional (3D) organoids (referred to as "enteroids" when specific to the intestinal epithelium) generated from adult Lgr5+ stem cells. Single intestinal stem cells or intestinal crypts can be used to generate 3D cellular structures, using an extracellular matrix along with a defined set of growth factors such as Wnt3a, EGF, R-spondin, and Noggin, that resemble the structure and mimic the functionality of the intestinal epithelium *in vivo*. Enteroids can be generated from different regions throughout the mouse and human intestinal tract, perpetuated in culture, and genetically modified using CRISPR/Cas 9 genome editing. As such, enteroid technology has shown great promise in helping to advance the understanding of disease pathogenesis such as norovirus and other enteroviral infections as well as new physiologic pathways such as the coupling of a subset of enteroendocrine cells to sensory neural pathways providing evidence for the role of the gut epithelium as a chemosensor.

Secretory cells in mucosal epithelia

Although the small intestinal epithelium contains differentiated lineages of specialized cells, virtually all mucosal epithelial cells are able to secrete water, ions, or macromolecules that contribute to luminal fluid and are relevant to barrier function. These epithelial cell products, as well as antimicrobial peptides produced by specialized cells in the epithelium, and mucin glycoproteins produced by goblet cells, form a barrier that limits contact between luminal microorganisms and the epithelium. Thus, nonimmune secretory cells play a critical role in host protection against the external environment.

4.5 Nonspecialized epithelial cells display remarkable plasticity

The plasticity of certain mucosal cells enables such cells to markedly increase their secretory capabilities in response to specific stimulation. In the respiratory tract, for example, club cells were not recognized to be mucin-producing cells because they lack mucin storage granules. However, club cells continuously produce small amounts of respiratory mucins under steady-state conditions. During lung inflammation, particularly helper T-cell (T_H2) dominated inflammation, inflammatory cytokines induce goblet cell hyperplasia, that is, increased numbers of mucin-producing cells via proliferation, and transdifferentiation of club cells into a classical goblet cell phenotype that secretes large amounts of respiratory mucins characteristic of T_H2 lung inflammation. Thus, although we describe here the classical differentiated cells in mucosal tissues, it is important to recognize that secretory phenotypes of mucosal epithelial lineages can be modified by infection or inflammation.

4.6 Mucus-producing cells are abundant in the gastrointestinal epithelium

The differentiated cells that produce large amounts of mucin glycoproteins are referred to as goblet cells (due to the goblet shape formed by stored mucin granules in a structure referred to as a theca) or mucus cells, depending on the location of the mucosal tissue. Ultrastructurally, these cells are characterized by copious perinuclear rough endoplasmic reticulum (ER) in the base of the cell, a well-developed supranuclear Golgi apparatus, large numbers of granules packed together occupying a large volume in the apical region of the cell, and reduced numbers of apical microvilli compared with adjacent nongoblet cells (Figure 4.4). In standard hematoxylin and eosin (H&E) stained sections, the stored mucin granules are unstained, but they are revealed using carbohydrate stains such as periodic acid-Schiff or Alcian blue. In different tissues, these cells are found within the surface epithelium, crypt structures, and/or submucosal glands. Mucin-producing cells are most abundant in the gastrointestinal tract, particularly in the stomach and colon where the mucus layer is thickest, but can be increased in abundance in all mucosal tissues, particularly in response to infection and inflammation.

4.7 Antimicrobial peptides are produced by specialized cells throughout the gastrointestinal tract

All regions of the alimentary tract express antimicrobial peptides under certain conditions. Saliva delivers several antimicrobial peptides and proteins to the oral cavity, and the oral mucosa produces constitutive and inducible β-defensins. In addition, β-defensin expression occurs in the epithelium of the oropharynx, tongue, esophagus, stomach, and colon during inflammation and infection. In the small intestine, α-defensins secreted by Paneth cells constitute the major antimicrobial peptides. Antimicrobial

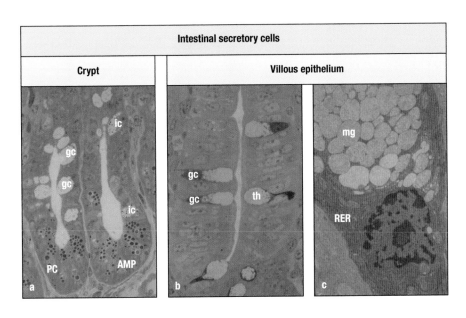

Figure 4.4 Intestinal secretory cells. Crypt (**a**) and villous (**b** and **c**) epithelium of the mouse small intestine shows Paneth cells (Pc), goblet cells (gc), and intermediate cells (ic) containing granules of antimicrobial peptides (AMP), pronounced rough endoplasmic reticulum (RER), and stored mucin granules (mg) in the goblet cell theca (th). Toluidine blue stained semithin resin embedded sections (**a** and **b**) and transmission electron micrograph (**c**).

peptides accumulate in Paneth cells within characteristic dense core secretory granules for subsequent release as components of regulated exocytotic pathways.

Biophysical features of the secreted mucus barrier

The biophysical properties of mucus are central to its protection of mucosal surfaces. Mucus is a highly hydrated viscous secretion with a complex macromolecular constituency. The rheologic properties of mucus vary between tissues and within tissues in response to different environmental and host-driven conditions. The major contributors to mucin viscosity are polymeric mucin glycoproteins. Although a barrier, mucus is permeable to macromolecules, as bidirectional diffusion of molecules through the mucus is frequently required. Nevertheless, thick mucus layers can establish concentration gradients that are functionally important, for example, in protecting the gastric mucosa from the low luminal pH. Similarly, mucus provides functional retention of some molecules, such as antimicrobial molecules and immunoglobulins. Retention in mucus is likely to increase the efficacy of these molecules in maintaining the barrier to infection. Many of these molecules may be retained in mucus due to noncovalent interactions with mucin glycoproteins, but this is poorly understood at present.

Molecules, and even viruses, can readily diffuse through mucus, although mucus is highly efficient at trapping particulate matter and cellular microbes. The surface properties of small particles determine their ability to diffuse through mucus, and binding to any component in mucus leads to highly efficient trapping. As an example, recent studies suggest that interactions between antibodies and mucin carbohydrates greatly enhance the capacity of antibodies to neutralize viruses. Integral to mucus physiology is the constant movement of mucus and its replacement by continuous secretion. In the gastrointestinal tract, the outer layer of mucus is moved by the flow of luminal contents, whereas in the lung, mucus is moved by the action of cilia, ensuring continuous clearing of the airway. In the lung, this system can be very rapidly increased in capacity by secretion of mucins by submucosal glands, such as after inhalation of particulate matter. Similarly, in the gut, the large number of goblet cells with stored mucin granules can be rapidly induced to secrete mucin granules in response to microbial, inflammatory, or neural signals. As an example, colonic goblet cells respond to a threshold concentration of microbial TLR ligands in the luminal environment at their apical surface by undergoing compound exocytosis; a system that can drive release of large amounts of mucus in response to microbes penetrating the mucus barrier.

4.8   Mucin glycoproteins have common structural features

The secreted polymeric mucins that form mucus share structural similarities, and most (MUC2, MUC5AC, MUC5B, and MUC6) are encoded by a family of genes in a cluster on chromosome 11, whereas the *MUC19* gene is on chromosome 12. These mucins have N- and C-terminal cysteine-rich domains that share homology with von Willebrand factor D domains and are involved in homo-oligomerization. Between these terminal domains are very large proline-, serine-, and threonine-rich domains that usually contain tandemly repeated sequences and are the scaffold for *O*-linked oligosaccharides.

The large mucin polypeptide chains are formed in the ER where the D domains are *N*-glycosylated and folded, and the *C*-termini are cross-linked via disulfide bonds to form dimers. The mucin dimers then move into the Golgi where the *O*-glycans are progressively assembled due to the action of Golgi-resident glycosyltransferases. Mass spectrometric analysis of mucin oligosaccharides shows that mucins are decorated by a diverse array of complex *O*-linked oligosaccharides and that these structures vary between species, within species between tissues/mucins, and within tissues between different conditions.

Following *O*-glycosylation, the dimers are further homo-oligomerized via N-terminal disulfide bond formation and packaged into granules for secretion. The true nature of the fully oligomerized mucin polymers remains clouded with conflicting data supporting the formation of linear polymers or more complex lattice-like structures, with current data suggesting that different mucin family members have differing mechanisms of N-terminal oligomerization. Although the details of packaging of the oligomerized mucin macromolecules into granules are not well understood, packaging involves association of Ca^{2+} ions with the negatively charged mucin oligosaccharides, which allows H_2O to be excluded and the dense granules to form.

4.9   Mucin release is a tightly regulated biological process

Secretion of mucin granules is a complex process involving movement of the granule to the apical cell surface, fusion with the cell membrane, and release of the granule contents. The process involves actin disruption and remodeling, Rab GTPases, and Munc proteins, which are essential mediators of exocytosis. In most mucosal tissues, both constitutive and inducible release of mucin granules occurs. Many different factors can promote mucin secretion, including the nucleotides ATP and UTP, which are potent mucin secretagogues via the $P2Y_2$-R purinoreceptor in the airway. The nervous system can also drive differential mucin release and, in the respiratory tract, differential innervation of submucosal glands versus the surface epithelium adds further regulation of mucin release. Once released into the luminal space, the granule contents are exposed to H_2O and the mucin complex rapidly expands as it is hydrated, increasing in volume by 100- to 1000-fold, indicating how such a large volume of mucus is maintained despite its continuous movement and degradation. Many other molecules are co-secreted with mucins into mucus, and proteomic approaches have revealed other host proteins present in mucus, some of which bind to mucins and thereby modulate the rheologic properties and other functions of mucus. The viscosity of mucus is also regulated by the concentration of ions such as Ca^{2+} and HCO_3^-, and by pH, which can be controlled by the epithelial cells, potentially allowing dynamic control of mucus properties.

The secreted mucins are expressed in a tightly regulated tissue-specific pattern. For example, MUC5B, MUC19, and the small nonpolymeric mucin MUC7 are the major constituents in saliva, and the stomach produces MUC6 deep in the glands and MUC5AC toward the surface. MUC5AC and MUC5B are the dominant respiratory tract mucins. In the duodenum, Brunner's glands produce MUC5AC, and the remainder of the goblet cells in the intestine produce mainly MUC2, although MUC6 is produced in the colon, and MUC5AC can be produced during inflammation.

4.10   Antimicrobial peptides play a key role in mucosal defense

Antimicrobial peptides collectively form a diverse population of gene-encoded protein effectors with selective microbicidal effects against bacteria and fungi. Varied epithelia release antimicrobial peptides onto mucosal surfaces, and evidence increasingly implicates them as components of a biochemical barrier against microbial challenges. Although antimicrobial peptide primary structures vary greatly (Figure 4.5), they have broad-spectrum microbicidal activities *in vitro*, generally at low micromolar concentrations, and most are 5 kDa or less, cationic at neutral pH, and amphipathic. Mammalian antimicrobial peptide structures range from linear,

Defensins		
α-defensins	HNP-2	CYCRIPACIAGERRYGTCIYQGRLWAFCC
β-defensins	BNBD-4	QRVRNPQSCRWNMGVCIPFLCRVGMRQIGTCFGPRVPCCRR
θ-defensins	RTD-1	GFCRCLCRRGVCRCICTR[1]
Cathelicidins		
human	LL-37	GKEFKRIVQRIKDFLRNLVPRTES
rabbit	CAP18	GLRKRLRKFRNKIKEKLKKIGQKIQGLLPKLAPRTDY
mouse	mCRAMP	GLLRKGGEKIGEKLKKIGQKIKNFFQKLVPQPE
rat	rCRAMP	GLVRKGGEKFGEKLRKIGQKIKEFFQKLALEIEQ
cattle	Bac5b	RFRPPIRRPPIRPPFYPPFRPPIRPPIFPPIRPPFRPPLGPFP-NH2
cattle	Bac7	RRIRPRPPRLPRPRPLPFPRPGPRPIPRPLPFPRPGPRPLPFPRPGPRPIPRPL
cattle	Indolicidin	ILPWKWPWWPWRR-NH2
cattle	Dodecapeptide	RLCRIVVIRVCR
sheep	SMAP-29	RGLRRLGRKIAHGVKKYGPTVLRIIRIA-NH2
sheep	SMAP-34	GLFGRLRDSLQRGGQKILEKAERIWCKIKDIFR-NH2
horse	eCATH-2	KRRHWFPLSFQEFLEQLRRFRDQLPFP
horse	eCATH-3	KRFHSVGSLIQRHQQMIRDKSEATRHGIRIITRPKLLLAS
pig	PR-39	RRRPRPPYLPRPRPPPFFPPRLPPRIPPGFPPRFPPRFP-NH2
pig	Protegrin-1	RGGRLCYCRRRFCVCVGR-NH2
pig	PMAP-23	RIIDLLWRVRRPQKPKFVTWVR
pig	PMAP-36	GRFRRLRKKTRKRLKKIGKVLKWIPPIVGSIPLGC-NH2

Figure 4.5 Representative members of the defensin and cathelicidin antimicrobial peptide families. To illustrate characteristic differences in the cysteine spacing in primary structures of the defensin subfamilies, HNP-2, a human neutrophil α-defensin, BNBD-4, a β-defensin from cattle neutrophils, and RTD-1, a θ-defensin from rhesus macaque neutrophils, are aligned in single-letter notation with Cys residues highlighted with red text. Because RTD-1 exists as a covalently closed macrocyclic molecule, assignment of its first residue position is arbitrary. The lower alignments are of cathelicidin peptides selected to illustrate the diversity of primary structures in the peptide family. Sequences correspond to characterized peptide products of exon 4 in the conserved peptide family genes.

disordered peptides that form α-helices in membrane-mimetic hydrophobic environments to molecules such as defensins which consist of antiparallel β-sheet-containing peptides that are constrained by up to four disulfide bonds. Certain antimicrobial peptides may be expressed constitutively or they may be inducible by exposure to bacteria or microbial antigens. Most antimicrobial peptides kill their target cells by peptide-mediated membrane disruption, creating defects that dissipate cellular electrochemical gradients, leading to microbial cell death. Despite their diverse primary, secondary, and tertiary structures, their amphipathicity enables the peptides to interact with and disrupt microbial cell membranes. There are two major families of antimicrobial peptides in mammals: cathelicidins and defensins (see Figure 4.5).

The cathelicidins are a family of highly diverse antimicrobial peptides. These peptides vary from cysteine-stabilized molecules (in swine and cattle), to linear peptides with unusually high arginine and proline content such as PR-39, to the tryptophan-rich tridecapeptide indolicidin (see Figure 4.5). These highly varied primary and secondary structures derive from a C-terminal cationic antimicrobial peptide domain. Although cathelicidins exhibit remarkable diversity across phylogenetic lines, the evolutionary relationship among cathelicidins is evident in the cathelin domain, which is conserved in the proregions of cathelicidin precursors. Although swine and cattle neutrophils contain numerous divergent cathelicidins, humans express a single cathelicidin, hCAP-18, which is processed to functional LL-37; mice also produce only a single cathelicidin, CRAMP. LL-37 and CRAMP are expressed and inducible at mucosal surfaces. hCAP-18 is expressed both in neutrophils and in epithelial cells, and it is processed to the antimicrobial peptide LL-37 by neutrophil proteinase 3 after release of secondary granules. hCAP-18 is reported to be activated in female reproductive tract mucosa by seminal pepsin C following intercourse. The low vaginal pH activates pepsin

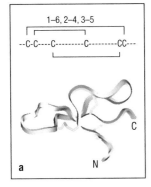

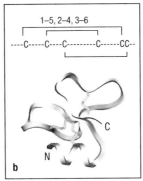

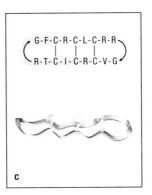

Figure 4.6 **Structural features of defensin peptides.** The general features of the three defensin peptide subfamilies are depicted. In each panel, the cysteine connectivities of the tridisulfide arrays are noted above the conserved cysteine residue positions of each subfamily displayed above ribbon diagrams of peptide solution structures. (**a**) Rabbit kidney α-defensin RK-1. (**b**) Human β-defensin-1. (**c**) Rhesus macaque neutrophil θ-defensin RTD-1. (Reprinted from Selsted, M., and Ouellette, A. *Nat. Immunol.* 2005, 6:551–557. With permission from Springer.)

C, enabling the processing of hCAP-18 molecule to ALL-38, the functional equivalent of LL-37. In mice, CRAMP is present in skin, eosinophils, salivary glands, and neutrophils. Disruption of mouse CRAMP results in susceptibility of mice to necrotic skin infection by group A streptococci. In addition to its antimicrobial activities, human LL-37 also is chemotactic for neutrophils, monocytes, mast cells, and T cells; induces degranulation of mast cells; alters transcriptional responses in macrophages; and stimulates wound vascularization and reepithelialization of healing skin.

Defensins comprise three families of cationic, cysteine-rich antimicrobial peptides, the α-, β-, and θ-defensins (Figure 4.6). The α-defensins are major granule constituents of mammalian phagocytic leukocytes and secretory granules of intestinal Paneth cells. α-Defensins are microbicidal against gram-positive and gram-negative bacteria, certain fungi, spirochetes, protozoa, and enveloped viruses in *in vitro* assays. Accounting for 5%–18% of total cellular protein in neutrophils, α-defensins are estimated to reach concentrations of ~10 mg mL^{-1} as granules dissolve in phagolysosomes following ingestion of microorganisms; Paneth cells secrete α-defensins at concentrations of 25–100 mg mL^{-1} at the point of release. The biosynthesis of α-defensins requires posttranslational activation by lineage-specific proteinases. α-Defensins derive from inactive 10 kDa precursors that contain canonical signal sequences, electronegative proregions, and a 3.5–4 kDa mature α-defensin peptide in the C-terminal portion of the precursor, and α-defensin bactericidal activity is dependent on proteolytic conversion of inactive 8.4 kDa pro-α-defensins to active forms. The details of human and mouse Paneth cell α-defensin processing differ in that human Paneth cells store unprocessed pro-α-defensin precursors that are processed coincident with or after secretion by trypsin, whereas mouse pro-α-defensins are activated intracellularly by Mmp7.

In humans, intestinal α-defensin-5 (HD5) is also expressed by female genital tract epithelia. HD5 mRNA is expressed in normal vagina, ectocervix, and by inflamed fallopian tube and in approximately 50% of normal endocervix, endometrium, and fallopian tube specimens. The HD5 peptide occurs in the upper half of the stratified squamous epithelium of the vagina and ectocervix, and immunoreactivity is more intense closer to the lumen. In positive endocervix, endometrium, and fallopian tube specimens, HD5 localizes to apically oriented granules and on the apical surface of some columnar epithelial cells, and HD5 peptide has been detected in cervicovaginal lavages, and levels peak during the secretory phase of the menstrual cycle. Endometrial HD5 occurs at highest levels during the early secretory phase of the menstrual cycle.

The β-defensins, first identified in neutrophils and in airway and lingual epithelial cells of cattle, are expressed by a greater variety of epithelial cell types than are α-defensins, and the gene family extends evolutionarily to birds and reptiles. For example, β-defensin peptides or transcripts have been detected in human kidney, skin, pancreas, gingiva, tongue, esophagus, salivary gland, cornea, and airway epithelium and in epithelial cells of a number of species. Unlike α-defensins, which are components of dense core granules in a regulated pathway, β-defensins expressed by epithelial cells are released via the constitutive pathway.

Human β-defensin-1 is expressed in genital and reproductive epithelium. Constitutively high human β-defensin-1 mRNA has been localized to epithelial layers of the vagina, ectocervix, endocervix, uterus, and fallopian tubes. Intracellular storage of human β-defensins has not been detected, suggesting that peptide release occurs via the constitutive pathway. Thus, the human female genital tract expresses both α- and β-defensins that could contribute to mucosal innate immunity. Also, inducible β-defensin human β-defensin-3 and human β-defensin-4 mRNAs have been detected in human endometrium, and exogenous steroid hormones alter levels of both β-defensins *in vivo*. In primary endometrial epithelial cells and in two endometrial cell lines, mRNA for human β-defensin-1 and human β-defensin-2 was detected in primary endometrial epithelium. In mice, the homolog of human β-defensin-1, mouse β-defensin-1, is expressed in female reproductive epithelium, and mouse β-defensin-1 transcripts have been detected in kidney, liver, and female reproductive organ tissues, and mRNA for the inducible peptide mouse β-defensin-3 was detected in ovary.

In primates, the atypical β-defensin DEFB126 has a conserved β-defensin domain and extended C-terminus containing several Ser and Thr residues. Those positions become *O*-glycosylated and are adsorbed onto spermatozoa as they migrate and mature in the epididymal duct. DEFB126 carbohydrates are major constituents of the sperm glycocalyx and are critical for efficient sperm transport in the female reproductive tract to deliver capacitated sperm to fertilization sites. Further exemplifying the diverse roles of β-defensins, canine β-defensin 103 is a determinant of coat color in dogs, and its ortholog in cattle, DEFB103, is responsible for the Variant Red phenotype in Holsteins. Accordingly, β-defensin biology spans innate immunity, reproductive fitness, and coat color phenotype.

The θ-defensins are unusual 2 kDa peptides found only in the neutrophils and monocytes of Old World monkeys and the only macrocyclic peptides in the Animal Kingdom. Like all defensins, they are stabilized by three disulfide bonds. θ-Defensins assemble from two hemi-precursors that derive from α-defensin genes that have stop codons that terminate the peptide at residue position 12. The ligation mechanisms that circularize the closed θ-defensin polypeptide chain remain unknown at this time. In addition to their direct microbicidal activities, θ-defensins are immunomodulatory, inhibiting the release of proinflammatory mediators by blood leukocytes responding to diverse TLR agonists and promoting survival of mice with polymicrobial peritonitis.

ROLE OF MUCOSAL CELL PRODUCTS IN MUCOSAL MICROBE HOMEOSTASIS

4.11   Microbes influence composition and structure of secreted mucosal barrier

Intestinal goblet cells continue to be produced in germ-free conditions. However, comparisons between normally exposed, germ-free, and germ-free conventionalized animals demonstrate decreased goblet cell number and reduced size of goblet cell thecae, together with a paradoxically thicker mucus layer, in the absence of microbes. The most likely explanations for this phenotype are that the microbial flora induce greater goblet cell production and greater mucin synthesis and that the mucus layer is thicker in germ-free animals because of substantially reduced mucin degradation. Mucosal epithelial cells increase production of cell-surface and secreted mucins after exposure to microbial products. Co-culture with probiotic bacteria increases expression of cell-surface mucins demonstrating how the host response to nonpathogenic bacteria may aid defense against pathogens. However, there are contrary examples where products of pathogens appear to reduce mucin synthesis, which is likely to facilitate infection. In addition to influencing mucin production, the microbial flora in the gut influences the glycosylation of mucins: conventionalization of gnotobiotic rodents and pigs results in substantial changes in mucin glycosylation, suggesting that the microbial flora either directly or indirectly influences expression of specific goblet cell glycosyltransferases. Nonmicrobial environmental factors can also influence expression of intestinal mucins. For example, dietary fructan (a prebiotic) alters goblet cell density, thecal size, and mucin glycosylation in rodents.

Two layers of mucus line the intestine: an inner layer that is resistant to removal and largely sterile, and an outer layer that can be readily removed and contains microbes (see Figure 4.1). The presence of two layers is likely due to progressive degradation of mucin polymers but could also occur as the mucus moves from an ionic environment regulated by the epithelium to one determined by the luminal contents. Importantly, in health, microbes colonize only the degraded mucus. Certain bacteria in the intestinal environment are mucolytic and can use mucin glycoproteins as an energy source. These bacteria are typically strict anaerobes, nonpathogenic, and do not encroach into the inner mucus layer. Combinations of multiple mucolytic bacteria result in more complete degradation of human MUC2 and more bacterial growth *in vitro*, suggesting that these bacteria collectively degrade mucins and utilize them as an energy source. An example of mucin-degrading bacteria occupying this niche is *Akkermansia muciniphila*, which is a prominent component of the colonic microbiota in health and often reduced in disease. Factors that prevent bacteria from penetrating the inner mucus layer likely include the higher pO_2, a higher concentration of antimicrobial molecules, and the continuous secretion of mucus by the epithelium. The thickness of the mucus layer varies throughout the gastrointestinal tract with a thinner layer in the small intestine than in the stomach and colon. The thinner mucus layer in the intestine may facilitate diffusion. In the respiratory tract, the mucus appears to lie over a thin aqueous layer covering the apical surface of the epithelial cells. Cilia beat back and forth in this layer, propelling the overlying mucus and any trapped microbes out of the airways.

In contrast to the strong influence of the microbiota on goblet cell and mucin biology, Paneth cell ontogeny is not dependent on luminal bacteria, reflected in the development of the lineage in germ-free mice and under aseptic conditions. Human Paneth cells appear in the first trimester of gestation and begin to express α-defensins coincident with the appearance of their differentiated morphology as early as 13.5 weeks of gestation. During postnatal crypt ontogeny in mice, newly induced Paneth cell gene products influence or respond to the microflora. In adult germ-free and conventionally reared mice, α-defensins, *Pla2g2a*, Mmp7, and lysozyme levels are similar. However, some Paneth cell granule constituents are inducible in response to microbial colonization or to changes in the microflora, at least in mice monocolonized with the anaerobe *Bacteroides thetaiotaomicron*, which have increased abundance of bactericidal ribonuclease Ang4 and the C-type lectin RegIIIγ compared with germ-free animals.

4.12 Secreted mucosal barrier influences composition of gut microbiome

Intestinal microbes can utilize mucins as an energy source, thereby influencing colonization. This was recently explored experimentally in germ-free mice with mono-association studies using *B. thetaiotaomicron*. These bacteria can feed on complex carbohydrates supplied in the diet or utilize mucin O-glycans for energy, switching to greater utilization of mucins when dietary sources are low. The bacterial genes involved in utilizing mucin O-glycans have been identified, and when disrupted the bacteria have diminished capacity to colonize the gut. Thus, some gut microbes depend on both host-derived and dietary factors. Further study will elucidate the relationship between the microbial community of the gut and host mucin glycosylation and production.

α-Defensins secreted by Paneth cells influence the composition of the small intestinal microbiome in mice, apparently by selecting peptide-tolerant microbial species in the microbial ecosystem. Such selection has been demonstrated in two mouse strains genetically modified to alter Paneth cell α-defensin production: the Tg-HD5 mouse, which expresses a human minigene in Paneth cells at levels comparable to those of endogenous mouse peptides, and the *Mmp7* null mouse, which is defective in the intracellular activation of pro-α-defensins in Paneth cells. Analysis of bacterial phyla and groups in the distal small intestine of *Mmp7*$^{-/-}$ mice and Tg-HD5 mice (which express HD5 in addition to murine Paneth cell α-defensins) showed that Firmicutes represented approximately 63% of the microbial composition in *Mmp7*$^{-/-}$ mice but only 26% in Tg-HD5 transgenic mice. In contrast, the percentage of Bacteroidetes was only 18% in the *Mmp7*$^{-/-}$ mouse but 69% in the Tg-HD5 mouse distal small bowel. Thus, the small bowel microflora of defensin-deficient mice compared with that of defensin-complemented mice was genotype dependent and showed reciprocal differences. Because a resident microflora is essential in IBD pathogenesis, the quantity and/or the repertoire of Paneth cell α-defensins and other gene products may influence gastrointestinal inflammation by shaping a healthy microbiome or one that promotes disease in genetically predisposed individuals.

4.13 Mucins and Paneth cell products contribute to protection against infectious pathogens

The central role of mucins during infection is reflected in changes in expression of multiple mucins during infection. In addition to forming a viscous biophysical barrier, mucin glycoproteins act as a barrier to microbes by providing carbohydrate decoy ligands, by having direct antimicrobial activity, and by facilitating retention of other antimicrobial molecules. Mucins display a complex array of O-linked glycans that reproduce many of the glycans present in the glycocalyx on the apical surface of mucosal epithelia. Pathogenic microbes that bind to the mucosal surface have evolved ligands for molecules in the glycocalyx; thus, the secreted mucins act as decoys for these microbial adhesins, facilitating entrapment of microbes within mucus. Mice lacking the Muc2 intestinal mucin develop spontaneous inflammation and intestinal cancers in the absence of pathogens, indicating that the mucus barrier is vital for preventing normal microbial flora from causing pathology. In addition, deficient Muc2$^{-/-}$ mice develop severe, usually fatal, pathology when infected with the attaching and effacing pathogen *Citrobacter rodentium*.

A family of transmembrane mucins highly expressed on the apical surface of mucosal epithelial cells has large, filamentous, heavily glycosylated, extracellular domains that can be released from the cell after engagement by bacteria. These mucins may represent a "final" defense against microbes that have penetrated the mucus barrier and reached the mucosal epithelial cells. Mice lacking individual cell-surface mucins appear more susceptible to infection by bacterial pathogens such as *Helicobacter pylori* and *Campylobacter jejuni*. Consistent with these findings, polymorphisms in the *MUC1* cell-surface mucin gene have been linked with susceptibility to *H. pylori* infection and the development of gastritis and gastric cancer. These transmembrane mucins also transmit signals into the epithelial cells that promote resistance to apoptosis and promote proliferation to aid wound repair during infection.

In addition to binding microbes, some mucin oligosaccharides have direct antimicrobial activity or can bind to other antimicrobial molecules. The gastric mucin oligosaccharide, α1-4-linked *N*-acetylglucosamine, inhibits synthesis of *H. pylori* cell-wall components and may limit *H. pylori* expansion in gastric mucus. The MUC7 mucin, a major component of saliva, has an N-terminal histatin domain with direct candidacidal activity and thus acts to limit growth of yeast in saliva. Mucins also directly bind antimicrobial molecules. For example, MUC7 binds statherin and histatin-1, and the other major mucin in saliva, MUC5B, binds histatin-1, -3, and -5 and statherin, retaining the antimicrobial molecules in the mucus.

α-Defensins contribute to innate mucosal protection in the small intestine. MMP7 gene disruption ablates processing of pro-α-defensins to their microbicidal form, resulting in impaired clearance of orally administered *Escherichia coli* and *Salmonella enterica* serovar *typhimurium* (*Salmonella typhimurium*). Conversely, the addition of human α-defensin to Paneth cell secretions augments innate defense in mice reflected in the ability of the Tg-HD5 mice to resist infection when challenged by *Salmonella typhimurium*. Resistance is likely due to increased killing of the bacteria, or possibly more complex HD5-related interactions in the intestinal lumen. The human Paneth cell α-defensin-6 (HD6) lacks bactericidal activity *in vitro*, but transgenic Tg-HD6 mice have enhanced innate mucosal immunity. As in Tg-HD5 mice, Tg-HD6 mice challenged with oral *Salmonella typhimurium* show improved survival relative to controls. However, rather than killing luminal *Salmonella typhimurium*, transgenic HD6 inhibits *Salmonella* translocation from the lumen to Peyer's patches and spleen without altering luminal *Salmonella* viability. The inhibition is due to the unique ability of HD6 dimers to produce stable tetramers that form higher-ordered self-assemblies, termed "nanonets," which trap bacteria both *in vitro* and *in vivo*. Although the mechanism superficially resembles neutrophil extracellular traps that occur at sites where neutrophils undergo cytolysis, nanonet formation only occurs in the presence of the HD6 peptide.

In mice that lack adaptor MyD88, small intestinal expression of RegIIIγ, RegIIIβ, CRP-ductin, and resistin-like molecule-β (RELM-β) fails to be induced by the normal flora or by oral bacterial infection with *Listeria monocytogenes*. In this model, cell-autonomous Paneth cell MyD88 activation, not activation in cells of myeloid origin, mediates signaling to maintain intestinal homeostasis to the infectious challenge. Also, oral infection of C57BL/6 mice with *Toxoplasma gondii* induced toll-like receptor 9 (Tlr9) and type I interferon (IFN) mRNA, and elevated levels of selected Paneth cell α-defensin mRNAs. In Tlr9$^{-/-}$ mice, the responses to *T. gondii* were replicated with IFN-β administration, and *T. gondii*–induced effects were eliminated in mice null for type I interferon receptors. The *T. gondii* effects are mediated by Tlr9-dependent production of type I IFNs, but whether the induction of certain Paneth cell α-defensin mRNAs is a direct effect on Paneth cells or a paracrine response to IFN is not known.

Regulation of secreted mucosal barrier by innate and adaptive immunity

Goblet cell differentiation, mucin glycosylation, and production rates of both secreted and cell-surface mucins are regulated in response to infection and inflammation. For example, release of T_H2 cytokines in response to parasitic infections is associated with goblet cell hyperplasia in both the intestine and lung. Responses are not restricted to T_H2 cytokines, however, as T_H1 and T_H17 cytokines also upregulate goblet cell mucin production. The ability of an array of cytokines and cell products including interleukin (IL)-1β, IL-4, IL-6, IL-9,

IL-13, interferons, tumor necrosis factor-α, nitric oxide, and granulocyte proteases to upregulate mucin secretion indicates that goblet cells are an integral component of innate immunity and inflammatory responses in the mucosa.

Paneth cells release secretory granules in a dose-dependent manner in response to bacterial antigens and pharmacologic agents. Paneth cell secretion induced by bacterial and carbamyl choline antigens is regulated by cytosolic Ca^{2+} mobilized from intracellular stores and an influx of extracellular Ca^{2+}. Inhibiting influx of Ca^{2+} by selective blockers of the Ca^{2+}-activated intermediate conductance K^+ channel KCa3.1 (also known as Kcnn4 and mIKCa1) attenuates Paneth cell secretion. KCa3.1 is expressed in Paneth cells and T lymphocytes, and enables Ca^{2+} influx to sustain the Paneth cell secretory response. However, whether defects in KCa3.1 impair Paneth cell secretion sufficiently to diminish levels of luminal α-defensins and other secreted components *in vivo* is not known.

Goblet cells as presenters of luminal antigens to dendritic cells

A fundamentally important role for goblet cells has emerged that is functionally distinct from their role in producing the secreted mucus barrier but is linked mechanistically to secretory events. Goblet cells in the small intestine sample soluble antigens from the luminal environment via exocytosis following granule release, and this antigen is then delivered by transcytosis to underlying dendritic cells closely opposed to the basal membrane of the goblet cell. This mechanism regulates antigen-specific controlled immune responses against the microbiota. In early neonatal life, this sampling function also occurs in goblet cells in the colon but is suppressed in adult life by exposure to microbial TLR ligands. The neonatal colonic goblet cell antigen transfer drives the establishment of regulatory T cells, and ablation of this function predisposes to subsequent inappropriate mucosal immune responses. Interestingly, colonic goblet cell antigen transfer is turned on in adults by suppression of the microbiota with antibiotics.

Defects in mucosal barrier secretion and pathogenesis of disease

Goblet cell pathology, particularly ER stress-related pathology, has been implicated in intestinal disease in murine models. Muc2$^{-/-}$- and Agr2$^{-/-}$-deficient mice switch off intestinal goblet cell mucin biosynthesis. Aberrant mucin biosynthesis results in accumulation of misfolded Muc2, ER stress, reduced Muc2 biosynthesis, premature goblet cell death, and spontaneous T_H17-dominated inflammation, leading to intestinal inflammation. Defects in elements of the unfolded protein response genes that are activated following misfolding, including *Xbp1*, *Ire1β*, *Mbpts1*, and *Chop*, also result in spontaneous inflammation or altered susceptibility to environmentally triggered inflammation. Information regarding goblet cell pathology in human disease is limited even though loss of goblet cells, smaller goblet cell thecae, and reduced thickness of the mucus barrier are hallmarks of ulcerative colitis.

Normally, Paneth cells are anatomically restricted to the small intestine. During conditions of inflammation, as in Barrett's esophagus, *H. pylori* gastritis, Crohn's disease, and ulcerative colitis, Paneth cells may appear ectopically. The "intermediate" or "granulo-goblet" cells (see **Figure 4.4**) are rare in healthy gut mucosa but may appear during episodes of inflammation. Reflecting the common progenitor of goblet cell and Paneth cell lineages, these cells produce mucins and contain small electron-dense cytoplasmic inclusions that are positive for Paneth cell-specific markers. Thus, pro-inflammatory conditions of diverse origin may induce the appearance of numerous intermediate cells that produce secretory proteins characteristic of both lineages.

A reduction in luminal α-defensins has been identified in cohorts of Crohn's disease patients. Such a reduction could impact the composition of resident microorganisms, promoting a more pro-inflammatory microbial population. In this connection, mouse models of cystic fibrosis have demonstrated undissolved Paneth cell secretory granules in mucus-occluded crypts, possibly contributing to the associated overgrowth of Enterobacteriaceae.

In premature infants with necrotizing enterocolitis, a leading cause of morbidity and mortality in neonates, Paneth cells appear to have reduced MD-2 expression. Premature infants appear to have a similar impaired expression of MD-2. Thus, the absence of a functional Paneth cell lipopolysaccharide-sensing apparatus in immature neonatal gut may predispose preterm newborns to necrotizing enterocolitis as the immature intestine is challenged by enteric bacteria.

Genetic defects in secretory cell pathology may become apparent under specific environmental conditions. For example, the Paneth cell phenotype in Atg16l1 hypomorphic mice is codependent on the genetic defect and infection by a particular strain of murine norovirus. Also, severe disruption of the unfolded protein response by deletion of *Xbp1* in epithelial cells induces massive Paneth cell apoptosis and fulminant ileitis. Germline and inducible gene deletions of *Agr2* result in goblet cell Muc2 deficiency, expansion of the Paneth cell compartment, and accumulation of intermediate cells (discussed earlier). Severe terminal ileitis and colitis are associated with ER stress induced by loss of Agr2. Thus, a diversity of genetic defects can disrupt Paneth cell function in mice, resulting in chronic ileitis that mimics adult human disease. When autophagy in Paneth cells is fully functional, ER stress-induced chronic inflammation of the intestine is averted, perhaps by restraining cytosolic IRE1α activity and NF-κB activation. Under normal conditions, autophagy controls Paneth cell unfolded protein responses (UPRs) to maintain homeostasis. However, when hypomorphisms in one or more autophagy components combine with environmental factors such as the microbiota or infectious agents, the UPR becomes chronically active in Paneth cells, thereby inducing enteric inflammation. In humans, familial and sporadic interstitial lung disease has been linked to misfolding mutations in the *SFTPC* gene that encodes the SP-C surfactant protein produced by lung epithelial cells. Interestingly, viral infection appears to precipitate the emergence of the phenotype in this disease, possibly through increased SP-C production and ER stress.

SUMMARY

Nonstructural components of the mucosal barrier, including mucus produced by goblet cells and antimicrobial peptides produced by Paneth cells, play a fundamental role in mucosal protection. Goblet and Paneth cells arise from stem-cell progenitors through intricate developmental pathways involving both Wnt and Notch signaling. Mucus, produced by goblet cells, protects the intestinal mucosal surface through its biophysical properties as well as its effects on luminal bacteria. The composition of mucus includes various mucins that are complex glycoproteins. Paneth cell products include antimicrobial peptides such as defensins, which influence the composition of the gut microbiota. Together, mucins and antimicrobial peptides form a secreted mucosal barrier that provides protection against an array of microbes. Components of both the innate and adaptive mucosal immune systems contribute to goblet cell and Paneth cell regulation in response to specific environmental challenges. Goblet cells are also involved in antigen transfer to antigen-presenting cells establishing appropriate regulatory and adaptive mucosal immune responses. Defects in either goblet or Paneth cells can lead to a disruption of the secreted mucosal barrier and thus contribute to the pathogenesis of mucosal diseases such as inflammatory bowel disease and necrotizing enterocolitis.

FURTHER READING

Barker, N., van Es, J.H., Kuipers, J. et al.: Identification of stem cells in small intestine and colon by marker gene Lgr5. *Nature* 2007, 449:1003–1007.

Davis, C.W., and Dickey, B.F.: Regulated airway goblet cell mucin secretion. *Annu. Rev. Physiol.* 2008, 70:487–512.

Dutta, D., Heo, I., and Clevers, H. Disease modeling in stem cell-derived 3D organoid systems. *Trends Mol. Med.* 2017, 23:393–410.

Gerbe, F., and Jay, P. Intestinal tuft cells: Epithelial sentinels linking luminal cues to the immune system. *Mucosal Immunol.* 2016, 9:1353–1359.

Linden, S.K., Sutton, P., Karlsson, N.G. et al.: Mucins in the mucosal barrier to infection. *Mucosal Immunol.* 2008, 1:183–197.

McGuckin, M.A., Eri, R., Simms, L.A. et al.: Intestinal barrier dysfunction in inflammatory bowel diseases. *Inflamm. Bowel Dis.* 2008, 15:100–113.

Mukherjee, S., Vaishnava, S., and Hooper, L.V.: Multi-layered regulation of intestinal antimicrobial defense. *Cell. Mol. Life Sci.* 2008, 65:3019–3027.

Knoop, K.A., McDonald, K.G., McCrate, S. et al.: Microbial sensing by goblet cells controls immune surveillance of luminal antigens in the colon. *Mucosal Immunol.* 2015, 8:198–210.

Porter, E.M., Bevins, C.L., Ghosh, D. et al.: The multifaceted Paneth cell. *Cell. Mol. Life Sci.* 2002, 59:156–170.

Salzman, N.H., Hung, K., Haribhai, D. et al.: Enteric defensins are essential regulators of intestinal microbial ecology. *Nat. Immunol.* 2010, 11:76–83.

Selsted, M.E., Brown, D.M., DeLange, R.J. et al.: Primary structures of MCP-1 and MCP-2, natural peptide antibiotics of rabbit lung macrophages. *J. Biol. Chem.* 1983, 258:14485–14489.

Selsted, M.E., and Ouellette, A.J.: Mammalian defensins in the antimicrobial immune response. *Nat. Immunol.* 2005, 6:551–557.

Thornton, D.J., Rousseau, K., and McGuckin, M.A.: Structure and function of the polymeric mucins in airways mucus. *Annu. Rev. Physiol.* 2008, 50:5.1–5.28.

Tomasinsig, L., and Zanetti, M.: The cathelicidins—Structure, function and evolution. *Curr. Protein Pept. Sci.* 2005, 6:23–34.

van der Flier, L.G., and Clevers, H.: Stem cells, self-renewal, and differentiation in the intestinal epithelium. *Annu. Rev. Physiol.* 2009, 71:241–260.

van Es, J.H., van Gijn, M.E., Riccio, O. et al.: Notchγ-secretase inhibition turns proliferative cells in intestinal crypts and adenomas into goblet cells. *Nature* 2005, 435:959–963.

Yang, Q., Bermingham, N.A., Finegold, M.J. et al.: Requirement of Math1 for secretory cell lineage commitment in the mouse intestine. *Science* 2001, 294:2155–2158.

PART TWO

CELLULAR CONSTITUENTS OF MUCOSAL IMMUNE SYSTEMS AND THEIR FUNCTION IN MUCOSAL HOMEOSTASIS

Immune function of epithelial cells

5

RICHARD S. BLUMBERG, WAYNE LENCER, ARTHUR KASER,
AND JERROLD R. TURNER

Epithelial cells lining mucosal surfaces essentially define the border between the environment and ourselves. At some mucosal surfaces, the epithelial border is only one cell thick, interfacing with the environment over huge surface areas (200 m^2 for intestine and lung). These vast and delicate epithelial barriers, typified by the intestinal, respiratory, and urogenital mucosa, have to distinguish between components of the outside world and selectively transport significant quantities of nutrients, ions, and water that are essential for life to the local tissues or systemic circulation. At the same time, the epithelial cell must also defend against invasion and absorption of unwanted, toxic, and pathogenic molecules or microbes, as well as producing and secreting factors required to provide for such host defense. From this pivotal position, the epithelial cell communicates dynamically with a multitude of targets that include solutes, macromolecules, microbes, and specific immune cells such as polymorphonuclear leukocytes within the lumen; all of the immunocompetent cells of the lamina propria that comprise the mucosal immune system; and cell types at a distance that affect systemic immunity, metabolism, and organ function. Therefore, the mucosal immune system cannot be understood without understanding the epithelial cell and the developmental and molecular biology that underlie its involvement in mucosal physiology and immunology. This chapter provides a brief introduction to these fields.

BARRIER FUNCTION

The epithelial barrier of mucosal surfaces separates a wealth of foreign antigens that consist of the commensal flora and ingested food, on the luminal side, and the mostly sterile environment harboring the mucosal immune system on the other side. This barrier function, which is as much physiologic as it is structural, is set up to efficiently deter the ingress of potential pathogens, including commensal organisms. Remarkably, the as yet only rudimentarily characterized commensal microbial communities including bacteria, fungi, viruses, bacteriophages, and archae associated with each mucosal surface and molecular components of the luminal (outside) space are required for the development of both the mucosal and systemic immune systems and can regulate assembly, maintenance, and repair of the barrier.

5.1   Stratified squamous and simple columnar epithelia form the major types of epithelial structures

Each of the "wet" epithelial surfaces that line mucosal tissues possesses the common feature of separating the internal sterile milieus of the host from the external environment. Because each of the different mucosal surfaces serves different functions (such as gas exchange in the lungs, or absorption of nutrient solutes in the intestine) and is uniquely colonized with microbes, the associated epithelium is highly adapted and differentiated by and within organ systems. For clarity in this chapter, we use the small intestinal mucosa as a paradigm for understanding all mucosal surfaces, because the intestine is perhaps the best-characterized and interfaces most intensely with highly dynamic and diverse microbial populations.

In general, there are two broad histological types of epithelia that line mucosal surfaces: stratified squamous epithelia and simple epithelia, which can be columnar or cuboidal (**Figure 5.1**). Stratified refers to the presence of multiple epithelial layers that are composed of squamous cells in the vagina, urethra, rectal canal, mouth, esophagus, pharynx, and nasal surfaces. Most often, the stratified squamous mucosa immediately abuts the skin, or is located close to it, and these surfaces do not participate in large-scale selective transport of materials necessary for life. The stratified squamous epithelium that is associated with mucosal tissues is distinguished from the squamous epithelium of the skin by the fact that the former is noncornified, i.e., there is not typically a dense layer of acellular keratin on the apical, i.e., luminal, surface. This can, however, develop in squamous mucosa as a response to chronic injury and is present in some neoplasms, e.g., esophageal verrucous (squamous cell) carcinomas.

The mucosal surfaces that support solute, ion, water, and gas transport are typically lined by a delicate single layer of epithelial cells, i.e., a simple epithelium. The epithelial cells present are typically columnar or cuboidal in shape, although they can assume flattened, squamous-like morphologies in the context of migration and wound repair. All epithelial cells lining mucosal surfaces are colonized with commensal organisms of different types and densities depending on the organ system and the location within that system.

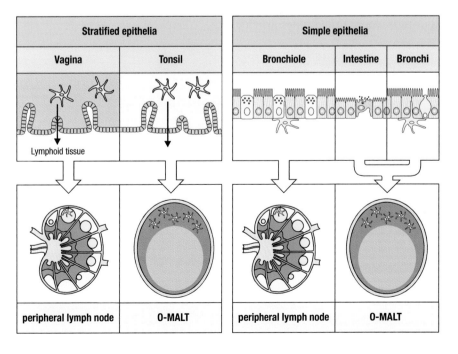

Figure 5.1 Simple and stratified epithelia form the barrier between the outside world and self. The architecture of epithelial cells has evolved according to the specific requirements of the body locale. Similarly, dendritic cells localize to places specific for the relevant locales and allow one specific route of antigen sampling, a process that is controlled by various factors, in particular, chemokine receptor signaling. O-MALT, organized mucosa-associated lymphoid tissue. (Adapted from Ogra, P. et al. [eds]: *Mucosal Immunology*, 2nd ed., New York: Academic Press, 1999.)

The commensal organisms have the common properties of resisting the invasion of potential pathogens, modifying luminal microbial composition, regulating mucosal immune cells, and permitting controlled sampling of antigenic materials for interpretation by the immune system.

5.2   Polarized simple epithelium permits vectorial transport

Simple columnar and cuboidal epithelial cells are structurally and functionally polarized, with a specialized apical membrane facing the mucosal surface that abuts the outside environment and a basolateral membrane facing the lamina propria and the internal environment of the host. The two membranes are composed of different protein and lipid compositions and structural features; the visually distinct microvilli present in the apical membrane of intestinal epithelial cells exemplify this dramatically. This polarity of structure and function is required for vectorial, i.e., directional, transport of solutes and water. Although apical and basolateral membrane domains are contiguous, they are separated by a functional "fence" that is presumed to prevent mixing of apical and basolateral lipids and proteins. Transport of membranes and cargo between the two surfaces occurs by vesicular traffic in a process termed "transcytosis."

Many luminal materials, including hydrophilic nutrients, are transported by distinct transport proteins within apical and basolateral domains. Most often, apical transporters take advantage of the steep, electrochemical Na^+ gradient (from extracellular to intracellular) to provide the driving force for absorption. The basolateral Na^+-K^+ATPase that maintains the Na^+ gradient and pumps apically transported Na^+ ions across the basolateral membrane is, therefore, essential to ongoing nutrient transport. Paracellular recycling of Na^+ ions from the lamina propria to the lumen (via the tight junction, as discussed later) is essential, as the diet does not otherwise contain sufficient Na^+ to support ongoing apical absorption. Solutes absorbed by apical transmembrane transport proteins cross the basolateral membrane via facilitated transporters that operate in a strictly concentration-dependent manner. This allows the basolateral transport proteins to drive nutrient absorption from the enterocyte cytoplasm toward the bloodstream when nutrients are being actively absorbed but to also operate in the reverse direction in order to bring nutrients into the enterocyte when none are present in the lumen, e.g., during fasting. Whether by vesicles or transmembrane transport proteins, transport of solutes, membranes, and cargo from one side, through the cell to the other side, is termed the "transcellular pathway" and is an energy-dependent process.

As with most eukaryotic cells, the epithelial cells lining mucosal surfaces maintain a dynamic endocytic network that continuously internalizes membranes and cargo from both sides of the cell, often in a regulated manner. The two apical and basolateral endosomal networks are distinct and can recycle membranes and proteins back to the cell surface of origin; however, some cargoes are sorted to a common endosome that receives and processes membranes from both sides of the cell. The common endosome can then sort and deliver cargo to the degradative lysosome (relevant to antigen processing), to the Golgi apparatus and endoplasmic reticulum (ER) via the retrograde pathways (relevant to autophagy, innate sensing by intracellular toll-like receptors [TLRs], and ER stress), or to the opposite cell surface via transcytosis (relevant to the transport of intact proteins and particles for sensing by the mucosal immune system, antigen presentation, or innate pattern recognition). The transcytotic pathway is one way in which luminal contents can breach the epithelial barrier to enter the host, and it is the pathway for the secretion of some intact proteins, such as secretory IgA, IgM, IgG, and IgE going in the opposite direction from basal to apical. Each of these immunoglobulin types are secreted by specific transport pathways to allow for mucosal defense and are discussed in more detail in the following sections.

5.3 Epithelial cells form a paracellular barrier via intercellular junctions

The other pathway for movement across epithelial barriers is to go around the cell, through the intercellular junctions that join adjacent cells together. This is termed the "paracellular pathway." Movement of solutes and water through the paracellular pathway can only be passive. There is no way to harness cell energy directly to this process, and paracellular transport is completely dependent on concentration gradients across the barrier epithelium, or indirectly harnessed to osmotic or electrical gradients established by other activities of the epithelial cell.

Intercellular junctions between adjacent epithelial cells are of course required for barrier function. Intercellular junctions are located along the lateral surfaces of the epithelial cells. They are composed of the most apically oriented tight junctions, and the subjacent adherens junction and desmosome (**Figure 5.2**). While the latter are required for physical integrity of the epithelial monolayer, movement across tight junctions is the rate-limiting barrier for solute and particle transport. Tight junctions are physiologically dynamic and regulate the access of small solutes, water, and macromolecules into the paracellular space. Movement across tight junctions can be considered to occur via two routes, the pore pathway and the transcellular pathway. The pore pathway is a high-capacity route that is exquisitely size selective, excluding molecules with Stokes radii greater than ~5 Å, and are also charge selective. Intestinal tight junctions allow cations to cross via the pore pathway at rates up to 10-fold greater than anions. While this cation selectivity is critical for normal physiologic function and can be reversed to anion selectivity by modifying tight junction protein expression, it is far less than that exhibited by transmembrane ion channels. Nevertheless, the paracellular channels that define the pore pathway are actively gated, i.e., they open and close, in a manner similar to transmembrane ion channels.

The second route across the tight junction is the leak pathway. This is a low-capacity route with an apparent channel diameter estimated to be ~125 Å, meaning that molecules up to this size can cross, although they do so inefficiently, and smaller molecules cross the leak pathway more readily. The leak pathway is not charge selective and, in healthy epithelial accounts for only

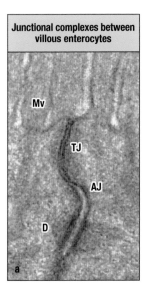
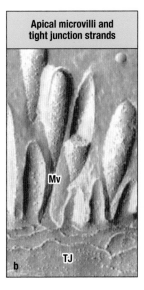
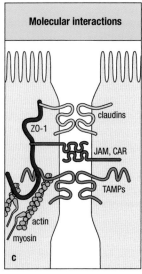

Figure 5.2 Intercellular junctions between adjacent epithelial cells. (**a**) Transmission electron micrograph showing junctional complexes between two villous enterocytes. The tight junction (TJ) is just below the microvilli (Mv), followed by the adherens junction (AJ). The desmosomes (D) are located basolaterally. (**b**) Freeze–fracture electron micrograph showing apical microvilli (Mv) and tight junction strands (TJ) in a cultured intestinal epithelial cell. (**c**) Interactions between F-actin, myosin II, zonula occludens-1 (ZO-1), claudins, tight junction–associated MARVEL proteins (TAMPs), and immunoglobulin superfamily members such as junctional adhesion molecules (JAM) and the Coxsackie–adenovirus receptor (CAR). (Adapted from Shen, L. et al.: *Annu. Rev. Physiol.* 2011, 73:11.1–11.27.)

a small fraction of overall paracellular transport. Flux across the leak pathway is, however, markedly upregulated in many disease states.

The structure and function of the tight junctions are established by the transmembrane claudin proteins, a family of 27 genes in mammals, the transmembrane tight junction–associated MARVEL proteins (TAMPs), which include occludin and tricellulin, peripheral membrane scaffolding proteins including zonula occludens-1 and -2, signaling molecules, the cytoskeleton, and the endocytic/vesicular trafficking machinery. The claudins direct assembly of the "molecular gasket" that seals the paracellular space and also define the capacity and charge selectivity of the pore pathway. While occludin and zonula occludens-1 are primarily associated with leak pathway maintenance and regulation, they also play other critical roles. For example, zonula occludens-1 and -2 are required for delivery of claudin proteins to tight junctions, and occludin is capable of regulating the function of some claudin proteins, e.g., claudin-2. Zonula occludens-1 and -2 also bind to and regulate the perijunctional actin filaments that are commonly involved in tight junction barrier regulation. Tight junctions also depend on adherens junctions (the cadherins are the primary component) and desmosomes (the cadherin-like desmoglein and desmocollin proteins are the primary components), both of which are also linked to and regulated by the cytoskeleton (actin and cytokeratin, respectively) but do not form significant barriers to diffusion.

Intercellular junctions are under the control of a variety of different host, dietary, and microbial factors. Dietary factors such as glucose and amino acids increase permeability through the actin cytoskeleton by activating myosin light-chain kinase. A similar, though quantitatively greater myosin light-chain kinase activation is responsible for increases in leak pathway flux induced by tumor necrosis factor (TNF)-α or interleukin (IL)-1β. IL-13 and IL-6 also reduce barrier function by increasing claudin-2 expression and pore pathway flux. Microbial factors such as the *Clostridium perfringens* enterotoxin may reduce barrier function by binding to barrier-enhancing claudin proteins and directing their removal from the tight junction. Other microbial factors, including repeats in toxin (RTX) from *Vibrio cholerae*, cytokines, or cytolytic T cells, may increase intestinal permeability in a tight junction–independent manner by causing epithelial damage, e.g., apoptosis, that enhances flux across the so-called unrestricted pathway, which is normally sealed by the presence of an intact epithelium. Similar factors can also affect cell polarity and the endocytic and transcellular pathways of membrane and cargo transport. Thus, barrier permeability is under the control of microbial and immune factors.

5.4   Various types of differentiated epithelial cells are derived from epithelial stem cell

Mucosal surfaces are rapidly self-renewing, and epithelial cells are shed on a regular basis. This is potentially beneficial both in terms of defense against invading pathogens and for the avoidance of neoplasia given the harsh environment in which they exist. Epithelial cell renewal depends on long-lived stem cells, from which all lineages of epithelial cells are derived. In the intestines, the common stem cell resides at the base of the crypt and is characterized by the expression of specific markers such as Lgr5 and prominin-1 (Figure 5.3). The Lgr5$^+$ stem cells, which are also known as crypt-base columnar cells, are mitotically active and, as a result, susceptible to injury from many sources, including radiation and chemotherapeutic agents that target dividing cells. A second, less well-defined, quiescent stem cell population is also present. These cells, which are located slightly above the crypt base, divide only occasionally in order to replenish the Lgr5$^+$ stem cell pool. Under the control of specific transcription factors, individual epithelial cell lineages differentiate from the common stem cell. In the small intestine,

Crypt-villus axis	Six types of differentiated epithelial cells	Epithelial shedding and renewal
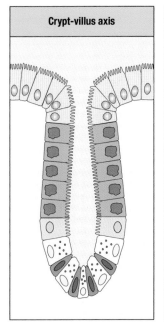	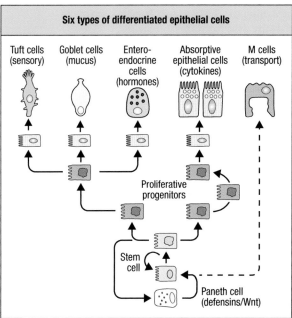	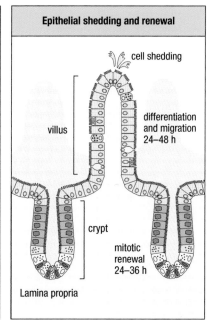

Figure 5.3 Intestinal epithelial cells arise from a common intestinal stem cell. The common stem cell is located at the base of small intestinal crypts and gives rise to epithelial cells, goblet cells, neuroendocrine cells, tuft cells, M cells, and Paneth cells. The last of these forms the niche for intestinal epithelial stem cells. (Adapted from Radtke, F., and Clevers, H. *Science* 2005, 307:1904–1909. With permission from AAAS.)

the common stem cell develops into six different lineages that migrate along the crypt villus axis; these are epithelial cells, goblet cells, Tuft cells, Paneth cells, M cells, and enteroendocrine cells (see **Figure 5.3**). Epithelial cells are a major component of innate and adaptive cellular immunity. Goblet cells and Paneth cells contribute significantly to innate immunity through the secretion of mucins and antimicrobial peptides, respectively. Enteroendocrine cells secrete neurohumoral factors such as substance P, which are immunologically active. As they are highly secretory cells, goblet cells and Paneth cells are highly dependent on the interrelated processes of autophagy (self-eating), which arises as a consequence of intracellular nutrient deprivation and the accumulation of abnormal intracellular organelles, and the unfolded protein response, which arises secondarily to the occurrence of misfolded proteins and the consequent elicitation of ER stress. Tuft cells are sensory cells that detect pathogens such as worm infections and secrete IL-25 to activate type 2 innate-like cells to secrete IL-13 that is involved in goblet cell activity and worm expulsion. M cells are specialized sampling cells that reside above lymphoid aggregates and transport luminal antigen to professional antigen-presenting cells (APCs) associated with Peyer's patches.

5.5 Non-immunologic and innate factors contribute to epithelial barrier function

The function of mucosal barriers is also provided by the factors that they secrete into the lumen, as well as by cell types that are associated with the epithelium, such as intraepithelial lymphocytes (IELs), macrophages, neutrophils, and dendritic cells (**Figure 5.4**). This section discusses each of these and emphasizes the barrier as a physiologic property rather than simply a structural function.

The intestinal epithelium constitutively secretes, or is induced to secrete, inflammatory mediators and, upon recognition of microbial components such as microbe-associated molecular patterns, a variety of host-protective soluble factors. These include those with antimicrobial activity, such as lactoferrin, lysozyme, peroxidase, cathelicidin, α-defensins, and β-defensins, which are largely provided by epithelial cells (β-defensins) and Paneth cells (α-defensins). These factors serve to regulate the composition of the commensal microbiota and protect the host from pathogenic invasion. The secretion of antimicrobial peptides is a major common property of all body surfaces. Epithelial cells also secrete a variety of different complement components into the lumen; consistent with this, epithelial cells express CD55 (decay accelerating factor), which protects the epithelium from inappropriate complement deposition. In other regions of the intestinal tract, a variety of other nonspecific factors serve a role in host defense, such as gastric acid from the stomach, bile salts from the biliary system, and urea in the urogenital system.

The mucosal barrier has both cell-associated (glycocalyx) and secreted carbohydrate components. The glycocalyx can extend up to 500 nm into the lumen and protects the epithelium from the direct adherence of bacteria and other microbes. The secreted mucus layer (or mucus blanket), which is demonstrable in all mucosal surfaces, is the major extracellular component of the nonspecific barrier, exhibits variable depths within mucosal tissues, and is organized as an attached layer and a more superficial detached layer in the colon. Only the detached layer is present in the small intestine. In the small intestine and colon, mucus is synthesized primarily by goblet cells and further modified by the enzymatic activity of the commensal microbiota. Through peristaltic activity (intestines) and ciliary function (respiratory tract), the mucus blanket is continuously renewed. The mucus layer includes not only a variety of mucin protein isoforms but also associated molecules such as intestinal trefoil factors. Mucins are important to homeostasis; loss of mucin-2 (*Muc2*) or intestinal trefoil factor in mice increases susceptibility to intestinal inflammation in the colon.

5.6   Adaptive immunologic factors contribute to epithelial barrier function

All human and nonhuman mucosal secretions contain variable quantities of a variety of different immunoglobulin (Ig) isotypes under physiologic and pathophysiologic conditions. As this topic is discussed in detail in Chapter 11, only a brief discussion is provided here. All Ig classes are potentially secretory, derive from mucosal lamina propria plasma cells, and are delivered into the secretions by either nonspecific or specific mechanisms. Nonspecific mechanisms can occur under homeostatic conditions but are more common during inflammation with disruption of the intestinal barrier. Specific mechanisms are those associated with transporting receptors that are capable of specific binding and transcytosis across the polarized simple epithelium; these are necessary given the exclusivity of the paracellular barrier (see earlier). Whereas the polymeric Ig receptor transports secretory IgA and IgM, with IgA being the dominant Ig in mucosal secretions in the basal-to-apical direction, the neonatal Fc receptor (FcRn) is responsible for the bidirectional transport of IgG. IgD is commonly found in association with tonsillar secretions, although the transporting receptor has not yet been defined. Secretory IgE is commonly observed during parasitic and allergic inflammation and is transported by CD23. Mucosal Igs serve to prevent microbial interactions with the immune system and/or regulate the uptake of cognate antigens through association as immune complexes and transport by their transcytosing receptors (see later).

5.7   Reduced barrier function increases susceptibility to mucosal pathology

Regulation of the physical epithelial barrier is important for maintenance of homeostasis, the normal composition of the mucosa-associated lymphoid tissue, and the avoidance of inflammation. In animal models, gross perturbation of the epithelial function, e.g., by disruption of E-cadherin interactions, leads to defective epithelial adherence, differentiation, and migration as well as inflammation, demonstrating the importance of these cellular structures. Expressed on IELs, $\alpha_E\beta_7$ is the receptor for cadherins on epithelial cells, which may link the latter to β-catenin and potentially to adenoma formation. Such disruptions of the physical barrier, which may be genetically determined, are possibly predisposing factors for the development of chronic intestinal inflammation as observed in inflammatory bowel disease. Proper nuclear factor κB (NFκB) signaling is required for maintaining the integrity of the epithelium and the functional linkage of the epithelium with subjacent immune cells (see later).

In contrast to E-cadherin disruption, less severe disruption of the barrier, by, for example, specifically upregulating junctional adhesion molecule (JAM) or myosin light-chain kinase activity within intestinal epithelia of mice, does not induce spontaneous disease. This likely reflects homeostatic immune activation, as upon challenge, these mice display increased susceptibility to disease. The chronic barrier defects present in a subset of healthy relatives of patients with inflammatory bowel disease (e.g., Crohn's disease) emphasize that barrier loss alone is insufficient to cause disease. It has been hypothesized but not demonstrated that these relatives are at greater risk of developing inflammatory bowel disease than those with normal barrier function. Normal physiologic epithelial barrier function is also necessary for intestinal homeostasis. Signals emanating from the microbiota and detected by cell-surface (TLRs) and intracellular (nucleotide-binding oligomerization domain [NOD]) pattern recognition receptors regulate antimicrobial peptide secretion. Consequently, loss of this activity may lead to alterations in the commensal microbiota and susceptibility to pathogens and pathobionts. A similar impairment in the ability to respond to ER stress through elicitation of a proper unfolded protein response may also lead to alterations in antimicrobial peptide production and barrier function leading to spontaneous or induced intestinal inflammation.

Antigen uptake and presentation

The functional aspect of epithelial cell barrier activity occurs at many different levels. Epithelial cells function as the sentinel cells to detect the presence of foreign antigen and, although still insufficiently understood, contribute to the immune hyporesponsiveness of the gut immune system. Epithelial cells interact with lymphocytes that abut against their basolateral cell surface, the IEL, as well as lymphocytes beneath the basement membrane—lamina propria lymphocytes—which stay in contact via soluble mediators as well as physically via basolateral projections of the epithelium through pores in the basement membrane. In a similar manner, epithelial cells are in contact with lamina propria macrophages, dendritic cells (DCs), and stromal cells. Notably, these interactions are two-way interactions with the epithelial cells instructing genuine immune cells and vice versa, as outlined in this chapter (**Figure 5.4**).

5.8   Three major mechanisms account for regulated antigen uptake across the epithelium

As discussed, the physicality of the epithelial cell barrier is formed by the polarity of a single layer of epithelial cells and their intercellular junctions, both of which are subject to a variety of homeostatic and inflammatory signals that alter composition and thus function. Inappropriate or dysregulated entry of antigens into the lamina propria may be associated with intestinal pathology (increased flux across the unrestricted pathway), but intact macromolecules are also often detected within the bloodstream of a seemingly normal

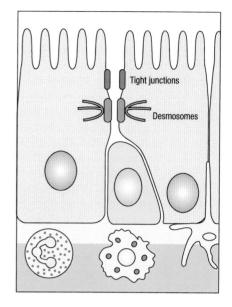

Figure 5.4 Myeloid and lymphoid cell types contribute to epithelial barrier function. Epithelial cells directly and indirectly interact closely with intraepithelial lymphocytes (IELs), neutrophils, macrophages, and dendritic cells.

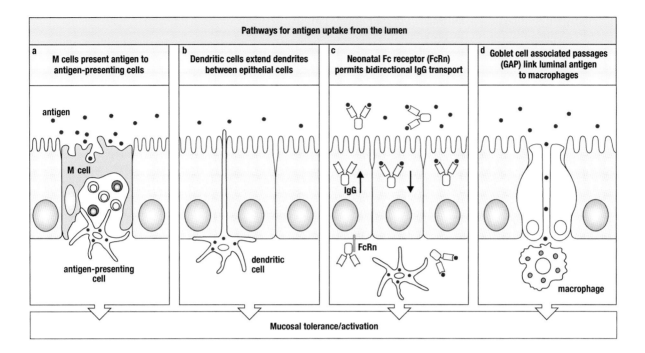

Figure 5.5 Epithelial cells sample luminal antigens through three general mechanisms. (**a**) Specialized epithelial cells (M cells) sample antigen and allow the antigen to access antigen-presenting cells localized immediately underneath. (**b**) Dendritic cells extend dendrites in between adjacent epithelial cells and thereby gain direct access to luminal contents. (**c**) The neonatal Fc receptor (FcRn) permits the bidirectional transport of IgG and thereby permits the retrieval of immune complexes from the lumen. (**d**) Goblet cell associated passages (GAP) link luminal antigen to specific types of macrophages that regulate immune responses.

epithelium. This may reflect flux across the leak pathway and is an important mechanism by which the mucosal immune system is exposed to food and microbial antigens.

There are four known regulated pathways for antigen uptake by the epithelium (see **Figure 5.5**). The first is that which is associated with M cells and the lymphoepithelial structures of Peyer's patches and isolated lymphoid follicles (covered in Chapter 15). A second pathway is by transcytosis and is associated with the ability of transporting Ig receptors within epithelial cells such as FcRn to retrieve and deliver IgG:antigen complexes to professional APCs in the lamina propria. Such antigen-laden APCs can migrate to regional lymphoid structures such as mesenteric lymph nodes and distant lymphoid tissues such as the liver and spleen. Through this mechanism, antigen exposures in the lumen can be linked to adaptive immune responses in systemic compartments. DCs that reside adjacent to the simple epithelium are capable of extending pseudopods containing tight junction proteins between epithelial cells in a chemokine-regulated pathway (such as CXCR3) for the purpose of retrieving luminal microbes and antigen. Finally, antigen can access the immune cells through goblet cell–associated passages (GAPs) that connect luminal antigen to specific subsets of macrophages that promote immune regulation.

As examples of one of the processes of antigen retrieval over mucosal barriers discussed earlier, IgG:antigen complexes can be efficiently and bidirectionally transported from the apical to the basolateral side of epithelial cells, and more importantly, basolateral IgG can retrieve antigen from the apical side as an immune complex that leaves the antigen intact, in an FcRn-dependent mechanism. Experimentally translating this observation *in vivo*, it was demonstrated that the basolateral to apical transport of IgG into the lumen is dependent on FcRn, and that antigen:IgG complexes form in the lumen as a consequence of such FcRn-dependent IgG transport. In turn, the transepithelially ported antigen:IgG complexes stimulate specific CD4$^+$ T cells in regional lymphoid structures such as the mesenteric lymph node. Further dissection of this mechanism revealed that intestinal lumen-derived antigen:IgG complexes can be transported into the lamina propria by FcRn and can be taken up by DCs, which in turn present antigen to CD4$^+$ T cells after they migrate to the regional lymphoid structures. Hence, luminal antigen can specifically be transported across the epithelial surface in an

FcRn-dependent manner; this has been applied experimentally to the development of vaccines such as for herpesvirus infection.

The relative contributions of the various transepithelial antigen transport mechanisms (FcRn, M cells, DC dendrite extension, GAP) have not yet been quantitatively assessed, and each of the mechanisms has been shown *in vivo* to have a role in the steady state as well as in experimental mucosal infection. With regard to the latter, the role of FcRn was addressed in an infectious model involving *Citrobacter rodentium. C. rodentium* is normally restricted in its localization to the epithelium, and its eradication is highly dependent on CD4$^+$ T-cell responses and IgG. FcRn selectively expressed in the intestinal epithelium renders mice less susceptible to *C. rodentium* infection in comparison with FcRn$^{-/-}$ mice. Anti-*C. rodentium* IgG protected from infection via FcRn expressed selectively on epithelial cells, and affected antigen-specific CD4$^+$ T-cell responses to *C. rodentium*. Remarkably, intestinal bacterial antigens can be transported from the lumen as immune complexes into the lamina propria by FcRn and received by CD11c$^+$ DCs.

5.9 Epithelium exerts important innate immune functions

The intestinal epithelium exerts innate immune function and is important in directly controlling the composition of the intestinal microbiota. Specifically, recognition of bacterial patterns via TLRs and downstream MyD88 signaling within Paneth cells regulates the release of bactericidal peptides (such as α-defensins and cryptdins) from Paneth cells and thereby controls intestinal barrier penetration by bacteria, both commensal and pathogenic. Moreover, intracellular pattern recognition receptors such as NOD2 have also been linked to the regulation of Paneth-cell α-defensin release. α-Defensins released from Paneth cells profoundly control the composition of the intestinal microbiota. It is further noteworthy that genetic deficiency in CD1d (a nonpolymorphic major histocompatibility complex [MHC] class I homolog presenting lipid antigens to CD1d-restricted natural killer T [NKT] cells; see later) alters the ultrastructure of Paneth-cell granules, and these Paneth cells exhibit a degranulation defect after bacterial colonization. This is likely due to decreased NKT-cell-derived interferon-γ which promotes Paneth-cell degranulation. Two further interrelated mechanisms that also control Paneth-cell granule morphogenesis and antibacterial function, and hence the innate immune functions of the epithelium, are autophagy and ER stress. However, innate immune functions of the epithelium are not confined to the Paneth cells. Goblet cells release resistin-like molecule-β, a molecule structurally linked to the adipocytokine resistin, which protects against worm infection. Specifically, resistin-like molecule-β not only propagates worm expulsion by acting on the host side via an IL-4-dependent pathway, but also by direct toxicity toward the worm through inhibition of the worm's ability to ingest nutrients.

5.10 Epithelium can present antigen via classical and nonclassical antigen-processing and antigen-presentation pathways

Epithelial cells constitutively express complex (MHC) class I molecules and are capable of presenting endogenous antigens, in the context of MHC class I, for defense against infections and neoplasia. Endogenous antigens from epithelial cells can also be captured by local dendritic cells and cross-presented on MHC class I for priming of CD8$^+$ T cells in draining lymph nodes. Epithelial cells also express MHC class II molecules, and do so preferentially in the small intestine. In response to pro-inflammatory stimuli, in particular, interferon-γ (IFN-γ), MHC II molecules are upregulated on epithelial cells. It has also been demonstrated that IFN-γ causes antigens to be sorted into MHC class II–positive late endosomes with presentation basolaterally in epithelial cell model systems. MHC class II is expressed constitutively by small intestinal epithelial cells and induced in colonic epithelium. Thus, epithelial cell expression of MHC class II can assist in the presentation of soluble antigens to CD4$^+$ T cells and the presentation of intracellular antigens via MHC class I to CD8$^+$ T cells during infection or neoplasia.

Epithelial cells express not only classical MHC class I and II molecules, but notably also nonpolymorphic MHC class I–related molecules such as CD1d, MR1, MHC class I chain–related gene A/B (MICA/B), UL-16 binding protein, and human leukocyte antigen E (HLA-E). CD1d expression on epithelial cells is regulated by IFN-γ and the heat-shock protein HSP-110 (in humans; Hsp105 in mice), which is derived in large quantities by the epithelium itself and found in rather high concentrations in the luminal content of the proximal small intestine and hence the feces of humans and mice. CD1d interactions on the intestinal epithelium with local NKT cells lead to IL-10 secretion, which in turn, protects epithelial cells from IFN-γ-induced leakage of the epithelial cell barrier function. MICA/B and UL-16 binding protein are ligands for natural killer group 2D (NKG2D) receptor, which is expressed by natural killer (NK) cells, type 1 innate like lymphoid cells and activated CD4$^+$ T cells and induce innate cytolysis of the NKG2D ligand bearing epithelial cells.

In a similar manner to the NKT-cell population that is selected by CD1d, another nonpolymorphic MHC class I–related molecule, MR1, selects another T-cell population that bears the canonical TCR-α chains hVα7.2-Jα33 or mVα19-Jα33. These T cells are preferentially localized to the gut lamina propria of humans and mice and are therefore genuine mucosa-associated invariant T cells (MAIT). Although the selection and expansion of these cells is dependent on MR1 expressed on B cells, it is notable that MR1 expression is found on epithelial cells as well. As the MR1 presents vitamin B metabolites from microbes to MAIT cells, the anatomic location of the epithelium is likely to be important to linking this important cell type to intestinal immunophysiology.

5.11 Epithelium can provide costimulatory and tertiary cytokine signals to lymphocytes

However, in contrast to professional APCs, epithelial cells lack the expression of the classical costimulatory molecules CD80 and CD86 but do express novel members of the B7 family, including inducible costimulatory ligand (ICOS) and programmed death-1 ligand (B7-H1). Further molecules that are relevant for epithelial cell–lymphoid cross talk and are constitutively expressed on epithelial cells are leukocyte function-associated antigen-3 (LFA-3; CD58), E-cadherin, the IL-7 receptor (IL-7R), carcinoembryonic antigen-related adhesion molecule 1 (CEACAM1), and CEACAM5 (CEA), as well as the common γ chain, which is required for signaling through the IL-2R, IL-4R, IL-7R, IL-9R, and IL-15R. CD86 may be expressed on epithelial cells during the inflammation associated with inflammatory bowel disease.

Furthermore, either constitutively or under certain conditions of specific stimulation, epithelial cells secrete multiple chemokines, cytokines, and other mediators relevant for orchestrating and regulating mucosal immune responses. Chemokines have a critical role in both homeostatic and inflammatory recruitment of myeloid cells, including dendritic cells, lymphocytes, and neutrophils. One excellent example of such chemokine-mediated recruitment of DCs to epithelial cells is CCL20 (MIP-3α) originating from epithelial cells, which recruits CCR6-expressing DCs to the subepithelial dome region of Peyer's patches just underneath M cells. In a *Salmonella typhimurium* infection model, it has been demonstrated that pathogen-specific T-cell responses are first activated in the Peyer's patch, and that DCs are required for this activation and expansion. CCR6+ DCs are restricted to the Peyer's patches and are rapidly recruited into the follicle-associated epithelium in this infection model. CCR6 expression on these DCs is required for the activation and expansion of *Salmonella*-specific T cells. Similarly, the CCR6 ligand CCL20 is expressed by the intestinal epithelium on exposure to *Salmonella*, indicating that a chemokine signal from the epithelium orchestrates the adaptive immune response via the recruitment and activation of macrophages and DCs at mucosal surfaces and hence the activation of pathogen-specific T-cell responses. Another chemokine ligand–receptor pair implicated in the homeostatic recruitment of APCs is CX_3CR1 and its ligand CX_3CL1, expressed in epithelial cells. The chemokine receptor CX_3CR1 on APC is required for the extension of their processes between epithelial cells, allowing these APC direct access to luminal contents and hence antigen uptake that circumvents the requirement for transepithelial transport of antigen (see earlier). The CX_3CR1 ligand CX_3CL1/fractalkine is a transmembrane chemokine expressed at the surface of epithelial cells and endothelial cells in the intestine. Although these CX_3CR1^+ cells are not required for T-cell activation in the aforementioned *Salmonella* infection model, it has been speculated that these resident CX_3CR1^+ cells may be involved in the steady-state antigen acquisition of luminal contents that contributes to tolerance induction under steady-state conditions. In line with this is the fact that orally administered labeled *Escherichia coli* can be cultured from mesenteric lymph nodes in wild-type mice but not in $CX_3CR1^{-/-}$ mice, implicating CX_3CR1 in the transport of commensals from the lumen to mesenteric lymph nodes via APCs. At the same time, similar numbers of *E. coli* can be recovered from Peyer's patches, implicating an unimpaired function of M cell–dependent transport mechanisms. Additional chemokines secreted by epithelial cells include IL-8, the related molecules ENA-78, MCP-1, GRO-α, and GRO-β, and RANTES.

Regulation of immune responses

As we have seen, epithelial cells are capable of exhibiting innate and adaptive immune functions. In this section, we discuss how the epithelium can integrate these functions into the fabric of the mucosa-associated lymphoid tissue (MALT). Specifically, the epithelium is strategically located as a sentry between the lumen with its associated commensal microbiota and the abluminal compartments, which are filled with the cellular components of the MALT. Thus, the polarized and stratified squamous epithelium can function as first responders to the environment (microbial and nonmicrobial) and consequently direct subsequent immunologic events and/or participate in responding to signals from abluminal immune and nonimmune cells. This section discusses these concepts.

5.12 Epithelial cells act as central organizers of immune responses

Insights obtained over the past few years have revealed a central role of epithelial cells in orchestrating the mucosal immune response (**Figures 5.6** and **5.7**). Mediators derived from epithelial cells can direct the type of immune response by acting distally on macrophages and DCs, with major consequences for T-cell and B-cell differentiation. Epithelial cells are typically the first cells to respond to luminal signals and thus as sentinels that guard the host. Thymic stromal lymphopoietin (TSLP) derived from epithelial cells is an instructive example of such a mediator. These concepts were initially envisioned on the basis of data derived from *in vitro* experiments involving epithelial cell lines and DCs and have been proven and further developed in mice with genetic targeting of specific genes in the intestinal epithelium by using Cre/*loxP* technology. Specifically, TSLP expression in epithelial cells is induced by the microbial flora of the intestine in a pathway that depends on NFκB signaling (in particular IKK2). TSLP binds to the TSLP receptor (TSLP-R) on DCs and renders them "tolerogenic" in that they are induced to produce high levels of IL-10 and IL-6, but not IL-12. Remarkably, epithelial cells genetically lacking *Ikbkb* (encoding IKK2) exhibit decreased TSLP expression, and such mice are unable to eradicate the worm *Trichuris muris* as a consequence of a failure to develop T_H2 immunity in a TSLP-dependent pathway. Instead, these IKK2- and TSLP-R-deficient mice develop a pathological helper T-cell response characterized by increased IFN-γ and IL-17 secretion, accompanied by increased TNF-α and IL-12/23p40 expression from DCs. Instead of eradicating *Trichuris muris*, these mice develop severe intestinal inflammation. Further underscoring the intricate network dictated by TSLP secretion from epithelial cells is the fact that blockade of IFN-γ and IL-12/23p40 signaling results in the restoration of resistin-like molecule-β expression from epithelial cells and the consequent expulsion of the worm. Further evidence for a key role of NFκB signaling in epithelial cells and the

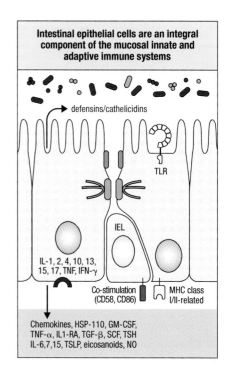

Figure 5.6 Epithelial cells direct numerous components of the innate and adaptive immune systems at various levels. Various mediators derived from myeloid, lymphoid, and stromal compartments act on epithelial cells and affect diverse biological processes including barrier function, the processing and presentation of antigens, the expression of MHC and costimulatory molecules, and the secretion of antimicrobial peptides. At the same time, epithelial cells secrete, and respond to, various mediators that profoundly affect the function of surrounding mucosal cell types. GM-CSF, granulocyte-macrophage colony-stimulating factor; HSP-110, heat-shock protein 110; IEL, intraepithelial lymphocyte; IFN-γ, interferon-γ; IL-1RA, interleukin-1 receptor antagonist; IL-6, interleukin-6; NO, nitric oxide; SCF, stem cell factor; TGF-β, transforming growth factor-β; TLR, toll-like receptor; TNF-α, tumor necrosis factor-α; TSH, thyroid-stimulating hormone; TSLP, thymic stromal lymphopoietin.

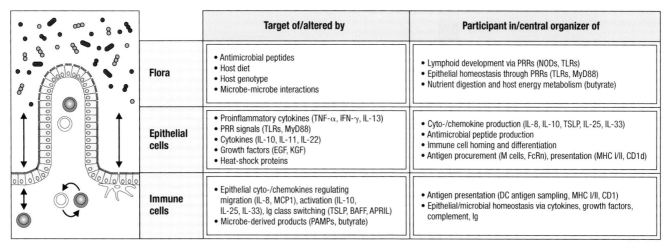

		Target of/altered by	Participant in/central organizer of
	Flora	• Antimicrobial peptides • Host diet • Host genotype • Microbe-microbe interactions	• Lymphoid development via PRRs (NODs, TLRs) • Epithelial homeostasis through PRRs (TLRs, MyD88) • Nutrient digestion and host energy metabolism (butyrate)
	Epithelial cells	• Proinflammatory cytokines (TNF-α, IFN-γ, IL-13) • PRR signals (TLRs, MyD88) • Cytokines (IL-10, IL-11, IL-22) • Growth factors (EGF, KGF) • Heat-shock proteins	• Cyto-/chemokine production (IL-8, IL-10, TSLP, IL-25, IL-33) • Antimicrobial peptide production • Immune cell homing and differentiation • Antigen procurement (M cells, FcRn), presentation (MHC I/II, CD1d)
	Immune cells	• Epithelial cyto-/chemokines regulating migration (IL-8, MCP1), activation (IL-10, IL-25, IL-33), Ig class switching (TSLP, BAFF, APRIL) • Microbe-derived products (PAMPs, butyrate)	• Antigen presentation (DC antigen sampling, MHC I/II, CD1) • Epithelial/microbial homeostasis via cytokines, growth factors, complement, Ig

Figure 5.7 Multiple inputs at the various levels of the epithelial barrier affect the mucosal immune compartment. Epithelial cells relay various signals emanating from the mucosal environment, including signals from the microbiota and immune cells. At the same time, epithelial cells act as central organizers or participants for multiple local and systemic immune functions and the regulation of microbial composition. APRIL, a proliferation-inducing ligand; BAFF, B-cell-activating factor; EGF, epidermal growth factor; FcRn, neonatal Fc receptor; IFN-γ, interferon-γ; KGF, keratinocyte growth factor; MCP-1, mast-cell precursor 1; NOD, nucleotide-binding oligomerization domain; PAMPs, pathogen-associated molecular patterns; PRR, pattern recognition receptor; TNF-α, tumor necrosis factor-α; TLR, toll-like receptor; TSLP, thymic stromal lymphopoietin. (Adapted from Kaser, A. et al. *Annu. Rev. Immunol.* 2010, 28:573–621. With permission from Annual Reviews.)

organization of mucosal immune responses comes from mice with a genetic deletion of NFκB essential modulator (NEMO), which acts upstream of IKK1 and IKK2. In those mice, epithelial NEMO deletion results in a MyD88- and TNF-α-dependent induction of severe inflammation of the large intestine. Interestingly, during the early stages of colonic inflammation in this model, innate immune cells dominate, whereas at later stages lymphocytes are predominantly involved.

Epithelial cells not only determine cell fate decisions in mucosal T cells via their actions on DCs, they also regulate IgG/IgA class switching in the B-cell compartment at mucosal surfaces. Epithelial cells lining tonsillar crypts form pockets that contain B cells actively performing class switching at that locale. When these epithelial cells sense microbial components via TLRs, they release soluble mediators such as B-cell activating factor (BAFF, encoded by *TNFSF13B*) that induce class switching through the induction of activation-induced cytidine deaminase (AID). This epithelial-cell-induced class switching is amplified by TSLP by means of a route that involves the "licensing" function of DCs. Remarkably, epithelial cells also secrete an inhibitor of this process, secretory leukocyte protease inhibitor, which inhibits AID function in B cells. A similar process is operative in the intestine, where bacteria induce the cytokine APRIL (a proliferation-inducing ligand, encoded by *TNFSF13*) in the epithelium and thereby induce T-cell independent IgA and IgG class switching.

5.13 Epithelial cells act as bystanders and participants in immune responses and inflammation

Although the functions described in the previous paragraphs have focused on mechanisms in which epithelial cells seem to be the central coordinators of mucosal immune functions, these central epithelial cell functions are influenced by various stimuli from outside compartments, be it from the microbial milieu or host-derived factors. In this section, we discuss factors for which this outside influence in regulating epithelial cell function seems more dominant, and in which the epithelial cells therefore fulfill a contributory role in mucosal physiology. However, such bystander or participant functions of epithelial cells can nevertheless become a central feature of pathophysiologic processes in the mucosa.

As discussed in other chapters, pro-inflammatory cytokines such as IFN-γ can upregulate MHC class II expression, and also upregulate the expression of costimulatory molecules such as CD86, which are not expressed in the baseline state. Epithelial cells express a wide variety of receptors for pro-inflammatory and anti-inflammatory cytokines, including IL-1, IL-2, IL-4, IL-10, IL-13, IL-15, IL-17, TNF-α, and IFN-γ. The mediators that bind to these receptors are derived from local myeloid, lymphoid, and nonimmune cells in the mucosal lamina propria and affect the epithelium. As examples, IL-13 and TNF-α can adversely affect epithelial cell barrier function via actions on tight junction expression, whereas anti-inflammatory IL-10 can counteract this. Although IELs may contribute to the maintenance of epithelial cell barrier function in various ways, they can also directly damage epithelial cells in the context of disease, thereby enhancing unrestricted pathway flux; this IEL function may be enhanced even further via their granzyme and FasL expression.

Moreover, epithelial cells themselves secrete a large panel of inflammatory mediators after receiving inductive signals from other immune and nonimmune cells or luminal stimuli, and thereby can amplify an evolving mucosal immune response. A particularly relevant mechanism involves the induction of neutrophil- and lymphocyte-recruiting chemokines in epithelial cell, as well as myeloid interferons such as IFN-$\alpha\beta\lambda$, which can protect from infections.

In addition to the contribution of epithelial cells to the recruitment of the inflammatory infiltrate, epithelial cells also respond to inflammatory mediators with increased chloride and mucus secretion, as well as with an increase in paracellular permeability. The clinical correlate of these alterations is the occurrence of diarrhea. Immune cell–secreted mediators that increase chloride secretion and mucus production are serotonin, prostaglandin E_2, leukotriene C_4, histamine, and vasoactive intestinal peptide, whereas IL-13, TNF-α, and other cytokines have important roles in increasing paracellular permeability.

$\gamma\delta$ T cells constitute an important subset of IELs that actively patrol the epithelium and are able to rapidly clear pathogens that overcome the tight junction barrier. $\gamma\delta$ T cells may also play a role in epithelial regeneration, likely due to their secretion of keratinocyte growth factor. For example, TCR$\delta^{-/-}$ mice were more susceptible to colitis induced by the damage-inducing agent dextran sulfate sodium (DSS), likely due to the absence of compensatory epithelial proliferation driven by $\gamma\delta$ T-cell–derived keratinocyte growth factor.

Cross-talk between epithelial cell and nonepithelial cell compartments further contributes to the regenerative response of the epithelium. Various pattern recognition receptors, prominently TLRs and NOD proteins, are expressed in the mucosa in both epithelial cells and professional APCs. The microbial ligands recognized by TLRs are not unique to pathogens and are produced by both pathogenic and commensal microorganisms. Commensal bacteria are recognized by TLRs under normal steady-state conditions; this interaction has a crucial role in the maintenance of intestinal epithelial homeostasis and is also important for protection against gut injury during experimental colitis. Remarkably, this colonic epithelial regenerative response involves colonic epithelial progenitors and requires the gut microbiota, because it is markedly reduced in germ-free animals. It seems that it is not TLR signaling in epithelial cells itself that mediates this regenerative response; instead, TLR ligation on mucosal macrophages relays a signal to epithelial cells that is required for this regenerative response of the epithelial compartment. During the regenerative response, macrophages in the pericryptal stem cell niche extend processes to directly contact colonic epithelial progenitors near the crypt base. Cyclooxygenase-2 (COX2) expression and function are required for the previously mentioned regenerative response after epithelial injury, and these functions are downstream of MyD88 signaling. Notably, COX2 expression is not regulated by injury and is present in MyD88$^{+/+}$ and MyD88$^{-/-}$ mesenchymal cells. However, in MyD88$^{+/+}$ mice, injury as a result of dextran sodium sulfate leads to the repositioning of COX2-expressing stromal cells from the mesenchyme surrounding the middle and upper parts of crypts to an area surrounding the crypt base adjacent to colonic epithelial progenitors.

A further factor critically involved in the regenerative response and wound healing of the epithelium is IL-22. IL-22 expression is induced in innate and adaptive immune cells, specifically a subtype of NK-like cells, as well as T_H17 helper T cells. IL-22-expressing NK-like cells, which are part of an increasingly recognized class of type 1 innate lymphocytes within the epithelium, are specifically found in tonsils, cryptopatches, and Peyer's patches; and the microbial flora is important for their induction. Remarkably, the transcription factor RORγt has a central role in the differentiation of both this subtype of mucosal NK cells and T_H17 cells. IL-22 induces the activation of STAT3 in epithelial cells and orchestrates a pro-regenerative, wound-healing program.

Important insights into the importance of epithelial cell–intraepithelial lymphocyte cross talk also come from mechanisms established in celiac disease, a disease precipitated by the exogenous antigen gluten. In the presence of gluten, epithelial cells secrete IL-15 and express nonclassical MHC class I molecules such as HLA-E and MICA. As a consequence of IL-15 secretion, NK receptors are upregulated on IELs such as the activating receptor NKG2D, which in turn leads to IEL activation through recognition of NKG2D ligands on the epithelial cell. Hence, low-affinity self-antigens lead to IEL activation as a result of a reduced activation threshold, and exhibit NK-cell-like activity, thereby destroying epithelial cells via cytolysis and IFN-γ secretion.

SUMMARY

Epithelial cells not only function as a physical barrier between the sterile environment of the host and the accumulation of foreign material including microbial constituents, but in fact they are also at the center of manifold immune functions of the mucosa. These involve antigen uptake and presentation, T-cell and NKT-cell interactions and reciprocal instruction, dendritic cell recruitment, and determination of the DC and T-cell response and the overall inflammatory milieu of the mucosa. Finally, epithelial cells have a central role in regulating the composition of the commensal microbiota and resisting pathogenic invasions by microbes.

FURTHER READING

Blander, J.M., Longman, R.S., Iliev I.D., Sonnenberg, G.F., Artis, D.: Regulation of inflammation by microbiota interactions with the host. *Nat. Immunol.* 2017, 18(8): 851–860.

Buckley, A., and Turner, J.R.: Cell biology of tight junction barrier regulation and mucosal disease. *Cold Spring Harb. Perspect Biol.* 2018, 10(1).

Clevers, H.: The intestinal crypt, a prototype stem cell compartment. *Cell* 2013, 154:274–284.

Gensollen, T., Iyer, S.S., Kasper, D.L., Blumberg, R.S.: How colonization by microbiota in early life shapes the immune system. *Science* 2016, 352:539–544.

Grootjans, J., Kaser, A., Kaufman, R.J., Blumberg, R.S.: The unfolded protein response in immunity and inflammation. *Nat. Rev. Immunol.* 2016, 16:469–484.

Gutzeit, C., Magri, G., Cerutti, A.: Intestinal IgA production and its role in host-microbe interaction. *Immunol. Rev.* 2014, 260:76–85.

Mukherjee, S., and Hooper, L.V.: Antimicrobial defense of the intestine. *Immunity* 2015, 42:28–39.

Johansson, M.E., Larsson, J.M., Hansson, G.C.: The two mucus layers of colon are organized by the MUC2 mucin, whereas the outer layer is a legislator of host-microbial interactions. *Proc. Natl. Acad. Sci. USA.* 2011, 108(Suppl 1): 4659–4665.

Kaser, A., Zeissig, S., Blumberg, R.S.: Inflammatory bowel disease. *Annu. Rev. Immunol.* 2010, 28:573–621.

Pasparakis, M., and Vandenabeele, P.: Necroptosis and its role in inflammation. *Nature* 2015, 517:311–320.

Pitman, R.S., and Blumberg, R.S.: First line of defense: The role of the intestinal epithelium as an active component of the mucosal immune system. *J. Gastroenterol.* 2000, 35:805–814.

Proud, D., and Leigh, R.: Epithelial cells and airway diseases. *Immunol. Rev.* 2011, 242:186–204.

Pyzik, M., Rath, T., Lencer, W.I., Baker, K., Blumberg, R.S.: FcRn: The architect behind the immune and nonimmune functions of IgG and albumin. *J. Immunol.* 2015, 194:4595–4603.

Ramanan, D., and Cadwell, K.: Intrinsic defense mechanisms of the intestinal epithelium. *Cell Host Microbe.* 2016, 19:434–441.

Rescigno, M.: Dendritic cell-epithelial cell crosstalk in the gut. *Immunol. Rev.* 2014, 260:118–128.

Turner, J.R.: Intestinal mucosal barrier function in health and disease. *Nat. Rev. Immunol.* 2009, 9:799–809.

Uhlig, H.H., and Powrie, F.: Translating immunology into therapeutic concepts for inflammatory bowel disease. *Annu. Rev. Immunol.* 2018, 36:755–781.

Wells, J.M., Rossi, O., Meijerink, M., van Baarlen, P.: Epithelial crosstalk at the microbiota-mucosal interface. *Proc. Natl. Acad. Sci. USA.* 2011, 108(Suppl. 1), 4607–4614.

Intraepithelial T cells: Specialized T cells at epithelial surfaces

6

HILDE CHEROUTRE

Intraepithelial lymphocytes (IELs) are small mononuclear cells interspersed between epithelial cells. The majority of these lymphocytes express a T-cell receptor (TCR) and belong to the T-cell lineage and are therefore called intraepithelial T cells (IETs). Some IELs lack a rearranged antigen receptor and are classified as innate lymphoid cells (ILCs). For the purpose of this chapter, we focus on the T-cell subsets. IETs are found in various epithelial tissues such as the skin and the epithelium of the small and large intestine, the biliary tract, the oral cavity, the lung, the upper respiratory tract, and the reproductive tract. The largest population of IETs resides within the columnar epithelial layer of the intestine (Figure 6.1), the single layer of cells that lines the lumen of the intestine and that forms the physical barrier between the core of the body and the environment. Complex innate and adaptive immune networks have developed at this prominent interface, and IETs are critical in maintaining intestinal homeostasis (see Figure 6.1). Although protection against pathogens is one of the major tasks of the mucosal immune system, it also needs to carry out these functions without excessive or unnecessary inflammation and tissue damage in order to preserve the integrity of the mucosal barrier. Furthermore, the intestine harbors trillions of microorganisms that establish a symbiotic relationship with their host. The fundamental role of the intestine as a digestive organ adds food to the antigenic load in the gut. Because of their location at this critical interface, IETs need to balance protective immunity while safeguarding the integrity of this major barrier, a *sine qua non* for the homeostasis of the organism.

UNIQUENESS AND HETEROGENEITY OF MUCOSAL INTRAEPITHELIAL T CELLS

Gut IETs are notably heterogeneous in phenotype and function and are distributed differently in the epithelium of the small and large intestines. The latter is probably a reflection of the distinct digestive functions and the unique adaptations to fight infections as well as maintain tolerance toward innocuous antigens derived from the diet (mainly in the small intestine) or resident noninvasive commensals (mainly in the large intestine) as well as numerous self-antigens derived from digestive enzymes, globulins, albumins, mucins, and shed epithelial cells.

Despite this, intestinal IETs share unique phenotypic and functional characteristics that distinguish them from systemic T cells in the periphery. First, there are more T cells residing within the gut epithelium than there are in other tissues. B cells are rare and only present in specialized epithelium overlying Peyer's patches (PP). Second, IETs include an abundance of γδ T-cell receptor (TCR) expressing cells. Third, IETs are antigen-experienced cells; however, they do not express markers of recently activated cells, such

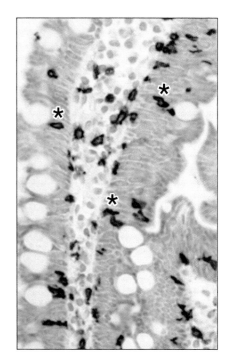

Figure 6.1 **Intraepithelial T cells (IETs) reside interspersed between the columnar epithelial layer of the intestine that forms the interface between the inner core of the body and intestinal lumen.** Shown is small intestine villus epithelium. Brown cells are CD8[+] T cells in epithelium (*) and lamina propria.

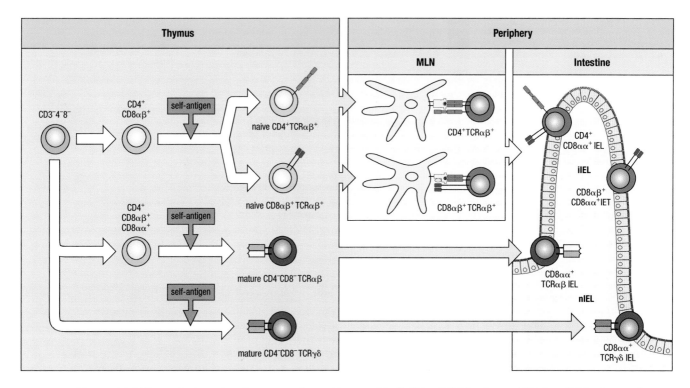

Figure 6.2 Differentiation of antigen-experienced intraepithelial T-cell (IET) subsets. TCRγδ and TCRαβ natural IET precursors acquire an antigen-experienced phenotype in the thymus in response to strong TCR signals and migrate directly to the gut epithelium where they further adapt and acquire CD8αα expression. TCRαβ induced IET precursors derive from conventional selected naive CD8αβ or CD4 single-positive T cells that encounter their cognate antigen in the periphery and subsequently home to the gut mucosae where they further adapt including expression of CD8αα. MLN, mesenteric lymph node.

as CD25. Fourth, the majority of IETs display a typical cytotoxic phenotype, but at steady state they secrete only low levels of effector cytokines. Fifth, and typical of stress-sensing cells, IETs also express innate activating and inhibitory natural killer (NK) cell-like receptors. Sixth, IETs express the $\alpha_E\beta_7$ integrin (CD103 indicates the α_E chain), which interacts with E-cadherin on the epithelial cells. Seventh, IETs express activation markers such as CD69 and CD8αα homodimers, a hallmark of activated cells. (The ligand for CD8αα, the nonclassical major histocompatibility complex [MHC] class I molecule thymus leukemia antigen, is also abundantly and constitutively expressed on gut epithelial cells.) Eighth, a large fraction of TCRαβ IETs are CD4⁻CD8⁻ double negative, typical of high-affinity T cells. They express a limited repertoire of TCR-α and TCR-β chains and are thus considered to be oligoclonal. Finally, IETs display a typical "restrained" phenotype hallmarked by the expression of molecules like CD160, CD244, CRTAM, and LAG3.

Although the antigen-experienced phenotype is typical of all IETs, the activation pathways they follow and the cognate antigens they recognize are very different. Consequently, IETs are divided into two major subtypes. The "natural" IETs (nIETs) acquire their activated phenotype during a self-antigen-induced differentiation process in the thymus, whereas "induced" IETs (iIETs) acquire their activated phenotype in response to foreign or self-antigens encountered in the periphery (**Figure 6.2**).

6.1 Natural IETs are enriched in the intestinal epithelium

Natural IET-type cells are most abundant in newborn and young mice and humans, and data concerning the phenotype, ontogeny, and function of nIETs have been obtained mainly from studies using mouse models.

Natural IETs, previously referred to as "unconventional" T cells, can be either TCRγδ or TCRαβ. They are typically CD2−, CD5−, CD28−, LFA-1−, and B220+ and mostly Thy1− (CD90−). In addition, most TCRαβ nIETs are CD69+B220+CD44low/int, a phenotype seldom found among peripheral T cells. Furthermore, although all TCRαβ nIETs depend on β_2-microglobulin, they are often restricted to "nonclassical" MHC class I molecules, whereas some TCRγδ IETs also respond to unprocessed antigens. nIETs commonly express CD3ζ-FcεRIγ heterodimers or FcεRIγ-FcεRIγ homodimers rather than CD3ζ-CD3ζ homodimers as part of their CD3 complex. In addition, many of them have the ability to respond in a TCR-independent fashion, and they characteristically share molecules expressed by natural killer (NK)–type cells, such as Ly49A, Ly49G2, Ly49E, NK1.1, DAP12, and CD122. nIETs expressing the γδ TCR make up approximately 50% of mouse epithelial lymphocytes and only approximately 10% of the human epithelial lymphocytes, but this number can be quite variable and increases in various conditions, including celiac disease. The factors controlling the numbers and composition of nIETs remain poorly defined.

6.2 Induced IETs are the mucosal counterpart of conventional T cells found in the periphery

iIETs are the progeny of CD4+ or CD8αβ+ TCRαβ MHC class II or class I restricted T cells. These TCRαβ+ T cells initially develop as naive cells in the thymus along the conventional selection pathway and then undergo further differentiation in response to their cognate antigens in the periphery (see Figure 6.2). Consistent with this, they express an "effector/memory-like phenotype" (CD2+, CD5+, Thy1+, LFA-1+, CD44+), often together with the activation markers, CD69 and CD8αα. Most iIETs are CD8+ TCRαβ cells that like the nIETs exhibit a typical cytotoxic effector memory phenotype. CD4+ IETs are less frequent, and especially in the small intestine, they display a cytotoxic phenotype similar to their CD8 counterparts (discussed later).

iIETs have much in common with other memory T cells, but they also have distinct features shared with the nIETs. Most iIETs are constitutively granzyme B^high, and they typically express CD103 and CD69, but they lack Ly6C and CD62L. Typical of effector or tissue memory cells, iIETs do not clonally expand, and at steady state, they produce only low levels of inflammatory cytokines. In addition, both inhibitory and activating NK receptors are found on most iIETs, giving them a phenotype typical of "stress-sensing" yet "highly regulated" or "restrained" T cells. Similar to the nIETs, the unique characteristics of iIETs might also have evolved in order to deal with the gut dilemma: to provide effective immediate immune protection without compromising barrier function.

6.3 Luminal factors control the numbers of IETs

Under normal physiologic conditions, there is one IET for every 3–10 epithelial cells in the small bowel, and about one IET per 40 epithelial cells in the colon. The different IET subtypes are present to varying extents in the small and large intestines, and the distribution depends in part on age, species, and environmental conditions. In contrast to the lamina propria, the "natural" type T cells are far more prevalent in the epithelium, especially of the small intestine.

In rodents, small intestine nIETs are the principal population at birth, and there is only a modest expansion early on (Figure 6.3). In contrast, iIETs largely expand after weaning, and they gradually accumulate with age in response to microbial commensalization.

Two important events shortly after birth have a profound impact on IETs: first, the microbial colonization of the gut, and second, the changes in the diet-related antigen load and composition. The increase in TCRαβ iIETs

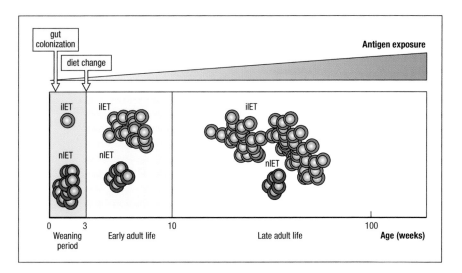

Figure 6.3 Intraepithelial T-cell (IET) subsets over a lifetime. Natural IETs are present during gestation and at birth and decrease with age, whereas induced IETs are few at birth but increase with antigen exposure and with age.

depends in part on the microbial colonization of the intestine. In addition, TCRγδ nIETs predominate in barrier-maintained specific pathogen-free adult mice, whereas in adult mice kept in nonbarrier facilities, TCRαβ iIETs prevail. The early migration of nIETs to the intestine is independent of exogenous antigens but crucially depends on CCR9/CCL25 and $\alpha_4\beta_7$. Once residing in the intestinal epithelium, recognition of gut microbiota by NOD2, an intracellular sensor for bacterial peptidoglycan, is essential for the homeostasis of both nIETs and CD8$\alpha\beta^+$ iIETs. Diet-derived factors also play a role in intestinal homeostasis as cruciferous vegetables contain ligands for aryl hydrocarbon receptor (AhR), a receptor critical for normal CD8$\alpha\alpha$TCR$\alpha\beta^+$ nIET numbers.

In human gut epithelium, there is extensive infiltration of T cells from around 12–14 weeks of gestation and the number continues to increase, reaching about half the density of IETs found in young children by 28 weeks of gestation. The absolute number increases dramatically during that time. The majority of IETs in humans, both at fetal and adult stages, express αβ TCRs. Very few (0.1%) human IETs are actively dividing, and in rodents, where survival can be measured, it appears that the majority of IETs are long lived. In addition, during homeostasis, there is only a limited contribution of circulating lymphocytes to the IET compartment, as shown in parabiotic mice.

T-cell receptor repertoire and specificity of IETs

In young and adult animals, apart from a few exceptions such as γδ T cells, the expressed T-cell repertoire is diverse, presumably to maximize the number of peptides capable of being recognized by T cells. One of the most striking features of IETs, however, is that in mice and humans, and in the αβ and γδ subsets, TCR usage is highly skewed toward a small number of clones.

6.4 T-cell receptor repertoire and antigen specificity of TCRγδ nIETs are different from T cells in rest of body

Regardless of species, the γδ TCRs expressed by nIETs are largely distinct from their peripheral counterparts. Mouse TCRγδ IELs predominantly express Vγ5 and Vγ1.1, and a few use Vγ2. With respect to TCRγδ, depending on the strain, 15%–60% of the TCRγδ IETs express Vδ4 and some express Vδ5 or Vδ6. For humans, most (approximately 70%) colonic TCRγδ IETs utilize Vδ1. Furthermore, human IETs express a non-disulfide-linked form of the γδ TCR unlike peripheral γδ T cells. The complexity of the γδ TCR is largely contributed by the complementarity-determining region 3 (CDR3) of the δ chain and by random joining of the variable (V), diversity (D), and joining (J) segments, random N-region nontemplate deletions, and additions of nucleotide palindromes at the coding ends of the V, D, and J segments (P additions). The composition of the TCRγδ IETs changes with development with changes in V region usage and junctional diversity. The TCRδ of fetal intestinal T cells is polyclonal, with different CDR3 lengths and several template-encoded P-region, but no N-region, additions. This polyclonal nature is maintained until birth, but at that time the junctional regions become more complicated with numerous N-region insertions. With age the TCRδ repertoire becomes increasingly restricted, and in young adults the TCRδ usage is comparable with that in older adults. Despite the diversity of nIET γδ TCRs, the oligoclonal expansions are distinct between individuals but appear to be stable over time.

The substantial junctional and combinatorial diversity implies that TCRγδ nIETs might recognize a diverse set of antigens. Identification of such ligands, however, remains a major deficit in our knowledge of TCRγδ nIET biology.

6.5   TCRαβ IETs are made up of a few dominant clones

At steady state, the TCRαβ repertoire of IETs is oligoclonal and comprises a few hundred clones that are quantitatively and qualitatively unique to the individual. In the case of iIETs, which increase with age and in response to environmental triggers and pathogen challenges (see **Figure 6.3**), the specificity and diversity of the TCRαβ repertoire are probably a direct reflection of the environmental antigen platform encountered by that particular individual. Although also oligoclonal and unique, the repertoire of nIETs shows very little overlap with that of the iIETs. Furthermore, αβ TCRs from nIETs of littermates in the same cage or from germ-free mice, show the same degree of oligoclonality. The oligoclonality of TCRαβ IETs is also seen in humans with, similarly to the γδ TCR nIETs, striking changes throughout life. Fetal intestinal lamina propria lymphocytes and IETs contain transcripts of only a few TCRβ members as early as 13 weeks. Diversity increases with age, and at birth all 24 Vβ families are detectable. Analysis of the CDR3 regions reveals variability with numerous nontemplated N-region additions. In contrast, rearranged TCRα chains are undetectable until 16 weeks of gestation, and only a few can be detected at later fetal stages. A large fraction of the fetal TCRα transcripts contain an immature precursor segment, termed T early α, which is also expressed in thymocytes prior to the onset of TCRVα recombinations. By 23 weeks of gestation, IETs containing TCRα transcripts with productively rearranged TCRV-J-C segments can be detected in the gut epithelium. IETs at the adult stage express an oligoclonal repertoire with dominant expression of the same αβ TCRs throughout the intestine.

6.6   TCRαβ nIETs exhibit higher affinity for self-antigens and unusual major histocompatibility restrictions

In addition to oligoclonality, some Vβ segments, which are deleted from the TCRαβ repertoire of naïve T cells in the periphery, are highly enriched in the repertoire of the CD8ααTCRαβ nIETs, including "forbidden" Vβ segments. For example, in mouse strains expressing I-E and Mls-1ᵃ antigen or mouse mammary tumor virus, forbidden Vβ6, Vβ8.1, or Vβ11 are overrepresented in the TCR repertoire of CD8ααTCRαβ nIETs. The presence of self-reactive TCRs among CD8ααTCRαβ⁺ nIETs indicates that their precursors that develop in the thymus do not follow the conventional selection process, which is thought to delete self-reactive TCRs.

This notion led to the discovery of an alternative selection process in the thymus, called "agonist selection." TCRαβ nIETs have long been associated with agonist TCR selection signals in TCR transgenic mouse models. Substantial numbers of transgenic CD8αα+TCRαβ+ nIETs were generated when a TCR and its cognate ligand were expressed as transgenes in the thymus. The TCR-instructive nature of nIET development was firmly demonstrated when TCRs were cloned from TCRαβ nIETs and retrogenically expressed in mouse models. These TCRs induced strong TCR activation and guided specific development to CD8ααTCRαβ+ nIETs. These nIET TCRs displayed unusual MHC restrictions as some clones recognized MHC class II or nonclassical MHC molecules, and those that were restricted to classical MHC class I appeared independently of transporter associated with antigen processing (TAP). This particular restriction pattern is confirmed at the polyclonal level as well, since CD8αα+TCRαβ+ nIETs are drastically reduced in β₂-microglobulin (β2m)–deficient mice but relatively normal in mice lacking TAP. In addition, some of the human TCRαβ IETs can be restricted to the MHC I–like CD1 molecules, CD1a, CD1b, CD1c, and CD1d.

Although thymus leukemia (TL) antigen does not serve as a restriction element or antigen-presenting molecule, TL antigen does bind with high affinity to CD8αα; this interaction does not lead to activation but to modulation of TCR reactivity and survival.

Development and differentiation of IETs

An idea dating back to the 1960s was that some IETs are T cells that are generated locally in the gut from bone marrow–derived precursors. However, the relative contribution of the thymus-dependent versus thymus-independent pathway has been controversial for many years. More recent studies have convincingly shown that under normal physiologic conditions, TCRαβ nIETs are thymus derived.

6.7   Thymic versus extrathymic development of IETs is controversial

Substantial efforts have been brought to bear on understanding the origin of IETs, with special emphasis on the CD8αα+ nIETs, in part due to their unusual phenotype, and on TCRαβ nIETs, for their restricted TCR repertoire highly enriched for self-reactive TCRs. Moreover, the intestine has been described as a primary lymphoid organ based on the demonstration that certain lymphocytes are present in the intestinal epithelium of neonatal-thymectomized mice and that IETs are present in primitive vertebrates. This concept appears reasonable considering that gut-associated lymphoid tissue arose prior to the thymus and may be the evolutionary antecedent. Nevertheless, in the absence of a functional thymus, as in nude mice, IETs are drastically reduced, and the residual lymphocytes chiefly express TCRγδ. Additionally, extrathymic development in the absence of irradiation results mainly in TCRγδ nIETs, suggesting that radiation promotes extrathymic T-cell development. Neonatal or fetal thymus grafting studies have demonstrated that all IETs are thymus derived, and consistent with this, neonatal thymectomy largely ablates all IET populations. These findings indicate that under normal circumstances, IETs are thymus derived but that, similar to the iIETs, further maturation may occur in the intestine. Lymphoid structures at the base of some small intestinal crypts in the mouse, termed *cryptopatches*, contain hematopoietic c-Kit+, interleukin-7 receptor α (IL-7Rα+), Thy1+, CD44+, Lin− precursors, and were proposed as a potential site where such "extrathymic" development could occur, although cryptopatches have not been described in humans or other species. The formation of mouse cryptopatches is interleukin-7 (IL-7) and IL-15 dependent and enterocyte-restricted IL-7 is crucial for cryptopatches and sufficient for extrathymic development of TCRγδ T cells. Cryptopatches, however, are not absolutely required for TCRγδ nIETs. Moreover, athymic RAG reporter mice show RAG expression in mesenteric lymph nodes (MLNs) and PP, but not in cryptopatches or IETs. Nevertheless, MLN and PP are also not required for extrathymic IETs, and RAG expression in mucosal tissues is negligible in euthymic animals, suggesting that the extrathymic pathway is a secondary consequence of a lack of thymus and thymus-derived cells and signals.

6.8 Under euthymic normal conditions, all IET subsets are progeny of bone marrow precursor cells that initially develop in thymus

Thymocyte differentiation and maturation do not compose a single streamlined process, and multiple pathways exist that ultimately lie at the basis of phenotypic and functional diversity of all mature T cells, including the IET subtypes. The γδ/αβ T-cell–lineage decision occurs early during thymic development, and both TCRαβ and TCRγδ T cells arise from a common precursor but diverge into separate lineages at the immature DN3 stage, just after T-cell commitment (Figure 6.4). Although the process of lineage differentiation is not fully understood, several underlying mechanisms have been proposed, including differential IL-7R expression and differences in signal strength propagated by pre-TCRα versus γδ TCR. Several observations indicate that strong signals at the immature double-negative stages favor the γδ lineage. Nevertheless, the constant association of pre-TCR with the αβ lineage choice points to unique qualities of the pre-TCR for αβ lineage commitment. The fact that nontransgenic CD8αα+TCRαβ+ nIELs are strictly dependent on pre-TCR expression indicates that these nIETs are not part of the γδ lineage but are genuine pre-TCR-dependent TCRαβ-lineage cells.

6.9 Natural IET precursor cells migrate directly from thymus to epithelium

The term "natural" refers to the lineage development of nIET precursor cells, which go through an "alternative" self (agonist) antigen-based thymic maturation process that results in the functional differentiation of mature double-negative, TCRγδ- or TCRαβ-expressing antigen-experienced T cells that directly migrate to the epithelium of the intestine (see Figure 6.2). In contrast to conventional thymocytes, nIET precursor cells are not deleted by agonist cognate self-antigens encountered in the thymus. Instead, such agonist signals are required to drive their antigen-experienced phenotype and functional differentiation (see Figure 6.4).

6.10 Development and differentiation of TCRγδ nIETs

In addition to strong and/or sustained TCR signals for γδ-lineage commitment, specific transcription factors promote and bias specific differentiation. In that aspect, mice deficient for the Wnt-induced β-catenin-controlled T-cell factor-1 showed drastic reduction in all TCRαβ+ subsets and also in the TCRγδ+ nIETs but not in TCRγδ+ splenocytes. Also, signals transduced through IL-7R play nonredundant roles during survival and expansion of T-cell progenitors and the rearrangement of the TCRγ locus. Consequently, IL-7-deficient (IL-7−/−) mice lack TCRγδ cells in all tissues.

Although genetic programming is involved in γδ gene rearrangements, selective TCRγδ-associated signals may also control the specific gut-homing

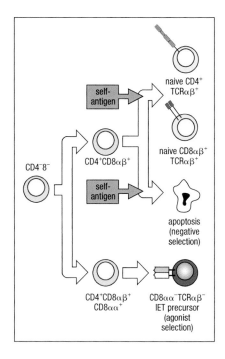

Figure 6.4 Thymic selection of TCRαβ precursor cells. Immature double-positive thymocytes that receive a strong selection signal undergo activation-induced death during a process of conventional negative selection. Double-positive thymocytes that receive intermediate signals are positively selected and exit the thymus as conventional naive CD4+ or CD8αβ+ T cells. Triple-positive thymocytes that receive a strong (agonist) selection signal are positively selected and functionally differentiate during an agonist selection process.

ability of the TCRγδ$^+$ precursors. For example, TCRγδ transgenic G8 thymocytes specific for the MHC class Ib molecules, T10^b/T22^b, differentiate in the presence of their cognate antigen, and directly home to the gut but nowhere else. Consistent with this, functional sphingolipid receptor, S1P1, is not required for egress of TCRγδ$^+$ nIET precursor thymocytes. Thymic programming of TCRγδ progenitors is imprinted at the DN3 stage and requires a full γδ TCR together with CD3 components and functional linker for activation of T cells. Whether thymic ligand engagement is required for nIET precursors expressing an endogenous TCRγδ is still unclear, but it was shown for fetal TCRVγ5$^+$ thymocytes that interaction with Skint1 on thymic epithelial cells induced a strong TCR signal that mediated induction of Tbx21 (or T-bet). The absence of ligand engagement resulted in the maintenance of Sox13 and induction of RORγt, mediating differentiation into distinct functional subsets.

Gene-expression analysis of TCRγδ nIETs identified an antigen-experienced profile that distinguishes TCRγδ$^+$ nIETs from conventional selected T cells. However, rather than being unique for TCRγδ nIETs, this profile is shared with the CD8αα$^+$TCRαβ$^+$ nIETs, which also require strong selection signals, induce Tbx21, and also directly home to the intestine in a S1P1-independent manner.

The specific loss of TCRVγ5$^+$ IETs in mice with defects in IL-15R-mediated signals and the increase of Vγ5$^+$ IETs in IL-15 transgenic mice suggest a central role for IL-15 in the generation and/or maintenance of TCRγδ$^+$ nIETs. In addition, the selective reduction of Vγ5$^+$ thymocytes in IL-15$^{-/-}$ or IL-15 signal transducer Stat5$^{-/-}$ mice indicates that IL-15 directly influences development and/or survival of Vγ5$^+$ thymocytes. Consistent with this, IL-15 is able to bias Vγ5 transcription by directing chromatin domain modifications and promoting the accessibility in the exclusive vicinity of the Vγ5 gene. Nevertheless, the absence of Vγ5 transcripts in IL-7R$^{-/-}$ thymocytes, in contrast to their presence in the gut, suggests that the IL-15 pathway is not a major mechanism during thymic differentiation, but it might operate in the gut when thymically selected IETs are limiting.

6.11 Agonist selection of self-reactive CD8αα$^+$TCRαβ$^+$ nIETs

The TCR repertoire of CD8αα$^+$TCRαβ$^+$ nIETs contains self-reactive TCRs that have encountered a high-affinity ligand during their thymic agonist selection process. It was shown that precursors to TCRαβ nIETs express high levels of the immediate early gene Nr4a1 (Nur77), reflecting strong TCR activation signals. The thymic agonist selection process is dependent on an intact store-operated calcium entry (SOCE) pathway to sustain intracellular calcium levels required for TCR-induced maturation. In contrast, positive selection of conventional naïve T cells occurs in the absence of SOCE. Accordingly, thymocytes engineered to express TCRs derived from TCRαβ nIETs show signs of strong TCR activation and mature to coreceptor negative double-negative (DN) T cells that exit the thymus and populate the small intestinal epithelium.

Although both the CD8ααTCRαβ$^+$ and the γδTCR$^+$ nIETs depend on strong signals and acquire their activated phenotype "naturally" in the thymus, they follow different pathways. Precursors of the CD8αα$^+$TCRαβ$^+$ nIETs pass through an additional checkpoint that involves the expression of the pre-TCR, and many of these transition to a CD4 and CD8αβ double-positive stage before they undergo the full TCR-based agonist selection process (see **Figure 6.4**). Also, *in vitro* differentiation of double-positive thymocytes exposed to their cognate antigen leads to CD8ααTCRαβ$^+$ T cells that express the typical gene transcription profile of CD8ααTCRαβ$^+$ nIETs. Fate-mapping studies, using expression of green fluorescent protein driven by the CD4 promoter, further confirmed that the majority of CD8ααTCRαβ+ nIET progenitors in adult mice transition through the double-positive checkpoint, implying that double-positive thymocytes are heterogeneous and contain precursors for conventional and CD8ααTCRαβ+ T cells. This led to the identification of a new subset of immature thymocytes that coexpress CD8αα together with CD4 and CD8αβ, called triple-positive (TP) cells (see Figure 6.4). Intrathymic injection of TP, but not double-positive, thymocytes led to the generation of CD8ααTCRαβ+ nIETs *in vivo*, and when exposed to their cognate antigen *in vitro*, TP thymocytes survived, expanded, and acquired an antigen-experienced phenotype as well as innate-like features and gut-specific homing receptors. These observations further underscore the importance of the thymic agonist selection process as a central mechanism for the unique differentiation and migration of CD8ααTCRαβ+ nIET thymic precursor cells. Agonist-selected TP thymocytes mature to CD5+TCRαβ+ double-negative thymocytes that egress the thymus and directly migrate to the intestinal epithelium (see Figures 6.2 and 6.4). The reexpression of CD8αα and the downregulation of CD5 are postthymic events that occur locally in the intestine and that are actively promoted by IL-15. Similar to TCRγδ+ nIETs, Tbx21 is induced by agonist TCR signals and intestinal IL-15, and this transcription factor is essential for the maturation and maintenance of CD8ααTCRαβ+ nIETs.

Studies in transgenic mice have shown that the development of CD8ααTCRαβ+ nIET is favored when a pre-rearranged transgenic TCRαβ with high affinity for a cognate self-antigen is expressed early, before conventional selection at the DP stage. These findings indicate that in addition to the strength of TCR signaling, the timing of TCRα rearrangement also influences the outcome of selection, since postponing TCRα expression until after the CD4 promoter is activated abolishes the selection of TCRαβ nIETs in the HY-TCR transgenic model. This was recently confirmed for a polyclonal population of natural human agonist-selected thymocytes expressing CD8αα. When analyzing TCRα rearrangements in human T-cell subsets, agonist selected CD8T cells showed enrichment for early TCRα rearrangements. It appears that TP thymocytes are early post-β-selection precursors enriched for early TCRα rearrangements that can survive and differentiate upon strong TCR engagement, whereas later post-β-selection double-positive (DP) precursors either undergo negative selection by strong TCR signaling or differentiate to naïve conventional T cells in response to weak TCR interactions.

6.12 iIETs are peripheral-antigen-driven progeny of conventional CD4+ or CD8αβ+TCRαβ T lymphocytes

The intestinal epithelium is particularly enriched for CD8αβ+TCRαβ+ iIETs, and *in vivo* priming experiments in mice demonstrate that intestinal primary and memory CD8αβ+ T cells are readily generated in response to infections. For example, oral reovirus or rotavirus infection results in the accumulation of virus-specific cytolytic iIETs. However, cytotoxic T lymphocytes (CTLs) are present in the lamina propria but not in the intestinal epithelium following footpad immunization, indicating discordance of the response within mucosal effector sites. Analysis of CTL induction in response to a human immunodeficiency virus-1 (HIV-1) combinatorial peptide indicated that intrarectal immunization leads to primary and memory CTL within the lamina propria, intestinal epithelium, and spleen. In contrast, subcutaneous immunization with the peptide leads to CTL activity in spleen but not mucosa. Interestingly, intranasal immunization targeting lung dendritic cells induced migration of protective T cells to the intestinal mucosa. This implies that induction of mucosal primary and memory CTL iIETs depends on the route of infection and the form of the immunogen.

The formation of pathogen-induced effector memory CD8αβ$^+$ nIETs was shown to depend on affinity-based induction of CD8αα on the CD8αβ TCRαβ memory precursor T cells. Further selective maintenance of high-affinity effector memory iIET at the mucosal border is mediated by TL, which is constitutively expressed on epithelial cells, and which eliminates weak affinity CD8αβ effector cells that fail to induce CD8αα.

Migration to and local adaptation in the gut

IETs are part of the effector arm of mucosal immune responses, but they differ markedly from many of the paradigms that control the migration of peripheral T cells into the gut. This is most dramatically seen in parabiotic mice, where there is rapid mixing of lamina propria T cells but not of the IETs from either parabiont.

6.13   Natural IETs are induced to home to the gut in the thymus

In addition to their antigen-experienced phenotype, nIET precursors also acquire gut-homing receptors in the thymus, including $α_E β_7$ and CCR9. The specific expression of their ligands, E-cadherin and CCL25 respectively, by the small intestine enterocytes, recruits the mature nIET precursors directly to the intestine epithelium. Consistent with this, the seeding of these cells is independent of S1P1. The direct and early population of the mucosal barrier with these stress-sensing self-reactive effector cells provides a layer of preexisting immunity, even before birth and during the weaning period, prior to the formation of an exogenous antigen-induced defense line and before the potential for exposure to pathogens becomes significant (see Figure 6.3).

6.14   Natural intraepithelial T cells undergo local adaptation in the gut

Additional local factors produced by the epithelial cells, such as IL-7 and IL-15, further expand and adapt the nIETs. In the absence of IL-15 or IL-15Rα, there is a preferential loss of CD8αα$^+$ nIETs. IL-15 operates by an unusual mechanism of trans-presentation in which IL-15 is bound to IL-15Rα intracellularly, transported to the cell surface, and presented to responding cells expressing the IL-15β and common γ-receptor chains. This mechanism is important for IET development since both IL-15 and IL-15Rα expression by the intestinal epithelium is essential for nIET development and maintenance. Notch receptor signaling and activation of c-Myc also contribute to the homeostasis of CD8αα$^+$ nIETs, at least in part through the IL-15 receptor pathway. Stem cell factor (SCF), produced by epithelial cells, also plays an important role in their homeostasis, and most nIETs, but not iIETs, express c-Kit. While early TCRγδ nIET development is normal in SCF or SCF-receptor (c-Kit) mutant mice, this population is gradually lost with time.

6.15   Only activated T cells migrate to the gut epithelium

Although some conventional CD8αβ$^+$ recent thymic emigrants have gut-homing receptors, in general, conventional thymically selected precursor cells and naive CD8 or CD4 T cells are not detected within the intestinal epithelium. However, organized lymphoid tissue, such as the PP and the MLN, form the main priming sites for peripheral naive T cells that acquire gut-homing capacity upon activation with their cognate antigens. In that aspect, specialized epithelial cells or M cells (discussed in Chapter 15), resident CX3CR1$^+$ macrophages, and migratory CD103$^+$ dendritic cells (DC) (discussed

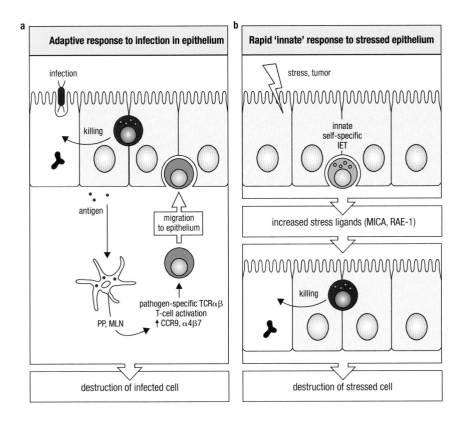

Figure 6.5 Adaptive and innate responses in the intestinal epithelium. (**a**) Conventional CD8αβ TCRαβ T cells are activated in the mesenteric lymph nodes (MLN) or Peyer's patches (PP) by dendritic cells that have acquired antigen following infection of the epithelium. This process may take days. Homing molecule expression is "imprinted" in the PP and MLN resulting in migration of activated CD8 T cells into the epithelium and destruction of infected cells through cytotoxic mechanisms. (**b**) Environmental stress or tumors induce expression of stress ligands (e.g., MICA, RAE-1) by intestinal epithelial cells. NKG2D+ TCRγδ IETs interact with these ligands resulting in destruction of infected cells through cytotoxic mechanisms. This process can occur in minutes to hours since the responding cells are resident in the epithelium.

in Chapter 12) all have the unique capacity to directly or indirectly sample the antigen content of the gut lumen, allowing for constant reporting on the condition of the intestinal compartment. Upon priming in these mucosal sites, activated naive T cells are induced to express gut-homing receptors that drive migration to the intestine (Figure 6.5). The ability of the mucosal DC to confer these properties is mediated at least in part by the release of a diet-derived factor, the vitamin-A metabolite retinoic acid (RA), which preferentially promotes migration to the small intestine, through induction of CCR9. The RA-mediated homing process greatly promotes migration to the intestine, but it is not absolutely required, and TCRαβ T cells activated in the periphery can also migrate to the gut, although under conditions of an intestinal challenge gut-specific homing may be significantly enhanced through RA-mediated imprinting.

The multistep paradigm of lymphocyte migration applies to T-cell movement into both the inductive and effector sites of the intestinal mucosa (discussed in Chapters 1 and 16). Integrins are also important players in lymphocyte migration and retention in the mucosa and the β_2 integrin $\alpha_L\beta_2$, whose primary ligand is intercellular adhesion molecule 1 (ICAM-1), is required for establishment of IETs as is the $\alpha_1\beta_1$ integrin (very late antigen-1; VLA-1) that binds to collagen. In the case of $\alpha_L\beta_2$, it is not clear whether integrin activity is needed for initial T-cell activation, migration to the intestine, or both. Another β_2 integrin CD11c ($\alpha_X\beta_2$), expressed by most IETs and induced by activation, is linked to lytic activity, but the function of CD11c *in vivo* is unknown.

The binding of $\alpha_4\beta_7$ to the mucosal addressin cell adhesion molecule-1 (MAdCAM-1), expressed by vascular endothelium in lamina propria, MLN, and PP blood vessels, promotes translocation into the respective tissues. After local or systemic infection, $\alpha_4\beta_7$ is transiently induced, which provides a finite window of time to migrate to the intestinal epithelium. Following entry into the mucosa, $\alpha_4\beta_7$ is lost; concomitant with this $\alpha_E\beta_7$ is induced and appears to be important for retention in the epithelium via interaction with its ligand, E-cadherin, expressed by epithelial cells. Induction of $\alpha_E\beta_7$ is mediated by transforming growth factor-beta (TGF-β), and the transcription factor Runx3 and is enhanced by signaling through CCR9. Similarly, reinduction of CD69 occurs when activated T cells enter the mucosa; however, its function on these mucosal T cells remains unknown.

6.16   Induced IET precursors undergo further adaptation in the gut

Activation of transferred OT-I transgenic cells by feeding soluble ovalbumin does not induce lytic activity in nonintestinal OT-I cells, but upon migration into the intestinal mucosa these cells become highly lytic. Moreover, optimal activation of mucosal T cells requires B7.1, and CD40L expressed by CD8αβ+ T cells is important for optimal expansion of mucosal CD8αβ+ T cells in response to virus infection. Similarly, CD70–CD27 and CD40–CD40L costimulation appears to be important for expansion of CD8αβ+TCRαβ+ iIETs after oral *Listeria monocytogenes* infection, presumably for "secondary" costimulation, since lymphoid tissue is required for initial priming. These observations indicate that tissue-specific cellular interactions further condition the iIETs resulting in tuning of their effector phenotype and function. In this light, and similarly to the nIETs, a large fraction of iIETs express CD8αα as part of their specific adaptation to the epithelial compartment. Although most memory CD8αβ iIETs retain an activated phenotype (e.g., CD69 expression and a heightened cytolytic state) that enables them to rapidly respond to antigen reexposure, they also display reduced ability to proliferate and to produce inflammatory cytokines typical of effector memory T cells. In this respect, intestinal CD8αβ and also CD8αα T cells express high levels of the P2X$_7$ purinoreceptor, which is induced by retinoic acid in the gut environment. Similar to the TL-mediated selective maintenance of only the high-affinity CD8 effector memory iIETs, intestinal effector T cells can also be deleted by P2X7 activation-dependent apoptosis. The retinoic acid-P2X7 axis limits excessive buildup of activated T cells and thereby controls intestinal homeostasis.

Overall, these results illustrate that (1) conventional naive T-cell activation results in migration to the gut epithelium, (2) secondary activation in the mucosa occurs upon secondary encounter with antigen, (3) environmental cues in the mucosa drive iIET activation even in response to tolerogenic forms of antigen, and (4) certain costimulatory elements preferentially regulate intestinal T-cell responses, including iIETs.

Beneficial functions of IETs

The location of IETs at the intersection between the outside milieu and the core of the body prompts teleological predictions about their function. They are geared to provide immediate immune protection in order to avoid the penetration of pathogens and also likely to prevent excessive infiltration of inflammatory immune cells from the periphery. Consistent with this, most IETs display constitutive cytolytic activity driven by various pathways, including serine proteases, perforin release, and Fas–FasL-mediated programmed cell death and natural-killer receptor-dependent killing.

6.17   TCRγδ+ nIETs maintain epithelial homeostasis

Although the exact functions and behavior of nIETs remain enigmatic, their primary role seems to be directed toward insuring the integrity of the epithelium and maintaining a local immune balance. TCRγδ nIETs respond to resident bacterial pathobionts that penetrate the epithelial lining by producing antimicrobial factors. Intestinal epithelial cells sensing invasion by bacteria activate resident γδ T cells specifically during the first hours of mucosal penetration. TCRγδ nIETs appeared crucial in preventing commensal bacteria from crossing the mucosal barrier during DSS-induced intestinal damage. This highlights the role for TCRγδ nIETs in maintaining host-microbial homeostasis whenever the intestinal lining has been breached. Additionally, TCRγδ nIETs may produce keratinocyte growth factor and express the junctional adhesion molecule known to induce keratinocyte growth factor upon ligation to its receptor, the Coxsackie virus and adenovirus receptor, thereby linking this costimulatory pathway directly to epithelial repair and maintenance function. In mice lacking all TCRγδ T cells, there are fewer enterocytes and crypt cells, and the enterocytes express less MHC class II. Intestinal infection of TCRδ$^{-/-}$ mice with the protozoan parasite *Eimeria vermiformis* results in more severe pathology and overt intestinal bleeding. This effect appears to be due to uncontrolled interferon-γ (IFN-γ) production by TCRαβ iIETs. The absence of TCRγδ cells also results in more severe injury in mouse models of colitis, and additionally, keratinocyte growth factor null mice are more susceptible to dextran sulfate sodium colitis suggesting that, in mice, TCRγδ cells control intestinal homeostasis in part by the production of keratinocyte growth factor. TCRγδ nIETs express other regulatory factors such as TGF-β1, TGF-β3, and prothymosin β4, which may aid in epithelial maintenance or repair. TCRγδ nIETs are also equipped with innate and adaptive cytotoxic mechanisms that could contribute to protective immunity. In a model of *Toxoplasma gondii* infection, TCRγδ nIETs greatly enhanced the protective function of pathogen-specific CD8αβ$^+$ T cells. Furthermore, γδ TCRs may recognize unprocessed antigens such as stress antigens expressed by enterocytes or molecular patterns generated by bacterial nonpeptide antigens, such as phosphorylated nucleotides and prenyl pyrophosphates. In humans, TCRγδ nIETs recognize the highly polymorphic MHC class I chain-related genes MICA and MICB, induced on stressed epithelium, which leads to death of damaged or infected epithelial cells. The invariant NKG2D receptors expressed by TCRγδ nIETs also bind to MICA and MICB, thereby inducing killing in a TCR-independent fashion (see Figure 6.5). Together with their cytotoxic capacity, TCRγδ nIETs produce protective cytokines, such as tumor necrosis factor-α and IFN-γ. Therefore, TCRγδ nIETs not only control the production of IFN-γ by other inflammatory cells, but they can also make IFN-γ that contributes to host protection. Nevertheless, much remains to be learned regarding the function of TCRγδ nIETs, not the least of which is the gap in our knowledge of their ligands for γδ TCRs.

6.18 Functions of double-negative TCRαβ nIETs remain largely unknown

The CD4 and CD8αβ double-negative (DN) TCRαβ nIETs are the most enigmatic subset of IETs. Similar to all other IET subsets, they express an activated phenotype and show high cytolytic potential, but they also display a highly constrained phenotype. In a lymphocytic choriomeningitis virus (LCMV) infection model using mice double-transgenic for an LCMV-epitope-specific TCR and an LCMV-specific antigen, LCMV-reactive DN TCRαβ nIETs displayed signs of virus-induced activation; however, unlike the conventional virus-specific CD8αβ T cells, they did not induce LCMV-specific cytotoxicity and they did not promote an inflammatory response.

This effector yet restrained dual nature of DN TCRαβ nIETs was confirmed by gene array analysis, which showed that they (1) express high levels of regulatory or inhibitory molecules including LAG-3, CTLA-4, PD-1, CD160, TGF-β, and the NK receptors 2B4 and Ly49 A, E, and G; and (2) express molecules associated with protective immunity, including granzymes, Fas ligand, RANTES, and CD69. Effector cytokines such as IL-2, and IL-4, and cytokine receptors (IL-2Ra, IL-12-R) are not abundantly expressed.

Therefore, similarly to other IET subsets, DN TCRαβ nIETs may function in the epithelium as gatekeepers, with the main emphasis on promoting homeostasis and maintaining barrier integrity, and upon the appropriate signals, they may gain full effector functions to quickly eliminate the threat before excessive damage by the pathogen or inflammatory immune cells can occur at the fragile epithelial barrier of the intestine.

Potential self-reactivity together with inherent cytotoxic capability and their close proximity to enterocytes suggest important roles for nIETs not only in sensing and eliminating infected cells but also transformed epithelial cells. Although self-reactive, nIETs are not self-destructive, and as discussed earlier, this self-based beneficial activity seems to be in part preprogrammed during agonist selection in the thymus. These observations underscore the importance of the thymic agonist selection process as a central mechanism to drive the antigen-experienced effector yet regulate differentiation of these unique nIETs. The agonist-selected DN thymic precursors that migrate to the intestine induce CD8αα expression driven by IL-15 from enterocytes. In contrast to the conventional TCR co-receptors, CD8αα does not function to enhance TCR activation, and although induced in response to strong TCR signals, CD8αα in turn suppresses TCR activation. This may be due to its ability to bind signalling molecules such as p56^{lck} and linker for activation of T cells, but its inability to join lipid rafts that contain the TCR activation complex (**Figure 6.6**). Therefore, CD8αα expression on high-affinity nIETs and iIETs may reset the threshold for activation, reducing the potential for unnecessary, or excessive, immune responses. CD8αα-mediated inhibition is not irreversible and increased antigen stimulation or cross-ligation of TCR coreceptors and TCR can readily take precedence over the CD8αα repressor function.

Other molecules associated with immune regulation, such as LAG-3, CTLA-4, PD-1, and TGF-β, together with the SLAM-related receptor 2B4 and several members of the killer-cell immunoglobulin-like receptors, KIR or Ly49 (Ly49 A, E, and G), may further regulate the activation status of high-affinity autoreactive nIETs. The NK-inhibitory receptors suppress cytotoxicity and cytokine secretion; however, 2B4 can also function as an activation receptor

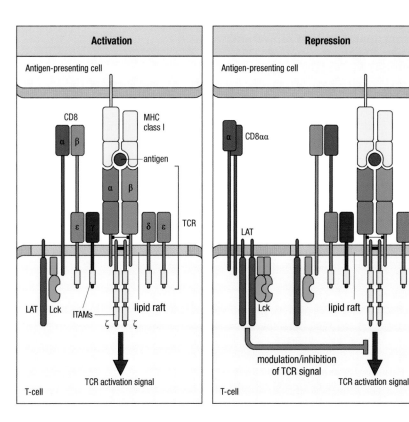

Figure 6.6 CD8αβ functions as a TCR coreceptor. CD8β moves CD8α into the lipid raft adjacent to the TCRαβ, whereas CD8α recruits signaling molecules, such as lymphocyte-specific protein tyrosine kinase (Lck) and linker for activation of T cells (LAT) that promote downstream signaling. CD8αα functions as a repressor. CD8αα does not move into the lipid raft and negatively influences TCR activation by sequestering signaling molecules such as LAT and Lck away from the TCR activation complex.

and under certain conditions even promote cytotoxicity. The dual expression of activating and inhibitory receptors is consistent with their "activated yet quiescent" state, which indicates that nIETs may gain full activation status for self-based protective and/or regulatory functions in response to self-antigen-based induction signals. In support of this, in a model of induced colitis, TCRαβ+ DN nIETs were able to prevent inflammation mediated by non-self specific naive CD4+CD45RB^hi donor T cells. In addition, their agonist self-antigen-based thymic functional differentiation, their cytolytic phenotype, and their early seeding and strategic location in the gut epithelium all suggest that they might serve an important role in maintaining and protecting the mucosal barrier at the time of gut colonization and before an exogenous-antigen-induced defence layer of iIETs has been formed (see Figure 6.3).

6.19 TCRαβCD8αβ iIETs have a protective function

Pathogen-induced IETs line the intestinal interface with a protective immune layer that gradually increases upon exposure to pathogens and which provides pre-existing immunity adapted toward those pathogens that are most likely to rechallenge (Figures 6.5 and 6.7). In the small intestine, iIETs

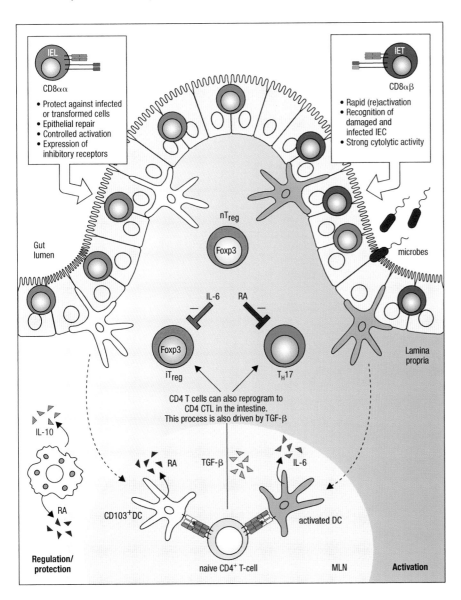

Figure 6.7 **The functionally diverse T-cell populations of the intestine that are shaped by the gut environment are important players in sustaining the delicate immune balance between activation and regulation.** The CD8αα+ intraepithelial cells (IETs) of the intestine play a crucial role in protection of the mucosal barrier. They are involved in maintaining and restoring barrier homeostasis by stimulating epithelial cell turnover. Upon pathogen entry, rapid activation and high cytolytic activity of the CD8αβ+ IETs contribute to the prevention of pathogen spreading by killing infected epithelial cells. The activation of IET is highly controlled through the expression of inhibitory receptors that may alter the threshold for activation. Under the influence of epithelial cells and IET, lamina propria dendritic cells (DCs) acquire the ability to produce retinoic acid (RA), thereby inducing gut-homing receptors during CD4 T-cell priming in the mesenteric lymph nodes (MLN). Additionally, gut-derived CD103+ DC and lamina propria macrophages favor the conversion of Foxp3+ induced regulatory T cells (iT_reg) in a RA, TGF-β, and IL-10 (in the case of lamina propria macrophages) mediated fashion, whereas activated DCs promote the differentiation of IL-17-producing T_H17 cells via a combination of IL-6 and TGF-β. This pro- and anti-inflammatory immune deviation of iT_reg and T_H17 is reciprocally controlled by RA and IL-6. Note that not shown in this slide is that TGF-β can also participate in converting CD4 T cells into cytolytic cells. Finally, the lamina propria is also home to agonist-selected Foxp3-expressing naturally occurring regulatory T cells (nT_reg). (Adapted from van Wijk, F., and Cheroutre, H. *Semin. Immunol.* 2009, 21:130–138. With permission from Elsevier Ltd.)

are mostly CD8αβ$^+$TCRαβ$^+$ cytotoxic effector memory cells. Compared with splenic memory T cells, they can be rapidly activated and provide immediate cytotoxic function. Their TCR repertoire is more restricted than that of peripheral memory CD8αβ T cells suggesting that repeated re-stimulation in the intestine may lead to survival and expansion of selected clones of T cells. In addition, the TL-mediated selective maintenance of only high affinity effector memory T cells also contributes to the oligoclonal repertoire of the iIETs. Locating high affinity effector memory cells in mucosal tissues makes teleological sense because a rapid sensing and response may be important to mediate an immediate attack on incoming pathogens and eliminating them before pathogen- or immune-induced pathology can occur. Early work showed that TCRαβ IETs can be primed against alloantigens, and against viral, bacterial, and parasite antigens. In mice the intestinal CD8αβ T-cell response has been characterized following systemic vesicular stomatitis virus, *L. monocytogenes*, and rotavirus infections. Systemic vesicular stomatitis virus infection results in CD8αβ T-cell activation followed by migration of effector cells to all non-lymphoid tissues. The magnitude of the response in the IET compartment is substantially larger and more sustained, and the contraction phase is more prolonged, compared with the spleen. Repeated challenge infections result in enhanced and sustained granzyme-B expression in secondary and tertiary antigen-specific IETs. The CD8αβ T-cell response to enteric infection with *L. monocytogenes* largely parallels that observed after systemic infection, but with some significant differences. Nevertheless, either route of infection generates antigen-specific CD8αβ iIETs, and enteric infections induce larger numbers of *L. monocytogenes*-reactive cells in the mucosa than do intravenous infections. Moreover, the TCR Vβ repertoire of splenic versus intestinal *L. monocytogenes*-specific CD8αβ T cells is distinct, indicating either differential migration and/or expansion of CD8αβ effectors.

Whether iIET memory cells are directly involved in this type of protection remains to be proven. However, rhesus macaques inoculated with rhesus cytomegalovirus vectors containing simian immunodeficiency virus (SIV) epitopes are protected from progressive wild-type SIV infection after repeated limiting-dose intrarectal challenges. Moreover, the protected animals showed no evidence of systemic infection, suggesting that infection was contained locally, well before extensive viral replication or systemic spreading of the virus occurred.

6.20 TCRαβ CD4 iIETs are also involved in protective immunity

Though less prevalent, the loss of mucosal CD4 T cells in SIV/HIV-1 infections, which results in translocation of enteric bacteria and increased local and systemic infections, indicates important participation of CD4 iIETs for local protective immunity.

Recently, a lineage-switch process has been shown *in vivo* that is characterized by the conversion of CD4 T helper cells (T$_H$) to CD4 CTL-like iIETs. Chronic exposure to luminal antigens in the absence of inflammation drives CD4 iIETs to lose expression of the master transcription factor ThPOK which controls the helper T-cell program. Chronic activated CD4 T cells that lose ThPOK expression acquire the CD8 lineage transcription factor Runx3 in a manner dependent on TGF-β and retinoic acid. The lineage-switch mechanism leads to the generation of protective cells but also represents a new form of immune regulation that prevents activated CD4 T cells from either becoming suppressive T$_{reg}$ or pro-inflammatory T$_H$17 iIETs. The conversion of T$_H$ effector CD4 T cells while preserving their protective function makes good sense at this mucosal interface where effective protective immunity has to coincide with the least inflammatory damage. These CD4 CTL iIETs are capable of dealing with infected class II$^+$ enterocytes and they might also provide protection against pathogens that have developed mechanisms to escape surveillance by the MHC class I restricted CD8αβTCRαβ iIETs. Recently, intravital microscopy was used to show that intestinal lamina propria T_{reg} cells lose Foxp3 expression and convert to CD4 CTLs upon migration into the epithelium. This demonstrates that microenvironmental cues guide different regulatory programs to establish and control immunity at distinct anatomical sites.

Potential aberrant function of IETs in inflammation

The dynamic interaction between the gut environment and the mucosal adaptive immune system is normally geared toward maintaining a stable equilibrium and sustaining the barrier function. Nevertheless, their heightened activation status, together with their close proximity and intense interactions with the vulnerable epithelial cells, also provide IETs with the opportunity to become potent destructive pathogenic cells that may fuel chronic intestinal inflammation or promote cancers.

6.21 TCRγδ nIETs may be involved in pathology

Even though TCRγδ nIETs may serve beneficial regulatory and protective roles, other evidence suggests a more pathogenic side of these nIETs. There are increased numbers of TCRγδ cells in the epithelium of patients with active celiac disease, and in this context, gliadin-mediated deregulation of IL-15 secretion may lead to overexpression of MICA/MICB on enterocytes and enhanced cytotoxicity by TCRγδ nIETs.

6.22 Self-specific DN TCRαβ nIETs may be destructive to the epithelium

Autoreactive αβTCRs are preserved and enriched in the repertoire of the DN TCRαβ nIETs. It thus remains possible that under inflammatory conditions or upon recognition of modified foreign antigens, these autoreactive T cells might overreact against self and induce autoimmune pathology. Furthermore, increased or uncontrolled inflammatory conditions, such as excessive production of IL-15, may trigger their heightened autoreactive cytotoxic effector function, directly jeopardizing the integrity of the mucosal barrier.

6.23 CD8αβTCRαβ iIETs may be involved in the pathogenesis of inflammatory bowel disease

Cytotoxic CD8αβTCRαβ iIETs have been implicated in the progression or even the induction of inflammatory bowel disease (IBD). The destruction of epithelial cells by pathogenic CD8$^+$ T cells may initiate the loss of epithelial barrier integrity, consequently leading to exposure to luminal antigens and bacteria, which may then attract other immune cells that further exacerbate the inflammatory pathology.

In support of this, in a hapten-induced mouse model of colitis, previously hapten-sensitized CD8$^+$ CTL are the earliest inducers of uncontrolled inflammation. Furthermore, although epithelial-cell-specific CD8$^+$ iIET may not mediate persisting inflammation, some hapten-specific CD8$^+$ effector cells survive long-term as effector memory T cells that initiate relapse disease upon secondary challenge. Although the pathogenic CTL may direct their responses toward epithelial-cell-expressed antigens, they are conventional selected cells, showing that a breakdown in self-tolerance in the periphery can lead to uncontrolled immune responses and self-destruction.

6.24 Aberrant functions of CD4 TCRαβ iIETs driving gut inflammation

Aberrant CD4$^+$ T cells in the epithelium together with their lamina propria counterparts are probably the most significant contributors to immunopathology in the intestine, and their roles in various inflammatory diseases such as celiac disease and IBD are extensively discussed in Chapters 31 and 34.

Overall, although all effector T cells, including the various IET subsets, have the potential to initiate or propagate gut inflammation, aberrant CD4 T$_H$ effector cells are almost always involved and play a big part in the destructive immune pathology that targets and jeopardizes the barrier function of the mucosal epithelium.

SUMMARY

In many ways, IETs are tightly regulated warriors and peacekeepers (see **Figure 6.7**). On the one hand, the self-antigen-based functional differentiation of nIETs and the expression of inhibitory and activating NK receptors endow them with natural antigen-experienced ability and innate-like properties to sense stressed and transformed epithelial cells, and to respond quickly and in a protective yet controlled fashion especially in the absence of pathogen-induced preexisting iIETs. The constant dilemma to prevent excessive and/or unwanted immune responses while at the same time provide rapid optimal protection against invading pathogens, presents truly unique challenges that undoubtedly shape the complexity and sophistication of the IETs, and their multidirectional interactions with the surrounding epithelial cells. Because the intestinal mucosa is the major primary target for many pathogens, including HIV-1, and because a major branch of the immune system is formed by the IETs, it is critical that we elucidate and identify all the components and interactions involved in their networking, and that we understand the rules governing their migration, functional differentiation, activation, and regulation. This knowledge will aid in the eventual intelligent design of vaccines able to direct appropriate effector functions to the desired locations in order to prevent the entry and spreading of pathogens, and in the generation of strategies to regulate immunity and avoid detrimental inflammation or the development of cancers.

FURTHER READING

Cheroutre, H.: IELs: Enforcing law and order in the court of the intestinal epithelium. *Immunol. Rev.* 2005, 206:114–131.

Cheroutre, H., and Lambolez, F.: The thymus chapter in the life of gut-specific intra epithelial lymphocytes. *Curr. Opin. Immunol.* 2008, 20:185–191.

Cheroutre, H., Mucida, D., Lambolez, F.: The importance of being earnestly selfish. *Nat. Immunol.* 2009, 10:1047–1049.

Hayday, A.C.: γδ T cells and the lymphoid stress-surveillance response. *Immunity* 2009, 31: 184–196.

Henderson, P., van Limbergen, J.E., Schwarze, J. et al.: Function of the intestinal epithelium and its dysregulation in inflammatory bowel disease. *Inflamm. Bowel Dis.* 2011, 17:382–395.

Huang, Y., Park, Y., Wang-Zhu, Y. et al.: Mucosal memory CD8$^+$ T cells are selected in the periphery by an MHC class I molecule. *Nat. Immunol.* 2011, 12(11):1086–1095.

Ismail, A.S., Severson, K.M., Vaishnava, S. et al.: γδ Intraepithelial lymphocytes are essential mediators of host-microbial homeostasis at the intestinal mucosal surface. *Proc. Natl. Acad. Sci. USA.* 2011, 108(21):8743–8748.

Jabri, B., and Ebert, E.: Human CD8$^+$ intraepithelial lymphocytes: A unique model to study the regulation of effector cytotoxic T lymphocytes in tissue. *Immunol. Rev.* 2007, 215:202–214.

Jabri, B., and Sollid, L.M.: Tissue-mediated control of immunopathology in coeliac disease. *Nat. Rev. Immunol.* 2009, 9:858–870.

Jiang, W., Wang, X., Zeng, B., et al.: Recognition of gut microbiota by NOD2 is essential for the homeostasis of intestinal intraepithelial lymphocytes. *J. Exp. Med.* 2013, 210(11):2465–2476.

Kronenberg, M., and Havran, W.L.: Frontline T cells: γδ T cells and intraepithelial lymphocytes. *Immunol. Rev.* 2007, 215:5–7.

Kunisawa, J., Takahashi, I., Kiyono, H.: Intraepithelial lymphocytes: Their shared and divergent immunological behaviors in the small and large intestine. *Immunol. Rev.* 2007, 215:136–153.

Laky, K., Lefrançois, L., Lingenheld, E.G., et al.: Enterocyte expression of interleukin 7 induces development of γδ T cells and Peyer's patches. *J. Exp. Med.* 2000, 191:1569–1580.

Lefrançois, L., and Puddington, L.: Extrathymic intestinal T-cell development: Virtual reality? *Immunol. Today* 1995, 16:16–21.

Lefrançois, L., and Puddington, L.: Intestinal and pulmonary mucosal T cells: Local heroes fight to maintain the status quo. *Annu. Rev. Immunol.* 2006, 24:681–704.

Li, Y., Innocentin, S., Withers, D.R. et al.: Exogenous stimuli maintain intraepithelial lymphocytes via aryl hydrocarbon receptor activation. *Cell* 2011, 147(3):629–640.

Mayans, S., Stepniak, D., Palida, S.F., et al.: αβT cell receptors expressed by CD4⁻ CD8αβ⁻ intraepithelial T cells drive their fate into a unique lineage with unusual MHC reactivities. *Immunity* 2014, 41(2):207–218.

McDonald, B.D., Bunker, J.J., Ishizuka, I.E. et al.: Elevated T cell receptor signaling identifies a thymic precursor to the TCRαβ⁺CD4⁻CD8β⁻ intraepithelial lymphocyte lineage. *Immunity* 2014, 41(2):219–229.

Mucida, D., Husain, M.M., Muroi, S. et al.: Transcriptional reprogramming of mature CD4⁺ helper T cells generates distinct MHC class II-restricted cytotoxic T lymphocytes. *Nat. Immunol.* 2013, 14(3):281–289.

Probert, C.S., Saubermann, L.J., Balk, S., et al.: Repertoire of the αβ T-cell receptor in the intestine. *Immunol. Rev.* 2007, 215:215–225.

Shires, J., Theodoridis, E., Hayday, A.C.: Biological insights into TCRγδ⁺ and TCRαβ⁺ intraepithelial lymphocytes provided by serial analysis of gene expression (SAGE). *Immunity* 2001, 15:419–434.

Sujino, T., London, M., Hoytema van Konijnenburg, D.P. et al.: Tissue adaptation of regulatory and intraepithelial CD4⁺ T cells controls gut inflammation. *Science* 2016, 352(6293):1581–1586.

Turchinovich, G., and Hayday, A.C.: Skint-1 identifies a common molecular mechanism for the development of interferon-γ-secreting versus interleukin-17-secreting γδ T cells. *Immunity* 2011, 35(1):59–68.

van Wijk, F., and Cheroutre, H.: Intestinal T cells: Facing the mucosal immune dilemma with synergy and diversity. *Semin. Immunol.* 2009, 21:130–138.

Verstichel, G.,Vermijlen, D., Martens L., et al.: The checkpoint for agonist selection precedes conventional selection in human thymus. *Sci. Immunol.* 2017, 2(8).

T lymphocyte populations within the lamina propria

7

THOMAS T. MACDONALD AND ANTONIO DI SABATINO

There are more T cells in the gut-associated lymphoid tissue (GALT) and intestinal lamina propria and epithelium than in the rest of the body. That they are absolutely essential for health is demonstrated when they are not present, as in children with severe combined immunodeficiency or adults with untreated human immunodeficiency virus-1 (HIV-1) infection: intestinal infections caused by cryptosporidia, *Isospora*, cytomegalovirus, and other low-grade pathogens result in chronic diarrhea, wasting, and eventually death (Table 7.1). The advent of highly active antiretroviral therapy that helps maintain gut T-cell numbers has seen gut infections and the enteropathy of HIV-1 infection diminish in clinical significance. Gut T cells are also at the forefront of protective responses to more aggressive pathogens such as *Salmonella* and *Shigella*. Identifying the pathways that allow gut T cells to react appropriately to commensal microbiota, yet respond to gut pathogens, remains one of the major challenges in biology. At the same time, overreactivity of gut T cells to harmless foods, or the commensal microbiota, underpins food-sensitive enteropathies and inflammatory bowel disease (IBD).

In healthy individuals, T lymphocytes are a natural component of the cell types present at mucosal surfaces. The largest numbers are found in the small intestinal mucosa, primarily in the epithelium, the lamina propria cores of the villi, Peyer's patches (PPs), and isolated lymphoid follicles (Figure 7.1). They are somewhat less common in normal colon epithelium and lamina propria, but abundant in isolated colonic lymphoid follicles. There are also many T cells in the appendix. Normal gastric mucosa only contains a few T cells. At other mucosal surfaces, such as the upper airways and the upper respiratory tract, T cells are rather sparse in the connective tissue, but they are abundant in nasopharynx-associated lymphoid tissue in rodents and in tonsils and adenoids in humans. It is important to emphasize that the organized mucosa-associated lymphoid tissues (MALTs) are the inductive sites for mucosal immune responses, and the connective tissue between MALT is the site of expression of the mucosal immune response (Chapter 1). The majority of T cells in MALT are small, nonactivated T cells, largely part of the recirculating pool, whereas T cells in the lamina propria are activated cells, the progeny of T cells responding to antigen in the MALT.

ORIGIN AND PHENOTYPE OF MUCOSAL T CELLS

In all mammalian species studied so far, CD4 T cells predominate in the lamina propria and CD8 cells in the gut epithelium. Intraepithelial lymphocytes are discussed in detail in Chapter 6; here we focus on CD4 T cells. Innate lymphoid cells lacking a T-cell receptor are covered in Chapter

TABLE 7.1 MUCOSAL INFECTIONS IN T-CELL–DEFICIENT PATIENTS

Mouth/esophagus

Candida albicans
Human papillomavirus

Intestine

Cryptosporidium parvum
Cystoisospora belli
Giardia lamblia
Microsporidia
Enterocytozoon bieneusi
Cytomegalovirus
Coronavirus, rotavirus, adenovirus
Mycobacterium avium complex

Airways

Pneumocystis carinii
Cryptococcus neoformans
Nontuberculous mycobacteria
  (e.g., *Mycobacterium avium* complex)
Cytomegalovirus
Bacterial pneumonia

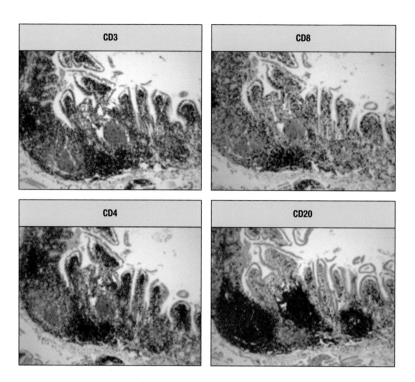

Figure 7.1 Immunostaining of human distal ileum for CD3, CD4, CD8, and CD20. The section contains three B-cell follicles (CD20). Note that surrounding the B-cell follicles are the T-cell zones, containing mostly CD4 cells.

8. Embedded in the connective tissue matrix, lamina propria T cells receive signals from epithelial cells, stromal cells, and the matrix itself via integrin receptors, and are closely associated with dendritic cells and macrophages. These cells and signals likely control the life span and function of T cells in the gut, but there is also cross talk between the T cells and other cell types; and importantly, the presence of the microbiota, signaling via pathogen-associated molecular patterns and producing short-chain fatty acids in the distal bowel, produces an extremely complex series of interactions, which are only now being unraveled.

7.1 Mucosal T-cell traffic from mucosa-associated lymphoid tissue to mucosal lamina propria

It has been known for 40 years that the gut-homing patterns of small lymphocytes and immunoblasts differ. Small resting lymphocytes from mesenteric lymph nodes (MLNs) do not migrate to the lamina propria when transferred into normal recipients. However, in animals, T- and B-cell blasts from the mesenteric nodes or dividing cells in thoracic duct lymph readily migrate to the lamina propria. While this is more difficult to show in humans, analysis of T-cell receptor rearrangements in GALT and nearby lamina propria has shown the same clones of cells at each site, strongly supporting the notion that in humans, GALT T cells traffic via blood to lamina propria. The molecular basis for gut homing was initially elucidated 20 years ago when it was shown that gut homing of blasts was conferred by the expression of the $\alpha_4\beta_7$ integrin on T- and B-cell blasts and mucosal addressin cell adhesion molecule-1 (MAdCAM-1) on intestinal vascular endothelium. Later studies showed that chemokines were also involved. The small bowel epithelium secretes CCL25, and gut-homing cells express the CCL25 receptor CCR9. CCL25 is also tethered to vascular endothelium in the gut. In contrast, CCR9 and CCL25 interactions do not seem to be important in T-cell migration to the colon, and in fact, colonic epithelium does not express CCL25. Instead, colon homing may be controlled by CCL28 made by colon epithelium, $\alpha_4\beta_7$ integrin, and perhaps CCR10. It was, however, only relatively recently that the molecular basis for imprinting of gut homing by PP T cells was elucidated, and it came from an unexpected angle, namely, vitamin A.

Vitamin A is stored in the liver, so to make vitamin A–deficient mice, pregnant mice are given a vitamin A–deficient diet. There is enough vitamin A in the mother to allow the mouse pups to be born. However, maintaining the pups on a vitamin A–deficient diet for 3 months has dramatic effects on mucosal T cells, in that these mice have marked depletion of small intestinal CD4 T cells and $CD8^+$ intraepithelial lymphocytes. Subsequent studies showed that in the PP and mesenteric nodes, dendritic cells (DCs) express a retinal dehydrogenase (RALDH), which can convert all-*trans*-retinol to all-*trans*-retinoic acid. When added to activated T cells and B cells, all-*trans*-retinoic acid is a potent inducer of both CCR9 and $\alpha_4\beta_7$. Thus, the preferential expression of RALDH by antigen-presenting cells (APCs) in PPs and MLNs accounts for gut homing.

Interestingly, vitamin A deficiency had no effect on T cells in the airways, an observation of relevance to a notion that has gone out of fashion, namely, the common mucosal immune system. This hypothesis suggested that immunization in the gut could protect other mucosal sites due to activated B cells and perhaps T cells migrating into all mucosal sites after enteric immunization. However, since respiratory tract vascular endothelium does not express MAdCAM, there is no longer a physiologic basis for the hypothesis. A possible explanation for the homing of mesenteric lymph node cells to the airways may lie in the fact that the rules that govern homing become disrupted during inflammation. For example, gut-homing cells lodge in inflamed skin but not normal skin. In the 1970s when the common mucosal immune system was suggested, virtually all mouse and rat colonies were infected with *Mycoplasma pulmonis*, a low-grade pathogen that induces lung inflammation.

7.2 Lamina propria T cells have characteristics of activated lymphocytes

Lamina propria mononuclear cells can be isolated relatively easily from human and mouse small bowel and colon. Many studies in humans have confirmed that the CD4 cells in normal gut lamina propria have the phenotype of activated effector T cells. Thus, they are $CD45RO^+$, $CD62^{low}$, $CD69^{high}$, $CD25^+$, Fas^+, and $FasL^+$, and as expected from earlier, $\alpha_4\beta_7^+$ and $CCR9^+$.

It has also been established for many years that MALT are on the major route of lymphocyte recirculation. Small recirculating naive T cells ($CD45RA^+$, L-selectin^{high}) can enter PPs because glycosylation of the mucin-like domains of the gut-specific addressin MAdCAM-1 allows it to bind L-selectin on naive T cells to allow them to cross high endothelial venules into the tissue. If these small naive T cells become activated by gut antigen in PPs, they lose L-selectin and express CD45RO. The presence of $CD45RO^+$ cells coexpressing high levels of the $\alpha_4\beta_7$ integrin and L-selectin^{low} near microlymphatics in PPs suggests that these cells are leaving the PPs and are on their way to the lamina propria via the blood. The other site in PPs at which activated T cells are seen is next to M cells. These cells are also $CD45RO^+$, L-selectin^{low}, and $CD69^+$ and are dividing.

These data are consistent with the idea that the gut-associated lymphoid tissues are the inductive sites of mucosal CD4 immune responses and that cells traffic via the thoracic duct and blood toward the lamina propria. The length of time that $CD4^+$ T cells persist in the lamina propria is not understood. In healthy human gut, many cells are terminal deoxynucleotidyl transferase dUTP nick end labeling (TUNEL) positive, suggesting they are undergoing apoptosis. However, in parabiotic congenic mice, only about 25% of lamina propria mononuclear cells are from the opposite mouse after almost 6 weeks of parabiosis, suggesting a long-lived population.

7.3 Cytokine production by mucosal T cells in healthy animals is T_H1/T_H17 dominated

Germ-free mice have virtually no T cells in their gut lamina propria, and the PPs in these mice have only primary follicles. Following colonization with bacteria, there is rapid immune activation, and mucosal T cells rapidly reach the same levels as normal mice. Thus, we can conclude that mucosal T cells are activated by gut microbial antigens, and it would be expected that as lamina propria T cells are activated effector cells, they would be actively secreting cytokines. In humans, enzyme-linked immunospot (ELISPOT) analysis of freshly isolated PPs and small intestinal lamina propria T cells shows a response dominated by interferon-γ (IFN-γ), with very few T cells secreting interleukin-4 (IL-4), IL-5, or IL-10. When human PP T cells are activated with mitogens or food antigens, the response is again IFN-γ dominated. These studies, however, pre-dated the discovery that IL-17A and IL-17F, made largely by a specialized helper T-cell (T_H) subset (T_H17 cells), seem to be particularly highly expressed at mucosal surfaces. Thus, CD4 T cells from healthy gut secrete equivalent amounts of IL-17A and IFN-γ when activated with anti-CD3/CD28. Mucosal T cells are therefore directed toward effector T_H1 and T_H17 responses; and in mice, in the absence of intestinal helminths, there are virtually no T_H2 cells in the small intestine. The ability of CD4 cells in the gut to make IFN-γ is probably very important in the control of gut infections such as cytomegalovirus and cryptosporidia, because, in the latter case, there is compelling evidence that immunity to this parasite depends on the ability of IFN-γ to render gut epithelial cells resistant to infection.

7.4 IL-17 and IL-22 may play an important role in protecting mucosal surfaces

The strategic positioning of T_H17 cells at barrier surfaces reflects their importance in the neutralization of pathogens and commensal microbiota alike, both by coordination of neutrophilic inflammation and by maintenance or restitution of epithelial barrier integrity. Mucosal epithelia express receptors for IL-17 and/or IL-22, T_H17 cytokines that promote tight junction formation, mucus production, antimicrobial peptide production, and epithelial regeneration following injury. Elevation of IL-17 and IL-22 is characteristic of intestinal inflammation in murine colitis and IBD. Whereas IL-17 contributes to both neutrophilic inflammation and aspects of epithelial barrier function, IL-22 functions appear more limited to epithelial barrier function in the gut. Accordingly, IL-22-deficient mice have more severe intestinal inflammation and epithelial injury in mouse models of colitis and are highly susceptible to colonic infection with the attaching-effacing natural mouse pathogen, *Citrobacter rodentium*.

Unlike the genes that encode IL-17A and IL-17F, expression of the gene encoding IL-22 is strictly dependent on the aryl hydrocarbon receptor, a ligand-activated transcription factor that binds a wide array of environmental and endogenous aromatic hydrocarbons, including tryptophan metabolites. Since tryptophan makes up 1%–2% of the total protein in many foods, there is the intriguing possibility that dietary proteins may regulate immunity by the AroA pathway. In humans, CD4 T cells that express IL-22 in the absence of IL-17 cytokines have been identified, supporting the possibility that these cells are regulatory in nature and are dedicated to the maintenance and repair of epithelial integrity. However, their role in normal and inflamed gut has not been established.

7.5   Different types of gut microbiota appear to induce different cytokine responses in mucosal T cells

In animals, where it is possible to experimentally control the gut flora, remarkable progress has been made on the texture of cytokine responses and their regulation by specific members of the microbiota (Figure 7.2). Germ-free mice monocolonized with the human gut bacterium, *Bacteroides fragilis*, show an expansion of T-cell numbers in the spleen and a shift in T-cell cytokine responses from a predominantly T_H2 response to a T_H1 response. Remarkably, this effect is not seen if mice are mono-associated with *B. fragilis* lacking polysaccharide A, a highly unusual capsular polysaccharide that can be processed by APCs and induce CD4 T-cell activation. Polysaccharide A

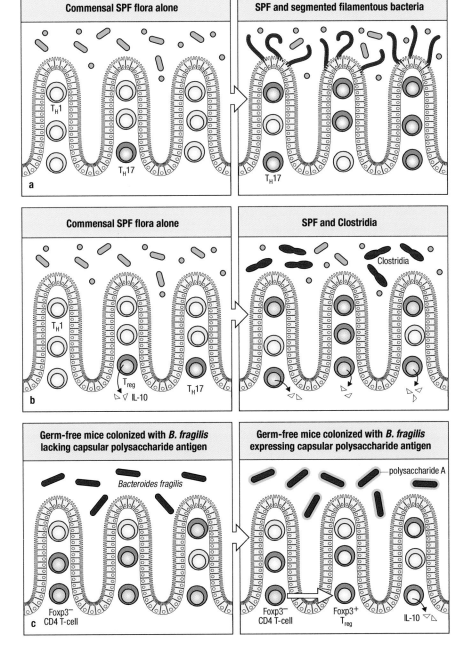

Figure 7.2 **Three examples of how specific microbes can influence T-cell subsets in the intestine.** (**a**) Colonization of mice with a specific pathogen free (SPF) flora containing a segmented filamentous bacteria (SFB) increases the numbers of T_H17 cells in the small intestine compared with mice colonized with the SPF flora alone. (**b**) Mice colonized with an SPF flora containing a variety of clostridial species have many more IL-10-secreting T_{reg} cells in the colon than mice colonized with an SPF flora alone. (**c**) Mice colonized with *Bacteroides fragilis* expressing a capsular polysaccharide antigen (polysaccharide A) have many more IL-10-secreting T_{reg} cells than mice colonized with a bacterium lacking polysaccharide A, because polysaccharide A appears to be able to convert $Foxp3^-$ CD4 T cells into $Foxp3^+$ regulatory cells in the gut.

seems to induce IL-10-secreting regulatory T cells (T_{reg}) in the colon, which can limit the extent of chemical colitis, both prophylactically and therapeutically.

Further support for the idea that specific types of bacteria can induce markedly different types of T-cell responses comes from the observations that mice from different suppliers have markedly different numbers of T_H17 cells in their small intestine, despite having identical numbers of T cells in their gut. It was shown that adult germ-free mice had virtually no T_H17 cells in their gut; however, on colonization with a microbial flora, T_H17 cells became abundant. If adult normal mice were treated with antibiotics to lower the flora, then T_H17 cell numbers were reduced by 50%. When mice from different vendors were examined, most had abundant T_H17 cells in their intestine, apart from mice from the Jackson Laboratory, in which there were essentially none. Likewise, reconstitution of germ-free mice with the altered Schaedler flora, for many years considered to be the gold-standard specific pathogen free (SPF) flora, also did not result in mucosal T_H17 cells. Detailed analysis of the components of the microflora in mice from different vendors showed that the key organism that drives T_H17 responses was an unculturable segmented filamentous bacterium (SFB) related to clostridia. Mice harboring SFB are somewhat more resistant to *C. rodentium* infection than mice lacking SFB.

The final example of commensals that appear to induce regulatory responses are gut clostridial species. Colonization of mice with a flora restricted to clostridia induces somewhat more IL-10-secreting Foxp3$^+$ inducible T_{reg} cells in the colonic lamina propria than in mice colonized with standard SPF flora. Increase in T_{reg} cells in the colon is associated with increased resistance to chemical colitis.

These studies are conceptually important because they indicate that different types of microbes may influence the generation of helper T-cell lineages. They are also interesting in a historical context because the components of the Schaedler flora were chosen by Russell Schaedler in the 1960s on the basis of their lack of immunogenicity, and so a whole generation of immunologists who used animals colonized with the Schaedler flora were working on highly unusual and atypical mice. It will be intriguing to identify which bacterial components skew immune responses. At the same time, however, care has to be taken in extrapolating from a mouse monocolonized with a single bacterial strain to the complex microbiota (1000 species) of humans.

7.6   Developmental pathways of T_H17 cells are not well defined, especially at mucosal surfaces

The developmental origins of T_H17 cells in mice and humans have stirred considerable controversy, and a clear understanding of potential differences between the species is not resolved. Akin to the earlier debate over the presence of T_H1 and T_H2 subsets in humans after their discovery in mice, it may be that many of the apparent species differences in T_H17 biology reflect more the differences in the cell types investigated (blood T cells in humans, spleen T cells in mice), and experimental conditions, than intrinsic species differences, although this remains to be determined. In fact, there is no *a priori* reason why different species should not adopt different strategies to generate effector T-cell responses appropriate for the lifestyle of the species, and the pathogens they face in their particular environments.

Whereas there is agreement over the requirement for pro-inflammatory cytokines in the differentiation of T_H17 cells, whether or not the same cytokines act similarly in both species is contentious. Even more contentious is the role of transforming growth factor-beta (TGF-β). In mice, where T_H17 cells were first described, IL-6 is indispensible for the induction of T_H17 cells and acts in concert with TGF-β to initiate T_H17 cell development from antigen-stimulated naive CD4 T cells. IL-23 acts downstream of IL-6 and TGF-β to reinforce T$_H$17 development in mice but is not needed for the development of IL-17-producing T cells in mouse intestinal mucosa, despite the requirement for IL-23 in some mouse models of intestinal inflammation. IL-1β has been found to amplify T$_H$17 development in mouse T cells, and along with IL-23, elicits the production of T$_H$17 cytokines from mature T$_H$17 cells without a requirement for T-cell receptor stimulation, providing an antigen-independent mechanism for T$_H$17 cell responses.

In humans, initial studies that examined the generation of the T$_H$17 cells using naive (CD45RA$^+$) CD4 T cells derived from peripheral blood found that IL-1β and IL-23 induced production of T$_H$17 cells without a requirement for IL-6 or TGF-β, contrasting greatly with studies from mice. However, it was subsequently found, using naive CD4 T cells derived from human umbilical cord blood, that T$_H$17 development did require low levels of TGF-β as well as IL-1β and a STAT3-inducing cytokine—IL-6, IL-21, or IL-23—more closely resembling findings from mice. These studies emphasize that differences in the previous activation history of T-cell precursors in human and mouse studies may influence T$_H$17 cell development, although fundamental differences in the development of human and mouse T$_H$17 cells cannot be excluded. Despite these differences, there are major similarities in some aspects of T$_H$17 cells in both species, such as the expression of the transcription factor retinoic acid–related orphan receptor RORγt, IL-23 receptor, and CCR6.

In mice, T$_H$17 cells can be derived from naive precursors by the concerted actions of IL-6, IL-1β, and IL-23 in the absence of TGF-β, in agreement with some human studies. Importantly, the T$_H$17 cells generated under these conditions have a gene expression profile distinct from T$_H$17 cells generated in the presence of TGF-β, raising the possibility of further T$_H$17 subsets that might have different functions. While further studies will be needed, there appears to be different pathways involved in the generation of subtypes of T$_H$17 cells, the precise functions of which are yet to be defined. It is plausible that cells derived via the TGF-β-dependent pathway of T$_H$17 development may play a more important role in maintenance of barrier integrity at mucosal sites, particularly in the intestine where there is abundant active TGF-β made by epithelial cells and stromal cells and the greatest number of T$_H$17 cells are found in health; whereas a TGF-β-independent pathway plays a more prominent role under conditions of pathogen-induced or autoimmune inflammation.

7.7 Role of local differentiation or selective migration in T$_H$1 and T$_H$17 responses is unclear

As for other CD4 T-cell subsets, the priming of T$_H$17-cell development is thought to occur in the GALT, where on exposure to antigen there is rapid downregulation in T cells of L-selectin and expression of CCR6 and CCR9, which favors trafficking to the intestinal lamina propria. As for T$_H$1 and T$_H$2 cells, the transfer of T$_H$17 cells into lymphopenic hosts has established that, at least in the setting of an "empty" T-cell compartment, migration of T$_H$17 cells into the intestinal lamina propria can readily be seen. Whether this occurs similarly in a T-cell–replete repertoire is not known. In fact, there is remarkably little known about the factors that control effector CD4 polarization in the MALT of experimental animals, and essentially no data for humans.

There is no doubt that the gut lamina propria is enriched for effector T$_H$1 cells, T$_H$17 cells, and T cells that make both IFN-γ and IL-17A. The critical question is whether the imprinting of gut homing in GALT following T-cell activation has another layer of complexity, with differences in homing of different CD4 cell lineages. Added to this is the question of how is plastic development in the lamina propria; in other words, is lineage commitment settled in the lamina propria, or is it committed prior to extravasation, perhaps imprinted in GALT? At the heart of this debate is whether there is a defined T_H17 cell precursor, separate from virgin CD4 T cells that can become T_H1, T_H2, or T_{reg} cells. In humans it has been suggested that CD161 along with IL-23R, CCR6, and the transcription factor retinoic acid–related orphan receptor C isoform 2 are markers of T_H17 cells. CCR6 is of particular interest in the gut because it is the only receptor for CCL20. Normal colonic epithelial cells express CCL20, but expression is highest in inflamed epithelium and, constitutively, in the follicle-associated epithelium overlying PPs and cryptopatches. CCR6 is also expressed on some lymphoid tissue inducer (LTi) cells, and mice lacking CCR6 have small PPs. It is unclear if the effects that different types of bacteria have on mucosal T_{reg}, T_H1, and T_H17 responses are controlled at inductive sites or the lamina propria.

When a T-cell extravasates into the lamina propria, the local environment contains molecules known to affect T_H17 cell development. Epithelial cells produce active TGF-β, and latent TGF-β is abundant in the matrix. Lamina propria DCs also secrete retinoic acid that may inhibit T_H17 generation and promote T_{reg} development. Mice lacking TGF-β have few T_H17 cells in their gut and spleen, suggesting that TGF-β has a global effect on T_H17 cell generation rather than a local effect in the gut lamina propria. A slightly different mouse that can make TGF-β but cannot activate the cytokine in the matrix also has a reduced number of lamina propria T_H17 cells. Confounders in these experiments, however, are that mice lacking TGF-β have extremely strong T_H1 responses that inhibit T_H17 cell generation, and there is the possibility that even at inductive sites, matrix-bound TGF-β may be needed for T_H17 cell generation. Recent studies have shown that the type of bacterium which colonises the gut can affect TH17 responses. In mice colonised with segmented filamentous bacteria (see below), TH17 cells have little ability to make pro-inflammatory cytokines. In contrast, in mice colonised with the pathogen Citrobacter rodentium TH17 cells are highly pro-inflammatory.

7.8 T_H17 lineage displays considerable flexibility

With the emergence of better technical approaches to track the fate of defined T-cell populations *in vivo*, it has become apparent that there is substantial flexibility, or plasticity, in the developmental programs of regulatory and effector T-cell subsets. CD4$^+$ T cells that coexpress lineage markers of regulatory (Foxp3) and T_H17 (RORγt) T cells have been detected in the normal gut of both mice and humans, and the frequencies of these cells increase in the setting of inflammation. At present, it is unclear whether these "dual-expressors" represent early precursors from which T_H17 or T_{reg} cells emerge, represent Foxp3$^+$ T_{reg} cells that are in transition to T_H17 cells in response to pro-inflammatory cytokines, are a functionally distinct T-cell subset, or are a combination thereof. However, there are growing data from studies in murine models of infectious disease that T_{reg} cells can transition to progeny with effector properties, such as T_H17 or T_H1, under the influence of inflammatory cytokines. Reciprocal programming (i.e., effector T cells into T_{reg} cells) has not been well demonstrated, suggesting that the conversion of T_{reg} cells to T effectors might represent an evolutionary adaptation to override the dominance of homeostatic T_{reg} function at mucosal sites in the face of a pathogenic threat. It is currently unclear to what extent similar transitions might occur in the context of immune dysregulation during the evolution of IBD. However, the possibility that T_{reg} cells may have the potential to become effector T cells needs to be considered when thinking of adoptive T_{reg} therapy for chronic inflammatory conditions, or even strategies to boost T_{reg} numbers *in vivo*.

T-cell subset plasticity is not, however, limited to T_{reg} cells. The transition of CD4$^+$ T cells of one effector subset to progeny with features of a distinct effector subset has been increasingly reported. This has been most clearly demonstrated for T_H17 cells, which can give rise to T_H1-like cells contingent on the hierarchy of cytokine signals received following T_H17 cell differentiation. In particular, T_H17 cells exposed to the T_H1-polarizing cytokine IL-12 extinguish expression of genes that encode IL-17 cytokines in favor of IFN-γ. The transition of T$_H$17 cells into IFN-γ-producing T$_H$1 cells is dependent on STAT4 signaling and induction of the T$_H$1 lineage transcription factor, T-bet. Similar to the transition of T$_{reg}$ cells to T effector cells, T$_H$17–T$_H$1 transitions appear to be unidirectional. That is, T$_H$17 cells can transition to T$_H$1 cells, but T$_H$1 cells do not transition to T$_H$17 cells. In intestinal inflammation in mice, the transition of T$_H$17 cells into T$_H$1-type cells is associated with the development of colitis, suggesting that the mix of T$_H$17 and T$_H$1 cells typically found in the intestinal lamina propria, both at homeostasis and in IBD, might be due to diversion of some T$_H$17 precursors locally in the intestines (**Figure 7.3**). An important contribution to the fate of Th17 cells is the observation using fate mapping that in the gut Th17 cells can become ex-Th17 cells and begin to make IL-10, similar to T$_{reg}$-1 cells. The relevance of this to gut inflammation is that IL-10R mutations in children lead to early onset gut inflammation, and the source of IL-10 may be exTh17 cells.

The plasticity of T$_H$17 cells appears to be shared in humans. CD4$^+$ T-cell isolates from the intestine of patients with Crohn's disease contain a mix of distinct subsets of T$_H$17 and T$_H$1 cells, as well as IL-17$^+$ IFN-γ$^+$ T cells (**Figure 7.4**). Similar to findings in mice, T$_H$17 and "T$_H$17/T$_H$1" cells cloned from intestinal T-cell isolates can be diverted to a T$_H$1 phenotype in response to IL-12. Interestingly, CD161, which was initially proposed as a marker for human T$_H$17 cells (mice do not express a CD161 homolog) is also expressed by a substantial fraction of IFN-γ$^+$ T cells in normal and inflamed intestines. Further, CD161$^+$ human T cells give rise to a distinct subpopulation of IFN-γ$^+$ cells under T$_H$17-polarizing conditions. Thus, as in studies in mice, human T$_H$17 cells may be phenotypically unstable and give rise to T$_H$1-like cells, whereas the converse has not been demonstrated. The implications of T$_H$17 plasticity for normal intestinal immune regulation and the development of IBD remain to be defined.

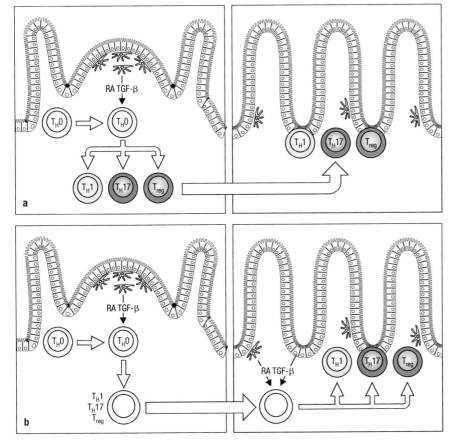

Figure 7.3 In the gut, the control of CD4 T-cell differentiation into T$_H$1, T$_H$17, and T$_{reg}$ cells, and their evolution into cells with other cytokine profiles is still not clear. In one model (**a**), lineage commitment and imprinting of gut homing occurs in gut-associated lymphoid tissue and T$_H$1, T$_H$17, and T$_{reg}$ cells migrate as distinct subsets into the lamina propria. In an alternative model (**b**), terminal differentiation into distinct subsets occurs in the lamina propria under the influence of local cytokines made by dendritic cells and epithelial cells. In health, however, where the cytokine milieu is inhibitory for further differentiation and survival, T cells die and are replaced by cells from the blood. RA, retinoic acid; TGF-β, transforming growth factor beta.

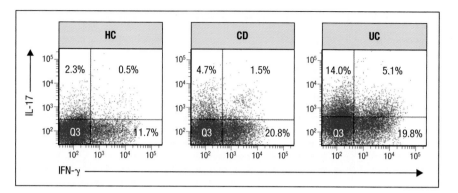

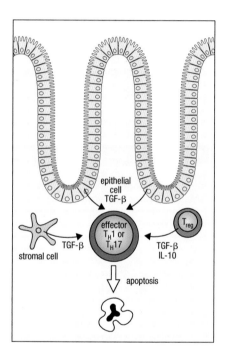

Figure 7.4 **Flow cytometric analysis of CD4 lymphocytes from normal human colon (HC), Crohn's disease colon (CD), and ulcerative colitis (UC) colon.** Cells making IFN-γ dominate, but in ulcerative colitis there are many cells making IL-17A only. In all groups, there are cells making both cytokines. Lamina propria mononuclear cells were stimulated with phorbol myristate acetate (PMA) and ionophore for 4 hours prior to analysis. (Adapted from Rovedatti, L. et al. *Gut* 2009, 58:1629–1636. With permission from BMJ Publishing Group Ltd.)

Regulatory T cells and control of effector T-cell responses

It is very well established that reconstitution of mice lacking T or B cells with small numbers of naive CD4$^+$ T cells leads to colitis. If the recipients are germ free, colitis does not develop. It is also well known that coinjection of natural CD4$^+$ T$_{reg}$ cells ameliorates the colitis induced by naive T cells. Thus, it is often claimed that pathogenic effector T-cell responses against the gut microflora are controlled or dampened by T$_{reg}$ cells, and a corollary of this statement, namely that IBD is somehow due to a failure of T-cell regulation, is often mooted. Suppression appears to be mediated by TGF-β, although other cytokines such as IL-10 and even IL-17A may be suppressive in the correct context. In reality, control of mucosal effector and regulatory T cells is likely to be multilayered and multifaceted, and there are almost certainly many different mechanisms that control immune responses in the gut, including the accessibility of antigen to the lamina propria. For example, patchy defects in the intestinal barrier leading to increased local intestinal permeability cause a patchy Crohn's-like lesion in mice with a normal immune system. The colonic epithelial cells in mice with a gut epithelial knockout of nuclear factor κB essential modulator (NEMO) undergo apoptosis and the barrier is broken, leading to colitis. Finally, in patients with neutrophil defects such as chronic granulomatous disease or glycogen storage disease type 1b, the normal flora enters the lamina propria and persists and elicits a disease almost identical to Crohn's disease. In normal lamina propria, there is abundant active TGF-β made by epithelial cells and stromal cells and as relatively little gut antigen crosses the intact epithelial/mucus barrier, it is entirely possible that effector T-cell regulation in health is due to "neglect," rather than active suppression (**Figure 7.5**).

7.9 Human diseases due to single gene defects are informative in understanding gut inflammation and immune regulation

Immunodysregulation, polyendocrinopathy, enteropathy, X-linked (IPEX) syndrome is a very rare X-linked disease where mutations in Foxp3 lead to defective regulatory T-cell activity. It is often used as an example of how regulatory T-cell defects cause colitis. Virtually all of these children have small bowel inflammation, and some have colitis. The gut manifestations, however, go alongside chronic inflammation in the pancreas and other endocrine organs, kidneys, liver, and skin; there is often anemia and/or thrombocytopenia, autoantibodies, and a generalized lymphadenopathy. It is therefore possible that children with IPEX have a general autoreactivity to self-antigens, with the gut tissue as a target of autoimmune attack because the gut is also an endocrine organ. The scurfy mouse also has mutations in Foxp3 but does not appear to have colitis; instead, the mouse dies at about 3–4 weeks of age showing lymphohistiocytic infiltration of many tissues and

Figure 7.5 **Different pathways probably downregulate effector CD4 T$_H$1 and T$_H$17 cell responses in the intestine in health and induce apoptosis of activated T cells.** Epithelial cells are a major source of bioactive TGF-β. Stromal cells also make TGF-β, which may be activated by the low levels of proteases in normal gut. CD103$^+$ dendritic cells may also induce T$_{reg}$ cells, which by making TGF-β and IL-10 dampen potentially damaging T$_H$1 and T$_H$17 responses. However, a major factor is probably the fact that very little antigen will cross the mucus layers and epithelium, and subepithelial macrophages (not shown) are highly efficient at killing and degrading bacterial antigens without evoking a pro-inflammatory response.

lymphadenopathy. Both of these situations differ markedly from human IBD, where, especially for Crohn's disease, pathologic manifestations in tissues other than the gut are generally thought to be a consequence of the gut lesions. It is also worth noting that prior to the development of the transfer model of colitis in lymphopenic mice, there was an extensive literature in the rat where transfer of naive T cells into athymic animals caused generalized autoimmune disease, including colitis.

Over 25 years ago it was shown that IL-10 null mice developed spontaneous enteritis, dependent on the microbiota. Mice with T-cell deficiency of IL-10 also develop a spontaneous colitis. Mice lacking IL-10 in the T_{reg} population also develop spontaneous colitis, but it is less severe than in mice with global loss of IL-10, suggesting that other sources of IL-10 may be important. Consistent with this, IL-10-producing Foxp3$^+$ cells are present in the intestinal lamina propria, and there are also data to suggest that T_H1 cells, as they differentiate, also begin to make IL-10. Extremely compelling evidence that IL-10 controls responses to the flora in humans comes from rare cases where there are defects in IL-10R as mentioned earlier. Very young children with IL-10R loss-of-function mutations develop a severe early onset inflammatory bowel disease with many characteristics of Crohn's disease. Successful therapy has been achieved using bone marrow transplantation.

7.10 Regulation of mucosal immune responses is complex and depends on whether effector T-cell can be regulated

It is worth considering whether potentially pathogenic T_H1 or T_H17 responses in the gut are controlled by regulatory T cells. First, in PPs and mesenteric lymph nodes, there is evidence that feeding protein antigens in intact mice leads to conversion of naive T cells into Foxp3$^+$ T_{reg} cells. However, the largest numbers of antigen-specific Foxp3$^+$ T_{reg} cells are found in the small bowel lamina propria. CD103$^+$ DCs from lamina propria can also convert antigen-specific T cells into T_{reg} cells, and this process is dependent on retinoic acid production. In mice, where Foxp3 is a good marker for T_{reg} cells, these experiments suggest that there is generation of T_{reg} cells in response to gut protein antigens in GALT, and cells then migrate to the lamina propria. In the small intestine, there is evidence for heterogeneity in the lamina propria in that macrophages appear to be able to induce T_{reg} cells by secreting retinoic acid, IL-10, and TGF-β, whereas dendritic cells activate T_H17 cells. It is not known, however, whether the same type of response occurs to the antigens of the microbiota.

In humans there are difficulties in using markers such as Foxp3 and CD25 to examine T_{reg} cells in the gut. Both markers are expressed on activated effector T cells, so they are not particularly good tools to analyze T_{reg} function. In humans there are few data on GALT; however, there have been some functional studies on lamina propria mononuclear cells. CD4$^+$CD25^{high} cells from normal lamina propria are less proliferative and produce fewer cytokines than CD4$^+$CD25$^-$ lamina propria mononuclear cells. When the former population is added to mitogen-activated CD4$^+$CD25$^-$ blood T cells, there is a marked reduction in proliferation. This suggests that there are regulatory T cells in the lamina propria; however, regulatory activity is partially abolished by exogenous IL-2, which does suggest that at least some of the effect is due to cytokine deprivation, rather than active suppression. There does not appear to be any reduction in T_{reg} numbers in the gut in IBD.

An alternative approach to understanding CD4 T_H1 or T_H17 responses in the gut in idiopathic inflammation is to assume that the problem is an anti-pathogen response gone awry, directed at harmless microbes. In pathogen responses, it is important to prevent T_{reg} cells dampening effector cell function until the pathogen has been eliminated. Thus, there is an advantage for effector CD4 cells in the gut to be resistant to suppression, regardless of the numbers of T_{reg} cells in the tissue. One way that effector T cells achieve this is by overexpressing Smad7, the intracellular inhibitor of TGF-β signaling. Smad7 functions in two main ways: by blocking the binding of Smad2/3 to ligand-activated TGF-β receptor; and by targeting the receptor itself for ubiquitination and degradation. As might be expected, T_{reg} cells and TGF-β protein are ineffective in dampening pathogenic T-cell responses from inflamed IBD tissues. However, knocking down Smad7 with an antisense oligonucleotide allows endogenous TGF-β to dampen T-cell responses. Overexpression of Smad7 in CD4 T cells makes animals more susceptible to experimental colitis because endogenous TGF-β and T_{reg} cells cannot dampen inflammation. Similarly, in the transfer model of colitis, T cells overexpressing Smad7 cannot be controlled by transferred T_{reg} cells. Paradoxically, however, in mouse models of colon cancer, overexpression of Smad7 in T cells is highly beneficial because the exaggerated T-cell responses translate into exaggerated antitumor responses, even though the mucosa remains highly inflamed (**Figure 7.6**). The importance of Smad7 in controlling gut inflammation is demonstrated by the recent findings that pills containing a smad7 antisense, when given orally to patients with right-sided Crohn's disease, induced remission in approximately two-thirds of patients. Presumably because endogenous TGFβ and T_{reg}s are free to inhibit T effectors.

One of the main difficulties in this area, especially when looking at the effects of the microflora, is the difficulty in generating T-cell responses to the flora. It was demonstrated many years ago that T cells from normal gut responded to microbial antigens from the flora of different individuals but were tolerant to their own flora, and that in IBD autologous gut T cells responded vigorously to an individual's own microflora. More recently CD4 T cells from blood and intestine of patients with Crohn's disease have been shown to have specificity for gut bacteria. Interestingly, these cells are predominantly Th17 cells. Their role in pathology must be carefully considered given that anti-IL-17A therapy exacerbates disease in Crohn's. In mice, T-cell lines reactive to gut bacterial antigens cause colitis when transferred into lymphopenic mice, and the colitogenic response can be inhibited *in vivo* by *in vitro*-generated T_{reg} cells with the same specificity.

T-cell–mediated gut diseases

Although IBD and celiac disease are covered in other chapters, it is worth considering these conditions as situations where the "balance" between CD4 T-cell activity in the gut and luminal antigens becomes skewed, and chronic T-cell activation causes disease.

Celiac disease is of interest because there is no doubt that a substantial part of the pathology is driven by CD4 T cells in the lamina propria responding to gliadin peptides presented in the context of HLA-DQ2 or DQ8. However, DQ2 is an extremely common haplotype in the general population, and it is still not known what additional factors lead to only a few percent of DQ2+ individuals becoming gluten reactive. Environmental factors are clearly important. There are cases in the literature where identical twins have developed celiac disease years apart, and when the proband was diagnosed, the healthy sibling has been investigated thoroughly and shown to be normal. One notion is that viral infections in the gut not only break the barrier to allow antigen to enter the lamina propria but can induce the local production of type I interferons, which are potent stimulators of T_H1 responses. Once sensitized, however, celiac disease is a lifelong condition, which implies T-cell memory. It is highly unlikely that long-lived gluten-specific memory T cells persist in the lamina propria; instead they are probably part of the recirculating memory T-cell pool, in which case they may encounter gluten in GALT. One feature of celiac disease, however, that has not been explained is the time to relapse

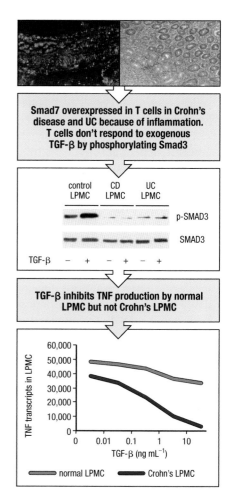

Figure 7.6 Smad7 overexpression in activated T cells in Crohn's disease (CD) and ulcerative colitis (UC) renders mucosal T cells resistant to TGF-β. In humans, the addition of exogenous TGF-β rapidly phosphorylates Smad3 (p-SMAD3) in lamina propria mononuclear cells (LPMCs) from normal human gut but has no effect on lamina propria mononuclear cells from IBD patients. Consistent with this, TGF-β can inhibit the production of tumor necrosis factor (TNF) transcripts by superantigen-activated normal lamina propria mononuclear cells but has no effect on lamina propria mononuclear cells from Crohn's patients. Collectively these data show that expression of Smad7 in effector T cells regulates whether T cells are susceptible to regulation. (Adapted from Monteleone, G. et al. *J. Clin. Invest.* 2001, 108:601–609. With permission from Elsevier Ltd.)

when returning to a gluten-containing diet. This is highly variable between individuals and implies that environmental factors may also be important at all stages of the disease.

Recently, genome-wide association studies have attempted to fill the gap between the high frequency of DQ2$^+$ individuals and the low numbers of these who develop disease. Virtually all of the polymorphisms identified are associated with genes that control the tone and texture of the T-cell responses, such as IL-2 and IL-21.

In Crohn's disease, there is overwhelming evidence that the disease is mediated by activated CD4 T$_H$1 and T$_H$17 cells. Genome-wide association studies have identified many dozens of polymorphisms related to innate immune function as being important risk factors (Chapter 31). It needs to be remembered, however, that the penetrance of Crohn's disease is almost completely environmental. Thus, Crohn's disease is very uncommon in India, but when families migrate to Westernized countries with high incidences of Crohn's disease, the incidence in the children of immigrants is the same as that of the indigenous population.

The unifying factor in both celiac disease and Crohn's disease is the persistence of activated CD4 T cells in the intestinal lamina propria. In celiac disease, the T cells are antigen driven, as shown by a gluten-free diet restoring normality of gut structure. In Crohn's disease, however, the situation is much more complex. The contribution of antigen is not known, and there are very many cytokines overexpressed in Crohn's disease mucosa, such as IL-1, IL-2, IL-6, IL-7, IL-12, IL-15, IL-18, IL-21, and IL-23, which can signal to activated T cells, prevent apoptosis, and allow the cells to continue to produce IFN-γ, IL-17A, and tumor necrosis factor-α (TNF-α). Identifying which, if any, of these cytokines are critical, and thus worthy of the expense of therapeutic intervention, represents a significant challenge. It may well be that so many factors are involved in driving CD4 T-cell survival in Crohn's disease with considerable variation between individuals that targeting individual cytokines may not be the best way to go, and that strategies to target the T cells themselves may be more productive. Another approach would be to use combination therapy. The caveat, however, remains that because CD4 T cells in the gut and airways are vital to maintain freedom from low-grade infections, any T-cell–directed therapy has to be effectively monitored.

The role of CD4 T cells in ulcerative colitis remains opaque. There are some data that suggest overexpression of IL-13; however, in absolute terms IL-13, or IL-5 for that matter, is dominated by IL-17A and IFN-γ, which are present in at least 10-fold excess compared with IL-13. Severe ulcerative colitis responds well to cyclosporin A, which clearly suggests a role for T-cell–mediated damage in fulminant disease; however, trials of tolerizing anti-CD3 antibodies in ulcerative colitis have not been a success.

Mucosal-associated invariant T cells, a cell type with growing importance at mucosal surfaces

Mucosal-associated invariant T (MAIT) cells are unconventional T cells, which are usually CD4$^-$ and CD8$^-$, although some expresss CD8. They express an invariant T-cell receptor (TCR)α chain, i.e., Vα19-Jα33 in mice and Vα7.2-Jα33/12/20 in humans, which is usually associated with a limited number of TCRβ chains, namely, Vβ6 and Vβ8 segments in mice and Vβ2, Vβ13, or Vβ22 segments in humans. MAIT cells are selected in the thymus by the evolutionarily conserved major histocompatibility complex class Ib-related molecule (MR)1, and enter the bloodstream with a naïve phenotype. They express CD161 in the thymus, and either CD161 or IL-18 receptor-α in cord blood. More recently, it has been demonstrated that the TCRα chain can also be heterogeneous in human MAIT cells, as the invariant TCRα chain Vα7.2 may be joined to Jα20 or Jα12 rather than to Jα33 in some individuals. MAIT cells are rare in most strains of mice but in man make up 5%–50% of blood T cells. The thymic selection of MAIT cells is followed by their expansion in the gut mucosa dependent on B cells and the commensal flora.

MAIT cells are located along the entire gut with a higher frequency in jejunum than in ileum, colon, and rectum. MAIT cells from healthy volunteers can be activated by B cells previously exposed to intestinal microbes, including commensals such as *Escherichia coli* Nissle, and pathogens such as *S. typhi*, enteropathogenic *E. coli*, and enteroinvasive *E. coli*. MAIT cells respond to bacterial compounds derived from the bacterial riboflavin pathway dependent on MR1 and antigen-presenting cells. MAIT cells from healthy volunteers are also activated by MR1-expressing epithelial cells infected with the invasive bacteria *Shigella Flexneri*, thus leading to killing of the infected cells. As part of the innate immunity response against *Vibrio cholerae*, MAIT cells become activated in and depleted from the periphery. MAIT cells thus might have a role in response to enteric pathogens by limiting the bacterial infections by producing cytokines, such as IFN-γ, TNF-α, and IL-17.

There is evidence that MAIT cells are reduced both in the blood and gut mucosa of patients with celiac disease, and this decrease could be secondary to bacterial overgrowth that in turn induces MAIT cell death. Likewise, MAIT cells have been found to be reduced in the peripheral blood and inflamed gut of IBD patients due to increased proapoptotic features, including activated caspases. Circulating MAIT cells from UC patients, which are characterized by high expression of IL-18 receptor α, are activated *in vitro* by IL-18. Activated MAIT cells from IBD patients release large amounts of IL-17 and IL-22, and their expression of the activation marker CD69 correlates with endoscopic scores in UC patients. There is evidence that MAIT cells infiltrate colon adenocarcinoma, but their ability to produce IFN-γ is reduced compared to unaffected colonic areas, suggesting that tumor microenvironment dampens local immune response sustained by this cell type.

MAIT cells were for many years identified using antibodies against their conserved T-cell receptors. However, an important technologic advance has been the development of MR1 tetramers to both identify and functionally characterize MAIT cells and their subsets in health and disease.

Natural killer T cells, a relatively uncommon T-cell type at mucosal surfaces which may have a role in inflammation

As the name says, natural killer T (NKT) cells share phenotypic and functional features with both conventional NK cells and T cells, such as the expression of NK1.1/CD161, Ly49 and NKG2, CD3 and an $\alpha\beta$ TCR. NKT cells are characterized by a relatively limited TCR usage. Type I NKT cells in mice express the evolutionarily conserved invariant TCRα chain Vα14-Jα18 associated with a limited set of β chains, Vβ8.2, Vβ7, and Vβ2 segments. Human type I NKT cells express the invariant TCRα chain Vα24-Jα18 paired with Vβ11. Type II (nonclassical) NKT cells in mice and man tend to use diverse α and β chains. Type I NKT cells can be double negative or CD4$^+$ in mice, and in humans, can be CD4$^+$, CD8$^+$, or double negative. All murine and human type I NKT cells but not type II cells bind the glycolipid α-galactosylceramide, NKT cells recognize glycolipids via the MHC-I like molecules CD1. In humans there are five CDI molecules (CD1a–e) with CD1d the best investigated. Mice only express CD1d.

Although type II NKT cells respond to sulfatide as ligand, sulfatide-loaded CD1d tetramers have not been widely used to investigate type II NKT cells due to poor stability of this tetramer complex and high background staining. CD1d$^{-/-}$ mice and Jα18$^{-/-}$ mice, which lack both the subgroups of NKT cells and only type I, respectively, have been used to assess type II NKT cells. Notwithstanding, much less is known regarding the biological function of type II NKT cells compared with type I NKT cells.

The frequency of type I NKT cells is 0.6% and 0.1% in lamina propria lymphocytes and intraepithelial lymphocytes, respectively.

Upon activation, type I NKT cells produce TNF-α, IFN-γ, IL-4, IL-5, and IL-13, IL-10, and TGF-β. Additionally, a subgroup of type I NKT cells, which are CD4$^-$ and NK1.1$^-$, release IL-17. Similarly, type II NKT cells release both pro-inflammatory (i.e., IFN-γ, TNF-α, IL-17A, IL-6) and anti-inflammatory cytokines (i.e., IL-10). In several murine models of colitis, NKT cells have been shown to play both protective and pathogenic roles, based on the inflammatory stimuli and the lipid antigens recognized. The protective function of NKT cells has been reported in three mouse models of Th1-mediated colitis, namely dextran sodium sulfate, adoptive transfer model of T cells into lymphopenic hosts and trinitrobenzene sulfonic acid. A pro-inflammatory role for type I NKT cells has been described in the oxazolone-induced colitis. High mucosal levels of IFN-γ and IL-17A are produced in a transgenic murine model overexpressing CD1d and a TCR from an autoreactive type II NKT cell. A number of studies have reported a reduced proportion of type I NKT cells in the peripheral blood and gut of IBD patients, while type II NKT cells accumulate in the lamina propria of patients with UC. The role of NKT cells appears to be complex in IBD, and it might be regulated by various local factors, including the amount of CD1d, the nature of the antigens presented by CD1d, the type of stimulatory antigen-presenting cells, the cytokine milieu, and the intestinal microbiota. Recently, administration of a monoclonal antibody blocking NKG2D, a receptor expressed on NKT cells, has been tested in the treatment of Crohn's disease, but unfortunately success was limited.

SUMMARY

The mucosal surfaces are the battleground between the immune system, pathogens, and dietary and environmental antigens. Terms such as *balance* and *homeostasis* are used to describe the situation in health. These words, however, are descriptors and give no insight into mechanisms. The fact that many different transgenic and knockout mice, many with normal numbers of T cells and B cells, develop chronic gut inflammation, mediated in most cases by CD4 T cells, strongly suggests that control of mucosal T cells is complex. In health, it would appear that activated mucosal T cells are generated in GALT (hence the large numbers of T_H1 and T_H17 cells in normal gut) in response to, and perhaps controlled by, the microbiota and perhaps even diet. When effector cells migrate to the lamina propria in health, they are short lived and die by apoptosis. An unpredictable coalition of antigen(s) (perhaps due to a break in the epithelial barrier or a genetically determined inability to break down the cell walls of the microbiota) and perhaps infection changing the local cytokine milieu may result in the rescue of CD4 T cells from the apoptotic path, and drive them toward survival. The inflammation generated by the activated CD4$^+$ T cells then tips the microenvironment into self-perpetuating inflammatory milieu, since the barrier is then broken by T-cell–derived cytokines, and inflammatory cells producing survival cytokines move into the tissues. Nonspecific agents such as corticosteroids may dampen the inflammation, and anticytokine agents such as anti-TNF may also produce transient relief. However, disease persists at low levels or returns, presumably because of CD4 T-cell memory, about which we know very little in mucosae. Exciting recent developments concerning MAIT cells and NKT cells in the gut add a new dimension to the complexity of gut immunity and inflammation, and it will be interesting to determine the role of these cells subsets at mucosal surfaces in humans.

FURTHER READING

Atarashi, K., Tanoue, T., Shima, T. et al.: Induction of colonic regulatory T cells by indigenous *Clostridium* species. *Science* 2011, 331:337–341.

Balasubramani, A., Mukasa, R., Hatton, R.D. et al.: Regulation of the *Ifng* locus in the context of T-lineage specification and plasticity. *Immunol. Rev.* 2010, 238:216–232.

Fantini, M.C., Rizzo, A., Fina, D. et al.: Smad7 controls resistance of colitogenic T cells to regulatory T cell-mediated suppression. *Gastro.* 2009, 136:1308–1316.

Gagliani, N., Vesely, M.C.A. et al.: Th17 cells transdifferentiate into regulatory T cells during resolution of inflammation. *Nature* 2015, 523:223–229.

Glocker, E.O., Kotlarz, D., Boztug, K. et al.: Inflammatory bowel disease and mutations affecting the interleukin-10 receptor. *N. Engl. J. Med.* 2009, 361:2033–2045.

Godfrey, D.I., Uldrich, A.P., McCluskey, J. et al.: The burgeoning family of unconventional T cells. *Nat. Immunol.* 2015, 16:1114–1123.

Hovhannisyan, Z., Treatman, J., Littman, D.R. et al.: Characterization of interleukin-17-producing regulatory T cells in inflamed intestinal mucosa from patients with inflammatory bowel diseases. *Gastroenterology* 2011, 140:957–965.

Ivanov, I.I., and Littman, D.R.: Segmented filamentous bacteria take the stage. *Mucosal Immunol.* 2010, 3:209–212.

Iwakura, Y., Ishigame, H., Saijo, S. et al.: Functional specialization of interleukin-17 family members. *Immunity* 2011, 34:149–162.

Murphy, K.M., and Stockinger, B.: Effector T cell plasticity: Flexibility in the face of changing circumstances. *Nat. Immunol.* 2010, 11:674–680.

Omenetti, S., and Pizarro, T.T.: The Treg/Th17 axis: A dynamic balance regulated by the gut microbiome. *Front Immunol.* 2015, 6:639.

Round, J.L., and Mazmanian, S.K.: Inducible Foxp3+ regulatory T-cell development by a commensal bacterium of the intestinal microbiota. *Proc. Natl Acad. Sci. USA.* 2010, 107:12204–12209.

Rubtsov, Y.P., Rasmussen, J.P., Chi, E.Y. et al.: Regulatory T cell-derived interleukin-10 limits inflammation at environmental interfaces. *Immunity* 2008, 28:546–558.

Sakaguchi, S., Miyara, M., Costantino, C.M. et al.: FOXP3+ regulatory T cells in the human immune system. *Nat. Rev. Immunol.* 2010, 10:490–500.

Spits, H., and Di Santo, J.P.: The expanding family of innate lymphoid cells: Regulators and effectors of immunity and tissue remodeling. *Nat. Immunol.* 2011, 12:21–27.

Innate lymphoid cells

8

GERARD EBERL AND NICHOLAS POWELL

Innate lymphoid cells (ILCs) have recently come to the forefront of mucosal immunology, being recognized as prompt effectors against infection and injury, and as orchestrators of immune responses. While it was understood that such orchestration was controlled by antigen-presenting cells (APCs)—dendritic cells (DCs), macrophages, and B cells—and helper T (T$_H$) cells, it came as a surprise that ILCs function like T$_H$ cells early in the unfolding of an immune response.

THREE MAJOR SUBSETS OF INNATE LYMPHOID CELLS

Three subsets of ILCs have been identified that mirror T$_H$ cells and their functions, namely, ILC1, ILC2, and ILC3 (Eberl et al., 2015). Although found in the blood in low numbers, ILC are enriched in tissue and have a tendency for expansion at barrier surfaces, including the gut, skin, and lung. The relative distribution of the ILC subsets across the different barrier surfaces varies. For instance, ILC2 are more abundant in lung (and adipose tissues), whereas the frequency of ILC3 is proportionally expanded in the gut (Figure 8.1).

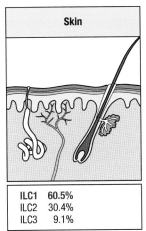
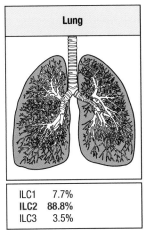
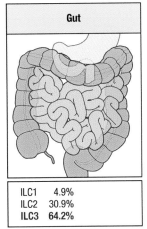

Skin		Lung		Gut	
ILC1	**60.5%**	ILC1	7.7%	ILC1	4.9%
ILC2	30.4%	**ILC2**	**88.8%**	ILC2	30.9%
ILC3	9.1%	ILC3	3.5%	**ILC3**	**64.2%**

Figure 8.1 Innate lymphoid cell (ILC) distribution at the barrier surfaces. The proportion of ILCs at different epithelial barrier surfaces varies. In the epithelial compartment, the predominant subset are ILC1s, whereas in the lungs the most common subset is ILC2. In the gut ILC3 are the most common subset.

8.1 Different ILC subsets respond to different kinds of pathogens

The ILC subsets react to different cues in their local environment to mount qualitatively different effector responses, depending on the type of infection or tissue injury present, which in turn dictates the orientation of the developing immune response. When a tissue is infected by intracellular microbes or viruses, or develops a tumor, stromal cells and APCs react by producing IL-12, IL-15, or IL-18. These inducer cytokines promote the differentiation of T_H1 cells and ILC1s and activate their expression of the effector cytokines IFN-γ and TNFα, which in turn induce the production of oxygen radicals and cytotoxic responses by macrophages and CD8$^+$ T cells. T-bet is the signature transcription factor of ILC1s and T_H1 cells and is required for their generation.

In contrast, when a tissue is infected by extracellular microbes such as bacteria and fungi, APCs react by producing IL-1β and IL-23. These inducer cytokines promote the differentiation of T_H17 cells and ILC3s and activate their expression of IL-17, IL-22, GM-CSF and lymphotoxin. These cytokines lead to the recruitment of phagocytes, such as neutrophils, that clear the microbes, and also induce the production of antimicrobial proteins (AMPs) by epithelial cells and stromal cells. The transcription factor RORγt is required for the generation of ILC3s and T_H17 cells, and its expression defines these cells.

When tissues are infected with large multicellular organisms such as worms, immune defenses do not develop necessarily with the aim to destroy the invader, but rather with the more practical aim of preventing tissue penetration and destruction of internal organs. Epithelial cells and stromal cells respond by producing IL-25, IL-33, or thymic stromal lymphopoietin (TSLP), which then promote the differentiation of ILC2s and T_H2 cells, which produce IL-4, IL-5, and IL-13. These effector cytokines lead to vasodilation, the production of mucus and deposition of collagen, which are all involved in the construction of a barrier against large parasites. The transcription factor Gata3 is required for the development of ILC2s and T_H2 cells, and high expression of Gata3 defines these cells.

8.2   ILCs have some similarities with T cells

ILCs exhibit important similarities with T cells in phenotype and function, with the difference, by definition, that they do not express an antigen-specific receptor (T-cell receptor [TCR] in the case of T cells and B-cell receptor [BCR] in the case of B cells). Therefore, even though they develop from a precursor common with B cells and T cells, the common lymphoid progenitor (CLP), they do not undergo thymic selection, clonal selection, and clonal expansion. As a consequence, ILCs react promptly (within less than hour) to tissue infection and injury but do not provide antigen-specific memory, whereas T cells require several days to be selected and expanded before being able to significantly contribute to the immune response and carry antigen-specific memory. Thus, ILCs are placed upstream of the immunologic cascade and, acting as helper cells, are the prime orchestrators of the immune response that is best suited to face a specific threat.

Of note, particular subsets of T cells exist that behave like ILCs. For example, mucosa-associated invariant T (MAIT) cells, invariant NKT cells, subsets of $\gamma\delta$ T cells, and tissue-resident memory (T_{RM}) cells react promptly to tissue infection and injury in specific organs and tissue niches, through TCR-mediated or inducer cytokine-mediated activation, and secrete a defined set of effector cytokines.

8.3   Lymphoid tissue inducer cells control lymphoid tissue development

A subset of ILC3s, termed *lymphoid tissue inducer* (LTi) cells, induce the development of lymph nodes and Peyer's patches during fetal development (Mebius, 2003). Starting during the second week after conception in mice, specialized stromal cells termed lymphoid tissue organizer (LTo) cells aggregate at specific locations around developing lymph vessels and produce chemokines, such as CXCL13, that recruit LTi cells generated in the fetal liver. LTi cells activate LTo cells through the expression of the membrane form of lymphotoxin, LT$\alpha_1\beta_2$, which bind to its receptor LTβR on stromal cells. The stromal cells increase the production of chemokines and also express the adhesion molecules ICAM-1 and VCAM-1, thereby establishing a positive feedback loop for the development of lymphoid tissues. A few days before birth, B and T cells are produced and are recruited to developing lymph nodes and Peyer's patches. B- and T-cell zones form through a mechanism of stromal differentiation. After birth, LTi cells are still found in the cortex of lymph nodes and Peyer's patches, in the T-cell zone between B-cell follicles, but their role in this location, for the structure or the function of the lymphoid tissue, remains unknown. Remarkably, the programmed fetal development of lymphoid tissue occurs only in mammals, whereas other vertebrates develop lymphoid tissues, termed tertiary lymphoid tissues (tLTs), which are only induced by chronic inflammation after birth (see later).

Shortly after birth, LTi cells cluster in the intestinal lamina propria into "cryptopatches." During colonization of the intestine by commensal bacteria, the innate receptor NOD-1 in epithelial cells is activated by peptidoglycans shed by gram-negative bacteria, which leads to the production of CCL20 by epithelial cells. CCL20 then activates CCR6$^+$ LTi cells in cryptopatches, which leads to the recruitment of B cells. This process leads to the formation of hundreds of isolated lymphoid follicles (ILFs) in the mucosa. ILFs are an important source of intestinal IgA-secreting B cells, mostly through a T-cell–independent pathway. ILFs thereby play a major role in the containment of the symbiotic microbiota and the host-microbe equilibrium in the intestine. It is possible that this process of microbe-induced ILF formation from LTi cells clustered in cryptopatches is the evolutionary predecessor of lymph node and Peyer's patch development. Mammals may therefore have programmed the activation of LTi cells in the sterile environment of the fetus to deal with the massive influx of bacteria after birth.

8.4   New lymphoid tissues can develop in adult in presence of chronic inflammation

In the adult, chronic inflammation induces the formation of tertiary lymphoid tissues that are not dependent on the activation of LTi cells. Instead, the activation of stromal LTo cells is induced by other types of lymphoid cells that express $LT\alpha_1\beta_2$, such as activated B cells, T cells, or NK cells. Notably, the formation of tLTs recapitulates the fetal development of lymph nodes and Peyer's patches. Conversely, the fetal development of lymphoid tissues can be viewed as the result of a programmed and stabilized local inflammation. Tertiary lymphoid tissues significantly contribute to chronic inflammation and severity of pathology during infection and autoimmunity. Perhaps the best example in the gut of tertiary lymphoid tissue is seen in Crohn's disease where the chronic transmural inflammation leads to generation of lymphoid tissue in the submucosa and muscle layers (Figure 8.2).

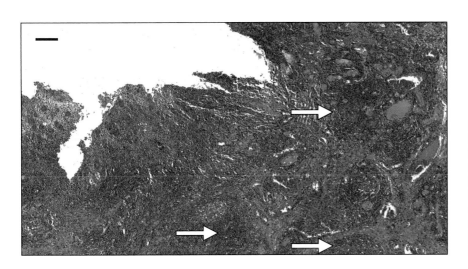

Figure 8.2 Tertiary lymphoid tissue forms in the deeper layers of the gut in Crohn's disease. The figure depicts extremely ulcerated mucosa where the epithelium and normal structure of the gut has been destroyed. Numerous ectopic (tertiary) lymphoid tissues (arrowed) have formed in the deeper layers. (Image courtesy of Dr. Adrian Bateman.)

8.5   ILC lineages are plastic

Plasticity is the process whereby an apparently terminally differentiated cell subset can be reprogrammed to adopt another phenotype/function. The best example of this is when T_H17 cells transdifferentiate into T_H1 cells. The notion of plasticity is conceptually attractive from an immunological perspective, allowing the host to adapt immune phenotype in the face of shifting threats. ILC2 can transdifferentiate into ILC1 when exposed to a sequence of different cytokines. IL-1β plays a key early role in priming ILC2 for transdifferentiation to ILC1, by inducing expression of the IL-12R, which is not normally expressed by ILC2 (Figure 8.3). IL-12 present in the local tissue environment binds to the IL-12R, which in turn drives induction of the master ILC1 transcription factor T-bet. Then T-bet orchestrates the ILC1 transcriptional program, including transactivation of interferon-γ and the IL-18R, which further promotes the transition toward ILC1. Finally, IL-1β also enhances the production of IL-5 and IL-13 (by inducing expression of the receptors for IL-25 and IL-33). It is possible that ILC2-ILC1 plasticity occurs in some human diseases. In patients with chronic obstructive pulmonary disease, a destructive lung disease caused by cigarette smoking, an increase in local production of IL-1β and IL-12 are associated with a marked loss of ILC2 in diseased lung tissue and an expansion of ILC1.

Bidirectional plasticity has been reported for ILC1 and ILC3. Under the influence of IL-23 and IL-1β, and in the presence of CD14$^+$ DCs, ILC1 can adopt an ILC3 phenotype. Conversely, IL-12 and CD14$^-$ DCs direct the transition of ILC3 into ILC1. An additional aspect of plasticity to consider is adoption of "hybrid" phenotypes (see Figure 8.3). In IBD, intestinal ILC3 from both man and mouse, express the effector cytokines IL-17, IL-22, and IFN-γ, which are functional attributes of both ILC3s and ILC1s. These cells resemble the T_H1/T_H17 hybrid T cells that are also expanded in IBD, which are regarded as pathogenic cells in IBD. Finally, during *Salmonella enterica* infection, a subset of ILC3s downregulates RORγt, upregulates T-bet, and expresses IFN-γ. Such cells are termed ex-ILC3 ILC1s.

ILC plasticity has also been reported in cultures of blood-derived human ILCs. Lines of ILC3s express type 2 cytokines, such as IL-5 and IL-13, when stimulated with TLR2 ligands (Crellin et al., 2010). These data suggest that ILC may adapt to tissue signals and adopt the effector phenotype best adapted to the local perturbation. However, the significance of ILC plasticity remains to be clearly demonstrated *in vivo* during infection and injury.

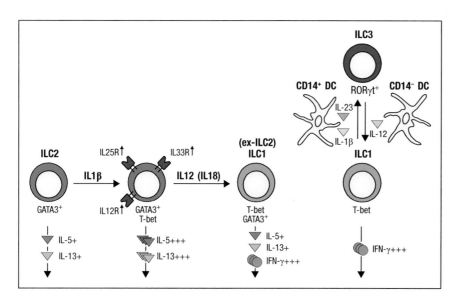

Figure 8.3 Innate lymphoid cell (ILC) plasticity. An important concept in immunology is plasticity, where apparently terminally differentiated cell lineages transition into phenotypically and functionally distinct cells, often resembling alternative cell lineages. ILCs transdifferentiate to ILC1 under the influence of environmental cytokines. IL-1β induces expression of cytokine receptors, including IL-25R and IL-33R (which augment ILC23 function), but also key ILC1 receptors (e.g., IL-12R and IL-18R), allowing ILC2 to respond to the canonical activating cytokines of the ILC1 response, allowing these ex-ILC2 to adopt ILC1 phenotype and function (induction of the ILC1 transcription factor T-bet and production of IFN-γ). There is also bidirectional plasticity of ILC1 and ILC3. Under the influence of IL-1β, IL-23, and CD14$^+$ mononuclear phagocytes ILC1 transition to ILC3, whereas increased exposure to IL-12 and CD14$^-$ mononuclear phagocytes, promotes transition of ILC3 toward ILC1. ILC plasticity allows the ILC family to adapt and tailor their functional responses according to local environmental cues.

ILCs in host defenses and pathology

There is accumulating evidence that ILCs are important in host defenses to infection and in mediating immunopathology.

8.6 ILC3 can be protective and cause tissue damage

In the intestine, the presence of the commensal microbiota induces a vigorous type 3 immune response, at the steady state and in the absence of pathogens. The highest density of ILC3s and T_H17 cells is found in the ileal lamina propria, where these cells play a critical role in the containment of the microbiota through the production of IL-17, IL-22, and lymphotoxins. Upon infection with pathogens, such as enteropathogenic *Escherichia coli* and related *Proteobacteria* species, ILC3s are critical effectors as they produce the bulk of IL-22 early in infection, which in turn induces the expression of protective AMPs by epithelial cells. However, during infection with *Salmonella enterica*, which is an invasive organism and elicits a strong inflammatory response, a subset of ILC3s expressing the NK marker NKp46, coexpresses IFN-γ and causes tissue damage.

8.7 ILC1 may have an important role in gut epithelium

The intestine, and in particular the intestinal epithelial compartment, is home to ILC1s, which produce IFN-γ. In the epithelial niche, ILC1s may interact with IEL in the control of epithelia-tropic viruses. ILC1s are also associated with chronic pathologic inflammation and are found in the intestine of IBD patients where large amounts of IFN-γ are produced. ILC1s include the oldest known of ILC subsets, NK cells, sometimes considered to be the innate homologue of cytotoxic CD8$^+$ T cells. The role of NK cells is best characterized in the context of antiviral and antitumor responses, during which the role of noncytotoxic ILC1s remains to be determined. ILC1s, in particular "intraepithelial ILC1s," have phenotypic and functional overlap with NK cells.

8.8 ILC2s play a major role in defense against helminths

Infection of the intestine with the nematode *Nippostrongylus brasiliensis* induces IL-25 production by a subset of epithelial cells, namely, Tuft cells, which leads to a strong ILC2-mediated response, both through expansion and the production of IL-4, IL-5, and IL-13. IL-4 and IL-13 induce the expansion of goblet cells and the production of mucus, an effective barrier to worm invasion. As prompt producers of type 2 cytokines, ILC2s also play a major role in the development of allergic responses in the lung. Epithelial cells react to allergens by the production of TSLP and IL-33, which activates ILC2s to release IL-5, a potent growth factor for eosinophils.

ILCs in immune-mediated inflammatory diseases

Immune-mediated inflammatory diseases (IMIDs) consist of a broad range of chronic, inflammatory diseases affecting different organ systems, including the joints (e.g., rheumatoid arthritis, ankylosing spondylitis), skin (e.g., psoriasis, eczema), lungs (e.g., asthma), and gut (e.g., celiac disease and inflammatory bowel disease). These incurable diseases are an important cause of morbidity, disability, and in some cases, premature death. Many patients with inflammatory conditions require drugs to suppress the immune system, including expensive biological therapies (such as anti-TNF-α or anti-IL-17A monoclonal antibodies). Exaggerated CD4$^+$ T-cell responses are strongly implicated in these diseases, but increasingly ILCs are also being implicated as important effector cells.

In IBD, interferon-γ and TNF-α producing ILC1 are expanded in diseased mucosa (**Figure 8.4**). Depletion of ILCs in a preclinical model of IBD (induced by administering agonistic anti-CD40 antibodies) alleviates disease. IL-23 and IL-1β are important proximal drivers of intestinal ILC activation. IL-23 and IL-1β responsive ILCs are pathogenic in several mouse models of IBD but seem to be especially harmful in chronic intestinal inflammation driven by intestinal microbiota dysbiosis. Neutralization, or genetic ablation, of IL-23 or IL-1β prevents ILC-mediated chronic colitis in animal models of disease. Early phase clinical trials evaluating the impact of blocking the p19 specific subunit of IL-23 look promising as a new treatment for IBD. IL-23 and IL-1β stimulate production of IL-17A, IL-22 and interferon-γ, by ILC3 and ILC1. Another way of alleviating harmful chronic inflammatory ILC responses in the intestine is to deprive them of IL-7, a key cytokine involved in sustaining ILC survival. Antibody-mediated neutralization of IL-7 alleviates chronic ILC-mediated colitis.

Different ILC subsets have been implicated in other inflammatory conditions, and similar to the situation with chronic inflammatory diseases

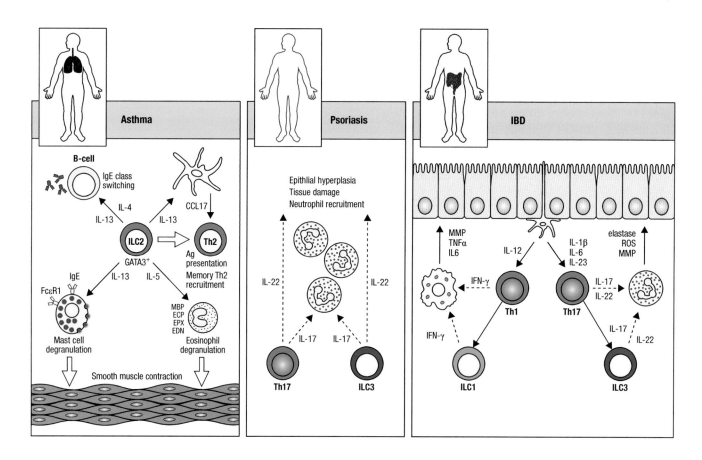

Figure 8.4 Innate lymphoid cells (ILCs) in immune-mediated inflammatory diseases (IMIDs). IMIDs are chronic diseases affecting many different organ systems. ILCs have been implicated as important cellular mediators of inflammation in these diseases. Asthma is a chronic disease of the airways, characterized by airflow limitation, wheezing, and difficulty in breathing. ILC2 contribute to disease by promoting recruitment, survival, and activation of eosinophils (IL-5/IL-13), augmenting mast cell degranulation and supporting immunoglobulin class switching to proallergic IgE antibodies. IL-13 derived from ILC2 also promote production of the chemokine CCL17, which recruits allergen-reactive T$_H$2 T cells. Psoriasis is an inflammatory skin disease. ILC3s act in conjunction with T$_H$17 cells to promote neutrophil recruitment and hyperproliferation of skin epithelial cells (epidermis). IBD, including Crohn's disease and ulcerative colitis, is a severe IMID affecting the gut. ILC1 may perform a similar role to T$_H$1 T cells. Activated by IL-12, they support mononuclear phagocyte recruitment/activation, including production of proinflammatory cytokines (e.g., TNF) and tissue damaging enzymes, such as matrix metalloproteinases (MMPs). Similar to the situation in psoriasis, ILC3 may share pro-inflammatory activity with T$_H$17 cells, including production of IL17 and IL22, which in turn promotes recruitment and activation of neutrophils.

of the gut, these cells seem to be especially important in immune-mediated damage at other epithelial barrier surfaces, such as the lung and skin. Atopic asthma is characterized by expansion of CD4$^+$ T$_H$2-cells reacting to inhaled aeroallergens, such as house dust mite, pollen, and animal dander. However, ILC2, which share many characteristics with T$_H$2 T cells, also contribute to pathology (see Figure 8.4). ILC2 are more abundant in the lining of the lungs than any other mucosal surface. As well as being an early source of type 2 cytokines, including IL-4, IL-5, and IL-13, which have proallergic properties, such as activation of mast cells, activation and recruitment of eosinophils, and promoting IgE class switching, ILC2 contribute to allergic inflammation of the airways in a different fashion from conventional T$_H$2 cells. In a model of allergen-induced airway inflammation, early depletion of ILC2 results in markedly impaired T$_H$2 responses in the lung. This activity of ILC2 was not mediated through direct cross talk with T$_H$2 cells, but instead through dialogue with lung dendritic cells. At very early time points following challenge with allergen (prior to activation of memory T cells), ILC2 produce IL-13, which in turns stimulates lung resident DCs to produce the T$_H$2 chemoattractant CCL17. Neutralization of IL-13 or CCL17 prevented accumulation of memory T$_H$2 cells in the lung. Therefore, ILC2 are potentially important proallergic effector cells acting upstream of T$_H$2 T cells, playing an important role in the recruitment and activation of memory T$_H$2 responses in the lung. Comparable responses were observed in a model of allergen-induced skin inflammation, indicating that ILC2 might also be involved in key early signals that set up recruitment of T$_H$2 memory cells to the skin, as occurs in atopic eczema.

ILCs have also been implicated as important effector cells in other skin diseases (see Figure 8.4). In psoriasis, there is expansion of IL-17A and IL-22 producing ILC3 in skin lesions. In keeping with their tendency for enrichment in tissue, all ILC populations were marked expanded in skin in comparison with blood. However, circulating ILC3 frequently expressed the skin homing receptor CLA enabling them to migrate to skin. CLA-expressing ILC3 were enriched in the circulating blood of psoriasis patients.

ILCs are involved in fat metabolism and repair

Intriguingly, ILC2s are enriched in adipose tissue. Upon cold shock, IL-33 is released in adipose tissues, which leads to the activation of ILC2s that make IL-5. This in turn recruits eosinophils, which release IL-4 and lead to the differentiation of blood-borne monocytes into alternatively activated macrophages (AAMs). These AAMs then respond to epinephrines produced by cold-sensing neurons to produce epinephrines themselves, thus amplifying the signal directed at adipocytes to generate heat instead of ATP. Whether this process is related to mucosal immunity is a matter of investigation. ILC2s also express methionine-enkephalin peptides that directly activate adipocytes to generate heat.

ILCs can function as antigen-presenting cells

The effector cytokines produced by ILCs regulate the expression of inducer cytokines by macrophages and DCs, which in turn regulate T-cell activity. For example, IL-12 produced by DCs activates NK cells to produces IFN-γ, which further promotes the expression of IL-12 by DCs and thereby the differentiation of T$_H$1 cells. Unexpectedly, ILCs also directly regulate lymphocyte function and differentiation. ILCs express MHC class II, as well as the antigen-processing machinery required to generate class II peptide antigens. In the intestine, ILC3s process and present microbial peptides on MHC class II and regulate the activity of T$_H$17 cells. In the spleen, the presentation of peptide antigen on MHC class II by ILC3 leads to the activation of T cells. In the lung, ILC2s promote the activation of T$_H$2 cells directly through MHC class II–restricted antigen presentation, as well as through the activation of DCs. In return, T$_H$2 cells produce IL-2, required for the activation of ILC2s.

In the intestinal lamina propria, the production of IL-22 by ILC3s leads to the activation of epithelial cells to produce serum amyloid A, which promotes the activation of T_H17 cells and their production of IL-17. In contrast, ILC3s also produce granulocyte-macrophage colony-stimulating factor (GM-CSF), which activates macrophages. In the steady state, GM-CSF leads to macrophages promoting the generation of regulatory T (T_{reg}) cells, demonstrating once more how myeloid cells, ILCs, and T cells exchange information to set the inflammatory tone of the tissues.

ILCs can regulate B cells

As previously discussed, ILCs in cryptopatches activate the recruitment of B cells and the generation of ILFs through their expression of $LT\alpha_1\beta_2$. In ILFs, B cells express IgA. The soluble form of lymphotoxin, $LT\alpha_3$, also produced by ILC3s, promotes the production of IgA through the recruitment of T cells. $LT\alpha_3$ thereby promotes the T-cell–dependent production of IgA. In the spleen, ILC3s appear enriched in the marginal zones, home to the B cells that produce T-cell–independent antibodies. At this site, ILC3s activate stromal cells through $LT\alpha_1\beta_2$ to produce BAFF and other factors that stimulate B cells.

SUMMARY

ILCs have been found to play critical roles in the development of lymphoid tissues, in the containment of the symbiotic microbiota at mucosal surfaces, in the early defense against pathogens, in the regulation of adaptive immunity, and in tissue repair responses and fat energy management. However, ILCs are also implicated in inflammatory pathology, during infection, IBD, and intestinal cancer, and in allergy in the lungs. Therefore, ILCs carry great promises as vehicles for enhanced immune responses, repair responses, and maintenance of homeostasis, or as targets against inflammatory pathology.

FURTHER READING

Bouskra, D., Brezillon, C., Berard, M., Werts, C., Varona, R., Boneca, I.G., Eberl, G.: Lymphoid tissue genesis induced by commensals through NOD1 regulates intestinal homeostasis. *Nature* 2008, 456:507–510.

Buonocore, S., Ahern, P.P., Uhlig, H.H., Ivanov, I.I., Littman, D.R., Maloy, K.J., Powrie, F.: Innate lymphoid cells drive interleukin-23-dependent innate intestinal pathology. *Nature* 2010, 464:1371–1375.

Crellin, N.K., Trifari, S., Kaplan, C.D., Satoh-Takayama, N., Di Santo, J.P., Spits, H.: Regulation of cytokine secretion in human CD127+ LTi-like innate lymphoid cells by Toll-like receptor 2. *Immunity* 2010, 33:752–764.

Eberl, G., Colonna, M., Di Santo, J.P., McKenzie A.N.: Innate lymphoid cells: A new paradigm in immunology. *Science* 2015, 348:aaa6566.

Halim, T.Y., Steer, C.A., Matha, L., Gold, M.J., Martinez-Gonzalez, I., McNagny, K.M., McKenzie, A.N., Takei, F.: Group 2 innate lymphoid cells are critical for the initiation of adaptive T helper 2 cell-mediated allergic lung inflammation. *Immunity* 2014, 40:425–435.

Hepworth, M.R., Monticelli, L.A., Fung, T.C. et al.: Innate lymphoid cells regulate CD4+ T-cell responses to intestinal commensal bacteria. *Nature* 2013, 498:113–117.

Lee, M., Odegaard, J.I., Mukundan, L., Qiu, Y., Molofsky, A.B., Nussbaum, J.C., Yun, K., Locksley, R.M., Chawla, A.: Activated type 2 innate lymphoid cells regulate beige fat biogenesis. *Cell* 2015, 160:74–87.

Mebius, R.E.: Organogenesis of lymphoid tissues. *Nat Rev Immunol.* 2003, 3:292–303.

Moro, K., Yamada, T., Tanabe, M., Takeuchi, T., Ikawa, T., Kawamoto, H., Furusawa, J., Ohtani, M., Fujii, H., Koyasu, S.: Innate production of T(H)2 cytokines by adipose tissue-associated c-Kit+Sca-1+ lymphoid cells. *Nature* 2009, 463:540–544.

Satoh-Takayama, N., Vosshenrich, C.A., Lesjean-Pottier, S. et al. : Microbial flora drives interleukin 22 production in intestinal NKp46+ cells that provide innate mucosal immune defense. *Immunity* 2008, 29:958–970.

Role of regulatory T cells in mucosal immunity

9

KENYA HONDA AND SHOHEI HORI

The mucosal surfaces are the first portals of entry for numerous infectious agents. Therefore, mucosal tissues contain the largest immune cell population in the body, tasked with preparing for, or reacting to, a broad range of infectious threats. At the same time, mucosal tissue has to maintain tolerance or quiescence to harmless environmental substances, such as food proteins and commensal microbes. This important task is achieved by a number of mechanisms, among which forkhead box P3 (Foxp3)–expressing CD4$^+$ regulatory T (T$_{reg}$) cells play a major role.

T$_{REG}$ CELLS ARE CRITICAL MEDIATORS OF MUCOSAL IMMUNE HOMEOSTASIS

9.1 There are 2 phenotypically different types of T$_{reg}$ in the mucosa

T$_{reg}$ cells were initially discovered as the T-cell subsets existing in healthy animals that play essential roles in constitutively preventing the development of autoimmune disease and inflammatory bowel disease (IBD). Several cell surface molecules were identified as markers for such disease-protective, "naturally occurring" T$_{reg}$. For example, CD25$^+$CD4$^+$ T cells were identified as T$_{reg}$ that prevent the development of organ-specific autoimmune diseases, such as the gastritis, thyroiditis, and oophoritis that develops in athymic *nude* mice transferred with CD25$^-$CD4$^+$ T cells, or in mice thymectomized around day 3 of life (Figure 9.1). CD45RB^{low}CD4$^+$ cells were also identified as T$_{reg}$ cells capable of preventing the development of mouse IBD. For example, adoptive transfer of CD45RB^{high}CD4$^+$ naïve T cells into SCID or RAG-deficient recipient mice resulted in the development of colitis, while cotransfer of CD45RB^{low}CD4$^+$ T cells prevented disease. The immune regulatory activity of CD45RB^{low}CD4$^+$ T cells was later ascribed to the CD25$^+$ subset, although some regulatory activity was also detected in the CD25$^-$CD45RB^{low} subset. The relationships among those naturally occurring T$_{reg}$ subsets remained unclear because of the lack of a more specific molecular marker that unequivocally identifies *bona fide* T$_{reg}$ cells. For example, CD25, the interleukin (IL)-2 receptor (IL-2R) α-chain, is expressed on activated T cells, and low-level expression of CD45RB is also a marker for effector T cells or memory T cells.

Those early studies culminated in the identification of the transcription factor Foxp3 as a specific molecular marker for T$_{reg}$ enriched within CD25$^+$ and/or CD45RB^{low}CD4$^+$ T-cell subsets. Foxp3 was originally identified as the gene mutated in the fatal autoimmune, inflammatory, and allergic disease that develops in patients with the immune dysregulation, polyendocrinopathy,

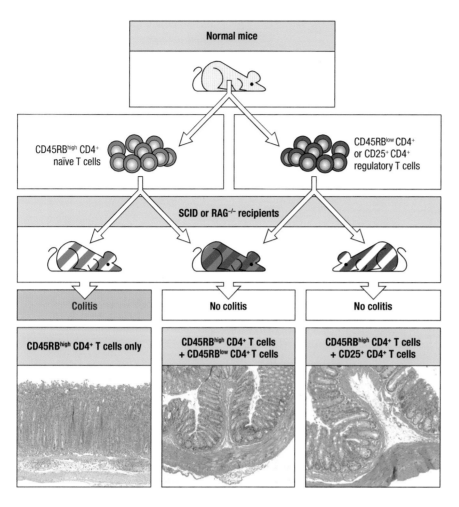

Figure 9.1 The naïve T-cell transfer model of colitis. When immunodeficient mice (e.g., SCID or RAG-deficient mice) are transferred with CD45RB^{high} naïve CD4$^+$ T cells sorted from the periphery of syngeneic normal unmanipulated mice, they develop severe colitis in 1–2 months. The development of colitis can be prevented by cotransferring CD45RB^{low} or CD25$^+$ CD4$^+$ T cells, which contain T$_{reg}$ cells. Adoptive transfer of CD45RB^{low} or CD25$^+$ CD4$^+$ T cells alone does not lead to disease development.

and enteropathy (IPEX) syndrome, and in a naturally arising mutant mouse strain, the *scurfy* mice. Foxp3 not only serves as a specific marker for T$_{reg}$ cells but also functions as a "master" transcription factor for their development and function. Ectopic expression of Foxp3 confers a T$_{reg}$ phenotype and function on conventional T cells, whereas loss-of-function Foxp3 mutations lead to defective development of functional T$_{reg}$ cells. The deficiency of functional T$_{reg}$ cells caused by Foxp3 mutations or inducible ablation of Foxp3$^+$ T cells in genetically modified mice results in the development of severe inflammation in multiple organs including the mucosal tissues. Notably, autoimmune enteropathy, which manifests as watery diarrhea, villous atrophy, and extensive lymphocytic infiltrates of the bowel mucosa, is one of the most common symptoms in IPEX patients. Moreover, in mice, CD4$^+$CD45RB^{high} T-cell–mediated colitis can be inhibited by cotransfer of Foxp3$^+$ T$_{reg}$ cells or Foxp3-transduced CD4$^+$CD45RB^{high} T cells. These findings have led to the notion that Foxp3$^+$ T$_{reg}$ cells are central to the maintenance of tolerance and homeostasis in the systemic and mucosal immune systems.

9.2   T$_{reg}$ cells regulate immune homeostasis in the gut by multiple pathways

Foxp3$^+$ T$_{reg}$ cells regulate immune responses at multiple cellular levels through various molecular mechanisms. These include the production of inhibitory cytokines (i.e., transforming growth factor [TGF]-β, IL-10, IL-35), cytolytic enzymes (i.e., granzyme and perforin), consumption/degradation of adenosine triphosphate (ATP) and IL-2 (CD39, CD73, IL-2R), and expression of inhibitory surface molecules (i.e., lymphocyte activation gene-3 [LAG-3] and cytotoxic T-lymphocyte antigen 4 [CTLA-4]) (Figure 9.2). With respect to mucosal immune regulation, IL-10, TGF-β, and CTLA-4 play particularly important roles.

At mucosal interfaces, Foxp3$^+$ T$_{reg}$ cells express IL-10 at much higher frequencies than in other tissues. IL-10 has particularly strong effects on myeloid cells, γδ T cells, and CD4$^+$ T cells to suppress their excessive activation. Indeed, T$_{reg}$-specific disruption of the *Il10* gene in mice results in the development of spontaneous colitis. IL-10 expression in T$_{reg}$ cells is controlled by STAT3, and Foxp3$^+$ cell-specific *Stat3*-deficient mice also develop spontaneous colitis. Additionally, mice with T$_{reg}$ cells lacking the *Il10ra* gene (encoding the IL-10R α-chain) develop colitis mediated by helper T 17 (T$_H$17) cells. Thus, IL-10 is also required for autocrine stimulation of T$_{reg}$ cells to activate their capacity to suppress T$_H$17 cell responses. The critical role for IL-10 in human gut homeostasis has been shown by the discovery that a rare disease, early onset IBD, previously termed *intractable enterocolitis of infancy*, is due in many babies to mutations in IL-10R.

TGF-β is another molecular mediator of T$_{reg}$-mediated immune suppression, particularly in the intestine. While TGF-β is critical for the extrathymic differentiation of T$_{reg}$ cells, it is also required for suppressive effector functions of T$_{reg}$ cells. In the CD45RB^{high}CD4$^+$ T-cell transfer model of colitis, T$_{reg}$-mediated suppression is abrogated when TGF-β is neutralized by specific antibody or when CD45RB^{high}CD4$^+$ T cells are rendered insensitive to TGF-β signals due to transgenic expression of a dominant negative form of the TGF-β receptor. However, T$_{reg}$ cells do not need to secrete TGF-β, because TGF-β1-deficient T$_{reg}$ cells are fully functional and because TGF-β can be produced from a variety of immune and nonimmune cells. Instead, T$_{reg}$ cells appear to be necessary for activation of TGF-β. This cytokine is produced in a latent form associated with latency-associated peptide that has to be proteolytically cleaved from TGF-β before the cytokine has biological activity. Activated (or effector) T$_{reg}$ cells are capable of activating TGF-β efficiently because they preferentially bind the latent TGF-β on the cell surface via the expression of leucine-rich repeat containing 32 (LRRC32, also known as GARP) and express αvβ8 integrin, which enzymatically activates the latent TGF-β. Active TGF-β binds to the TGF-β receptor that is expressed on many types of immune and nonimmune cells, leading to phosphorylation of Smad2/3. Phosphorylated Smad2/3 then forms a complex with Smad4 and translocates to the nucleus, where the Smad2/3/4 complex influences the transcriptional activity of multiple genes, including those encoding inflammatory molecules. In the gut of patients with IBD, mucosal T cells express elevated levels of Smad7, an intracellular protein that binds to TGF-β receptor, thereby preventing phosphorylation of Smad2/3. Inhibition of Smad7 with a specific antisense oligonucleotide restores TGF-β1/Smad3 signaling, resulting in the suppression of inflammatory cytokine production, supporting the concept that Smad7 constitutes a molecular target for direct therapeutic interventions in IBD.

CTLA-4, one of the most prominent immune inhibitory receptors, is constitutively expressed by T$_{reg}$ cells, in part through the transcriptional action of Foxp3. It is structurally similar to the costimulatory molecule CD28 but shows a much higher affinity for the common ligands CD80 and CD86 expressed by dendritic cells (DCs). CTLA-4 thus effectively competes with CD28 for binding CD80/86. In contrast to CD28, which exists predominantly on the cell surface, CTLA-4 shows a high turnover rate, being rapidly internalized from the cell surface independent of ligand binding and is either degraded or trafficked back to the cell surface via recycling endosomes. Upon CD80/CD86 binding, CTLA-4 induces transendocytosis of these costimulatory ligands from the surface of antigen-presenting cells, reducing their expression, minimizing costimulatory interactions between DCs and naïve T cells. The finding that Foxp3$^+$ cell-specific CTLA-4-deficient

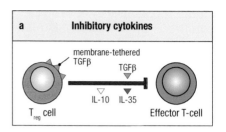

a Inhibitory cytokines

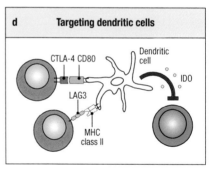

b Cytolysis

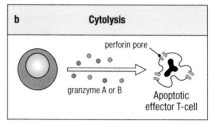

c Metabolic disruption

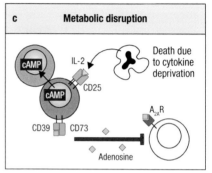

d Targeting dendritic cells

Figure 9.2 Basic mechanisms used by T$_{reg}$ cells. (a) Inhibition of effector T cells by inhibitory cytokines. **(b)** Cytolysis of effector T cells by granzyme-A and -B and perforin. **(c)** Metabolic disruption through high-affinity IL-2Rα-mediated IL-2-deprivation, and CD39/CD73-mediated ATP degradation and generation of immunosuppressive adenosine and cAMP. **(d)** Targeting dendritic cells including CTLA4-CD80/CD86-mediated suppression of maturation. (Adapted from Dario A.A. Vignali et al.: *Nat. Rev. Immunol.* 2008, 8:523–532. With permission from Springer Nature.)

mice spontaneously develop autoimmune inflammation in multiple tissues indicates an indispensable role for CTLA-4 in T_{reg} function. CTLA-4 also plays a critical role in mucosal immune regulation, as evidenced by the development of colitis in patients with cancer receiving anti-CTLA-4 antibody therapy. Moreover, T_{reg} expression of CTLA-4 is strongly enhanced by the gut microbiota.

T_{reg} cells also suppress immune responses through passive or competitive mechanisms. *In vitro*, T_{reg} cells can outcompete conventional T cells in aggregating around DCs due to high-level expression of the adhesion molecule LFA-1, which binds to ICAM on DCs. T_{reg} cells are also capable of depriving activated conventional T cells of IL-2, which is a critical T-cell growth factor. T_{reg} cells constitutively express the *Il2ra* (encoding CD25) and *Il2rb* (CD122) genes and express the high-affinity IL-2R complex at much higher levels than conventional T cells. In contrast to conventional T cells, T_{reg} cells do not secrete IL-2 upon activation and are thus "anergic" because Foxp3 actively represses IL-2 transcription. Thus, T_{reg} cells efficiently consume IL-2, thereby limiting clonal expansion of activated T cells. This "IL-2 consumption" mechanism plays an important role in the regulation of CD8 T-cell responses. For example, T_{reg} cells that cannot bind IL-2 (due to CD122 deficiency) but can transduce IL-2 signals (due to transgenic expression of a constitutively active STAT5 mutant) are impaired in their ability to inhibit the activation of $CD8^+$ T cells but not $CD4^+$ T cells. T_{reg} cells also express CD39 and CD73, both of which are positively regulated by Foxp3 and sequentially degrade immune-stimulatory ATP into immune-regulatory adenosine. It remains unclear, however, whether IL-2 and ATP deprivation are essential for mucosal immune regulation.

MUCOSAL T_{REG} CELLS ARE HETEROGENEOUS IN ORIGIN

9.3 Thymic versus peripherally derived T_{reg} cells

T_{reg} cells in mucosal tissues contain two developmentally distinct subsets, namely, natural thymus-derived T_{reg} (tT_{reg}) cells that differentiate and mature intrathymically, and peripherally induced T_{reg} cells (pT_{reg}) that differentiate from conventional naïve $CD4^+$ T cells in extrathymic tissues (**Figure 9.3**). Compared to other tissues, the intestinal microenvironment favors pT_{reg} cell generation.

Adoptive transfer of T-cell receptor (TCR) transgenic naïve $Foxp3^-CD4^+$ T cells and subsequent oral exposure to cognate antigen results in differentiation of donor-derived $Foxp3^+$ pT_{reg} cells in the intestine. These pT_{reg} cells play a critical role in mediating tolerance against dietary antigens. The nuclear protein Helios and cell-surface protein Neuropilin 1 (Nrp1) are constitutively expressed by tT_{reg} but not by pT_{reg} cells. Therefore, these molecules are used as markers to distinguish tT_{reg} and pT_{reg} cells, although they can be upregulated in pT_{reg} cells under some conditions. $Helios^-Nrp1^-Foxp3^+$ pT_{reg} cells are most abundantly present in the intestine. In the colon of germ-free mice, the frequency of $Helios^-Nrp1^-Foxp3^+$ cells is significantly reduced, suggesting that colonic pT_{reg} cells are likely to be induced by commensal microbes. Supporting this notion, TCR usage among colonic $Foxp3^+$ pT_{reg} cells differs greatly from other tissues. Moreover, analysis of transgenic mice expressing TCRs cloned from colonic pT_{reg} cells demonstrates that differentiation of transgenic T cells into pT_{reg} cells preferentially occurs in the colon in the presence of gut commensal bacteria but not in the thymus. Therefore, a subset of pT_{reg} cells in the colon is likely to differentiate extrathymically as a result of an encounter with bacterial antigens. It is also possible that tT_{reg} cells are selected by self-antigen recognition in the thymus, followed by expansion through recognition of cross-reactive microbial antigens in the intestine.

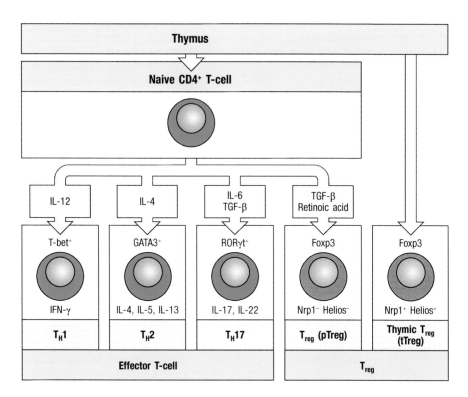

Nevertheless, whatever their origin, the TCR repertoire of intestinal T_{reg} cells is heavily influenced by the microbiota. In the small intestine, in addition to commensal microbe-dependent pT_{reg} cells, there is another distinct subset of Helios⁻Nrp1⁻Foxp3+ pT_{reg} cells that are absent in mice fed an antigen-free diet and appear to be generated in response to food antigens. Therefore, intestinal pT_{reg} cells recognize and suppress immune response against antigens derived from commensal microbes or ingested foods.

There are multiple upstream pathways for Foxp3 induction in pT_{reg} cells, some shared with, but some distinct from, those in tT_{reg} cells. Whereas TCR and IL-2/STAT-5 signals are required for Foxp3 induction in both tT_{reg} and pT_{reg} cells, TGF-β/Smad signals are uniquely required for Foxp3 induction in pT_{reg} cells. In addition, the retinoic acid signaling pathway enhances TGF-β-dependent Foxp3 induction in pT_{reg} cells. Importantly, both retinoic acid and TGF-β are found at high concentrations in the intestine. In line with the distinct molecular requirements for tT_{reg} versus pT_{reg} differentiation, their lineages depend on distinct *cis*-regulatory elements within the *Foxp3* locus, namely, its intronic enhancers termed *conserved noncoding sequences 1–3* (CNS1-3) (**Figure 9.4**). CNS3 plays an essential role in tT_{reg} cells and regulates the initial induction of Foxp3 by recruiting the transactivating factor, c-Rel, to the *Foxp3* locus. CNS3-deficient mice as well as *c-Rel*-deficient mice have a drastic reduction in the numbers of tT_{reg} cells, whereas the pT_{reg} cell compartment remains intact. In contrast, CNS1 serves as the response element for the TGF-β/Smad and retinoic acid signaling pathways such that ablation of CNS1 impairs pT_{reg} induction, particularly in the gut and lung, whereas tT_{reg} differentiation is unaffected. Importantly, CNS1-deficient mice spontaneously develop T_H2-type pathologies in the gut and lung, highlighting the nonredundant role of pT_{reg} cells in the mucosa.

STABILITY OF MUCOSAL T_{REG} CELLS

Foxp3+ T_{reg} cells are capable of maintaining Foxp3 expression and suppressive function even in the face of a variety of stimuli from the extracellular environments and thus constitute a stable cell lineage. Although Foxp3 is the most critical transcription factor that delineates T_{reg} cells from other

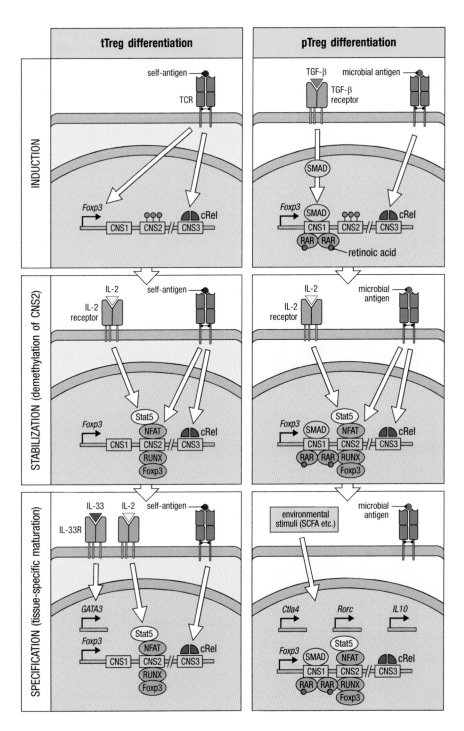

Figure 9.4 Stepwise differentiation of tT_reg and pT_reg cells. The tT_reg cell differentiation program is initiated by strong T-cell receptor (TCR) signals induced by self-antigens in the thymus, which results in nuclear entry of c-Rel. c-Rel, together with several other transcription factors (not depicted), bind to CNS3 and the Foxp3 promoter to induce Foxp3 gene expression. These cells then become a stable T_reg lineage when CNS2 is demethylated, and Stat5 activated by IL-2 contributes to this demethylation. Demethylated CNS2 allows binding of NFAT, Foxp3, and Runx1, resulting in stable Foxp3 expression. tT_reg cells migrate into mucosal tissues and further undergo functional maturation in response to environmental stimuli, including IL-33-mediated GATA3 expression. pT_reg cell differentiation from uncommitted naïve CD4+ T cells occurs in the intestine but progresses in a stepwise manner like tT_reg cells. The differentiation is initiated by TCR signaling induced by microbial antigens together with TGF-β and retinoic acid–mediated signal. These signals promote c-Rel binding to CNS3, as well as Smad3 and RAR-RXR binding to CNS1 of the Foxp3 gene. Similar to tT_reg cells, pT_reg cells establish their T_reg character and lineage commitment by demethylation of CNS2, which is induced by IL-2 and Stat5 signaling. Microbiota-derived short-chain fatty acids contribute to chromatin remodeling and lineage commitment by their histone deacetylase inhibitory activity. pT_reg cells further acquire functionally specialized features such as expression of RORγt, IL-10, and CTLA-4 to adapt to the intestinal environment. (Adapted from Tanoue, T. et al.: *Nat. Rev. Immunol.* 2016, 16: 295–309. With permission from Springer Nature.)

conventional CD4+ T cells, Foxp3 expression alone is not sufficient to ensure their phenotypic and functional stability. Transcriptome analyses have revealed substantial differences between Foxp3+ T_reg cells isolated from normal individuals and Foxp3-transduced conventional T cells. Foxp3 can be transiently upregulated in conventional T cells upon TCR stimulation without inducing a stable T_reg cell phenotype. Furthermore, a fraction of Foxp3+ T cells downregulates Foxp3 expression under lymphopenic or inflammatory conditions and differentiates into Foxp3− T_H cells. For example, in T-cell–deficient recipients of Foxp3+ T cells, extensive accumulation of such "ex-Foxp3" cells is observed in the Peyer's patches, and many of them have a T-follicular-helper (T_FH) cell phenotype and promote IgA production.

Epigenetic mechanisms are central for the phenotypic and functional stability of T$_{reg}$ cells. In particular, complete DNA demethylation of the CpG-rich CNS2 region (or "T$_{reg}$-specific demethylation region") in the *Foxp3* locus is important for the stability of both tT$_{reg}$ and pT$_{reg}$ cells (see Figure 9.4). CNS2 demethylation enables bindings of multiple transcription factors including runt-related transcription factor 1 (RUNX1), NFAT, STAT5, and Foxp3 to the CNS2 region, which cooperatively stabilize Foxp3 transcription. NFAT is activated by TCR signals and facilitates the interaction between CNS2 and Foxp3. STAT5 is activated by IL-2 and counteracts the effects of STAT3 or STAT6 signaling, activated by pro-inflammatory cytokines (e.g., IL-4 or IL-6) that can potentially destabilize Foxp3 expression. Therefore, CNS2 demethylation is required for protection of TCR-stimulated T$_{reg}$ cells from losing Foxp3 expression during cell division in the presence of destabilizing signals such as pro-inflammatory cytokines.

Although a fraction of Foxp3$^+$ T cells can lose Foxp3 expression and differentiate into effector T cells under lymphopenic or inflammatory conditions, such ex-Foxp3 T cells may not result from reprograming of mature stable T$_{reg}$ cells (i.e., remethylation of demethylated CNS2) but from selective downregulation of Foxp3 in T cells that have transiently and promiscuously upregulated Foxp3 expression without engaging CNS2 demethylation. Foxp3 expression can be induced *in vitro* in activated conventional CD4$^+$ T cells in the presence of TGF-β without inducing CNS2 demethylation. Moreover, both in mice and humans, the peripheral Foxp3$^+$ T-cell compartment contains nonregulatory Foxp3$^+$ T cells with methylated CNS2. Importantly, Foxp3$^+$ pT$_{reg}$ cells generated from TCR transgenic naïve CD4$^+$ T cells by oral administration of the cognate antigen show a largely, albeit not completely, demethylated CNS2. Moreover, intestinal ROR-γt$^+$Foxp3$^+$ cells, which are presumably pT$_{reg}$ cells generated in response to commensal bacteria (see later), also show fully demethylated CNS2 locus. Therefore, intestinal pT$_{reg}$ cells belong to a stable T$_{reg}$ lineage.

EFFECTOR T$_{REG}$ CELL SUBSETS EXIST THAT EXPRESS ROR-γt$^+$ AND GATA3$^+$

Foxp3$^+$ T$_{reg}$ cells show a significant degree of phenotypic plasticity to adapt to specific microenvironments and to suppress different types of immune responses ("specification" in Figure 9.4). T$_{reg}$ cells accomplish this functional adaptation by employing the same transcriptional machinery that directs the differentiation of effector T$_H$ cells. For example, interferon regulatory factor IRF4, and an AP-1 transcription factor BATF, can be activated in response to TCR signals and environmental cues and cooperatively direct the differentiation of multiple T$_H$ cell subsets; they can also drive T$_{reg}$ cells to differentiate into various effector states. Conventional effector CD4$^+$ T cells can differentiate into T$_H$1, T$_H$2, T$_H$17, or T$_{FH}$ cell subsets, defined by expression of their signature transcription factors T-bet, GATA3, RAR-related orphan receptor γ-t (ROR-γt), and Bcl6. In a similar fashion, effector T$_{reg}$ cells are also heterogeneous and consist of subsets distinguished by expression of T$_H$ cell signature transcription factors, chemokine receptors, and cytokine receptors. For example, T-bet-expressing T$_{reg}$ cells, which are induced under conditions of T$_H$1 inflammation, express CXCR3, accumulate at sites of T$_H$1 inflammation, and exhibit the ability to specifically regulate T$_H$1 responses. Bcl6-expressing T$_{reg}$ cells, which are termed *follicular regulatory T* (T$_{FR}$) cells, express CXCR5, and migrate into lymphoid follicles where they control T$_{FH}$-cell-dependent germinal center reactions. In the mucosa, T$_{reg}$ cells contain high frequencies of the ROR-γt$^+$ and GATA3$^+$ subsets. By differentiating into those diverse effector subsets with distinctive migratory, functional, and homeostatic characteristics, T$_{reg}$ cells can influence a broad range of

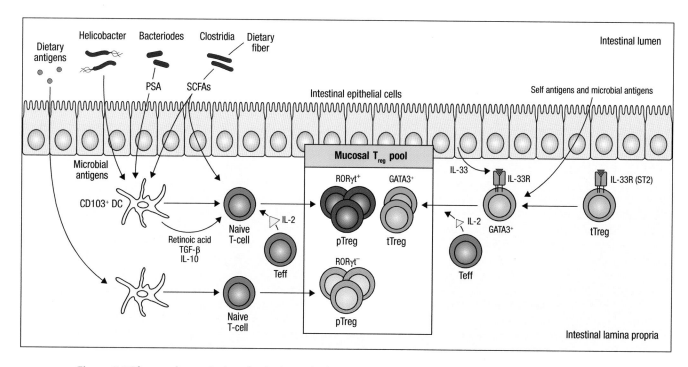

Figure 9.5 Three subpopulations in the intestinal T$_{reg}$ cells. The intestinal T$_{reg}$ pool can be divided into at least three subsets on the basis of RORγt and GATA3 expression. RORγt$^+$ T$_{reg}$ cells are considered to be pT$_{reg}$ cells that derive from naïve T cells in response to microbiota-derived antigens and metabolites/molecules including SCFAs and PSA. Induction of RORγt$^+$ pT$_{reg}$ cells in the intestine depends on CD103$^+$ DCs, which are activated in the intestinal environment, particularly by the gut microbiota and produce retinoic acids, IL-10 and TGF-β. IL-2 derived from effector T cells likely contributes to stabilizing T$_{reg}$ differentiation. RORγt$^-$ T$_{reg}$ cells are considered to be pT$_{reg}$ cells that are induced by dietary antigens and have a critical role in the suppression of immune response to dietary antigens. The third subset is GATA3$^+$ T$_{reg}$ cells, which are thought to be tT$_{reg}$ cells. GATA3$^+$ T$_{reg}$ cells express ST2 (a component of IL-33 receptor complex) enabling them to quickly respond to IL-33. IL-33 is released from epithelial cells in steady state, and its expression is greatly increased under inflammatory conditions. IL-33 acts together with IL-2 to induce the expression of GATA3 in T$_{reg}$ cells. GATA3 upregulates the expression of Foxp3 and contributes to T$_{reg}$ proliferation and maintenance. (Adapted from Tanoue, T. et al.: *Nat. Rev. Immunol.* 2016, 16:295–309.)

cell populations and exert suppressive activities in a variety of anatomical locations and immune environments.

RORγt$^+$ T$_{reg}$ cells are found predominantly in the intestine and are absent in germ-free mice. They are negative for Helios and Nrp1 expression, suggesting that they are a subpopulation of pT$_{reg}$ cells induced by microbial stimuli (**Figure 9.5**). In contrast, RORγt$^-$Nrp1$^-$ T$_{reg}$ cells, abundant in the small intestine, are present in germ-free mice but severely reduced in numbers when germ-free mice are maintained on an antigen-free diet (see **Figure 9.5**). Several lines of evidence suggest that RORγt$^+$ T$_{reg}$ cells are major producers of IL-10 and play a critical role in the regulation of microbiota-driven T$_H$17 cell responses in the intestine. Inactivation of STAT3 in the T$_{reg}$ compartment or impaired TGF-β signaling in CD4$^+$ T cells results in a selective deficiency of RORγt$^+$ T$_{reg}$ cells without affecting GATA3$^+$ T$_{reg}$ cells, and the mice also develop T$_H$17 cell-driven spontaneous colitis. A similar phenotype is also observed in mice with Foxp3$^+$ cell-specific deficiency of the transcription factor c-Maf. The deficiency of c-Maf selectively impairs the generation of bacteria-specific IL-10-producing RORγt$^+$ pT$_{reg}$ cells in the intestinal lamina propria and of T$_{FR}$ cells with unknown antigen-specificity in the Peyer's patches. Because c-Maf expression is synergistically promoted by STAT3-activating pro-inflammatory cytokines (e.g., IL-6 or IL-27) and TGF-β, c-Maf may integrate anti-inflammatory TGF-β/Smad signaling with microbe-induced pro-inflammatory STAT3 signaling to induce RORγt$^+$ pT$_{reg}$ cells.

A significant proportion of Foxp3$^+$ T$_{reg}$ cells in the colon, dermis of the skin, and visceral adipose tissues express GATA3. The skin T$_{reg}$ pool is dominated by GATA3$^+$ T$_{reg}$ cells. GATA3$^+$ T$_{reg}$ cells are distinct from RORγt$^+$ T$_{reg}$ cells, in that they are mostly Helios$^+$Nrp1$^+$ and unaffected by the absence of the microbiota, suggesting that the vast majority of them are tT$_{reg}$ cells (see Figure 9.5). GATA3 expression in T$_{reg}$ cells is upregulated by TCR or IL-2 signaling but unaffected by deficiency of T$_H$2 cytokines (i.e., IL-4 and IL-13) or STAT6, which contribute to GATA3 upregulation by conventional CD4$^+$ T cells. *GATA3*-deficient T$_{reg}$ cells exhibit an impaired ability to accumulate at inflamed sites and to maintain high levels of Foxp3 expression. GATA3$^+$ T$_{reg}$ cells express ST2, a subunit of the IL-33 receptor. IL-33 represents one of the damage-associated molecular patterns (DAMPs) or "alarmins," produced by damaged or inflamed epithelial cells. IL-33 signals activate and phosphorylate GATA3, and the phosphorylated GATA3 in turn upregulates expression of ST2 and Foxp3 to provide a necessary signal for T$_{reg}$ proliferation and their maintenance at the sites of inflammation. *ST2*-deficient T$_{reg}$ cells are unable to suppress the development of colitis in the CD45RB^{high} CD4$^+$ T-cell transfer model of colitis due to their impaired accumulation and reduced expression of Foxp3 in the intestine. In the colon, ST2$^+$ T$_{reg}$ cells have a more activated phenotype and express higher levels of OX40 than ST2$^-$ T$_{reg}$ cells. OX40 is required for the accumulation of T$_{reg}$ cells in the colon and the T$_{reg}$-mediated suppression of colitis. Thus, the IL-33-ST2-GATA3 axis functions under inflammatory conditions to activate and maintain T$_{reg}$ cells to prevent excessive inflammation. Furthermore, recent studies have demonstrated that GATA3$^+$ T$_{reg}$ cells also exert tissue-repair activity in part through IL-33-ST2-dependent production of the tissue-remodeling factor amphiregulin. The mutually exclusive expression of RORγt and GATA3 in colonic T$_{reg}$ cells suggests that microbe-specific RORγt$^+$ T$_{reg}$ cells and self-reactive GATA3$^+$ T$_{reg}$ cells complementarily facilitate mucosal homeostasis.

ENVIRONMENTAL STIMULI ARE NEEDED FOR THE DEVELOPMENT OF MUCOSAL T$_{REG}$ CELLS

9.4 Cell-derived signals

Intestinal DCs express CD103 and are migratory, potent antigen-presenting cells that can extend their dendrites through the intestinal epithelial cell layer and capture luminal antigens, in a manner similar to that of CX3CR1$^+$ macrophages. CD103$^+$ DCs also obtain luminal antigens through a unique transport system via goblet cells. CD103$^+$ DCs express $\alpha v \beta 8$ integrin, which convert latent TGF-β into its active form, as well as retinaldehyde dehydrogenase (RALDH) that metabolizes vitamin A into retinoic acid, which cooperates with TGF-β to induce pT$_{reg}$ cell differentiation (see Figure 9.5).

DCs can be further divided into two subsets, DC1 and DC2, based on CD11b expression. CD103$^+$CD11b$^+$ DCs (DC2) are the most prominent DCs in the small intestine but are less abundant in the colon and uncommon in other tissues. The development of CD103$^+$CD11b$^+$ DCs is dependent on Notch2 and IRF4. Although CD103$^+$CD11b$^+$ DCs are a major constituent of the tolerogenic CD103$^+$ DC population, depletion of this subset does not affect the number of intestinal T$_{reg}$ cells. CD103$^+$CD11b$^-$ DCs (DC1) are functionally related to splenic CD8α^+ DCs. Their development is dependent on the basic leucine zipper transcription factor ATF-like 3 (Batf3) and IRF8, and they can efficiently cross-present internalized soluble or cell-associated antigens to naïve CD8$^+$ T cells. CD103$^+$CD11b$^-$ DCs express high levels of RALDH and as such contribute to intestinal pT$_{reg}$ development together with the CD103$^+$CD11b$^+$ DCs. Mice lacking both CD103$^+$CD11b$^+$ and CD103$^+$ CD11b$^-$ DC subsets have reduced numbers of intestinal T$_{reg}$ cells.

9.5   Microbiota-mediated T_{reg} accumulation

The gut microbiota affects the number, function, and TCR repertoire of intestinal T_{reg} cells (see **Figure 9.3**). Germ-free mice have a striking reduction in the frequency of pT_{reg} cells in the intestine, particularly the RORγt⁺Helios⁻Foxp3⁺ pT_{reg} cell subpopulation. Among the intestinal microbiota, one of the best-characterized examples of pT_{reg}-inducing bacteria is *Bacteroides fragilis*. *B. fragilis*–derived polysaccharide A (PSA), a zwitterionic capsular carbohydrate, drives pT_{reg} differentiation and induces production of IL-10. PSA is incorporated into outer membrane vesicles (OMVs) released by *B. fragilis*, which are delivered to DCs through an autophagy-dependent pathway. PSA stimulates IL-10 production by DCs. The DC-derived IL-10 then acts on pT_{reg} cells, causing them to produce IL-10. Besides *B. fragilis*, several *Bacteroides* species including *B. caccae* and *B. thetaiotaomicron* are also able to induce accumulation of Nrp1⁻RORγt⁺ pT_{reg} cells in the colon following monocolonization of germ-free mice.

Another example of bacteria that induce pT_{reg} cells are *Clostridium* species within clusters IV and XIVa. Consortia of *Clostridium* species can promote the differentiation of colonic pT_{reg} cells. *Faecalibacterium prausnitzii* is a bacterial species found in man and belonging to *Clostridium* cluster IV. It can promote the production of IL-10 by peripheral blood monocytes. *Clostridium ramosum* is also a strong inducer of pT_{reg} cells. *Clostridium* species induce accumulation of RORγt⁺Helios⁻ pT_{reg} cells, rather than GATA3⁺ tT_{reg} cells. *Clostridium* strains can also facilitate the expression of IL-10 and CTLA-4 by pT_{reg} cells. Mice with abundant *Clostridium* strains in their intestines exhibit resistance to experimental colitis or food allergy.

Clostridia stimulate the induction of pT_{reg} cells by a mechanism distinct from that of PSA. One suggested mechanism is the production of short-chain fatty acids (SCFAs), which in the gut are mainly acetate, butyrate, and propionate produced by fermentation of dietary fibers. SCFAs promote intestinal pT_{reg} responses via several mechanisms. SCFAs activate signaling through G-protein-coupled receptors (GPRs) to induce anti-inflammatory genes in DCs and directly stimulate pT_{reg} cell proliferation. SCFAs also drive differentiation of naïve CD4⁺ T cells into pT_{reg} cells, presumably through histone H3 acetylation of CNS1 of the *Foxp3* gene by histone deacetylase inhibition (HDACi). SCFAs also upregulate GPR15 expression, which promotes accumulation of pT_{reg} cells in the colon.

SUMMARY

Having adapted to the barrier sites, mucosal T_{reg} cells are phenotypically distinct from those found in other organs. Their development is affected by antigens and metabolites derived from both the diet and the microbiota. Induced mucosal T_{reg} cells then contribute to the constitutive suppression and prevention of aberrant immune reactions to innocuous environmental factors, such as commensal bacteria and dietary components. T_{reg} cells play a nonredundant role in the maintenance of mucosal immune homeostasis, as illustrated by the fact that defects in the development and/or functional maturation of T_{reg} cells lead to mucosal inflammation. In contrast to the intestine, the pathways that drive T_{reg} development at other mucosal sites, such as the skin and lungs, are less well defined. Obtaining a deeper and more comprehensive understanding of the cellular and molecular mechanisms responsible for the mucosal tissue-specific development of T_{reg} cells is critical, especially in the context of designing targeted therapies for inflammatory diseases.

FURTHER READING

Atarashi, K. et al.: Induction of colonic regulatory T cells by indigenous *Clostridium* species. *Science* 2011, 331:337–341.

Fontenot, J.D., Gavin, M.A., and Rudensky, A.Y.: Foxp3 programs the development and function of CD4$^+$CD25$^+$ regulatory T cells. *Nat. Immunol.* 2003, 4:330–336.

Hori, S., Nomura, T., and Sakaguchi, S.: Control of regulatory T cell development by the transcription factor Foxp3. *Science* 2003, 299:1057–1061.

Josefowicz, S.Z., Lu, L.F., and Rudensky, A.Y.: Regulatory T cells: mechanisms of differentiation and function. *Annu. Rev. Immunol.* 2012, 30:531–564.

Josefowicz, S.Z. et al.: Extrathymically generated regulatory T cells control mucosal TH2 inflammation. *Nature* 2012, 482:395–399.

Lathrop, S.K. et al.: Peripheral education of the immune system by colonic commensal microbiota. *Nature* 2011, 478:250–254.

Monteleone, G., Boirivant, M., Pallone, F., and MacDonald, T.T.: TGF-β1 and Smad7 in the regulation of IBD. *Mucosal Immunol.* 2008, 1(Suppl 1), S50–53.

Mucida, D. et al.: Oral tolerance in the absence of naturally occurring Tregs. *J. Clin. Invest.* 2005, 115:1923–1933.

Ohkura, N., Kitagawa, Y., and Sakaguchi, S.: Development and maintenance of regulatory T cells. *Immunity* 2013, 38:414–423.

Powrie, F., Carlino, J., Leach, M.W., Mauze, S., and Coffman, R.L.: A critical role for transforming growth factor-β but not interleukin 4 in the suppression of T helper type 1-mediated colitis by CD45RB(low) CD4$^+$ T cells. *J. Exp. Med.* 1996, 183:2669–2674

Powrie, F., Leach, M.W., et al.: Phenotypically distinct subsets of CD4+ T cells induce or protect from chronic intestinal inflammation in C. B-17 SCID mice. *Int. Immunol.* 1993, 5:1461–1471.

Rubtsov, Y.P. et al.: Regulatory T cell-derived interleukin-10 limits inflammation at environmental interfaces. *Immunity* 2008, 28:546–558.

Sakaguchi, S., Sakaguchi, N., Asano, M., et al.: Immunologic self-tolerance maintained by activated T cells expressing IL-2 receptor α-chains (CD25). Breakdown of a single mechanism of self-tolerance causes various autoimmune diseases. *J. Immunol.* 1995, 155:1151–1164.

Tanoue, T., Atarashi, K., and Honda, K.: Development and maintenance of intestinal regulatory T cells. *Nat. Rev. Immunol.* 2016, 16:295–309.

Vignali, D.A., Collison, L.W., and Workman, C.J.: How regulatory T cells work. *Nat. Rev. Immunol.* 2008, 8:523–532.

Mucosal B cells and their function

10

JO SPENCER, EDWARD N. JANOFF, AND PER BRANDTZAEG

Mucosal epithelia separate the external environment from the tissues of the body. These surfaces provide specific immune protection through antibodies released into the external exocrine secretions such as tears, nasal fluids, saliva, intestinal juice, and breast milk. The protection of the approximately 400 m^2 of internal epithelial surface area in healthy adult humans involves antibody production by approximately 6×10^{10} mucosal plasma cells, comprising 80% of the total plasma cell population in the body. Thus, the largest and most active compartment of the immune system is dedicated to defense of mucosal surfaces.

Mucosal tissue can be subdivided into structures and cells associated with the inductive or the effector arms of the mucosal immune response. The inductive compartment consists of mucosa-associated lymphoid tissue (MALT), classically Peyer's patches in the intestinal mucosa (gut-associated lymphoid tissue [GALT]) and bronchus-associated lymphoid tissue (BALT) (Chapter 23), which resemble lymph nodes and are highly organized and dynamic in terms of lymphocyte traffic and proliferation. However, because MALT is constantly exposed to antigens coming directly from mucosal surfaces, MALT structures are uniquely adapted to generate diverse precursors of mucosal effector cells. In contrast, the GALT effector compartment is diffusely located in the subepithelial lamina propria. The lamina propria contains plasma cells and their immediate precursors, together with effector T cells, macrophages, dendritic cells, granulocytes, and mast cells. The most abundant effector molecule produced in the lamina propria is IgA, which is among the best understood mediators of mucosal protection.

In this chapter, we describe B cells and antibodies associated with the mucosal immunity and the progression of mucosal B-cell development from the inductive phase in organized follicles in GALT to the production of immunoglobulins by plasma cells in effector mucosal tissues.

CELLS AND PROTEINS INVOLVED IN HUMORAL IMMUNITY AT MUCOSAL SURFACES

Fundamental differences distinguish the structure of the lymphoid tissues in MALT from that of other lymph nodes scattered throughout the body. Likewise, there are major differences in the control of their development and the molecules that drive their formation. In particular, peripheral lymphoid tissue develops in a largely antigen-free environment, whereas mucosal lymphoid tissues are exposed to the lumenal microbiota or their components from the earliest moments after birth and face constant antigenic stimulation throughout life.

10.1 Mucosa-associated lymphoid tissue has major differences from conventional lymphoid tissue

The unencapsulated organized lymphoid tissue in MALT (both GALT and BALT) is separated from the lumen by a specialized follicle-associated epithelium (FAE) that contains microfold or membrane (M) cells. FAE is infiltrated by B cells that express high levels of costimulatory molecules CD80/CD86, CD4$^+$ T cells, and dendritic cells, and therefore is distinct from ordinary mucosal surface epithelium such as villus epithelium, which contains mostly CD8$^+$ T cells. Lymphoepithelial FAE is a portal for the entry and sampling of luminal antigen, and its presence is a defining characteristic of MALT.

Exogenous stimuli from the gut lumen are transported into GALT via M cells in the FAE, probably aided by dendritic cells, which have been shown in mice to be capable of sending processes through the epithelium into the lumen to sample lumenal antigens. Antigens that reach the intraepithelial and subepithelial compartments of GALT encounter a mixed lymphocyte population of B and T cells in a microenvironment that also is rich in macrophages and dendritic cells. Macrophages and dendritic cells in the dome region form a critical first line of defense beneath the epithelium that can influence downstream immunologic events and B-cell function in adjacent organized lymphoid follicles and beyond.

The structure of GALT is similar throughout the gastrointestinal tract with prominent B-cell follicles with intervening T-cell zones. CD4$^+$ T cells in the outermost zone intermingle and interact with antigen-exposed B cells. GALT contains a marginal zone that resembles the splenic marginal zone and is surrounded by and merged with the memory B-cell population (**Figure 10.1**). Splenic and GALT marginal zone B cells are similar morphologically and are medium-sized cells with cleaved nuclei. These B cells are CD27$^+$IgM$^+$ and do

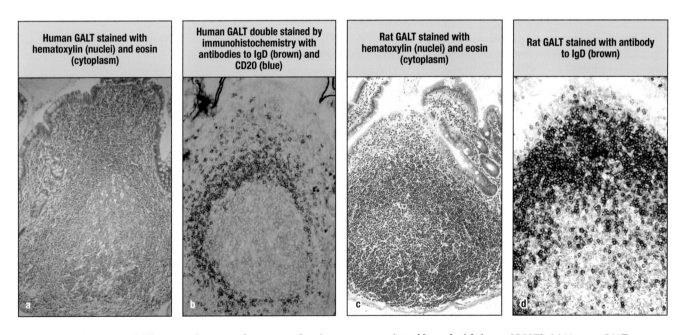

| Human GALT stained with hematoxylin (nuclei) and eosin (cytoplasm) | Human GALT double stained by immunohistochemistry with antibodies to IgD (brown) and CD20 (blue) | Rat GALT stained with hematoxylin (nuclei) and eosin (cytoplasm) | Rat GALT stained with antibody to IgD (brown) |

Figure 10.1 Differences between human and rodent gut-associated lymphoid tissue (GALT). (**a**) Human GALT stained with hematoxylin (nuclei) and eosin (cytoplasm). (**b**) Human GALT double stained by immunohistochemistry with antibodies to IgD (brown) and CD20 (blue). IgD-negative B cells in blue are present in the germinal center and on the periphery of the B-cell follicle. (**c**) Rat GALT stained with hematoxylin and eosin for comparison with human GALT in (**a**). (**d**) Rat GALT stained with antibody to IgD (brown). This section has a blue nuclear counterstain. In contrast to human GALT, the IgD-expressing zone of B cells in rats extends into the follicle-associated epithelium.

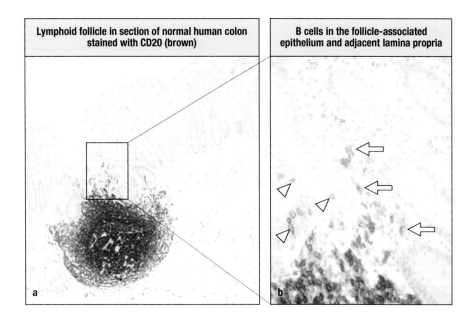

| Lymphoid follicle in section of normal human colon stained with CD20 (brown) | B cells in the follicle-associated epithelium and adjacent lamina propria |

Figure 10.2 Paraffin section of normal human colon stained with anti-CD20. (**a**) An isolated lymphoid follicle is composed mostly of B cells that express CD20 (brown). In contrast, the lamina propria contains rare cells that express CD20. (**b**) The area inside the rectangle in (**a**) illustrated at higher magnification shows B cells in the follicle-associated epithelium (arrowheads) and the infiltration of B cells into the adjacent lamina propria at the periphery of gut-associated lymphoid tissue (arrows).

not express high levels of IgD. Splenic marginal zone B cells have mutations in their IgHV genes that are acquired in the germinal centers of GALT. The GALT marginal zone also contains B cells with mutations in IgHV. Internal to and overlapping with the marginal zone is the mantle zone of IgD high naive B cells. The broadest aspect of the often narrow, crescent-shaped mantle zone faces the antigen-exposed FAE. Memory B cells expressing predominantly IgA or IgM (but not IgD) are located on the periphery of the B-cell area of GALT. At the center of the follicle is the germinal center. B cells proliferate and mutate in the cell-dense dark zone of the germinal center and are selected for further antigen-driven mutation and maturation by T-follicular-helper (T_{FH}) cells in the light zone.

Organized MALT structures do not have a distinct boundary in the form of a fibrous capsule. Instead, they diffuse gradually into the adjacent lamina propria; consequently, the boundary between the inductive site of GALT and the effector tissue of the lamina propria may be indeterminant. A useful indicator of the margins of MALT in tissue sections is to stain with antibodies to mature B cells such as anti-CD20. Most B cells in MALT express CD20, but $CD20^{+}$ cells are very rare in the intestinal lamina propria (**Figure 10.2**), where $CD20^{-}$ plasma cells and their immediate precursors predominate. Awareness of the microanatomy of GALT and the lamina propria reveals that B cells identified in the lamina propria may localize in the periphery of GALT if the cells express mature B-cell antigens. The B cells on the periphery of GALT are often large cells that may have connections with adjacent cells via cytoplasmic processes. Expression of J chain by IgA-producing plasma cells is reduced at the GALT–lamina propria boundary region compared with the more distant lamina propria. This difference is consistent with the idea that the extreme margins of GALT have distinct properties related to B-cell function or the stage of differentiation.

In the small intestine, the induction and regulation of the mucosal B-cell response occurs primarily in Peyer's patches, the clusters of organized lymphoid tissue concentrated in the terminal ileum in most mammalian species. In contrast, individual isolated lymphoid follicles are dispersed throughout the intestine in the lamina propria. In mice, Peyer's patches and isolated lymphoid follicles develop sequentially. Whereas Peyer's patches are constitutive and induced to develop before birth by lymphoid tissue inducer (LTi) cells, isolated lymphoid follicles develop after birth in response to the microbiota that induces the accumulation of T and B cells around clusters

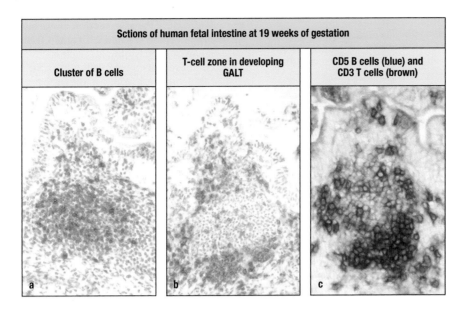

Figure 10.3 Frozen sections of human fetal intestine at 19 weeks of gestation. (**a**) Section stained with antibody to IgD (brown) shows a cluster of B cells. (**b**) Section stained with anti-CD3 shows surrounding T-cell zone in developing GALT. (**c**) Section double-stained with anti-CD3 (brown) and anti-CD5 (blue). Blue CD5 B cells are visible in the central B-cell zone of fetal GALT. The follicle-associated epithelium in this section is infiltrated by CD3-expressing T cells and CD5-expressing B cells.

of LTi-like cells, termed *cryptopatches*. The epithelial pattern-recognition receptor nucleotide-binding oligomerization domain 1 (NOD1) is crucially involved in the maturation of murine isolated lymphoid follicles through recognition of bacterial peptidoglycans as the gut becomes colonized with the microbiota.

The Peyer's patch component of human GALT is constitutive in humans, as in mice. In humans, constitutive GALT, where clusters of B cells express CD5, can be observed in fetal intestine from around 18 weeks of gestation. Although used to identify the B-1 B-cell lineage in mice, CD5 can be expressed by immature transitional B cells in humans and can be an activation marker on human B cells (**Figure 10.3**). Indeed, fetal intestinal B cells are large activated cells, although there is no evidence of germinal center formation in the healthy fetal intestine. Like GALT in the postnatal intestine, B cells in fetal GALT infiltrate between the epithelial cells in the FAE, suggesting that recognition of microbial antigens is not required for this localization. However, results in originally germ-free then colonized rodents highlight the role of the microbiota in driving the magnitude of epithelial homing.

Whether isolated lymphoid follicles in human small intestine and colon are constitutive or acquired is not known. However, GALT can be acquired in humans. For example, the normal stomach mucosa contains few lymphoid or plasma cells and no GALT structures. However GALT develops in the stomach of patients infected with the gastric bacterium *Helicobacter pylori*. These structures provide the background for the evolution of "MALT lymphomas," tumors of GALT marginal-zone B cells that in many cases are dependent on *H. pylori* for growth.

MALT is generally associated with B-2 (conventional) B-cell function. However, in mice, B-1 B cells may also contribute significantly to the intestinal plasma cell population, estimated to include up to 50% of mouse gut plasma cells. B-1 B cells are a self-renewing and self-sustaining B-cell subset that reside predominantly in the peritoneal cavity and originate from B cells in the omentum, the fatty, blanket-like structure covering the intestine in the peritoneal cavity. B-1 B cells contribute to innate immunity through their production of polyspecific antibodies. In mice, B-1 B cells are suggested to differentiate into plasma cells in the lamina propria rather than in GALT. This function may be related to the ability of dendritic cells to extend processes through the intestinal epithelium and sample antigens outside GALT. However, IgA responses in general are dependent on LTi cells that are founders and orchestrators of organized lymphoid tissues. Thus, even the B-1 B-cell axis of plasma cell production in mice may be dependent on GALT.

B-1 cells and B-2 cells are not well-distinguished in humans, nor are peritoneal or omental B cells known to contribute to human mucosal immunity. Nevertheless, human IgA includes low-affinity cross-reactive antibodies comparable to those produced by the B-1 B-cell lineage in mice, but these antibodies may also be induced by the continuously shifting antigen repertoire of the large commensal microbiota.

The structure of GALT in humans shows interesting similarities and differences with other species. In rats and mice, Peyer's patches are bulbous structures clearly visible from the serosal surface of the intestine. In contrast, Peyer's patches in humans are more difficult to detect, even from the mucosal surface in unfixed intestine. Histologically, in rodents and humans, the most prominent features of Peyer's patches are the secondary lymphoid follicles with germinal centers and the FAE. In rat and mouse Peyer's patches, the follicles often abut directly onto the muscularis mucosae, whereas in humans, where the mucosa of the intestine is thicker, the T-cell zone tends to mark the boundary on the serosal side (see Figure 10.1). The FAE appears similar in structure in humans and rodents and is infiltrated by a mixed lymphoid population that includes medium-sized B cells that are phenotypically diverse. However, in rats this lymphocyte subset expresses levels of IgD similar to mantle-zone B cells (see Figure 10.1), implying a fundamental difference in the lineages of mucosal B cells between species. The species differences in the Peyer's patch function are in fact more extreme in other species. In sheep, for example, the large ileal Peyer's patch is a site of B-cell development with gene rearrangement in developing B cells.

10.2 IgA and IgM are the dominant mucosal immunoglobulins

Mucosal epithelia are protected by secretory immunoglobulin A (SIgA) and, to a lesser extent, secretory IgM (SIgM) antibodies produced by plasma cells and/or plasmablasts in the lamina propria. These tissue plasmablasts and plasma cells are identified in humans by expression of CD38, CD27, lower levels of CD19, and the absence of the B-cell marker CD20 and constitute the body's largest population of antibody-secreting cells. CD138 (syndecan-1) is reliably expressed on plasma cells, but epithelial cells also express CD138. Although IgA is the major isotype produced at mucosal surfaces, its production varies among different regions variations of the intestinal tract. Mucosal plasma cells produce mainly dimeric IgA and some larger polymers of IgA (collectively called polymeric IgA, pIgA).

In addition to the light and heavy chains, pIgA and pentameric IgM contain a 15 kDa polypeptide termed the *joining* or *J chain*, which facilitates the spontaneous noncovalent interaction of these ligands with the pIg receptor (pIgR) expressed basolaterally on secretory epithelial cells. As a basis for secretory immunity, MALT must therefore favor the development and dispersion of B cells with J-chain expression, a prerequisite for production of pIgA and pentameric IgM that can be exported to the lumen by pIgR. The detailed mechanism for J-chain induction is not known, at least not in humans.

Notably, systemic IgA may be induced, in part, at mucosal sites. Circulating IgA in plasma and IgA-secreting cells in blood appear to preferentially share reactivity with mucosal pathogens (e.g., *Giardia lamblia*, *Vibrio cholerae*, rotavirus), commensals and other antigens (e.g., tissue transglutaminase in celiac disease). However, IgA in serum is not likely produced directly at mucosal sites. In rodents, but not humans, IgA is transported from serum to breast milk.

Of the two IgA subclasses (IgA1 and IgA2) in humans, IgA1 comprises 80%–85% of total serum IgA. In the upper respiratory tract (tonsils and nasal mucosae) and peripheral lymph nodes, plasma cells predominantly secrete

IgA1 (75%–93%). In the gut, however, a relatively large proportion of plasma cells secrete IgA2 (29%–64%), particularly in the large bowel. SIgA2 (pIgA2 linked to the epithelial-transporting secretory component) may be particularly advantageous due to its resistance to several IgA1-specific bacterial proteases. Compared with monomeric IgA (mIgA), pIgA also enhances agglutination of organisms to limit mucosal infection and has greater avidity for antigens and organisms, thereby facilitating clearance and phagocytosis. The proportions of the two SIgA subclasses in various secretions correspond to the IgA1$^+$:IgA2$^+$ plasma cell ratios in the local secretory tissues, consistent with the notion that both subclasses of pIgA are transported equally from lamina propria to lumen by the epithelial pIgR.

The antigenic and immunologic molecular events underlying preferential IgA1 or IgA2 responses remain unclear. Secretory antibodies to lipopolysaccharide (LPS) are generally SIgA2, whereas protein antigens stimulate predominantly SIgA1. IgA$^+$ plasma cells are mainly of the IgA1 subclass (approximately 71%) in the jejunum, whereas IgA2 dominates in the colon (64%), possibly reflecting differences between responses to food antigens versus bacterial antigens, respectively. Consistent with this hypothesis, bacterial overgrowth in bypassed jejunal segments produces a change in the plasma cell isotype proportions, with an increase of IgA2 and a decrease of IgM, suggesting LPS-induced isotype switching from C_μ to $C_{\alpha 2}$ or progressive sequential downstream switching of the IgC$_H$ genes. Thus, preferential subclass switching may derive from differential induction of chemokines and chemokine receptors in MALT in different locations that guide cells to different effector sites.

IgM$^+$ plasma cells constitute a substantial, albeit variable, fraction of antibody-secreting cells in the adult human gut. The relatively high proportion (approximately 18%) of IgM plasma cells in the proximal small intestine may be related to the low levels of LPS (discussed earlier), whereas the frequency of these cells is much lower in the upper airways and nasopharyngeal mucosa. Importantly, despite a lack of mucosal IgA, persons with selective IgA deficiency do not typically develop gut infections, likely due to the compensatory presence of plasma cells that secrete pentameric IgM, which also is transported across the epithelium by the pIgR.

IgG$^+$ plasma cells comprise only 3%–4% of the antibody-secreting cells in the normal human intestine but a considerably larger proportion in gastric and nasal mucosae. Similar to its distribution in serum, the majority of plasma cells at multiple mucosal sites are of the IgG1 subclass (56%–69%). However, IgG2$^+$ plasma cells (20%–35%) are generally more frequent than IgG3$^+$ plasma cells (4%–6%) in the distal gut, whereas the reverse is often true in upper-airway mucosa. Such IgG subclass differences support the idea that C$_H$ gene switching may take different pathways in various mucosal regions as discussed later. Interestingly, the $C_{\gamma 2}$ and $C_{\alpha 2}$ genes are located on the same duplicated DNA segment, and many carbohydrate and bacterial antigens preferentially induce an IgG2 and, in part, an IgA2 response, whereas proteins (which are usually T-cell-dependent antigens) primarily generate IgG1 responses and IgA1. In inflammatory bowel disease, the number of mucosal IgG$^+$ plasma cells is dramatically increased, and the predominant IgA$^+$ plasma cells in the colon are atypically preferentially IgA1 and show decreased J-chain expression, reflecting a more systemic type of plasma-cell pattern.

IgD$^+$ plasma cells are rarely encountered in the human gut, but they make up a significant fraction (3%–10%) in the upper airways, particularly the sinuses. In IgA deficiency, this disparity may be even more striking for IgD$^+$ plasma cells than for IgM$^+$ plasma cells. No active epithelial transport has been identified for short-lived IgD, but this isotype may facilitate antimicrobial activity through binding specific antigens on organisms such as *Haemophilus influenzae* and *Moraxella catarrhalis*.

IgE$^+$ plasma cells are virtually absent from normal human mucosae. However, IgE plasma cells and especially IgE-bearing mast cells are commonly present in the oropharyngeal mucosa and sinuses in patients with allergic diseases and asthma.

MUCOSAL B-CELL RESPONSES

Interaction between dendritic cells and luminal contents sampled directly or transported into the MALT by M cells results in a change in gene expression that has several potential consequences. Dendritic cells activated by microbe-associated molecular patterns (MAMPs) directly affect B-cell function through modification of T-cell-dependent and -independent production of IgA and the imprinting of homing properties.

10.3 Most IgA responses in MALT germinal centers are T-cell dependent

Dendritic cells in MALT are exposed to antigens such as bacterial products for presentation to T cells and B cells and may be modified by bacterial products via pattern recognition receptors. T-cell activation induces the upregulation of CD278 (ICOS) and CD154 (CD40L), essential for cognate B-cell–T-cell interactions. This T-cell-dependent mechanism of B-cell activation initiates the formation of germinal centers. The importance of cognate interactions between B and T cells for germinal center formation is supported by the fact that no germinal centers are formed when the CD40–CD154/CD40L interaction is disabled in humans (CD154 mutation in X-linked hyper-IgM syndrome) or in CD40 knockout mice.

The importance of T-cell help for germinal center formation in mice has been documented by immunization with polysaccharide-based T-cell-independent antigens. Although germinal centers are temporarily induced, the level of V-region mutation remains low, and the germinal centers involute by massive B-cell apoptosis at the initiation of B-cell centrocyte selection. Thus, germinal centers are "designed" to generate conventional B-2 memory/effector cells for the production of antibodies to T-cell-dependent antigens; in this manner, MALT structures can mount adaptive pathogen-directed responses of high affinity and specificity.

When B-cell activation occurs in a primary follicle where there is no existing germinal center, the proliferation of B cells activated with cognate T-cell help generates a cluster of large cells, precursors of a mature germinal center. The cluster of dividing cells nudges the naive B-cell component of a primary follicle aside to form the mantle component of a secondary follicle. This effect has been demonstrated to occur within the first few days of life in human Peyer's patches, which invariably contain germinal centers throughout life. This scenario may differ with colonic isolated lymphoid follicles, which frequently lack germinal centers.

In germinal center responses in peripheral lymphoid organs and, likely in GALT, newly activated B cells can be recruited into existing germinal centers that contain dividing B cells with other specificities, providing that the cognate T-cell help has developed prior to their colonization to sustain specific germinal center B-cell growth. Germinal center B cells (centroblasts) undergo proliferation (expressing Ki67) in the germinal center "dark zone" (Figure 10.4). Here, activation-induced cytidine deaminase (AID) supports somatic hypermutation of their surface antibody's antigen-binding variable (IgV) regions and class-switch (e.g., IgM to IgA) of the effector constant region. These cells then relocate within the germinal center to compete as centrocytes for antigen with B cells bearing surface antibodies of related specificities but with different mutations in the germinal center "light zone." The cells bearing the antibody with the highest avidity compete successfully

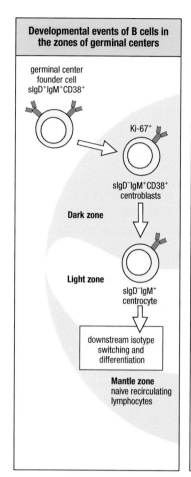

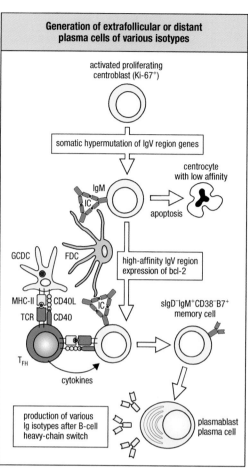

Figure 10.4 Developmental events of B cells in the dark and light zones of germinal centers, leading to the generation of extrafollicular or distant plasma cells of various isotypes. The germinal center founder cell is activated in the extrafollicular compartment (see **Figure 10.7**) and then migrates to the dark zone where it proliferates and differentiates. The molecular cell–cell interactions and immune events are schematically depicted on the right, and further details are discussed in the text. FDC, follicular dendritic cell; GCDC, germinal center dendritic cell; IC, immune complex; Ki-67, proliferation marker; MHC-II, major histocompatibility complex class II molecule; TCR, T-cell receptor; T$_{FH}$, T-follicular-helper cell. (Adapted from Brandtzaeg, P. *Immunol. Invest.* 2010, 39:1–53. With permission from Taylor & Francis Group.)

for antigens presented on the follicular dendritic cells (FDCs) that extend throughout the germinal center light zone. Thus, under the influence of local T-follicular-helper (T$_{FH}$) cells, germinal center B cells are selected for sustained expansion, further mutation and differentiation into memory cells, and antibody-secreting cells that exit the germinal center and potentially undergo further differentiation in the lamina propria. In this setting, BCL6 is a germinal center-targeted transcription factor that directs T cells to become T$_{FH}$ cells and protects DNA in B cells from excessive damage but allows somatic hypermutation and class-switch recombination.

10.4 High level of somatic hypermutation characterizes mucosal B-cell responses

Germinal center B cells undergo modification of DNA sequences through the activity of AID, the B-cell–specific enzyme that catalyzes the mutation of target DNA sequences (**Figure 10.5**). AID is upregulated in B cells following engagement of the B-cell receptor or toll-like receptors by microbial antigens to effect somatic hypermutation and class-switch recombination. Mutations in the functional domains of AID in Type 2 hyper-IgM syndrome lead to the absence of germinal centers, IgA and other class-switched immunoglobulins, and somatic hypermutation, underscoring the importance of AID.

AID deaminates cytidine in targeted segments of DNA in rapidly dividing centroblasts in the dark zones of germinal centers (see **Figure 10.5**). The deamination of cytidine generates uridine that is a normal component of RNA, but not DNA. A subsequent balance in activity between cell division, removal of uridine by uracil glycosylase, and lesional repair by error-prone polymerases generates a highly diverse spectrum of mutational outcomes.

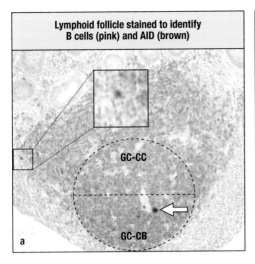

Lymphoid follicle stained to identify
B cells (pink) and AID (brown)

GC-CC

GC-CB

a

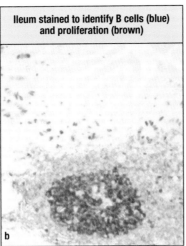

Ileum stained to identify B cells (blue)
and proliferation (brown)

b

Figure 10.5 **Activation-induced cytidine deaminase (AID) and cell division in human gut.** (a) Lymphoid follicle in human colon tissue stained for B cells (CD20 pink) and AID (brown). The germinal center is identified by a dotted line and divided into the centroblast zone (GC-CB), where AID-mediated somatic hypermutation occurs, and the centrocyte zone (GC-CC), where selection for immunoglobulin specificity takes place. Note the double staining for CD20 and AID in the centroblast zone. Large interfollicular B cells containing cytoplasmic AID are also apparent (black arrow). The very dark brown staining identified by a white arrow is a macrophage. This is a previously described artifact of the AID staining. (b) Human ileum stained to identify B cells (CD20 in blue) and proliferation (Ki-67 in brown). B-cell proliferation is observed in GALT inside and outside of the germinal center. B-cell proliferation is not observed in the adjacent lamina propria.

Somatic hypermutation introduces mostly point mutations but also some insertions and deletions in the immunoglobulin heavy (IgH)- and light-chain (IgL) variable (V) region genes. Mutations tend to cluster in AID-targeted hotspots in the complementarity determining regions (CDRs) that form the antigen-binding loops upon protein folding, rather than in the more structural framework regions. Both murine and human Peyer's patches demonstrate a very high frequency of somatic mutations in IgV genes, indicating the antigen-driven specificity of most antibodies produced at mucosal sites. The very high mutation frequencies of Peyer's patch germinal center cells in mice were helpful in elucidating the mechanisms of somatic hypermutation. However, some intestinal plasma cells, including those with high specificity for transglutaminase 2 in humans with celiac disease, do not show high mutation frequencies.

Somatic hypermutation changes on average over 5% of the variable region sequence of human mucosal immunoglobulins. Mutation frequencies in plasma cells in the lamina propria of the gut, salivary gland, and other mucosal tissues exceed those from plasma cells from the systemic compartment, including the splenic red pulp, independent of isotype. Mutation frequencies in the 250–350 nucleotide IgV sequence also provide an estimate of the number of cell divisions with approximately one mutation per two cell divisions. A plasma cell with the average number of 15 IgHV mutations is likely to have undergone approximately 30 divisions, with the resulting greater than 10^9 potential progeny from a single cell, providing tremendous capacity for generating plasma cell precursors. That the estimated progeny (up to 10^{18} cells) may exceed the number of cells in the human body by a million-fold predicts that intestinal plasma cells show diverse antigen reactivity, are turning over rapidly, and undergo extensive cell death. Like plasma cells in bone marrow, a subset of intestinal plasma cells is very long-lived. Antibody diversity may also be facilitated by receptor editing in which the initial κ light chain is replaced by a λ chain, a process that also limits initially autoreactive B cells.

A constant flow of lymph from the gut into the thoracic duct contains immunoblast precursors of plasma cells derived from Peyer's patches. Nevertheless, when this lymph flow is interrupted in rats, following an initial decline, the intestinal IgA⁺ plasma cell density is remarkably well maintained, perhaps implying sideways migration of precursor cells from GALT structures. Similar observations of highly localized B-cell responses to topically applied antigen have been made in a multiple-intestinal-loop model in lambs. Production of IgA antibodies in the gut lamina propria, therefore, probably depends on the potential of GALT to generate a continuous supply of plasma cell precursors, both laterally and by homing through lymph and peripheral blood. The cells at the periphery of GALT illustrated in **Figure 10.2** may be involved in this process of lateral migration prior to differentiation.

The junctional sequences between the V, D, and J regions created in the CDR3 region of IgH genes by RAG-mediated recombination during B-cell development are unique to an individual B-cell or B-cell clone. This unique "clonal signature" can therefore be used to track the distribution and migratory potential of B cells and their plasma cell progeny. This methodology demonstrated that B cells in the germinal centers of human Peyer's patches and lamina propria plasma cells can be developmentally linked and that clones of plasma cells can be widely distributed along the gut, consistent with dissemination of GALT-derived plasma cell precursors via the blood. The identification of clonally related cells in the intestinal lamina propria at multiple sites also is consistent with the idea that GALT produces large numbers of progeny from a single founder B-cell. Whether chronic exposure to an excess of an antigen in the gut affects this process is not known. The iterative processes of B cells competing for interactions with follicular dendritic cell ligands then cognate, and soluble T_{FH} interactions result in selection of optimized B-cell clones for maturation into memory and plasma cells that exit the germinal centers for recirculation and tissue homing (see **Figure 10.4**).

10.5 Class-switch recombination generates IgA

Somatic hypermutation in germinal centers may be accompanied by another AID-mediated DNA modification, class-switch recombination (**Figure 10.6**). This DNA deletional mechanism results in switching from Cμ to downstream CH genes on chromosome 14 in humans, while maintaining the same antigen-binding variable region gene sequence. In most mammals, there is a single germline gene encoding the α constant heavy chain (CHα) domains at the 3′ end of the Ig heavy-chain locus. In humans and higher primates, however, there has been a duplication of the γ→α CH region, resulting in two functional CHα genes encoding the IgA1 and IgA2 subclasses. In both mice and humans, class switch to IgA can be directed in Peyer's patches and isolated lymphoid follicles by predominantly T-cell–dependent as well as T-cell–independent mechanisms.

Figure 10.6 Class-switch recombination from IgM to IgA1 at the Ig heavy-chain locus. The variable region of the Ig heavy chain is assembled from the variable (V), diversity (D), and joining (J) gene segments by VDJ recombination during B-cell development. Transcription across the locus is driven by a promoter upstream of the rearranged VDJ segment (bent black arrow), which facilitates the synthesis of a μ heavy chain. The latter associates with a light chain, thereby forming a complete IgM molecule, which is displayed on the surface of the B-cell as part of the B-cell receptor. Secondary isotypes are produced by class-switch recombination (CSR), which is initiated by germline (GL) transcription of the C_{μ} and the $C_{\alpha}1$ genes (for switch to IgA1; bent red arrows). Products of recombination at the C_H locus are the rearranged chromosome and an episomal circle, from which circle transcripts are derived. Cytokines stimulate transcription (bent red arrows) through the μ gene and determine the Ig isotype to which the B-cell switches. The constant regions of downstream isotypes are denoted by their corresponding Greek letters and the 3′ enhancers (3′E), which influence GL transcription and thereby the CSR, are indicated below the schematic chromosome. (Adapted from Brandtzaeg, P., and Johansen, F. *Immunol. Rev.* 2005, 206:32–63. With permission from John Wiley & Sons.)

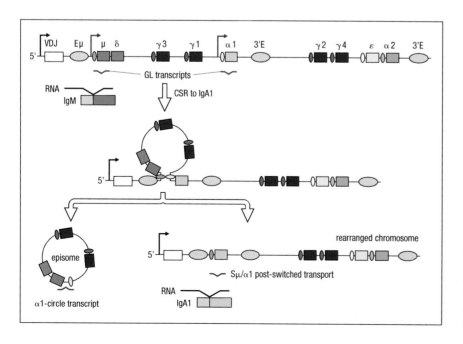

In these inductive sites, T-cell–dependent class-switch recombination to IgA follows cognate engagement of CD40 on B cells with CD40 ligand (CD40L) from T_{FH} cells and upregulation of AID in B cells. T-helper cells also provide noncognate help by secreting cytokines such as TGF-β, a potent switch factor, along with IL-10, IL-4, IL-5, IL-6, and IL-21 that augment B-cell class-switch recombination, differentiation, and survival. Mice deficient in the TGF-β receptor TGF-βRII are almost completely deficient in IgA. Mice deficient in SMAD7, the intracellular negative regulator of TGF-β signaling, have increased class-switch recombination to IgA but reduced proliferation in response to LPS, whereas mice deficient in SMAD2, a component of the TGF-β signal pathway, lack IgA. Mice have a single IgA isotype and require TGF-β signaling for IgA class switching, whereas humans, with two subclasses (IgA1 and IgA2) do not. IL-4-deficient mice have normal total IgA levels but an impaired specific response to mucosal immunization with cholera toxin, whereas IL-6-deficient mice have defective IgA responses in some systems. Dendritic cells enhance TGF-β receptor-mediated induction of IgA class switch by upregulating TGF-β receptor expression on B cells through the production of inducible nitric oxide synthase.

T-cell–independent mechanisms of IgA class-switch recombination in the intestinal tract in mice, and possibly in humans, are mediated by interactions of B cells with relatively homologous CD40L-like molecules, a proliferation-inducing ligand (APRIL), and B-cell–activating factor (BAFF). Both factors favor class-switch recombination to IgA1 and IgA2 in humans when combined with IL-4 and cell division. These molecules are produced locally by GALT dendritic cells, epithelial cells, macrophages, and neutrophils and interact directly with surface receptors on B cells (B-cell maturation antigen, transmembrane activator and calcium-modulator and cyclophilin ligand interactor, or BAFF-R) to induce the upregulation of AID and IgA class switching (see Figure 10.6). TLR-induced activation of dendritic cells enhances APRIL production. Mice deficient in APRIL and humans lacking APRIL receptor exhibit substantially reduced IgA production. However, mice lacking T cells, CD4, or CD40L still develop relatively normal intestinal IgA responses, highlighting the importance of both T–independent and T-dependent class-switch recombination in the generation of intestinal IgA. The microbiome may drive class-switch recombination though both T-dependent and T-independent pathways by activation of CD40L on T cells and engagement of the antigen-specific B-cell receptor and toll-like receptors (e.g., TLR9), respectively.

IgA switch factors initiate transcription through switch regions of C_α in cells expressing sIgM. The RNA produced is noncoding and may be referred to as a sterile transcript. Experimentally, the presence of sterile transcripts in a population of B cells reflects the potential for class-switch recombination. The function of sterile germline transcript production is to open up the DNA double helix in a transcription bubble, allowing the molecular mediators of class-switch recombination, including AID, access to single-stranded DNA substrate (see Figure 10.6). The switch regions are long DNA sequences composed largely of repeated nucleotide motifs that include the DNA target substrate for AID. Deamination of cytidine on both DNA strands generates uracil that can be removed by uracil glycosylase. The abasic site can then be recognized by apurinic/apyrimidinic endonuclease that generates nicks in the DNA. The subsequent repair (by a nonhomologous end-joining pathway related to repair of DNA strand breaks during RAG-mediated recombination) replaces the C_μ constant region with C_α, leaving a circular deleted episomal fragment (see Figure 10.6). RNA transcripts of the circular DNA fragment (noncoding circle transcripts) are produced for a short time after the recombination event. Circle transcript production is therefore a feature of class-switch recombination in the recent past history of a B-cell—a feature that can be exploited experimentally when investigating class-switch recombination events.

Activation of class-switch recombination with expression of transcription factors PAX5 and BLC6 is a negative signal for terminal B-cell differentiation within germinal centers, during which the former are downregulated and XBP1, IRF4 and BLIMP1 are expressed. The decision whether activated B cells should continue down the memory pathway or differentiate along the effector pathway remains elusive, but interactions between CD38+CD27+ germinal center B cells and CD70 (CD27L)+ T cells may be a decisive event.

10.6 Chemokines and chemokine receptors are involved in the organization of MALT and germinal centers

Several chemokines contribute to lymphocyte trafficking and determinants of lymphocyte positioning in lymphoid tissue (**Figure 10.7**). The human B-cell-attracting chemokine (BCA-1, CXCL13) is produced in human lymph node follicles. CXCL13 and its receptor CXCR5 are involved in the formation of organized lymphoid tissue by upregulating $LT_{\alpha1\beta2}$ on B cells, establishing a positive-feedback loop to generate follicle formation and attracting B cells into the tissue. The follicular expression of murine CXCL13 is reportedly stronger in murine Peyer's patches than in peripheral lymph nodes. Other chemokines acting on B cells include stromal cell-derived factor 1 (SDF-1, CXCL12), which is produced by cells lining tonsillar germinal centers and attracts naive and memory B cells that express CXCR4, at least *in vitro*.

Low level expression of CXCR5 in follicular mantle zones is consistent with moderate CXCL13 attraction of naive B cells *in vitro*, whereas strong CXCR5 expression occurs where scattered T cells are located. These CD4+CXCR5+ cells, which also express PD-1 and often CD57 are functionally defined as T_{FH} cells (see **Figure 10.7**), are distinct from T_H1 and T_H2 cells, and are particularly efficient in providing B-cell help required for the production of high-affinity antibodies (see **Figure 10.4**).

The partial overlap in immunostaining for CXCL13 in human lymphoid tissue and several follicular dendritic cell markers suggests that CXCL13 is

Figure 10.7 Adhesion molecule- and chemokine-regulated steps in T- and B-cell migration to, and positioning within, gut-associated lymphoid tissue (GALT) compartments. Antigens are sampled from the gut lumen by M cells, and mesenteric lymph nodes receive antigens via draining lymph (either in soluble form or carried by dendritic cells; not shown). Naive T and B cells enter both GALT and mesenteric lymph nodes (left) via high endothelial venules (HEV) by interactions principally between CD62L/L-selectin (L-sel) and endothelial mucosal addressin cell adhesion molecule-1 (MAdCAM-1) or peripheral lymph node addressin (PNAd), distributed as indicated. Activated (memory/effector) T and B cells may to some extent reenter these sites by leukocyte-integrin $\alpha_4\alpha_7$-MAdCAM-1 interactions. The chemokines involved (right) at the level of HEVs are CCL21 (SLC) and CCL19 (ELC), provided by stromal cells and redistributed to the HEV endothelium to preferentially attract CCR7+ naive T cells and, at a less extent, B cells (broken arrow). SLC also may be involved in the exit of lymphoid cells from GALT via draining lymphatics. Naive B cells are CXCR5+ and extravasate mainly via modified HEVs presenting CXCL13 (also called BCA-1 in humans) juxtaposed to, or inside of, the lymphoid follicles. The cells next are attracted to the mantle zone where BCA-1 is deposited on dendritic elements such as the follicular dendritic cell tips. Also, CXCR5+CD4+CD57+ follicular T-helper cells (T_{FH}) are attracted to the follicle by similar interactions. B cells are primed just outside of the lymphoid follicle by interaction with cognate T cells and antigen-presenting cells and then reenter the follicle and become CCR7+ germinal center cells after interactions with follicular dendritic cells and T_{FH} cells (see **Figure 10.4**). The B cells thereafter may leave the follicle as memory or effector cells. (Adapted from Brandtzaeg, P., and Johansen, F. *Immunol. Rev.* 2005, 206:32–63. With permission from John Wiley & Sons.)

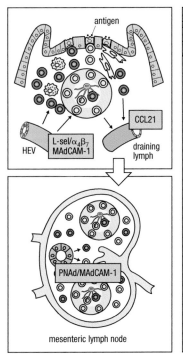

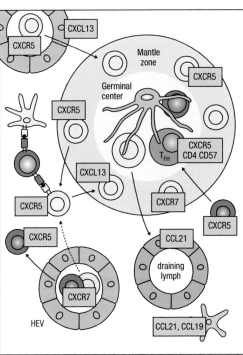

deposited on peripheral extensions of follicular dendritic cells after secretion by another cell type (see Figure 10.7), most prominently, germinal center dendritic cells (see Figure 10.4). However, both germinal center dendritic cells and large CXCL13-producing cells in inflammatory bowel disease-associated B-cell aggregates and lymphoid neogenesis exhibit a phenotype consistent with macrophage derivation.

The germinal center reaction induced by the constant challenge of the commensal gut microbiota may involve BCR-independent polyclonal stimuli via innate receptors to generate a restricted IgA repertoire with broad reactivity but relatively low affinity. Such a dual model for anti-microbial IgA responses may be adequate for the host's coexistence with the indigenous bacteria while providing protection against pathogens. Ongoing polyclonal activation of mucosal memory B cells in GALT also might help maintain a diverse SIgA antibody memory repertoire throughout life. Recent work highlights the reactivity of circulating IgG and IgA in humans with their intestinal microbiome. Sorting of antibody-bound intestinal bacteria and high throughput 16S DNA sequencing has resulted in the detection of distinct populations of bacteria recognized by some of these antibodies. Microbial antigen arrays have also been used to identify prominent microbial antigens differentially recognized by persons with inflammatory bowel disease. Some bacteria such *Faecalibacteria* may be more likely to promote humoral responses, whereas others such as *Bifidobacteria* spp. in breastfed infants and *Bacteroides fragilis* may limit local inflammation and enhance responses to mucosal and systemic vaccines. The mechanisms by which specific microbiota modulate local host responses is an area of active investigation.

10.7   Exit of activated B cells from germinal centers is controlled by specific factors

The exit of the activated B cells from the germinal centers is likely controlled by chemokines, and the actual cues, particularly ligands for CCR7, appear to be extrafollicular (see Figure 10.7). Thus, germinal center B cells downregulate CXCR5 and upregulate CCR7, which in experiments with mice has profound consequences for cellular positioning. Most MALT-induced sIgD⁻IgM⁺CD38⁻ putative memory B cells continuously emigrate from germinal centers to extrafollicular sites such as the tonsillar crypt epithelium or M cell pockets in Peyer's patches, where they presumably present recall antigens to cognate memory T cells. Likewise, most plasma cell precursors (CD20⁻CD38^{hi}) become rapidly dispersed in extrafollicular compartments or migrate via lymph and blood to distant effector sites, where they undergo terminal differentiation to plasma cells (see Figure 10.4), a process that depends on downregulation of the transcription factor Bcl-6 and upregulation of B lymphocyte-induced maturation protein-1 (Blimp-1). The egress of lymphoid cells to draining lymph is regulated by sphingosine-1-phosphate but apparently not by a chemotactic process.

10.8   IgA responses can occur in the absence of cognate B-cell–T-cell interaction

Antigens from the gut are rich in potential T-cell independent antigens, such as the polysaccharides abundant in the Gram-positive and Gram-negative bacteria. Mouse B cells can be activated by the type 1 T-cell-independent bacterial antigen LPS that can induce polyclonal B-cell proliferation through ligation of TLR4 in the absence of any form of T-cell help. Unlike mice, human B cells do not express TLR4 or TLR5, and thus do not respond to bacterial LPS or flagellin, respectively. However, human B cells express TLR9, binding of which by its ligand hypomethylated bacteria DNA elicits proliferation and antibody production, as does binding of viral nucleic acid ligands to TLR7/8. However, engagement of lumenal bacterial antigens with repeating subunits may cross-link the BCR and also promote B-cell differentiation with other local soluble factors.

The germinal center response is dependent on signaling between CD154 (CD40L) on T cells and CD40 on B cells. Despite the absence of germinal centers in individuals with X-linked hyper-IgM syndrome who have mutations in CD154, these individuals do have IgA$^+$ plasma cells in the intestinal lamina propria. CD40 knockout mice also have IgA plasma cells in the gut. In CD40 knockout mice, IgA class switch occurs only in MALT, but is independent of the germinal center response. Human and mouse GALT are known to express extrafollicular AID outside the context of the germinal center. In humans, AID expression has been observed in the cytoplasm of large interfollicular B cells in GALT, where it is considered to be in a storage compartment, but AID function requires access to the nucleus for DNA deamination for somatic hypermutation and class-switch recombination. The location of cells expressing nuclear AID outside the germinal center remains unknown. Thus, T-cell independent B-cell activators and IgA switch factors have been identified, but the nature of the B-cell proliferation stimuli that could underpin and support extrafollicular recombination event are poorly defined.

INSTRUCTIONS INVOLVED IN INDUCING GALT B CELLS TO MIGRATE TO EFFECTOR SITES

Migration to different anatomical destinations is a feature of mucosal B-cell responses, allowing cells to migrate or home from local inductive to appropriate effector sites. The ability of the GALT-derived B cells in blood to home via the lymphatics to the intestinal lamina propria is imprinted on B cells through binding of the vitamin A derivative retinoic acid to the B-cell retinoic acid receptor (RAR). Retinoic acid is generated by dendritic cells and macrophages exposed to pathogen-associated molecular patterns from the intestinal lumen via the activity of induced retinal dehydrogenase 1 (RALDH1). RAR activation upregulates the expression of the integrin $\alpha_4\beta_7$ that mediates lymphocyte homing to the gut lamina propria by specific binding to its ligand mucosal addressin cell adhesion molecule-1 (MAdCAM-1), constitutively expressed by mucosal postcapillary endothelial cells. Binding of retinoic acid to B-cell RAR also results in the induction of chemokine receptor CCR9 which mediates preferential homing and retention in the small intestinal mucosa, because epithelial cells produce the CCR9 ligand CCL25. Migration of plasma cells to different destinations involves complex combinations of chemokines and their receptors to guide the plasmablasts to the effector sites. For example, B-cell migration to the colon is favored by the expression of CCR10 on B cells controlled by the production by colonic epithelium of CCL28 but is also dependent on high expression of the homing receptor $\alpha_4\beta_7$ to bind to colonic endothelial MAdCAM-1 to gain entry. These motifs allow plasmablasts to migrate to the colonic microenvironment.

10.9 Activation of mucosal B cells and their maturation to IgA-producing plasma cells outside of GALT is controversial

The notion that the gut lamina propria could be a site of induction of B-cell responses gained support from three observations in mice. First, B-1 cells not associated with follicular B-cell structures contribute to the intestinal plasma cell population. Also, mice that lack Peyer's patches and isolated lymphoid follicles do not have IgA$^+$ plasma cells in the gut, implying that follicular structures are required for all B-cell lineage routes to IgA production. Second, lamina propria dendritic cells can acquire antigen by extending their dendrites between epithelial cells to sample lumenal contents without involving the FAE. Third, isolated lamina propria cells include B cells that express AID (AID would be required for IgA class-switch recombination).

10.10 Evidence indicates that the switch event to human IgA1 or IgA2 production may occur outside of GALT

The mucosa along the intestinal tract varies in the percentage of plasma cells that produce IgA1 and IgA2. The proportion of IgA2-producing plasma cells is greatest in colon, likely the consequence of one of the following processes. First, B cells may class switch to IgA1 and IgA2 to different degrees in MALT, and then home differentially to mucosal sites. In this connection, colonic and small intestinal epithelium produce different cytokine profiles, possibly guided by developmental programming or induction by factors such as MAMPs in the local microenvironment. Second, the switch to IgA2 may occur once the plasmablasts have extravasated into their effector sites. In this context, signaling through TLR5 in epithelium induces the local production of IgA switch factor APRIL. APRIL may mediate sequential IgA class switch to IgA2 in the lamina propria of areas rich in TLR5 ligands such as the colon. This model is elegant and has considerable support but has been disputed because class-switch recombination requires cell division, which has rarely been convincingly shown in lamina propria cells, and the local expression of AID remains controversial.

10.11 B cells in gut lamina propria do not divide before plasma cell differentiation

B-cell division has been proposed to occur at effector sites prior to terminal plasma cell differentiation. Early animal studies suggested that local amplification of relevant B-cell responses occurred in response to local challenge, but isolated lymphoid follicles might be the source of clonal proliferation, since they would still be present after Peyer's patches had been dissected from mouse gut, thus falsely attributing local expansion to gut lamina propria cells. Local clones of plasma cells can be detected in human lamina propria by polymerase chain reaction (PCR)-based methods that identify clones of cells by sequence identity in the junctional regions of the immunoglobulin variable region genes generated by gene rearrangement. However, methods based on clonal amplification of DNA or RNA by PCR may be unreliable in this context because identical PCR products may be derived by PCR rather than from adjacent related dividing cells.

In addition, clonally related plasma cells in lamina propria may have disseminated laterally from local GALT structures, or they may be widely disseminated clonal progeny of distant GALT germinal center cells. However, arguing against the notion that plasmablasts terminally differentiate in the lamina propria is the absence of Ki-67$^+$ (dividing) IgA$^+$ cells in the lamina propria. These results are extremely powerful because Ki-67 is expressed by all dividing cells, and importantly, all the tissue sections contain positive cells in the crypts of Lieberkühn or colonic glands. Although plasma cells do not divide, human B cells may home from the germinal center to the lamina propria and without further division differentiate into mature plasma cells under the influence of activated CD4$^+$ T cells in the effector site.

SUMMARY

The extensive mucosal plasma cell population is derived from B-cell responses in MALT initiated by antigens taken up directly from the external environment. The antigens are acquired from the mucosal surface via the FAE and stimulate B-cell responses through T–dependent and T–independent mechanisms that initiate extensive cell division, diversification of antigen-binding capacity by somatic hypermutation, and commitment to IgA production by class-switch recombination. The subsequent dissemination and localization of the IgA responses to the mucosae are dependent on regional or compartmentalized migration regulated by lymphocyte homing receptors with affinity for local endothelial ligands and chemokines secreted by mucosal or glandular epithelia. The mucosal B-cell responses are orchestrated by antigen-activated dendritic cells in MALT that support T-cell–dependent and T-cell–independent routes to IgA class-switch recombination in mucosal B cells and induce molecular mediators of regional homing. The interdependent regulation of the microbiome and mucosal B cells and antibodies is emerging as a key paradigm in understanding immune development and primary protection against disease at these sites.

FURTHER READING

Barone, F., Patel, P., Sanderson, J.D. et al.: Gut-associated lymphoid tissue contains the molecular machinery to support T-cell-dependent and T-cell-independent class switch recombination. *Mucosal Immunol.* 2009, 2:495–503.

Brandtzaeg, P.: Mucosal immunity: Induction, dissemination, and effector functions. *Scand. J. Immunol.* 2009, 70:505–515.

Brandtzaeg, P., and Johansen, F.E.: Mucosal B cells: Phenotypic characteristics, transcriptional regulation, and homing properties. *Immunol. Rev.* 2005, 206:32–63.

Brandtzaeg, P., Kiyono, H., Pabst, R. et al.: Terminology: Nomenclature of mucosa-associated lymphoid tissue. *Mucosal Immunol.* 2008, 1:31–37.

Chorny, A., Puga, I., Cerutti, A.: Innate signaling networks in mucosal IgA class switching. *Adv. Immunol.* 2010, 107:31–69.

Gustafson, C.E., Higbee, D., Yeckes, A.R., Wilson, C.C., DeZoeten, E.F., Jedlicka, P., and Janoff, E.N.: Limited expression of APRIL and its receptors prior to intestinal IgA plasma cell development during human infancy. *Mucosal Immunol.* 2014, 7:467–477.

Hapfelmeier, S., Lawson, M.A., Slack, E. et al.: Reversible microbial colonization of germ-free mice reveals the dynamics of IgA immune responses. *Science* 2010, 328:1705–1709.

Hooper, L.V., and Macpherson, A.J.: Immune adaptations that maintain homeostasis with the intestinal microbiota. *Nat. Rev. Immunol.* 2010, 10:159–169.

Landsverk, O.J., Snir, O., Casado, R.B. et al.: Antibody-secreting plasma cells persist for decades in human intestine. *J. Exp. Med.* 2017, 214:309–317.

Lycke, N.Y., and Bemark, M.: The regulation of gut mucosal IgA B-cell responses: Recent developments. *Mucosal Immunol.* 2017, 10:1361–1374.

Macpherson, A.J., McCoy, K.D., Johansen, F.E. et al.: The immune geography of IgA induction and function. *Mucosal Immunol.* 2008, 1:11–22.

Maul, R.W., and Gearhart, P.J.: AID and somatic hypermutation. *Adv. Immunol.* 2010, 105:159–191.

Reboldi, A., Arnon, T.I., Rodda, L.B. et al.: IgA production requires B cell interaction with subepithelal dendritic cells in Peyer's patches. *Science* 2016, 352:aaf4822.

Spencer, J., Barone, F., Dunn-Walters, D.: Generation of immunoglobulin diversity in human gut-associated lymphoid tissue. *Semin. Immunol.* 2009, 21:139–146.

Yeramilli, V.A., and Knight, K.L.: Requirement for BAFF and APRIL during B cell development in GALT. *J. Immunol.* 2010, 184:5527–5536.

Zhao, Y., Uduman, M., Siu, J.H.Y. et al.: Spatiotemporal segregation of human marginal zone and memory B cell populations in lymphoid tissue. *Nat. Commun.* 2018, 9(1):3857.

Secretory immunoglobulins and their transport

11

CHARLOTTE S. KAETZEL, JIRI MESTECKY, AND JENNY M. WOOF

The early demonstration of the presence of antibodies in external secretions of the gastrointestinal tract and later in milk, saliva, tears, respiratory and genital tract secretions, and urine stimulated extensive studies of their structure, function, and induction following immunization. In contrast to plasma, external secretions contain lower levels and a markedly different immunoglobulin (Ig) isotype distribution of antibodies with characteristic structural and functional properties. The dominant Ig isotype in tears, nasal secretions, saliva, milk, and intestinal fluid is IgA, mainly in the form of secretory IgA (SIgA) (Table 11.1). By contrast, IgG predominates in secretions of the lower respiratory tract and genitourinary tracts of both females and males. Immunoglobulin levels vary between different external secretions and reflect the dominant source of Ig (local production or plasma). In addition, there can be pronounced variability in Ig levels within the same secretion, related to the method of collection, processing of the sample, expression of relevant receptors on epithelial cells involved in Ig transport, and distinct regulatory mechanisms that affect Ig class switching to different isotypes. The Ig content of mucosal secretions is also influenced by hormones (e.g., during the menstrual cycle in the female genital tract), nutritional status, age of the individual, damage of the mucosal barrier due to infection and inflammation, and other factors, including interactions of host tissues with the microbiota that colonize mucosal surfaces.

FEATURES OF SECRETORY IMMUNOGLOBULINS

All isotypes of Igs are present in external secretions, albeit at markedly different concentrations. Thus, all Ig isotypes can be considered as secretory, because their entry into the lumen is largely determined by active transport mechanisms across epithelia. In health, paracellular movement of Igs is negligible because antibodies are too large to passively diffuse into the lumen. However, in inflammation when barrier function breaks down, macromolecules such as serum IgG and albumin are readily detectable in the lumen overlying all epithelia.

11.1 Relative distribution and molecular forms of Ig isotypes in external secretions display marked differences in comparison with plasma

Immunoglobulins of all isotypes are present in external secretions; however, their relative distribution and molecular forms display marked differences in comparison with plasma. IgA occurs in plasma in a monomeric,

TABLE 11.1 LEVELS OF IMMUNOGLOBULINS IN HUMAN EXTERNAL SECRETIONS

Fluid	IgM (μg mL^{-1})	IgG (μg mL^{-1})	IgA (μg mL^{-1})	IgA1 (%)	IgA2 (%)
Tear	0–18	Trace–16	80–400	63	37
Nasal secretions	0	8–304	70–846	71	29
Parotid saliva	0.4	0.4 2–5	15–39 120–319	ND	ND
Whole saliva	60–64	26–42	99–206	56	44
Bronchoalveolar fluid	0.1	13	3	67	33
Colostrum	610	100	12,340–53,800	40–77	23–59
Milk	50–340	40–168	470–1632	52–65	35–48
Duodenal fluid	207	104	313	ND	ND
Jejunal fluid	2	4–340	32–276	ND	ND
Colonic fluid	Trace–860	1	240–827	40	60
Intestinal fluid	8	4	166	70	30
Rectal fluid	30 49	0.9 379 297	143 3624[a] 2317[b]	30 34	70 66
Urine	ND	0.06–0.56	0.1–1.0	ND	ND
Pre-ejaculate	ND	0.6	1.7	ND	ND
Ejaculate	0–8	16–33	11–23	83	17
Uterine cervix	5–328	1–1200	3–330	50	50
Vaginal fluid	16	10–467	21–118	42	58

[a] Dry.
[b] Wet.
Source: From Jackson, S. et al. in Mestecky, J. et al. (eds), *Mucosal Immunology*, San Diego, CA: Elsevier, 2005:1829–1839; Moldoveanu, Z. et al. *J. Immunol.* 2005, 175:4127–4136; Pakkanen, S. et al. *Clin. Vaccine Immunol.* 2010, 17:393–401.
Abbreviation: ND, Not determined.

four-chain form with two heavy (H) α chains and two light (L) κ or λ chains (**Figure 11.1a–d**). Only a small proportion (approximately 1%–5%) of IgA in plasma consists of polymeric (p) forms with two to four monomeric IgA molecules covalently linked by disulfide bridges between the α chains of monomeric IgA and an additional joining (J) chain (**Figure 11.1e**). In typical external secretions such as intestinal fluid, milk, tears, and saliva, the dominant form of IgA is SIgA, which comprises two or four monomeric IgA molecules, one molecule of J chain, and an additional glycoprotein, secretory component, acquired during the receptor-mediated transport of J chain-containing pIgA through the epithelial cells (see later). The proportion of dimeric (two monomeric IgA molecules) to tetrameric (four monomeric IgA molecules) SIgA in saliva and milk is approximately 3:2. The structure of

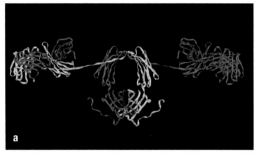

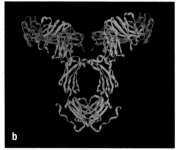

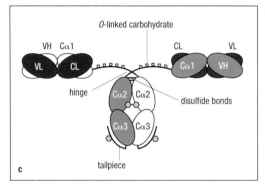

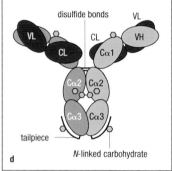

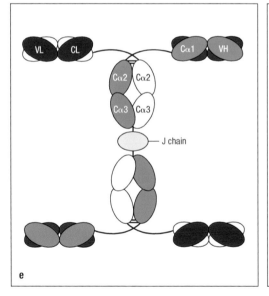

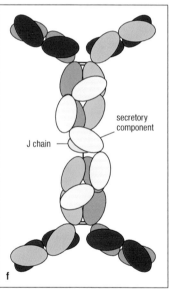

Figure 11.1 **Structure of human IgA molecules.** (**a** and **b**) Ribbon diagrams of molecular models based on solution scattering studies showing the course of the backbones of the polypeptide chains. (**a**) IgA1. (**b**) IgA2. Schematic representations of these structures, (**c**) IgA1 and (**d**) IgA2, illustrate the four-chain composition and the separate domains comprising each chain. Domains in each polypeptide are similarly colored. *N*-linked and *O*-linked carbohydrates are shown in turquoise, as hexagons and small squares, respectively. (**e**) Dimeric IgA (in this case of the IgA1 subclass) where the IgA monomers are linked via their tailpieces to J chain. (**f**) Secretory IgA (in this case of the IgA2 subclass). The five globular domains of secretory component, a cleavage product of the polymeric immunoglobulin receptor, form an integral part of this molecule.

secretory IgA is shown in Figure 11.1f. A variable fraction of IgA in external secretions, particularly in bile, male and female genital tract secretions, and urine, occurs as monomeric IgA and pIgA devoid of secretory components. Studies of the structures of SIgA, pIgA, and monomeric IgA have revealed the dimensions and shapes of these molecules as shown in Figure 11.1.

Many species have a single IgA gene. However, humans and most hominoid primates have two IgA subclasses (IgA1 and IgA2) encoded by two distinct constant region genes. IgA1 and IgA2 display characteristic structural differences on their α1 and α2 H chains and exhibit distinct biological activities (Figures 11.1 and 11.2). In plasma, approximately 84% of the molecules belong to the IgA1 and approximately 16% to the IgA2 subclasses. The proportions of IgA subclasses in external secretions display a characteristic distribution with, for example, nasal fluid dominated by IgA1, while large intestinal fluid contains more IgA2 than IgA1 (see Figure 11.2).

In addition to IgA of various molecular forms, external secretions contain IgG, IgM, and trace amounts of IgD and IgE, depending on the

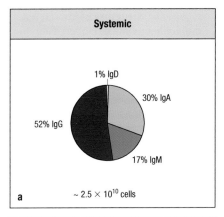

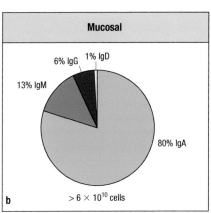

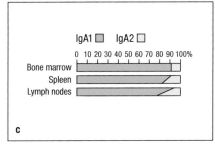

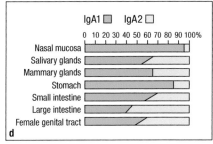

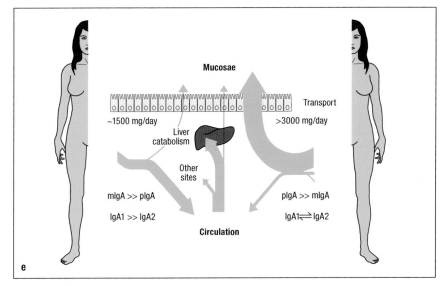

Figure 11.2 Comparative distribution and properties of IgA-producing cells in systemic and mucosal compartments and metabolism of IgA. (**a** and **b**) Relative distribution of cells producing IgA, IgG, IgD, or IgM in the systemic compartment, which is dominated by IgG, and the mucosal compartment, which is dominated by IgA. Note that the total number of Ig-producing cells is higher in the mucosal compartment ($>6 \times 10^{10}$ cells) than in the systemic compartments ($\sim2.5 \times 10^{10}$ cells). (**c** and **d**) Relative distribution of cells producing the IgA1 and IgA2 subclasses in systemic and mucosal tissues. (**e**) Biosynthesis, properties, mucosal transport, and sites of catabolism of monomeric IgA (mIgA), polymeric IgA (pIgA), and IgA1 and IgA2 in humans.

fluid examined. IgM appears in secretions in its secretory form (SIgM) with secretory component acquired during the transepithelial transport mediated by the polymeric Ig receptor (pIgR), shared by both pIgA and IgM (see later). However, the level of SIgM is substantially lower than that of SIgA with the exception of individuals with selective IgA deficiency; in these individuals, SIgM may replace and functionally compensate for deficient SIgA. The level and relative proportion of IgG to IgA varies among secretions, with low levels of IgG in the distal intestinal fluid of the small intestine and colon, milk, tears, and saliva, but dominance of IgG over IgA in the urine, semen, and cervicovaginal secretions collected at various phases of the menstrual cycle. Monomeric IgG is actively transported into external secretions by the neonatal Fc receptor (FcRn) and can also diffuse paracellularly following damage to the epithelial barrier. IgE, which is transported across mucosal epithelial cells by FcεRII (CD23), has been detected in saliva and respiratory and intestinal secretions in extremely low concentrations. Nevertheless, it may participate in immune defense in parasitic infections and in local Type 1 hypersensitivity reactions in allergic individuals and in these circumstances can achieve very high levels. Low levels of IgD are found in human milk, but a specific transport mechanism has yet to be defined.

11.2 Distribution of mucosal immunoglobulins differs between species

Significant differences in the isotypes, their levels, and their molecular forms exist among various vertebrate species, and these must be considered in the design of experiments, their interpretation, and their relevance to the mucosal immune system of humans. Common laboratory animals such as mice and rats have only a single IgA isotype that structurally resembles human IgA2 (in that it lacks a hinge region cleavable by bacterial IgA1 proteases—see later), and in their plasma almost all IgA occurs in the polymeric form. In contrast, the rabbit genome encodes more than 13 IgA subclasses but only one IgG class. Early milk (colostrum) and milk of species such as rodents, pigs, cows, horses, sheep, and goats contain IgG as the dominant isotype. Epithelial cells in the mammary glands of these species express FcRn, which mediates transport of IgG from the circulation into colostrum and milk. The presence of IgG in milk is of enormous functional significance for survival. Unlike humans and many other species, IgG is not transported through the placenta antenatally at significant levels in rodents, pigs, cattle, and horses, and consequently, the newborns of these latter species lack maternally acquired IgG in their circulation. Thus, without colostrum and milk, they succumb within several days to infections with common environmental microorganisms. Ingestion of milk and selective, receptor-mediated transfer of IgG from the gut lumen into the newborns' circulation is essential for their survival. Such lumen-to-blood transport of IgG across the polarized epithelial cell is mediated by FcRn via a process called transcytosis. During the neonatal life of rodents (mice and rats), the lumen of the intestine is highly acidic, and FcRn is expressed at extremely high levels by the intestinal epithelium. As discussed later, these acidic conditions are perfect for FcRn-mediated uptake of IgG and vectoral transport from an apical-to-basal direction across the simple, polarized epithelium. At the time of weaning (approximately 2 weeks of life), FcRn expression is downregulated nearly 1000-fold. Such developmentally regulated, functional expression of FcRn in rodents and other animals has led to naming this IgG receptor as neonatal. However, as discussed later, the expression of this receptor is not in the least restricted to neonatal life or to the epithelium of the mucosa, such that active low-level transport can be present throughout life as seen in humans, nonhuman primates, pigs, and likely other animals.

IgM, which is capable of contributing to mucosal defense, is present in all vertebrate species except primitive Agnathans. An evolutionary precursor of IgD is present in cartilaginous and bony fishes, and a true homolog to mammalian IgD is found in all higher vertebrates except birds. Other mucosal Ig isotypes have appeared independently during vertebrate evolution, including IgT in teleost fishes and IgX in amphibians. The IgY isotype present in amphibians, reptiles, and birds is homologous to both mammalian IgG and IgE, and homologs of mammalian IgA are present in mucosal secretions of crocodilians (alligators and crocodiles) and birds.

11.3 Biosynthesis of mucosal IgA is distinct from circulating IgA

In humans and several other species, the production of IgA ($\sim$66 mg kg^{-1} body weight/day) far exceeds the combined synthesis of all other isotypes of immunoglobulins. This reflects both the distribution into mucosal secretions and markedly different mechanisms of catabolism. The lower serum concentration of IgA in comparison with IgG is the result of the distribution of IgA produced for the mucosal versus systemic pools: approximately two-thirds of IgA is produced mainly in its polymeric form in mucosal tissues, especially the gut, and is selectively transported into external secretions; approximately one-third is produced as monomeric IgA in the bone marrow, lymph nodes, and spleen and enters the circulation. Furthermore, the half-life of IgA in the circulation is considerably shorter than that of IgG (approximately 5–6 days for IgA versus approximately 20–25 days for IgG in humans) due to the much faster catabolism that takes place mainly in the liver, with the hepatocyte-expressed asialoglycoprotein receptor playing the dominant role in IgA binding. IgG is in fact the longest lived of the serum proteins due to its protection from catabolism by FcRn expressed in the endothelium and hematopoietic system (mostly monocytes, dendritic cells, B cells, and macrophages), which recycles IgG internalized by pinocytosis away from a degradative fate in lysosomes. The numbers of lymphoblast and plasma cells in mucosal tissues also exceed the numbers of cells in the bone marrow, spleen, and lymph nodes, with a characteristic distribution of Ig isotypes (see Chapter 1).

In the bone marrow, IgA is produced almost exclusively as monomeric IgA with a pronounced dominance of IgA1-producing cells. In the lymph nodes and spleen, cells producing a small proportion of pIgA are present. The α and L (κ or λ) chains are produced on separate sets of polyribosomes and assembled into monomeric IgA molecules through several possible pathways. The glycosylation of the α chains is initiated with the attachment of N-acetylglucosamine (GlcNAc) on the polyribosome, continues in the Golgi apparatus, and is terminated during the intracellular passage from the Golgi apparatus to the cell surface before secretion. In cells engaged in pIgA production, the J chain is complexed with the monomeric IgA molecules to generate IgA dimers or higher molecular forms. It is not certain whether the J chain initiates or terminates Ig polymerization; pIgA and IgM molecules may be assembled in the absence of J chain. The incorporation of J chain and generation of polymers appear to occur as a last step, shortly before secretion. The absolute majority of intracellular IgA in cells producing pIgA is present as monomeric IgA and IgA half-molecules. However, the incorporation of J chain into pIgA is essential in the subsequent binding of pIgA to the epithelial pIgR (see later). Thus, almost all pIgA-producing cells in mucosal tissues are positive for intracellular J chain, while in the bone marrow they are negative. Secretory component, the cleaved extracellular domain of pIgR, remains attached to the J chain-containing pIgA following transepithelial transport to form the SIgA complex (see later). Thus, SIgA is a terminal product of two structurally and functionally distinct cell types: plasma cells producing pIgA with J chain, and epithelial cells producing secretory component. Several findings provide evidence for a relative independence of the systemic and mucosal compartments of the IgA system. Levels of SIgA in external secretions reach adult levels within the first year of life, while the plasma IgA levels reach adult levels at puberty. The distribution of cells producing IgA1 and IgA2 in mucosal tissue parallels the distribution of these two subclasses in corresponding external secretions. Mucosal, especially oral, immunization with microbial antigens induces SIgA responses in external secretions but low levels of IgA responses in plasma. In contrast, systemic immunization with antigens that induce initially dominant IgA responses in plasma (e.g., bacterial polysaccharide vaccines) does not induce parallel vigorous responses in external secretions. Finally, intravenous injection of radiolabeled pIgA or pIgA myeloma proteins present in high levels in plasma of patients with IgA myeloma does not result in an efficient appearance of IgA in external secretions. The latter point is of considerable importance because it indicates that pIgA specific for a desired antigen (e.g., human immunodeficiency virus-1 [HIV-1] or influenza virus) administered by a systemic route may be unlikely to lead to elevated levels of SIgA in external secretions.

11.4   IgA exists as two subclasses, IgA1 and IgA2

As noted, structural and genetic studies have shown that IgA exists as two subclasses in humans and hominoid primates. Comparative structures of IgA from human and other species indicate that IgA2 represents the phylogenetically older form and that IgA1 was generated by the insertion of a gene segment encoding the IgA1 hinge region. The structural difference between $\alpha 1$ and $\alpha 2$ H chains is minimal except for 13 amino acid residues in the IgA1 hinge region. In addition, IgA1 and IgA2 display differences in the number, composition, and types of glycosidic bonds, and in the sites of attachment of their glycan side chains (see Figure 11.1). The proportion of IgA1 to IgA2 in various body fluids mirrors the distribution of IgA1- and IgA2-producing cells in corresponding tissues. Furthermore, studies relating antibody responses to different types of antigens to their IgA subclass have revealed marked differences. For example, antibodies specific for the influenza virus, HIV-1, or common protein food antigens are associated predominately with the IgA1 isotype, while antibodies to bacterial polysaccharide, lipopolysaccharide (LPS), and lipoteichoic acid antigens are mainly present in the IgA2 subclass.

The longer hinge of IgA1 results in a greater distance between the antigen-binding sites formed by the pairing of VH and VL domains than in IgA2 (see Figure 11.1a and b). Hence, monomeric IgA1 is equipped for bivalent binding to antigens spaced at greater distances apart, which may offer advantages over IgA2 and other Ig isotypes in terms of high-avidity recognition of certain antigens. In turn, increased avidity is likely to result in longer residency of IgA1 on the antigenic surface, providing increased opportunities for the triggering of effector functions that will eliminate the antigenic target being recognized. Other functional differences between IgA1 and IgA2 include their susceptibility to certain bacterial proteases (see later), inhibition of bacterial adherence mediated by IgA-associated glycans, and the ability to bind other proteins (e.g., fibronectin, lactoferrin, and a broad variety of enzymes) and interact with lectins. Structural aberrations in IgA1 molecules (altered glycan moiety or aggregation) may result in the conversion of IgA from a non-complement-activating Ig into a complement activator (lectin and alternative pathway) and in the formation of pathogenic immune complexes as demonstrated in the common glomerulonephritis known as IgA nephropathy (see Chapter 32).

11.5   IgA is heavily glycosylated with important functional consequences

Analyses of IgA-associated glycans indicate that 6%–10% of total molecular mass is contributed by *N*- and *O*-linked side chains that display a remarkable heterogeneity in their number and composition (see Figure 11.1c and d). The most striking difference is the presence of three to five short *O*-linked glycans in the IgA1 hinge region composed of *N*-acetylgalactosamine (GalNAc), galactose (Gal), and sialic acid. The total glycan content of SIgA is even higher than that of serum monomeric IgA due to the presence of glycans on J chain and especially on secretory component, which is particularly rich in *N*-linked glycans (approximately 22% of the total molecular mass). Glycans appear to play an important role in IgA catabolism, the ability to activate the lectin pathway of complement, and binding to bacteria with consequent inhibition of their adherence to epithelial cells (see later).

11.6   J chain is a unique protein that links together monomeric IgA to create polymeric IgA

The human gene for J chain lies on chromosome 4. It encodes a 15–16 kDa polypeptide, unrelated to any other known protein. J chain is rich in acidic amino acids and contains eight Cys residues, six of which form conserved intrachain disulfide linkages. It is very highly conserved (approximately 70%) across a wide range of species, from mammals through birds and reptiles to amphibians and fishes. J chain is produced in lymphocytes of the B-cell lineage from early stages of their differentiation (pro-B and pre-B cells). While a detailed three-dimensional structure is still awaited, a number of models of J chain conformation have been proposed. These include a two-domain structure, with *N*-terminal β-sheets and *C*-terminal α-helical segments, a single-domain β-barrel structure reminiscent of an Ig VL domain, and an alternative two-domain model with an *N*-terminal β-barrel domain and a *C*-terminal domain combining both α-helices and β-strands that takes into account the disulfide bond pairing arrangement.

Early electron micrographs revealed dimeric IgA to have a double-Y shape. The dimensions were consistent with an arrangement where the two IgA monomers were linked via the tips of their Fc regions, and sedimentation and viscosity experiments as well as more recent mutagenesis and molecular modeling studies support such an "end-to-end" arrangement of the monomers with J chain interposed (see **Figure 11.1e**). The arrangement is stabilized by disulfide bridges linking the so-called tailpiece regions of the two IgA monomers to J chain. Each tailpiece, a highly conserved stretch of 18 amino acids at the extreme *C*-terminus of the α H chain, carries a cysteine as its penultimate residue. One of these cysteines in each IgA monomer forms a disulfide bridge to one of the two Cys residues in J chain not involved in intrachain disulfide bonds (Cys14 and Cys68). *N*-linked sugars on both the tailpiece and J chain appear to influence dimer formation, and the domains of IgA Fc also seem to play some role in determining the efficiency of dimerization. The fact that some J chain epitopes are not recognized in polymeric IgA suggests that parts of the Fc may partially obscure J chain. However, J chain is easily released from polymeric IgA by cleavage of interchain disulfide bonds, indicating that any noncovalent interactions between J chain and the Fc region are relatively weak. J chain is also produced by mucosal plasma cells that express IgM, IgD, and IgG, and is linked by disulfide bonds to two of the μ heavy chains in polymeric IgM. However, J chain does not form disulfide bonds with IgG or IgD due to the lack of penultimate Cys residues in γ and δ heavy chains.

EPITHELIAL TRANSCYTOSIS OF SECRETORY IMMUNOGLOBULINS

The single layer of columnar epithelial cells that form the linings of mucous membranes and exocrine glands are connected by tight junctions and other barrier structures that prevent paracellular transport of most solutes, especially large multisubunit protein complexes such as immunoglobulins (see Chapter 5). Locally synthesized polymeric IgA and IgM as well as monomeric IgG and IgE are therefore actively transported across mucosal epithelial cells into external secretions by specialized receptors. The receptor for polymeric IgA and IgM is cleaved at the luminal surface, and its extracellular domain becomes a functional part of the SIgA or SIgM complex. The receptors for IgG and IgE mediate multiple rounds of Ig transport in both directions (so-called bidirectional transcytosis) across mucosa epithelia.

11.7 Polymeric immunoglobulin receptor delivers polymeric IgA and IgM into mucosal secretions

The transport of polymeric immunoglobulins (pIgA and pIgM) across mucosal epithelial cells is mediated by a transmembrane glycoprotein called the polymeric immunoglobulin receptor (pIgR; **Figure 11.3**). The pIgR binds to the Fc region of polymeric antibodies and can thus be classified as a type of

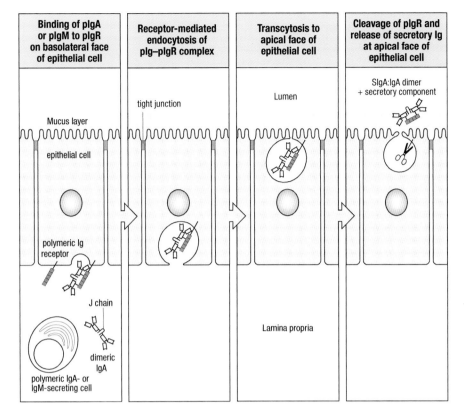

Figure 11.3 Transcytosis of polymeric IgA and IgM across epithelia is mediated by the polymeric Ig receptor. Polymeric IgA and IgM immunoglobulins (pIgs) are synthesized by plasma cells in the lamina propria of the gut, the respiratory epithelia, exocrine glands, and the lactating mammary gland. These J chain-containing pIgs, mainly dimeric IgA, diffuse across the basement membrane and are bound by the polymeric Ig receptor (pIgR) on the basolateral face of the epithelial cell. The pIg–pIgR complex is internalized by receptor-mediated endocytosis and undergoes transcytosis through a series of intracellular vesicles. At the apical face, proteolytic cleavage of pIgR releases SIgA or SIgM bound to secretory component, the soluble extracellular domain of pIgR. The residual cytoplasmic domain of pIgR is nonfunctional and is degraded. Oligosaccharide side chains on the secretory component moiety bind to mucins in the mucus layer, thus anchoring SIgA or SIgM to the epithelial surface. (Adapted from Murphy, K., and Weaver, C. *Janeway's Immunobiology*, 9th ed, New York, Garland Science, 2016.)

Fc receptor. The single gene encoding pIgR, which is located on chromosome 1 in humans and mice, first emerged in bony fishes and is present in all higher vertebrates. The requirement for pIgR for epithelial transcytosis of SIgA was demonstrated by the finding that pIgR-deficient mice have markedly reduced IgA in external secretions and elevated levels of serum IgA. The magnitude of pIgR-mediated antibody transport is greatest in the gut, resulting in delivery of approximately 3 g per day of SIgA into the intestinal fluids of the average adult. Lesser amounts of SIgM are transported into gut secretions due to the far greater abundance of IgA-producing plasma cells, although SIgM can partially compensate for loss of SIgA in IgA-deficient humans. The pIgR is synthesized as an integral membrane protein in the rough endoplasmic reticulum and is further modified in the Golgi apparatus, including extensive *N*-linked glycosylation. In the trans-Golgi network, pIgR is sorted into vesicles that deliver it to the basolateral surface of the epithelial cell, where it binds pIgA and pIgM secreted by plasma cells in the lamina propria underlying the epithelium. With or without bound pIg, pIgR is endocytosed and delivered through a series of sorting vesicles to the apical (luminal) membrane of the epithelial cell. During transcytosis, a disulfide bond forms between pIgR and one of the two subunits of dimeric IgA, resulting in a permanent association between the receptor and antibody. Once the pIg–pIgR complex reaches the apical plasma membrane, the extracellular domain of pIgR is proteolytically cleaved to form soluble secretory component. Cleavage of unoccupied pIgR results in release of free secretory component, while cleavage of pIgR bound to pIgA or pIgM results in release of SIgA or SIgM. Many external secretions such as colostrum contain significant amounts of free secretory component, which contributes to mucosal homeostasis by inhibiting the binding of certain microbes and modulating the activity of pro-inflammatory factors. The seven *N*-glycan chains of human secretory component have a unique structure that is heavily fucosylated and sialylated, similar to those of the antibacterial milk protein lactoferrin. These *N*-glycans bind a variety of host-, pathogen-, and environment-derived substances with lectin-like activity.

Carbohydrate-mediated binding of secretory component to intestinal mucins anchors free secretory component, SIgA, and SIgM to the mucus layer overlying the epithelium, thereby creating an immunologic barrier against infection. Secretory component also stabilizes SIgA by inhibiting the access of microbial proteases to the vulnerable IgA hinge region (see later).

In some animals, such as rodents and rabbits, SIgA can also be delivered into intestinal secretions via the hepatobiliary route. In these species, unlike humans, most of the circulating IgA is polymeric rather than monomeric. Much of this pIgA appears to originate from plasma cells in the gut lamina propria, where pIgA that is not transported across intestinal epithelial cells enters the blood circulation through the portal vein. In the liver, pIgR on the sinusoidal surface of hepatocytes binds pIgA and transports it to the bile canaliculus, analogous to the apical surface of mucosal epithelial cells. Proteolytic cleavage of pIgR causes release of SIgA into bile, which drains into the proximal small intestine. Because the pIgR in these species does not bind pIgM, antibodies of this isotype are not transported from blood to bile and remain in the circulation. Humans and other primates have evolved a distinct pathway for transport of secretory antibodies into bile. In these species, pIgR is expressed by epithelial cells lining the bile ducts but not by parenchymal hepatocytes. Plasma cells in the lamina propria underlying the bile duct epithelium secrete pIgA (and to a lesser extent, IgM), which is then delivered into bile by pIgR-mediated epithelial transcytosis.

Several unique features of the pIg–pIgR transcytosis pathway contribute to the immune functions of secretory antibodies (see **Figure 11.6**). During transport, intracellular vesicles containing pIg–pIgR complexes can fuse with endocytotic vesicles internalized from the apical surface of epithelial cells. This intracellular colocalization allows antibody-mediated intracellular neutralization of endocytosed microbes and antigens. Another important feature of pIgR is its ability to transport pIgA or pIgM bound to antigens or whole microbes from the lamina propria to the luminal surface of mucosal epithelial cells. The ability of pIgR to bind and transport large pIgA immune complexes demonstrates that cross-linking of the Fab fragments with antigen does not appreciably alter binding of pIgR to the Fcα-J chain segments of pIgA. Locally produced pIgA antibodies within the mucosa can therefore trap and excrete antigens derived from the environment, diet, commensal microbiota, or infectious microorganisms.

11.8 Expression of pIgR by mucosal epithelial cells is regulated by microbial and host factors

Because membrane-bound pIgR is proteolytically cleaved to a soluble secretory component with each round of SIgA or SIgM transport, epithelial expression of pIgR must be maintained at high levels to support the demand for delivery of mucosal antibodies into external secretions. The single polypeptide chain of pIgR is encoded by the *PIGR* gene on chromosome 1 in humans and mice. Expression of pIgR in epithelial cells is regulated by a variety of microbial and host factors (Table 11.2). In the gut, cross-talk between the commensal microbiota and host cells is crucial for maintenance of pIgR expression. Expression of pIgR is eight- to tenfold higher in the colon than in the small intestine of humans and mice, consistent with the much greater bacterial content in the colon. Germ-free mice have been shown to have underdeveloped intestinal lymphoid structures (see Chapter 19), and also have lower than normal expression of pIgR in intestinal epithelial cells. Colonization of germ-free mice with *Bacteroides thetaiotaomicron*, a prominent commensal of the normal mouse and human gut, restores expression of pIgR. Expression of pIgR is regulated by direct effects of microbes and their products on intestinal epithelial cells, as well as by

TABLE 11.2 FACTORS THAT REGULATE EXPRESSION OF THE POLYMERIC Ig RECEPTOR

Microbes/microbial factors	Constituents of the commensal microbiota, including but not limited to *Escherichia coli, Bacteroides thetaiotaomicron, Saccharomyces boulardii* Pathogenic microbes such as *Salmonella* and reovirus Microbial factors that signal through pattern recognition receptors, such as LPS and double-stranded RNA
Cytokines	IFN-γ, TNF-α, IL-1, IL-4, IL-17
Hormones	Estrogen, progesterone, androgens, glucocorticoids, prolactin

Abbreviations: IFN-γ, interferon-γ; IL, interleukin; TNF-α, tumor necrosis factor-alpha.

indirect effects of host factors produced in response to microbial stimulation. LPS shed by bacteria of the family Enterobacteriaceae, including both commensals and pathogens, upregulates transcription of the *PIGR* gene via signaling pathways initiated through an innate receptor for LPS termed Toll-like receptor 4 (TLR4) and activation of the transcription factor nuclear factor κB (NF-κB). Mice with a selective deficiency in the TLR adaptor protein MyD88 in epithelial cells have reduced pIgR expression and IgA transport in the colon, demonstrating the importance of TLR signaling for regulation of pIgR. Expression of pIgR has also been shown to be enhanced by butyrate, a bacterial fermentation product that has anti-inflammatory properties. Introduction of double-stranded RNA into epithelial cells, as a result of infection with intestinal viruses such as reovirus, upregulates *PIGR* gene transcription via signaling through TLR3, another Toll-like receptor, and activation of NF-κB as well as the interferon regulatory factor family of transcription factors. Thus, the rate of transport of secretory antibodies into intestinal fluids can be regulated by constituents of the normal gut microbiota as well as pathogens, and epithelial pIgR serves as a link between innate and adaptive immunity.

Cytokines produced by host cells, in response to colonization with commensal microbes or infection with pathogenic microbes, regulate expression of pIgR in mucosal epithelial cells. The pro-inflammatory cytokines interferon-γ (IFN-γ) and tumor necrosis factor-α (TNF-α) upregulate *PIGR* gene transcription via JAK-STAT and NF-κB signaling, respectively. Interleukin-1 (IL-1), which activates a MyD88-dependent signaling pathway that shares many elements with the TLR pathway, synergizes with IFN-γ and TNF-α. Another pro-inflammatory cytokine, IL-17, has been shown to enhance pIgR expression in respiratory epithelial cells, and likely contributes to pIgR regulation in the gut. A unique aspect of pIgR regulation is cooperativity between the helper T-cell T_H1-type cytokine IFN-γ and the T_H2-type cytokine IL-4, the effects of which are usually antagonistic. The ability of pIgR to be regulated by a wide range of host cytokines facilitates optimal transport of secretory antibodies in response to the broad spectrum of microbial and environmental stimuli encountered at mucosal surfaces.

Paradoxically, expression of pIgR in intestinal epithelial cells is downregulated during intestinal inflammation associated with inflammatory bowel disease (IBD) in humans and experimental colitis in mice. This loss of pIgR expression is associated with accumulation of IgA in the intestinal lamina propria and increased levels of circulating IgA, which may lead to increased exposure of the systemic immune system to intestine-derived antigens. Decreased expression of pIgR is also seen in dysplastic epithelium and carcinomas of the gastrointestinal and respiratory tracts. A common feature of these conditions is rapid proliferation of relatively undifferentiated epithelial cells, which may not respond properly to microbial and host factors that maintain pIgR expression under homeostatic conditions and enhance pIgR expression during infections.

Expression of pIgR is regulated by hormones in a cell-type specific manner. The effects of estrogen and progesterone are antagonistic in endometrial epithelial cells, and thus the expression of pIgR and transport of IgA varies during the estrous cycle. Androgens have been shown to upregulate pIgR expression in both male and female reproductive tissues. The lamina propria underlying the epithelium of the genitourinary tract contains IgA-secreting plasma cells, albeit at lower density than in nasal and intestinal mucosae, and SIgA in genital secretions has been shown to contribute to immunity against sexually transmitted diseases. The polypeptide hormone prolactin upregulates pIgR expression in mammary gland epithelial cells during pregnancy and lactation, thus enhancing the delivery of free secretory components and SIgA into colostrum and milk. The lamina propria of the mammary gland is populated by pIgA-secreting plasma cells that were originally stimulated by antigens in the gut-associated and nasopharynx-associated lymphoid tissues. Newborn mice that are the offspring of pIgR-deficient dams, which lack SIgA antibodies in breast milk, have an altered composition of the commensal gut microbiota compared with the offspring of pIgR-sufficient dams, and these changes persist after weaning and into adulthood. Thus, SIgA antibodies transported by pIgR across the mammary gland epithelium into breast milk provide protection against infectious agents and exogenous antigens in the shared environment of mother and child and also shape the development of the intestinal microbiota in the breastfed infant.

11.9   Bidirectional transport of IgG and IgE across mucosal epithelia is mediated by specialized Fc receptors

IgG is transported across polarized mucosal epithelial cells by the neonatal FcRn, so named for its high level of expression by intestinal epithelial cells of neonatal rats and mice. Initially characterized for its role in the metabolism of circulating IgG, FcRn (then known as the "Brambell receptor" for its discoverer) was shown to protect internalized IgG from intracellular degradation in a variety of cell types. FcRn was subsequently shown to mediate transfer of IgG from the maternal to the fetal circulation in pregnant women, via transport across placental endothelial cells. The importance of this receptor for mucosal immunity was suggested by the discovery that FcRn could transport maternal milk-derived luminal IgG antibodies across the intestinal epithelium of neonatal rodents and into the systemic circulation. Whereas FcRn is expressed in intestinal epithelial cells of rats and mice only during the neonatal period, in humans it is expressed throughout life in enterocytes and a variety of other cell types, including epithelial cells of the respiratory tract, mammary gland, skin, kidneys, and eyes, as well as endothelial cells and hematopoietic cells (monocytes, macrophages, dendritic cells, neutrophils, and B cells). FcRn is a major histocompatibility complex (MHC) class I-like molecule with a transmembrane α heavy chain containing three extracellular domains that is noncovalently bound to soluble β_2-microglobulin (**Figure 11.4a**). The α heavy chain is encoded by the *FCGRT* gene outside of the MHC gene locus, on chromosome 19 in humans and chromosome 7 in mice. A major difference between FcRn and classical MHC class I proteins is that the α_1 and α_2 domains do not fold to form a peptide-binding groove, rather forming a platform that contacts the Fc region of IgG at the interface of the Cγ2 and Cγ3 domains (**Figure 11.4b**). Evidence from *in vitro* binding studies suggests that high-affinity binding of IgG requires dimerization of two FcRn molecules.

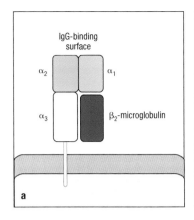

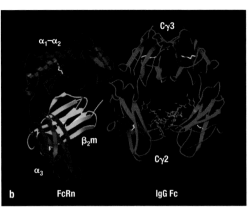

Figure 11.4 **Structure and IgG binding of FcRn.** Shown schematically in (**a**) FcRn is an MHC class I–like molecule with a membrane-spanning α chain bound noncovalently to β_2-microglobulin. (**b**) A ribbon diagram of a molecule of FcRn (blue and green) bound to one chain of the Fc region of an IgG molecule (red, note that the hinge region and Fab arms of IgG are not shown). The α_1 and α_2 domains of the FcRn α chain fold to form a surface that binds at the interface of the Cγ2 and Cγ3 domains of IgG. The dark blue structures attached to the Cγ3 domain of IgG are oligosaccharide side chains. Although only one molecule of FcRn is shown, high-affinity binding of IgG requires dimerization of FcRn such that FcRn binds to IgG with a 2:1 stoichiometry. β_2m, β_2-microglobulin. (Courtesy of P. Björkman.)

In polarized epithelial cells, FcRn directs IgG through a complex intracellular itinerary that can result in either apical-to-basolateral or basolateral-to-apical transcytosis (Figure 11.5). This pathway differs in three important ways from pIgR-mediated epithelial transcytosis of polymeric IgA and IgM. First, whereas pIgR-mediated transport is strongly biased in the basolateral-to-apical direction, FcRn-mediated transport can deliver IgG to both surfaces, with important functional consequences for mucosal immunity (see later). Second, FcRn is not cleaved at the plasma membrane like pIgR and can mediate multiple rounds of IgG transport. Third, binding of FcRn is optimal at a slightly acidic pH of around 6.0, thus favoring association with IgG within acidic intracellular compartments and release of IgG at the neutral pH of the plasma membrane. An exception to this rule is the luminal plasma membrane of neonatal rodent enterocytes, where the naturally acidic environment favors binding of IgG to FcRn and uptake by receptor-mediated endocytosis. It is possible that this cell-surface FcRn–IgG binding also may occur at the surface of other epithelial cells within localized acidic microenvironments such as may occur during inflammation. However, for most cell types, IgG is primarily internalized via fluid-phase pinocytosis and delivered to acidic endosomal compartments where binding to FcRn occurs. Consistent with this, the majority of FcRn that is expressed within the epithelium is contained intracellularly within endosomes. Once bound, FcRn diverts monomeric IgG away from lysosomes and into a series of transport and recycling vesicles for delivery to the basolateral and apical surfaces. In nonepithelial cells such as endothelium and hematopoietic cells, recycling vesicles carry FcRn–IgG complexes back to the plasma membrane for release into interstitial fluids or the blood circulation. The ability of FcRn to protect internalized IgG from lysosomal degradation is the mechanism through which this receptor inhibits IgG catabolism and increases the half-life of circulating IgG. Interestingly, FcRn also binds serum albumin and appears to play a similar role in regulating albumin catabolism.

In contrast to the downregulation of pIgR expression and SIgA transcytosis observed in human and experimental IBD, the pro-inflammatory environment of the gut in IBD results in increased IgG production and absorption of luminal IgG-bound antigens by FcRn-mediated transcytosis. The combination of reduced SIgA-mediated immune exclusion and increased IgG-mediated antigen absorption may account for the increased levels of circulating IgG antibodies against self and bacterial antigens that are a common feature of human IBD and which may further promote the intestinal inflammation.

Compared with IgA, IgM, and IgG, the levels of IgE are normally very low in mucosal secretions. However, numbers of IgE-producing plasma cells in the lamina propria and levels of IgE in secretions of the respiratory and gastrointestinal tracts are elevated in individuals with allergic diseases such as asthma, allergic rhinitis, and food allergy, and during intestinal parasite

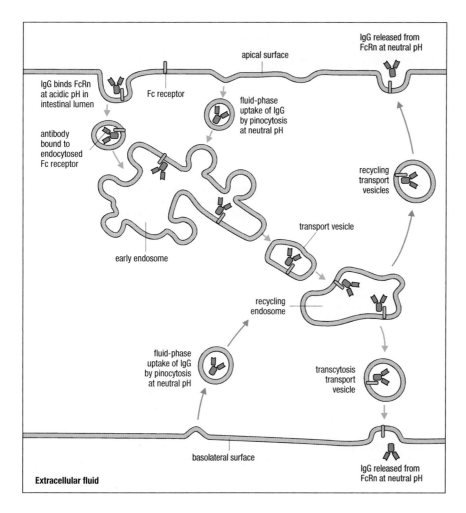

Extracellular fluid

Figure 11.5 Bidirectional transcytosis of IgG across polarized cells by FcRn. Both apical-to-basolateral (green arrows) and basolateral-to-apical (blue arrows) transcytosis of IgG are mediated by FcRn. Most of the FcRn molecules are found in acidic intracellular endosomes and transport vesicles, where low pH promotes binding of FcRn to IgG. Some FcRn molecules reach the plasma membrane via recycling vesicles. In the acidic environment of the neonatal rodent intestinal lumen, IgG binds FcRn at the apical plasma membrane and is delivered to early endosomes via receptor-mediated endocytosis. In cells where fluids bathing the plasma membrane are closer to neutral pH, most of the IgG is internalized via fluid-phase pinocytosis. Fusion of pinocytic vesicles with acidic endosomes results in binding of soluble IgG to membrane-bound FcRn. FcRn–IgG complexes that are taken up by the basolateral and apical surfaces are delivered to the apical and basolateral surfaces of polarized epithelial cells, respectively, through a network of transport and recycling vesicles and endosomes in a process called transcytosis. At the neutral pH of the plasma membrane, IgG is released from FcRn into external secretions and interstitial fluids. Membrane-associated FcRn molecules then reenter the endocytic pathway, where they can mediate many rounds of IgG transport to the apical and basolateral cell surfaces. In humans, FcRn mediates transport of IgG across endothelial cells in the placenta, thus delivering IgG from the maternal to the fetal circulation. Expression of FcRn by a variety of cell types, including endothelial and immune cells, prevents degradation of internalized monomeric IgG by directing these FcRn–IgG complexes away from the lysosome and into recycling vesicles. (Adapted from Alberts, B. et al. *Molecular Biology of the Cell*, 5th ed, New York, Garland Science, 2008.)

infections. The high-affinity receptor for IgE, FcεRI, is expressed by a variety of immune cells and plays an important role in effector functions of IgE (see later). It has recently been appreciated that the low-affinity IgE receptor, FcεRII (CD23), mediates endocytosis and bidirectional transcytosis of IgE and IgE–antigen complexes in several cell types, including enterocytes. CD23-mediated IgE transcytosis may enhance transport of potential allergens across the intestinal epithelial barrier, with important implications for mucosal defense and allergy (see later).

FUNCTIONS OF MUCOSAL IMMUNOGLOBULINS

The characteristic distribution and biological properties of Igs of various isotypes in external secretions reflect functional requirements for optimal protection at diverse mucosal surfaces. SIgA and SIgM antibodies generally have a lower intrinsic affinity for antigens than do IgG antibodies, but due to the bonus effect of multivalency (e.g., four or eight antigen-binding sites for SIgA dimers and tetramers, respectively), SIgA and SIgM have a higher avidity for and enhanced ability to cross-link antigens. Furthermore, the presence of bound secretory component protects SIgA and SIgM from proteolytic degradation, resulting in a longer half-life compared with IgG antibodies in protease-rich secretions such as those of the gastrointestinal tract. IgG antibodies may thus be more important for immune defense in secretions of the respiratory and genital tracts, where they are more stable. The direct protective role of mucosal antibodies was demonstrated *in vivo* in animal models of germ-free piglets and in mice with passively administered

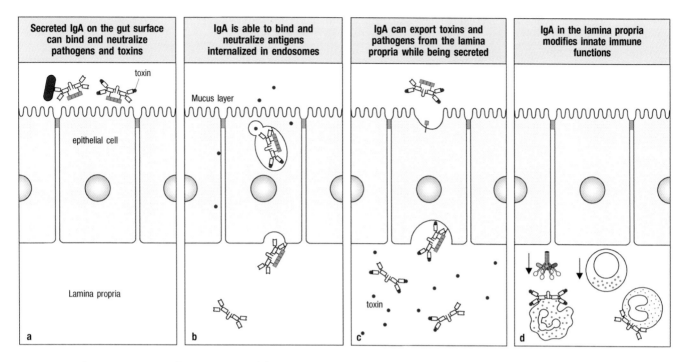

Secreted IgA on the gut surface can bind and neutralize pathogens and toxins	IgA is able to bind and neutralize antigens internalized in endosomes	IgA can export toxins and pathogens from the lamina propria while being secreted	IgA in the lamina propria modifies innate immune functions

Figure 11.6 Mucosal IgA has several functions in epithelial surfaces. (**a**) IgA binds to the layer of mucus covering the epithelium, where it can neutralize pathogens and their toxins, preventing their access to tissues and inhibiting their functions. (**b**) Antigen internalized by the epithelial cell can meet and be neutralized by IgA in endosomes. (**c**) Toxins or pathogens that have reached the lamina propria encounter pathogen-specific IgA in the lamina propria, and the resulting complexes are reexported into the lumen across the epithelial cell as the dimeric IgA is secreted. (**d**) IgA in the lamina propria can inhibit (↓) complement activation, decrease (↓) natural killer cell activity, promote opsonization and phagocytosis of antigens by neutrophils, and trigger degranulations of eosinophils. (Adapted from Murphy, K., and Weaver, C. *Janeway's Immunobiology*, 9th ed, New York, Garland Science, 2016.)

IgA specific for bacterial and viral pathogens or their products. Mucosal Igs effectively diminish absorption of antigens from mucosal surfaces, a process called immune exclusion. In the case of biologically active antigens, such as bacterial or plant toxins and viruses, mucosal Igs can neutralize their activity or block binding to the surface of epithelial cells (**Figure 11.6**). The antibody isotype profoundly influences the protective effect of mucosal Igs. SIgA, due to its inability to activate complement cascades, excludes antigens without generating local inflammation. In contrast, activation of complement by IgM– or IgG–antigen complexes may induce local inflammation, resulting in increased nonspecific uptake of antigens and pathogens due to the altered mucosal barrier. For example, antigen challenge of individuals with a selective IgA deficiency results in a higher level of circulating immune complexes than is seen in healthy individuals.

11.10 Secretory Igs protect mucosal surfaces against microbial invasion

Mucosal Igs utilize a variety of mechanisms to enhance barrier function of mucosal surfaces (see **Figure 11.6**). Coating of an enormously diverse range of mucosal microorganisms with SIgA via antigen-specific binding and glycan-mediated interactions results in inhibition of their attachment to receptors expressed on the surfaces of epithelial cells. *In vitro* experiments have indicated that glycans associated with the α heavy chains and the secretory component moiety of SIgA act as decoys, effectively preventing the attachment of gram-positive and gram-negative microorganisms to glycan-dependent receptors expressed on epithelial cells in the gastrointestinal and respiratory tracts. Such interactions confine Ig-coated bacteria to their natural biological niches and contribute to the formation of physiologic "biofilms" essential for their continued survival and establishment of a mutually beneficial mucosal microbiota. IgG antibodies also participate in immune exclusion, particularly in the lower respiratory tract and the genitourinary tract where they are more stable. These antibody activities may be further enhanced through synergy with innate factors of humoral immunity such as lactoferrin, lysozyme, and the peroxidase system. SIgA antibodies in breast milk appear to be particularly important in protection of newborn mammals from microbial invasion, as suckling offspring of pIgR-deficient dams have been found to have several species of invasive bacteria in the normally sterile mesenteric lymph nodes draining the gut epithelium.

11.11 Epithelial transcytosis of mucosal Igs enhances immune defense

In addition to interactions with antigens on mucosal surfaces, SIgA antibodies can contribute to immune protection during the process of epithelial transport. Biologically active antigens such as toxins can be internalized from the apical surface of epithelial cells via receptor-mediated endocytosis and delivered to intracellular sorting vesicles. At the same time, transcytotic vesicles carrying pIgA–pIgR complexes from the basolateral surface fuse with the apical sorting vesicles, allowing the IgA antibodies to neutralize their specific antigens (see **Figure 11.6**). For example, IgA antibodies have been shown to neutralize bacterial LPS within epithelial cells, thus dampening a potential pro-inflammatory response in the mucosa and limiting access of LPS to the systemic circulation. IgA antibodies have also been shown to interfere with the assembly of influenza virus and Sendai virus by binding to newly synthesized envelope glycoproteins within intracellular sorting vesicles. Mucosal IgA antibodies also contribute to clearance of antigens by pIgR-mediated transport of antigen–antibody immune complexes to the luminal surface of mucosal epithelial cells. In addition to immune complexes containing soluble protein antigens, pIgR has been shown to mediate transcytosis of whole viruses and bacteria complexed to pIgA. This excretory function of mucosal IgA serves to limit access of environmental, dietary, and microbial antigens to the systemic immune system, thus minimizing the generation of potentially pro-inflammatory IgG antibodies. Excretion of infectious microbes within epithelial cells or in the lamina propria can prevent their spread to other regions of the body.

FcRn-mediated transcytosis of IgG from the basolateral to the apical surface of mucosal epithelial cells delivers protective IgG antibodies into external secretions. In turn, transport of IgG–antigen complexes from the luminal to the abluminal surface of the epithelium associated with the intestines, lung, and genitourinary tract leads to absorption of intact antigens. Importantly, these IgG–antigen complexes can be phagocytosed by mucosal dendritic cells, which traffic to regional lymph nodes and present antigenic peptides to mucosal and systemic T cells. In experimental models, this FcRn-driven pathway of antigen absorption has been shown to induce tolerance to mucosal antigens in both the gut and the respiratory tract. This mechanism is likely to be particularly important in newborns, where IgG antibodies in maternal milk can promote absorption of intestinal antigens to which the mother is immune. Thus, IgG antibodies, in addition to their roles in mucosal defense, may promote physiologic tolerance to the commensal gut microbiota and regulate the development of allergic airway inflammation. Although not yet formally demonstrated, it is possible that CD23-mediated absorption of IgE-bound antigens contribute to mucosal tolerance and/or inflammation, particularly in allergic individuals with relatively high levels of mucosal IgE.

11.12 SIgA and secretory component possess innate immune functions through their carbohydrate modifications and regulate mucosal inflammation

In addition to antigen-specific immune functions of IgA, SIgA and free secretory component also contribute to the regulation of innate, "nonspecific" responses to pathogens. Many of these functions are mediated by binding of the unusual *N*-glycans of secretory component to bacterial and host factors. Free secretory component has been shown to limit infection or reduce morbidity by binding to bacterial components such as *Clostridium difficile* toxin A and fimbriae of enterotoxigenic *Escherichia coli*. During both innate and antigen-specific immune responses, secretory component and SIgA enhance mucosal homeostasis by limiting potentially pro-inflammatory immune responses. Because IgA, unlike IgG, is a poor activator of complement, IgA can neutralize antigens and excrete immune complexes without eliciting an inflammatory response that could cause collateral damage to mucous membranes and disrupt barrier function. Although artificially aggregated IgA or IgA with altered glycan moieties may under some circumstances activate the complement cascade by the alternative or lectin pathways, this mechanism does not appear to occur in mucosal tissues. Although external secretions contain complement components, they typically do so at levels that would not promote effective complement activation. Furthermore, the abundance of IgA relative to IgG antibodies in these secretions leads to competition for antigen binding and limitation of IgG-mediated activation of the classical complement pathway. Free secretory component can also prevent inflammation by forming a high-molecular-weight complex with the chemokine IL-8, limiting its activity as a neutrophil chemoattractant.

11.13 Mucosal Igs also interact with Fc receptors on immune cells

By activating isotype-specific Fc receptors on phagocytes and other immune cells, mucosal Igs may facilitate protective immune responses against mucosal pathogens. Some of these immune cells, such as macrophages, are normal residents at the mucosal surface, while others tend to infiltrate mucosal sites only during times of infection or other insult. In humans, three classes of IgG-specific Fc receptors (FcγRI, FcγRII, and FcγRIII), an IgA-specific Fc receptor (FcαRI), and the high-affinity IgE Fc receptor (FcϵRI) are expressed on various types of immune cells in lamina propria of mucous membranes. Ligation of these Fc receptors triggers elimination of Ig-coated targets through mechanisms such as phagocytosis, degranulation and release of antimicrobial factors, and release of activated oxygen species. Intriguingly, the mode of interaction of IgG with FcγR is very different from that of IgA with FcαRI. All of the FcγRs interact with a site on IgG lying at the *N*-terminal end of the Fc region in close proximity to the hinge region. In contrast, the interaction site for FcαRI lies close to the midpoint of the IgA Fc, at the interface between the Cα2 and Cα3 domains (see **Figure 11.1**). Targeting of this receptor site on IgA by microbial proteins appears to offer an effective means for certain pathogens to evade elimination via FcαRI-mediated clearance mechanisms (see later). Intestinal macrophages, dendritic cells, and B cells also express FcRn. In these cell types, although FcRn mediates the recycling of monomeric IgG, it participates in directing polymeric IgG-containing antigen–antibody complexes into antigen processing compartments associated with MHC class II-restricted presentation that may be important in defending against mucosal pathogens or promoting intestinal inflammation as in inflammatory bowel disease. Binding of IgE–antigen complexes to FcεRI on mucosal antigen-presenting cells such as dendritic cells, macrophages, and basophils likely regulates IgE-mediated adaptive immunity. In the effector phase, cross-linking of cell-surface FcεRI by IgE–antigen complexes results in degranulation of mucosal mast cells, an important mechanism for elimination of parasitic infections.

STRATEGIES EMPLOYED BY PATHOGENS TO EVADE IGA-MEDIATED DEFENSE

The relationship between IgA and the microbiota is complex. Commensal microbes, as well as pathogens, regulate the production and transport of IgA, and their colonization is, in turn, affected by IgA secretion. The importance of this relationship and the effectiveness of IgA as a form of immune defense are highlighted by the fact that a number of pathogens have targeted IgA as a means of evading the immune response.

11.14 Pathogens express IgA1 proteases that interfere with IgA function

Certain pathogenic bacteria, known to cause clinically important and life-threatening infections, secrete proteolytic enzymes that cleave specifically in the hinge region of IgA1. These IgA1 proteases are produced by bacteria that colonize and invade mucosal sites to cause diseases such as acute meningitis (*Neisseria meningitidis*, *Haemophilus influenzae*, and *Streptococcus pneumoniae*), oral cavity disease (*Streptococcus sanguis*, *Streptococcus oralis*), and sexually transmitted infections (*Neisseria gonorrhoeae*). The proteases appear to be associated with virulence because closely related nonpathogenic strains do not produce them. Diversity in the structural and mechanistic features of IgA1 proteases suggests that they have arisen by convergent evolution. Presumably, their ability to disrupt IgA function affords the bacteria concerned with advantageous opportunities for colonization and invasion of mucosal surfaces.

IgA1 proteases cleave post-proline peptide bonds within the extended hinge region of IgA1, each protease cleaving at a particular proline–serine or proline–threonine site (**Figure 11.7**). IgA2 lacks the susceptible sequence and therefore resists cleavage. Proteolysis of the IgA1 hinge essentially disengages the antigen-recognition function of the Fab regions from the effector function capabilities of the Fc region, allowing any bacteria recognized by the free Fab fragments to escape the elimination processes normally triggered through the Fc region. Moreover, Fab binding blocks access of intact Igs, enabling the bacterium to evade the protective functions of mucosal Igs.

The efficient recognition and cleavage by IgA1 proteases is governed both by the hinge amino acid sequence and the structural context of the hinge within the antibody as a whole. Thus, mutagenesis experiments have shown that susceptible bonds must be positioned at a suitable distance away from the Fc region, and that motifs in the lower part of the Fc region (Cα3 domain) are required for substrate recognition by several IgA1 proteases. The solution of the first X-ray crystal structure of an IgA1 protease, the type 1 protease from *Haemophilus influenzae*, suggested a binding mechanism in keeping with these findings. In this model, the binding of the Fc is postulated to stabilize the protease in an open conformation, thereby allowing access of the hinge peptide to the active site so that cleavage may ensue.

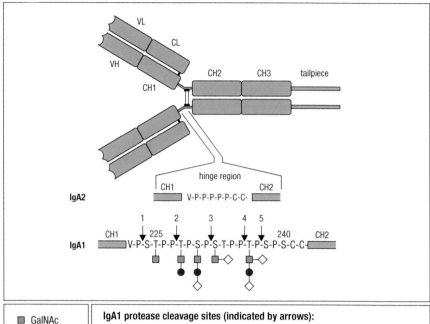

Figure 11.7 Differences between the hinge regions of human IgA1 and IgA2. The amino acid sequences of the hinge regions are shown. Arrows above the IgA1 hinge sequence indicate the cleavage sites of particular IgA1 proteases. The *O*-linked oligosaccharides carried by the IgA1 hinge are indicated below the sequence.

■ GalNAc
● Gal
◇ Sialic acid

IgA1 protease cleavage sites (indicated by arrows):

1 *Prevotella* species, *Capnocytophaga* species
2 *S. pneumoniae, S. sanguis, S. mitis, S. oralis*
3 *H. influenzae* type 1
4 *N. meningitidis* type 2, *H. influenzae* type 2, *N. gonorrhoeae* type 2
5 *N. meningitidis* type 1, *N. gonorrhoeae* type 1

11.15 Pathogenic bacteria express binding proteins specific for IgA and secretory component that are involved in virulence

Another bacterial strategy to evade the mucosal IgA response employs surface proteins, termed IgA-binding proteins, that bind specifically to both serum and secretory forms of human IgA. Such proteins are expressed by strains of group A streptococcus (*Streptococcus pyogenes*), an important pathogen causing acute infections sometimes resulting in compromised heart and kidney function; group B streptococcus, a major cause of septicemia, meningitis, and pneumonia in newborn babies; and *Staphylococcus aureus*, responsible for skin infections, abscesses, pneumonia, bacteremia, and other illnesses, many of which are life-threatening. Remarkably, although the IgA-binding proteins produced by these organisms are structurally unrelated, they all interact with broadly the same region of IgA Fc. Their sites of interaction on the Fc region of IgA overlap with that bound by FcαRI, and their binding has been shown to block binding of IgA to this receptor and inhibit triggering of FcαRI-mediated responses. This inhibitory capability suggests that these bacteria have evolved IgA-binding protein expression as a means to evade elimination mechanisms elicited by IgA through interaction with FcαRI.

Streptococcus pneumoniae, which causes diseases ranging from relatively mild otitis media to potentially fatal sepsis, pneumonia, and meningitis, expresses a surface protein called CbpA (also known as SpsA and PspC) that binds specifically to human pIgR and secretory component. *S. pneumoniae* uses this protein to enhance colonization and invasion of the nasopharynx by binding to pIgR on the surface of epithelial cells. Strains of *S. pneumoniae* that lack CbpA have reduced ability to colonize the nasopharynx, suggesting that its ability to bind pIgR may represent an important virulence factor. However, free secretory component and SIgA secreted by nasopharyngeal epithelial cells can bind to CbpA and inhibit its ability to bind to cell-surface pIgR. The balance between this bacterial virulence factor and host defense mechanism may contribute to variations in susceptibility among humans to nasopharyngeal carriage of *S. pneumoniae.*

SUMMARY

All isotypes of antibodies can function as secretory immunoglobulins, and as such play a critical role in mucosal defense. Secretory IgA is the chief antibody class at most mucosal sites, whereas IgG predominates in secretions of the lower respiratory tract and genitourinary tract. IgM is found at lower concentrations but can compensate for IgA in selective IgA deficiency. Mucosal IgE can play a protective role in parasitic infections and a pathogenic role in food allergies and asthma. Most of the mucosal IgA and IgM form polymers through covalent interactions with the J chain polypeptide. Polymeric IgA and IgM are transported across mucosal epithelial cells in the basolateral-to-apical direction by the polymeric immunoglobulin receptor, which is cleaved at the apical surface to form SIgA and SIgM. The extracellular domain of pIgR, a five-domain polypeptide also known as secretory component, remains bound to SIgA and SIgM and provides protection against proteolysis and enhances immune effector functions. SIgA is highly glycosylated, which increases its stability and some aspects of its protective function. IgG is transported across mucosal epithelial cells in both the apical-to-basolateral and basolateral-to-apical directions by the neonatal Fc receptor, and IgE is transported bidirectionally by FcɛRII (CD23).

SIgA in mucosal secretions neutralizes biologically active antigens, prevents absorption of food antigens, and inhibits adherence of mucosal microbes to epithelial cells. SIgA also neutralizes pathogens and antigens during pIgR-mediated epithelial transcytosis and clears antigen–antibody immune complexes from the lamina propria via transport to the luminal surface of epithelial cells. SIgA antibodies in breast milk are important for protection of the neonatal gut against microbial invasion and shape the composition of the developing gut microbiota. IgG antibodies facilitate uptake and presentation of mucosal antigens to link mucosal antigen exposures to systemic induction of immune responses and promote immune tolerance via FcRn-mediated transcytosis. IgG, IgA, and IgE antibodies engage specific Fc receptors on phagocytes and immune cells in mucosal tissues, triggering elimination of pathogens and antigens by mechanisms such as phagocytosis and release of antimicrobial substances. Some pathogenic bacteria inhibit SIgA defense mechanisms by secreting IgA-binding proteins or proteolytic enzymes that cleave IgA1.

FURTHER READING

Conley, M.E., and Delacroix, D.L.: Intravascular and mucosal immunoglobulin A: Two separate but related systems of immune defense? Ann. *Int. Med.* 1987, 106:892–899.

Johansen, F.-E., Massol, R., Baker, K. et al.: Biology of gut immunoglobulins, in Johnson, L.R. (ed): *Physiology of the Gastrointestinal Tract*, 5th ed. Burlington, MA, Elsevier Academic Press, 2012:1089–1118.

Kaetzel, C.S.: Cooperativity among secretory IgA, the polymeric immunoglobulin receptor, and the gut microbiota promotes host-microbial mutualism. *Immunol. Lett.* 2014, 162:10–21.

Kuo, T.T., Baker, K., Yoshida, M. et al.: Neonatal Fc receptor: From immunity to therapeutics. *J. Clin. Immunol.* 2010, 30:777–789.

Li, H., Nowark-Wegrzyn, A., Charlop-Powers, Z. et al.: Transcytosis of IgE-antigen complexes by CD23a in human intestinal epithelial cells and its role in food allergy. *Gastroenterology* 2006, 131:47–58.

Mattu, T.S., Pleass, R.J., Willis, A.C. et al.: The glycosylation and structure of human serum IgA1, Fab, and Fc regions and the role of *N*-glycosylation on Fcα receptor interactions. *J. Biol. Chem.* 1998, 273:2260–2272.

Mestecky, J., Strober, W., Russell, M.W. et al. (eds): *Mucosal Immunology*, 4th ed. Burlington, MA, Elsevier Academic Press, 2015.

Oettgen, H.C.: Fifty years later: Emerging functions of IgE antibodies in host defense, immune regulation, and allergic diseases. *J. Allergy Clin. Immunol.* 2016, 137:1631–1645.

Pyzik, M., Rath, T., Lencer, W.I. et al.: FcRn: The architect behind the immune and nonimmune functions of IgG and albumin. *J. Immunol.* 2015, 194:4595–4603.

Rogier, E.W., Frantz, A.L., Bruno, M.E. et al.: Lessons from mother: Long-term impact of antibodies in breast milk on the gut microbiota and intestinal immune system of breastfed offspring. *Gut Microbes.* 2014, 5:663–668.

Royle, L., Roos, A., Harvey, D.J. et al.: Secretory IgA *N*- and *O*-glycans provide a link between the innate and adaptive immune systems. *J. Biol. Chem.* 2003, 278:20140–20153.

Woof, J.M., and Burton, D.R.: Human antibody-Fc receptor interactions illuminated by crystal structures. *Nat. Rev. Immunol.* 2004, 4:89–99.

Woof, J.M., and Russell, M.W.: Structure and function relationships in IgA. *Mucosal Immunol.* 2011, 4:590–597.

Ye, L., Zeng, R., Bai, Y. et al.: Efficient mucosal vaccination mediated by the neonatal Fc receptor. *Nat. Biotechnol.* 2011, 29:158–163.

Yoshida, M., Claypool, S.M., Wagner, J.S. et al.: Human neonatal Fc receptor mediates transport of IgG into luminal secretions for delivery of antigens to mucosal dendritic cells. *Immunity* 2004, 20:769–783.

Role of dendritic cells in integrating immune responses to luminal antigens

12

BRIAN L. KELSALL AND MARIA RESCIGNO

Mucosal tissues constitute the major lymphoid compartments of the body. Containing more lymphocytes than all nonmucosal lymphoid tissues combined, these tissues generate complex and unique immune responses that protect the body against invading microbes and prevent untoward responses to common mucosal antigens and symbiotic bacteria. This chapter focuses on the role of dendritic cells (DCs) in the induction and regulation of immune responses in the intestine.

DCs are divided into follicular and nonfollicular DCs. Follicular DCs (FDCs) develop from a non-bone-marrow-derived precursor, express high levels of receptors for immunoglobulin and complement, and are present in germinal centers, where they contribute to B-cell responses. Nonfollicular DCs comprise a family of mononuclear phagocytes that develop from bone marrow–derived stem cells under the influence of FMS-like tyrosine kinase-3 (Flt-3) ligand, granulocyte-macrophage colony-stimulating factor (GM-CSF) and other cytokines, and include conventional DCs (cDCs) and plasmacytoid DCs (pDCs). Three immunologic compartments contain these cells: (1) nonlymphoid tissues, including internal organ interstitium, mucosal lamina propria, and skin dermis and epidermis (nonlymphoid or tissue DCs, Langerhans cells, and dermal DCs, respectively); (2) lymphoid tissues such as the T-cell regions of draining lymph nodes (resident or interdigitating DCs); and (3) lymph (often referred to as "veiled cells"). Focusing on intestinal cDCs, we discuss DC phenotype, localization, antigen uptake and trafficking, interaction with T and B cells, and role in the pathogenesis of inflammatory bowel disease (IBD).

DEFINING CHARACTERISTICS OF DENDRITIC CELLS

12.1 DCs capture antigens and select and activate naïve T-cell clones *in vivo*

In the early 1970s, Ralph Steinman discovered a population of "tree-like" cells in cultures of splenic accessory cells (cells required to generate antibody responses *in vitro*), which he called "dendritic cells" derived from *dendron*, the Greek word for "tree." These cells displayed distinct appearance, motility, and function. Initially, DCs were shown to capture antigens, direct T- and B-cell responses *in vitro*, and select T cells in the thymus. Subsequently, unique characteristics of DCs were identified that distinguish them from other members of the mononuclear phagocyte family. These characteristics include the ability to capture and process antigens for major histocompatibility complex (MHC) presentation, migrate to T-cell zones in lymphoid tissues, and attract and activate naïve T cells, as well as direct the nature of lymphocyte response.

Immature DCs phagocytose antigen through scavenger, Fc, and C-type lectin-like receptors and capture antigens by endocytosis and micropinocytosis. In addition, DCs have a distinct endocytic system that promotes efficient capture, processing, and presentation of antigens via MHC I (cross-presentation) and MHC II antigens to CD8 and CD4 T cells, respectively.

Immature DCs in peripheral tissues capture tissue-specific cells and proteins. In barrier sites (skin and mucosal tissue), DCs capture innocuous environmental and food proteins, take up microbial antigens from commensal and pathogenic microbes, and migrate to lymph nodes. In internal organs, DCs capture tissue antigens from apoptotic cells during normal cell turnover or after tissue injury. Resident DCs in lymph nodes and the spleen capture and process blood-borne antigens, including injured and dying cells and pathogens, and can acquire lymph-borne antigens, possibly including those transferred from lymph-node macrophages.

DCs that have captured antigens undergo a process of maturation, either constitutively through unclear mechanisms that may involve changes in homeotypic adhesion or through activation by pattern-recognition receptors (PRRs) or inflammatory cytokines (interleukin [IL]-1α/β, tumor necrosis factor [TNF]-α), resulting in increased maturation and migration. DC maturation is characterized by a decreased capacity to take up antigen (to prevent processing of additional self-antigens), increased processing of antigen into MHC I and II and peptide complexes, and expression of costimulatory molecules. Expression of MHC peptide complexes and costimulatory molecules is markedly increased on the expanded cell surface that results from formation of extended dendritic processes. This maturation process coincides with DC migration from nonlymphoid tissues or from the marginal zones to T-cell zones of draining lymph nodes due to expression of CCR7, which allows DCs to enter lymphatics. cDCs localize along connective tissue fibers and present the antigen to circulating naïve T cells attracted by chemokines such as DC-derived CCL18. In addition, mature DCs express adhesion molecules, including LFA-1 (CD11a/CD18), LFA-3 (CD58), and DC-SIGN (CD209). DC-SIGN binds transiently and with high avidity to ligands expressed by naïve T cells, thereby enhancing T-cell sampling of DC-expressed MHC–peptide complexes. LFA-1 and LFA-3 expressed on mature DCs bind naïve T-cell ICAM-3 and CD2, respectively, promoting T-cell activation.

In contrast to DCs, tissue macrophages lack dendrites, migrate poorly to draining lymph nodes, and inefficiently activate naïve T cells. The latter is likely due to the production of antigen-degrading enzymes such as lysozyme and cathepsins, the decreased expression of appropriate adhesion molecules, and the production of inhibitory cytokines such as IL-10 and transforming growth factor (TGF)-β, resulting in inefficient and shortened presentation of antigen–MHC complexes. Under inflammatory conditions, however, monocytes can differentiate into cells with a higher capacity to prime naïve T cells in lymph nodes and tissues; these monocytes are variably referred to as "inflammatory monocytes," "inflammatory macrophages," or "monocyte-derived DCs."

12.2 DCs direct the nature of lymphocyte responses

In addition to their ability to prime naïve T cells, DCs determine the nature of the resulting T-cell activation. Thus, DCs are responsible for driving either a homeostatic T-cell response to maintain immunologic "tolerance" to environmental and self-antigens under steady-state conditions, or an effector T-cell response to pathogens that promotes host defense and inflammation.

Presentation of antigens by peripheral DCs under steady-state conditions normally results in homeostatic responses that are appropriate for the tissue of origin. Such responses are often skewed toward the differentiation or expansion of CD4+Foxp3+ regulatory T cells (T_{reg} cells) or nonpathogenic T cells such as nonpathogenic helper T cells T_H17 that provide tissue protection. In contrast, in the setting of infection or inflammation and in response to activating signals from PRRs and inflammatory cytokines, DCs rapidly mature and drive responses skewed toward effector CD4+ and CD8+ T cells. DC induction of T-cell differentiation occurs through a combination of MHC antigen presentation; engagement of costimulatory molecules, including CD80, CD86, OX40L, PDL1, and PDL2; and the production and/or activation of soluble factors, including IL-12, IL-23, IL-6, TGF-β, retinoic acid (RA), indolamine 2,3-dioxygenase (IDO), and factors in the draining lymphoid microenvironment.

DCs also direct the differentiation of B cells in lymphoid tissues by presenting repetitive carbohydrate antigens on their surface to B cells, inducing the differentiation of T-follicular and other helper T cells, or producing cytokines that influence B-cell isotype switching.

FACTORS AFFECTING DC FUNCTION

DC induction of T-cell and B-cell responses are complex, as schematically depicted in **Figure 12.1**. This complexity is reflected in the subpopulations of DCs that arise from distinct precursors, the array of stimuli to which DCs are exposed during maturation and T-cell priming, local microenvironmental factors that influence DC migration, maturation, survival, and the production of cytokines that can directly influence T-cell differentiation.

12.3 DCs are a family of cells with distinct subpopulations

Subpopulations of DC are defined on the basis of unique surface marker expression, localization, gene expression, and function. Importantly, surface markers that define populations of DCs in mice and humans may differ, and additional markers have been identified that distinguish DC populations in both species.

Myeloid populations are now more clearly defined on the basis of ontogeny (**Figure 12.2**). According to this scheme, the three primary circulating myeloid precursors, monocytes, pre-cDCs, and pDCs, differentiate from hematopoietic stem cells in the bone marrow. In mice, monocytes can be further distinguished by their expression of Ly6C. In the steady state,

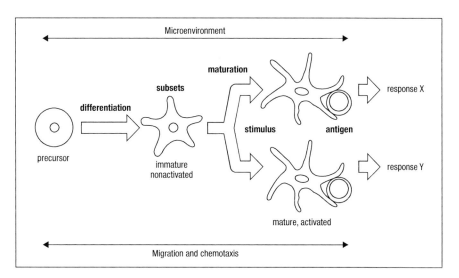

Figure 12.1 Factors that affect dendritic cell (DC) function. Bone marrow–derived precursor cells differentiate into distinct subsets of DCs that mature and differentiate under the influence of microenvironmental factors. In steady-state conditions, the factors that stimulate maturation have not been defined, but in infected, inflamed, or injured mucosa, DCs mature after activation by microbial products, inflammatory cytokines, and/or products of damaged tissue. The signals that induce maturation also affect the functional phenotype of the DC population.

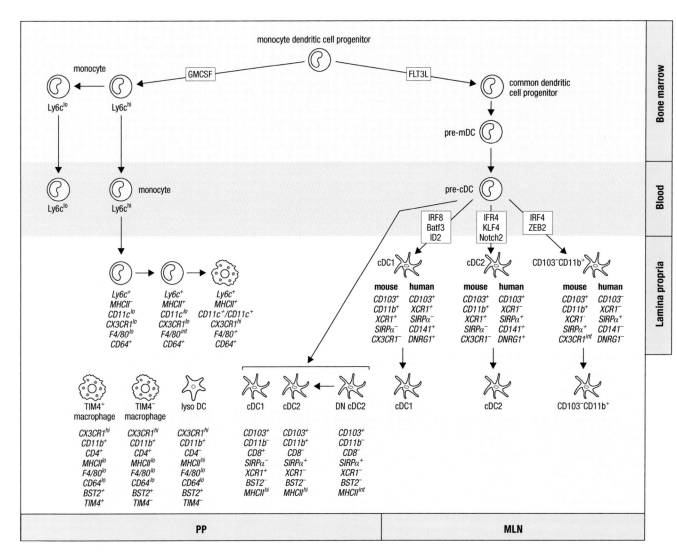

Figure 12.2 Mononuclear phagocyte populations in the intestine. Dendritic cell (DC) and monocyte/macrophage populations arise from circulating bone marrow–derived precursors. Ly6C^{hi} monocytes give rise to macrophages and LysoDCs in the Peyer's patches (PPs) and macrophages in the intestinal lamina propria (LP), the latter through a well-described maturation process, whereas pre-cDCs (conventional DCs) give rise to cDC populations. Differentiation of different cDC populations is dependent on individual sets of transcription factors, and homogenous populations can be identified with common and unique surface markers. LP cDC migration to the mesenteric lymph node (MLN) in the steady state increases during inflammation, and additional populations of resident cDCs are present in the MLN (not depicted). Resident cDC populations in the PP also differentiate from pre-cDCs, and while cDC1s are dependent on Batf3, the transcription factors responsible for the other populations have not been addressed experimentally.

circulating Ly6C^{hi} monocytes give rise to CD11c$^+$ and CD11c$^-$ mature intestinal macrophages in a process in which expression of Ly6C is lost and expression of MHC II, CX3CR1, F4/80, and CD64 is increased. A population of resident self-renewing macrophages that are not dependent on blood precursors also has been identified that expresses CD4 and Tim-4 (see later and Chapter 13).

Pre-cDCs differentiate in the bone marrow in response to Flt3L and mature into two primary cDC populations, cDC1 and cDC2 cells, in tissues and lymphoid organs under the influence of the hematopoietic cytokines Flt3L, GM-CSF, and unique transcription factors. cDC1 cells are CD103$^+$CD8$^+$CD11b$^-$ (SIRPα$^-$, XCR1$^+$) and are dependent on the transcription factors BATF3, IRF8, and Id2, whereas cDC2 cells are CD103$^+$CD11b$^+$ (SIRPα$^+$, XCR1$^-$) and dependent on IRF4, KLF4, and Notch 2. These populations also have been identified in other species, including humans, the definitions of which depend on the expression of CD103, XCR1, and SIRPα, which have emerged as more reliable markers than CD11b and CD8 across species. In the mouse intestine, an additional population of pre-cDC-derived CD103$^-$CD11b$^+$ XCR1$^-$ DCs uniquely shares developmental dependence on the transcription factor Zeb2 with monocyte-derived cells and expresses intermediate levels of CX3CR1 that is present at high levels on macrophages. This population is considered a unique DC lineage, as it is derived from pre-cDCs and dependent on FLT3L for its differentiation.

12.4   DC function is influenced by activation signals

DCs drive a particular T-cell response that is influenced by signals generated during antigen uptake and activation. These signals are dependent on the form of antigen (e.g., soluble, apoptotic body-associated, or pathogen-associated) and tissue factors (e.g., inflammatory cytokines or products of damaged cells), which in turn affect the type of T-cell response induced by the DCs. For example, the microbial products cholera toxin, lipopolysaccharide (LPS) from *P. gingivalis*, soluble egg antigens from *S. mansoni*, hyphae from *C. albicans*, and the mediators histamine and thymic stromal lymphopoietin (TSLP, IL-50), can drive DCs, at least in part, through suppressed IL-12 production, to induce T_H2 responses. In contrast, toll-like receptor (TLR) ligands generally induce high levels of IL-12 production by DCs to drive T_H1 responses. Furthermore, certain cytokines, including IL-10 or TGF-β, pathogens such as *P. falciparum*, or pathogen products such as *B. pertussis* fimbrial hemagglutinin, induce DCs that inhibit effector T cells and drive T_{reg} differentiation. Finally, the T-cell phenotype (e.g., T_H1 versus T_H2) and the level of homing molecule expression on naïve T cells differentiated with DCs is influenced, at least *in vitro*, by the antigen dose and the ratio of DCs to T cells. High antigen doses and DC/T-cell ratios favor T_H1 responses, whereas low antigen doses and DC/T-cell ratios favor T_H2 responses.

The DC "plasticity" described earlier and illustrated *in vitro* by bone marrow (mouse) DCs and blood monocyte (human)–derived DCs correlates with responses during infection and immunization. However, the *in vivo* effects are more complex and depend on individual DC subpopulations, for which plasticity has been demonstrated, but which may be more limited. This is due to different intrinsic capacities of DC subpopulations to respond to PRR-ligands and cytokines and to produce cytokines that affect lymphocyte differentiation.

INTESTINAL DC POPULATIONS

Intestinal cDC populations display an array of specialized functions. These functions include the imprinting of lymphocytes with homing receptors for their recirculation to intestinal tissues, driving the differentiation of Treg cells involved in tolerance to soluble oral antigens, providing signals for the differentiation of IgA-producing B cells, and inducing appropriate T-cell response to commensal bacteria to maintain homeostasis and to intestinal pathogens for host defense.

12.5   Distinct populations of DCs present in mucosal inductive and effector sites capture antigens by different mechanisms

Primary induction sites for intestinal T- and B-cell responses are the gut-associated lymphoid tissues (GALT), which include Peyer's patches (PPs) in the small intestine, cecal patches in the cecum/appendix, colonic patches in the colon, and isolated lymphoid follicles (ILFs) in the small and large intestines that develop in response to bacterial signals, and mesenteric lymph nodes (MLNs). In contrast, the primarily effector sites are the diffuse lamina propria and the intraepithelial cell compartments. DC populations in these tissues capture antigens by different mechanisms, have different abilities to migrate to the MLNs, and have distinct functional capacities (**Figure 12.3**).

Luminal antigens, including macromolecules, bacteria, and viruses, gain access to PPs and ILFs through specialized epithelial cells called M (microfold) cells in the follicle-associated epithelium (FAE) (see Chapter 15). cDC2 cells are present in the subepithelial dome (SED), the region just below the FAE, together with two populations that do not express F4/80 or CD64, normally present on mouse macrophages. One population expresses high levels of CD11b and MHC II, has a rapid turnover, is located just beneath the FAE, and can extend dendrites into the intestinal lumen through M cell–specific transcellular pores. Called "LysoDCs" because of their ability to present antigens to naïve T cells, these cells present antigen at much lower efficiency than cDCs. Another population of macrophages expresses low

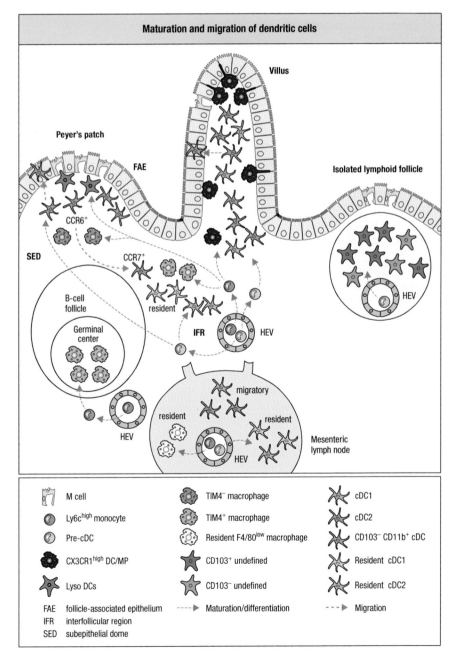

Figure 12.3 **Maturation, localization, and migration of intestinal mononuclear phagocytes.**
Conventional dendritic cells (cDCs), monocyte-derived macrophages, and LysoDCs are derived from different precursors and are localized to specific sites. DCs in Peyer's patches (PPs) and villus LP can migrate into the intraepithelial cell space to capture antigens. DCs in PPs migrate to the interfollicular region (IFR), while DCs in the villus lamina propria (LP) and migrate to the MLNs. DCs in the PPs and isolated lymphoid follicles likely remain in these tissues. See text for further details.

levels of CD11b and MHC II and does not present antigens to naïve T cells, similar to other macrophage populations. The expression of high levels of lysozyme by both LysoDCs and T-cell immunoglobulin mucin receptor 4 (Tim-4)$^-$ macrophages suggests that these cells have a role in innate defense against incoming microbes.

cDC2 cells in the SED express low levels of CD103, and low to intermediate levels of CD11b, which increases with cell maturation. PP cDC2 cells express high levels of CCR6 and CCR1, which mediate the cells' localization in the SED in response to CCL20 and CCL9 expressed by the FAE and cells in the SED matrix. cDC2s form clusters with T cells in the SED, where they may activate naïve T cells in response to bacteria and TLR ligands. cDC2s then migrate to the T-cell zone (interfollicular region, IFR) after a switch in chemokine receptor expression to CCR7, which directs migration to the CCR7 ligands CCL19 and CCL21 constitutively expressed in the T-cell–rich IFR. Activation of cDC2 in the intestinal villi located between follicles also results in their migration to the IRF, the functional meaning of which is not yet clear.

Microparticles that enter the SED via M cells are taken up by cDC2s, LysoDCs, and Tim-4$^-$ macrophages, although pathogens such as *S. typhimurium* and *Listeria monocytogenes* appear to be taken up primarily by macrophages. Both cDC2s and LysoDCs can interact with M cells and take up apoptotic epithelial cells in steady-state conditions, but LysoDCs may be more efficient in this setting. Following reovirus infection of the FAE, CD11c$^+$CD8$^-$CD11b$^-$ cells take up apoptotic infected FAE cells in the SED, which are likely macrophages, due to their poor ability to present viral antigens to primed T cells *ex vivo*. In addition, translocated *E. coli*, SIgA coated *S. flexneri*, *C. albicans*, β-glucan, and prion particles are captured by CD11c$^+$ cells that remain poorly defined.

In the steady state, cDCs and macrophages also are present in the IFR but are different from populations in the SED. CD103$^+$CD11b$^-$CD8$^+$ BATF3-dependent DCs (cDC1 cells) likely represent a blood-derived population that enters this region through high-endothelial venules. These DCs can process viral antigens for T-cell activation, but the mechanism by which IFR cDCs acquire antigens has not been elucidated. Furthermore, macrophages present in the IFR and at the base of the follicles express the phosphatidylserine receptor Tim-4, suggesting they may play a role in the uptake of apoptotic T cells. Following systemic activation with TLR7/8 agonist resiquimod/R848, cDC1 cells in the IFR increase in number and cDC2 cells from the SED and overlying villi migrate to the IFR, and both populations display an activated phenotype.

cDCs also are present within ILFs of the small intestine and colon. ILFs develop from rudimentary structures called cryptopatches, collections of lymphoid-tissue inducer (LTi) ILC3 cells and lymphoid tissue organizing (LTo) stromal cells that are present at birth. Upon bacterial colonization, cDCs are recruited to cryptopatches and produce CXCL13, which attracts B cells to form an ILF. Intestinal bacteria enter the ILFs via M cells similar to PPs and induce the production of IgA. ILF cDCs appear to be both CD103$^+$ and CD103$^-$ cells, the latter likely including CD103$^-$CD11b$^-$ cDCs.

A second site for antigen entry into the intestine is the nonfollicular absorptive epithelium overlaying the lamina propria. Food antigens, microbial products, and pathogenic bacteria can traverse the intestinal epithelium by multiple mechanisms. Such mechanisms include endocytosis, transcytosis, paracellular "leak" through intercellular junctions, goblet cell–associated passages (GAPs), Fc receptor-mediated transport of antibody-bound antigen, and active uptake by CX3CR1$^+$ cells that extend transepithelial dendrites (TEDs) across epithelial cell tight junctions. In addition, villous M cells can sample bacteria similar to FAE M cells.

CX3CR1^{hi} CD64$^+$ macrophages are highly represented in the lamina propria (LP), where they account for 70%–80% of MHCII^{hi} CD11c$^+$ cells in the small and large intestines. cDC2s and cDC1s account for the majority (85%–90%) of cDCs in the small intestinal LP, whereas CD103$^-$ CD11b$^+$ cDCs that express intermediate levels of CX3CR1 account for 10%–15%. cDC2 cells are more prevalent than cDC1s in the small intestine; the opposite proportion is present in the colon.

CD103$^+$ cDCs and CX3CR1$^+$ cells acquire soluble antigens and bacteria from the intestinal lumen by several mechanisms. CX3CR1$^+$ cell TEDs sample luminal bacteria and antigen and are more prevalent in the proximal intestine (duodenum and jejunum) than the terminal ileum. The presence of TEDs is dependent on the intestinal microbiota and expression of CXCR1 and CXCL1 on the basolateral surface of epithelial cells. TEDs have not been identified in colonic epithelium. In the terminal ileum, the number of TEDs increases in response to *Salmonella* and TLR ligands, indicating that constitutive and inducible mechanisms contribute to TED formation. TEDs have been shown to capture both invasive and noninvasive *S. typhimurium* and nonpathogenic *E. coli.*

CX3CR1$^+$ cells also are the major cells that capture soluble antigens in the small intestine, even in the absence of TEDs, indicating they capture soluble antigens transported into the LP by other mechanisms. Whether these cells represent true macrophages or CD103$^-$CD11b$^+$CX3CR1^{int} cDCs is not yet clear.

cDCs in the LP can acquire antigens by several mechanisms. Soluble antigens captured by CX3CR1$^+$ cells can be transferred to CD103$^+$ cDCs. This occurs in the small intestine, particularly in the duodenum, and is dependent on connexin-43, a major component of gap junctions that form between CX3CR1$^+$ cells and CD103$^+$ cDCs. CD103$^+$ cDCs also can migrate into the epithelium above the basement membrane and return to the LP. Within the epithelium, CD103$^+$ cDCs capture soluble proteins (e.g., ovalbumin) but not as efficiently as CX3CR1$^+$ cells in the LP. CD103$^+$ DCs migrate into the epithelium following *Salmonella* exposure, where they send TEDs into the lumen to capture invasive and noninvasive *S. typhimurium.*

In addition, CD103$^+$ cDCs can capture Ag-IgG immune complexes transported from the lumen by FcRn, a receptor expressed in the small intestine in the human neonate, but throughout life in rodents. CD103$^+$ cDCs also can capture soluble antigens transported by GAPs. GAP-dependent antigen capture by LP CD103$^+$ DCs occurs throughout the small intestine in the steady state. Underscoring the importance of antigen capture by GAPs, LP CD103$^+$ DCs from antigen-fed mice without goblet cells are not able to stimulate T cells *ex vivo*. Following antibiotic treatment, GAPs are induced in the proximal colon, where they can transiently transport both intestinal bacteria and soluble antigens for capture by CX3CR1$^+$ cells.

Finally, CD103$^+$ cDCs in the LP appear capable of capturing fragments of intestinal epithelial cells, which have been identified in cDCs in the MLN. This suggests that luminal antigens taken up by, or bound to epithelial cells, are carried to the MLN with epithelial cell fragments by CD103$^+$ DCs.

cDCs in the MLN consist of migratory cells from the afferent lymph draining intestinal tissues and resident cDCs that arise from blood pre-cDCs. Resident MLN cDCs can be distinguished from migratory cDCs by their relatively lower expression of MHCII and higher expression of CD11c. Resident cells contain both cDC1(CD8α^+) and cDC2 (CD11b$^+$) populations similar to other lymph nodes, but detailed phenotypic and functional data for these cells are lacking.

In contrast, migratory cDCs in the MLN in the steady state are composed of cDC populations and proportions similar to those of the intestinal LP, with a majority (80%) CD103$^+$ with equivalent percentages of CD11b$^+$ and CD11b$^-$ cells, together with smaller percentages of CD103$^-$CD11b$^+$ cells, and even fewer CD103$^-$CD11b$^-$cDCs. All cDC populations migrate to the MLNs in the steady state, but differential migration may contribute to T-cell responses in particular settings, such as helminth infection.

In addition, lymph from the small intestine and colon drains to different MLNs, and migratory cDC populations reflect the cDC population differences in these tissues. cDCs from the colon also drain to caudal and iliac lymph nodes, where CD103$^-$CD11b$^+$ cells represent the majority of migratory cDCs in the steady state.

Importantly, migratory cDCs carry both soluble antigens and bacteria to the MLN in a CCR7-dependent manner. CCR7-deficient mice lack migratory cDCs in the MLNs, are deficient in the development of oral tolerance to ovalbumin, and have reduced numbers of noninvasive *S. typhimurium* in the MLNs compared with WT mice. Noninvasive *Salmonella* is carried in the lymph by CD103$^+$ cDC1 and cDC2 cells and, to lesser degree, by CD103$^-$CD11b$^-$ cDCs,

Most information regarding intestinal DC phenotype and function is derived from mouse studies. However human intestinal cDC populations analogous to the three major cDC populations in the mouse have been identified using common cell surface markers SIRPα, in addition to DNRG1 and CD41. Thus, human cDC1s are CD103$^+$ SIRPα^-XCR1$^+$CD141$^+$ DNRG1$^+$, and cDC2s are CD103$^+$ SIRPα^+XCR1$^-$CD141$^-$DNRG1$^-$. Furthermore, human CD103$^+$SIRPα^{hi} cDCs express IRF4, similar to intestinal cDC2s in mice, and mouse and human intestinal CD103$^+$ DC populations have conserved transcriptional programs. Use of these novel markers will provide a more accurate profile of human intestinal DC phenotypes and function in the future.

12.6 DCs drive intestinal homing receptors on T cells and B cells

Intestinal T and B cells are initially activated and undergo selection and maturation in lymph nodes, PPs, and ILFs, where they are induced to express tissue-specific homing receptors. Lymphocytes that home to the small intestinal mucosa express CCR9, the receptor for CCL25, which is highly expressed in the intestinal mucosa, and $\alpha 4 \beta 7$, an integrin that binds to mucosal vascular addressin cell adhesion molecule 1 (MAdCAM-1) expressed on high endothelial venules of intestinal tissues.

CD103$^+$ DCs isolated from PPs and MLNs imprint gut homing properties on both CD4$^+$ and CD8$^+$ T cells, as well as IgA$^+$ B cells. Such imprinting is dependent on retinoic acid (RA) released by PP and MLN DCs. CD103$^-$CD11b$^+$ cDCs also express the vitamin A (retinol) metabolizing enzyme RALDH1 for the conversion of vitamin A to RA and induce $\alpha 4 \beta 7$ and CCR9 on CD8 T cells. However, since CD103$^+$ cDCs constitute the majority of cDCs migrating to the MLN, the effect of RA metabolism *in vivo* has been ascribed to CD103$^+$ cDCs. RA signaling confers on CD103$^+$ DCs the ability to produce RA, thereby providing autocrine signals for RA production by cDCs. RA also is produced by intestinal epithelial cells from retinol present in intestinal bile, and by stromal cells in the MLN.

12.7 Oral tolerance is dependent on *de novo* induction of T$_{reg}$ cells by CD103$^+$ DCs in the mesenteric lymph node

Oral administration of soluble protein antigens results in systemic tolerance to subsequent immunization and is referred to as oral tolerance. Oral tolerance is mediated by T-cell anergy, T-cell deletion, and the differentiation of CD4$^+$ T$_{reg}$ cells (**Figure 12.4**). Following single high doses of oral soluble protein antigen such as ovalbumin, immunologically intact proteins or peptides are promptly detected in the circulation, and when transferred to an unchallenged host can confer tolerance to subsequent antigen challenge. Further studies with oral peptide administration demonstrated the induction of peripheral antigen-specific T-cell anergy and deletion. Such anergy is likely due to systemic dissemination to resident DCs in LNs and the spleen, followed by activation of T cells in the absence of sufficient costimulation, similar to that shown for soluble proteins administered intravenously.

The *de novo* induction of CD4$^+$Foxp3$^+$ T$_{reg}$ cells (iT$_{reg}$) is essential for oral tolerance. Following oral antigen administration, CD4$^+$Foxp3$^+$ T$_{reg}$ cells are

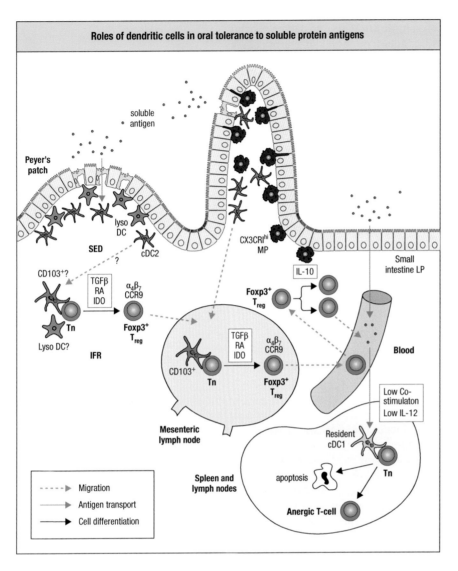

Roles of dendritic cells in oral tolerance to soluble protein antigens

Figure 12.4 Role of conventional dendritic cells (cDCs) in oral tolerance to soluble protein antigens. Oral tolerance is dependent on the induction of CD4+Foxp3+ regulatory T (T$_{reg}$) cells in the mesenteric lymph node. Antigens are presented by cDCs, most importantly CD103+ cDCs, in the presence of transforming growth factor (TGF)-β, retinoic acid (RA), and indolamine 2,3-dioxygenase (IDO), which results in the breakdown and consumption of tryptophan into active metabolites. TGF-β and RA are produced by cDCs, but stromal cells are also an important source of RA; TGF-β also may be produced by other cells. Latent TGF-β is activated by αvβ7 expressed by cDCs. Induced T$_{reg}$ cells migrate to the lymphopoietin guided by the addressins α4β7 and CCR9, driven by RA signaling, where they proliferate in response to interleukin-10 produced by CX3CR1+ cells, most likely macrophages. Expanded T$_{reg}$ cells appear to reenter the circulation for dissemination to systemic sites. T$_{reg}$ cells also are induced efficiently in the Peyer's patches (PPs) and are likely induced by CD103+ cDCs at this site; however, PPs are not required for oral tolerance induction, and T$_{reg}$ cells induced in this site may be more important in controlling responses to particulate antigens and commensal bacteria. High doses of soluble oral antigens can enter the systemic circulation, where they have been shown to induce anergy and deletion of antigen-specific CD4+ T cells, and possibly T$_{reg}$ cells in systemic lymphoid tissue, but the contribution of this pathway to systemic tolerance induction is not clear.

induced in PPs and MLNs. T$_{reg}$ induction in the MLNs is mediated primarily by CD103+ cDCs that have migrated from the intestinal LP. iT$_{reg}$ induction is dependent on the production of TGF-β and RA by CD103+ DCs. CD103+ cDC expression of αvβ8 integrin, important for the activation of latent TGF-β, also is essential for iT$_{reg}$ induction by CD103+ cDCs. RA is a nonessential cofactor for TGF-β-signaling, and in the absence of active TGF-β, RA can contribute to the induction of T$_H$17 cells in the steady state and the induction of T$_H$1 and T$_H$17 cells in inflammation, infectious conditions, and following vaccination.

CD103+ DCs also express indoleamine 2,3-dioxygenase (IDO), an enzyme important for DC tolerogenic functions. IDO participates in tryptophan catabolism, and its immunoregulatory effects may be due to either the reduction of local tryptophan concentration or to the production of immunomodulatory tryptophan metabolites such as kynurenine, which activates the aryl-hydrocarbon receptor (AHR) to suppress T$_H$17 development and enhance T$_{reg}$ functions by suppressing their expression of IFN-γ.

After differentiation in the MLN, iT$_{reg}$ cells migrate in the blood to the intestinal LP, where they undergo expansion required for oral tolerance induction. This has been attributed to CD11c+CD11b+ cells and CX3CR1+ macrophages and their production of IL-10.

Evidence is not currently available to reconcile the effects of systemic antigen dissemination on T-cell anergy and deletion with the essential role of *de novo* induction of T$_{reg}$ cells in mucosal tissues after oral antigen administration. Although further study is required to resolve this issue, "anergic" T cells may represent iT_{reg} cells, reflected in their poor capacity for proliferation.

12.8 DCs are involved in T-cell–dependent IgA B-cell responses in the intestine

In addition to neutralizing toxins and providing protection against infectious agents, a major role for secretory IgA (SIgA) in the intestine is to limit growth and entry of symbiotic bacteria (**Figure 12.5**).

Bacteria, viruses, and particulate antigens that enter PPs and ILFs through M cells, and bacteria taken up by DCs in the terminal ileum, are transported to MLNs within CD103+ DCs. In the MLNs, naïve B cells can undergo IgA class switching and differentiation. IgA switching in the MLNs is limited in the steady state but can be induced in response to systemic and local bacterial challenge. PPs are the primary site for both T-cell–dependent and T-cell–independent IgA B-cell responses in the small intestine. ILFs contribute to T-cell–independent IgA responses to commensal bacteria. Lymphoid structures in the cecum (cecal patches) may be more important for IgA responses in the colon.

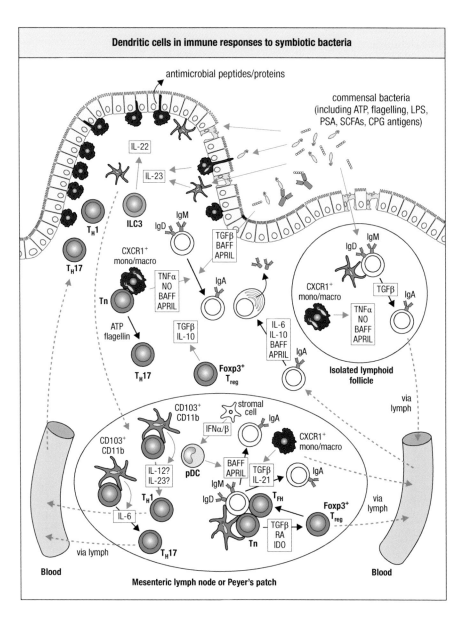

Figure 12.5 Mononuclear phagocyte populations in immune responses to commensal microbiota. Commensal intestinal bacteria and their products can directly or indirectly induce T_H1, T_H17, and Foxp3+ T_{reg} cell responses and drive the differentiation of IgA B cells. CD103+ dendritic cells (DCs) drive the differentiation of CD4+Foxp3+ T_{reg} cells and contribute to IgA B-cell differentiation and homing through the production of retinoic acid, transforming growth factor (TGF)-β, and indolamine 2,3-dioxygenase (IDO). CD4+Foxp3+ T_{reg} cells limit expansion of potentially pathogenic T_H1 and T_H17 cells and can contribute to IgA production by providing TGF-β. T_{reg} cells and T_H17 cells convert into T-follicular-helper cells that provide help (e.g., CD40, IL-21) for T-cell–dependent B-cell IgA isotype switching and affinity maturation. Commensal bacteria also can drive T-cell–independent B-cell differentiation that involves pDC production of BAFF and APRIL, bacteria-induced TGF-β, BAFF and APRIL production by epithelial cells, and CX3CR1+ mononuclear phagocyte release of TNF-α, nitric oxide, BAFF, and APRIL. T_H1 and T_H17 cells are induced by DCs, the latter more specifically by CD103+CD11b+ DCs, and the former possibly more efficiently by CD103−CD11b+ DCs. Through the blood circulation, Peyer's patch and mesenteric lymph node derived IgA+ B cells, CD4+Foxp3+ T_{reg} cells, T_H1, and T_H17 cells reach the lamina propria. IgA+ B cells further differentiate into plasma cells that release dimeric IgA antibodies that are transported to the lumen by the polymeric immunoglobulin receptor. Some specific bacteria (e.g., segmented filamentous bacteria, some *Clostridia* species, *Escherichia coli*, and *Proteus* species) and bacterial products (e.g., LPS and flagellin, CpG oligonucleotides, short-chain fatty acids, adenosine triphosphate, and *Bacteroides fragilis* polysaccharide) have been identified to be important in inducing particular T-cell and IgA responses. See text for further details.

Both T-cell–dependent and T-cell–independent IgA B-cell differentiation is highly dependent on TGF-β, which induces germline Ig α-transcripts in B cells, initiating IgA class-switch recombination (CSR), although RA, LPS, and IL-10 also can contribute to IgA B-cell CSR. In addition, nitric oxide (NO) plays a critical role in IgA CSR and the generation of antibody-producing cells. NO induces the expression of TGF-β receptors and SMAD proteins important for TGF-β-signaling in B cells and expression of B-cell activation and proliferation factors BAFF and APRIL for T-cell–independent responses.

The precise contribution of cDC populations in T-cell–dependent and T-cell–independent IgA B-cell differentiation is currently difficult to discern, as most studies have not defined the involved cDC populations. However, available information suggests that both cDC populations and monocyte-derived cells are important for IgA B-cell differentiation in the intestine.

cDCs are clearly implicated in T-cell–dependent IgA responses in PPs (see Figure 12.5). Initial signals for IgA CSR in response to commensal bacteria and cholera toxin, a highly T-cell–dependent response, occur in the PP SED. CD40/CD40L interaction induces CCR6 expression on pregerminal center B cells to position B cells in the SED. Here CD11c$^+$ cells, including CD11b$^+$ and CD11b$^-$CD8$^-$ cDCs, have intimate prolonged contact with naïve B cells, where they express low levels of activation-induced cytidine deaminase (AID), an enzyme that promotes CSR and affinity maturation through somatic hypermutation (SHM) but at low levels required for CSR.

B cells activated in the SED may also be dependent on T-follicular-helper (T$_{FH}$) cells to provide CD40, IL-21, or other signals for full B-cell differentiation and SHM. T$_{FH}$ cells also have been shown to be important for T-cell–dependent IgA B-cell responses, where they are activated in T-B cell borders. T$_{FH}$ cells in PPs may arise from T$_H$17 cells and/or Foxp3$^+$ T$_{reg}$ cells. PP T$_{FH}$ cells from both sources express FoxP3, and some may express TGF-β, which paradoxically limits T$_{FH}$ function (T$_{FH}$ regulatory cells), rather than acting as a source of TGF-β. cDCs also likely have a role in T$_{FH}$ cell differentiation, as mice with cDC-specific deletion of MHCII have significantly fewer T$_{FH}$ cells (by 50%) and GC B cells (by 70%) compared with *Cre-* littermates. However, when made germ-free, the mice did not generate these cells after reconstitution with commensal bacteria. Furthermore, Batf3$^{-/-}$ mice lacking cDC1s have normal IgA responses in the steady state and after rotavirus infection or immunization with cholera toxin. Together, these findings indicate an important role for cDC2s, or possibly CD103$^-$CD11b$^+$ cDCs, but not cDC1s, in the induction of T$_{FH}$ and T-cell–dependent responses in PPs.

T-cell–independent IgA B-cell CSR relies on the B-cell activators APRIL, and to a lesser degree BAFF substitution for CD40 signaling, to drive AID expression by binding to transmembrane activator and CAML interactor (TACI) on B cells. In PPs, FDCs respond to TLR signals and RA to produce TGF-β, as well as BAFF. BAFF and APRIL in PPs also are produced by pDCs and CD11c$^+$ cells, which may include DCs and macrophages.

Additional potential sources of TGF-β in the PPs are epithelial cells, T cells, B cells, cDCs, and possibly LysoDCs and macrophages. MHCII$^+$CD11c$^+$ PP cells produce RA that contributes to IgA B-cell differentiation, likely by enhancing TGF-β signaling, and inducing α4β7 expression on B-cell blasts that promote homing to the intestinal LP. NO in the PPs is produced in response to TLR signals largely by CD11c^{lo} cells, and in the LP by cells expressing MHC II and low levels of Ly6c, suggesting monocyte-derived populations.

The role of cDCs versus macrophages is less clear in T-cell–independent B-cell CSR in the ILFs, which occurs in response to stimulation with commensal bacteria, as these populations have been poorly defined. Bacteria at these sites stimulate epithelial cells to produce BAFF, APRIL, and TSLP through TLR activation. The production of TSLP can condition CD11c$^+$ cells to produce BAFF, APRIL, and NO, as well as TNF-α, in response to TLRs. TNF-α drives the production of proteinases that activate latent TGF-β produced by stromal cells, whereas NO drives the production of BAFF and APRIL from CX_3CR1^+ cells. Furthermore, $CD11c^+$ $TLR5^+$ cells release RA, which contributes to IgA CSR. Both LP $CD11c^+$ cells and stromal cells produce IL-6 and IL-10, which may contribute to B-cell differentiation into plasma cells. Thus, DCs and macrophages play important roles in T-cell–independent IgA B-cell CSR and differentiation through their production of TGF-β, BAFF, APRIL, TNF-α, NO, IL-6, and IL-10 in response to bacterial signals.

12.9   Specialized functions of DC populations drive T-cell responses in the intestine

Intestinal homeostasis in the setting of approximately 10^{14} commensal bacteria is achieved through the protective mucosal barrier, cytokines, and chemokines that regulate the function of resident innate immune cells, LP macrophages that eliminate bacteria that breach the epithelium, and coordinated T-cell responses to commensal bacteria that prevent pathogenic inflammation.

Commensal bacteria drive $CD4^+Foxp3^+$ T_{reg}, T_H1, and T_H17 cell accumulation and function in the intestinal LP (see **Figure 12.5**). Specific cDC populations are involved in driving these T-cell responses. In early mouse studies, different PP $CD11c^+$ cells were shown to induce unique T-cell responses *ex vivo*: $MHCII^{hi}CD11c^+CD11b^-$ cells induced T_H1 cells, whereas $MHCII^{hi}CD11c^+CD11b^+$ cells induced T_H2 and IL-10-producing cells. Since then, additional functions have been identified for intestinal cDC populations. cDC1s in the small intestine drive T_H1 responses and cross-presentation of soluble antigens to $CD8^+$ T cells; cDC2s drive T_H2 and T_H17 cell differentiation; and $CD103^-CD11b^+$ cDCs drive T_H1- and T_H17-cell differentiation. $CD103^+$ populations can induce T_{reg} differentiation, and all three populations induce the intestinal homing receptor CCR9.

cDC1 cells have been shown in mouse models to have specific effects on the induction, differentiation, and survival of T_H1 cells *in vivo*. The induction of MLN T_H1 cells is reduced, and small intestinal LP $CD4^+$ T cells express lower levels of IFN-γ mRNA in *Xcr1*-DTA mice treated with diphtheria toxin (DT) to selectively eliminate cCD1 cells. In *Cd11c-cre.Irf8*$^{fl/fl}$ mice, T_H1 cells are significantly reduced in the intestinal and colonic LP. cDC1 cell IL-12 contributes to intestinal T_H1 responses during *Toxoplasma gondii* infection and to intestinal T-cell release of IFN-γ, which protects epithelial cells against damage induced by dextran sodium sulfate (DSS).

cDC1 cells also contribute to intestinal T-cell homeostasis, as *Cd11c-cre.Irf8*$^{fl/fl}$ and *Xcr1*-DTA mice treated with DT have reduced numbers of $CD4^+$ and $CD8^+$ LP T cells, conventional $CD8αβ^+TCRαβ^+$ and unconventional $CD8αα^+TCRαβ^+$, and $CD8αα^+TCRγδ^+$ intraepithelial lymphocytes in the small intestine. These effects may be due to poor homing of T cells to the small intestine, as these cells express the highest level of RA in the MLN compared with other cDCs, and their deficiency results in lower numbers of $CCR9^+α_4β_7^+$ small intestinal homing T cells in the MLNs. In addition to decreased homing, low numbers of $CD8^+$ T cells in the LP also may be due to reduced priming in the MLN. High levels of αvβ7 expression by cDC1s compared with other cDCs in the LP may result in higher local concentration of active TGFβ, which contributes to the conversion of $CD4^+$ T cells into $CD4^+CD8αα^+$ intraepithelial lymphocytes in the LP.

In contrast to cDC1s, the *in vivo* function of cDC2 cells and $CD103^-CD11b^+$ cDCs, both of which express SIRPα, have been evaluated in *Cd11c-cre.Irf4*$^{fl/fl}$, *Cd11c-cre.Notch2*$^{fl/fl}$, *Cd11c-cre.Klf4*$^{fl/fl}$, *Clec4a4*-DTR, *Sirpα*$^{-/-}$, as well as *huLangerin*-DTA mice. The loss of SIRPα$^+$ cDCs in these strains reflects both cDC2s and $CD103^-CD11b^+$ cDCs, except *huLangerin*-DTA mice lose primarily cDC2 after DT administration.

cDC2s and $CD103^-CD11b^+$ cDCs have a specific role in T_H17 responses *in vivo*. *Cd11c-cre.Irf4*$^{fl/fl}$, *Cd11c-cre.Notch2*$^{fl/fl}$, and *Sirpα*$^{-/-}$ mice, which have a significant reduction in the number of SIRPα$^+$ cDCs but not cDC1 cells, display a selective loss of T_H17 cells in the intestinal and colonic LP in the steady state. A similar reduction in LP T_H17 cells in DT-treated *huLangerin*-DTA mice and *Cd11c-cre.Notch2$^{fl/fl}$* mice with a deficiency of cDC2 development indicates that cDC2s participate in T_H17 cell responses. Thus, the absence of T_H17 cells in the steadys tate in these cDC-deficient mice implies a role for cDC2 cells in T_H17 responses to commensal bacteria.

The loss of T_H17 cells in the absence of cDC2s and CD103$^-$CD11b$^+$ cDCs may be due to poor T-cell priming in the MLN. *Cd11c-cre.Irf4$^{fl/fl}$* and *Cd11c-cre.Notch2$^{fl/fl}$* mice have low numbers of cDC2s, and *Cd11c-cre.Irf4$^{fl/fl}$* mice have fewer MLN T_H17 cells in the steady state and generate reduced MNL T_H17 responses following systemic immunization with ovalbumin, αCD40, and LPS. Furthermore, the induction of T_H17 responses in the MLN, as well as T_H17 differentiation by cDC2s, was dependent on IL-6, implying a role for IL-6 in T_H17 differentiation in response to commensal bacteria. In addition to IL-6, IL-23 and TGF-β may contribute to cDC-driven T_H17 responses to commensal bacteria, as small intestinal LP CD103$^-$CD11b$^+$ cDCs express IL-12/23p40 mRNA, and LP cDC2s express mRNA for IL-23, TGF-β $\alpha v \delta$ IL-6 in the steady state. Together, these studies provide strong evidence that SIRPα^+ cDCs contribute to driving T_H17 cells in the intestine.

SIRPα^+ cDC also have been shown to be important for T_H2 cell responses in the intestinal mucosa. Although little is known about T_H2 responses to commensal bacteria, *Cd11c-cre.Irf$^{fl/fl}$* do not develop a protective T_H2 response to intestinal helminth infections in mice. Furthermore, cDC2s are important for small intestinal T_H2 responses to *S. mansoni* egg challenge, whereas CD103$^-$CD11b$^+$ cDCs are important for colonic challenge, indicating different cDC populations are involved in the induction of T_H2 cells in different sites.

In addition to driving the induction of CD4$^+$Foxp3$^+$ T_{reg} cells following oral antigen administration, CD103$^+$ cDCs also are implicated in driving T_{reg} induction to commensal bacteria. *huLangerin-DTA x BatF3$^{(-/-)}$* mice deficient in all CD103$^+$ DC populations have fewer T_{reg} cells in the intestinal LP, but the specific cDC populations involved are not clear.

huLangerin-DTA and *Cd11c-cre.Irf4$^{fl/fl}$* mice have lower numbers of cDC2s and CD103$^-$CD11b$^+$ cDCs, and normal numbers of T_{reg} cells in the LP, suggesting a role for cDC1s. Furthermore, *de novo* induction of T_{reg} cells does not occur in the MLN of *Cd11c-cre.Notch2$^{fl/fl}$* mice lacking cDC2s following adoptive transfer of naïve commensal-specific CD4$^+$ TCR-transgenic T cells, whereas T_{reg} induction readily occurs in wild-type mice. Thus, cDC1, CD103$^-$CD1b$^+$ cDCs, and cDC2s may have redundant functions in driving T_{reg} responses to commensal bacteria.

LOCAL FACTORS AFFECT THE DIFFERENTIATION AND FUNCTION OF DCs

Commensal bacterial and tissue-specific factors may influence the differentiation of distinct circulating precursor cells into specific cell phenotypes and functions in the intestine in steady-state conditions (Figure 12.6). These factors include epithelial cell-, macrophage-, neuron-, and stromal cell-derived cytokines, and bacteria- and cell-derived metabolites and neuropeptides.

12.10 Intestinal epithelial cells produce cytokines that condition DCs

Intestinal epithelial cells (IECs) produce TSLP, TGF-β, and RA, which impact DCs in different experimental systems (see Figure 12.6). TSLP is an IL-7-like cytokine that signals through a heterodimer of TSLPR and IL-7Rα to activate STAT5. TSLP potently drives cDC maturation and the capacity of DCs to induce T-cell proliferation and polarization toward T_H2 and T_{reg} responses. In the

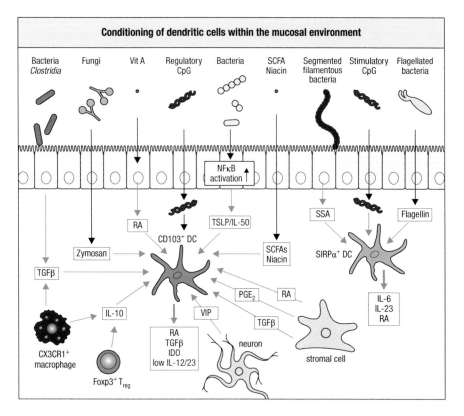

Figure 12.6 Factors affecting conventional dendritic cells (cDCs) within the mucosal environment. Bacterial, epithelial cell, stromal cell, and neuronal cell products, in addition to dietary and microbial metabolites, affect cDC functions in the lamina propria (LP). Circulating pre-DCs can acquire a tolerogenic phenotype in response to epithelial cell production of retinoic acid (RA), transforming growth factor (TGF)-β, and thymic stromal lymphopoietin (TSLP) (IL-50), to active bacterial metabolites such as short-chain fatty acids and niacin, as well as to specific microbial products, such as zymosan (β−glucan) and regulatory CpGs. Immune and nonimmune cells produce interleukin (IL)-10, TGF-β, prostaglandin E2, RA, and vasoactive intestinal peptide (VIP) that can influence CD103$^+$ cDC function. SIRPα$^+$ (cDC2 and CD103$^-$CD11b$^+$) cDCs are influenced by bacterial products, such as flagellin, and stimulatory CpGs, as well as by epithelial cell-derived serum amyloid A. In response to these factors, CD103$^+$ cDCs and SIRPα$^+$ produce different cytokines and factors that control T$_{reg}$ and T$_H$17 cell differentiation. Little is known regarding factors that affect the phenotype of cDC1 cells in the intestine. See text for further details.

mouse, TSLP is constitutively produced largely by colonic, but not intestinal, IECs in response to commensal bacteria, where it restrains T$_H$17 cells and promotes the expansion of iT$_{reg}$ cells. *Tslpr*$^{-/-}$ mice are skewed toward the development of effector T-cell responses in the setting of infection such as helminth infection, which induces NF-κB-dependent TSLP production by small intestinal IECs. In humans, IEC lines and low doses of TSLP can inhibit IL-12 production by monocyte-derived DCs in response to bacteria, driving the differentiation of T$_H$2 cells and T$_{reg}$ cells while inhibiting T$_H$1 differentiation. In addition, TSLP induces human blood DC precursors to potently drive naïve CD4$^+$ T-cell proliferation and the differentiation of T$_{FH}$ and T$_H$2 cells *in vitro*. Human IECs express short and long isoforms of TSLP, the former genetically absent in mice, and may influence TSLP functions in the human intestine. IECs in healthy persons express predominantly the short isoform, which has immunoregulatory activity in a STAT5-independent manner. When administered systemically to mice, it inhibits DSS colitis severity. In contrast, the long isoform is expressed at higher levels in tissue from patients with ulcerative colitis and can drive the production of TNF-α and the T$_H$2 chemokines CCL17 and CCL22 in monocyte-derived DCs. Together, these data indicate that TSLP is produced in the colon by IECs in response to commensal bacteria and in the small intestine in response to infection and contributes to homeostasis by balancing effector and T$_{reg}$ responses, at least in part by limiting cDC production of IL-12, IL-23, and other inflammatory cytokines.

IECs also produce TGF-β and RA, which drive the development *in vitro* of mouse CD103$^+$ DCs from CD103$^-$ cells that are capable of inducing CD4$^+$Foxp3$^+$ T$_{reg}$ cells. Adoptive transfer of CD4$^+$Foxp3$^+$ T$_{reg}$ cells suppresses experimental colitis. As previously mentioned, TGF-β activation by CD11c$^+$ cells is essential for normal induction and maintenance of CD4$^+$Foxp3$^+$ T$_{reg}$ cells by DCs and intestinal homeostasis *in vivo*.

The ability of IECs to condition DCs may be largely influenced by their interactions with commensal bacteria. For example, IEC expression of TGF-β is induced by *Clostridia* species, which have a particular propensity for inducing LP T$_{reg}$ cells. IECs also produce TSLP and TGF-β1 *in vitro* when exposed to certain species of bacteria, and exposure of IECs to a probiotic strain of *Lactobacillus paracasei* induces factors that block the production of pro-inflammatory cytokines by activated DCs. Probiotic-treated epithelial cell supernatant can reduce IL-12p70 release and the capacity of DCs to drive the development of T_H1 T cells in response to *Salmonella*. These data indicate that intestinal bacteria can regulate DC function by indirect effects on epithelial cells. DCs may also be influenced directly by microbial-derived metabolites that can freely cross the intestinal epithelium.

12.11 Stromal cells and macrophages also influence the phenotype of DCs

Stromal cells also play an important role in conditioning cells in the intestine. Stromal cells can produce TGF-β, RA, and prostaglandin E2 (PGE_2). Stromal cell–derived TGF-β induces the differentiation of blood monocytes into intestinal macrophages, and stromal cell–derived RA is important in conditioning MLN DCs to then confer homing properties on T cells. In addition, stromal cells constitutively express Cox-2 and produce PGE_2. PGE_2 inhibits IL-12 production by human cDCs, impairs production of type I IFNs by mouse PP pDCs, and promotes oral tolerance.

Mouse intestinal macrophages also have the capacity to regulate DC function through their production of IL-10, which can inhibit DC production of IL-12 and other pro-inflammatory cytokines. Although IL-10 appears to be essential for preventing spontaneous inflammation in mice, an essential source of IL-10 for protecting mice in the steady state appears to be T_{reg} cells, rather than macrophages. When cultured with intestinal macrophages, LP DCs acquire a reduced capacity to induce T_H17 T cells independent of IL-10 production by unclear mechanisms. However, macrophages also can release GM-CSF, which has been shown to cooperate with RA and IL-4 in inducing RALDH2 expression in bone marrow–derived DCs and to favor Foxp3$^+$ T_{reg} development.

INFLAMMATORY BOWEL DISEASE

Crohn's disease and ulcerative colitis, the two major forms of IBD in humans, result from an inappropriate and exaggerated mucosal immune response to gut bacteria in genetically susceptible persons. DCs play a key role in the immunopathogenesis of these diseases (**Figure 12.7**).

12.12 Intestinal DCs in mice with experimental colitis display impaired induction of T_{reg} cells and enhanced induction T_H1 and T_H17 cell differentiation

In mouse models of IBD, an influx of neutrophils and monocytes that become inflammatory cells produce high levels of IL-6, IL-1, and TNF-α, as well as chemokines that attract additional neutrophils, monocytes, and lymphocytes. In addition, DC precursors are recruited, rapidly mature into cells with pro-inflammatory features, and migrate to MLNs to drive T-cell responses dominated by pathogenic T cells.

In the transfer model of mouse colitis in which naïve Foxp3$^-$ CD45RB^{hi} CD4$^+$ T cells are transferred into lymphocyte-deficient RAG$^{-/-}$ or SCID mice, CD11c$^+$ cDCs have an activated phenotype in colonic mucosa and MLNs. CD103$^+$ DCs isolated from the MLNs of mice with induced colitis are deficient in their ability to induce Foxp3$^+$ T_{reg} cells and instead induce the differentiation of T_H1 and T_H17 effector T cells. Activated MLN CD103$^+$ cDCs produce IL-23, which can enhance T_H1 and T_H17 differentiation by blocking the differentiation of Foxp3$^+$ T_{reg} cells and expanding the T_H17 cell

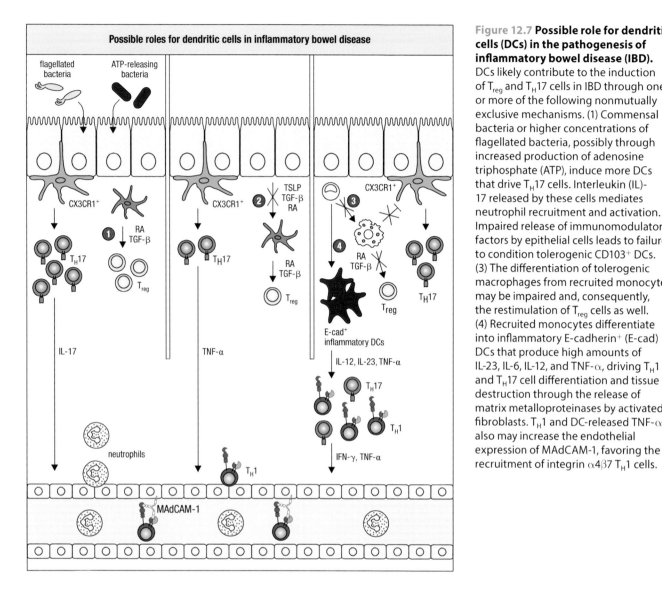

Figure 12.7 Possible role for dendritic cells (DCs) in the pathogenesis of inflammatory bowel disease (IBD). DCs likely contribute to the induction of T_{reg} and T_H17 cells in IBD through one or more of the following nonmutually exclusive mechanisms. (1) Commensal bacteria or higher concentrations of flagellated bacteria, possibly through increased production of adenosine triphosphate (ATP), induce more DCs that drive T_H17 cells. Interleukin (IL)-17 released by these cells mediates neutrophil recruitment and activation. (2) Impaired release of immunomodulatory factors by epithelial cells leads to failure to condition tolerogenic CD103$^+$ DCs. (3) The differentiation of tolerogenic macrophages from recruited monocytes may be impaired and, consequently, the restimulation of T_{reg} cells as well. (4) Recruited monocytes differentiate into inflammatory E-cadherin$^+$ (E-cad) DCs that produce high amounts of IL-23, IL-6, IL-12, and TNF-α, driving T_H1 and T_H17 cell differentiation and tissue destruction through the release of matrix metalloproteinases by activated fibroblasts. T_H1 and DC-released TNF-α also may increase the endothelial expression of MAdCAM-1, favoring the recruitment of integrin $\alpha4\beta7$ T_H1 cells.

population. Migratory CD11b$^+$ and CD11b$^-$ cDCs that accumulate in the MLNs in T-cell transfer colitis also may drive T_H1 responses. Finally, following adoptive transfer of CD4$^+$Foxp3$^+$ T_{reg} cells into colitic mice to treat the colitis, CD4$^+$Foxp3$^+$ T_{reg} cells associated with LP DCs form tight clusters around the DCs, suggesting that CD4$^+$Foxp3$^+$ T_{reg} cells interact directly with DCs to promote their expansion and regulatory function or block interaction with effector cells.

12.13 DCs are activated in the intestinal mucosa in human IBD

Study of DCs and macrophages in humans with IBD has been limited largely to inflamed tissue and blood cells, resulting in incomplete characterization of cDCs with nonspecific surface markers. Despite limited information, possible roles of DCs in IBD are shown in Figure 12.7. In humans, activated cDCs accumulate along with macrophages at sites of mucosal inflammation. In Crohn's disease tissue, DCs that express CD83, a glycoprotein associated with DC activation, are present in association with numerous CD83$^-$ CD80$^+$ DC-SIGN$^+$ DCs, producing IL-12 and IL-18. The expression of TLR2, TLR4, and CD40 is enhanced in DCs isolated from inflamed mucosa that also overproduce IL-6 and IL-12. Furthermore, mature DCs recruited to the lamina propria form clusters with proliferating T cells in the affected colonic tissue. In ulcerative colitis, an increase in activated CD83$^+$ cells produces macrophage inhibitory factor (MIF), which is thought to contribute to neutrophil recruitment and activation.

In Crohn's disease, activated DCs may migrate from the mucosa to MLNs. DC populations have been described in MLNs from patients with Crohn's disease that initiate potent T$_H$1 responses and have an enhanced ability to drive T$_H$17 differentiation. In addition, MLN DCs from patients release significantly higher amounts of IL-23 but lower amounts of IL-10 compared with MLN DCs from patients with ulcerative colitis or healthy subjects. Although CD103$^+$ DCs capable of driving homing receptor expression and Foxp3$^+$ T$_{reg}$ differentiation are present in the MLNs of both normal humans and patients with CD, DCs from Crohn's MLNs express higher levels of the activation markers CD83 and CD40 than DCs from noninflamed MLN. Thus, MLN DCs may be involved in Crohn's disease pathogenesis through the suppression of T$_{reg}$ differentiation and the initiation of potent inflammatory T$_H$1 T cells. Together, these findings are consistent with the concept that DC populations in humans with IBD are highly activated and produce high levels of inflammatory cytokines that skew T cells toward pathogenic T-cell populations, and suggest that MLNs are a primary site of T-cell differentiation, similar to mouse models of IBD.

SUMMARY

DCs play a fundamental role in the maintenance of homeostasis of the intestinal mucosa in steady-state conditions and in the induction of T-cell responses to pathogenic microbes in the setting of infection. These functions are accomplished by specialized DC populations that are either resident within lymphoid and peripheral tissues or that migrate from tissues to draining lymph nodes. DCs are conditioned by local environmental factors, many of which are induced by symbiotic bacteria, and in turn, DCs contribute to the maintenance of a symbiotic relationship with the intestinal microbiota. Unraveling the biology of DC activation and response to the local microenvironment will further our understanding of intestinal immune disorders.

FURTHER READING

Allaire, J.M., Crowley, S.M., Law, H.T. et al.: The intestinal epithelium: Central coordinator of mucosal immunity. *Trends Immunol.* 2018, 39(9):677–696.

Bekiaris, V., Persson, E.K., and Agace, W.W.: Intestinal dendritic cells in the regulation of mucosal immunity. *Immunol. Rev.* 2014, 260(1):86–101.

Bernardo, D., Chaparro, M., and Gisbert, J.P.: Human intestinal dendritic cells in inflammatory bowel diseases. *Mol. Nutr. Food. Res.* 2018, 62(7):e1700931.

Bonnardel, J., Da Silva, C., Wagner, C. et al.: Distribution, location, and transcriptional profile of Peyer's patch conventional DC subsets at steady state and under TLR7 ligand stimulation. *Mucosal Immunol.* 2017, 10(6):1412–1430.

Da Silva, C., Wagner, C., Bonnardel, J. et al.: The Peyer's patch mononuclear phagocyte system at steady state and during infection. *Front. Immunol.* 2017, 8:1254.

Erkelens, M.N., and Mebius, R.E.: Retinoic acid and immune homeostasis: A balancing act. *Trends Immunol.* 2017, 38(3):168–180.

Honda, K., and Littman, D.R.: The microbiota in adaptive immune homeostasis and disease. *Nature* 2017, 535(7610):75–84.

Joeris, T., Muller-Luda, K., Agace, W.W. et al.: Diversity and functions of intestinal mononuclear phagocytes. *Mucosal Immunol.* 2017, 10(4):845–864.

Knoop, K.A., and Newberry, R.D.: Goblet cells: Multifaceted players in immunity at mucosal surfaces. *Mucosal Immunol.* 2018, 11(6): 1551–1557.

Lamas, B., Natividad, J.M., and Sokol, H.: Aryl hydrocarbon receptor and intestinal immunity. *Mucosal Immunol.* 2018, 11(4):1024–1038.

Magnusson, M.K., Brynjolfsson, S.F., Dige, A. et al.: Macrophage and dendritic cell subsets in IBD: ALDH+ cells are reduced in colon tissue of patients with ulcerative colitis regardless of inflammation. *Mucosal Immunol.* 2016, 9(1):171–182.

Mowat, A.M., and Agace, W.W.: Regional specialization within the intestinal immune system. *Nat. Rev. Immunol.* 2014, 14(10):667–685.

Nakamura, Y., Kimura, S., and Hase, K.: M cell-dependent antigen uptake on follicle-associated epithelium for mucosal immune surveillance. *Inflamm. Regen.* 2018, 38:15.

Pabst, O., and Mowat, A.M.: Oral tolerance to food protein. *Mucosal Immunol.* 2012, 5(3):232–239.

Sun, M., Wu, W., Liu, Z. et al.: Microbiota metabolite short chain fatty acids, GPCR, and inflammatory bowel diseases. *J. Gastroenterol.* 2017, 52(1):1–8.

Intestinal macrophages in defense of the mucosa

LESLEY E. SMYTHIES AND PHILLIP D. SMITH

The gastrointestinal mucosa, the largest body surface to interact with the external environment, contains the most abundant population of macrophages in the body (Figure 13.1). The three major functions of intestinal macrophages include protection against pathogens and foreign substances that breach the epithelium, contribution to tolerance to commensal bacteria and food antigens, and clearance of apoptotic and dead cells in the lamina propria. Intestinal macrophages mediate these functions through powerful phagocytic and bactericidal capabilities, but, in contrast to macrophages in other tissues, are profoundly restricted in pro-inflammatory capability in healthy mucosa in the steady state. Specifically, intestinal macrophages display markedly downregulated pro-inflammatory cytokine production, decreased antigen presentation, and diminished oxygen radical production—a unique macrophage phenotype that serves to maintain low-level inflammation in normal intestinal mucosa. Focusing on human intestinal macrophages, here we discuss intestinal macrophage effector functions in the context of their evolutionary development and differentiation from pro-inflammatory monocytes recruited from the circulation into uniquely inflammation anergic resident intestinal macrophages.

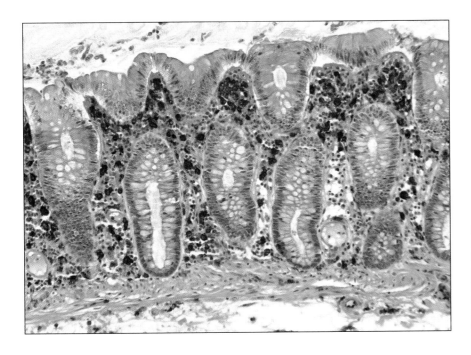

Figure 13.1 **Human intestinal macrophages are distributed throughout the subepithelial lamina propria.** Section of normal human small intestine shows CD68+ macrophages distributed throughout the lamina propria but concentrated in the subepithelial region (×20).

ORIGIN OF LAMINA PROPRIA MACROPHAGES

Lamina propria macrophages are part of the system of first-line defense mechanisms referred to collectively as the innate immune system, which orchestrates initial responses to invading microorganisms and their products. Primitive macrophage-like cells arose a billion years ago in early unicellular organisms such as protozoa, evolved in multicellular invertebrates, including insects such as the *Drosophila*, then evolved further in primitive vertebrates such as the lamprey, and finally vertebrates. Macrophages are most numerous in the gastrointestinal mucosa, where they reside in close proximity to the diverse multitude of microbial species that colonize the gut mucosa. Through complex cross-talk pathways involving the microbiota, epithelium, and stroma, intestinal macrophages contribute to the regulation of mucosal homeostasis. In contrast to the innate immune system, the acquired immune system is composed of second-line mechanisms of adaptive or acquired immune defense and arose approximately 400–500 million years ago, during the separation of vertebrates from invertebrates. Thus, the acquired immune system is present in fish, amphibians, reptiles, birds, and mammals, but not invertebrates.

The mucosal innate and acquired immune systems have two goals in common: the recognition and elimination of invading pathogens, and the recognition and tolerance of non-harmful (and often beneficial) commensal organisms. Contributing to these goals at the level of the mucosa, intestinal and colonic epithelial cells form a structural barrier consisting of a single layer of cells that separates luminal microorganisms from the subepithelial lamina propria, where intestinal macrophages reside. Goblet cells in the epithelium secrete a protective coat of mucus that contains antimicrobial peptides, including defensins and other proteins such as serine leukocyte protease inhibitor, that restrict microbial translocation. In the setting of mucosal injury or infection, microbes that translocate directly into the lamina propria are contained by intestinal macrophages located strategically in the subepithelial space (see Figure 13.1). In normal or healthy mucosa, microbes enter via dendritic cell (DC) para-epithelial extensions, M cells, and epithelial cell transcytosis to initiate contact with subepithelial effector defense cells, especially lamina propria macrophages.

Effector cells of the innate immune system include macrophages, DCs, neutrophils, eosinophils, and natural killer cells. Effector T cells and B cells of the acquired immune system act in concert with the innate immune system. Innate responses are initiated within minutes and are directed toward conserved carbohydrate, lipid, and nucleic acid structures, referred to collectively as pathogen-associated molecular patterns (PAMPs), which are expressed by microbes and are essential for microbe survival. Macrophages recognize PAMPs through predetermined repertoires of pattern-recognition receptors (PRRs) that include the germline-encoded transmembrane toll-like receptors (TLRs), cytosolic nucleotide-binding oligomerization domain (NOD)–like receptors, retinoic acid inducible gene-1 (RIG)–like receptors, and endocytic C-type (calcium-requiring) lectin receptors (CLRs). The predetermined nature of PRRs facilitates rapid innate responses to conserved microbial antigens but limits the diversity of ligands to which macrophages can respond. Indeed, the innate immune system cannot easily respond to novel molecules because memory for antigen recall is not a feature of macrophages or other innate immune cells. Adaptive responses to infectious agents by effector T and B cells are slower (hours to days) and use antigen-specific receptors, T-cell receptors, and immunoglobulins (Igs) to recognize antigenic epitopes on protein antigens. In contrast to the evolutionarily conserved genes for innate response receptors, the genes encoding antigen-specific receptors undergo somatic recombination, thereby conferring on the acquired immune system the ability to identify and recall novel antigens. Thus, the innate and acquired immune systems complement and support one another to provide broad, rapid responses while developing specific, albeit delayed, responses to foreign cells and antigenic epitopes.

13.1 Mucosal macrophages are derived from pluripotent stem cells in bone marrow

Intestinal macrophages are derived from bone marrow hematopoietic stem cells, which originate from fetal liver hematopoietic stem cells. Elegant mutant mouse and fate-mapping studies in mice have shown that after infancy, non-replicating intestinal macrophages are derived and maintained by blood monocytes, although recent studies have also identified a population of locally maintained resident intestinal macrophages in mice (see later). In contrast, brain macrophages (microglia) originate exclusively from yolk sac progenitors and liver, lung, splenic, and peritoneal macrophages originate from varying contributions of yolk sac and fetal liver progenitors; macrophages in these tissues self-maintain in the adult through longevity and limited self-renewal.

In the bone marrow, pluripotent stem cells are exposed sequentially to combinations of growth factors, hormones, and cytokines that regulate cell maturation and differentiation into common lymphoid progenitor cells and common myeloid progenitor cells. The cytokines interleukin (IL)-1, IL-3, and IL-6 together stimulate pluripotent stem cells to lose their capacity for self-renewal and become myeloid progenitor cells capable of differentiation into monocytic, granulocytic, megakaryocytic, or erythroid lineages. Myeloid progenitor cells become committed to a granulocyte/monocyte lineage and lose their ability to differentiate into the other myeloid lineages under the continued influence of IL-1 and IL-3. Further exposure to this combination of cytokines plus granulocyte-macrophage colony-stimulating factor (GM-CSF) induces the proliferation of both granulocyte and monocyte precursors. Exposure to IL-1, IL-3, GM-CSF, and then macrophage CSF (M-CSF), induces the sequential differentiation of monocyte precursors, monoblasts, promonocytes, and finally monocytes, which are released from the bone marrow and enter the circulation. Under homeostatic conditions, the same precursors can differentiate into monocytes, DCs, or polymorphonuclear neutrophils, but differentiation can be redirected in appropriate circumstances. In addition to the cytokine growth factors mentioned earlier, each stage of this differentiation pathway may be influenced by constitutive and inducible transcription factors, including PU.1 and AML1, which direct the expression of myeloid-specific genes involved in monocyte differentiation. PU.1 is particularly important because it regulates the expression of the receptor for M-CSF, which is critical for M-CSF-dependent differentiation. The transcription factors GATA-2, SCL, and Myb regulate myeloid cell survival. Additional transcription factors, including CCAAT-enhancer-binding protein-β (C/EBP-β), HOXB7, and c-Myc, regulate the intermediate stages of myeloid differentiation, whereas C/EBPβ, EBR-1, IRF-1, NF-Y, and some Jun/Fos and STAT proteins regulate monocyte maturation.

13.2 Blood monocytes continually populate lamina propria of uninflamed healthy intestinal mucosa

After leaving the bone marrow, monocytes circulate in the blood compartment for approximately 3–4 days, during which time a proportion of monocytes migrate into the lamina propria, where they embed in the extracellular matrix (stroma) and differentiate into tissue macrophages (Figure 13.2). Monocytes recruit into the lamina propria in response to local chemotactic cytokines (chemokines), which are divided into lymphoid and inflammatory chemokines. The lymphoid chemokines recruit cells into noninflamed tissue, whereas inflammatory chemokines recruit cells into inflamed tissue. In this

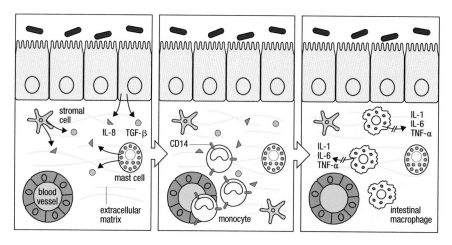

Figure 13.2 Monocytes recruit to the lamina propria and differentiate into inflammation anergic intestinal macrophages in the stroma. In the homeostatic conditions of normal mucosa, TGF-β and IL-8 produced by epithelial cells, mast cells, and stromal cells bind to and are released from the stromal extracellular matrix (first panel). Mucosal factors, including TGF-β and IL-8, promote the recruitment of resting but potentially pro-inflammatory CD14$^+$ blood monocytes from the circulation into the mucosa (center panel). Newly recruited monocytes rapidly differentiate into inflammation-anergic intestinal macrophages as they embed in the lamina propria stroma (last panel). TNF-α, tumor necrosis factor-α.

connection, circulating monocytes express abundant chemokine receptors and adhesion molecules that facilitate their recruitment to, and retention in, the tissues. Reflecting the remarkable plasticity of mononuclear phagocytes, the differentiation of newly recruited monocytes into tissue macrophages, rather than DCs, is regulated by local cytokines, including IL-6 and M-CSF. During differentiation, monocytes enlarge as they increase their content of lysosomal and hydrolytic enzymes and their number and size of mitochondria to become terminally differentiated resident macrophages. Distinct functional roles have been reported for macrophages from different tissues on the basis of macrophage cytokine and chemokine release and enzyme production. Thus, macrophages (particularly mouse macrophages) have been characterized as classically activated (M1) or alternatively activated (M2) macrophages along the lines of the CD4$^+$ helper T-cell T_H1 and T_H2 paradigm. Human blood monocytes also polarize into M1/M2 macrophages, but human intestinal macrophages in the steady state appear to be terminally differentiated and do not undergo polarization. Importantly, macrophages from different tissues may be distinguished phenotypically and functionally, suggesting that local tissue microenvironments influence macrophage differentiation and contribute to macrophage heterogeneity. For example, blood monocytes express high levels of CD14, the receptor for lipopolysaccharide (LPS) and LPS-binding protein, and CD89, the receptor for IgA, but promptly lose these receptors during differentiation in the intestinal mucosa.

In healthy intestinal mucosa, epithelial cells, mast cells, and stromal cells (fibroblasts, myofibroblasts, and pericytes) produce and release transforming growth factor-β (TGF-β), the most potent of all monocyte chemokines, and IL-8, which is chemotactic for monocytes as well as neutrophils. These chemokines bind to, and are released from, the lamina propria stroma (see **Figure 13.2**). *In vitro* studies of human intestinal tissues have shown that stromal TGF-β and IL-8 recruit blood monocytes, which express both TGF-β receptors I and II (TGF-β RI and RII) and IL-8 receptors (CXCR1 and CXCR2). Once recruited to the lamina propria, monocytes take up residence in the stroma and differentiate into resident macrophages. In mice, blood monocytes express CX_3CR1, the receptor for fractalkine (CX_3CL1), which

Figure 13.3 **Resting human blood monocytes and intestinal macrophages are phenotypically and functionally distinct.** *Blood monocytes, but not intestinal macrophages, upregulate expression of CD25 in response to stimulation. (Adapted from Smith, P.D. et al. *Mucosal Immunol.* 2011, 4:31–42, with permission from Nature Publishing Group, and Dennis, E. et al. *Curr. Protoc. Immunol.* 2017, 118:14.3.1–14.3.14.)

Phenotype	Blood monocyte	Intestinal macrophage
CD4	+	low
CD11a, b, c	+	–
CD13	+	+
CD14	+	–
CD16, 32, 64	+ (subset)	–
CD18/integrin β_2	+	–
CD25/IL-2Rα*	–	–
CD33	+	+
CD36	+	+
CD40/TNFRSF5	+	–
CD69	– (transient)	–
CD80/B7-1	+	–
CD86/B7-2	+	–
CD88/C5aR	+	low
CD89/FcαR	+	–
CD123/IL-3Rα	+	–
HLA-DR	+	+
TGF-βRI	+	+
TGF-βRII	+	+
CX$_3$CR1	–	–
IL-10R	–/+	–
TREM-1	+	–
TLR1,2, 4, 6	low	low
TLR3, 5, 7–9	+	+
Function		
Phagocytosis	+	+
Killing	+	+
Chemotaxis	+	–
Respiratory burst	+	–
Antigen presentation	+	?
Cytokine production	+	
Co-stimulation	+	–
Polarization capacity	+	–

likely facilitates mucosal localization of the recruited monocytes. However, human blood monocytes and intestinal macrophages do not express CX$_3$CR1 (Figure 13.3). The constitutive expression of these, and probably other, chemokines by mucosal cells promotes the continuous recruitment of blood monocytes to the mucosa, which, together with the long lifespan of resident macrophages, makes the gastrointestinal tract lamina propria the largest reservoir of macrophages in the body. Importantly, human lamina propria macrophages do not proliferate. After terminal differentiation, the macrophages survive for weeks to months as resident macrophages in the lamina propria of the small intestine and colon, after which they undergo programmed cell death and are replaced by newly recruited blood monocytes.

13.3 Blood monocytes are predominant source of macrophages in inflammatory lesions in intestinal mucosa

In studies of macrophage accumulation in inflamed gastrointestinal mucosa, immunohistochemical analysis has shown that the endothelial cells lining small blood vessels in the mucosa of patients with Crohn's disease display high levels of CD34, a ligand that promotes the rolling of L-selectin$^+$ monocytes in high endothelial venules. More recently, antibody blocking studies have revealed that P-selectin glycoprotein ligand-1, P-selectin, and vascular cell adhesion molecule-1 (VCAM-1) promote CD14$^+$ monocyte rolling and adherence in the intestinal mucosa, particularly ileal mucosa, in a mouse model of spontaneous ileitis. Increased levels of intercellular adhesion molecule-1 (ICAM-1) and CD31, which facilitate the transendothelial migration of monocytes, are also present in Crohn's disease lesions. Thus, endothelial cells in mucosal vessels express molecules that promote the rolling, adherence, and subsequent transendothelial migration of circulating blood monocytes into inflamed gastrointestinal mucosa. In addition, recruitment factors, including CCL2 and CCL4, which selectively induce monocyte transendothelial migration and accumulation, may be released in inflamed and/or infected mucosa, enhancing the migration of monocytes into the mucosa. The interdiction of such recruitment is an attractive therapeutic strategy, but localization of such therapy to inflammatory sites will be difficult to achieve.

In the 1990s, macrophages in normal human colonic and intestinal mucosa were shown to lack CD14 (see Figure 13.3). However, the expression of this receptor on a substantial proportion of macrophages in inflammatory lesions in the mucosa of patients with inflammatory bowel disease (IBD) suggested that the lesions were populated by newly recruited blood monocytes, which express CD14. In experiments designed to explore the origin of CD14$^+$ macrophages in human inflamed intestinal mucosa, investigators harvested CD14$^+$ blood monocytes from subjects with IBD, labeled the cells with technetium Tc 99m, inoculated the cells back into the subjects, and then examined the lesions: the labeled monocytes recruited to the inflamed mucosa, confirming that the CD14$^+$ macrophages in the inflammatory lesions were newly recruited pro-inflammatory blood monocytes. The absence of respiratory burst activity in macrophages in normal gut mucosa, but the presence of this function in macrophages in the inflamed mucosa of patients

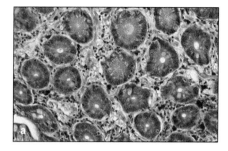
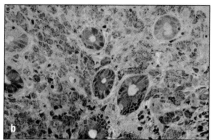

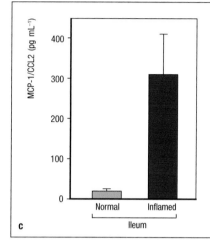
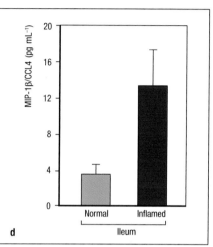

Figure 13.4 Monocytes migrate into the lamina propria in response to infection or inflammation. (**a**) Normal intestinal mucosa in which CD68+ macrophages are distributed throughout the lamina propria. (**b**) Infection of intestinal mucosa by *M. avium* (identified by red acid-fast stain), which results in striking increases in macrophage accumulation and phagocytosis of the bacteria in the lamina propria. Conditioned medium generated from inflamed (inflammatory bowel disease) intestinal lamina propria contains increased levels of monocyte-targeted chemokines, including MCP-1/CCL2 (**c**) and MIP-1β/CCL4 (**d**), which contribute to the recruitment of monocytes into inflamed or infected intestinal mucosa. (Reprinted from Smith, P.D. et al. *Mucosal Immunol.* 2011, 4:31–42. With permission from Macmillan Publishers Ltd.)

with Crohn's disease, is consistent with the recruitment of blood monocytes to sites of mucosal inflammation. In addition, triggering receptor expressed on myeloid cells (TREM-1) is present on blood monocytes and mucosal macrophages in inflamed IBD mucosa, but not on mucosal macrophages in normal intestinal mucosa. Furthermore, in contrast to monocytes, resident intestinal macrophages do not undergo chemotaxis. Together, these findings support the concept that inflammatory lesions are populated largely by newly recruited monocytes.

In addition to mucosal inflammatory diseases, the recruitment of monocytes to the gastrointestinal mucosa is accelerated during mucosal infections. In *Mycobacterium avium*-infected mucosa, for example, the number of mucosal macrophages, including those that contain internalized mycobacteria, increases in response to bacterial invasion (**Figure 13.4**). Mucosal infection by cytomegalovirus also is accompanied by macrophage accumulation, even though the majority of such macrophages do not contain the virus. Whether the mucosal macrophages associated with or infected by opportunistic pathogens in immunosuppressed subjects exhibit a CD14+ phenotype has not been defined, but the cells probably represent monocyte-derived macrophages that have exited from the circulation at sites of infection in response to the elaboration of chemokines induced by the intracellular pathogen.

Role of lamina propria macrophages in intestinal homeostasis

For millions of years, a highly contaminated environment challenged evolving host defense mechanisms of early vertebrates, nonhuman primates, and eventually humans, with a spectrum of colonized microorganisms and infectious pathogens likely very different from the "typical" enteric microbiota of today. In this inhospitable environment, a disrupted intestinal epithelium resulting from frequent infections was probably common. Responding to this formidable immunostimulatory challenge, the intestinal innate immune system developed host defense mechanisms to respond to microorganisms that breach the disrupted epithelium. Because the exposure of myeloid effector cells to bacteria or their products can trigger a potent, and potentially life-threatening, inflammatory response, the gastrointestinal mucosa evolved the capacity to downregulate inflammatory, but not host defense, responses to luminal microorganisms. Thus, resident lamina propria macrophages emerged, at least in humans, that were unique for their capacity to phagocytose and digest microorganisms and innate material without a concomitant inflammatory component, a distinct selective advantage to the human host. Extended to immune surveillance, non-inflammatory host defense also enabled intestinal macrophages to scavenge apoptotic mononuclear cells without the release of pro-inflammatory cytokines.

In the gastrointestinal mucosa, macrophages are strategically located adjacent to the epithelial basement membrane in close proximity to luminal bacteria, as well as throughout the lamina propria (see Figure 13.1). The number of luminal bacteria in the duodenum and jejunum is relatively low compared with that of the colon (10^3–10^4 versus 10^{10}–10^{12} bacteria per milliliter luminal contents) but sufficient to activate monocyte-derived macrophages *in vitro*. Nevertheless, normal intestinal tissue is characterized by strikingly low levels of inflammation, designated "physiological inflammation," suggesting that the large number of effector cells, including lamina propria macrophages, resident in the intestinal mucosa are actively downregulated by the mucosal microenvironment. We next discuss how the unique phenotype and profoundly downregulated pro-inflammatory function of intestinal macrophages, in concert with powerful defense mechanisms, contribute to the regulation of inflammation (homeostasis) in the normal intestinal mucosa.

13.4 Intestinal macrophages are uniquely inflammation anergic

Circulating monocytes typically express surface molecules, including chemokine receptors, adhesion molecules, TLRs, the LPS receptor, and Fc receptors, which contribute to cell recruitment, retention, and PAMP recognition at sites of infection and/or inflammation. In addition to these key innate receptors, blood monocytes have powerful and inducible pro-inflammatory capabilities. Together, these features equip monocytes for rapid, pro-inflammatory defense after they take up residence in most organ tissues. In the gut lamina propria, however, macrophages display a unique innate receptor phenotype and potently downregulated pro-inflammatory capabilities. This phenotypic and functional profile reflects both the plasticity of mononuclear phagocytes and the unique microenvironment of the intestinal mucosa and is termed *inflammation anergy*.

Intestinal macrophages display many, but not all, of the receptors expressed by blood monocytes. For example, intestinal macrophages and monocytes express high levels of the major histocompatibility complex (MHC) class II molecule human leukocyte antigen (HLA)-DR and the myeloid marker aminopeptidase N (CD13). Both populations also express surface receptors involved in the recognition of, and interaction with, potentially harmful microbes; these receptors include TLR1 and TLR3–9 (Figures 13.3 and 13.5a), as well as TGF-β RI and RII, which mediate recruitment (see earlier) and active Smad signaling (see later). However, in sharp contrast to blood monocytes, resident macrophages in healthy mucosa do not express the receptors for LPS (CD14), IgA (CD89), IgG (CD16, 32, and 64), CR3 (CD11b/CD18), CR4 (CD11c/CD18), growth factor receptors for IL-2 (CD25) and IL-3 (CD123), the integrin leukocyte function-associated antigen-1 (LFA-1), CD11a, and TREM-1. Intestinal macrophages express very low levels of

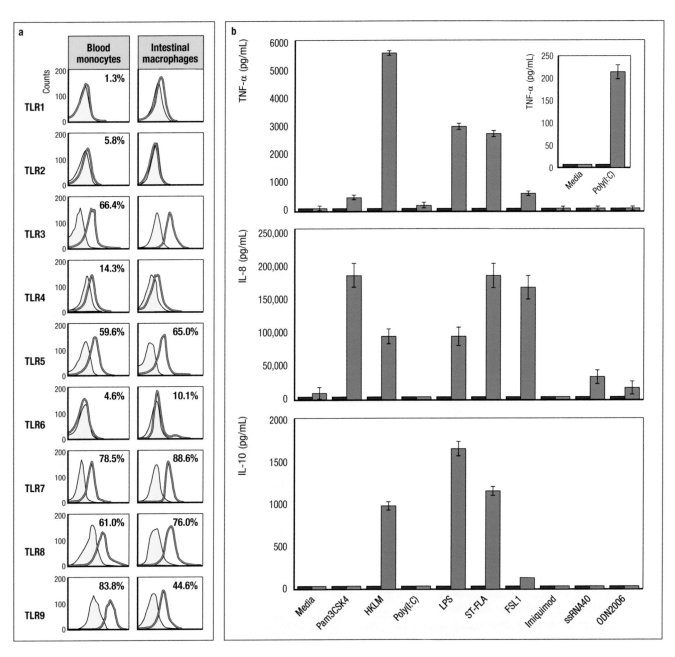

Figure 13.5 Intestinal macrophages express toll-like receptors (TLRs) but do not produce inflammatory cytokines. (a) Comparison of the expression of levels of TLRs in intestinal macrophages and blood monocytes as determined by flow cytometry. Open blue histograms indicate TLR expression on autologous intestinal macrophages and blood monocytes; yellow histograms indicate isotype control staining. (b) Blood monocytes, but not intestinal macrophages, produce tumor necrosis factor-α, interleukin (IL)-8, and IL-10 in response to exposure to ligands for TLRs1-9. (Adapted from Smythies, L.E. et al. *J. Biol. Chem.* 2010, 285:19593–19604. With permission from the American Society for Biochemistry and Molecular Biology.)

chemokine receptors CCR5 and CXCR4 and receptors for the chemotactic ligands f-Met-Leu-Phe and C5a. This unusual phenotype has profound functional implications. In particular, the absence of CD14 is consistent with the inability of intestinal macrophages to respond to LPS, a feature well suited to macrophages residing in a microenvironment potentially rich in immunostimulatory LPS. The absence of the receptors for IgA and IgG limits macrophage recognition and uptake of IgA- and IgG-opsonized microorganisms and particles, thereby reducing macrophage integration of innate and adaptive immune responses. The striking downregulation of CCR5 and CXCR4 contributes to the non-permissiveness of intestinal macrophages to CCR5-tropic (R5) and CXCR4-tropic (X4) human immunodeficiency virus 1 (HIV-1), the causative agent of the acquired immunodeficiency syndrome (AIDS). This mucosa-specific phenotype disables the interaction between intestinal macrophages and an array of inflammatory signals.

Intestinal macrophages also are functionally a unique population of mononuclear phagocytes. Their downregulation for the production of pro-inflammatory cytokines is not solely dependent on absent or low-level expression of CD14, because intestinal macrophages express TLR3 and TLR5–9 (see Figure 13.5a), yet the cells do not release detectable cytokines, including tumor necrosis factor-α (TNF-α), IL-8, and IL-10 (see Figure 13.5b) or IL-1, IL-6, and IFN-β, in response to TLR-specific ligands. However, intestinal macrophages retain potent host defense function, reflected in their powerful phagocytic and bactericidal activities. Thus, despite profound inflammation anergy, intestinal macrophages retain antimicrobial activities through phagocytic and bactericidal function, enabling the cells to rapidly eliminate bacteria that penetrate the epithelium without promoting local inflammation.

13.5   Inflammation anergy of intestinal macrophages is due to inactive NF-κB signaling

A network of mechanisms downregulates nuclear factor-κB (NF-κB) activation to cause inflammation anergy in intestinal macrophages. In the mucosa, latent TGF-β is produced by an array of cells, including epithelial cells, mast cells, stromal cells, regulatory T cells, and T cells undergoing apoptosis. Released into the lamina propria, TGF-β binds to the extracellular matrix, establishing a "TGF-β reservoir." TGF-β released from mucosal cells or the extracellular matrix engages the cognate receptors TGF-β RI and RII expressed on blood monocytes (and intestinal macrophages), promoting monocyte recruitment into the lamina propria and the cells' rapid differentiation into inflammation-anergic macrophages.

The inability of intestinal macrophages to release pro-inflammatory cytokines in response to TLR-specific ligands, despite the expression of TLRs, is consistent with defective downstream NF-κB signaling. Molecular studies have shown that intestinal macrophages express markedly lower (or undetectable) levels of TRIF, MyD88, and TRAF6 proteins, leading to a failure to phosphorylate NF-κB p65 (Figure 13.6a and b). However, intestinal macrophages express increased levels of mRNA for both suppressor of cytokine signaling (SOCS1), which promotes the degradation of MAL (MyD88 adaptor-like protein), and sterile and Armadillo motif-containing protein (SARM) (Figure 13.7), which inhibits TRIF signaling (see Figure 13.6b). MyD88 is a critical element in the NF-κB activation pathway of all TLRs, except TLR3, and TRIF mediates TLR3-induced RANTES and IFN-γ production, as well as TLR4-mediated MyD88-independent signaling. Thus, the inhibition of MyD88-dependent and MyD88-independent NF-κB signaling powerfully decreases TLR-mediated pro-inflammatory responses in intestinal macrophages. Intestinal macrophages are similarly unable to activate NF-κB through mitogen-activated protein kinase (MAPK) pathways involving phosphorylated (p) p38, p-ERK, or p-JNK, pathways that are dependent on TRAF6. These dysregulations lead to the inability of intestinal macrophages to activate NF-κB and produce NF-κB-dependent cytokines.

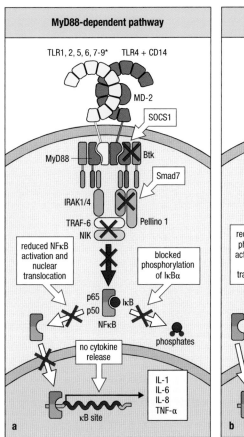

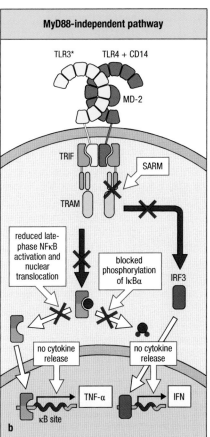

Signal molecule	Macrophage:monocyte fold difference in mRNA	
CD14	−194	
TLR4	−31.4	
MD-2	−64.9	
MyD88	−16.6	
TRAM	−17.6	
IRAK1	−15.3	
IRAK4	−3.9	
TRAF6	4	
NFκB p105	−9.1	
NFκB p65	−4.6	
IκBα	−5.9	
TOLLIP	2.1	
SOCS1	5	
SARM	9.1	
BTK	−4.1	
Smad6	17.7	
Pellino 1	−13.9	
CEPBα	−6.7	
CEPBβ	−28.8	
Sp1	−7.4	

Expression of intestinal macrophage mRNA relative to monocyte mRNA		
	−199	to −100
	−99	to −50
	−49	to −5
	−4	to +5
	+6	to +50

Figure 13.6 MyD88-dependent and MyD88-independent signaling pathways are inhibited in intestinal macrophages. Inhibition of MyD88-dependent (**a**) and MyD88-independent (**b**) signaling pathways through blocked phosphorylation (and thus ubiquitination) of IκBα, reduced signaling through Bruton agammaglobulinemia tyrosine kinase (BTK) and Pelle-interacting protein (Pellino), subsequent reduced activation and nuclear translocation of nuclear factor-κB (NF-κB), and inhibition of interferon regulatory factor 3 (IRF3) signaling, resulting in blockade of pro-inflammatory cytokine release from intestinal macrophages. (TLR1, 2, 4–6 are expressed on the outer cell membrane; *TLR3, 7–9 are intracellular.)

Figure 13.7 Intestinal macrophages and blood monocytes express markedly different levels of mRNA for signal proteins. mRNA for signal proteins is downregulated and mRNA for signal inhibitors is upregulated in intestinal macrophages compared with autologous blood monocytes.

13.6 Active TGF-β/Smad signaling contributes to inflammation anergy of intestinal macrophages

A second line of investigation has implicated active TGF-β signaling in the regulation of IκBα expression in intestinal macrophages, amplifying downregulated NF-κB signaling in these cells. TGF-β mediates its functions through signaling via TGF-β receptors and the intracellular effectors of TGF-β signaling, the Smad proteins, which translocate into the nucleus and regulate the transcription of target molecules. In this respect, intestinal macrophages express TGF-β RI and RII and display constitutive Smad signaling (**Figure 13.8**). Thus, intestinal macrophages express nuclear Smad4, a critical component of the TGF-β signal cascade that associates with the phosphorylated heterodimeric Smad2/3 complex and then translocates into the nucleus to initiate gene transcription. In addition, intestinal macrophages do not express cytoplasmic Smad7 (see **Figure 13.8**), the inhibitor of Smad2/3 phosphorylation, which prevent the nuclear translocation of the Smad2/3–Smad4 complex. Moreover, active Smad signaling in the macrophages promotes constitutive expression of IκBα, which sequesters NF-κB in the cytoplasm and, together with the cells' inability to phosphorylate NF-κB, blocks NF-κB signal transduction, thereby inhibiting NF-κB-mediated activities. Consistent with these findings, TGF-β also has been shown to block NF-κB activation in response to TLR2, 4, and 5 signaling by facilitating the ubiquitination and proteasomal degradation of MyD88. Resting blood monocytes also express low levels of inhibitor Smad7 but, unlike intestinal macrophages, upregulate the expression of these inhibitors after exposure to LPS or other TLR-specific ligands, thereby permitting NF-κB translocation into the nucleus and the subsequent production of pro-inflammatory cytokines. Thus, stromal extracellular matrix-derived TGF-β induces downregulation in NF-κB signaling and NF-κB-mediated function in blood monocytes *in vitro*, recapitulating the abrogated NF-κB signaling in intestinal macrophages.

Taken together, ineffective NF-κB signaling in intestinal macrophages and TGF-β-mediated Smad-induced IκBα upregulation provide a mechanism for the inflammation anergy of intestinal macrophages, promoting the absence of inflammation in intestinal mucosa despite the close proximity of lamina propria macrophages to luminal bacteria and food antigens. In this way, a breach in mucosal (epithelial) integrity would be met by potent host defense activity, but not an inflammatory response, in the normal intestinal mucosa.

13.7 Macrophages do not present antigen in normal intestinal mucosa

As a consequence of their strategic location in the subepithelial lamina propria and their avid phagocytic activity, intestinal macrophages are assumed to interact with antigens translocated across the epithelium and to participate in accessory cell function. Consistent with this notion, intestinal macrophages express high levels of HLA-DR, whose major function is the presentation of antigenic peptides to T-cell receptors. However, human small intestinal macrophages from noninflamed mucosa lack constitutive and inducible expression of the costimulatory molecules CD40, CD80 (B7.1), and CD86 (B7.2), which have a key role in accessory cell stimulation (see Figure 13.3). The cells also lack the ability to produce the costimulatory cytokines IL-1, IL-10, IL-12, IL-21, IL-22, and IL-23, which participate in the priming and expansion of T-cell populations. Not surprisingly, therefore, intestinal macrophages pulsed with antigen (or mitogen) are markedly less efficient at stimulating T-cell proliferation and cytokine release than similarly pulsed autologous blood monocytes. Recent reports have suggested that colonic macrophages may have some antigen-presenting capability, particularly in inflamed colonic mucosa; however, because these cells are CD14+, they are probably recently recruited blood monocytes rather than resident macrophages. Together, these unique features of human intestinal macrophages may provide an indirect tolerogenic effect on the normal mucosal response to luminal antigens and the microbiota. An active tolerogenic role for intestinal macrophages has been identified in mouse macrophages, as discussed next.

13.8 Mouse intestinal macrophages have immunoregulatory function

In contrast to human resident intestinal macrophages, which are CD11b⁻CD11c⁻ and do not secrete the downregulatory cytokine IL-10 required to maintain immune tolerance (see Figures 13.3 and 13.5b), mouse intestinal macrophages are CD11b+CD11c+/− and secrete constitutive and inducible IL-10. IL-10 released by mouse intestinal macrophages inhibits T_H1 polarization through the blockade of IFN-γ production by CD4+ T cells and blocks T_H17 polarization through the blockade of IL-17 production by co-cultured DC-T cells. IL-10 signaling, mediated through the IL-10R, inhibits

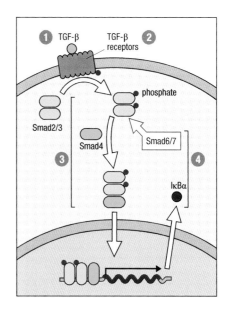

Figure 13.8 Active TGF-β/Smad signaling promotes inflammation anergy in intestinal macrophages. (1) Intestinal epithelial cells, mast cells, and stromal cells in normal mucosa produce TGF-β. (2) Intestinal macrophages express both TGF-βRI and TGF-βRII. (3) TGF-β signaling promotes the phosphorylation of Smad2/3, which complexes with Smad4, translocates into the nucleus, and drives the increased production of IκBα. (4) Absence of Smad7 inhibitor in resident intestinal macrophages permits active Smad signaling and increased constitutive expression of cytoplasmic IκBα, causing cytoplasmic sequestration of NF-κB, resulting in inhibition of NF-κB nuclear translocation and blockade of NF-κB-mediated pro-inflammatory cytokine production.

cytokine production by blocking NF-κB-dependent signals, and inhibiting canonical activation of the inflammasome and production of IL-1 through transcriptional and post-translational regulation of *NLRP3*. These findings suggest an explanation, at least in part, for the early onset IBD in infants with IL-10R deficiency, where mutational loss of IL-10 sensing in innate immune cells, including macrophages, is critical for regulating mucosal homeostasis. Data implicating diet in the regulation of intestinal macrophage function has also recently emerged showing that deprivation of enteral nutrition in mice results in impaired IL-10-producing lamina propria macrophages. Precisely how macrophage-derived IL-10 promotes and maintains intestinal homeostasis in mice appears to involve complex cellular networks, including direct effects on macrophages themselves, as well as on intestinal epithelial cells leading to enhanced barrier integrity. However, additional mechanisms are likely involved since human resident intestinal macrophages in the steady state do not produce IL-10 or express IL-10R (see **Figure 13.3**), despite originating from blood monocytes capable of both.

Intestinal macrophages in host defense

Consistent with their central role in host defense against microbes and noxious molecules, intestinal macrophages are avidly phagocytic for microorganisms and inert material. *In vitro* studies have shown that they phagocytose bacteria (*Salmonella enterica* serovar *typhimurium*, *Escherichia coli*, and *Mycobacterium*), fungi (*Candida albicans*), apoptotic cells (Jurkat cells), and latex beads. The phagocytic and ensuing bactericidal activity is generally equivalent to or greater than that of autologous blood monocytes. An exception to the potent bactericidal activity of intestinal macrophages is their inability to kill *Tropheryma whipplei*, at least in patients with Whipple's disease. Some investigators have hypothesized that Whipple's disease is due to a genetic defect in macrophage processing. Importantly, macrophage phagocytosis and killing of engulfed bacteria do not induce inflammatory responses, whereas such activities in monocytes cause strong release of pro-inflammatory cytokines. Thus, in contrast to blood monocytes and macrophages in other tissues, human resident intestinal macrophages perform host defense activities without inducing an inflammatory response, providing an evolutionary advantage that enables gut macrophages to protect against enteric microorganisms without contributing to inflammation in the mucosa. The mechanism by which intestinal macrophages kill bacteria is not known but seems to be independent of reactive oxygen intermediates.

13.9 Intestinal macrophages scavenge apoptotic cells

As previously discussed, intestinal macrophages phagocytose microorganisms, inert material, and apoptotic cells. Intestinal macrophages also express the scavenger receptor CD36, a receptor for lipoprotein and phospholipid, endowing intestinal macrophages with phagocytic activity for apoptotic cells. The ability of intestinal macrophages to phagocytose apoptotic mononuclear cells (and apoptotic cell debris) and clear this potential source of inflammatory material from the lamina propria is critical for the maintenance of mucosal homeostasis. Recent studies in mice indicate that an abundant resident population of Tim-4$^+$CD4$^+$ intestinal macrophages express the receptor for phosphatidylserine, promoting efficient uptake and removal of apoptotic cells in the lamina propria. This population appears to be developmentally dependent on the microbiome, locally maintained and not supported by recruited blood monocytes, which do not express Tim-4.

13.10 Intestinal macrophages participate in host defense against mucosal pathogens

Macrophages contribute to host defense against a wide array of bacterial, viral, parasitic, and fungal pathogens. In the intestinal mucosa, however, the stringent downregulation of innate response receptors and pro-inflammatory signal transduction in lamina propria macrophages influences the cells' response to an array of microorganisms, including viruses such as HIV-1 and cytomegalovirus (CMV). After HIV-1 is inoculated onto the intestinal mucosa and subsequently translocated into the subepithelial lamina propria, the virus encounters the largest reservoir of macrophage and lymphocyte target cells in the body. However, the permissiveness of intestinal macrophages to HIV-1 is profoundly downregulated due to markedly reduced levels of (a) CD4, the primary receptor, (b) CCR5, the coreceptor for CCR5-trophic (R5) HIV-1 virus entry, and (c) NF-κB activation, which is required for HIV-1 replication. Consequently, lamina propria lymphocytes, not macrophages, are the early target cells for infection and replication by HIV-1 R5 viruses, which dominate in acute HIV-1 infection. Thus, the unique phenotype and signal transduction profile of intestinal macrophages restricts their permissiveness to HIV-1 in primary infection.

In sharp contrast to the direct mucosal entry of HIV-1, CMV enters the intestinal mucosa through the recruitment of CMV-infected blood mononuclear cells, especially monocytes, into the lamina propria. Recent evidence indicates that CMV infection of blood monocytes enhances expression of CD14 and TLR5, the receptors for LPS and flagellin, and upregulates expression of the TGF-β antagonist Smad7, promoting monocyte resistance to stromal TGF-β inactivation of NF-κB. Thus, CMV infection of monocytes blocks inflammation anergy when the cells differentiate into intestinal macrophages, leading to inducible pro-inflammatory cytokine production and macrophage-driven inflammation characteristic of CMV mucosal inflammatory disease in immunosuppressed subjects.

SUMMARY

The gastrointestinal mucosa, the largest body surface to interface with the external environment, contains the major tissue reservoir of macrophages, recruited to the lamina propria stroma by extracellular matrix–associated TGF-β and IL-8 in normal mucosa and by inflammatory chemokines and microbial products in inflamed and/or infected mucosa. Reflecting a remarkable evolution and dynamic plasticity, the newly recruited monocytes undergo downregulation of some, but not all, innate response receptors and TGF-β-induced inactivation of NF-κB and Smad-induced IκBα sequestration of NF-κB. The resultant blockade of NF-κB-mediated pro-inflammatory function promotes noninflammatory host defense in normal mucosa. In infected intestinal mucosa, however, recruited monocytes may retain pro-inflammatory function, promoting the containment of microbes and foreign molecules that have entered the lamina propria. In addition, the massive numbers of intestinal macrophages resident in the mucosa promptly phagocytose and digest invading microorganisms as well as apoptotic and dead host cells. Thus, the intestinal mucosa is host to continuous mononuclear phagocyte activity as huge numbers of blood monocytes recruit to the lamina propria, embed into the extracellular matrix, quickly undergo induction of inflammation anergy in normal mucosa, but retain many pro-inflammatory activities in infected mucosa, thereby providing long-lived and highly efficient protection against harmful enteric microbes and potentially inflammatory apoptotic cells.

FURTHER READING

Denning, T.L., Wang, Y.-C., Patel, S.R. et al.: Lamina propria macrophages and dendritic cells differentially induce regulatory and interleukin 17-producing T cell responses. *Nat. Immunol.* 2007, 8:1086–1094.

Dennis, E.A., Robinson, T.O., Smythies, L.D. et al.: Characterization of human blood monocytes and intestinal macrophages. *Curr. Protoc. Immunol.* 2017, 118:14.3.1–14.3.14.

Dennis, E.A., Smythies, L.E., Grabski, R. et al.: Cytomegalovirus promotes intestinal macrophage-mediated mucosal inflammation through induction of Smad7. *Muc. Immunol.* 2018, 11:1694–1704.

Geissmann, F., Gordon, S., Hume, D.A. et al.: Unravelling mononuclear phagocyte heterogeneity. *Nat. Rev. Immunol.* 2010, 10:453–460.

Hausmann, M., Kiessling, S., Mestermann, S. et al.: Toll-like receptors 2 and 4 are up-regulated during intestinal inflammation. *Gastroenterology* 2002, 122:1987–2000.

Ochi, T., Feng, Y., Kitamoto, S., Nagao-Kitamoto, H. et al.: Diet-dependent, microbiota-independent regulation of IL-10-producing lamina propria macrophages in the small intestine. *Sci. Rep.* 2016, 5:27634.

Schenk, M., Bouchon, A., Seibold, F. et al.: TREM-1-expressing intestinal macrophages crucially amplify chronic inflammation in experimental colitis and inflammatory bowel diseases. *J. Clin. Invest.* 2007, 117:3097–3106.

Shaw, T.N., Houston, S.A., Wemyss, K. et al.: Tissue-resident macrophages in the intestine are long lived and defined by Tim-4 and CD4 expression. *J. Exp. Med.* 2018, 215: 1507–1518.

Shen, R., Meng, G., Ochsenbauer, C. et al.: Stromal down-regulation of macrophage CD4/CCR5 expression and NF-κB activation mediates HIV-1 non-permissiveness in intestinal macrophages. *PLOS Pathogens* 2011, 7:e1002060.

Shouval, D.S., Biswas, A., Goettel, J.D. et al.: Interleukin-10 receptor signalling in innate immune cells regulates mucosal immune tolerance and anti-inflammatory macrophage function. *Immunity* 2014, 40:706–719.

Smith, P.D., Saini, S.S., Raffeld, M. et al.: Cytomegalovirus induction of tumor necrosis factor-α by human monocytes and mucosal macrophages. *J. Clin. Invest.* 1992, 90:1642–1648.

Smith, P.D., Smythies, L.E., Shen, R. et al.: Intestinal macrophages and response to microbial encroachment. *Mucosal Immunol.* 2011, 4:31–42.

Smythies, L.E., Sellers, M., Clements, R.H. et al.: Human intestinal macrophages display profound inflammatory anergy despite avid phagocytic and bactericidal activity. *J. Clin. Invest.* 2005, 115:66–75.

Smythies, L.E., Shen, R., Bimczok, D. et al.: Inflammation anergy in human intestinal macrophages is due to Smad-induced IκBα expression and NF-κB inactivation. *J. Biol. Chem.* 2010, 285:19593–19604.

Varol, C., Mildner, A., and Jung, S.: Macrophages: Development and tissue specialization. *Annu. Rev. Immunol.* 2015, 33:643–675.

Mucosal basophils, eosinophils, and mast cells

14

EDDA FIEBIGER AND STEPHAN C. BISCHOFF

Mast cells, which are abundantly found in the skin and mucosal tissues, play an important role as regulators of multiple mucosal and mucosa-associated functions such as epithelial secretion, smooth muscle contraction, and local activation of nerves. Moreover, mast cells are involved in host defense against bacteria, viruses, and helminth parasites, and contribute to the progression of allergic and other types of inflammation at mucosal sites. In contrast, eosinophils and basophils are blood leukocytes, which can enter mucosal tissues under particular circumstances. In particular, eosinophils are found at mucosal sites, either under normal conditions (e.g., intestine) or under pathologic conditions (e.g., esophagus and bronchial mucosa). Eosinophils, like mast cells, are found at elevated numbers during the course of allergic inflammation and helminth infection. They have multiple pro-inflammatory properties but are also involved in immunity to parasites and other infectious agents. Basophils, which are a relatively rare granulocyte population found primarily in the blood and spleen, were traditionally thought to have redundant roles to mast cells in immunity to parasites or the pathogenesis of allergic inflammation. However, recent studies have demonstrated that basophils express effector molecules distinct from mast cells and are found in the inflamed lung tissue of humans following exposure to allergens or lethal asthmatic attack. In addition to their effector functions, recent murine studies have demonstrated that basophils contribute to the development and maintenance of CD4$^+$ T-cell responses and allergic inflammation at multiple mucosal sites. Collectively, these results suggest that mast cell, eosinophil, and basophil populations are critical regulators of mucosal immunity and inflammation. The focus of this chapter is to discuss the molecules and pathways that control the development, activation, and effector functions of these three cell populations in mucosal tissues.

BIOLOGY OF BASOPHILS

Basophils are the least abundant granulocyte population and account for less than 1% of the leukocytes found in the blood, spleen, and bone marrow. Although originally discovered in 1879 by the German scientist Paul Ehrlich, almost a century passed before basophils were demonstrated to bind IgE and release histamine. Even after these findings, basophils were considered to be a redundant cell population with the same effector functions as mast cells. However, later studies that compared and contrasted the functions of basophils and mast cells discovered that these two populations differ in their gross phenotype, signal transduction pathways, and release of inflammatory mediators. More recently, studies have identified several previously unrecognized functions of basophils in multiple models of helper

T-cell (T_H)2 cytokine-dependent immunity and allergic inflammation. In addition to their known role as effector cells in inflamed tissues, findings now indicate that basophils express major histocompatibility complex (MHC) class II and costimulatory molecules, can migrate into draining lymph nodes, act cooperatively with dendritic cells to present antigen to naive CD4+ T cells, and promote T_H2 cell differentiation. In this context, basophils have been shown to contribute to the optimal induction and propagation of T_H2 cytokine responses in murine models of helminth infection and allergic inflammation. In this section, we discuss our current understanding of basophil biology in the context of immunity and inflammation and discuss the factors that regulate basophil development, activation, and effector functions and the role that they have during helminth infection, allergic inflammation, and some forms of autoimmunity.

14.1 Basophils arise from common granulocyte-monocyte precursor in the bone marrow

Basophils arise from a common granulocyte-monocyte precursor in the bone marrow that has the capacity to differentiate into eosinophils, basophil–mast cell precursors, mast-cell precursors, and basophil precursors *in vitro* (**Figure 14.1**). Both mast-cell precursors and basophil precursors generally remain in the bone marrow and give rise to mast cells and the majority of mature basophils in the bloodstream. Another source of basophils includes basophil–mast cell precursors, which migrate to the spleen where they mature into basophils. Basophils exhibit a relatively short life span of 60–70 hours. It has thus been suggested that basophils are continually replenished by precursor populations in the bone marrow in order to maintain their presence in the periphery. Although relatively small in number and short in life span, basophils have the capacity to perform multiple effector mechanisms that are introduced and discussed later.

14.2 Basophils provide effector functions by releasing preformed mediators as consequence of IgE-dependent and IgE-independent stimuli

Basophils are best known as effector cells that release preformed mediators in response to activation via surface-bound IgE. Circulating basophils bind IgE through the high-affinity IgE receptor FcεR1α and degranulate upon FcεR1α cross-linking. Basophils activated via surface-bound IgE produce histamines, leukotrienes, cytokines, and chemokines. However, basophils can also be activated by an array of stimuli in both antibody-dependent and antibody-independent fashions including via antibodies (IgE, IgG1, and IgD), cytokines (interleukin [IL]-3, IL-18, and IL-33), antigens, proteases, pathogen-associated molecular patterns (PAMPs), and complement components (**Figure 14.2**). Activated basophils are capable of secreting a variety of effector molecules including histamines, leukotrienes, IL-4, IL-6, IL-13, tumor necrosis factor (TNF)-α, and thymic stromal lymphopoietin (TSLP). In the following sections, we discuss the stimuli that activate basophils and the bioactive molecules that are secreted as a consequence of this activation.

14.3 Effector functions of basophils can be activated by IgE, IgG1, and IgD antibody-antigen complexes

As previously described, basophils rapidly produce histamines, leukotriene C_4 (LTC$_4$), and cytokines in response to the cross-linking of FcεR1α via IgE. The rapid production of effector molecules by basophils in response to activation

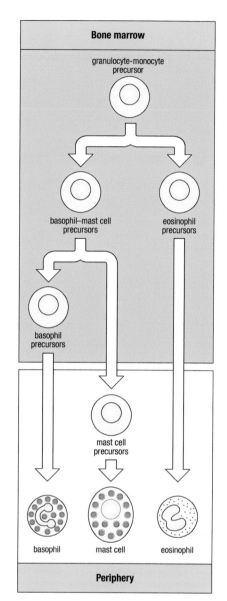

Figure 14.1 Pathways of basophil, eosinophil, and mast-cell development. The bone marrow resident granulocyte-monocyte precursor has the capacity to develop into eosinophils, mast cells, and basophils. Granulocyte-monocyte precursors can become eosinophil precursors and ultimately mature eosinophils that will enter the periphery. Granulocyte-monocyte precursors can also become basophil–mast cell precursors that have the capacity to become mast cells and basophils. Some basophil–mast cell precursors will develop into mast-cell precursors that will enter the periphery and ultimately develop into mature tissue-resident mast cells, while others will develop into basophil precursors and ultimately become mature basophils that will then enter the periphery.

through surface-bound IgE supports their role as contributors to systemic anaphylaxis as a consequence of blood-borne antigen exposure in humans. Although similar studies have not identified a role for mouse basophils in IgE-mediated anaphylaxis, they are capable of producing platelet-activating factor (PAF) as a consequence of IgG1 binding. The IgG1-mediated release of PAF by mouse basophils after sensitization to penicillin V leads to a pathway of alternative anaphylaxis that is dependent on functional expression of FcγRII and FcγRIII. Thus, basophils can be activated by both IgE and IgG1 antibody-antigen complexes leading to release of preformed mediators that contribute to the development of systemic anaphylaxis in both mice and humans.

In addition to the antibody-mediated immediate activation of basophils described earlier, basophils can be activated by IgE during certain types of chronic inflammation. Such IgE-mediated chronic allergic inflammation (IgE-CAI) is independent of mast cells and T cells but dependent on basophil populations. Interestingly, while basophils account for only 1%–2% of the cellular infiltrate at the site of such lesions in the skin, their depletion results in a dramatic reduction in inflammation. Pathologically, depletion of basophils results in decreased infiltration with eosinophils and neutrophils and a marked reduction in skin thickness. The loss of cellular infiltrates suggests that basophils produce chemokines and/or other factors that result in cellular recruitment in this model of chronic skin inflammation. Thus, basophils have a role in the initiation and maintenance of chronic IgE-mediated inflammatory responses.

In addition to IgE- and IgG1-containing immune complex activation of basophils via Fc receptor ligation, basophils can also be activated by IgD antibody-antigen complexes. In this case, human basophils can selectively bind IgD, a class of antibody produced early in B-cell development. Although the biological function of IgD remains enigmatic, it is interesting that IgD is highly expressed in the human upper respiratory tract where it can bind organisms such as *Haemophilus influenzae* and *Moraxella catarrhalis* and activate basophils to produce antimicrobial peptides that inhibit bacterial replication. IgD-mediated activation of basophils stimulates a distinct effector phenotype from that elicited by IgE-mediated activation. In comparison to cross-linking with IgE, IgD cross-linking of basophils results in enhanced expression of IL-4 and B-cell activating factor (BAFF), which facilitates B-cell class switching, and the production of antimicrobial peptides, which inhibits bacterial replication, but not the induction of histamine release.

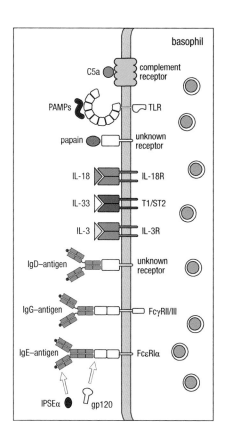

Figure 14.2 Mechanisms of basophil activation. Basophils can be activated by many signals including antigen-antibody complexes (IgE, IgG1, IgD), cytokines (IL-3, IL-18, IL-33), proteases, viral antigens, toll-like receptor (TLR) agonists, and complement components. gp120, glycoprotein of human immunodeficiency virus; IPSEα, interleukin-4-inducing principle from *Schistosoma mansoni* eggs-α.

14.4 Basophils are activated by cytokines such as IL-3, IL-18, and IL-33

Basophil development and effector functions can also be regulated by cytokines. For example, IL-3 is crucially important for the expansion, survival, and activation of basophils. IL-3 promotes the differentiation of basophils from bone marrow cells *in vitro*, and administration of IL-3 *in vivo* induces their generation. Consistent with this, IL-3 is necessary for infection-induced basophilia in the bone marrow, blood, spleen, and liver following infection with organisms such as *Nippostrongylus brasiliensis* and *Strongyloides venezuelensis*. In addition, IL-3 can augment basophil functions induced by other factors such as IgE-dependent stimulation of IL-4 and IL-13 secretion.

IL-18 and IL-33, members of the IL-1 cytokine family, are also capable of activating basophils. IL-18 is produced by innate cells such as macrophages and Kupffer cells and is known to play a role in allergic disease and immunity to helminthic parasites. IL-33 is expressed by dermal fibroblasts, airway epithelial cells, and bronchial smooth muscle cells and is important in the induction of IL-4, IL-13, and IgE production *in vivo*. Both IL-18 and IL-33 are closely linked to T_H2 cytokine-mediated inflammation. Consistent with their ability to enhance T_H2 cytokine-associated immune responses, IL-18 and IL-33 are capable of activating basophils. Murine bone marrow–derived basophils stimulated with IL-3 and IL-18 exhibit enhanced secretion of IL-4 and IL-13 compared with stimulation with IL-3 alone. Furthermore, administration of IL-18 *in vivo* increases the production of basophil-dependent IL-4 and histamine in mice. IL-18 also enhances the survival of murine basophils. Interestingly, although human blood basophils express the IL-18R, a role for IL-18 in the activation of human basophils remains to be established. Similarly to IL-18, IL-33 can stimulate the production of IL-4 and IL-13 from murine bone marrow–derived basophils. IL-33 also induces the production of IL-4, IL-5, IL-6, and IL-13 from human blood-derived basophils.

14.5 Basophils can sense microbial products to provide innate immune resistance

In addition to antibody- and cytokine-mediated activation, basophils are activated by allergens, parasite antigens, and other pathogen-associated molecules. Specifically, secretions of the hookworm parasite *Necator americanus* and the house dust mite antigen Derp1, which possess active protease activity, can induce the production of IL-4, IL-5, and IL-13 by human basophils. These activities are eliminated by protease inhibitors, suggesting that basophils may express protease-activated receptor-like factors that are capable of the proteolytic activation of such antigens (or organisms) and activate innate and adaptive immune cells.

In addition to proteases, basophils can be activated by a class of "superantigens" that function independently of any known receptors. For example, the gp120 glycoprotein of the human immunodeficiency virus is able to interact with the VH3 region of IgE in a pathway that induces IL-4 and IL-13 production from human basophils. Similarly, murine basophils are activated by the schistosome-derived glycoprotein IPSE/α-1, which induces the production of IL-4 from basophils in a nonspecific manner.

14.6 Basophils are activated by toll-like and complement receptors for innate immune function

Basophils express toll-like receptors (TLRs) and complement receptors (CRs). On human basophil populations, these include TLR1, TLR2, TLR4, TLR6, TLR9, CR1, CR3, CR4, and CD88. This predicts that basophils may be capable of recognizing PAMPs and complement components. Consistent with this, there is evidence that TLR2 ligands can activate basophils and induce the production of IL-4 and IL-13 while the complement component C5a can induce histamine production. Despite these observations, this is likely an underestimate of the full innate immune capacities of basophils as later discussed.

14.7 Basophils are recruited to inductive and effector sites throughout an immune response

As we have seen, basophils are potent producers of type 2 cytokines such as IL-4 and IL-13 in response to a wide variety of stimuli. In addition, they are able to do so throughout all phases of an immune response and consequently accumulate at sites of inflammation during an ongoing inflammatory response. Basophils were originally considered to be late-phase effector cells and thought to be excluded from lymph nodes and for the induction and early phases of inflammation. However, it is now appreciated that basophils may serve as liaisons between the innate and adaptive immune response and may even promote the latter. For example, MHC class II$^+$ basophils can migrate to draining lymph nodes following exposure to papain, *Schistosoma mansoni* eggs, or *N. brasiliensis* infection. Although basophils appear to be only transiently present in the lymph nodes, there is evidence that they are capable of directly interacting with lymph-node-resident CD4$^+$ T cells, B cells, and dendritic cell (DC) populations. Basophils may directly contribute to the induction of T$_H$2 cytokine-mediated inflammation. Consistent with this, depletion of basophils prior to papain challenge can prevent the subsequent development of an allergen-induced T$_H$2 cell response. Finally, it should be noted that the mechanisms by which basophils enter lymph nodes are poorly defined, although they appear to involve IL-3. For example, basophils are induced to accumulate in mediastinal lymph nodes during the first week of infection with *N. brasiliensis*, and this recruitment is dependent on IL-3 and IL-3 receptor signaling. Whether this relates to basophil recruitment in other circumstances is unknown but demonstrates that recruitment of basophils is doubtless directed.

14.8 Basophils possess antigen-presenting functions and cooperate with dendritic cells in the induction of T$_H$2 immunity along mucosal surfaces

Recent studies have introduced the notion that basophils may be capable of providing the initial cellular and soluble stimuli for the development and induction of T$_H$2 cytokine-mediated immunity and inflammation independently of dendritic cells. For example, on the one hand, when MHC class II expression is restricted to DCs or when DCs are depleted, T$_H$2 cytokine-dependent immune responses may be impaired, but not eliminated, following exposure to mucosal helminth infection or mucosal exposure to allergens such as those associated with papain or allergic airway sensitization. This suggests that other cells may provide the adaptive signals necessary for induction of T$_H$2 cells. Consistent with this, there is some direct evidence that basophils may function as antigen-presenting cells in the context of mucosal exposure to helminths or allergens. In these studies, basophils have been shown to endocytose soluble antigens and IgE-allergen complexes, express MHC class II and costimulatory molecules, migrate to draining lymph nodes, and promote T$_H$2 cell differentiation in response to *S. mansoni* egg antigens. Depletion of basophils results in impaired protective immunity to the mucosal whipworm *Trichuris muris*, together with elimination of T$_H$2 cell development after mucosal challenge with allergens.

As a whole, multiple studies now indicate that both basophils and DCs are important for the optimal induction of T$_H$2 cytokine-mediated immunity and inflammation including those associated with mucosal tissues. The apparent identification of DC-dependent and DC-independent pathways of T$_H$2 cell differentiation suggests that there are multiple pathways by which T$_H$2 cell responses may develop, as summarized in Figure 14.3. For example, in some models of helminth infection or allergic inflammation, T$_H$2 cell differentiation may involve priming of naive T cells by DCs augmented with IL-4 or other soluble factors produced by basophils (Figure 14.3a). However, other models of antigen presentation by IL-4 and MHC class II–expressing basophils in mice may be sufficient to promote T$_H$2 cell differentiation in the absence of DC populations (Figure 14.3b). Alternatively, basophils and DCs may present antigen to naive T cells cooperatively to initiate the propagation of optimal T$_H$2 cell responses (Figure 14.3c and d). Consistent with a cooperative model of antigen presentation, T$_H$2 cell development in response to immunization with papain is dependent on the presence of both

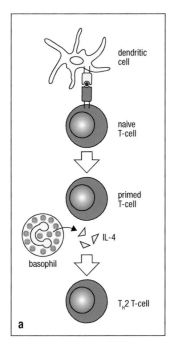
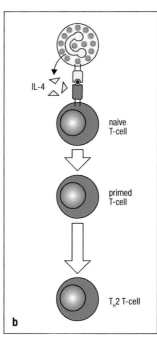
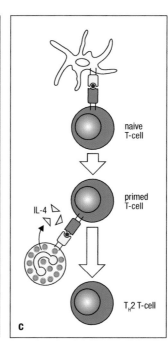
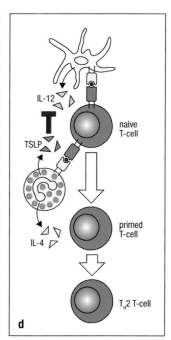

Figure 14.3 Basophils regulate T$_H$2 cell differentiation. There are several potential pathways by which basophils may interact with T cells to augment T$_H$2 cell differentiation. First, as shown in (**a**), basophils may function as accessory cells that provide a source of IL-4, which helps to polarize T cells that were previously activated by DC-mediated antigen presentation. Second, in (**b**), basophils may activate T cells by directly presenting antigen while simultaneously providing an innate source of IL-4. Third, basophils and DCs may act cooperatively to present antigen either in series (**c**) or in parallel (**d**). In the context of the model shown in (**d**), basophils would augment DC-mediated antigen presentation and provide the innate IL-4 needed for T$_H$2 cell differentiation. In addition, basophils can produce thymic stromal lymphopoietin (TSLP) that is capable of inhibiting IL-12 production by DCs, thereby further promoting T$_H$2 cell differentiation.

DC and basophil populations. In this context, immunization with papain and soluble antigen results in DC populations releasing reactive oxygen species, which promote the production of oxidized lipids that are capable of triggering epithelial cells to produce TSLP, which suppresses the production of IL-12 by DCs. This cascade of events also results in the induction of DCs to produce CCL7, which mediates the recruitment of basophil populations to the lymph nodes, actions that together show how DCs and basophils cooperate in the induction of T$_H$2 cytokine-mediated responses (**Figure 14.3d**). It is likely that the nature of these responses will depend on the characteristics of the antigen being studied (e.g., helminth infection or allergens), the duration and site of antigen exposure, and whether the differentiation of T$_H$1, T$_H$17, or regulatory T (T$_{reg}$) cells is occurring simultaneously.

Biology of mucosal eosinophils

Eosinophil granulocytes are bone marrow–derived leukocytes involved in inflammatory and immunoregulatory processes. This dual function of eosinophils is similar to that of basophils and mast cells. Obviously, inflammatory and immunoregulatory functions are frequently combined in the same cell type. Like basophils and mast cells, which are discussed later in this chapter, eosinophils exert many biological functions by releasing humoral mediators that preferentially function locally in a paracrine manner. Both the factors that regulate eosinophil development and activation and the mediators that are released on activation are discussed. Consistent with the pleiotropic functions of eosinophils in health and disease, eosinophils are capable of releasing a wide range of granule-associated proteins and cytokines, upregulating a variety of cell-surface receptors and adhesion molecules, and potentially providing antigen-presentation functions for T cells. Eosinophils participate in parasitic and helminth infections, virus infections, asthma, gastrointestinal diseases, and neoplastic disorders of the leukocyte lineage and are thus an important component of mucosal immunity.

14.9 Development of eosinophils depends on lineage-specific transcription factors and cytokines

Eosinophils are derived from pluripotent bone marrow cells, which first develop into a common precursor of basophils and eosinophils, and then into a separate eosinophil lineage. Eosinophil lineage specification is dictated by the interplay of at least three classes of transcription factors including GATA-1 (a zinc finger family member), PU.1 (an Ets family member), and c/EBP members (CCAAT/enhancer-binding protein family). The specificity of these factors for eosinophils is conserved across species. Of these transcription factors, GATA-1 is clearly the most important for eosinophil lineage specification as revealed by loss of the eosinophil lineage in mice harboring a targeted deletion of the high-affinity GATA-binding site in the GATA-1 promoter. Three cytokines, IL-3, IL-5, and granulocyte-macrophage colony-stimulating factor (GM-CSF) are particularly important in regulating eosinophil development. These eosinophilopoietins likely provide permissive proliferative and differentiation signals following the instructive signals specified by the transcription factors GATA-1, PU.1, and c/EBPs. Interestingly, these cytokines are encoded by closely linked genes on chromosome 5q31. They bind to receptors that share a common β-chain and possess unique α-chains. Of these three cytokines, IL-5 is the most specific to the eosinophil lineage and is responsible for selective differentiation of eosinophils. IL-5 also stimulates the release of eosinophils from the bone marrow into the peripheral circulation. The critical role of IL-5 in the production of eosinophils is best demonstrated by genetic manipulation of mice. Overproduction of IL-5 in transgenic mice results in profound eosinophilia, and deletion of the IL-5 gene causes a marked reduction of eosinophils in the blood and lungs after allergen challenge. The overproduction of one or a combination of these three cytokines occurs in humans with eosinophilia, and diseases with selective eosinophilia are often accompanied by overproduction of IL-5. The critical role of IL-5 in regulating eosinophils in humans has been demonstrated by several clinical trials using a humanized anti-IL-5 antibody that lowers eosinophil levels in the blood and, to a lesser extent, in the inflamed lung.

14.10 Eosinophil survival and recruitment are largely determined by IL-5 and eotaxin

The recruitment and survival of eosinophils are regulated by cytokines and chemokines. IL-3, IL-5, and GM-CSF support the survival of eosinophils; however, IL-5 has the most potent and most eosinophil-specific regulatory properties. IL-5 promotes the proliferation and maturation of eosinophils in the bone marrow and their release into the circulation. IL-5 also promotes the survival of eosinophils by diverting them from an apoptotic fate, primes them for responses to chemoattractant signals that are responsible for their mucosal recruitment, and promotes their degranulation in response to triggering agents.

The recruitment of eosinophils to mucosal sites is also critically dependent on the local production of chemoattractants, especially chemokines. Eotaxin-1 (CCL11) is the most important and selective eosinophil chemoattractant. Eotaxin-1 is responsible for the physiologic recruitment of eosinophils to the intestines in healthy individuals, and is expressed constitutively in the intestinal lamina propria. In the absence of eotaxin-1, or its eosinophil receptor, CCR3, eosinophils fail to home to the gastrointestinal tract.

Eosinophils also express the mucosal integrin $\alpha_4\beta_7$ on their cell surface, which is responsible for the selective recruitment of these leukocytes toward the mucosa of the intestines rather than other tissue compartments. Integrin $\alpha_4\beta_7$ binds to its specific ligand, MAdCAM-1 (mucosal vascular addressin cell adhesion molecule-1), which is preferentially expressed on the vascular endothelium of intestinal lamina propria venules.

14.11 Eosinophils are activated by a broad array of signals and function primarily through their ability to secrete a wide range of soluble mediators

The granules of eosinophils contain major basic protein, eosinophil cationic protein, eosinophil-derived neurotoxin, eosinophil peroxidase, and other enzymes of uncertain significance, as well as proteins that include a broad range of preformed cytokines and chemokines (**Figure 14.4**). Major basic protein is cytotoxic for helminthic parasites and mammalian cells, activates the complement cascade, and leads to increased smooth muscle reactivity by causing dysfunction of vagal muscarinic M2 receptors. Moreover, major basic protein has been shown to stimulate substance P release from neonatal rat dorsal root ganglia neurons. Eosinophil cationic protein, eosinophil peroxidase, and eosinophil-derived neurotoxin induce cytotoxic effects in helminthic parasites and mammalian cells by exerting ribonuclease activity or generating unstable oxygen radicals. Among the cytokines and chemokines produced by eosinophils are IL-1α, IL-2, IL-3, IL-4, IL-5, IL-6, IL-9, IL-10, IL-12, IL-16, interferon-γ (IFN-γ), GM-CSF, TNF-α, eotaxin (CCL11), IL-8 (CXCL8), macrophage inflammatory protein-1α (MIP-1α), RANTES (CCL5), nerve growth factor, stem cell factor (SCF), platelet-derived growth factor (PDGF), and transforming growth factor (TGF)-α and TGF-β1, among others.

Most of the studies that have investigated the ability of human eosinophils to produce cytokines have been performed using cells isolated from peripheral blood, making it difficult to know whether these properties also relate to tissue-residing eosinophils. Eosinophils do indeed change when they migrate into mucosal tissues, as revealed by the induction of activation markers such as CD69, intercellular adhesion molecule-1 (ICAM-1), and CD25 on the cell surface when this occurs. In the few studies that have been performed, *in situ* expression of IL-3, IL-5, GM-CSF, IL-16, and TGF-β1 has been detected in eosinophils residing within the intestinal mucosa. In addition, human blood and intestinal eosinophils have been found to store vasoactive intestinal polypeptide, substance P, calcitonin-gene related peptide, and somatostatin. When activated, these preformed mediators are released and additional mediators generated, for example, reactive oxygen species and lipid mediators such as leukotrienes, platelet-activating factor, and prostaglandins.

Under homeostatic conditions, eosinophils are present in a state of rest in order to avoid tissue damage from eosinophil-derived mediators. In contrast to the central role of IgE receptor cross-linking for activation of mast cells, as discussed later, eosinophil activation does not seem to depend on a dominant mediator but seems to respond to a broad range of stimuli, most of which also mediate survival or chemotaxis. Activation of eosinophils involves initial priming and subsequent triggering of the cells for effector functions. Triggering is mediated by cytokines such as IL-5, IL-3, GM-CSF, or TNF-α; by chemokines such as RANTES, MIP-1α, monocyte chemotactic protein 3 (MCP-3), MCP-4, eotaxin, or IL-8; and by complement components, aggregated immunoglobulins, lipid mediators, or histamine. Interestingly, many of these agonists are also known as mast-cell secretory products suggesting that mast-cell activation might be involved in the rapid activation of eosinophils

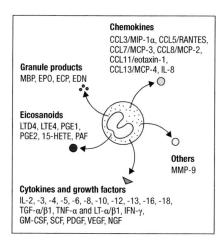

Figure 14.4 Eosinophil effector functions. Mediators of human eosinophils include cationic granule proteins, pro-inflammatory lipid mediators (leukotrienes D$_4$ and E$_4$, prostaglandins E$_1$ and E$_2$), cytokines (interleukins, IL; tumor necrosis factor, TNF-α; transforming growth factor, TGF; interferon, IFN; nerve growth factor, NGF; stem cell factor, SCF; platelet-derived growth factor, PDGF; vascular endothelial growth factor, VEGF), chemokines, and matrix metalloproteinases (MMP). 15-HETE, 15-hydroxyeicosatetraenoic acid; ECP, eosinophil cationic protein; EDN, eosinophil-derived neurotoxin; EPO, eosinophil peroxidase; LT, lymphotoxin; MBP, major basic protein; MCP, monocyte chemotactic protein; MIP, macrophage inflammatory protein.

as observed during allergic reactions. Human eosinophils also have functions in phagocytosis and antigen presentation, but these are poorly characterized. With regard to the latter function, studies in mouse models support a role for eosinophils in antigen presentation in view of their expression of MHC class II and accessory costimulatory molecules, and their ability to induce expansion of antigen-specific T cells. However, the biologic significance of this function remains to be firmly established.

14.12 Mucosal eosinophils collaborate with epithelial cells, T cells, mast cells, and nerve cells in mucosal immunophysiology

A major function of eosinophils is through their contribution to the maintenance of mucosal barrier function and protection against invasion by mucosal pathogens. This is consistent with the general role of innate immunity in barrier function. For example, following direct exposure to bacterial products derived from the intestinal microbiota *ex vivo*, eosinophils release toxic mediators. Furthermore, in contrast to blood eosinophils, intestinal eosinophils exhibit an activation phenotype (expression of CD69, for example) in noninflamed mucosa in the absence of obvious degranulation, suggesting that they are poised for immediate responses to commensal bacteria or pathogens in the event of loss of epithelial integrity.

The physiologic functions of eosinophils also depend on their capacity for coordinated interactions with other innate and adaptive immune cells. As noted, eosinophils likely receive stimulatory signals from mast cells. Eosinophils also interact with T cells in a bidirectional manner. At sites of mucosal allergic inflammation, mast cells and T_H2 cells secrete eosinophil-activating cytokines, including IL-5, which promote local survival and degranulation of eosinophils. Eosinophils are capable of presenting antigens to T cells through their expression of MHC class II and costimulatory molecules and as such stimulate or modulate T-cell function (Figure 14.5).

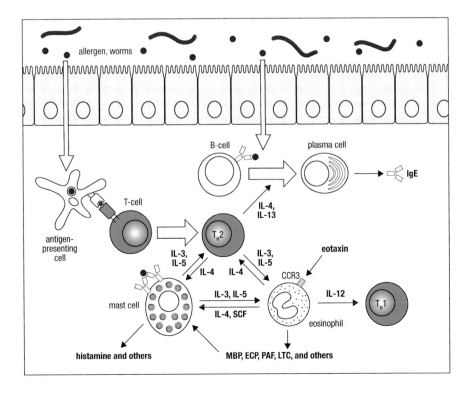

Figure 14.5 Interaction between eosinophils, mast cells, and T lymphocytes. The potential mechanisms that lead to immunologic hypersensitivity and T_H2-type inflammation as a result of cross talk between mast cells, eosinophils, and T cells in the intestinal mucosa are shown. LTC, leukotriene C. (Adapted from Bischoff, S., and Crowe, S., *Gastroenterology* 2005, 128:1089–1113. With permission from Elsevier Ltd.)

Eosinophils also interface directly and indirectly (through cross talk with mast cells) with the enteric nervous system. Eosinophil-derived granule proteins, particularly major basic protein and other soluble mediators such as stem cell factor, induce human mast-cell activation, which leads to the release of tryptase, histamine, and prostaglandin D_2 (PGD_2). Eosinophils bind to nerve cells in a specific fashion, which results in nerve remodeling, changes in neurotransmitter activity, and the induction of genes involved in neurotransmitter metabolism.

Another physiologic role of eosinophils could be in epithelial repair. Epithelial cell damage or necrosis is a potent signal for eosinophil chemotaxis, which prompts secretion of cytokines that possess important tissue repair and regulatory properties, such as TGF-β1 and fibroblast growth factors. Therefore, depending on local signals, eosinophils may be incited to mediate either epithelial cell damage or repair.

Mucosal mast cells

Unlike basophils and eosinophils, mature mast cells are only found within tissues, not in the blood. Although all three cell types (basophils, eosinophils, and mast cells) are derived from myeloid progenitors, mast cells leave the bone marrow in an immature state, with the tissue environment such as the mucosa necessary for full maturation. Mast cells have a general property of being strategically located at sites of the host-environment interface such as the skin and the mucosa of the respiratory, gastrointestinal, and urogenital tracts. This location suggests that a major function of mast cells is as immunologic "gatekeepers" or "watchdogs," and indeed functional data support this idea as discussed later.

The normal human gastrointestinal tract contains numerous mast cells. The largest numbers are found in the lamina propria, where 2%–3% of the cells are mast cells. There are two types of mast cells within mucosal tissues. The mast cells contained within the human lamina propria are considered to be the mucosal mast-cell subtype (MC_T) because they are tryptase positive and chymase negative and resemble a similar subtype in rodents. The second subtype of mast cells within human mucosal tissues is contained within the submucosa. Here, the mast-cell density is lower than in the lamina propria (about 1% of all cells) with the majority of cells being tryptase and chymase positive. This subtype of double-positive mast cells (MC_{TC}) is also observed in rodents.

Human mast cells are recognized as key effector cells in allergic inflammation, consistent with their strategic location within the skin and mucosal barriers and close relationship with the vascular system. Moreover, they bind IgE on their surface by virtue of their expression of the high-affinity IgE receptor that induces the release of histamine and other mediators on cross-linking by surface-bound IgE in the context of an allergen. Not only are mast cells cellular mediators of allergy, they regulate many tissue functions such as blood flow and coagulation, smooth muscle contraction and peristalsis of the intestine, mucosal secretion, wound healing, and innate and adaptive immune responses, including that associated with immune tolerance. This explains why mast cells have been found to be involved in many different types of human disease in addition to allergic disorders. These include inflammatory diseases, neurologic diseases, and functional diseases such as irritable bowel syndrome, functional dyspepsia, and fibromyalgia. Recent studies in rodents, and to some extent also in humans, have shown that mast cells also have a central role in host defense against bacteria and parasites, through the release of cytokines and other mediators that serve to recruit neutrophils, eosinophils, and T_H2 cells to the site of infection. These findings are consistent with the notion that mast cells are uniquely important in the defense of mucosal barriers.

14.13 Mast cells develop from immature bone marrow–derived precursors in mucosal tissues

Mast cells originate from immature, bone marrow–derived CD34$^+$ hematopoietic stem cells and circulate in the peripheral blood as committed progenitors before homing to tissues. In mice, homing to mucosal tissues of the intestinal lamina propria is dependent on immature mast-cell expression of $\alpha_4\beta_7$ integrin and MAdCAM-1 expression on high endothelial venules. In humans, the regulation of homing and the stage of maturation at which mast cells migrate from the blood into the tissue remain largely unknown. Electron-microscopic studies have revealed that mast-cell progenitors are not only found in peripheral blood but also in tissues such as the intestine where they represent 5%–15% of the total mast-cell numbers. This suggests that mast-cell densities in the intestine are regulated by the influx of early mast-cell progenitors and their growth-factor-dependent survival and proliferation.

Human mast cells develop from myeloid-cell progenitors under the influence of particular growth factors such as stem cell factor and IL-4, cytokines that also regulate the development of mast-cell subtypes. In addition to SCF and IL-4, other cytokines such as IL-6 and IL-9 promote mast-cell survival and proliferation. Consistent with a common progenitor, human mast cells exhibit significant similarities to basophils and have been considered as their tissue equivalent. Both cells contain basophilic intracellular granules, release histamine, and express the high-affinity IgE receptor. Morphologic and functional analyses, including gene expression studies, also support similarities between mast cells and monocytes or macrophages more than similarities with basophils (Figure 14.6). This is supported by the fact that although some murine mast-cell populations and human basophils respond well to IL-3, human mast cells either lack the IL-3 receptor or are poorly responsive. Thus, the developmental origin of human mast cells is incompletely known and may coincide with that of monocytes and/or basophils.

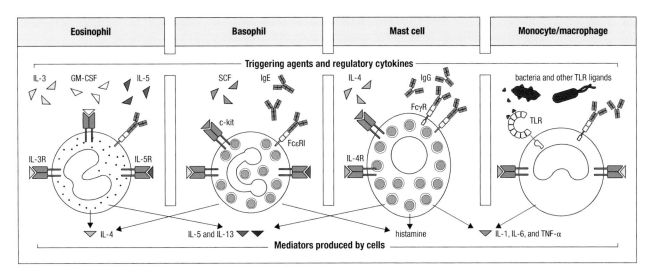

Figure 14.6 **Comparison between human mast cells, eosinophils, and basophils.** Mast cells exhibit similarities to, and differences from, eosinophil and basophil granulocytes and monocytes/macrophages. Mast cells share FcɛRI expression and histamine release with basophils. Mast cells share responsiveness to cytokines and bacterial products as well as nuclear morphology with monocytes and macrophages. However, mast cells generally do not express CD14 like monocytes, or IL-3 or IL-5 receptors like basophils or eosinophils. Mast cells express almost exclusively the stem cell factor (SCF) receptor c-Kit. Triggering agents and regulatory cytokines of the four cell types are shown above each cell; mediators released from the cells after stimulation are shown below each cell. (Adapted from Bischoff, S., *Nat. Rev. Immunol.*, 2007, 7:93–104. With permission from Macmillan Publishers Ltd.)

14.14 Mast-cell effector function is initiated by IgE-dependent and IgE-independent stimuli

The classical, and possibly most effective, mast-cell stimulus is cross-linking of cell-surface-bound IgE by allergen in sensitized individuals. This mechanism, which was first described shortly after the discovery of IgE in the late 1960s, is central to type I hypersensitivity reactions. Such responses can be readily modeled by stimulating mast cells with antibodies that cross-link IgE or the type 1 IgE receptor (FcεRI).

FcεRI-independent mast-cell regulators are typically growth factors or cytokines that either promote mast-cell development from progenitor states, or function as regulators of mediator release, or do both. For example, SCF acts not only as a mast-cell growth factor, but also as a regulator of mast-cell mediator release by either enhancing IgE-dependent mediator release or by directly inducing mediator release in mast cells maintained in a SCF-deprived milieu. The mechanisms of SCF effects in human mast cells are well understood (**Figure 14.7**). Binding of SCF induces autophosphorylation of its receptor CD117 (c-Kit) and subsequent activation of several signaling molecules including phosphatidylinositol-3-kinase (PI3K) and mitogen-activated protein kinase (MAPK). SCF-mediated signaling of mast cells derived from human CD34$^+$ peripheral blood cells involves activation of a series of kinases, including the Src-related kinase Lyn and BTK, and phosphorylation of signal transducer of activation 5 (STAT5) and STAT6 that link c-Kit activation to specific transcriptional programs. c-Kit signaling also involves a transmembrane adaptor (NTAL), which is not only linked to c-Kit signaling but also to that associated with signaling by FcεRI. Both c-Kit and FcεRI signaling enhance IgE-dependent degranulation.

IL-4 is another important human mast-cell regulator. In contrast to SCF, IL-4 does not affect mast cells by itself but acts synergistically with SCF on mast-cell survival, proliferation, and IgE-dependent mediator release. Moreover, IL-4 alters the cytokine profile released by mast cells by reducing pro-inflammatory cytokines such as TNF-α and IL-6, and in turn enhancing secretion of type 2 cytokines such as IL-5 and IL-13. Considering that IL-4 also induces the development of T$_H$2 cells and IgE switching in B cells, this cytokine is a key mediator of allergic inflammation.

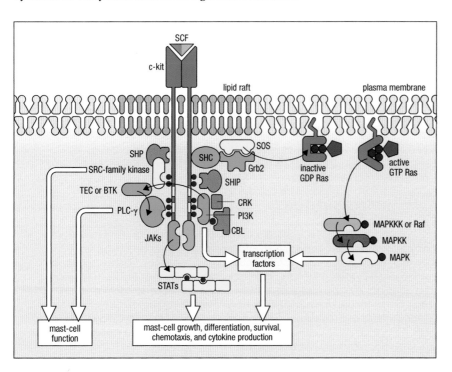

Figure 14.7 Stem cell factor (SCF) signaling pathways in mast cells. SCF binding to its receptor, c-Kit, results in its dimerization and autophosphorylation within the cytoplasmic tail. This results in recruitment of numerous adaptors and signaling modules to the phosphorylated cytoplasmic tail. Four general signaling pathways are activated, including those associated with mitogen-activated protein kinases (MAPK); phosphatidylinositol-3-kinase (PI3K); Janus kinase-signal transducer and activator of transcription (JAK/STAT); and Src-family kinases that regulate mast-cell function, growth, differentiation, survival, mediator production, and chemotaxis. BTK, Bruton's tyrosine kinase; GRB2, growth factor-receptor-bound protein 2; MAPKK, MAPK kinase; PLCγ, phospholipase Cγ; SHC, Src homology 2 (SH2)-domain-containing transforming protein C; SHIP, SH2-domain-containing inositol-5-phosphatase; SHP2, SH2-domain-containing protein tyrosine phosphatase; SOS, son-of-sevenless homolog; TEC, tyrosine kinase expressed in hepatocellular carcinoma. (Adapted from Gilfillan, A.M., and Tkaczyk, C., *Nat. Rev. Immunol.*, 2006, 6:218–230. With permission from Nature Publishing Group.)

Little is known about other FcεRI-independent triggers of mediator release from human mast cells. The types of IgE-independent mast-cell agonists vary not only between human and rodent mast cells but also between human mast cells from different body sites. Human mast cells challenged with IFN-γ express FcγRI at sufficient quantities that allow them to become activated for mediator release on FcγRI aggregation. This mechanism could be of relevance for the otherwise poorly understood IgE-independent allergic reactions as well as for nonallergic mast-cell activation during type III hypersensitivity reactions or infections. Mediators such as C3a, C5a, substance P and other neuropeptides, IL-8, nerve growth factor, SCF, bacterial products, and ultraviolet light have all been suggested to be mast-cell activators. However, most of these mediators only affect human skin mast cells and not mucosal mast cells and may require prior stimulation with SCF or IL-4. SCF and IL-4 may thus be considered as primary, costimulatory mediators since they not only enhance FcεRI-mediated signals but also induce the upregulation of secondary stimulatory receptors. One example is the receptor for substance P (NK-1) which is upregulated by SCF and IL-4. Such costimulation draws comparisons between mast-cell and T-cell signaling because both require the cooperation of two signals for optimal activation: an antigen-dependent signal such as the T-cell receptor or IgE bound to the cell surface, and a costimulatory molecule, for example, CD80 in T cells or SCF/IL-4 in mast cells.

Mast cells are also subject to regulation by inhibitory receptors that counter the activity of the aforementioned activating ligands and their receptors. These inhibitors include ligands of receptors that contain an immune tyrosine inhibitory motif such as FcγRIIB, gp49B1, SIRPα, and the human analogs LIR-5 and LILR B4. They also include ligands of the anti-inflammatory cytokines TGF-β1 and IL-10; CD200; intracellular signaling molecules such as NTAL or RabGEF1; and several other molecules including retinol, β2-adrenoceptor agonists, and extracellular matrix proteins binding to CD63.

14.15 Mast cells exert their biological functions almost exclusively by secretion of humoral factors

Although there are a few reports describing mast-cell phagocytosis and other nonsecretory mast-cell functions, the biological activities of mast cells are mainly associated with the secretion of mediators. The array of mediators released by human mast cells is enormous and further explains how mast cells can be involved in so many different physiologic and pathophysiologic functions (Figure 14.8). The mast-cell mediators can be classified into small-molecule mediators (histamine, serotonin), protein mediators (cytokines, proteases), lipid mediators (leukotrienes, prostaglandins), and proteoglycans (heparin). Some of the mediators are stored in granules (histamine, proteases, proteoglycans, small amounts of TNF-α) and therefore can be released within seconds or minutes. Others are newly synthesized within minutes to hours upon stimulation of the cells (lipid mediators and most cytokines) and often require transcription.

14.16 Mucosal mast cells respond to extrinsic and intrinsic signals and thus regulate function of mucosal barrier and smooth muscle motor function

The function of the mucosal barrier is under the influence of mast cells in response to extrinsic (e.g., infections and toxins) and intrinsic (e.g., acute

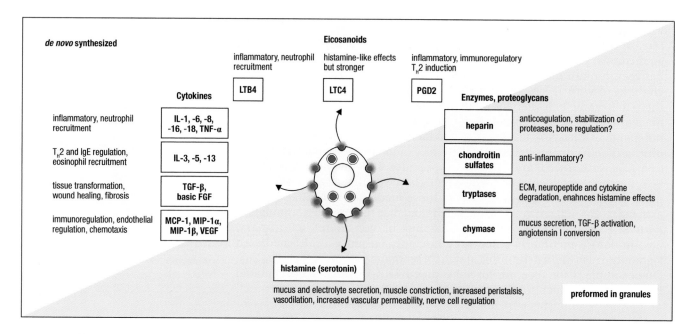

Figure 14.8 Mast cell effector functions. Mediators of human mast cells include small molecules that primarily function by promoting inflammation (histamine, leukotriene B$_4$ and C$_4$, prostaglandin D$_2$), cytokines (interleukins, IL; tumor necrosis factor, TNF-α; transforming growth factor-beta, TGF-β; basic fibroblast growth factor, FGF; monocyte chemotactic protein 1, MCP-1; macrophage inflammatory protein, MIP; vascular endothelial growth factor, VEGF), proteases (tryptases, chymase), and proteoglycans (heparin, chondroitin sulfate). The mediators displayed in the yellow section are mediators synthesized *de novo* upon stimulation of mast cells, and those in the blue section exist within intracellular granules. The major biological effects of each mediator are indicated. ECM, extracellular matrix. (Adapted from Bischoff, S., *Semin. Immunopathol.*, 2009, 31:185–205. With permission from Springer.)

or chronic stress) events. For example, in response to parasitic infection or stress, mast-cell proteases are released that increase epithelial paracellular permeability by inducing a redistribution of epithelial tight junction proteins. Psychological stress can also lead to enhanced release of mast-cell mediators such as histamine and tryptase in the proximal small intestine both in healthy individuals and to a larger extent in individuals who have food allergies.

The basis for this likely resides in the fact that mast cells are in close contact with epithelial cells and nerve endings. Histamine, when released at low concentrations by mast cells, can bind histamine receptors on polarized epithelial cells resulting in increased transepithelial ion transport and mucus secretion. These effects on the epithelium are blocked by antihistamines and cyclooxygenase inhibitors, which in the latter case disrupt production of PGD$_2$ and LTC$_4$. This is clinically associated with diarrhea, which although abnormal in the context of food allergy would be beneficial in the case of exposure to microbes, toxins, and other harmful substances.

These physiologic activities of mast cells are further enhanced by the relationship between mast cells and nerve cells contained within all mucosal tissues. In intestines, these nerve cells are highly organized and contained within the enteric nervous system. Information that is received by the enteric nervous system is derived from local sensory receptors, the central nervous system, and immune cells including mast cells. Specific IgE or IgG antibodies attach to the mast cells, enabling the mast cell to detect sensitizing antigens when they appear in the gut lumen, and are transcytosed by transporting immunoglobulin receptors. This allows for the mast cell to signal the presence of the antigen to the enteric nervous system. The enteric nervous system interprets the mast cell signal as a threat and releases neurotransmitters that drive the propulsive motor behavior of the gut tube, which is in turn organized to rapidly and effectively eliminate the threat. Although operation of this alarm program protects the individual, it is at the expense of symptoms that include cramping abdominal pain, fecal urgency, and diarrhea in the intestines or bronchial constriction in the lungs. Mast cells use paracrine signaling, such as the release of histamine, proteases, and lipid mediators, to transfer the necessary chemical information to the neural networks. In addition to such humoral mediators, adhesion molecules such as ICAM-1 appear to be involved in mast cell–nerve interactions. It can be surmised that the regulation of epithelial secretion and smooth muscle motor function are due to the properties of MC_T and MC_{CT}, respectively.

14.17 Mast cells provide host defense against microbes

Human mast cells are capable of recognizing a large number of PAMPs and other bacterial products by TLRs and other pattern recognition receptors. Consistent with this, mast cells respond to a variety of microbially derived pathogenic factors. For example, *Escherichia coli* α-hemolysin induces calcium influx in human intestinal mast cells leading to the release of histamine, sulfidoleukotrienes, and pro-inflammatory cytokines. Furthermore, the type 1 fimbrial protein FimH, cholera toxins, and glucopeptides derived from parasites have been characterized as capable of activating mast cells. These factors are important in the host's response to infections with organisms such as *Trichinella spiralis* or *Nippostrongylus brasiliensis*, which are accompanied by substantial accumulation of mast cells in infected tissues. It is not clear whether this mast cell accumulation is triggered by the infectious agents or by the associated T_H2 response. Nonetheless, mast-cell proteases, such as monocyte chemotactic protein 1 (MCP-1), are critical mediators of the immune response to helminths. In the absence of MCP-1, for example, expulsion of *Trichinella* sp. and *Nippostrongylus* sp. is significantly delayed.

The mechanism by which mast cells sense parasitic infection likely involves not only pattern recognition receptors but also activation by FcεRI aggregation through cross-linking of IgE directed against parasitic antigens. This is consistent with the generation of parasite-specific IgE antibodies during a parasite infection and the fact that blocking mast-cell function leads to impairment in host defense against parasites despite the presence of antiparasite IgE antibodies. Of the mast-cell mediators that are induced by FcεRI cross-linking, IL-5 (for eosinophil recruitment) and IL-13 (for B and T_H2 cell immunity) are of particular importance.

Mast cells also participate in host defense against bacterial pathogens. This has been demonstrated in mouse models of *Klebsiella pneumonia*–induced infections of the lung and in models of peritoneal sepsis. In response to bacterial infections, mast cells produce TNF-α and leukotrienes, which serve to recruit neutrophils and as such result, along with other mechanisms, in increased bacterial clearance. Mast-cell activation during bacterial infection also induces dendritic cell migration, supporting a role for mast cells in linking innate to adaptive immune responses. In response to bacterial pathogens, mast cells are activated by both TLR and non-TLR (e.g., complement) dependent mechanisms.

Mast cells may also provide host defense against viruses. Stimulation of mast cells through TLR3 using polyinosinic/polycytidylic acid leads to the recruitment of $CD8^+$ T cells. Human mast cells also express TLR3, and stimulation through this receptor leads to decreased mast-cell adhesion to extracellular matrix proteins and increased production of IFN-γ consistent with a role in host defense against viral infections.

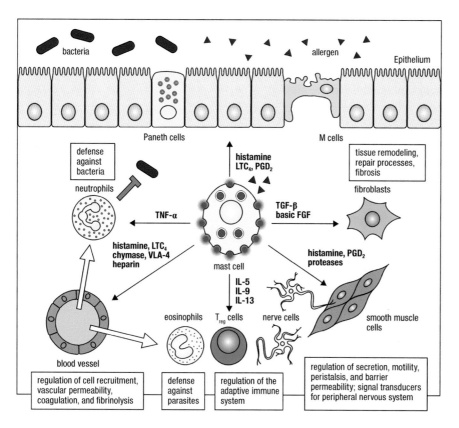

Figure 14.9 Physiologic functions of mast cells at the mucosal barrier. Mucosal mast cells regulate tissue homeostasis (epithelial secretion and permeability, blood flow and vascular permeability, smooth muscle functions and peristalsis, wound healing, and fibrosis), immune functions (recruitment and activation of neutrophils, eosinophils, and lymphocytes; induction of T_{reg} cells; and defense against microbes), neuronal functions (neuroimmune interactions, peristalsis, bronchoconstriction, and pain), and inflammation (allergy, asthma, inflammatory bowel disease, irritable bowel syndrome, infection). FGF, fibroblast growth factor; IL-5, interleukin 5; LTC_4, leukotriene C_4; PGD_2, prostaglandin D_2; TGF, transforming growth factor; TNF-α, tumor necrosis factor; T_{reg}, regulatory T cells; VLA-4, very late antigen 4. (Adapted from Bischoff, S., *Nat. Rev. Immunol.*, 2007, 7:93–104. With permission from Macmillan Publishers Ltd.)

14.18 Mast cells not only induce an immune response and inflammation but also regulate immune response and resultant processes of tissue repair

Mast cells are also involved in effecting peripheral tolerance induced by $CD4^+CD25^+Foxp3^+$ T_{reg} cells. Release of IL-9, a mast-cell growth and activation factor, by activated T_{reg} cells is associated with mast-cell recruitment and activation during the induction of tolerance. Finally, mast cells regulate wound healing, fibrosis, blood flow and coagulation, and protection against neoplasms (**Figure 14.9**). As such, mast cells are important in both the maintenance and reestablishment of tissue homeostasis.

Mucosal basophils, eosinophils, and mast cells in human disease

Given their similarities in lineage development, sources of stimuli to which they respond, localization to mucosal tissues, and association with allergy and T_H2-dependent immune responses, it is not surprising that basophils, eosinophils, and mast cells exhibit overlapping but distinct relationships with mucosal diseases. The human diseases with which these cell types exhibit a unique relationship are discussed next.

14.19 Mucosal basophils are involved in allergies of the skin and intestines, asthma, and autoimmunity

The ability of basophils to be activated by multiple stimuli, coupled with their ability to migrate to draining lymph nodes with their many effector functions, suggests that they may be critical regulators of inflammation in multiple tissue sites. Accordingly, elevated basophil populations are correlated with the inflammation associated with allergies and asthma. For example, basophils are elevated in the airways of asthmatics, increase upon allergen challenge, and are observed at elevated numbers in the lungs during severe asthma. In addition, blood basophil counts are increased in asthma, and elevated basophil populations directly correlate with airway hyperresponsiveness and decreased lung function.

Increased basophil populations are also associated with multiple allergic disorders of the skin. For example, elevated numbers of basophils are found in skin biopsies associated with atopic or allergic contact dermatitis and are reported to increase on allergen challenge. Basophils are also implicated in chronic urticaria (hives), a common skin disease involving mast cells and basophil recruitment to the lesional sites. Although basophils are most commonly associated with allergic diseases, they likely play a role in autoimmune disorders as revealed by observations in systemic lupus erythematosus (SLE), an autoimmune disease characterized by autoantibody production that may affect mucosal tissues. More specifically, individuals suffering from SLE possess self-reactive IgE and activated circulating basophils as determined by increased levels of HLA-DR and CD62L. HLA-DR-expressing basophils are also detectable in the lymph nodes of SLE patients but not in healthy controls.

14.20 Mucosal eosinophils are associated with asthma, parasitic and viral infections, mucosal allergy, and several hypereosinophilic disorders characterized by mucosal infiltration with eosinophils

The respiratory mucosa, in contrast to the gastrointestinal mucosa, does not contain significant numbers of eosinophils in healthy conditions. However, eosinophil influx into the respiratory mucosa is observed during the course of several inflammatory lung diseases and especially asthma. As such, eosinophils have been considered as major inflammatory cells involved in the pathogenesis of asthma. This paradigm was challenged when clinical trials revealed that treatment of asthma patients with anti-IL-5 antibodies substantially decreased the number of eosinophils in the lungs but failed to provide symptomatic benefit. More recent studies have, however, suggested that anti-IL-5 treatment benefits patients with sputum eosinophilia, as shown by a reduction in prednisone requirements or asthma exacerbations and improvement of asthma patients' quality of life. Therefore, the eosinophil is still viewed as a terminal effector cell in subgroups of allergic airway diseases. But this multifunctional cell could be more involved in the initial stages of allergic disease development than was previously thought, particularly with regard to the ability of the eosinophil to modulate T-cell responses. Alternatively, eosinophils might play a role not only in initiating and perpetuating inflammation, but also in repair and remodeling processes. There is a well-documented association of tissue eosinophilia and eosinophil degranulation with certain fibrotic syndromes, and the eosinophil is the source of several fibrogenic and growth factors, including TGF-α, TGF-β1, fibroblast growth factor 2, vascular endothelial growth factor, matrix metalloproteinases, and selected cytokines.

Eosinophilia is commonly observed during helminth infection and is typically associated with a strong T_H2 immune response. Eosinophils are able to effectively kill or damage larvae and adult worms. However, eosinophils are only partly effective in the control of helminth infection. This suggests that eosinophils have other functions during parasitic infection. Eosinophils, as described earlier, possess a range of immunomodulatory effects such as increased production of anti-inflammatory cytokines (IL-10 and TGF-β) and stimulation of regulatory T cells or alternatively activated macrophages. Therefore, as described for asthma, immunomodulation that allows parasite survival but reduces pathology may be a dominating function of eosinophils in association with parasitosis. Whether this is a primary function of the eosinophil or an immune-evasive function that is induced by the parasite is not known. However, it can be conjectured that such helminth (parasite)-induced immunomodulation may contribute to protection from allergic and autoimmune responses, as proposed by the "hygiene hypothesis."

Eosinophils are also important in viral infections. Human respiratory syncytial virus causes respiratory tract infection, primarily among infants and toddlers. The severity of infection can extend from mild upper respiratory symptoms to severe bronchiolitis and pneumonia, and may progress to acute respiratory distress syndrome and death. Human respiratory syncytial virus infections are associated with significant eosinophilia in the bronchoalveolar fluid. This is most likely because eosinophils are attracted to the site of inflammation by chemokines such as RANTES and MIP-1α. The role of eosinophils in this disease is uncertain, as there is no clear evidence from human studies as to whether they promote host defense or enhance pathology. Among these pathologies, there is a clear association between severe human respiratory syncytial virus infection, particularly among young infants, and the development of postinfection asthma.

Eosinophilic gastrointestinal diseases (EGIDs) comprise primary and secondary subtypes. Primary EGIDs are defined as gastrointestinal symptoms occurring in the context of increased numbers of mucosal eosinophils, and in the absence of other recognized causes of tissue eosinophilia, such as drugs, allergies, autoimmunity, malignancy, and parasitic infestation, which would be considered secondary EGIDs as discussed later. They include eosinophilic esophagitis, eosinophilic gastritis, eosinophilic gastroenteritis, eosinophilic enteritis, and eosinophilic colitis. The prevalence, or at least recognition, of EGIDs is increasing. The majority of patients (50%–80%) with EGIDs are atopic. The close link with atopy, in combination with the increased likelihood of food-allergen-specific IgE in the serum or positive food allergen skin prick tests, suggests that there is a pathologic role for an allergic inflammatory process in EGIDs. Allergic inflammation is characterized by the activation of T_H2 cells, which elaborate specific profiles of proallergic cytokines, including IL-4, IL-5, and IL-13, that promote the formation of IgE antibodies and support eosinophil and mast-cell recruitment, maturation, and activation. The potential importance of IL-5 has been highlighted by preliminary studies in which treatment of patients with eosinophilic esophagitis using anti-IL-5 antibody is associated with marked reductions in the numbers of eosinophils in the blood and esophageal mucosa.

Secondary EGIDs are inflammatory diseases that have recognized causes of gastrointestinal eosinophilia such as parasite infection, gastroesophageal reflux disease (GERD), inflammatory bowel disease (IBD), irritable bowel syndrome (IBS), allergic disease, drug reactions, and malignancy. The mild eosinophilia typically occurring in the distal esophagus in about 50% of patients with GERD has to be separated from the more pronounced and ubiquitous eosinophilia in patients with eosinophilic esophagitis.

Finally, hypereosinophilic syndrome comprises a heterogeneous group of disorders characterized by excessive numbers of circulating and tissue-infiltrating eosinophils resulting in organ dysfunction. Some hypereosinophilic syndrome subtypes are myeloproliferative disorders, whereas the etiology of other variants remains undefined. Involvement of the gastrointestinal tract is common, and about 20% of patients experience diarrhea. Hypereosinophilic syndrome is a well-recognized cause of eosinophilic gastritis, gastroenteritis, and colitis, but in contrast to the other primary EGIDs, eosinophilic infiltration of other organs including the heart, skin, and nerves is characteristic. Consequently, hypereosinophilic syndrome might be considered to be a secondary EGID.

14.21 Mucosal mast cells are important mediators of inflammation associated with allergy, asthma, celiac disease, inflammatory bowel disease, and diseases of unknown origin such as irritable bowel syndrome and systemic mastocytosis

The inflammatory mediators produced by mast cells, basophils, and eosinophils are responsible for the clinical symptoms and the organ dysfunction that occur during IgE-mediated type I reactions. Type I allergic reactions, which are also taken as a basis for many cases of bronchial asthma, food allergy, and other atopic diseases, are divided into an immediate phase and a late phase, which occurs commonly 6–8 hours after the immediate phase (see Chapter 21). The immediate phase is characterized by the IgE-dependent activation of mast cells and basophils and the release of pro-inflammatory mediators—such as histamine, proteases, leukotrienes, and cytokines—from these cells. At the same time, the cells start to synthesize mediators leading to a more sustained release of mediators such as eicosanoids and cytokines. The late phase is characterized by the infiltration of the tissue with further inflammatory cells such as neutrophils, eosinophils, and lymphocytes. These cells are attracted by mediators such as TNF-α, IL-5, IL-4, and IL-3 released by mast cells and basophils upon IgE-dependent immediate-type activation. Most importantly, human mast cells induce the recruitment and local activation of eosinophils by expressing factors such as IL-5 upon IgE-dependent activation, and induce the recruitment of neutrophils by releasing IL-8 and TNF-α.

Human mast cells also likely participate in regulating lymphocyte functions during the course of allergic inflammation. Upon IgE cross-linking, mast cells produce IL-13, a cytokine that supports the production of allergen-specific IgE by B cells. The release of IL-13 can be further increased by the presence of IL-4, which is known to shift the cytokine profile produced by human mast cells away from pro-inflammatory cytokines—such as TNF-α, IL-1β, and IL-6—to T_H2 cytokines, including IL-13. Human mast cells can also regulate T-cell functions through other mediators, such as PGD_2 that almost exclusively derives from activated mast cells, is released during allergic reactions, and is particularly important at the onset and in the perpetuation of asthma in young adults. This lipid mediator evokes airway hypersensitivity and chemotaxis of T cells, basophils, and eosinophils through interaction with two receptors: the prostanoid DP receptor (PTGDR) on granulocytes and smooth muscle cells; and CRTH2 (chemoattractant receptor-homologous molecule expressed on T_H2 cells) on T_H2 cells. Furthermore, *PTGDR* has been identified as an asthma-susceptibility gene. Apart from PGD_2, other human mast-cell mediators, such as LTB_4, CCL3, and CCL4; OX40 ligand; and TNF-α, are involved in recruiting T cells and triggering T-cell–mediated adaptive immune responses, including memory induction, that enhance and perpetuate allergic reactions.

Elevated levels of histamine, its metabolite methylhistamine, tryptase, eosinophil cationic protein, eosinophil-derived protein X (EPX), IL-5, and TNF-α are detectable in serum, urine, gut or bronchial lavage fluid, and stool from patients with allergy. This is further evidence that activation of mast cells, eosinophils, and to some extent basophils is involved in allergy. This is supported by histologic studies showing degranulation and cytokine production by these cell types after allergen provocation tests. Mast cells stimulated by IgE cross-linking also trigger local nerve responses resulting in bronchial hypersecretion, pain, and diarrhea.

Mucosal mast cells also play an important role in specific mucosal diseases such as celiac disease and IBD. In celiac disease, the histamine content and numbers of MC_T cells are increased. Moreover, gliadin challenge in celiac disease results in decreased mast cells consistent with mast-cell degranulation. Increased numbers of mast cells are also observed in the mucosa of the ileum and colon in IBD. Mast cells in IBD also exhibit significant increases in expression of histamine, tryptase, TNF-α, IL-16, and substance P. Such mast cells are found in close proximity to the basal lamina of the epithelial cell glands and in other layers of the gut wall. Studies with tryptase-deficient animal models suggest that mast cells function in the promotion of inflammation.

Irritable bowel syndrome (IBS) is a disorder of unknown cause that is largely defined by clinical criteria and is thus a diagnosis of exclusion in the absence of any known gastrointestinal disease such as infection, IBD, celiac disease, food allergy, or malignancy. There is increasing evidence that IBS may represent a subclinical inflammatory disorder. A low-grade mucosal inflammation can be detected in a subset of patients with IBS. This inflammatory infiltrate is mainly represented by increased numbers of T lymphocytes and mast cells that are located in the lamina propria. The close apposition of immunocytes to gut nerves supplying the mucosa provides a basis for neuroimmune cross talk, which may explain the gut sensorimotor dysfunction and related symptoms in patients with IBS. Consistent with this, duodenal mast-cell hyperplasia is associated with IBS. However, the mechanisms that induce mast-cell hyperplasia and the presumed mast-cell mediator release in IBS still need to be defined.

The final disease to be discussed is systemic mastocytosis. Systemic mastocytosis is a disorder of unknown origin in which mast cells accumulate in multiple tissues, especially those in association with mucosal tissues. Consequently, gastrointestinal symptoms are common and consist of abdominal pain or diarrhea in the majority of patients. Systemic mastocytosis is also associated with gastric acid hypersecretion due to the hyperhistaminemia. This can result in ulcer disease that in turn can cause small intestinal mucosal damage due to the effects of the secreted acid and consequently disruption of digestion and absorption of nutrients. The malabsorption that occurs is similar to that observed in another disease associated with acid hypersecretion by the stomach—Zollinger-Ellison syndrome, which is a disorder due to tumors that secrete excessive gastrin, a hormone that stimulates acid secretion.

Systemic mastocytosis is characterized by the accumulation of neoplastic mast cells not only in the gastrointestinal tract but also in the bone marrow and other organs. Consequently, enlargement of the liver and the spleen, portal hypertension, and ascites occur frequently. Diagnosis of systemic mastocytosis is based on histologic and immunohistologic examination of gastrointestinal tissue specimens using antibodies directed against tryptase, CD117, and CD25.

SUMMARY

Basophils, eosinophils, and mast cells are immunoregulatory cells with inflammatory potential that are derived from bone marrow myeloid progenitors. They are found in the blood and in tissues, especially mucosal tissues, under normal conditions but accumulate during the course of many diseases. These cells intercommunicate with each other and many other hematopoietic and parenchymal cells (e.g., epithelial cells and nerve cells) and contribute to protection from infections (e.g., parasitic infections) and pathologies associated with allergy and related conditions such as asthma. Their function during the course of these diseases is not always clear because they may either initiate and maintain the inflammatory process or be involved in tissue remodeling and healing. These cells have the common property of responding to IgE-complexed antigens and characteristically exert their biological functions by releasing inflammatory, cytotoxic, and immunomodulatory mediators that are tightly controlled by numerous agonists and antagonists.

FURTHER READING

Abraham, S.N., and St John, A.L.: Mast cell-orchestrated immunity to pathogens. *Nat. Rev. Immunol.* 2010, 10:440–452.

Bischoff, S.C.: Role of mast cells in allergic and non-allergic immune responses: Comparison of human and murine data. *Nat. Rev. Immunol.* 2007, 7:93–104.

Bischoff, S.C.: Food allergy and eosinophilic gastroenteritis and colitis. *Curr. Opin. Allergy Clin. Immunol.* 2010, 10:238–245.

Bochner, B.S., and Gleich, G.J.: What targeting eosinophils has taught us about their role in diseases. *J. Allergy Clin. Immunol.* 2010, 126:16–25.

Brightling, C.E., and Bradding, P.: The re-emergence of the mast cell as a pivotal cell in asthma pathogenesis. *Curr. Allergy Asthma Rep.* 2005, 5:130–135.

Finkelman, F.D., Khodoun, M.V., Strait, R.: Human IgE-independent systemic anaphylaxis. *J. Allergy Clin. Immunol.* 2016, 137:1674–1680.

Guilarte, M., Sala-Cunill, A., Luengo, O. et al.: The mast cell, contact, and coagulation system connection in anaphylaxis. *Front. Immunol.* 2017, 26:(8):846.

Hogan, S.P., Rosenberg, H.F., Moqbel, R., et al.: Eosinophils: Biological properties and role in health and disease. *Clin. Exp. Allergy.* 2008, 38:709–750.

Liu, C., Liu, Z., Li, Z., et al.: Molecular regulation of mast cell development and maturation. *Mol. Biol. Rep.* 2010, 37:1993–2001.

Obata, K., Mukai, K., Tsujimura, Y., et al.: Basophils are essential initiators of a novel type of chronic allergic inflammation. *Blood* 2007, 110:913–920.

Ohnmacht, C., and Voehringer, D.: Basophil effector function and homeostasis during helminth infection. *Blood* 2009, 113:2816–2825.

Perrigoue, J.G., Saenz, S.A., Allenspach, E.G., et al.: MHC class II-dependent basophil-CD4+ T cell interactions promote T(H)2 cytokine-dependent immunity. *Nat. Immunol.* 2009, 10:697–705.

Powell, N., Walker, M.M., and Talley, N.J.: Gastrointestinal eosinophils in health, disease and functional disorders. *Nat. Rev. Gastroenterol. Hepatol.* 2010, 7:146–156.

Schroeder, J.T.: Basophils beyond effector cells of allergic inflammation. *Adv. Immunol.* 2009, 101:123–161.

Stone, K.D., Prussin, C., and Metcalfe, D.D.: IgE, mast cells, basophils, and eosinophils. *J. Allergy Clin. Immunol.* 2010, 125:S73–80.

15

M cells and the follicle-associated epithelium

HIROSHI OHNO, MARIAN NEUTRA, AND IFOR R. WILLIAMS

The follicle-associated epithelium (FAE) overlying organized mucosal lymphoid tissues (Figure 15.1a) plays a major role in mucosal immunity. Representing a very small fraction of the total mucosal surface area of the gastrointestinal tract mucosa, the FAE contains a unique epithelial cell type, the microfold (M) cell, whose primary function is to translocate luminal material across the epithelial barrier to dendritic cells (DCs) and lymphocytes within and below the epithelium. The FAE is separated from the underlying lymphoid follicle by a subepithelial "dome" region filled with T and B cells, as well as DCs, that efficiently capture materials transported by M cells (see Figure 15.1b). Some DCs and lymphocytes migrate into intraepithelial pockets formed by the M cells, as described in more detail later. FAE and M cells are associated with the organized Peyer's patches in the small intestine, and with isolated follicles in the small and large intestine, appendix, rectum, crypts of the adenoids and tonsils, airways, and even the conjunctivae of the eyes.

FUNCTION OF THE FOLLICLE-ASSOCIATED EPITHELIUM

15.1 FAE promotes antigen sampling

In contrast to the well-defended villus and surface epithelium of the intestine, the FAE lacks many defensive features and appears designed to allow macromolecules, particles, and microorganisms access to the apical surface. First, the secretions emanating from the FAE lack key protective molecules. Whereas the villus epithelium is composed primarily of absorptive enterocytes and mucin-secreting goblet cells, the FAE contains few or no goblet cells and secretes little or no mucus (see Figure 15.1a). Instead, it contains M cells along with enterocytes that display an FAE-specific phenotype that differs from villus enterocytes. Enterocytes of the FAE produce very low levels of digestive enzymes and alkaline phosphatase. The crypts surrounding the FAE lack defensin- and lysozyme-producing Paneth cells and do not secrete IgA into the lumen, as they lack polymeric immunoglobulin receptors.

FAE enterocytes have closely packed apical microvilli and a thick brush border glycocalyx that provides a barrier to large molecules and particles. Unlike M cells, FAE enterocytes do not participate in transport but appear to facilitate transport activities of neighboring M cells through distinct glycosylation patterns that allow recognition and adherence of certain microorganisms to the FAE. Gene expression studies indicate that the enterocytes of the FAE express specific extracellular matrix and

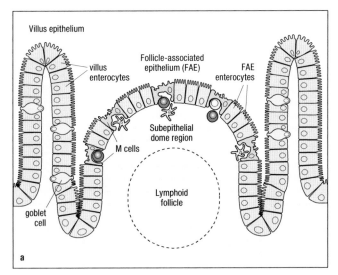

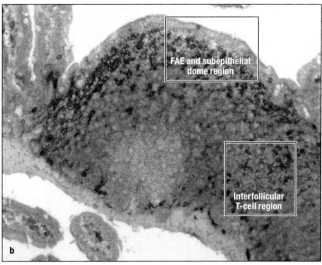

Figure 15.1 A specialized follicle-associated epithelium (FAE) covers organized mucosal lymphoid follicles. (a) FAE cells and villus epithelial cells. The FAE consists of special follicle-associated enterocytes and M cells. The villus epithelium is composed primarily of absorptive enterocytes and mucus-secreting goblet cells. Cells in the crypts located between a villus and a lymphoid follicle supply cells to both villus epithelium and FAE, differentiating along distinct phenotypic pathways. **(b)** A section of mouse Peyer's patch with labeled antibodies specific for the DC marker CD11c (brown) shows immature DCs congregated in the subepithelial dome region, where the cells capture incoming microbes and macromolecules transported by M cells. The DCs later migrate down the sides of the follicle to the interfollicular region to present antigen to naive T cells. ([a] Adapted from Neutra, M., and Kozlowski, P. *Nat. Rev. Immunol.* 2006, 6:148–158. With permission from Macmillan Publishers Ltd. [b] Adapted from Chabot, S. et al. *Vaccine* 2007, 25:5348–5358. With permission from Elsevier Ltd.)

matrix-interacting proteins. For example, the basal lamina of the FAE lacks laminin-2 subunits and contains large pores that presumably reflect the migration of cells into and out of the M cell pockets in the epithelium.

15.2 FAE attracts specific populations of dendritic cells and lymphocytes into organized mucosal lymphoid tissues

FAE cells differ from villus cells in their ability to release certain chemokines that attract immune cells toward the FAE and thus to sites of organized lymphoid tissue. In the small intestine of mice and humans, for example, chemokine CCL20, also designated macrophage inflammatory protein-3α (MIP-3α), is constitutively expressed in the FAE but not in the villus epithelium (**Figure 15.2**). CCL20 attracts subpopulations of DCs and lymphocytes that express the chemokine receptor CCR6. Mice that lack CCR6 have lower numbers of CD11c$^+$ DCs in the subepithelial dome regions of Peyer's patches and have an impaired humoral immune response to orally administered antigen and certain enteropathogenic viruses. Cells of the mouse FAE also express CCL9 (analogous to CCL23 in humans), which attracts CCR1-expressing myeloid DCs, and CXCL16, which attracts CXCR6-expressing B and T lymphocytes into Peyer's patches.

Gene expression in the FAE may be modulated by microorganisms that contact the FAE or are transported by M cells into the mucosa. In the small intestine, CCL20 gene expression is upregulated in the FAE by exposure to pathogenic bacteria or bacterial components. In the colon, where microbial populations are more numerous, CCL20 is expressed on FAE and non-FAE epithelial cells. Like other intestinal epithelial cells, FAE cells express toll-like receptors (TLRs), the microbial pattern recognition receptors that transmit intracellular signals and induce expression of pro-inflammatory

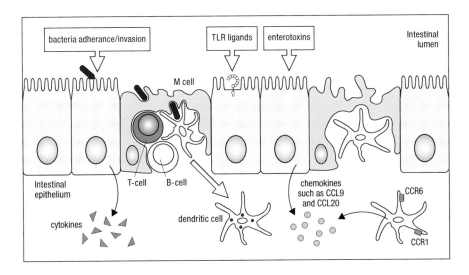

Figure 15.2 Follicle-associated epithelium (FAE) mediates cross talk between the luminal flora and the mucosal immune system. M cells conduct endocytosis and rapid transepithelial transport of intact antigens and microorganisms into intraepithelial pockets that contain B and T cells and dendritic cells (DCs). The majority of FAE cells are enterocytes that do not transport antigens. FAE enterocytes may contribute to antigen sampling by sensing luminal pathogens and their products via toll-like receptors and then releasing cytokine and chemokine signals that attract and activate DCs. (Adapted from Neutra M., and Kozlowski, P. *Nat. Rev. Immunol.* 2006, 6:148–158. With permission from Macmillan Publishers Ltd.)

and immunomodulatory cytokines and chemokines. In villus cells, TLRs are generally confined to basolateral cell membranes, allowing activation when microbes breach the epithelial barrier. In the FAE, however, transported microbes routinely contact both sides of the epithelium as well as subepithelial TLR-bearing DCs. Constitutive activation of TLRs in the FAE may contribute to the constitutive expression of chemokines.

M CELLS AS GATEWAYS TO THE MUCOSAL IMMUNE SYSTEM

The cardinal feature of the FAE is the presence of M cells. The major function of M cells is to deliver samples of macromolecules, particulate foreign material, and microorganisms by transepithelial transport from the lumen to organized mucosal lymphoid tissues. M cells, like all the epithelial cells of the gastrointestinal and respiratory tracts, are joined to adjacent epithelial cells by junctional proteins that seal the paracellular pathway. However, M cells provide functional openings through the epithelial barrier by their transepithelial vesicular transport activity. M cell differentiation is largely restricted to the FAE; consequently, transport of foreign material and microbes across the epithelial barrier is targeted to the organized, inductive sites of the mucosal immune system. M cells were first recognized microscopically by their unique morphology, especially their intraepithelial "pocket" that provides a sequestered space for activated or memory B and T lymphocytes and DCs. The pocket shortens the transcytotic pathway and provides for rapid delivery of luminal samples to intraepithelial and subepithelial cells.

15.3 M cells are polarized epithelial cells with unique architecture

M cells, like all epithelial cells lining the intestine, are highly polarized. The tight junctions that seal their apical poles prevent the passage of most molecules between cells and the lateral diffusion of integral membrane glycolipids and proteins between apical and basolateral domains of the plasma membrane. The basolateral domain is further subdivided into a lateral subdomain that adheres to the neighboring cell through homotypic interaction of adhesion molecules such as E-cadherin and a basal subdomain that adheres to the basal lamina through heterotypic interaction of integrins with extracellular matrix components. These interactions transmit signals internally, resulting in complex cytoskeletal organization and membrane polarization. The apical domain of typical intestinal epithelial cells is highly organized, with long, actin-supported microvilli, between which are small microdomains that include clathrin-coated pits and caveolae capable of limited endocytosis. Maintenance of the apical and basolateral domains involves sorting of newly synthesized membrane components in the trans-Golgi network and budding of vesicles destined for either the apical or basolateral domain.

As in other epithelial cells, M cell tight junctions define the apical and basolateral plasma membrane domains. However, in M cells, the basolateral domain is expanded and modified to form a large, invaginated subdomain that lines a novel space, the intraepithelial pocket. Little is known about the molecular composition of the pocket domain, although it contains adhesion molecules that promote docking of DCs and intraepithelial lymphocytes, as described later.

15.4 M cell apical surface designed for easy access and efficient endocytosis

Importantly, M cell apical membranes lack the organized brush border and thick, protective glycocalyx of integral membrane mucins that blankets the microvillus tips on enterocytes (**Figure 15.3a**). M cells instead have widely spaced microvilli or microfolds, interspersed with large plasma membrane subdomains that are organized for endocytosis. Thus, microorganisms and particles can readily contact and adhere to the areas involved in endocytosis; this is important because adherence increases the likelihood of endocytosis. For example, polystyrene or latex beads measuring up to 1 μm in diameter can adhere nonspecifically to many cell surfaces, but they adhere avidly to M cells in Peyer's patches and are rapidly taken up into the mucosa.

Although M cells lack a thick glycocalyx, their apical membranes contain glycoproteins and glycolipids that can serve as binding sites. Many bacteria and viruses adhere to target host cells using specific adhesion molecules that recognize specific carbohydrate structures on host-cell glycoproteins or glycolipids. Carbohydrate-based recognition mechanisms contribute to M cell–specific uptake into the mucosa in several ways. Most glycolipids and

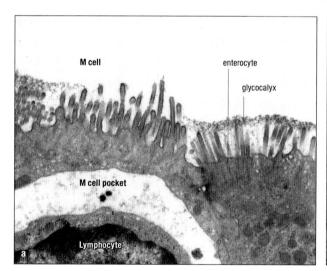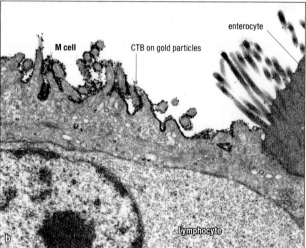

Figure 15.3 The M cell apical surface is designed for easy access and efficient endocytosis. (a) The apical poles of two FAE cells in a Peyer's patch that have been processed for electron micrographic visualization of membrane glycoproteins. The M cell (left) has irregular microvilli with a very thin surface coat. The microvilli of the enterocyte (right) are blanketed by a thick glycocalyx. A lymphocyte is present in the M cell pocket. Cholera toxin B subunit (CTB) is a soluble protein that binds to a glycolipid receptor on all cell surfaces in the intestine. **(b)** Explanted rabbit Peyer's patch mucosa exposed to cholera toxin coupled to 15 nm colloidal gold particles. Electron microscopic analysis revealed that CTB-gold adhered to the apical surfaces of M cells but not to adjacent enterocyte microvilli. In M cells, gold particles were present in clathrin-coated pits and vesicles, indicating endocytosis and transport. (Adapted from Frey, A. et al. *J. Exp. Med.* 1996, 184:1045–1060. With permission from Rockefeller University Press.)

glycoprotein-associated carbohydrate structures are not unique to M cells but are ubiquitous, as they also are present in the mucin-like molecules of the brush border glycocalyx, thereby promoting microbe and particle adherence to intestinal epithelium. Particles that adhere to the enterocyte glycocalyx, however, do not interact with endocytic membrane domains and are eventually shed into the lumen. In contrast, particles that adhere to M cell membranes are endocytosed and transported. Other glycoproteins and glycolipids present on membrane molecules of intestinal epithelial cells are masked by the brush border glycocalyx. Large macromolecules or microbes that cannot diffuse through the glycocalyx are unable to adhere to these enterocyte epitopes but can bind M cells. This was demonstrated using cholera toxin, whose pentameric B subunit binds to the ganglioside GM1, a membrane glycolipid present on all intestinal epithelial cells. Soluble cholera toxin diffuses through the glycocalyx to bind to the enterocyte membrane, triggering a cascade of events that leads to severe diarrhea. However, toxin immobilized on microparticles and administered into the intestine in particulate form adheres only to M cells (see Figure 15.3b). Similarly, certain viruses are able to gain access to their receptors only on M cells, as described next.

15.5 M cells display cell type–specific membrane molecules that may serve as receptors

M cells show a unique "face" to the lumen by displaying surface molecules that are not present on the apical surface of neighboring epithelial cells. For example, β-1 integrins and intercellular adhesion molecule-1 (ICAM-1) are located on the apical surfaces of M cells but on the basolateral surfaces of neighboring epithelial cells, where they function in lymphocyte and matrix interactions. Structures expressed only on M cells include specific carbohydrates identified by lectin or antibody binding, reflecting M cell–specific glycosyltransferase expression. These M cell–specific carbohydrate epitopes vary among species, among different mucosal regions, and even within the same FAE. Also present on the apical surfaces of M cells are highly glycosylated, lipid-linked proteins that bind bacterial type 1 pili of gram-negative bacteria, and proteins that bind proteoglycans of gram-positive bacteria. Thus, M cell–specific oligosaccharides and glycoproteins likely serve as receptors for immune sampling of broad classes of microorganisms, but such receptors may be exploited by pathogens to invade the mucosa and cause disease, as discussed later.

The fine balance of host-microbe symbiosis in the intestine is crucial for normal mucosal immune function as well as maintenance of a healthy epithelium. Consequently, an important function of M cells in the absence of pathogens is uptake of commensal microorganisms that normally inhabit mucosal surfaces, especially in the intestine. Nonpathogenic bacteria are regularly delivered by M cells into Peyer's patches, and some may live within mucosal follicles without causing disease. In this regard, M cells express a putative IgA receptor that recognizes IgA and secretory IgA, potentially promoting the uptake of IgA-coated microbes and IgA-antigen complexes. Thus, through IgA-mediated uptake and selective microbial adherence, M cells participate in host interaction with normal luminal microflora, as well as with certain pathogens.

M cells display distinct proteins on their basolateral membrane to attract lymphocytes into the intraepithelial pocket. For example, M cells express on their basolateral membranes a membrane-bound form of the chemokine CXCL16, a molecule that interacts with CXCR6 on T and B cells. M cells also express CD137, an integrin family member that is recognized by certain B cells. Mice deficient in CD137 have Peyer's patches, FAE, and M cells with typical apical markers, but their M cells fail to form pockets and do not harbor intraepithelial lymphocytes.

15.6 Adherent macromolecules and particles are efficiently endocytosed and transported across M cells

Epithelial cells move macromolecules and microbes from the apical cell surface to the basolateral surface by a process called transcytosis (**Figure 15.4**). Transcytosis is mediated by a complex series of events, including formation and fusion of endosomes and polarized recycling of membrane vesicles, with directional information provided by G proteins and adaptors and the highly polarized cytoskeleton. The first step in the transcytotic process is invagination of apical membrane microdomains to form intracellular compartments called early endosomes. The mechanisms by which M cells take up cargo are diverse. Adherent macromolecules, small particles, and viruses are taken up via clathrin-coated or noncoated pits and vesicles (see **Figure 15.4**); soluble macromolecules are captured in the fluid phase. Large adherent particles are internalized in a process resembling phagocytosis that involves the assembly of organized submembrane actin networks. Electron-micrographic and immunocytochemical data suggest that certain proteases are delivered into M cell endosomes, and that the endosomal lumen is acidified to pH 6–6.2, perhaps allowing some ligands to be released from their receptors.

In model epithelial cells, membrane proteins and lipids are sorted in the early endosomes or in the trans-Golgi network for transport to multiple locations. Some are recycled to the cell surface and others transported along the degradative pathway via "late endosomes" or "multivesicular bodies" (that contain small internal vesicles) to lysosomes. Only a small amount of selected material enters vesicles directed to the basolateral membrane domain via transcytosis. In contrast, M cell transcytosis traffics the majority of vesicles containing endocytosed material to a single destination: the specialized membrane domain lining the invaginated pocket. The short distance between the apical and basal cell surfaces, 1–2 μm, allows transcytosis to be completed in as little as 15 minutes. Fusion of vesicles with this membrane releases vesicle content into the intraepithelial space for uptake by DCs and lymphocytes. The extent to which the brief exposure to proteases in M cell transport vesicles results in degradation or modification of content is not clear, but

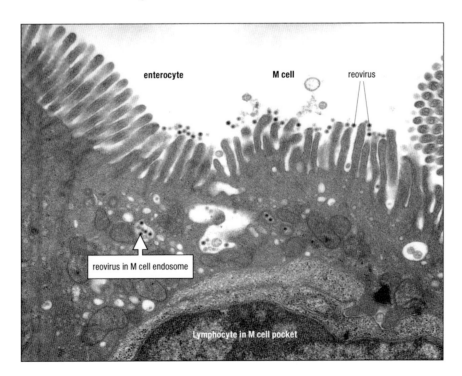

Figure 15.4 Reovirus adheres selectively to M cells in mouse Peyer's patch. This electron micrograph of mouse FAE shows reovirus bound to and taken up into the transepithelial transport pathway of an M cell. Some adherent virus particles have been taken up into endosomes. After transcytosis through the cell, endocytosed virus will be released at the modified basolateral membrane into the intraepithelial pocket that contains lymphocytes.

microorganisms arrive in the pocket intact and alive. Transcytosis by M cells is accelerated when the FAE is exposed to bacteria or TLR ligands, possibly due in part to release of macrophage inhibitory factor from subepithelial DCs.

15.7 Transported cargo is delivered to cells in the M cell pocket and subepithelial dome region

Antigens and pathogens released into the M cell pocket immediately contact the intraepithelial DCs, B cells, and CD4$^+$ T cells located in the pocket. The lymphocytes display markers typical of activated or memory cells. The B cells express major histocompatibility complex (MHC) class II, suggesting that they are capable of antigen presentation. DCs also migrate into the pocket from the subepithelial dome region and quickly take up transported material such as viruses and particles. DC migration toward and into the FAE is enhanced when the mucosa is exposed to TLR ligands or enterotoxins, likely due to release of chemokines such as CCL20 from the epithelium (see Figure 15.2). Thus, the intraepithelial pocket containing subepithelial DCs and lymphocytes permits efficient antigen exposure, possibly without the influence of preexisting systemic antibodies, which appear incapable of diffusion into the pocket.

Mucosal lymphoid follicles consist primarily of a cluster of immature B cells, often including a germinal center and a few T cells, supported by a network of specialized follicular DCs. The follicles are flanked by interfollicular T-cell areas that contain a distinct population of antigen-presenting DCs, naive and antigen-sensitized T cells, and high endothelial venules that allow entry and exit of migrating cells (Figure 15.5). Between each mucosal lymphoid follicle and the FAE is a subepithelial dome region filled with lymphocytes and DCs. Antigens and pathogens transported across the FAE by M cells are efficiently captured by immature DCs in the pocket and subepithelial dome region. Thus, the site of initial antigen entry via M cells and capture by DCs occurs in close proximity to organized T- and B-cell zones.

Antigens and pathogens are endocytosed by immature DCs in the subepithelial dome region and translocated by DCs the short distance to the adjacent interfollicular T-cell zones (see Figure 15.5), where DCs express maturation markers and antigen presentation occurs. Some DCs enter lymphatic vessels, migrating to the nearest draining lymph node, such as a mesenteric node, to present antigen in the context of information from broader mucosal areas and interface with the systemic immune system. DC migration is accelerated by uptake of live bacterial pathogens and enterotoxins, materials that are known to induce local cytokine and chemokine "alarms," but not by uptake of inert nonliving particles. DC migration and the location and nature of subsequent intercellular interactions are important determinants of the location and quality of the resulting mucosal immune response. DC migration can also facilitate dissemination of infectious agents that exploit the M cell transport pathway to enter the mucosa.

FORMATION, DIFFERENTIATION, AND MAINTENANCE OF THE FAE AND M CELLS

15.8 FAE is continuously renewed

Epithelial cells emerging from multiple adjacent crypts migrate onto the surface of the underlying lymphoid follicle to form the FAE. The epithelial cell kinetics and sloughing at the crest of the dome resemble that of villus epithelial cells. Crypt cells destined for the FAE follow a distinct differentiation program, becoming FAE enterocytes and M cells (see Figure 15.1a). Thus, a crypt located between a mucosal lymphoid follicle and a villus must provide cells to the villus on one side and the FAE on the other. In the crypt, cells on the villus side express the polymeric Ig receptor for export of IgA and differentiate

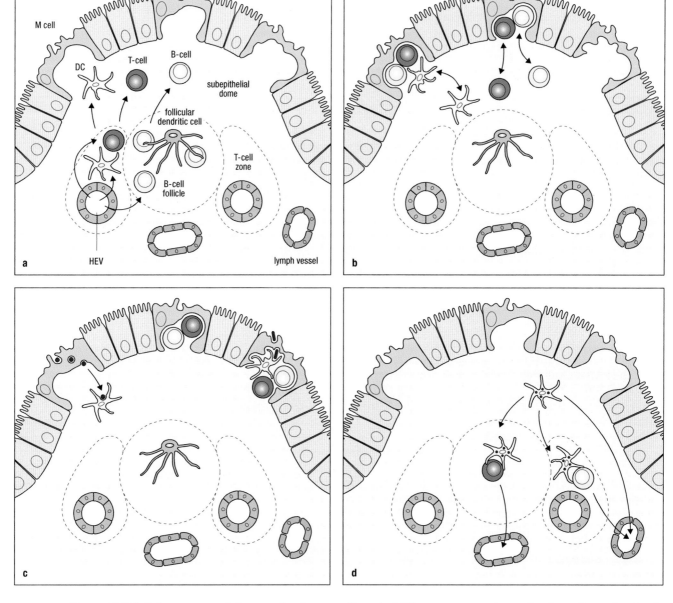

Figure 15.5 Cellular movements in organized mucosal lymphoid tissues are complex and only partly understood. (**a**) Naive and memory lymphocytes (B and T cells) and immature dendritic cells (DCs) enter the mucosa through high endothelial venules (HEV). Many of these cells are attracted to the subepithelial dome region by chemokines released from the FAE. (**b**) Some B and T cells migrating into M cell pockets where they express maturation or memory markers; the ultimate fate of these lymphocytes is not known. Most immature DCs remain in the subepithelial dome region, but a few migrate into the FAE. (**c**) Antigens and microorganisms transported by M cells are captured by DCs. (**d**) Antigen capture along with other signals induces DC maturation and movement into interfollicular T-cell areas where they process and present antigen to naive T cells. DCs may also enter the B-cell follicle or migrate into draining lymphatics. (Adapted from Neutra, M., and Kozlowski, P. *Nat. Rev. Immunol.* 2006, 6:148–158. With permission from Macmillan Publishers Ltd.)

into the cell types typical of the villus (enterocytes and goblet cells), whereas cells on the opposite wall of the same crypt do not secrete IgA and gradually acquire the features of M cells and follicle-associated enterocytes. M cells in the crypts can be identified by their distinct glycosylation patterns, but they do not yet have intraepithelial lymphocytes or pockets. As they emerge from the crypt, differentiating M cells form a pocket, acquire resident lymphocytes, and begin endocytic activity.

15.9 Signals from organized mucosal lymphoid tissues are required for induction of FAE and M cells

Considerable evidence indicates that the distinctive patterns of FAE gene expression and M cell differentiation are dependent on cells of the underlying follicle. New sites of FAE and M cells appear *in vivo* where mucosal lymphoid follicles assemble, as observed when new isolated lymphoid follicles form in response to microbial challenge and experimentally by local injection of Peyer's patch cells into the mucosa. Factors produced by mucosal follicles appear to act very early in the differentiation pathway, inducing crypt cells to commit to FAE phenotypes. Evidence also indicates that during pathogen challenge from the lumen, some FAE enterocyte-like cells can be converted within hours to antigen-transporting M cells. Factors that signal crypt cells to follow the FAE differentiation program are discussed next.

The concept that lymphocytes play a role in differentiation of the FAE originated from indirect evidence. Immunodeficient *SCID* (severe combined immunodeficiency) mice, which lack both B and T lymphocytes, have no mucosal B-cell follicles, as well as no FAE or identifiable M cells. However, mucosal follicles with FAE appeared after adoptive transfer of Peyer's patch cells, especially when the transfer was enriched in B cells. B-cell–deficient mice have T-lymphocyte clusters in the mucosa and markedly reduced numbers of M cells. In contrast, T-cell–deficient mice have small but recognizable Peyer's patches with follicles, FAE, and M cells. The apparent requirement of lymphocytes and their products for the induction of FAE has been elucidated *in vitro* with the Caco-2 epithelial cell line. When grown on permeable filters, Caco-2 cells form polarized monolayers that resemble normal villus enterocytes with tight junctions, microvilli without a brush border glycocalyx, and variably expressed enterocyte enzymes and other molecules. When mouse Peyer's patch lymphocytes or cloned human B cells were added to the basolateral chamber of these cultures, some lymphocytes migrated through the holes in the filter and into the epithelium; some of the Caco-2 cells showed M cell–like changes, including sparse microvilli, loss of the surface enzyme sucrase-isomaltase, and increased endocytic activity for the transport of latex particles and bacteria from the apical to the basolateral compartment. The addition of lymphotoxin (LT), TNF-α, and other factors to the culture medium also induced M cell–like changes in endocytic activity. Although the cells induced in this way do not resemble normal FAE M cells in all respects, the system has proven useful for elucidating the factors that differentiate the FAE and for identifying M cell–specific genes.

Some of the cytokine signals required for differentiation of FAE have been identified in mice with defined genetic disruptions. For example, LT single-knockout and TNF/LT double-knockout mice do not form Peyer's patches, indicating a role for LT and possibly TNF-α in Peyer's patch assembly.

The cytokine signals and intracellular signaling pathways that control enterocyte differentiation into M cells are beginning to be elucidated, particularly using enteroids. The addition of receptor activator of nuclear factor kappa B ligand (RANKL) to mouse enteroids rapidly increases the expression of M cell–associated transcripts such as Sp1B and CCL9; the effect is enhanced if TNF-α is added along with RANKL. Similar results occur when RANKL is added to monolayers of human small intestinal crypts. When RANKL binds to RANK, a signaling pathway is initiated via TNF receptor associated factor (TRAF6). Gut epithelial knockout of TRAF6 in mice completely abolishes the formation of M cells in FAE.

A population of CD11b$^+$CD8$^-$ DCs in the subepithelial dome of the Peyer's patch of mice express high levels of IL-22 binding protein (IL-22bp). Deletion of IL-22bp leads to enhanced expression of genes associated with villus epithelium, including muc-2. Although numbers of M cells in IL-22bp knockout mice are not reduced, these mice display defects in the uptake of bacterial antigens from the lumen.

15.10 New organized mucosal lymphoid tissue and FAE form in response to antigenic challenge

Solitary or isolated lymphoid follicles and associated T-cell clusters occur in many mucosal locations and are common throughout the gastrointestinal tract (Figure 15.6). The formation of these structures depends on exposure to antigens and microorganisms, as they appear only after birth and increase in number after microbial challenge. In the human digestive tract, isolated lymphoid follicles are most frequent where microbial populations are abundant, such as the large intestine, cecum, and rectum. Isolated lymphoid follicles are common at mucocutaneous transitions such as the anal-rectal junction and near the ducts of secretory glands that empty onto mucosal surfaces. Isolated lymphoid follicles also form in the trachea and bronchi under conditions of antigenic challenge. Isolated lymphoid follicle formation involves a distinct sequence of events, including differentiation of local stromal cells, entry of migratory DCs as well as distinct lymphocyte populations, and induction of a FAE. The "mature" isolated lymphoid follicle structure is functionally analogous to Peyer's patches for antigen sampling and initiation of immune responses.

EXPLOITATION OF M CELLS BY MICROORGANISMS

In addition to M cell delivery of microorganisms into Peyer's patches for induction of IgA production and maintenance of a normal microflora in the intestine, M cell delivery of nonpathogenic bacteria into the rudimentary Peyer's patches of neonates contributes to maturation of the entire mucosal immune system. The potential harm of local infection is minimized by the translocation of microorganisms across the epithelial barrier via M cells at immune inductive sites, where phagocytes are abundant. Usually bacterial and viral pathogens transported by M cells are immediately killed by phagocytes and other innate components of the organized lymphoid tissues or are cleared from the mucosa by the ensuing adaptive mucosal immune response. However, certain bacterial and viral pathogens successfully exploit this system to establish a mucosal infection and/or spread systemically and cause disease. In this section, we discuss M cell and pathogen interactions involved in the initiation or control of disease.

15.11 M cell transport of noninvasive, surface-colonizing bacterial pathogens can limit duration of diarrheal disease

Some pathogenic bacteria such as *Vibrio cholerae* and enterotoxigenic *Escherichia coli* cause severe diarrheal disease without invasion. *V. cholerae*, for example, adhere to the epithelial glycocalyx using adhesive toxin-coregulated pili, form dense colonies, and then upregulate production of cholera toxin that diffuses through the glycocalyx to bind to a common glycolipid in enterocyte apical membranes. Endocytosis of the toxin initiates complex intracellular events that result in passage of the enzymatic toxin subunit into the cytoplasm, production of cyclic AMP, and apical secretion of chloride ions that drive massive loss of water into the intestinal lumen. Some vibrios, however, adhere to M cells and are rapidly endocytosed, transported into Peyer's patches, and processed by DCs (Figure 15.7). The

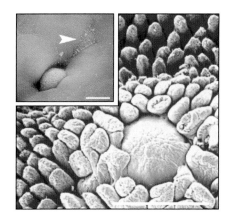

Figure 15.6 Isolated lymphoid follicles are covered by the FAE and M cells. This scanning electron micrograph of mouse intestine shows a single organized mucosal follicle surrounded by villi, as viewed from the lumen. (Scale bar: 500 μm.) The inset is a higher-magnification view, showing the distinctive surface of an M cell (arrowhead). (Scale bar: 5 μm.) (Adapted from Lorenz, R., and Newberry, R. *Ann. NY Acad. Sci.* 2004, 1029:44–47. With permission from John Wiley & Sons.)

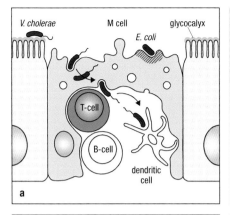

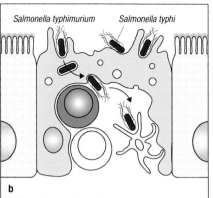

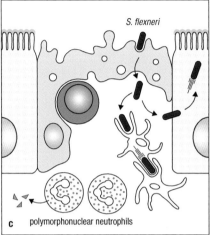

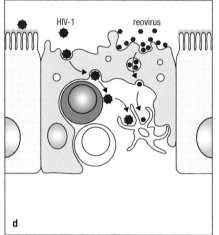

Figure 15.7 Microorganisms exploit M cells to cross the intestinal epithelial barrier. (a) *Vibrio cholerae*, a noninvasive pathogen, adhering to the periphery of the glycocalyx of enterocytes (left) to form toxin-secreting colonies. In contrast, *V. cholerae* that adhere to M cells make close contact with M cell apical membranes and are endocytosed and delivered to antigen-presenting cells. *Escherichia coli* RDEC-1 induces stable, actin-rich pedestals on M cells. **(b)** *Salmonella enterica* serovar *typhimurium* and *Salmonella enterica* serovar Typhi adhere preferentially (but not exclusively) to M cells and initiate signal transduction events that alter the M cell apical cytoskeleton, causing active ruffling of the M cell surface and uptake of bacteria by macropinocytosis. **(c)** *Shigella flexneri* exploits the transport activity of M cells to enter the intestinal mucosa and invade the basolateral surfaces of epithelial cells. This invasion may trigger recruitment of inflammatory cells and local cytokine release, which, in turn, result in disruption of the epithelial barrier. **(d)** Reovirus adheres selectively to M cells in mice, and transcytosis delivers the virus to the basolateral side where it can infect other epithelial target cells. Human immunodeficiency virus-1 (HIV-1) applied to mouse or rabbit mucosa adheres to epithelial cell surfaces. Virus adherent to M cells may be transcytosed into the intraepithelial pocket and subepithelial region, where it may infect target cells. (Adapted from Neutra, M. et al. *Cell* 1996, 86:345–348. With permission from Elsevier Ltd.)

resultant adaptive immune response leads to the secretion of IgA antibodies to *V. cholera* pili, lipopolysaccharide, and cholera toxin. These antibodies participate in the clearance of the adherent colonies and protection against subsequent challenge.

15.12 Invasive bacterial pathogens can exploit rapid M cell transport to establish local and systemic infection

M cells transcytose most particles that adhere to their surface. In the transcytotic vesicles, content may be acidified and undergo protease activity, but transport is rapid, and some microbes survive to infect the epithelium, underlying phagocytes, and other mucosal cells (see Figure 15.7).

Salmonella species are able to infect many cell types, including epithelial cells and phagocytes. In humans, *Salmonella enterica* serovar *typhimurium* causes a mucosa-limited diarrheal disease, but in mice the bacteria spread systemically and cause a lethal septicemia that closely mimics typhoid fever caused by *Salmonella enterica* serovar Typhi in humans. After adhering to the host-cell membrane, the bacteria use a type III secretion system to deliver proteins into the host cell, inducing massive actin reorganization and macropinocytosis, a process that disrupts normal cell architecture as bacteria are internalized. Because the FAE and M cells are relatively accessible, intestinal *Salmonella* preferentially adhere to M cells in both mice and humans, rapidly invade Peyer's patches, and are taken up by subepithelial DCs. Uptake of *Salmonella* has cytotoxic effects that can result in M cell death. In mice, some *Salmonella* may also be taken up from villus surfaces, captured by DCs that send extensions through tight junctions.

Salmonella enterica serovar *typhimurium* survive in special vacuoles in phagocytes and are carried by these migrating cells via the lymphatic vessels to the spleen, causing murine typhoid fever.

Shigella infections cause widespread mucosal damage to the colonic mucosa, but the bacterium does not spread systemically. Entry into the colonic mucosa is highly efficient, as only a few *Shigella* organisms ingested orally are required to initiate infection. Studies in experimental animals showed that *Shigella* initially use M cell transport to gain access to the basolateral side of the FAE and adjacent epithelium. Other studies of epithelial cells *in vitro* showed that *Shigella* are unable to infect enterocytes at the apical surface but readily attach and are endocytosed at the basolateral membrane. The bacteria rapidly spread from cell to cell, usurping the host-cell cytoskeleton and signaling networks. Initial uptake into phagosomes occurs through a macropinocytic process that is induced by secretion of bacterial proteins through a type III secretory apparatus. *Shigella* then lyse the phagosome membrane and escape into the cytoplasm, where they initiate a complex actin-based motility process that propels the bacterium through the host-cell cytoplasm, form torpedo-like extensions on the lateral cell surface, and induce endocytosis by the adjacent cell. While spreading from cell to cell, *Shigella* induce activation of nuclear factor κB (NF-κB), resulting in interleukin-8 production by epithelial cells and the rapid influx of polymorphonuclear leukocytes into the lamina propria. The resultant inflammatory process leads to disruption of the epithelial barrier, facilitating widespread *Shigella* invasion.

15.13 Endocytosis of viruses by M cells results in mucosal and/or systemic disease

Viruses that adhere to cell surfaces may invade the host cell through endocytic pathways. Viruses adherent to M cell surfaces are efficiently transcytosed through the cell. Thus far, only a few viruses are known to adhere specifically to M cells. Other viruses adhere to many epithelial cell surfaces, but only M cells mediate their transcytosis. Still other enteric viruses exploit epithelial receptors that are located basolaterally, necessitating entry via epithelial breaks or capture by protruding intraepithelial DCs.

Indirect evidence suggests that M cells may be involved in HIV-1 and simian immunodeficiency virus (SIV) entry. In the rectum and recto-anal junction, organized lymphoid tissues and M cells are numerous, possibly contributing to rectal transmission of HIV-1 and SIV. SIV administered orally to monkeys was detected first in organized lymphoid tissues of the tonsils. HIV-1 applied to mouse and rabbit Peyer's patch explants adhered to M cells and was rapidly endocytosed and transported across the epithelial barrier. Cells with morphologic and functional features of M cells generated by the coculture of Caco-2 cells and B cells transported HIV-1 in a galactosylceramide- and chemokine receptor–dependent manner. In rabbit Peyer's patch explants, some HIV-1 particles adhered to the glycocalyx on enterocytes but did not contact the cell membrane and were not endocytosed. Epithelial cells are not infected by these viruses, but HIV-1 entry via M cells would provide rapid access to DCs and activated T cells in the M cell pocket and would facilitate DC delivery of virus to T cells. However, whether HIV-1 enters the human rectum through M cells *in vivo* is not known.

Two closely related viral pathogens, reovirus in mice and poliovirus in humans, have evolved to exploit selective adherence to M cells as an invasion strategy (see **Figure 15.7**). Reovirus is a nonenveloped virus that adheres only to M cells and uses this pathway to enter the Peyer's patch mucosa of mice (see **Figure 15.4**). Reovirus also binds to M cells in the colon and airways. The virus uses the pancreatic proteases trypsin and chymotrypsin to process its outer capsid and extend its trimeric attachment

SUMMARY **237**

protein, sigma 1, a distance of 45 nm from the viral surface. Reovirus type 1 binding to M cells is mediated by interaction between the extended sigma 1 protein and a specific sialic acid–containing trisaccharide on the M cell apical membrane. Although this determinant is present on epithelial cells, viral particles cannot pass through the overlying glycocalyx. On M cells, adherent reovirus is taken up by clathrin-coated vesicles and is released into the intraepithelial pocket and subepithelial dome region where it is endocytosed by DCs. Although reovirus is unable to adhere to the apical membranes of enterocytes in adult mice, it binds to a basolateral component near tight junctions called "junction adhesion molecule-1" and proceeds to infect the entire epithelium from the basolateral side.

Poliovirus causes paralytic disease by infecting neurons, but it enters the body by the oral route. The ability of the virus to proliferate in Peyer's patches before spreading systemically suggested an M cell entry strategy. When explants of human Peyer's patch mucosa were exposed *in vitro* to wild poliovirus type 1, electron microscopy showed that the virus adhered selectively to M cells and then was endocytosed. The receptor for poliovirus on target cell membranes is CD155, a member of the immunoglobulin superfamily that mediates interactions between cells of the immune system. CD155 is expressed on human intestinal epithelial cells, including M cells, primarily on basolateral but also on apical domains. That M cell expression of CD155 is required for intestinal uptake of poliovirus is supported by studies in which transgenic mice that expressed CD155 on many cell types, but not epithelial cells, could be infected with injected but not orally administered poliovirus. The ability of poliovirus to exploit M cell transport for uptake into Peyer's patches has led to the testing of poliovirus-based vaccine vectors, including recombinant viral particles or pseudovirus particles, for mucosal delivery of foreign antigens.

15.14 M cells can be exploited for vaccine antigen delivery

Given the importance of organized mucosal lymphoid tissues for the initiation of mucosal immune responses, there is now great interest in the delivery of vaccine antigens to these sites. However, vaccines administered orally or through other mucosal routes encounter host defenses, including mucus, proteases, and nucleases at the epithelial barrier, and dilution by gastrointestinal secretions. In addition, the marked infrequency of M cells reduces the likelihood of contact between an M cell and vaccine antigen. Consequently, much larger vaccine doses are usually required for oral immunization compared with injection. Also, endocytic and transcytotic activity in the FAE is efficient only for macromolecules and particles that adhere to M cell apical membranes. In contrast, soluble, nonadherent antigens are taken up at low levels in the intestine and generally induce immune tolerance. Live or attenuated bacteria and attenuated viruses that adhere to mucosal surfaces, especially those that normally use M cells to invade organized mucosal lymphoid tissues, could theoretically make ideal mucosal vaccines or vaccine vectors. The biology of these vaccines and vectors, including the advantages and disadvantages of each, is discussed in Chapter 30.

SUMMARY

A specialized FAE covers mucosal lymphoid follicles. The FAE contains a unique epithelial cell type, the M cell. M cell differentiation is induced by cytokines from cells in organized mucosal lymphoid tissues. The major function of M cells is to transport samples of macromolecules, particles, and microorganisms across the epithelial barrier from the lumen to organized mucosal lymphoid tissues. A key feature of M cells is the invagination of the basal surface to form an intraepithelial "pocket" that contains activated lymphocytes and DCs. M cells display apical surface features that promote adherence of microorganisms, and basolateral membrane proteins that promote lymphocyte docking in the intraepithelial pocket. After M cell transport, antigens and pathogens are endocytosed by DCs that subsequently migrate to the adjacent interfollicular T-cell zones where antigen presentation occurs. The transepithelial transport activity of M cells is a "double-edged sword," playing a key role in initiating protective mucosal immune responses but also providing microbes with a rapid entry route into the mucosa. Thus, M cells are important in the pathogenesis of certain bacterial and viral diseases and could be exploited in mucosal vaccine delivery systems.

FURTHER READING

Anderle, P., Rumbo, M., Sierro, F. et al.: Novel markers of the human follicle-associated epithelium identified by genomic profiling and microdissection. *Gastroenterology* 2005, 129:321–327.

Debard, N., Sierro, F., Browning, J. et al.: Effect of mature lymphocytes and lymphotoxin on the development of the follicle-associated epithelium and M cells in mouse Peyer's patches. *Gastroenterology* 2001, 120:1173–1182.

Frey, A., Lencer, W.I., Weltzin, R. et al.: Role of the glycocalyx in regulating access of microparticles to apical plasma membranes of intestinal epithelial cells: Implications for microbial attachment and oral vaccine targeting. *J. Exp. Med.* 1996, 184:1045–1060.

Fukuoka, S., Lowe, A.W., Itoh, K. et al.: Uptake through glycoprotein 2 of FimH$^+$ bacteria by M cells initiates mucosal immune response. *Nature* 2009, 462:226–230.

Helander, A., Silvey, K.J., Mantis, N.J. et al.: The viral σ1 protein and glycoconjugates containing α2-3-linked sialic acid are involved in type 1 reovirus adherence to M cell apical surfaces. *J. Virol.* 2003, 77:7964–7977.

Herbrand, H., Bernhardt, G., Förster, R. et al.: Dynamics and function of solitary intestinal lymphoid tissue. *Crit. Rev. Immunol.* 2008, 28:1–13.

Kanaya, T., Sakakibara, S., Jinnohara, T. et al.: Development of intestinal M cells and follicle associated epithelium is regulated by TRAF6-mediated NF-κB signalling. *J. Exp. Med.* 2018, 215:501–519.

Kraehenbuhl, J.P., and Neutra, M.R.: Epithelial M cells: Differentiation and function. *Annu. Rev. Cell Dev. Biol.* 2000, 16:301–332.

MacPherson, A.J., McKoy, K.D., Johansen F.E. et al.: The immune geography of IgA induction and function. *Mucosal Immunol.* 2008, 1:11–22.

Neutra, M.R., and Kozlowski, P.A.: Mucosal vaccines: The promise and the challenge. *Nat. Rev. Immunol.* 2006, 6:148–158.

Neutra, M.R., and Kraehenbuhl, J.P.: Immune defense at mucosal surfaces, in Kaufmann, S.H.E., Sher, A., Ahmed, R. (eds): *Immunology of Infectious Diseases.* Washington, DC, American Society for Microbiology, 2010.

Neutra, M.R., Mantis, N.J., and Kraehenbuhl, J.P.: Collaboration of epithelial cells with organized mucosal lymphoid tissues. *Nat. Immunol.* 2001, 2:1004–1009.

Neutra, M.R., Sansonetti, P., and Kraehenbuhl, J.P.: Interactions of microbial pathogens with intestinal M cells, in Blaser, M., Smith, P.D., Ravdin, J.I. et al. (eds): *Infections of the Gastrointestinal Tract*, 2nd ed. New York, Lippincott Raven, 2002:141–156.

Rouch, J.D., Scott, A., Lei, N.Y. et al.: Development of functional microfold (M) cells from intestinal stem cells in primary human enteroids. *PLOS ONE* 2016, 11(1):e0148216.

Wood, M.B., Rios, D., and Williams, I.R.: TNF- augments RANKL-dependent intestinal M cell differentiation in enteroid cultures. *Am. J. Physiol. Cell Physiol.* 2016, 311:C498–C507.

Lymphocyte trafficking from inductive sites to effector sites

16

VALERIE VERHASSELT, WILLIAM AGACE, OLIVER PABST, AND ANDREW STAGG

The initiation, maintenance, and resolution of innate and adaptive immune responses are critically dependent on immune cell migration, not only within tissues but often over long distances between organs. This process is highly dynamic and requires that immune cells interact with multiple vascular, lymphatic, and tissue environments in a tightly controlled and organized fashion.

Most infections will initially constitute a small number of pathogens and be localized to a small tissue area. If naive lymphocytes were tissue-resident cells dispersed at random throughout the body, the chances that a given antigen-specific lymphocyte would be at the right site—that is, at the site of pathogen entry—would be very slim. In a rough calculation, we might estimate that a T lymphocyte with a diameter of 10 μm and a volume of about 0.5×10^{-18} m^3 occupies less than $10^{-15}\%$ of the body volume. Even if we suppose that several hundred T lymphocytes recognize a distinct antigen, the odds of an antigen-specific lymphocyte being in the same location as a pathogen remain extremely low. Lymphocyte recirculation solves this problem by enabling the entire T-lymphocyte population to scan antigen-presenting cells within lymphoid compartments.

Perhaps nowhere is this more obvious than during the induction of adaptive immune responses. Adaptive immunity is initiated when antigen-presenting cells, primarily dendritic cells (DCs), present antigen to lymphocytes in inductive immune compartments, such as lymph nodes and Peyer's patches. During mucosal immune responses, antigen-bearing DCs migrate from mucosal tissues through draining afferent lymph vessels to regional lymph nodes and into the lymph node T-cell zone. Conversely, to find a DC presenting relevant cognate antigen/major histocompatibility complex (MHC), naive lymphocytes continually traffic from the bloodstream into lymph nodes and back via efferent lymph to the venous blood. In combination, these migratory routes of DCs and lymphocytes allow frequent contact of both cell types and thus form the basis for the efficient induction of adaptive immune responses (Figure 16.1).

Besides the entry of DCs and constitutive recirculation of naive lymphocytes into lymph nodes, large numbers of immune cells enter peripheral tissues in the setting of infection and inflammation. This holds true for cells of the innate immune system, including monocytes, granulocytes, and innate lymphoid cells, as well as for effector T cells and plasma cells that are generated during an adaptive immune response.

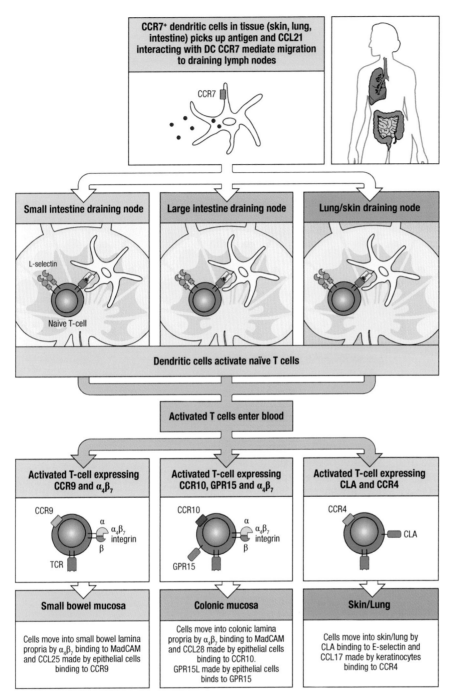

Figure 16.1 The journey of a circulating naive T-cell to a tissue effector T-cell. Naive T lymphocytes circulating in the blood and tissue dendritic cells both migrate to lymph node as the result of their expression of CCR7, which binds to CCL21 expressed on high endothelial venules. T cells are activated by dendritic cells and, according to location of lymph node, will express specific tissue homing receptors. After leaving the lymph node by lymph vessels and joining the thoracic duct, effector lymphocytes reach the systemic circulation and are dispatched into tissue according to their expression of homing receptors.

BASIC CONCEPTS IN IMMUNE CELL MIGRATION

Cell adhesion molecules (CAMs) mediate intercellular interactions and cellular interactions with the extracellular environment. It is thus perhaps not surprising that CAMs also have essential roles in regulating immune cell function and, as described in detail later, form the basis for regulated immune-cell migration. One of the best-studied examples of such an interaction is that between immune cells and the endothelial cells that line blood vessels. We therefore briefly describe the CAMs and chemoattractants stabilizing the interaction of immune cells with the microvascular endothelium. As we discuss later in this chapter, the dynamic regulation of these molecules permits the directed migration of immune cells and offers strategies to modulate the migration of immune cells therapeutically.

16.1 Adhesion molecules and chemoattractants involved in immune cell migration

CAMs involved in immune cell migration fall into three main families: the selectin, integrin, and immunoglobulin gene superfamily of adhesion receptors. Although the precise molecules involved in immune cell adherence depend on the immune cell type, the tissue, and the inflammatory environment under study, generally each CAM needs to bind to its specific ligand expressed on the immune cell and endothelial cell, respectively (Figure 16.2).

Selectins are glycoproteins of the C-type lectin group that bind to specific carbohydrate determinants on selectin ligands. Mammals express three types of selectins: L-selectin, P-selectin, and E-selectin. L-selectin is expressed by many types of immune cells and interacts with ligands expressed by vascular endothelial cells. Conversely, E- and P-selectins are expressed by endothelial cells, and the interacting carbohydrate selectin ligands can be expressed by immune cells. Selectin ligands comprise a heterogeneous group of molecules that frequently, but not exclusively, carry the sulfated tetrasaccharide sialyl-Lewis^x. Thus, the generation of functional selectin ligands depends on the posttranslational modification of proteins by glycosyltransferases. Selectins bind their ligands with low affinity, facilitating the capture of immune cells freely flowing in the blood and resulting in their rolling along the vascular endothelium.

The subsequent strong adhesion of immune cells to vascular endothelial cells is mediated by integrins and their interaction with ligands of the immunoglobulin gene superfamily. Integrins control a wide array of cellular functions, including cell growth, migration, apoptosis, and differentiation. They are heterodimers comprising one α-chain and one β-chain. Both chains exist in multiple forms that, through dimerization, give rise to more than 20 distinct integrin-family members. All major integrins involved in immune-cell migration contain the β_1, β_2, or β_7 integrin chain and are expressed on immune cells, whereas their respective ligands are expressed on the vascular endothelium. Well-known integrin ligands include intercellular adhesion molecules-1 and -2 (ICAM-1, ICAM-2), vascular cell adhesion molecule-1 (VCAM-1), and mucosal addressin cell adhesion molecule-1 (MAdCAM-1) (see Figure 16.2).

Although ligation of selectins and integrins can activate intracellular signaling pathways, the chief molecules that mediate the critical signals driving immune cell migration and arrest on vascular endothelial cells belong to the chemokine family. Chemokines are a large family of low molecular weight proteins (about 8–12 kDa) that are classified into groups depending on the positioning of cysteine residues at their amino terminus. The two major subgroups of chemokines are the CC and CXC chemokines, containing adjacent cysteine residues or cysteine residues separated by a single amino acid (X), respectively. Chemokines can be produced by a broad spectrum of cells and signal through seven-transmembrane domain G-protein-coupled receptors. Many of these receptors are promiscuous in that they can bind more than one chemokine. However, such promiscuity does not extend across CC or CXC chemokine-family members. Thus, chemokine receptors binding CC chemokine ligands (CCLs) are termed CCRs, and those binding CXCL chemokines are termed CXCRs. A major consequence of chemokine receptor activation in immune cells is alterations in integrin structure that occur within seconds of chemokine binding, through a process termed "inside-out signaling," and lead to an increased affinity of targeted integrins for their cellular and extracellular ligands. Together with chemokine-induced integrin clustering, these conformational changes result in a stable binding of the immune cell to the vascular endothelium. An additional important property of chemokines is their ability to bind proteoglycans, and chemokines are

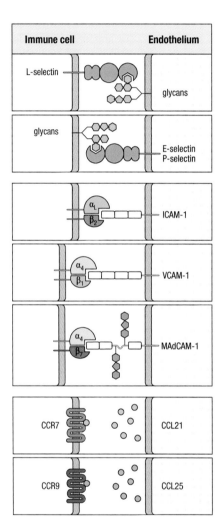

Figure 16.2 Adhesion molecules involved in immune cell migration. Cell-adhesion molecules involved in immune cell migration fall into three large groups: selectins, integrins, and members of the immunoglobulin superfamily of adhesion receptors. These adhesion molecules, together with chemokine/chemokine receptors, are expressed on immune cells and endothelial cells. Selectins bind sulfated glycans provided by a scaffold of glycoproteins. Integrins consist of an α- and a β-chain, forming heterodimers that bind to immunoglobulin superfamily members such as ICAM-1.

thought to be "presented" in this form on the surface of vascular endothelium to circulating immune cells.

16.2   Immune cell migration into tissues is a multistep process

A general model has been proposed to describe the entry of immune cells into tissue (**Figure 16.3**). This model is known as the "multistep adhesion cascade" and postulates the sequential involvement of several adhesion pathways mediated by selectins, chemokines, integrins, and their respective ligands. These molecular interactions enable the cells (1) to tether to the endothelial surface, resulting in a slow rolling of the cell on the endothelium; (2) to undergo an activation step enabling the cells to arrest stably on the endothelium; and (3) to traverse the endothelial layer in a process known as diapedesis. In what follows, we discuss first, as a paradigmatic example, the entry of naive T lymphocytes into lymph nodes, and subsequently, the variations of this process as they occur in the mucosal immune system.

The entry of T cells from the blood into lymph nodes occurs across high endothelial venules (HEVs). HEVs are specialized postcapillary venules that constitutively express the CAMs and chemokines needed for the recruitment of naive lymphocytes, namely, the molecules mediating the distinct steps of the multistep adhesion cascade. During the first step, "tethering and rolling," L-selectin expressed on naive T cells engages oligosaccharides containing sialyl-Lewis^x residues present on the luminal side of the HEV. This interaction is too weak to arrest the cells, but it reduces their speed and manifests in their rolling along the HEV. The second step, "arrest," is initiated by chemokines presented on the HEV. In the context of T-cell homing into lymph nodes, the most important chemokine is CCL21, which binds to the chemokine receptor CCR7 expressed on the surface of all naive T cells. Signaling through the CCR7 receptor induces a high-affinity conformation of the $\alpha_L\beta_2$ integrin (also referred to as lymphocyte function-associated antigen-1 [LFA-1]) on the T cells. Activated $\alpha_L\beta_2$ integrin binds to ICAMs constitutively expressed on the HEV and mediates a stable arrest of the cells on the HEV that initiates the step of transendothelial migration. Thus, the constitutive expression of L-selectin, $\alpha_L\beta_2$ integrin, and CCR7 by naive T lymphocytes and their interaction with glycans, CCL21, and ICAMs constitutively present on HEVs permits their entry into lymph nodes (**Figure 16.4**). At the same time, naive lymphocytes are largely excluded from peripheral tissues such as the skin, lung, or gut,

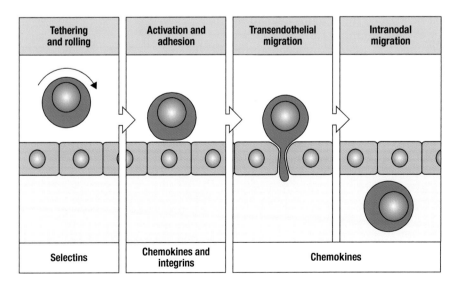

Tethering and rolling	Activation and adhesion	Transendothelial migration	Intranodal migration
Selectins	Chemokines and integrins	Chemokines	

Figure 16.3 The multistep adhesion cascade. Immune-cell entry into tissues is a multistep process coordinated by tissue-homing molecules. The sequential interaction of cell adhesion molecules expressed on immune cells and specialized high endothelial venules allows the cells to tether and roll along the endothelium, to bind stably to the vascular endothelium, to transmigrate, and eventually to follow chemotactic gradients to localize to distinct subcompartments inside the lymph node.

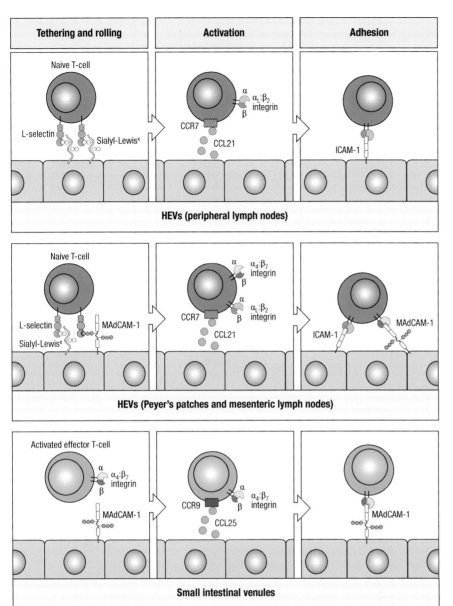

Figure 16.4 Immune cells can use tissue-specific homing receptors to direct their entry into different tissues. Tissue-specific and immune cell–specific expression of homing receptors enable rolling, activation, and adhesion. In peripheral lymph node high endothelial venules (HEVs), rolling is mediated by L-selectin interacting with sulfated glycans decorating CD34 and GlyCAM-1. In Peyer's patches and intestinal venules, rolling can be achieved by the interaction of nonactivated $\alpha_4\beta_7$ integrin with MAdCAM-1, whereas in mesenteric lymph nodes, $\alpha_4\beta_7$ integrin and L-selectin both contribute. Activation of integrins in HEVs is achieved by CCL21 signaling via CCR7, whereas in intestinal venules CCL25 signaling via CCR9 leads to integrin activation. Stable adhesion in HEVs is dominated by activated $\alpha_L\beta_2$ integrin interacting with ICAM, and $\alpha_4\beta_7$ integrin binding to MAdCAM-1 in intestinal venules.

simply because in those tissues the essential molecular interaction partners required for the homing of naive lymphocytes are not expressed. We discuss later in this chapter the distinct patterns of CAMs and chemokine/chemokine receptors that direct immune cells into mucosal tissues and how changes in CAM expression come about.

The CAMs and chemokines directing the immune response to distinct tissues have been studied in considerable detail. A particularly important technical approach in this regard had been the use of adoptive cell transfer, meaning the transfer of immune cells from donor animals to syngeneic or congenic recipients. Early experiments isolated immune cells from different compartments and studied their distribution after intravenous transfer into recipient animals. Results from such experiments revealed that lymphocytes derived from mucosal lymph nodes preferentially relocated to such lymph nodes in the recipient animals. Conversely, lymphocytes isolated from skin-draining lymph nodes retained a preference for the skin-draining lymph nodes of recipients. Similarly, plasma cells producing IgA but not IgG accumulated at mucosal sites such as the intestine and lung. These experiments demonstrated that the homing cues must differ between mucosal and other tissues and led to the concept of a distinct mucosal immune system, which was thought to differ from the systemic immune system principally in respect to cell homing properties, cellular composition, and function. In the light of the multistep adhesion cascade and with the forthcoming use of gene-deficient mice and selective CAM inhibitors, differential homing properties could be tracked down to the divergent expression of selectins, integrins, and chemokines by different immune cells and tissues. It is now apparent that immune cells are equipped with unique "address codes." Comparing mucosal and nonmucosal compartments, these codes show both differences and overlaps. Thus, the concept of a distinct mucosal immune system has been replaced with a refined definition of distinct homing codes, encoded chiefly by selectins, chemokines, and integrins. Still, we are far from understanding the mechanisms regulating the distinct homing properties of immune cells in detail, and it seems likely that in the future we will need to add further molecular players for an adequate description of the homing properties of immune cells.

16.3 Dendritic cells migrate from mucosal tissue to draining lymph nodes to initiate adaptive immune responses

All mucosal tissues contain large numbers of DCs. In the steady state, these cells continually sample their local environment, taking up any self-antigen (from, e.g., apoptotic epithelial cells) or innocuous antigen in their environment and transporting it to local draining lymph nodes for presentation to T cells. The steady-state migration of peripheral DCs to draining lymph nodes is dependent on the chemokine receptor CCR7, which is induced on DCs during their maturation, and the chemokine CCL21, which is constitutively expressed by lymphatic endothelial cells in peripheral tissues. In contrast to lymphocyte migration from the blood into lymph nodes, the migration of DCs into draining lymph nodes can occur in the absence of integrins. DCs migrating into draining lymph nodes in the steady state seem to have a central role in the induction of tolerance to peripheral self and innocuous foreign antigen. However, in the setting of mucosal inflammation and/or infection, there is a marked increase in CCR7-dependent DC migration into draining lymph nodes. Such increased migration of DCs can be mimicked in animal models by local or systemic administration of toll-like receptor ligands, and under these conditions, migratory DCs probably have a central role in the initiation of adaptive immune responses.

16.4 Naive lymphocytes recirculate through mucosa-associated lymphoid tissue

Mucosa-associated lymphoid tissue (MALT), such as Peyer's patches and the appendix in humans, or follicular structures present in inflamed lungs, together with the mucosa-draining lymph nodes, serve as the inductive sites of mucosal adaptive immune responses.

It is now clear that CAM expression on HEVs in certain MALT structures differs from that of other HEVs, for example, those in skin-draining lymph nodes, and that selective CAM expression has an important impact on the mechanisms by which naive lymphocytes gain entry into these sites (**Figure 16.5**). The best example of this is the immunoglobulin (Ig) superfamily member MAdCAM-1, which in adults is constitutively expressed on HEVs in the gut-draining mesenteric lymph nodes and Peyer's patches but not on the HEVs of most extraintestinal lymph nodes. MAdCAM-1 can serve as a scaffold for sulfated glycans, thereby enabling the engagement of L-selectin,

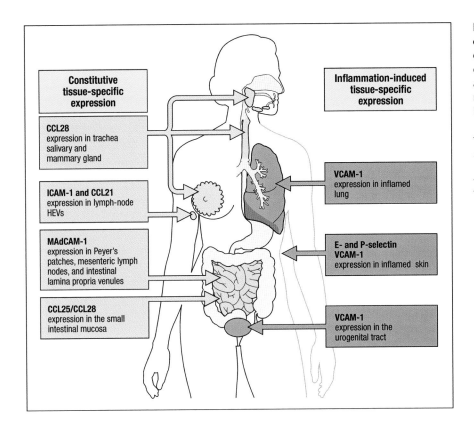

Figure 16.5 Tissue-specific expression of cell adhesion molecules and chemokines. Although immune cells can regulate the expression of cell adhesion molecules and chemokine receptors, their destination is determined by the divergent expression of the respective receptors and ligands in the target tissues. Constitutive tissue-specific expression of homing molecules is depicted in blue, and inflammation-triggered expression is depicted in red. The constitutive expression of ICAM and CCL21 in lymph nodes is critical in driving the homing of lymphocytes to lymph nodes, whereas the expression of MAdCAM-1 and CCL25 in the small intestine supports the entry of gut-homing immune cells. Under inflammatory conditions, VCAM-1, which serves as a ligand for $\alpha_4\beta_1$ integrin, becomes upregulated in numerous tissues including the lung and urogenital tract. In inflamed skin, upregulation of E- and P-selectin and VCAM on endothelium serves in the recruitment of skin-homing effector cells.

and can interact weakly with low-affinity $\alpha_4\beta_7$ integrin. Both interactions contribute to lymphocyte rolling on MAdCAM-1$^+$ venules. Subsequent to chemokine signaling, $\alpha_4\beta_7$ integrin binds MAdCAM-1 with high affinity, resulting in lymphocyte arrest on these HEVs. Interactions of $\alpha_4\beta_7$ integrin with MAdCAM-1 seem particularly important in mediating the entry of naive T lymphocytes into Peyer's patches, as demonstrated by the presence of hypocellular Peyer's patches in mice deficient for MAdCAM-1 or β_7 integrins, whereas mice with a deletion in both L-selectin and β_7 integrin display hypocellular gut-draining mesenteric lymph nodes. Thus, the constitutive expression of MAdCAM-1 and its binding to β_7 integrins are required for the efficient homing of naive lymphocytes into the mesenteric lymph nodes and gut-associated lymphoid tissue (GALT) (see Figure 16.4).

The homing of T and B cells to mucosal lymph nodes and Peyer's patches, as in peripheral lymph nodes, uses CCL21/CCR7 signaling (and can be helped by the chemokine receptor CXCR4). In addition to the CCL21-decorated HEVs that are localized in the interfollicular T-cell zone of all lymph nodes, the B-cell follicles of Peyer's patches contain CXCL13-decorated HEVs. Naive B cells express the CXCL13 chemokine receptor CXCR5 and can use CXCL13/CXCR5 to gain direct entry into B-cell follicles in Peyer's patches. Such specialized HEVs are absent from lymph nodes and might have developed in response to the particular need of a B-cell–rich compartment such as the Peyer's patches. B-cell–rich lymphoid tissues exist at various anatomical locations, including small follicular structures in the intestine, the rodent nasopharyngeal-associated lymphoid tissue, and the omentum (a unique streak of adipose tissue studded with numerous B-cell–rich follicular structures known as milky spots), and CXCL13/CXCR5 contribute to B-cell entry at all these sites. These findings illustrate that besides constitutive differences in the expression of integrin ligands on HEVs in intestinal inductive sites, the selective distribution of chemokines along HEVs can provide additional positional cues for migrating lymphocytes.

16.5 Lymphocytes traffic inside lymph nodes and leave to return to the blood

On entry into the lymph node, lymphocytes are guided by chemokines into distinct anatomical localizations, most prominently the paracortical T-cell zone and the B-cell follicles (Figure 16.6). The backbone of the lymph node is made up of nonhematopoietic stromal cells, fibers, and extracellular matrix. Stromal cells constitute a heterogeneous array of nonhematopoietic cells that can be subdivided into endothelial cells (including cells forming the HEVs), follicular dendritic cells (FDCs) in the B-cell follicles, and fibroblastic reticular cells in the T-cell regions. Fibroblastic reticular cells are distributed throughout the T-cell zone, forming a three-dimensional network of cells that surround highly organized extracellular channels termed *conduits*. The conduits are composed of a fibrous collagen core that is surrounded by a basement membrane, through which molecules of low molecular weight, including antigens, cytokines, and chemokines, can disperse. Of particular relevance to lymphocyte migration is the fact that conduits can channel chemokines to HEVs. Thus, HEVs can display chemokines produced at distant sites that are channeled to HEVs through the conduit system.

Pioneering technical advances in imaging and microscopy, and in particular the development of multiphoton microscopy, have allowed researchers to visualize the migratory paths of immune cells at depth within lymph nodes in living animals in real time. Such studies have provided fundamental insights into immune-cell migratory patterns and immune-cell interactions within lymph nodes. These techniques have demonstrated that naive T cells, after entering lymph nodes, are highly motile and actively scan antigen-presenting cells for their cognate antigens. Within lymph nodes, T cells migrate along the fibroblastic reticular cell network. Fibroblastic reticular cells express the CCR7 ligands CCL21 and CCL19, which seem to promote T-cell motility within the T-cell zone and thus the speed with which T cells interact with and scan the surface of resident DCs. In a similar fashion, naive B cells, although migrating on average more slowly than T cells, are thought to migrate along FDCs in the B-cell follicles in a process that is probably promoted through the CXCR5 ligand CXCL13.

In most cases, naive T cells will not find their cognate antigen:MHC on the surface of DCs and will leave the lymph nodes within a few hours via the efferent lymph. Depending on the particular situation, lymphocytes in lymph will directly reenter the circulation or enter the subcapsular

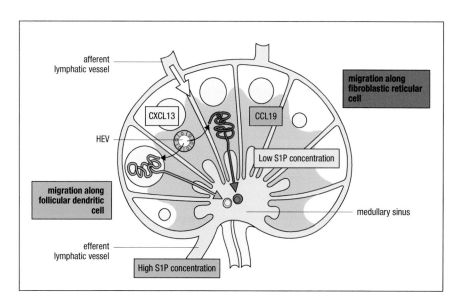

Figure 16.6 Intranodal migration of naive T and B cells. Naive lymphocytes enter lymph nodes via high endothelial venules. Within the lymph nodes, the chemokine CXCL13 guides B cells into the B-cell follicles, whereas T cells are guided by CCL19 into the T-cell zone. Within these subcompartments, lymphocytes migrate along stromal cells and scan resident antigen-presenting cells for their cognate antigens. In most cases, lymphocytes will not become activated and exit from the lymph nodes via sphingosine-1-phosphate (S1P) signaling.

region of higher-order lymph nodes. In contrast, if activated, T cells are initially retained within the lymph node, where they undergo rapid clonal expansion. A subset of these cells subsequently migrates via the efferent lymph back to the circulation, from where they can seed other organs. The underlying cellular and molecular mechanisms regulating the egress of lymphocytes from lymph nodes are only beginning to emerge. A key factor in the regulation of lymphocyte egress is the lysophospholipid sphingosine-1-phosphate (S1P). S1P concentrations are higher in blood and lymph than in lymph nodes, and these differential levels of S1P are needed to drive lymphocyte exit. S1P is produced by sphingosine kinases through the phosphorylation of sphingosine and is degraded by sphingosine lyase. Thus, S1P concentrations in tissues, lymph, and blood are regulated by the differential activity of the respective S1P-producing and/or -degrading enzymes as well as S1P release from intracellular stores. High S1P levels in lymph are maintained by the expression of sphingosine kinase in lymphatic endothelial cells, whereas high S1P concentrations in blood rely on S1P release from red blood cells.

S1P binds to five G-protein-coupled receptors designated $S1P_1$–$S1P_5$. Egress of naive T cells from lymph nodes depends on $S1P_1$, because $S1P_1$-deficient T cells fail to leave lymph nodes. $S1P_1$ is internalized on encountering high S1P concentration such as that found in blood and lymph but is maintained on the cell surface in areas of low S1P concentration such as the lymph nodes. Thus, shuttling of $S1P_1$ between the cell surface and cytosol seems to regulate lymphocyte egress from lymph nodes, and the reappearance of $S1P_1$ on the cell surface during lymph node transit might override retention signals, possibly including CCL21–CCR7 interactions, and initiate lymphocyte egress. After their activation, T cells express the early T-cell activation marker CD69, a transmembrane protein of the C-type lectin family. Association between CD69 and $S1P_1$ induces $S1P_1$ internalization, leading to reduced responsiveness to S1P and inhibition of lymphocyte exit. Later, after extensive proliferation in the lymph node, T cells lose their expression of CD69 and reexpress $S1P_1$; as a consequence, they regain their ability to leave via efferent lymph.

LYMPHOCYTE MIGRATION INTO MUCOSAL TISSUES: GENERATION OF TISSUE-TROPIC LYMPHOCYTE SUBSETS

After recognizing their cognate antigen/MHC on the surface of antigen-presenting cells, activated T cells undergo extensive clonal expansion. In the context of appropriate costimulatory signals, T cells expand and differentiate into effector T cells, the functions and migratory properties of which seem to be determined, at least in part, by the environmental context in which antigen presentation and recognition take place. During this process, T cells alter their expression of CAMs and chemokine receptors and thus their migratory potential.

Effector T cells can acquire distinct migratory properties. A large share of the effector T-cell population will start to express CAMs and chemokine receptors involved in lymphocyte migration to extralymphoid tissues and sites of inflammation. Changes in the expression of adhesion molecules allow effector T cells to leave lymph nodes, to enter peripheral tissues, and to coordinate immune and inflammatory responses in peripheral sites, including mucosal tissues. Yet other effector T cells will exert their function within lymph nodes. For example, follicular helper CD4 T cells downregulate CCR7 and acquire expression of the chemokine receptor CXCR5, which allows these cells to migrate toward B-cell follicles and provide B-cell help.

After resolution of the immune response, the expanded effector T-cell population will collapse, and a smaller population of memory T cells will be retained. Memory lymphocytes also display characteristic migratory behavior. Memory T cells in blood have been distinguished based on their expression of CCR7 and CD62L. Memory T cells expressing these two homing factors retain lymph node homing capacity and have been termed *central memory cells*. In contrast, memory cells lacking CD62L and CCR7 have been named *effector memory cells* and are suggested to recirculate through peripheral tissues. More recently, another migration pattern has been observed: memory T cells that remain in peripheral tissues and do not recirculate. These have been termed *tissue resident memory cells*.

16.6   Specific integrin–adhesion-molecule interactions and chemokines direct cell migration into mucosal tissues

As with naive T-lymphocyte entry into secondary lymphoid organs, the migration of effector/memory lymphocytes into peripheral tissues is coordinated through the interaction of CAMs and chemokine receptors on circulating lymphocytes with their respective ligands on vascular endothelial cells. Circulating effector/memory T cells and B cells are heterogeneous in their expression of CAMs and chemokine receptors. Thus, the subset of circulating effector lymphocytes that enter into any given peripheral tissue will depend on the array of receptor ligands expressed within that particular tissue. Homing receptor ligand expression can be overlapping as well as distinct between different mucosal tissues and ultimately determines which lymphocyte subsets are recruited into that tissue from the circulating effector T-cell pool. Homing receptor ligand expression in mucosal tissues can also change in the setting of infection or inflammation resulting in the recruitment of alternative or additional lymphocyte subsets.

One of the best-characterized examples of a common homing receptor usage is that of CCR10 for the localization of IgA plasma cells to mucosal tissues. Epithelial cells at a range of mucosal sites, including the intestinal tract, trachea, and salivary and mammary glands, constitutively express the CCR10 ligand, CCL28, and IgA-positive plasma cells at many of these sites have been shown to express CCR10. Furthermore, CCR10 has a redundant role (together with CCR9) in mediating IgA plasmablast recruitment to the small intestine, but has a nonredundant role in the recruitment of IgA plasmablasts into the lactating mammary gland and thus the passive transfer of IgA into the nursing neonate, and has been implicated in recruitment of these cells to the colon. Such common receptor usage allows for the broad dissemination of IgA plasma cells to a wide range of mucosal tissues. Similar to CCR10 for IgA plasma cells, effector and regulatory T-cell recruitment to the murine colon appears in part to be mediated by their expression of the "orphan" G-protein-coupled receptor GPR15; however, GPR15 has also been implicated in T-cell localization to extraintestinal sites and is, for example, expressed on a wide range of $\alpha_4\beta_7$ integrin + (gut homing) and $\alpha_4\beta_7$ integrin negative (extraintestinal homing) memory T cells in human blood. While the physiologic ligands for GPR15 remain to be characterized, it is of note that GPR15 is not expressed by human colonic FoxP3$^+$ T cells, indicating that inhibition of GPR15 may selectively block inflammatory T-cell influx to the colon in patients with ulcerative colitis. Finally, the chemokine receptor CCR4 has been implicated in effector T-cell localization to both the lung and skin, indicating a potential immunologic common cross talk between these sites. Nevertheless, according to the "multistep adhesion cascade" theory, which proposes that lymphocyte trafficking from the blood into peripheral tissues requires the concerted action of several CAMs, lymphocyte recruitment to distinct mucosal tissues probably involves the use of partly nonoverlapping CAM mechanisms.

The best-studied example of a mucosal organ that preferentially recruits specific subsets of effector T cells from the circulation is the small intestinal mucosa. Vascular endothelial cells that line the blood vessels within the intestinal mucosa constitutively express MAdCAM-1, and efficient migration of B and T lymphocytes into the small intestinal and colonic mucosa is dependent on interactions between MAdCAM-1 and $\alpha_4\beta_7$ integrin, which is expressed on a subset of circulating lymphocytes. Further specificity in effector T lymphocyte homing to the small intestinal mucosa is provided by the chemokine CCL25 which is constitutively expressed at high levels by epithelial cells of the small intestine but not other peripheral or mucosal tissues, including the colon. The CCL25 receptor CCR9 is expressed on a subset of circulating T lymphocytes that coexpress high levels of β_7 integrins, and on the majority of T cells in the small intestinal mucosa. T-cell adoptive transfer studies in mice have demonstrated an important and selective role for this chemokine/chemokine receptor pair in mediating the efficient recruitment of effector T cells into the small intestinal mucosa. Thus, expression of MAdCAM-1 and CCL25 in the small intestinal mucosa induces the selective recruitment of T cells coexpressing CCR9 and $\alpha_4\beta_7$ integrin from the circulating effector T-lymphocyte pool.

16.7 Lymphocytes usually remain in tissues, but some reenter the circulating pool

In contrast to naive lymphocytes, which as described earlier continually recirculate through secondary lymphoid organs, effector/memory cells that enter peripheral tissues in many cases never leave and may persist at these sites for long periods. The retention and survival of lymphocytes and their localization to distinct anatomical localizations within peripheral tissues probably require survival and retention signals provided by the local environment, including the involvement of CAMs and chemokine receptors. A striking example of divergent sublocalization of lymphocyte subsets in peripheral tissues is the differential localization of plasma cells and CD8+ lymphocytes to the lamina propria and intraepithelial compartments of the intestinal mucosa, respectively. Although the underlying mechanisms regulating the differential localization of lymphocyte subsets in peripheral tissues are incompletely understood, CD8 T cells within epithelial tissues, unlike most of their lamina propria counterparts, express the α_E (for epithelial) β_7 integrin. The cellular ligand for $\alpha_E\beta_7$ integrin is not an Ig superfamily member but E-cadherin that is expressed on the lateral and basolateral surface of epithelial cells. Heterotypic interactions between E-cadherin on the epithelial cell and $\alpha_E\beta_7$-integrin-expressing intraepithelial lymphocytes (IELs) are believed to help maintain T cells within the epithelium and probably modulate IEL and epithelial cell cross talk. The $\alpha_E\beta_7$ integrin, at least on effector CD8 T cells, is not a classical "tissue-homing" molecule, because it is induced on these cells once they have migrated into the epithelium in response to local signals. In addition to $\alpha_E\beta_7$, subsets of lymphocytes in mucosal tissues, including the lung and intestine, express the β_1 integrins $\alpha_1\beta_1$ and $\alpha_2\beta_1$, whose extracellular ligands include collagen IV and collagen I, respectively, the former being a major epithelial basement membrane constituent. Although a role for $\alpha_2\beta_1$ integrin in lymphocyte localization is currently unclear, $\alpha_1\beta_1$ integrin has been shown to be important in CD8 effector/memory T-cell survival and/or retention within the lung mucosa.

Despite the findings described earlier, there is also clear evidence that subsets of effector/memory lymphocytes may leave peripheral tissues. Early studies in sheep demonstrated the presence of effector/memory cells in afferent lymph (lymph draining from peripheral tissues to draining lymph nodes). Thus, some effector/memory T cells are capable of exiting from peripheral tissues via the lymph and migrating back to the draining lymph nodes. Some effector/memory T cells within peripheral tissues have been shown to express the chemokine receptor CCR7, as do most T cells in sheep afferent lymph, and studies in mice have demonstrated that effector/memory T-cell migration from the lung and skin to draining lymph nodes is largely dependent on CCR7, whose ligand CCL21 is constitutively expressed by lymphatic vessels within peripheral tissues. It remains unclear whether these migratory cells are a subset of effector/memory cells that expressed CCR7 before their entry into peripheral tissues, or whether they are a subset of cells that selectively upregulated CCR7 after they received signals within the tissue environment. A second receptor-ligand system that seems to regulate T-cell exit from peripheral tissues is the $S1P_1$-S1P system, described previously in regulating lymphocyte exit from secondary lymphoid organs. Tissue resident memory CD8 T cells downregulate expression of $S1P_1$ potentially preventing their entry into lymphatics.

16.8 Different lymph nodes support generation of distinct tissue-tropic T-cell subsets

Circulating effector/memory T-cell subsets are heterogeneous in their expression of CAMs and chemokine receptors, but the mechanism regulating the expression of these molecules on effector T-cell subsets and thus their homing potential to peripheral tissue has only recently become a focus of study. Certain chemokine receptors, including CXCR5, CXCR3, and CCR6, seem to be induced on T-cell subsets after their activation in all secondary lymphoid organs, whereas other tissue-homing receptors are preferentially induced on activated T cells in only a restricted set of lymph nodes. The best current examples of selective induction of tissue-homing receptors are found in the lymph nodes draining cutaneous tissues and the intestine-associated lymph nodes (mesenteric lymph nodes and Peyer's patches). By using T-lymphocyte adoptive transfer models, it has been demonstrated that T cells primed in mesenteric lymph nodes are induced to express higher levels of $\alpha_4\beta_7$ integrin and CCR9, and thus acquire enhanced "small intestinal homing" properties in comparison with T cells primed in nonintestinal lymph nodes. Conversely, T cells primed in skin-draining lymph nodes upregulate E-selectin ligands and CCR4, two homing receptors involved in T-cell migration to cutaneous and extraintestinal sites of inflammation. This does not mean that effector T cells generated in, for example, intestinal inductive sites are incapable of entering into extraintestinal tissues, but they do so utilizing "nonintestinal" homing receptors. It seems likely that this selective induction of gut- and skin-homing receptors in gut- and skin-draining lymph nodes, respectively, serves to enhance effector T-cell migration to the tissues in which infection initially takes place. However, the extent to which this paradigm holds true for other mucosal tissues, the impact of mucosal inflammation and infection on such selective imprinting capacity, and whether lymph nodes draining distinct compartments of the same mucosal tissue have similar imprinting capacities remain to be fully explored (see **Figure 16.1**).

16.9 Dendritic cells appear important in generation of tissue-tropic effector lymphocyte subsets

In addition to their role in initiating adaptive immune responses, DCs appear to contribute to the selective induction of tissue-homing receptors during T-cell priming in secondary lymphoid organs. For example, DCs isolated from the small intestinal lamina propria or draining mesenteric lymph nodes efficiently induce expression of the "small intestinal homing" receptors CCR9 and $\alpha_4\beta_7$ integrin on T lymphocytes *in vitro* compared with colonic DCs or DC subsets isolated from extraintestinal sites. This activity appears to be restricted to those mesenteric lymph node DCs that have migrated from the small intestine. Conversely, DCs isolated from the skin, or skin-draining lymph, induce enhanced levels of E-selectin ligand but not CCR9 or $\alpha_4\beta_7$ integrin. Finally, lung DCs appear to imprint T cells with enhanced lung homing capacity compared with DCs from other sites, partially through the upregulation of CCR4, a chemokine receptor implicated in T-cell trafficking to both the lung and skin. The expression of tissue-specific homing receptors seems to be reversible, because gut-homing T lymphocytes restimulated *in vitro* with skin DCs downregulate gut-homing receptors and adopt a skin-homing receptor profile, and vice versa. It seems likely that such plasticity in tissue-homing receptor expression serves to enhance protective immune responses to less tissue-specific pathogens on a second infection with the same pathogen. There is currently considerable interest in trying to understand the underlying molecular mechanisms regulating the tissue-homing receptor expression, because these pathways may provide interesting targets for modulating the tissue-homing potential of T cells for the treatment of inflammatory diseases and for vaccine development. Most progress in our understanding of tissue-homing receptor induction has been made with the gut-homing receptors CCR9 and $\alpha_4\beta_7$ integrin, and these are discussed in more detail later.

16.10 Vitamin A can confer generation of gut-specific migration signature

The specific capacity of small intestine-derived DC to induce CCR9 and $\alpha_4\beta_7$ integrin on T lymphocytes appears to result from their enhanced ability to metabolize vitamin A. Vitamin A is a dietary vitamin that is taken up in the small intestine and transported to the liver, where it is stored as retinyl esters. Retinol is released into the circulation, primarily from the liver, in association with retinol-binding protein 4 and is taken up by tissue cells, where it is further metabolized. The major active metabolite of vitamin A is retinoic acid (RA). RA is generated by the consecutive oxidation of retinol to retinal and from retinal to RA. The oxidation of retinol to retinal is catalyzed by a family of alcohol dehydrogenases, which seem to be ubiquitously expressed in a wide range of tissues, and the oxidation of retinal to RA is catalyzed through an irreversible reaction by a family of retinal dehydrogenases, which seem to have more tissue- and cell-restricted expression patterns. Once generated, RA can be used by the cell generating it or released into the local environment to regulate the function of neighboring cells. RA signaling is mediated through retinoic acid receptor (RAR)–retinoid X receptor (RXR) heterodimers that function as ligand-induced transcription factors bound to DNA target sequences. The addition of low (subnanomolar) concentrations of RA to stimulated T lymphocytes *in vitro* is sufficient to induce expression of the gut-homing receptors CCR9 and $\alpha_4\beta_7$ integrin on responding T cells; more importantly, the ability of intestinal DCs to induce gut-homing receptors on responding T cells is blocked by the addition of antagonists of retinoic acid receptors. In this regard, small intestinal-derived migratory DCs express higher levels of key enzymes involved in vitamin A metabolism and have been shown to induce enhanced RA signaling in responding T cells (**Figure 16.7**). Numerous additional cells in the intestinal mucosa, including epithelial cells of the intestinal wall and stromal cells in the mesenteric lymph nodes, also have the capacity to metabolize vitamin A and probably contribute to generating an RA-rich environment at these sites that helps drive the selective generation of gut-homing T cells in the mesenteric lymph node *in vivo*. The ability of RA to induce gut-homing receptors is also not restricted to T cells, because RA also induces the expression of gut-homing receptors on B cells. As one might expect from these findings, mice whose vitamin A stores have been depleted by being kept on a diet deficient in vitamin A for a long period have

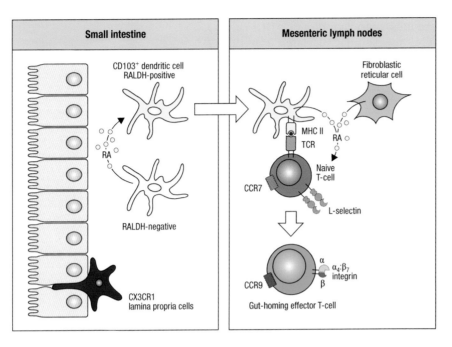

Figure 16.7 Small intestinal–derived migratory DCs induce the expression of gut-homing molecules. In the intestinal lamina propria under the influence of retinoic acid, DCs are imprinted to express increased levels of retinal dehydrogenase (RALDH) enzymes. On migration into the gut-draining mesenteric lymph nodes, these DCs, in cooperation with lymph-node-resident fibroblastic reticular stromal cells, induce the expression of the gut-homing molecules CCR9 and $\alpha_4\beta_7$ integrin on activated T lymphocytes. This pathway is restricted to migratory DCs and does not extend to nonmigratory lamina propria cells expressing CX3CR1. TCR, T-cell receptor.

markedly reduced numbers of T cells and B cells in the intestinal mucosa. These findings probably explain some of the beneficial effects of dietary supplementation with vitamin A in reducing childhood mortality from persistent diarrhea induced by infectious disease in developing countries.

The ability of small intestinal derived CD103$^+$ DCs to induce gut-homing receptors efficiently on responding T cells indicates that factors within the small intestinal mucosa are involved in selectively imprinting these cells with the enhanced ability to metabolize retinol. Although the nature of these factors remains to be fully elucidated, intestinal DCs from mice kept on a diet deficient in vitamin A express lower levels of retinal dehydrogenase (RALDH) enzymes and have lower aldehyde dehydrogenase activity, indicating that RA itself, either directly or indirectly, may be involved in small intestinal DC imprinting *in vivo*.

TARGETING LYMPHOCYTE MIGRATION FOR TREATMENT OF MUCOSAL DISEASE

Although the recruitment of lymphocytes to mucosal tissues is a critical component of the immune response to mucosal pathogens, in certain susceptible individuals, this process can contribute in a major way to the establishment and maintenance of mucosal disease, including celiac disease, asthma, and inflammatory bowel disease (Crohn's disease and ulcerative colitis). Given the dual role of lymphocytes in protecting mucosal surfaces and in driving mucosal inflammation, there is considerable interest in determining the CAMs and chemokine receptors that regulate immune cell influx as potential therapeutic targets for enhancing or preventing immune-cell recruitment in these settings. During mucosal infection and inflammation, the expression of CAMs on vascular endothelial cells and the chemokine milieu of mucosal tissues are markedly altered, with enhanced expression as well as *de novo* expression of multiple chemokine family members. In this context, defining key molecular targets that regulate lymphocyte trafficking to sites of infection and inflammation has proven difficult. In particular, different sets of CAMs and chemokines seem to have sometimes distinct and sometimes overlapping roles depending on the infection/inflammatory model, lymphocyte subset, and mucosal tissue under study. Nevertheless, there is evidence, at least in the intestinal mucosa, that some of the cellular adhesion receptors regulating lymphocyte recruitment in the steady state remain relevant in the setting of inflammation.

Inflammatory bowel disease (IBD), as discussed in detail in several chapters of this book, comprises two major disease groups: Crohn's disease, which can affect all areas of the gastrointestinal tract, and ulcerative colitis, which is restricted to the colon and rectum. Importantly, vascular endothelial cells of the intestinal lamina propria in patients with IBD express enhanced levels of MAdCAM-1, and preclinical studies demonstrated that neutralizing antibodies against $\alpha_4\beta_7$ integrin alleviates inflammation in several animal models of chronic intestinal inflammation. These findings have led to the intriguing possibility of targeting tissue-homing receptors or their endothelial ligands for the treatment of inflammation in a tissue-specific manner. As a consequence, an inhibitor of $\alpha_4\beta_7$ (vedolizumab) is now an approved therapy for the treatment of Crohn's disease and ulcerative colitis, and several additional drugs targeting the interaction of $\alpha_4\beta_7$ with MAdCAM-1 are in clinical trials. Similarly, CCL25 is expressed in the small intestinal mucosa in patients with ileal Crohn's disease, although the impact of inhibiting CCL25:CCR9 interactions in animal models of intestinal inflammation and in initial phase II and III trials in patients with ileal Crohn's disease have produced mixed results. Drugs that block egress of T cells by targeting $S1P_1R$ and $S1P_5R$ have shown some success in relapsing multiple sclerosis and ulcerative colitis and hopefully will be in the clinic soon. When successful, selective targeting of tissue-specific migration may offer an important advantage over many of today's current treatments by alleviating inflammation in a tissue-selective manner without broadly inhibiting systemic immune function.

SUMMARY

Although it has been recognized for almost 50 years that small lymphocytes recirculate from blood into lymphoid tissues and reenter the blood via the lymphatic system, and that lymphoblasts migrate into tissues, especially the gut, it is only in the past decade or so that the molecular basis for specific homing has been elucidated. In addition to being of fundamental interest, this knowledge is being exploited for health benefits. The idiopathic inflammatory diseases of the modern world such as IBD, rheumatoid arthritis, and multiple sclerosis are all consequences of the migration of immune cells into tissues and subsequent injury. Preventing lymphocytes from entering tissues is therefore a rational way of treating these serious conditions, and it is gratifying that antibodies against $\alpha_4\beta_7$ integrin are now routinely used in the clinic for the treatment of IBD.

FURTHER READING

Agace, W.W.: Tissue-tropic effector T cells: Generation and targeting opportunities. *Nat. Rev. Immunol.* 2006, 6(9):682–692.

Alvarez, D., Vollmann, E.H., and von Andrian, U.H.: Mechanisms and consequences of dendritic cell migration. *Immunity* 2008, 29:325–342.

Gorfu, G., Rivera-Nieves, J., and Ley, K.: Role of β7 integrins in intestinal lymphocyte homing and retention. *Curr. Mol. Med.* 2009, 9:836–850.

Habtezion, A., Nguyen, L.P., Hadeiba, H., and Butcher, E.C.: Leukocyte trafficking to the small intestine and colon. *Gastroenterology* 2016, 150:340–354.

Hart, A.L., Ng, S.C., Mann, E., et al.: Homing of immune cells: Role in homeostasis and intestinal inflammation. *Inflamm. Bowel Dis.* 2010, 16:1969–1977.

Kunkel, E.J., and Butcher, E.C.: Plasma-cell homing. *Nat. Rev. Immunol.* 2003, 3:822–829.

Mora, J.R., and von Andrian, U.H.: Role of retinoic acid in the imprinting of gut-homing IgA-secreting cells. *Semin. Immunol.* 2009, 21:28–35.

Mora, J.R., Iwata, M., Eksteen, B., et al.: Generation of gut-homing IgA-secreting B cells by intestinal dendritic cells. *Science* 2006, 314:1157–1160.

Sigmundsdottir, H., and Butcher, E.C.: Environmental cues, dendritic cells and the programming of tissue-selective lymphocyte trafficking. *Nat. Immunol.* 2008, 9:81–87.

Stagg, A.J., Kamm, M.A., and Knight, S.C.: Intestinal dendritic cells increase T cell expression of α4β7 integrin. *Eur. J. Immunol.* 2002, 32:1445–1454.

Zundler, S., and Neurath, M.F.: Novel insights into the mechanisms of gut homing and anti-adhesion therapies in inflammatory bowel diseases. *Inflamm. Bowel Dis.* 2017, 23:617–627.

17

Mucosal tolerance

CHARLES O. ELSON AND OLIVER PABST

Immunologic tolerance has been a subject of fascination and study for as long as the immune system has been known. Immunologic tolerance to exogenous antigen is the reduction or abrogation of an immune response due to previous exposure to a given antigen. For decades, this was largely accomplished by the intravenous injection of antigens such as soluble proteins. However, at the beginning of the twentieth century, there was a rebirth of the concept that tolerance can also be induced by feeding an immunogen prior to immunization, hence, the term *oral* or *mucosal* tolerance, which is the topic of this chapter. The intestine and other mucosal surfaces have features and a microenvironment that are distinctly different from those of systemic lymphoid tissues. First, the mucosal surfaces are colonized with microorganisms, which are collectively called the *microbiota*. The microbiota has profound effects on the host, particularly on the host immune system. In addition, mucosal surfaces are exposed to exogenous antigens in the form of food in the intestine and aeroantigens and dust in the nasal and respiratory tracts. The mucosal immune system has coevolved with these exogenous antigens and thus has developed mechanisms to respond to this continuous antigenic load. Considering the size of the antigen exposure, the dominant mucosal immune response mechanisms are those that dampen the immune and inflammatory responses to such antigens and thus limit inflammation that could injure the mucosal layer. Antigens transiently applied to a mucosal surface are encountered among a remarkably large number and variety of antigens associated with the continuously present microbiota. Multiple cell types are located in the mucosal surface, as shown in Figure 17.1. Among the possible outcomes of antigen exposure at the mucosal surface is mucosal tolerance. This chapter focuses on the mechanisms of these processes, which have profound implications for understanding the biological basis of many inflammatory disorders associated with the mucosal surfaces and for the development of vaccines and other types of therapies.

GENERAL FEATURES OF MUCOSAL TOLERANCE

Mucosal tolerance is a complex process that shows both important differences and similarities among mucosal tissues and types of antigens. Many studies have focused on the gut and the effects of orally applied exogenous antigen: Oral application of antigen that induces a striking reduction in mucosal and systemic immune responses is called *oral tolerance*. Many different types of antigens have been used to induce oral tolerance, including exogenous proteins, red blood cells, contact allergens, inactivated viruses, self-antigens, and alloantigens. A typical experiment demonstrating oral tolerance is shown in Figure 17.2. Oral tolerance has been characterized in some detail, and frequently the term *oral tolerance* has been used synonymously with *mucosal*

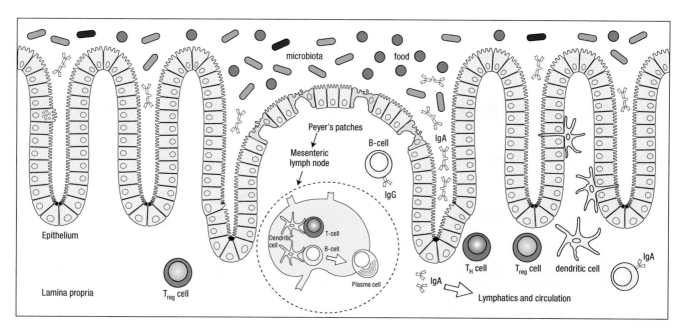

Figure 17.1 The mucosal microenvironment is complex, and an antigen encountered at the mucosal surface has multiple possible outcomes. The mucosal surfaces constitute very complex microenvironments. The intestine is colonized by a large variety of commensal microbes and exposed to a huge array of food antigens and digested proteins, whereas the respiratory and nasal mucosae are exposed to a variety of airborne foreign antigens. Cell types present in the mucosal surfaces include epithelial cells, myeloid cells such as dendritic cells, and macrophages, B cells, and T cells of the various subsets. Antigens encountered at the mucosal surface can result in a variety of different outcomes, including a local mucosal IgA response, induction of helper T (T_H) and/or regulatory T (T_{reg}) cells, systemic immune hyporesponsiveness (tolerance), and an active systemic immune response. Combinations of these outcomes are possible, depending on the microenvironmental conditions, nature of the antigen, and dose. The microbiota that colonizes mucosal surfaces is now being recognized as having a profound impact on both the mucosal and systemic immune systems. An antigen applied to a mucosal surface is encountered in the context of this large and ongoing response to the microbiota.

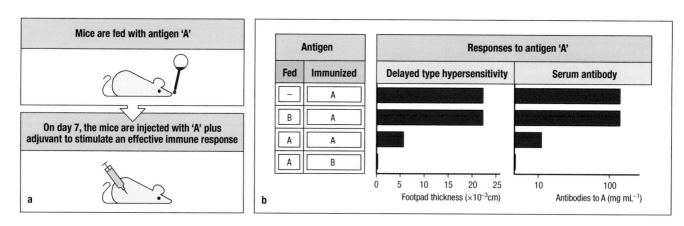

Figure 17.2 A generic oral tolerance experiment. (a) Antigen A is fed or applied to a mucosal surface prior to systemic priming with antigen A plus adjuvant, which is generally followed by a booster immunization. Following the immunization, the ensuing immune response to antigen A is measured. Two of the most common types of immune response are shown in (**b**). Delayed-type hypersensitivity is measured by intradermal injection of small amounts of antigen into the footpad or skin dermis; the degree of induration, or thickening, 24–48 hours later is measured. Delayed-type hypersensitivity is mediated by T cells. Assays of antigen-specific T-cell proliferation or T-cell cytokine production are also utilized. The humoral or B-cell response to the antigen is measured by various antibody assays or measurement of antibody-producing cells using the ELISPOT (enzyme-linked immunospot) technique. In experimental disease models, measurements of clinical signs of the disease or of inflammation in the relevant tissue have been used. Tolerance is specific for the antigen that was fed or applied mucosally, such as feeding a control antigen, denoted B in the figure, does not affect the systemic response to antigen A, and vice versa. As shown, the antibody response usually is reduced by one log10 unit, and the delayed-type hypersensitivity response is also reduced but not abolished because mucosal tolerance causes a reduction rather than abolition of the ensuing immune response.

tolerance. However, in a strict definition, the term *oral tolerance* should not be used to describe all forms of mucosal tolerance. For example, the mucosal immune system is tolerant to the microbiota, but exposure to the microbiota does not dampen systemic responses to the microbiota. This illustrates that mechanisms of mucosal tolerance differ depending on the type of antigen: Fed antigen triggers classical oral tolerance and dampens systemic immune responses, whereas tolerance to the microbiota remains local. Besides the oral route, inhalation, intravaginal, and intranasal applications have been investigated. Common to these routes, the tolerance induced by mucosal exposure is antigen-specific, partial rather than complete, and abrogates an initial immune response more effectively than reducing an established one.

17.1 Mucosal tolerance involves many mechanisms, including anergy, deletion, and induction of active regulatory pathways

The major mechanisms underlying immunologic tolerance generally include deletion of antigen-reactive T cells, clonal anergy of antigen-reactive T cells, and induction of regulatory T (T_{reg}) cells. Each of these mechanisms has been demonstrated in mucosal tolerance. The dose administered at the mucosal surface may be a crucial variable in this regard. In T-cell receptor (TCR) transgenic systems, large doses of antigens, such as 20 mg or more per dose, given to an animal result in clonal deletion and/or anergy. Modest or "low" doses such as 1 mg per dose, particularly if given repeatedly, appear to induce T_{reg} cells as their dominant mechanism (Figure 17.3). Most of the mucosal administrations are of fairly short duration, and whether the

Figure 17.3 Induction of T_{reg} cells by antigen feeding. (a) A typical experiment in which antigen A is fed to animal 1. Subsequently, mesenteric lymph node T cells from animal 1 are transferred to animal 2, which is immunized systemically with antigen A in adjuvant. Following the immunization, the immune response to the antigen is determined in animal 2, which had never been fed the antigen in question. The same types of assays as in **Figure 17.2** are performed to measure delayed-type hypersensitivity responses to antigen A after its injection into the dermis or footpad (**b**). In addition, *in vitro* cocultures of CD4$^+$ T cells from animals fed antigen A with CD4$^+$ T cells from animals parenterally immunized with antigen A have been performed to test whether the former inhibit the latter, as measured by changes in proliferation or cytokine production (**c**). This is an *in vitro* assay for antigen-specific T_{reg} cells induced by feeding antigen. Such T_{reg} cells are antigen-specific in that CD4$^+$ T cells from animals fed control antigen B do not suppress CD4$^+$ T cells of animals primed to antigen A. The dose of the antigen fed is a critical determinant of mucosal tolerance and largely determines the mechanism involved. In mice, very low doses (micrograms) of antigen do not induce oral tolerance, indicating that there is a threshold effect. Very high doses such as 20–25 mg, even given only once, induce mucosal tolerance due to anergy and/or deletion of antigen-specific T cells. Low doses such as 1 mg per dose given multiple times in mice appear to induce oral tolerance by inducing regulatory T cells specific for the fed antigen, although the specific dose varies with the antigen. The T_{reg} cells can be demonstrated by measurements of antigen-specific inhibition in *in vitro* assays, as shown, or by measurement of the production of inhibitory cytokines produced selectively by T_{reg} cells. T_{reg} cells induced by antigen feeding preferentially produce TGF-β; T_{reg} cells induced in the respiratory mucosa appear to be predominantly IL-10-producing T_{reg} cells.

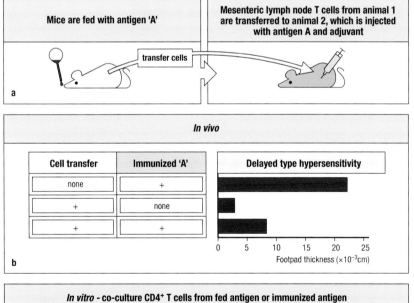

Mice are fed with antigen 'A'

transfer cells

Mesenteric lymph node T cells from animal 1 are transferred to animal 2, which is injected with antigen A and adjuvant

a

In vivo

Cell transfer	Immunized 'A'	Delayed type hypersensitivity
none	+	
+	none	
+	+	

0 5 10 15 20 25
Footpad thickness ($\times 10^{-3}$cm)

b

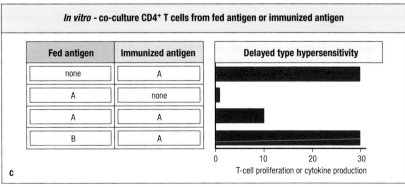

In vitro - co-culture CD4$^+$ T cells from fed antigen or immunized antigen

Fed antigen	Immunized antigen	Delayed type hypersensitivity
none	A	
A	none	
A	A	
B	A	

0 10 20 30
T-cell proliferation or cytokine production

c

same mechanisms would apply if the antigen exposure were more chronic is unclear. Although most experimental studies have been performed with model antigens such as ovalbumin and keyhole limpet hemocyanin, as discussed later, similar mechanisms may apply to "self" antigens.

17.2 Mucosal tolerance is likely induced in organized lymphoid structures associated with mucosal tissues and is disseminated widely throughout the mucosa-associated lymphoid tissues

The site of induction of mucosal tolerance has not been clearly defined. Candidate sites are the gut-associated lymphoid tissue, Peyer's patches, and equivalent structures in the respiratory tract (nasopharynx-associated or bronchus-associated lymphoid tissue). T_{reg} cells have been demonstrated in these structures shortly after mucosal exposure and subsequently in draining mesenteric lymph nodes. Mucosal epithelial cells also have been implicated in tolerance induction. Epithelial cells can express class I and II major histocompatibility complex (MHC) molecules, as well as nonclassical MHC class I molecules (e.g., CD1d, HLA-E, and HLA-G), process soluble antigen, produce cytokines, express costimulatory molecules, and appear to preferentially stimulate CD8$^+$ T cells, among which may be CD8$^+$ T_{reg} cells. A second possible site of tolerance induction is mesenteric lymph nodes, which contain dendritic cells (DCs) that induce tolerance. Mucosally applied antigen may cross the epithelium and reach these nodes via the lymph or "tolerogenic" dendritic cells. Migrating CD103$^+$ DCs take up the antigen in the lamina propria and then migrate to the mesenteric lymph nodes to induce T_{reg} cells. Subsequently, T_{reg} cells relocate to the intestinal mucosa to undergo local modification, potentially under the influence of intestinal antigen-presenting cells. Besides mucosa-associated lymphoid tissues and the draining lymph nodes, the liver and systemic lymphoid tissue may contribute to tolerance induction. After administration of large doses of antigen, intact antigenic material can escape the local mucosal immune system and shape immune responses systemically.

17.3 An array of immune effector functions are subject to the impact of mucosal tolerance

Mucosal tolerance can inhibit a wide variety of immune responses, but the major commonality among these different immune responses is the T-cell. Sensitivity to mucosal tolerance is a gradient with the T_H1 helper T-cell the most sensitive, T_H2 cell of intermediate sensitivity, and B cells being refractory. The reduction in antibody responses is largely due to the inhibition of T_H cells rather than direct reduction in B-cell activity. In fact, the feeding of T-cell independent B-cell antigens does not result in tolerance. The reduction in T-cell–dependent antibody responses is largely due to inhibition of the T-cell, rather than the B-cell.

As an active form of the immune response, mucosal tolerance can be enhanced or reduced by various molecules coadministered with the antigen to the mucosal surface (**Figure 17.4**). One example of an agent that can enhance oral tolerance is cholera toxin B subunit (CTB). To amplify mucosal tolerance, CTB must be covalently coupled to the antigen and applied to the nasal or respiratory mucosa. CTB-antigen conjugates applied to the nasal mucosa result in mucosal tolerance at surprisingly small doses of antigen. Simple mixtures of CTB and antigen applied in a similar fashion do not have this effect. Other substances that have been shown to either enhance or reduce mucosal tolerance when coadministered with antigen are shown in **Figure 17.4**.

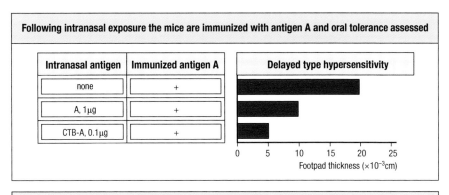

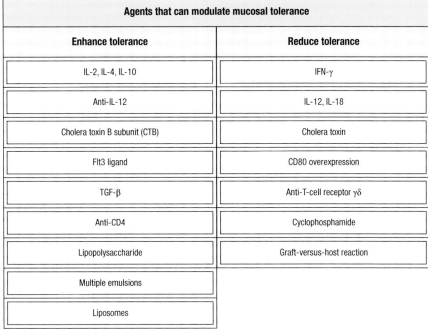

Figure 17.4 Mucosal tolerance can be enhanced or reduced by other compounds. A variety of compounds either enhance or reduce mucosal tolerance, demonstrating that mucosal tolerance is not the absence of an immune response, but an active immune response that itself can be enhanced or reduced. Enhancement of tolerance is shown by more profound inhibition of the subsequent immune response or by having equivalent tolerance at much lower doses of mucosally applied antigen. In this example, oral tolerance is enhanced by coupling cholera toxin B subunit (CTB) covalently to an antigen A and then applying it intranasally. Following intranasal exposure, the mice are immunized with antigen A, and oral tolerance is assessed in the same assays already described (top panel). In this figure, typical delayed-type hypersensitivity data are shown, demonstrating that the CTB-A conjugate induced greater tolerance than did intranasal antigen A alone, at one-tenth the dose. The lower panel shows agents that can either enhance or reduce mucosal tolerance.

Mucosal tolerance in experimental autoimmune and inflammatory disease

Organ-specific autoimmune disease can be induced in rodents by parenteral immunization with an autoantigen in a strong adjuvant. For example, immunization with myelin basic protein, a constituent of brain and spinal cord, results in inflammation in the nervous system and subsequent neurologic symptoms and deficits, which are similar to those that occur in multiple sclerosis in humans. This model is called experimental allergic encephalomyelitis (EAE). EAE, initially induced in the Lewis strain of rat, occurs in various susceptible mouse strains and can be induced by immunization with other nervous system proteins such as proteolipid protein. EAE has been used extensively to probe the mechanisms involved in this multiple sclerosis–like disease. Other organ-specific autoimmune diseases can be induced experimentally in mice by immunization with autoantigens, including collagen type II, which results in arthritis, retinal S-antigen in uveitis, thyroglobulin in thyroiditis, and the acetylcholine receptor in a myasthenia gravis–like condition. A spontaneous autoimmune disease that occurs in nonobese diabetic mice has been used extensively as a model of autoimmune diabetes in humans. Such models are useful for investigating the utility of mucosal (oral) tolerance in the treatment of autoimmune diseases.

17.4 Mucosal tolerance to autoantigens can be elicited in experimental model systems

Administration of the same autoantigen to the mucosal surface prior to the parenteral immunization reduces the frequency and severity of the subsequent autoimmune disease (Figure 17.5), as demonstrated with a diverse array of antigens and models (Table 17.1). The most common route of mucosal exposure has been by feeding the antigen, but the nasal/respiratory route also is an effective route of antigen exposure. The same mechanisms of mucosal tolerance regulate tolerance to exogenous nonantigens—namely, clonal deletion, clonal anergy, and induction of T_{reg} cells—but the generation of antigen-specific T_{reg} cells appears to be the predominant mechanism for the induction of mucosal tolerance. Antigen applied to the mucosa is most effective when delivered prior to the parenteral immunization. Depending on the model, the delivery of autoantigen to the mucosa after immunization has less or no effect on the disease.

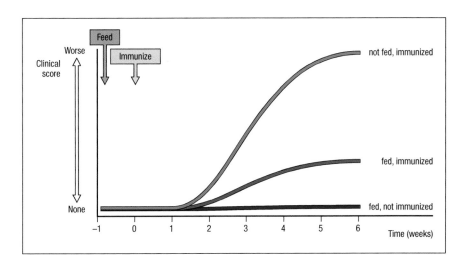

Figure 17.5 Abrogation of experimental autoimmunity by mucosal tolerance. Mucosal tolerance has been used to abrogate experimental autoimmunity in multiple models. Most models involve feeding the autoantigen prior to parenteral immunization with the tissue-specific autoantigen, but others involve spontaneous autoimmunity in a genetically susceptible host such as nonobese diabetic mice that develop type 1 diabetes. The sequence is similar to that in **Figure 17.2** in which feeding precedes immunization with the tissue-specific antigen. Various measurements are made after the feeding and immunization, including a clinical score, a score of inflammation in the relevant tissue, or, in the case of nonobese diabetic mouse autoimmune diabetes, the blood glucose level. In the case of the nonobese diabetic mouse, there is no systemic immunization as the animal spontaneously develops autoimmune diabetes. Feeding a tissue-specific antigen after disease is already established is significantly less effective than feeding before the disease is established. Examples of the different models in which mucosal tolerance has been effective are provided in **Table 17.1**.

TABLE 17.1 ORAL TOLERANCE AS THERAPY FOR EXPERIMENTAL AUTOIMMUNE DISEASE

Model	Antigen fed
EAE	MBP, PLP
Arthritis (CII)	Collagen type II
Uveitis	S-antigen, IRBP
Diabetes (nonobese diabetic mouse)	Insulin, GAD
Myasthenia gravis	Acetylcholine receptor
Thyroiditis	Thyroglobulin
Transplantation	Alloantigen (cells), MHC peptides

Abbreviations: CII, type II collagen-induced arthritis; EAE, experimental allergic encephalomyelitis; GAD, glutamic acid decarboxylase; IRBP, interphotoreceptor retinoid binding protein; MBP, myelin basic protein; MHC, major histocompatibility complex; PLP, proteolipid protein.

TABLE 17.2 EFFECTS OF MUCOSAL TOLERANCE IN EXPERIMENTAL ALLERGIC ENCEPHALOMYELITIS

Decreased severity of disease

Delayed onset and decreased frequency of disease

Decreased frequency of MBP-specific T cells

Altered T-cell receptor usage of MBP-specific T cells

Decreased T-cell epitope spreading

Increased TGF-β-producing T_{reg} cells

Decreased serum IgG anti-MBP

Increased salivary IgA anti-MBP

Abbreviations: MBP, myelin basic protein; TGF, transforming growth factor.

Many experimental models have a clinical scoring system that allows repeated measurement of disease features such as neurologic deficits in EAE, the number and size of swollen joints in arthritis, and the blood glucose level in nonobese diabetic mice. Increasing clinical severity correlates with increasing infiltration of the relevant organ by immune and inflammatory cells and increasing T-cell and B-cell reactivity to the autoantigen. Mucosal tolerance can reduce the clinical score, organ inflammation, and diverse aspects of the antigen-specific immune response. Examples of the many effects of mucosal tolerance on EAE induced by immunization with myelin basic protein are presented in Table 17.2. In the affected organ, T_H cell subsets and cytokines are generally altered, with a decrease in T_H1 cells and interferon-γ secretion, and an increase in T_{reg} cells and in local production of interleukin-10 (IL-10) and transforming growth factor-β (TGF-$\beta1$).

Not only can mucosal tolerance be elicited against orally administered autoantigens in the treatment of experimental models of autoimmunity, such tolerance also can be elicited to other types of antigens associated with immune-mediated inflammation. For example, oral administration of trinitrobenzene sulfonic acid (TNBS) hapten-conjugated colonic proteins, including proteins associated with the commensal microbiota, results in the induction of mucosal tolerance, which is associated with the prevention of TNBS-induced colitis. The mechanism of this tolerance involves induction of TGF-β-producing regulatory cells and indicates the potential broad range of clinically relevant antigens to which oral tolerance can be applied.

17.5 Mechanisms of mucosal tolerance induced by autoantigens are similar to those induced by model antigens

Most models involving mucosal immunization of a known autoantigen are followed by parenteral immunization with the same antigen. These models have provided important insights into mechanisms of disease and its amelioration, but tissues contain multiple autoantigens. Moreover, during an autoimmune disease, the immune system responds to many tissue autoantigens. Nonobese diabetic mice, for example, have T cells that react to more than a dozen autoantigens, including insulin from pancreatic β cells. The multiplicity of autoantigens may limit the potential usefulness of mucosal tolerance for therapy of spontaneously occurring autoimmune or inflammatory disease. However, T_{reg} cells bear T-cell receptors (TCRs) specific for a single antigen and must be activated via that antigen-specific TCR in

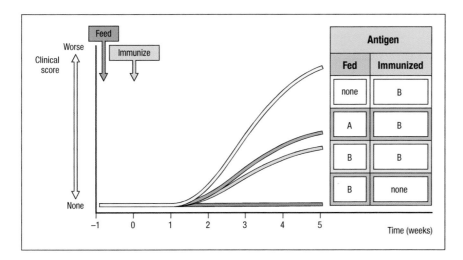

Figure 17.6 Bystander suppression of experimental autoimmune disease. In experimental, and presumably in human, autoimmune disease, more than one antigen contributes to the pathogenic immune response. In most cases, the antigen that initiates the disease is not known. Bystander suppression refers to a phenomenon in which activation of an organ-specific T_{reg} cell causes inhibition of the immune response to other, unlinked tissue-specific antigens. The typical protocol is shown in which animals are fed one organ-specific antigen A but are immunized with a different, unrelated antigen that is present in, and specific for, the same organ B. The concept is that the mucosal exposure to antigen A induces antigen-specific T_{reg} cells to antigen A. These T_{reg} cells are then activated *in vivo* by encounter with antigen A in the tissue or organ. Once activated, the T_{reg} cell specific for antigen A produces inhibitory cytokines that suppress T cells of other specificities, such as B, that are in the lesion or organ. Some examples in which bystander suppression has been demonstrated include (with the antigens used for feeding and immunization in parentheses) experimental allergic encephalomyelitis (myelin basic protein, proteolipid protein), arthritis (collagen type II, complete Freund's adjuvant), and autoimmune diabetes (insulin, lymphocytic choriomeningitis virus in an LCMV-islet cell transgenic mouse).

order to express their inhibitory program, including the release of inhibitory cytokines such as TGF-β1, IL-10, and IL-35. Once released, these cytokines will inhibit T cells of other specificities in the local environment in an antigen-nonspecific manner. Called *bystander suppression*, this inhibition has been demonstrated in multiple experimental models (**Figure 17.6**) in which mucosal tolerance is induced by one organ-specific antigen and inhibits autoimmune disease induced by a different, "unlinked," autoantigen. In one model, insulin feeding inhibited disease triggered by lymphocytic choriomeningitis virus (LCMV) infection in a mouse transgenically expressing an LCMV antigen in pancreatic islets. In these models, mucosal tolerance preceded disease initiation, which is unlikely to apply to spontaneous autoimmune disease. However, mucosal tolerance induced early in the course of disease might theoretically interrupt epitope spreading and broadening of the immune responses, thus ameliorating disease, although this remains to be demonstrated.

Oral tolerance in treatment of human immune-mediated diseases

The effectiveness of oral tolerance in preventing experimental autoimmune disease raised interest in using oral tolerance to treat human autoimmune diseases. Oral tolerance is clearly complex and can be demonstrated only *in vivo* and not in cells tested *ex vivo*. Data on mucosal tolerance in humans are sparse. Experiments performed in the 1960s with contact allergens exposed to the buccal mucosa in the mouth provided results compatible with the presence of mucosal tolerance to contact sensitizers in humans. Contact sensitizers, also known as contact allergens, are covalently reactive compounds that couple to host proteins to form a complex that provokes a cutaneous T-cell–mediated inflammation.

17.6   Mucosal (oral) tolerance is a property of normal human mucosal tissues

To determine whether oral tolerance to protein antigens exists in humans, keyhole limpet hemocyanin (KLH) has been fed to a group of human volunteers. KLH is a potent immunogen that has been used safely in humans to assess immunocompetence. The group fed KLH and a group not fed KLH were parenterally immunized, and the resulting systemic and mucosal immune responses were compared (**Figure 17.7**). KLH feeding resulted in significant reductions of the delayed-type hypersensitivity skin test response and of KLH-specific T-cell proliferation. Conversely, KLH feeding prior to

Figure 17.7 **Mucosal tolerance in humans.** Oral tolerance to protein antigens in humans was demonstrated using keyhole limpet hemocyanin (KLH), which is a strong immunogen, but not a food antigen. (**a**) Human volunteers ingested 10 doses of 0.5 g KLH over 10 days and then were immunized with KLH by subcutaneous injection; controls received only the KLH immunization (p.o., *per os* [orally]; s.c., subcutaneous). (**b**) shows that KLH feeding resulted in a significant decrease in delayed hypersensitivity reaction to KLH, as well as reductions in antigen-specific T-cell proliferation by peripheral blood T cells (not shown). Although KLH feeding by itself did not induce significant antibody responses, feeding did prime B cells at both systemic (**c**) and mucosal (**d**) sites as evidenced by an enhanced antibody response at these sites upon parenteral immunization with KLH. This dissociation between T-cell and B-cell tolerance, "split tolerance," has been observed after low-dose antigen feeding in animals. Subsequent studies showed that mucosal tolerance to KLH in humans was antigen specific.

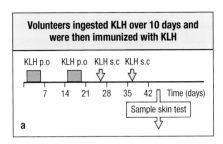

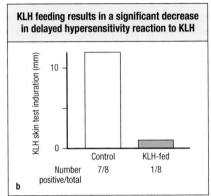

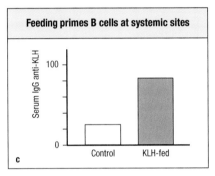

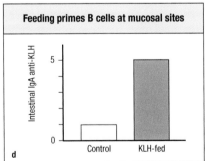

immunization resulted in increased B-cell responses reflected in higher anti-KLH antibodies in serum and secretions compared with individuals not fed KLH. Such "split tolerance" has been demonstrated in some experimental models as well. Split tolerance may be biologically relevant, since secretory antibodies (e.g., secretory IgA) might further promote tolerance by excluding the uptake of "allergic" or "disease-producing" antigen.

17.7 Mucosal tolerance may be amenable to therapeutic manipulation in human immune-mediated disease

The KLH feeding experiments confirmed that at least some aspects of experimental oral tolerance are achievable in humans. These studies, and the vast majority of animal models, used the oral administration of exogenous antigen not previously encountered, or, in mice, the application of self-antigen prior to the induction of disease. In human autoimmune disease, however, the immune response to the antigen has already been established. Nevertheless, the mucosal tolerance demonstrated in numerous experimental animal models, the noninvasive nature of the induction of mucosal tolerance, and the limited adverse effects led to clinical trials to address the use of mucosal tolerance as a therapy for human disease. Controlled clinical trials of oral tolerance have been performed in rheumatoid arthritis, multiple sclerosis, autoimmune uveitis, and autoimmune diabetes (Table 17.3). Unfortunately, the results of these clinical trials have been ambiguous at best and showed only a limited efficacy in the treatment of these diseases. Possible explanations for the lack of efficacy in these trials include the fact that the antigens responsible for autoimmune and inflammatory disease in humans are unknown, the lack of induction of bystander suppression by T_{reg} cells, the antigen in most of the trials was not of human origin, and the dose fed was based on animal studies rather than direct testing in humans. In each trial, the antigen was administered without an agent that might enhance mucosal tolerance. The results suggest that agents that enhance mucosal tolerance are necessary if oral tolerance is to succeed as a therapy for human immune-mediated inflammatory diseases.

For some human diseases, the failure of tolerance to food antigens or to the microbiota initiates a pathologic process. Celiac sprue (or gluten-sensitive enteropathy) and food allergies (peanut, egg, milk) are examples of the former, whereas inflammatory bowel disease (IBD) is an example of the latter. Studies of tolerance in IBD using KLH (described earlier) show that patients lack tolerance and that healthy family members share the same deficit. Thus, while defects in tolerance do not necessarily result in disease, they contribute to susceptibility to disease.

TABLE 17.3 CONTROLLED CLINICAL TRIALS OF ORAL TOLERANCE IN HUMANS

Disease	Antigen
Rheumatoid arthritis	Collagen type II, bovine/chicken
Multiple sclerosis	Myelin, bovine
Uveoretinitis	Retinal S-antigen, bovine
Diabetes mellitus type I	Insulin, human

17.8   Mucosal tolerance can be used in treatment of human allergic disorders

Although mucosal tolerance induction has been a relatively unsuccessful treatment for autoimmune disease, a voluminous body of literature has shown that such therapy is safe and effective for many forms of allergy. Early introduction of food allergens in the diet, such as peanuts, has been shown to prevent the development of food allergy. For established allergy, three primary routes of mucosal tolerance therapy have been utilized: oral immunotherapy in which antigen is swallowed, sublingual immunotherapy (SLIT) in which antigen is held under the tongue for a short period of time, and nasal immunotherapy in which the antigen is applied directly to the nasal mucosa typically as a dry powder. Oral immunotherapy is most commonly used for the treatment of food and contact allergies, whereas SLIT and nasal immunotherapy are used primarily to treat inhaled antigens, given the known preferences for mucosal homing to the site of antigen administration based on the induction of specific homing molecules. Regardless of the route of administration, most clinical allergy trials are conducted in a similar manner. Enrolled patients are subjected to a double-blind, placebo-controlled challenge that serves to determine the minimum dose of allergen that elicits a response in a given individual. The treatment phase is initiated using some fraction of this "eliciting" dose. Tolerance induction begins with a dose escalation phase in which mucosal application of the antigen is conducted in increasing concentrations over a period of time. The escalation may be performed over weeks or months at the patient's home or in a "rush protocol" with dose escalation performed over hours or days at a medical institution. Dose escalation continues until the "maintenance dose" is reached, which is continued for the length of the protocol, typically months to years. Treatment efficacy is usually based on clinical scores generated from patient self-evaluations or physician evaluations, rescue medication scores (scores assigned to medications taken to alleviate allergic symptoms), and allergen challenges conducted at the end of the trial.

The administration of an antigen orally may be more effective therapy for allergy than autoimmune disease for several reasons. First, oral immunotherapy for allergy is directed against the same antigen that is responsible for causing the inappropriate immune response, whereas the antigen used to induce tolerance in autoimmune disease may not be the antigen that caused the initial insult. Because tolerance is induced to a single antigen in allergic disease, clonal deletion or anergy of antigen-specific cells will have a significant impact on the subsequent immune response. Therapy for autoimmune disease requires the generation of antigen-specific regulatory cells to suppress an immune response directed against multiple antigens and consequently necessitates bystander suppression, which is less potent. In addition, inflammation may impair tolerance induction or responsiveness of inflammatory cells to the tolerance-inducing factors (e.g., regulatory cytokines). Furthermore, allergens, and thus inflammation, can be avoided during the induction of tolerance, which is not possible in autoimmunity when the antigen and inflammatory cells are always present. Moreover, oral tolerance protocols for allergy are carried out over significantly longer periods of time than those for autoimmune disease, as many tolerance protocols attempting to treat autoimmune disease are simply too short.

SUMMARY

Due to the constant exposure of the mucosal immune system to food and microbiota, avoidance of immune responses to innocuous antigens is necessary to maintain mucosal homeostasis. Mucosal tolerance is the result of many different mechanisms, including anergy, deletion and induction of T_{reg} cells, which specifically inhibits immune responses to a given antigen. One form of mucosal tolerance involves the induction of T_{reg} cells that secrete TGF-β, IL-10, and IL-35 and is associated with bystander suppression, which may be used in the treatment of autoimmune and inflammatory diseases in which the autoantigen is not known. However, initial successes in animal models have not been replicated in human diseases, despite the evidence that induction of mucosal tolerance can be elicited in humans. Whether these failures are due to the inability to shut down ongoing immune responses, especially in the context of ongoing inflammation, or to the dose and character of the antigen being incorrect, remains to be determined. Moreover, understanding mucosal tolerance has important implications for understanding diseases that are due to a loss of tolerance to environmental antigens such as those associated with food (e.g., gluten in celiac disease) or microbes (e.g., commensal microbiota in IBD). In addition, the tendency of the mucosal surfaces to elicit suppression in response to mucosally applied antigens represents a major challenge for the development of mucosal vaccines and indicates the need for the development of mucosal adjuvants capable of bypassing the restrictive nature of mucosal tolerance.

FURTHER READING

Burks, A.W., Laubach, S., Jones, S.M.: Oral tolerance, food allergy, and immunotherapy: Implications for future treatment. *J. Allergy Clin. Immunol.* 2008, 121:1344–1350.

du Pré, M.F., and Samsom, T.N.: Adaptive T-cell responses regulating oral tolerance to protein antigen. *Allergy* 2011, 66:478–490.

Hadis, U., Wahl, B., Schulz, O., Hardtke-Wolenski, M., Schippers, A., Wagner, N. et al.: Intestinal tolerance requires gut homing and expansion of FoxP3+ regulatory T cells in the lamina propria. *Immunity* 2011, 34:237–46.

Husby, S., Mestecky, J., Moldoveanu, Z. et al.: Oral tolerance in humans. T cell but not B cell tolerance after antigen feeding. *J. Immunol.* 1994, 152:4663–4670.

Knoop, K.A., Gustafsson, J.K., McDonald, K.G. et al.: Microbial antigen encounter during a preweaning interval is critical for tolerance to gut bacteria. *Sci. Immunol.* 2017, 2(18).

Mayer, L., and Shao, L.: Therapeutic potential of oral tolerance. *Nat. Rev. Immunol.* 2004, 4:407–419.

McDole, J.R., Wheeler, L.W., McDonald, K.G. et al.: Goblet cells deliver luminal antigen to CD103+ dendritic cells in the small intestine. *Nature* 2012, 483:345–349.

Pabst, O., and Mowat, A.M.: Oral tolerance to food protein. *Mucosal Immunol.* 2012, 5:232–239.

Prakken, B.J., Samudal, R., Le, T.D. et al.: Epitope-specific immunotherapy induces immune deviation of proinflammatory T cells in rheumatoid arthritis. *Proc. Natl Acad. Sci. USA.* 2004, 101:4228–4233.

Rezende, R.M., and Weiner, H.L.: History and mechanisms of oral tolerance. *Semin. Immunol.* 2017, 30:3–11.

Sun, J.B., Gerkinsky, C., Holmgren, J. et al.: Mucosally induced immunological tolerance, regulatory T cells and the adjuvant effect by cholera toxin B subunit. *Scand. J. Immunol.* 2010, 71:1–11.

Toit Du, G., Roberts, G., Sayre, PH., Bahnson, HT., Radulovic, S., Santos, AF. et al.: Randomized trial of peanut consumption in infants at risk for peanut allergy. *N. Engl. J. Med.* 2015, 372:803–813.

Tordesillas, L., and Berin, M.C.: Mechanisms of oral tolerance. *Clin. Rev. Allergy Immunol.* 2018, 55(2):107–117.

Vickery, B.P., and Burks, A.W.: Immunotherapy in the treatment of food allergy: Focus on oral tolerance. *Curr. Opin. Allergy Clin. Immunol.* 2009, 9:364–370.

Weiner, H.L., da Cunha, A.P., Quintana, F., Wu, H.: Oral tolerance. *Immunol. Rev.* 2011, 241:241–259.

Recognition of microbe-associated molecular patterns by pattern recognition receptors

18

ELKE CARIO

The immune system has two major components: the innate or nonspecific immune system and the adaptive or specific immune system. After structural barriers, the innate immune system is the first line of host defense against microorganisms, whereas the adaptive immune system represents the second line. Both systems contribute to self and nonself discrimination, to protect the host from pathogens and to eliminate modified or host altered cells. Both systems also contain cellular and humoral components and interact with each other. Innate immunity is readily available at birth, whereas adaptive immunity arises later as a consequence of pathogen exposure. The two phases of immune host defense are pathogen recognition followed by pathogen removal.

In contrast to adaptive immunity, innate immunity uses defense mechanisms that promote the immediate detection and rapid destruction of microorganisms independently of the clonal expansion of effector cells or the generation of immunological memory. Consequently, the innate immune system is not antigen specific and reacts to pathogen structures without previous exposure. Innate host defenses are limited and present in all multicellular organisms, in contrast to adaptive host defenses, which are diverse and present only in vertebrates. Tight control of innate immunity is critical to mucosal homeostasis in the intestine. After physical barriers (e.g., epithelia), chemical barriers (e.g., mucus), and biological barriers (e.g., competing microbiota) have been breached, innate immune responses are initiated.

Innate immunity relies on germline-encoded receptors that recognize structures common to pathogens and commensal microorganisms. These receptors are called pattern recognition receptors (PRRs) and are expressed by innate cells, including epithelial cells, macrophages, and dendritic cells. PRRs are grouped into toll-like receptors (TLRs), NOD-like receptors (NLRs), and retinoic acid-inducible gene-I (RIG-I)-like receptors (RLRs) (Figure 18.1). In this chapter, we review the contribution of PRRs and NLRs to innate immune defense of the intestinal mucosa, focusing on TLRs and NLRs. We also discuss the genetic imbalances in pattern recognition that may contribute to inflammatory bowel diseases.

PRINCIPLES OF PATTERN RECOGNITION AND SIGNALING

A key function of the mucosal innate immune system is to recognize invading pathogens and quickly mount defensive responses. Pathogen recognition is accomplished by PRRs that recognize highly conserved, pathogen-associated molecular patterns (PAMPs). PAMPs are expressed by microorganisms and not by mammalian host cells or tissues. The major features of innate cell PRRs are outlined in Table 18.1.

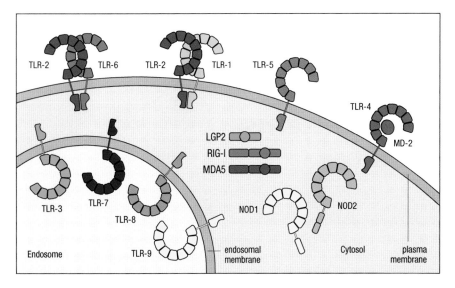

Figure 18.1 Pattern recognition receptors (PRRs) and their cellular location. PRRs are grouped into toll-like receptors (TLRs); NOD-like receptors (NLRs); and retinoic acid-inducible gene-I (RIG-I)-like receptors (RLRs). TLR1, 2, 4, 5, and 6 are cell-surface receptors, whereas TLR3, 7, 8, and 9 are located on intracellular endosomal membranes. The NLRs NOD1 and NOD2 are present exclusively in the cytosol, where they sense intracellular pathogens. The RLRs, including RIG-I, melanoma differentiation-associated gene 5 (MDA-5), and laboratory of genetics and physiology 2 (LGP2), are cytoplasmic helicases that recognize viral RNA.

TABLE 18.1 FEATURES OF PATTERN RECOGNITION RECEPTORS IN THE INNATE IMMUNE SYSTEM
Encoded within the genome (inherited)
Display limited diversity
Recognize broad, shared classes of conserved molecules on pathogens
Initiate responses within minutes
Expressed by all cells of a particular lineage (nonclonal)

18.1 Toll-like receptors comprise a family of conserved receptors that recognize specific pathogen-associated molecular patterns

TLRs comprise a major group of PRRs and have a key role in the innate cell recognition of PAMPs, leading to the induction of pro- and anti-inflammatory genes, phagocytosis, and control of adaptive immune responses. The *Toll* gene was originally discovered during the study of fruitfly (*Drosophila melanogaster*) embryogenesis. After the discovery of a role for the toll pathway in controlling host defense through the induction of potent antifungal factors, a human homolog of the *Drosophila* toll protein was identified. A constitutively active mutant of the toll homolog transfected into human cell lines mediated the activation of the transcription factor NF-κB and the expression of several pro-inflammatory genes. The first TLR, now designated TLR4, recognizes bacterial lipopolysaccharide (LPS). Since its discovery by Bruce Beutler, for which he received the 2011 Nobel Prize in Physiology or Medicine, 10 TLRs in humans and 13 in mice have been identified.

Mammalian TLRs are type I transmembrane glycoproteins containing three common structural features: (1) a large extracellular domain with 19–25 consecutive leucine-rich repeats and one or two cysteine-rich regions, (2) a short transmembrane region, and (3) a highly conserved cytoplasmic domain (**Figure 18.2**). The extracellular domain is responsible for the recognition of PAMPs. The intracellular domain is highly homologous among the individual TLRs and contains a toll/interleukin-1 (IL-1) receptor (TIR) domain that mediates homodimeric or heterodimeric interactions between TLRs and initiates downstream signaling cascades, as discussed

Figure 18.2 **Toll-like receptor (TLR) structure.** TLRs are type I transmembrane proteins with common structural features, including multiple leucine-rich repeats and one or two cysteine-rich regions in the large and divergent ligand-binding ectodomain, a short transmembrane region, and a conserved cytoplasmic domain that contains a toll/interleukin-1 receptor (TIR) domain. TLRs mediate their receptor function as homodimers or heterodimers.

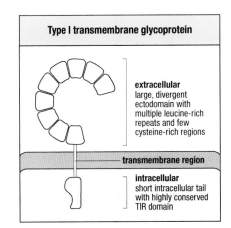

later. The TIR domain also is present in members of the closely related IL-1, IL-18, and IL-33 receptor families.

Different classes of PAMPs activate different TLRs; that is, each TLR recognizes a limited number of specific molecular signatures in different classes of microorganisms (Table 18.2). Although the diversity of pattern recognition is limited, TLRs collectively recognize most pathogenic or commensal microorganisms that may invade the mucosa. For example, TLR2 signals the presence of bacterial lipopeptides and lipoteichoic acid, which are cell-wall constituents of gram-positive bacteria. Select combinations of TLRs may act in concert to expand the repertoire of these PRRs; thus, TLR2 cooperates with TLR1 or TLR6. TLR4 is the major receptor for LPS signal activation but requires the presence of the accessory molecules MD-2 and CD14. Flagellins of flagellated bacteria are ligands for TLR5. Unmethylated CpG DNA present in prokaryotic genomes and DNA viruses is detected by TLR9. In addition, TLRs may recognize endogenous damage-associated molecular patterns such as heat-shock proteins, which are released by dying cells and function as alarm signals during inflammation.

TABLE 18.2 SPECIFIC PATTERN-RECOGNITION RECEPTORS BIND DIFFERENT MOLECULAR SIGNATURES OR STRUCTURES IN MICROBES

Ligand	Receptor	Signaling adaptor
Peptidoglycan Triacylated lipoproteins Diacylated lipoproteins Glycosylphosphatidylinositol anchors Zymosan (yeast)	TLR1–TLR2 TLR6–TLR2 TLR2 (with Dectin-1)	MyD88 MAL
Double-stranded RNA	TLR3	TRIF
LPS Lipoteichoic acids	TLR4 (with MD-2, CD14, LBP)	MAL MyD88 TRIF TRAM
Flagellin	TLR5	MyD88 TRIF
Single-stranded RNA	TLR7	MyD88
G-rich oligonucleotides	TLR8	MyD88
Unmethylated CpG DNA	TLR9	MyD88
Profilin	TLR11 (mouse)	MyD88
iE-DAP	NOD1	
MDP	NOD2	

Abbreviations: iE-DAP, γ-D-glutamyl-meso-diaminopimelic acid; MDP, muramyl dipeptide.

In general, TLRs are expressed predominantly on cells that are the first to encounter the invading threat. Phagocytic cells, including macrophages and dendritic cells, exhibit the broadest repertoire of TLRs, but TLR expression is not restricted to these cells. In the gastrointestinal tract mucosa, TLRs are constitutively or inducibly expressed by an array of cell types, including the four principal intestinal epithelial cell lineages, subepithelial myofibroblasts, as well as antigen-presenting cells in the lamina propria. TLR expression is influenced by the level of activation of the cell and/or surrounding tissue. In healthy intestinal mucosa, TLR expression is generally low, minimizing recognition of the omnipresent microflora. However, in the diseased intestinal mucosa, TLR expression may be increased, maximizing immune responsiveness to PAMPs.

TLRs are present in extracellular and intracellular forms (**Figure 18.1**). The TLRs that bind to surface bacterial antigens (TLR1, 2, 4, 5, and 6) are expressed on the outer membrane of the cell, whereas TLRs that bind to intracellular viral and bacterial structures (TLR3, 7, 8, and 9) are expressed on the endosomal membrane within the cell. Endosomes are phagolysosomes that degrade pathogens and their products. Another family of PRRs, the NLRs, exists exclusively in the cytosol and is discussed next.

18.2 NOD1 and NOD2 are NLR-family members that recognize peptidoglycan motifs

NLRs comprise a second group of evolutionarily conserved host defense PRRs encoded in both animal and plant genomes. NLRs act as intracellular surveillance molecules, sensing microbial PAMPs and endogenous damage-associated molecular patterns that enter the cytoplasmic compartment. NLR orthologs were first discovered in plants and are called resistance (R) proteins because they protect against fungal, viral, parasitic, and insect pathogens.

Typically, NLRs contain three distinct functional domains: (1) a structurally variable amino-terminal effector domain that comprises caspase recruitment domains (CARDs), a pyrin domain (PYR), and a baculovirus-inhibitor-of-apoptosis repeat domain (BIR); (2) a central nucleotide-binding domain; and (3) a leucine-rich repeat ligand recognition domain at the carboxy-terminal end (**Figure 18.3**). The NLR family is large and encompasses more than 23 human and 34 murine cytosolic members. NLRs are divided into five subfamilies on the basis of the domain at the amino terminus.

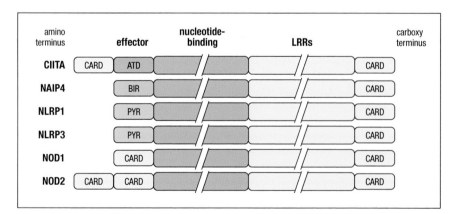

Figure 18.3 NOD-like receptor structure. NLRs are cytoplasmic proteins with conserved tripartite structures composed of an amino-terminal effector domain, a central nucleotide-binding domain, and a carboxy-terminal leucine-rich repeat (LRR) domain. The amino-terminal domain may contain caspase recruitment domains (CARDs), as in NOD1 and NOD2; a pyrin domain (PYR), as in NLRP1 and NLRP2; or a baculovirus-inhibitor-of-apoptosis repeat domain (BIR), as in NAIP4.

Certain NLRs play complementary and nonredundant roles to TLRs in the innate immune detection of PAMPs. The participation of NLRs in host defense against pathogens that invade the cell, and in the regulation of NF-κB and mitogen-activated protein kinase (MAPK) signaling and cell death, indicates that NLRs are important in the pathogenesis of a variety of inflammatory human diseases. Nucleotide-binding oligomerization domains 1 and 2 (NOD1 and NOD2) are the best-characterized NLR members in the intestinal mucosa.

NOD1 and NOD2 were first discovered as mammalian members of the Ced4/Apaf-1 family of apoptosis regulators. Structurally, NOD2 shares significant homology with NOD1 but contains two, instead of one, CARDs at its amino terminus. The CARD recruits caspases, a family of intracellular proteases that induce apoptotic cascades. NOD1 is encoded by the *CARD4* gene and NOD2 by the *CARD15* gene. Both NOD proteins recognize peptidoglycan fragments released from the cell wall of bacteria: NOD1 senses a gram-negative peptidoglycan derivative, γ-D-glutamyl-meso-diaminopimelic acid (iE-DAP), whereas NOD2 senses muramyl dipeptide (MDP), a minimal bioactive motif of peptidoglycan from both gram-negative and gram-positive bacteria. Thus, NOD2 acts as a general sensor of bacterial infection, whereas sensing via NOD1 is restricted to gram-negative microorganisms. The ligands that activate NOD1 and NOD2 probably enter the cell by endocytosis through clathrin-coated pits. After enzymatic processing in endosomes, the peptide ligands enter the cytosol. NOD1 and NOD2 are implicated in the intracellular recognition of pathogens, including enteroinvasive *Escherichia coli*, *Shigella flexneri*, *Campylobacter jejuni*, *Helicobacter pylori*, and *Listeria monocytogenes*.

NOD2 is constitutively or inducibly expressed in monocytes, macrophages, T and B cells, dendritic cells, epithelial cells, and Paneth cells. In contrast, NOD1 is ubiquitously expressed in many tissues and cells. NOD2 protein has been detected throughout the cytoplasm of cells, but redistribution to the cell membrane is required for the induction of downstream effects after ligand recognition. NOD1 expression is regulated by interferon-γ (IFN-γ), and NOD2 expression by LPS, tumor necrosis factor-α (TNF-α), and IFN-γ. On activation, NOD1 and NOD2 induce the downstream production and release of chemokines and pro-inflammatory cytokines, such as TNF-α, IL-1β, and IL-6. NOD2 activation critically induces autophagy and mediates bacterial clearance. Both NOD1 and NOD2 interact with the inhibitor of apoptosis proteins cIAP1 and cIAP2.

18.3 Inflammasomes may be involved in gut homeostasis and inflammation

In recent years, there has been considerable interest in the NOD-LRR-PYR (NLRP) family of molecules in gut homeostasis, immunity to infection in the gut, inflammation in the gut, and control of the microbiota. Various NLRP family members have been shown to be involved in recognition of bacteria such *Clostridium difficile, Helicobacter pylori, Citrobacter rodentium, Salmonella typhimurium, Listeria monocytogenes*, and *Yersinia enterocolitica*; the protozoan *Entamoeba histolytica*; and encephalomyocarditis virus that infects via the oral route. The range of ligands is extensive, ranging from dsDNA, toxins, MDP, PAMPS, and DAMPS, all thought to be recognized by the LRR regions of the NLRPs. Ligand binding in the canonical pathway results in the formation of the inflammasome through interaction of the pyrin domain and the adaptor protein apoptosis-associated speck-like protein containing a CARD (ASC), which recruits pro-caspase 1 to the inflammasome, which is then activated and cleaves IL-1β and IL-18 into their active forms. NLRP family members are highly expressed in gut epithelial cells and immune cells.

Studies on NLRP6 predominate in the published literature. NLRP6 null mice develop more severe inflammation in DSS colitis, as do mice deficient in ASC. It was claimed that the altered microbiota seen in these null mice could transfer increased susceptibility to colitis to wild-type mice. NLRP null mice are also more susceptible to *C. rodentium* infection, attributable to changes in colonic goblet cells. Subsequent studies, however, cast doubt on some of these results. NRLP6 null mice and their control littermates (rather than co-housed mice) show no change in the microbiota or susceptibility to DSS colitis. This does not appear to be the case with NLRP12 null mice, since null littermates have a less diverse microbiota than their wild-type littermates.

The key point, however, in terms of human gut inflammation is that despite the somewhat confusing literature on the role of the inflammasome on the gut microbiota and susceptibility to colitis, the end point of the pathway is the production of active IL-1β and IL-18, neither of which have tested in the clinic for inflammatory bowel disease (IBD) using either their receptor antagonists, binding proteins, or therapeutic antibodies.

18.4 TLR and NOD signaling pathways converge on downstream NF-κB and mitogen-activated protein kinase

The recognition of PAMPs by PRRs signals danger to host innate cells. Once PAMPs have been recognized, PRRs initiate intracellular signal transduction cascades, resulting in the activation of transcription factors (**Figure 18.4**). The transcription factor NF-κB is a major common end result of TLR and NOD signaling pathways in innate immune cells.

Individual TLRs differentially activate distinct signaling events through adaptor proteins, resulting in specific immune responses. The "classical" and first identified pathway involves the recruitment of the adaptor molecule MyD88 via the conserved TIR domain, and activation of the serine-threonine kinases of the IRAK family—leading ultimately to the degradation of IκB and the translocation of NF-κB into the nucleus. Importantly, all TLRs except TLR3 signal through the adaptor protein MyD88, and TLR4 uses both MyD88-dependent and MyD88-independent pathways. Subsequent transcriptional activation of TLR target genes that encode pro- and anti-inflammatory cytokines, chemokines, effector molecules, and type I interferons initiates the activation of antigen-specific and nonspecific adaptive immune responses. The resultant downstream effects promote mucosal homeostasis.

On ligand recognition, the NODs undergo conformational change and self-oligomerization. Both NOD1 and NOD2 enter into CARD-CARD interactions with the serine-threonine kinase RIP2, which leads to the activation of MAPKs (see **Figure 18.4**). NOD1 signals through TRAF2/5, whereas NOD2 signals through TRAF6. NOD1 and NOD2 signaling pathways converge downstream on the transcription factor NF-κB to induce the expression of many important

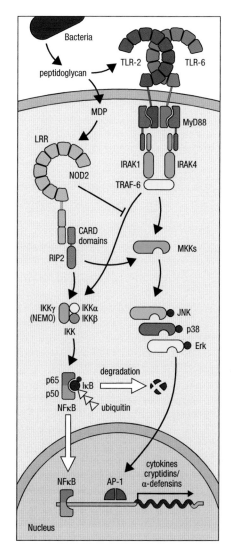

Figure 18.4 Schematic representation of NOD and toll-like receptor (TLR) activation pathways. Bacterial peptidoglycan activates intracellular NOD2 via the interaction of muramyl dipeptide (MDP) with the leucine-rich repeat (LRR) in the carboxy-terminal domain. This interaction causes the oligomerization of NOD2 and recruitment of the kinase RIP2 through CARD-CARD interactions. RIP2 activates the IKK complex through IKK γ/NEMO, resulting in phosphorylation of IκB and its degradation through ubiquitination and release of the sequestered NF-κB for translocation into the nucleus and initiation of gene transcription. Peptidoglycan also interacts with surface TLR2/6 to induce the MyD88 signal pathway, leading to activation of TRAF6, a key signal protein. TRAF6 can activate the IKK complex, leading to NF-κB activation, and/or activate MAP kinases, including JNK, p38, and ERK, which translocate into the nucleus to induce AP-1 transcription factors. NF-κB and AP-1 induce gene transcription for an array of pro-inflammatory cytokines and host defense proteins. MKKs, mitogen-activated protein kinase kinases. (Reprinted from B. Kelsall, *Nature Med.* 2005, 11:383–384. With permission from Macmillan Publishers Ltd.)

mediators of innate immune host defense, including cytokines, chemokines, and antimicrobial genes.

Function of pattern recognition molecules in healthy mucosa

The normal gut flora has an essential role in promoting local immune balance, called homeostasis, in the intestinal mucosa. PRRs contribute to the maintenance of this balance by promoting the avoidance of excessive and deleterious activation of host cell inflammatory responses.

18.5 Negative regulation prevents prolonged and detrimental TLR/NOD signaling

In healthy intestinal mucosa, tolerance is an essential defense mechanism for the maintenance of hyporesponsiveness to commensal microbiota and their products. However, when a pathogen disrupts or traverses the epithelial barrier and enters the underlying lamina propria, mechanisms that promote tolerance are switched off, and regulators that promote PRR signaling to induce mucosal immune defenses are switched on (Figure 18.5). To this end, control mechanisms in the healthy intestinal mucosa regulate TLR expression, localization, and signaling cascades. In the healthy intestine, TLR2 and TLR4 are expressed at low levels on the apical surface of intestinal epithelial cells, limiting microbe recognition but maintaining a basal state of activation toward luminal commensals. The differentiation and polarity of epithelial cells influence TLR localization and ligand responsiveness. In differentiated intestinal epithelial cells, for example, TLR2 and TLR4 are present at the apical surface; however, in undifferentiated intestinal

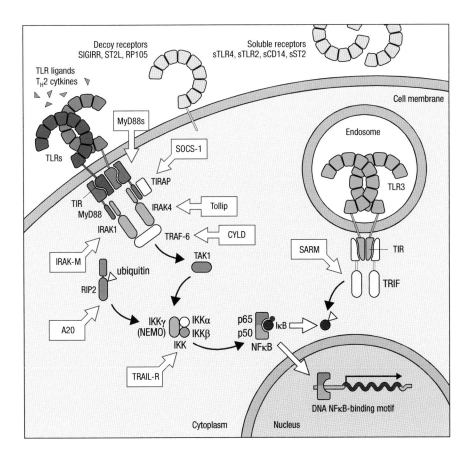

Figure 18.5 Toll-like receptor (TLR) signaling and inhibitors. TLR signaling is tightly regulated by signal proteins and inhibitors. TLR recognition of PAMPs on microorganisms initiates myeloid differentiating factor 88 (MyD88)-dependent or MyD88-independent cascades that culminate in the activation of NF-κB and MAPK, transition factors that initiate pro-inflammatory and host defense response gene transcription. Regulation is achieved through inhibitors of the various signaling steps. The main inhibitors include Tollip, which inhibits IRAK4 and IRAK1; IRAK-M, a member of the IL-1 receptor-associated kinases in monocytes that blocks the formation of IRAK1:TRAF6 complexes; SOCS1, which inhibits TIRAP; CYLD, an enzyme that interferes with TLR2 signaling through inhibition of TRAF6 and TRAF7; A20, a cytoplasmic zinc-finger protein that blocks receptor interacting protein (RIP) activation; and SARM, which blocks TRIF to inhibit TLR3 and TLR4 signaling. T$_H$2, helper T-cell type 2. (Adapted from O. Shibolet and D. Podolsky: *Am. J. Physiol. Gastrointest. Liver Physiol.* 2007, 292:G1469–G1473. With permission from the American Physiological Society.)

epithelial cells, both TLRs are expressed in cytoplasmic compartments. TLR5 is expressed exclusively on the basolateral surface of differentiated colonocytes and is thus positioned to detect translocated flagellin when the epithelial barrier has been injured. TLR9 is expressed on both the apical and basolateral membranes of intestinal epithelial cells, and signaling cascades seem to vary in a pole-dependent manner.

Other mechanisms that contribute to the hyporesponsiveness of intestinal mucosa include intracellular molecules that block TLR-induced signaling cascades that activate NF-κB and MAPKs. Examples of such molecules include Tollip, which inhibits IRAK activation, and A20, which downregulates pro-inflammatory NF-κB. The main inhibitor molecules are depicted in **Figure 18.5**. In contrast to TLRs, expression of NOD1 or NOD2 is not downregulated in innate cells in healthy intestinal mucosa. Thus, normal cells remain responsive to invasive bacteria via NOD1 or NOD2. However, inhibitors of NOD1- and NOD2-induced NF-κB activation that disrupt receptor oligomerization or block RIP2, regulating NOD-induced immune responses, have been identified. These NOD1 and NOD2 inhibitors include NOD2-S, an alternatively transcribed single-CARD-domain splice variant of NOD2; centaurin β1, a GTPase-activating protein; and erbin, a cell polarity protein.

TLRs and NODs may interact with each other through cross-tolerizing feedback loops. Thus, continuous exposure to TLR ligands can suppress exaggerated inflammatory responses after exposure to NOD ligands, and vice versa. IRAK has been identified as a central negative regulator of cross-tolerization between a variety of microbial components of both PRR families. In contrast, pro-inflammatory cytokines, including IFN-γ and TNF-α, may modulate PRR expression and downstream signaling to promote mucosal inflammation, as in IBDs.

18.6 TLR function is involved in the maintenance of mucosal barrier integrity

The interaction between commensal microbes and the intestinal epithelium induces TLR signaling that promotes barrier preservation, cell survival, and restitution. Aberrant TLR signaling may alter the response to commensal microorganisms, facilitating injury and inflammatory responses. For example, mice in which the TLR2, 4, 5, or 9 or MyD88 genes have been deleted exhibit delayed or diminished tissue repair responses during acute dextran sulfate sodium (DSS)-induced colonic inflammation. Conversely, treatment with TLR2, 4, 5, and 9 ligands prevents or delays the onset of acute DSS-induced colitis. In this connection, treatment with the TLR2 ligand Pam3-Cys-Ser-Lys4 has been shown to protect tight-junction-associated barrier integrity and decrease intestinal permeability, thereby reducing DSS-induced colonic inflammation during the recovery phase. TLR2 also controls terminal goblet-cell differentiation by selectively regulating trefoil factor 3 (TFF3) expression in the colon, promoting antiapoptotic protection of the intestinal mucosa. TLR-mediated epithelial cell proliferation in gut mucosa seems to be site specific and developmentally regulated, possibly as a result of differences in the commensal microbiota in different regions of the gastrointestinal tract and during different stages of development.

Activation of innate immune signaling in the intestinal epithelium by commensal bacteria influences adaptive immunity through cross talk between epithelial cells and immune cells in the underlying lamina propria. Resident or infiltrating dendritic cells, T cells, and B cells amplify antigen recognition and sorting, generate antibody production, and enhance host-protective inflammatory responses. In mice, bacteria-induced TLR signaling by epithelial cells drives subjacent dendritic cells to sample luminal bacteria. TLR-dependent secretion of mediators such as thymic stromal lymphopoietin

(TSLP) and B-cell activating factor belonging to the TNF family (BAFF) may provide an additional link between intestinal epithelial cells and both dendritic cells and B cells, modulating the secretion of mucosa-protective IgA$_2$ in the presence of IL-10. TLR ligation on intestinal epithelial cells also may promote IgG and IgA class switching via *BAFF*. Together, these TLR-dependent mechanisms contribute to bacterial clearance, limit deleterious pro-inflammatory responses, and thus promote mucosal homeostasis.

18.7 NOD2 modulates antimicrobial peptide secretion and bacterial clearance via autophagy

Mucosal cell-derived antimicrobial peptides have a key role in the host response to luminal microbes. One important group of such peptides, the α-defensins, is produced by Paneth cells, which are located in the crypts of Lieberkühn in the small intestine. These positively charged, hydrophobic peptides are induced after PRR-mediated activation of NOD2 in Paneth cells. The released α-defensins interact with negatively charged phospholipids to kill bacterial cells efficiently. Activation of NOD2 in Paneth cells exerts antibacterial activity by limiting bacterial invasion and expansion through α-defensin production. In studies with intestinal epithelial cells expressing a functional NOD2 protein, the *in vitro* bacterial clearance of pathogenic *Salmonella typhimurium* was strongly accelerated, whereas epithelial cells overexpressing a NOD2 mutant (L1007fsinsC) were unable to clear the pathogen. Mice that lack NOD2 or express a dysfunctional NOD2 show a decreased ability to clear intracellular bacteria, such as *Listeria monocytogenes*. A NOD-mediated decrease in bacterial clearance may lead to persistent immune activation, dysregulated cytokine production, and subsequent T-cell activation in the lamina propria.

Emerging evidence indicates that autophagy is induced by innate immune pathways to defend against bacterial invasion and maintain cellular homeostasis. Autophagy involves complex catabolic signaling pathways to sequester cytosolic regions within double-membrane-enclosed compartments, called *autophagosomes*, which fuse with lysosomes for cargo degradation. Autophagy is important in the removal or recycling of aged macromolecules and damaged organelles under stress conditions and in the elimination of pathogens or virulence factors delivered into the cell through type III or IV secretion systems. NOD2 recruits the autophagic regulator ATG16L1 to the plasma membrane at the bacterial entry site. NOD2 then activates the autophagy pathway and directs bacterial trafficking to the autophagosome, which on fusion with the lysosome forms the autophagolysosome. The degraded cargo is loaded onto major histocompatibility complex (MHC) class II molecules and is able to stimulate CD4 T cells. Thus, the autophagic machinery links NOD2 to antigen presentation and subsequent adaptive immune responses.

Genetic alterations in pattern recognition

Genetic variations in TLRs or NODs may influence ligand recognition, mucosal immune tolerance, and commensal microbe composition. Such variations may affect host susceptibility and disease progression in certain pathogenic processes (Table 18.3). For example, NOD2 mutations have been associated with increased susceptibility to ileal Crohn's disease. About 30% of patients of European ancestry have at least one of three NOD2 polymorphisms; in contrast, TLR variants are relatively rare. Although TLR polymorphisms may not predict overall disease risk, polymorphisms may influence phenotype severity in subgroups of patients with IBD.

TABLE 18.3   EFFECTS OF GENETIC DEFECTS IN PATTERN-RECOGNITION RECEPTORS

Defect	Association	Effect
Human genetic defects in NOD2		
NOD2-R702W NOD2-G908R NOD2-L1007fsinsC	Increased susceptibility to Crohn's disease (fibrostenosis/terminal ileal disease)	Loss of antibacterial activity as a result of impaired bacterial recognition and clearance, decreased defensin production and deregulated autophagy, TLR hypersensitivity, unbalanced T-cell activation with shift toward T_H1/T_H17 responses
Human genetic defects in TLR		
TLR2-R753Q	Ulcerative colitis pancolitis	Impairs intestinal epithelial cell restitution and communication
TLR4-D299G	Increased susceptibility to IBD	Interrupts LPS signaling, but phenotype in IBD so far unclear
TLR5-stop	Decreased susceptibility to IBD	Decreases adaptive immune responses to flagellin
Murine defects in IBD-associated genes, secondarily influencing TLR function		
IL-10 $^{-/-}$	In mice. No association	TLR4 limits propagation of colitic effector CD4 T cells, thus ameliorating disease course
NOD2 $^{-/-}$	In mice. No association	TLR4 limits propagation of colitic effector CD4 T cells, thus ameliorating disease course

18.8   NOD2 is a major susceptibility gene for Crohn's disease

In 2001, the first IBD susceptibility locus was found to involve NOD2. In some geographical regions, patients with Crohn's disease had a relatively high frequency of NOD2 variants. The major variants involving the leucine-rich repeat region include a frameshift mutation (L1007fsinsC) and two missense mutations (R702W and G908R), suggesting that a defect in peptidoglycan recognition may be associated with Crohn's disease. The most common mutation, L1007fsinsC, causes a truncated NOD2 protein that lacks the last 33 amino acids. Immune cells carrying such a mutant NOD2 have an impaired capacity to induce NF-κB activation in response to ligand stimulation and fail to produce the anti-inflammatory cytokine IL-10. In addition, the presence of a Crohn's disease–associated NOD2 mutation may impair cross-tolerization by disturbing homeostatic TLR signaling and shift signal outcome toward pro-inflammatory hypersensitivity in antigen-presenting cells. Thus, impaired signaling via a variant of the *NOD2* gene could perpetuate disease by uncontrolled and excessive actions of TLR pathways in normally quiescent cells within the intestinal mucosa.

In Crohn's disease, expression of NOD2 is increased in Paneth cells and macrophages, possibly inducing pro-inflammatory cytokines. The production of cryptidins, homologs of α-defensins, is significantly diminished in mice that lack a *Nod2* gene, increasing susceptibility to infection with bacterial pathogens. Similarly, Crohn's disease–associated NOD2 mutants may represent loss-of-function phenotypes. In this connection, transfection studies in humans have shown that Crohn's disease–associated NOD2 mutants fail to constrain the survival and overgrowth of pathogenic bacteria. In patients with NOD2-mutant Crohn's disease, Paneth-cell-derived expression of the human α-defensins HD5 and HD6 is significantly diminished. Paneth-cell-mediated antimicrobial signaling may also be compromised by dysregulated autophagy during colonic inflammation. Crohn's disease-associated NOD2 mutants fail to recruit the critical autophagy protein ATG16L1 to the plasma membrane during bacterial invasion. Consequently, autophagy is defective on bacterial infection or stimulation with the peptidoglycan fragment muramyl dipeptide, resulting in decreased bacterial killing and defective antigen presentation.

Thus, aberrant bacterial handling and immune priming may act as combined triggers for initiating and perpetuating intestinal inflammation in patients with Crohn's disease with NOD2 mutations. Genetic defects in NOD2 may lead to changes in the composition of the microbiota as a result of defensin deficiency, which may alter homeostatic TLR signaling and promote inflammatory disease. However, NOD2 polymorphisms alone are not sufficient to cause Crohn's disease, supporting the concept that multiple factors contribute to the pathogenesis of this disease.

18.9 TLR polymorphisms may modulate inflammatory bowel disease severity

Cohort studies of the incidence and prevalence of TLR polymorphisms in selected populations with IBD have identified TLR1, 2, and 6 genes as being associated with distinct disease phenotypes. Patients with ulcerative colitis and the polymorphisms TLR1-R80T and TLR2-R753Q seem to have an increased risk for pancolitis. The TLR2-R753Q variant represents a loss-of-function mutation that mediates intestinal epithelial dysfunction *in vitro*. Goblet cells with TLR2-R753Q fail to induce trefoil factor 3, a goblet cell–secreted protein important in mucosal defense, and enterocytes with TLR2-R753Q do not communicate properly with each other. Thus, TLR2-R753Q is associated with impaired epithelial cell restitution during wound healing and may contribute to more extensive disease in a subset of patients with IBD.

In active IBD, variant alleles in the *TLR4* gene could induce functional dysregulation of the LPS receptor. Two common variants in the human *TLR4* gene, D299G and T399I, occur in up to 20% of Caucasians and have been associated with increased susceptibility to IBD. Although the D299G variant has been shown to interrupt TLR4-mediated LPS signaling *in vitro*, the functional consequence of the mutation in the mucosa in IBD has not been determined.

Other TLR mutations have been identified in patients with IBD. A recently detected dominant-negative TLR5 polymorphism (TLR5-stop), which leads to a 75% loss of TLR5 function, decreases adaptive immune responses to flagellin and, in Jewish cohorts, protects against the development of Crohn's disease. A complete loss of TLR5 in mice results in the development of spontaneous colitis via aberrant TLR4 signaling in response to changes in the commensal composition.

18.10 C-type lectin receptors also contribute to microbe recognition

C-type lectin receptors (CLRs) are membrane proteins expressed on myeloid cells that recognize microbe carbohydrates, as well as self carbohydrates. CLRs also recognize many lipids and proteins. Although the role of CLRs in recognizing fungi and shaping the host response to fungal pathogens is well known, accumulating evidence indicates that CLRs also recognize bacterial, viral, and parasitic pathogens. Ligand recognition by CLRs triggers endocytic, phagocytic, inflammasome, inflammatory mediator, or anti-inflammatory activities, depending on the receptor and the downstream signaling pathway. Thus, CLRs expand the repertoire of innate receptors by functioning as PRRs for PAMPs, host damage-associated molecular patterns (DAMPs) and tumor-associated molecular patterns (TAMPs).

Structurally, CLRs contain a short cytoplasmic tail, a type II transmembrane protein, and one or more carbohydrate recognition domains (CRDs). The CRD is a structural unit of two protein loops stabilized by two disulfide bridges with the second loop generally containing the conserved residue motifs that determine the carbohydrate specificity. CLRs include more than 1000 proteins divided into 17 groups (I–XVII) based on structure and domain organization.

Similar to TLRs, intracellular signaling pathways are shared by multiple CLRs. CLR signaling pathways, however, have not been as well-defined as TLR pathways. The best-described CLR pathway involves DC-associated C-type lectin-1 and -2 (Dectin-1 and Dectin-2) and macrophage-inducible C-type lectin (Mincle), which recognize fungi. These activation CLRs trigger recruitment and phosphorylation of spleen tyrosine kinase (Syk) via immunoreceptor tyrosine-based activation motif (ITAM) that in turn binds CARD9, BCL10, and MALT1 into a complex that activates NF-κB, which induces an array of cell responses, including inflammatory mediator production, and T_H17 cell responses. Syk activation also initiates the MAPK pathway and the JNK, p38 and Erk kinase cascade, which regulate cell cycle progression and, consequently, cell proliferation, differentiation, and transformation, as well as inflammatory responses. In contrast to Syk-coupled activation signaling, the SHP-coupled inhibitory CLRs, e.g., DC inhibitory receptor (DCIR), trigger the immunoreceptor tyrosine-based inhibition motif (ITIM). Other CLRs, e.g., DC-specific intercellular adhesion molecule-3-grabbing non-integrin (DC-SIGN), mannose receptor (MR), and CD-205, do not have ITAM or ITIM domains and trigger signaling through Raf-1 or yet to be defined pathways.

Among the most studied CLRs are Dectin-1 (Group V, Ca^{++}-independent) and Dectin-2 (Group II, Ca^{++}-dependent). Dectin-1 plays a key role in antifungal immunity by initiating signals that lead to phagocytosis and killing of fungi. Expressed on macrophages, monocytes, DCs, neutrophils, microglia, and eosinophils, Dectin-1 binds β-glycans, a major cell-wall component of nearly all fungi. Deficiencies in Dectin-1 or CARD9 in mice and humans result in increased susceptibility to fungal infections. In mice, for example, Dectin-1 deficiency causes increased mortality in response to infection by fungal pathogens such as *Candida albicans*, *Aspergillus fumigatus*, and *Coccoidiodes podasii*. Some patients with familial chronic mucocutaneous candidiasis have a nonsense mutation in the Dectin-1 gene. Dectin-1 also appears to contribute to mucosal myeloid cell sensing of enteric commensal fungi, thereby contributing to the maintenance of "mucosal homeostasis." For example, mice with Dectin-1 deficiency (*Clec7a⁻/⁻*) show an increased mucosal inflammatory response to commensal fungi in dextran sodium sulfate-induced colitis. Further, a single nucleotide polymorphism in the human Dectin-1 *CLEC7A* gene in patients with ulcerative colitis is associated with medically refractory disease.

Dectin-2 is expressed by DCs, as well as macrophages and monocytes. The Dectin-2 receptor has a Glu-Pro-Asn motif in the extracellular CRD that is the putative binding site for mannose. The high mannose content of an array of fungal and parasitic pathogens makes such microbes targets for Dectin-2-bearing myeloid cells. Dectin-2-deficient mice have significantly reduced survival after systemic infection with *C. albicans*, which abundantly express α-mannans in their outer cell wall. Notably, α-mannan recognition by Dectin-2 induces T_H17 polarization, and *IL-17⁻/⁻* mice have reduced survival in response to systemic *C. albicans* infection, consistent with Dectin-2-associated IL-17 production in host defense against fungi. Although the link between T_H17 cytokines and antifungal immunity also has been demonstrated in humans, the involvement of Dectin-2 in mediating this association has not been fully explored.

An important feature of CLRs is their redundancy. This important feature of CLRs is best reflected in the detection of *Mycobacteria tuberculosis* by Dectin-1, Mincle, MR, and DC-SIGN. Receptor deletion models show that whereas loss of one of these receptors may not impact survival in response to *M. tuberculosis* infection, together they play a critical role in the detection and control of this life-threatening bacterial pathogen. Characteristic of PRRs, a single CLR may recognize components on multiple pathogens. For example, DC-SIGN plays a key role in the recognition and host cell entry of HIV-1 but also contributes to the recognition of cytomegalovirus, as well as dengue, ebola, hepatitis C, West Nile, and measles viruses. Not limited to virus recognition, DC-SIGN also recognizes glycan motifs in the soluble egg antigen of multiple schistosome species. Beside parasitic worm recognition, the CLRs contribute to the recognition of certain protozoa, evidenced by the critical role played by DCIR in the development of experimental cerebral malaria caused by *Plasmodium berghei.*

18.11 Retinoic acid gene-I-like receptors sense viral RNA, and cyclic GMP-AMP synthase and stimulator of interferon genes recognize bacterial and viral DNA and host DNA

Nucleic acid sensing in the cytosol is particularly important for the detection of viral and bacterial pathogens that enter this compartment. The downstream pathways are also important for immunoregulation and responding to host nucleic acids that enter the cytosol from dying cells or accumulate during normal cellular processing. Pathways therefore exist for the detection of RNA and DNA species, which may intersect with each other, depending on the initial stimulus.

RNA sensing occurs through the activity of retinoic acid gene-I (RIG-I)-like receptors (RLR). These classically serve to sense the presence of RNA species generated during viral infection and can detect both single (ss) and double-stranded (ds) RNA. Three RLRs have been described: RIG-I, melanoma differentiation associated gene 5 (MDA5), and laboratory of genetics and physiology 2 (LGP2). RIG-I and MDA5 consist of two amino-terminal caspase-associated recruitment domains (CARDs), which signal the presence of RNA species that bind a DEAD box (*N*-terminal) helicase/ATPase domain that is normally repressed by a carboxy-terminal regulatory domain (CTD). RIG-I detects short dsRNA species that possess 5′ end di- and tri-phosphorylated sequences generated by oligoadenylate synthetase (OAS) and RNaseL processing of RNA viruses or Pol III generated dsRNA species from dsDNA viruses. Upon RNA binding and ubiquitination by Riplet and TRIM25, RIG-I binds mitochondrial antiviral-signaling protein (MAVS) to activate two pathways, which together converge on the production of myeloid interferons: TANK-binding kinase-1 (TBK1) and IκB kinase epsilon (IKKe) phosphorylation and activation of IRF3 and IRF7 or induction of NF-κB on repression of IκB. The interferons induced are secreted and activate IFN receptor signaling and the induction of interferon-stimulated genes through the activation of STAT1, STAT2, and IRF9. MDA-5 senses long dsRNA species in a ubiquitin-independent pathway. The third member of the RLR family is less well understood; it binds RNA species but lacks a CARD domain.

RLRs likely have evolved for the sensing of and resistance to distinct viral pathogens. As viral infections typically invade through the mucosal surfaces, the expression of these pattern recognition receptors is extremely important for host resistance. For example, although both RIG-I and MDA5 are responsive to *Reoviridae*, only RIG-I senses the presence of influenza or norovirus.

TABLE 18.4 MAJOR C-TYPE LECTIN RECEPTORS

Microbe	Host receptors	Activation motif	Function
Fungi Bacteria	**Dectin-1** (Clec7a)	**Syk**	Phagocytosis killing
Fungi Bacteria	**Dectin-2** (Clecsf8)	**Syk**	Phagocytosis killing
Fungi Bacteria	**Mincle** (Clec4e)	**Syk**	Phagocytosis killing
Fungi Helminths Bacteria Viruses	**Mannose Receptor** (CD206)	?	Phagocytosis endocytosis
Viruses Bacteria Fungi	**DC-SIGN** (CD209)	**Raf-1**	Phagocytosis
Apoptotic and necrotic cells	**DEC-205** (CD205)	?	Endocytosis cross-presentation

Microbes and products of apoptosis and necrosis contain carbohydrates recognized by the indicated CLRs. These CLRs signal through spleen tyrosine kinase (Syk), a key immunoreceptor tyrosine activation motif (ITAM), and yet to be defined ITAMs that trigger host cell protective responses

DNA sensing in the cytosol mainly occurs through the activity of cyclin GMP-AMP (cGAMP) synthase (cGAS), which activates stimulator of interferon genes (STING) associated with the endoplasmic reticulum, which in turn stimulates the TBK1 and IKKe pathway to activate IRF3 and IRF7 that converge on the induction of myeloid interferons. In this and other ways, DNA and RNA sensing of pathogens and their signaling pathways intersect (Table 18.4).

These pathways are important for the host responses to viral infections, and thorough understanding of their pathways can be useful in the development of vaccines and other antiviral strategies. Moreover, in addition to the RLR/MAVS and cGAS/STING pathways sensing host nucleic acids, these circuits are important for immunoregulation and, when overexpressed, to disease phenotypes. Regarding the former pathway, RIG-I deficiency has been associated with colitis, consistent with the downregulation of RIG-I in human inflammatory bowel disease and the role of RLRs in the induction of regulatory T cells. The abnormal accumulation of nucleic acids in certain autoimmune diseases can activate myeloid interferons in a pathologic manner, such as in systemic lupus erythematosus or in the intestinal mucosa undergoing ER stress. For example, during ER stress, activation of the RNAase activity of inositol requiring enzyme 1 (IRE1) generates RNA species potentially capable of activating RLRs.

SUMMARY

In the intestinal mucosa, cells of the innate immune system express an array of germline-encoded PRRs that detect highly conserved microbial structures. Composed of surface TLRs plus intracellular NLRs and some intracellular TLRs, PRRs participate in host defense and tissue repair responses. Together, TLRs and NLRs ensure mucosal homeostasis in healthy mucosa, and rapid microbe elimination in infected mucosa. Conversely, defects in PRR genes lead to dysfunctional microbe recognition and may contribute to mucosal inflammation in subpopulations of patients with Crohn's disease.

FURTHER READING

Beutler, B.: Microbe sensing, positive feedback loops, and the pathogenesis of inflammatory diseases. *Immunol. Rev.* 2009, 227:248–263.

Cario, E.: Toll-like receptors in inflammatory bowel diseases: A decade later. *Inflamm. Bowel Dis.* 2010, 16:1583–1597.

Franchi, L., Warner, N., Viani K. et al.: Function of NOD-like receptors in microbial recognition and host defense. *Immunol. Rev.* 2009, 227:106–128.

Fukata, M., Vamadevan, A.S., and Abreu, M.T.: Toll-like receptors (TLRs) and NOD-like receptors (NLRs) in inflammatory disorders. *Semin. Immunol.* 2009, 21:242–253.

Hoving, J.C., Wilson, G.J., and Brown, G.D.: Signaling C-type lectin receptors, microbial recognition and immunity. *Cell. Microbiol.* 2014, 16:185–194.

Kawai, T., and Akira, S.: The role of pattern-recognition receptors in innate immunity: Update on Toll-like receptors. *Nat. Immunol.* 2010, 11:373–384.

Mamantopoulos, M., Ronchi, F., Van Hauwermeiren, F. et al.: Nlrp6- and ASC-dependent inflammasomes do not shape the commensal gut microbiota composition. *Immunity* 2017, 47:339–348.

O'Neill, L.A.: The interleukin-1 receptor/Toll-like receptor superfamily: 10 years of progress. *Immunol. Rev.* 2008, 226:10–18.

Ori, D., Murase, M., and Kawai, T.: Cytosolic nucleic acid sensors and innate immune regulation. *Int. Rev. Immunol.* 2017, 36:74–88.

Palm, N.W., and Medzhitov, R.: Pattern recognition receptors and control of adaptive immunity. *Immunol. Rev.* 2009, 227:221–233.

Philpott, D.J., and Girardin, S.E.: NOD-like receptors: Sentinels at host membranes. *Curr. Opin. Immunol.* 2010, 22:428–434.

Place, D.E., and Kanneganti, T.D. Recent advances in inflammasome biology. *Curr. Opin. Immunol.* 2018 50:32–38.

Ting, J.P.Y., Duncan, J.A., and Lei, Y.: How the noninflammasome NLRs function in the innate immune system. *Science* 2010, 327:286–290.

Vijay-Kumar, M., and Gewirtz, A.T.: Flagellin: key target of mucosal innate immunity. *Mucosal Immunol.* 2009, 2:197–205.

Zmora, N., Levy, M., Pevsner-Fischer, M., Elinav, E.: Inflammasomes and intestinal inflammation. *Mucosal Immunol.* 2017, 10:865–883.

Commensal microbiota and its relationship to homeostasis and disease

19

JONATHAN BRAUN, ELAINE Y. HSIAO, AND NICHOLAS POWELL

Vertebrates and bacteria have coevolved in intimate contact for more than 150 million years. This has created evolutionary pressures on both sides of the divide: on the one hand driving improvement and refinement of defense mechanisms employed by multicellular organisms to counter infection; and on the other hand prompting bacteria to develop more sophisticated evasive maneuvers to subvert the immune system. Selection pressures have also encouraged mutually beneficial relationships, and some bacterial species have adapted to coexist in harmony with the host, particularly at the epithelial barrier surfaces. Despite dwelling in such close proximity, these colonizing microbial communities, or microbiota, seldom cause disease in immunologically competent individuals and instead often confer survival advantage to the host. For instance, gut-residing bacteria liberate nutrients from otherwise indigestible dietary polysaccharides, synthesize essential vitamins, detoxify xenobiotics, and limit the growth of potential pathogens by competing for space and nutrients (Figure 19.1). For this symbiosis to flourish, the host must curtail its potent immune responses that might otherwise be directed against the microbiota. However, the barrier surfaces are vast and include the skin and mouth, and the respiratory, gastrointestinal, and genital tracts, providing an extensive target and potential portal of entry for pathogens. So at the same time as tolerating commensals at the barrier surfaces, the host immune system must also remain alert and poised to repel infections. The gravity of this task should not be underestimated, as worldwide the most common causes of human mortality are infections caused by organisms occupying or gaining entry to the body through the mucosal epithelia, including respiratory tract infections, intestinal infections, and human immunodeficiency virus-1 (HIV-1).

Mammals therefore provide a scaffold upon which numerous microbial communities are assembled, and subsequently provide residence to thousands of individual species. Almost every environmentally exposed surface of our bodies teems with microbes. These populations include bacteria, archaea, fungi, viruses, protozoans, and in some cases even multicellular helminths. These polymicrobial communities play important roles in the architecture and function of the tissues they inhabit. The simple but profound concept is that the properties of our mucosal surfaces involve a true integration with their resident microbiota to create what has been called a "supraorganism."

PRINCIPLES AND DEFINITIONS OF COMMENSAL MICROBIOTA

As a starting point, the important definitions and principles of such microbial communities need to be explained. Three definitions and principles will be helpful. *Microbiome* refers to the members of a microbial community found in

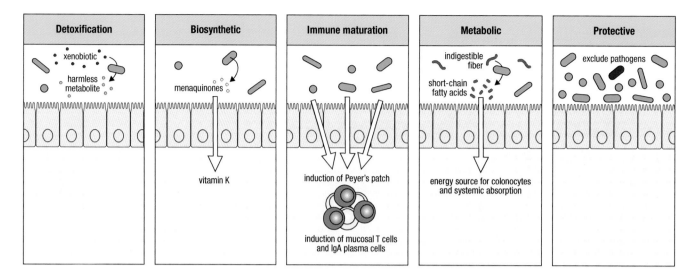

Figure 19.1 The intestinal microbiota confers many health benefits to the host. The intestinal microbiota performs a complex variety of functions that benefit the host. During detoxification, ingested xenobiotics are metabolized into inert by-products that can pass harmlessly in the feces. Biosynthetic functions involve the synthesis of biomolecules that are absorbed and utilized by the host. For example, microbial enzymes are important for the biosynthesis of vitamins such as vitamin K—a critical cofactor involved in hepatic biosynthesis of several key clotting factors. Immune maturation requires an intestinal microbiota: evidence from germ-free mice convincingly shows that in the absence of the intestinal microbiota, there is impaired mucosal and systemic immune maturation (see **Table 19.1**). For example, the microbiota facilitates the formation of Peyer's patches and isolated lymphoid follicles and increases the recruitment of lymphocytes to the lamina propria. Metabolic functions involve the fermentation of indigestible dietary fiber into energy-rich metabolites, such as short-chain fatty acids, which can be utilized by the host. These metabolites supply energy to colonocytes and can be utilized systemically. Protective functions are performed by excluding potential pathogens, both by competing for limited nutrients and by physically excluding them from the epithelial barrier surface, thus protecting the underlying mucosa.

a particular anatomical habitat. The microbiome is defined by a combination of traditional microbiology that is dependent on the ability to culture specific microorganisms under defined conditions and, more recently, the application of unbiased approaches based on the genetic composition of the specific microbe. These unbiased approaches largely rely on next-generation sequencing of conserved regions within the 16S rRNA subunit of microbes using region-specific primers and high-throughput DNA sequencing. Such approaches are changing the definitions of classical taxonomy such that an alternative definition of a species is as a phylotype as defined by 99% identity at the level of the conserved sequences within the 16S ribosomal subunit. Finally, it is important to note that the composition of a microbiome may change over time, even abruptly, due to physiologic, metabolic, or disease states, and be markedly different in different parts of the world. However, the composition of each anatomic microbiome is characteristic, reflecting the specialist microorganisms and their functions devoted to this habitat.

A second important definition is that of the *metagenome*, which refers to the aggregate of genes found in the microbiome. The metagenome is also defined by the application of high-throughput sequencing. However, in this case, rather than focusing on specific conserved regions of the 16S rRNA structure, next-generation sequencing is performed on random DNA fragments that are derived from the entire microbial community. Using bioinformatic approaches, a metagenome is assembled from the "shotgunned" sequenced DNA fragments that provide, in contrast to the high-throughput biased sequences of the microbiome that are focused on 16S rRNA, an unbiased assessment of the genetic ensemble of the community that can then be organized into functional metabolic repertoires. Distinct sets of genes are found in each microorganism, such that the composite of genes in a microbiome is huge, complex, and overlapping. However, the concept of an "aggregate" genome for

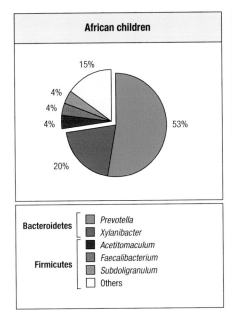

African children

15%
4%
4%
4%
53%
20%

Bacteroidetes
 Prevotella
 Xylanibacter

Firmicutes
 Acetitomaculum
 Faecalibacterium
 Subdoligranulum
 Others

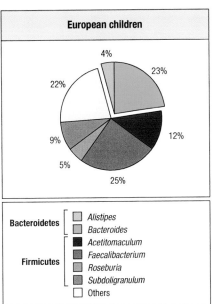

European children

4%
23%
22%
12%
9%
5%
25%

Bacteroidetes
 Alistipes
 Bacteroides

Firmicutes
 Acetitomaculum
 Faecalibacterium
 Roseburia
 Subdoligranulum
 Others

Figure 19.2 Culture-independent 16S ribosomal RNA gene sequencing is useful for mapping the composition of the intestinal microbiota. 16S ribosomal RNA was sequenced to disclose differences in the microbial composition in Italian children consuming a typical Western-style diet in comparison with children from rural villages in Burkina Faso who consume high-fiber diets resembling ancient human diets. The intestinal microbiome of children from Burkina Faso was enriched in *Bacteroidetes* and organisms from the genera *Prevotella* and *Xylanibacter* that are known to be a rich source of enzymes capable of degrading cellulose and xylose, consistent with the possibility that our intestinal microbiome has coevolved with our changing dietary patterns. (Adapted from De Filippo, C. et al., *Proc. Natl Acad. Sci. USA.* 2010, 107:14691–14696. With permission from National Academy of Sciences.)

the microbial community is important, because each microbiome operates in an integrated fashion, akin to the function of other multicellular organs of the body. A spectacular example of the use of this technique to assess the effect of environment comes from analysis of the microbiome in children in Italy and children in Burkina-Faso (Figure 19.2). In this case, Firmicutes make up 51% of the microbial community in Italy, but only 12% in Burkina Faso.

Interindividual variation is another important feature of the microbiome (see Figure 19.2). Because individuals are born germ-free, the composition is initially determined by microorganisms acquired in the local environment. As the intimate caretaker in most species and cultures, the mother's microbiome is the predominant microbial source for a newborn. Due to founder advantage, the composition of the maternally derived microbiota is preserved as a major fingerprint in the microbiome of her progeny even after they have reached adulthood. The genetics of an individual can also affect his or her microbiome. A substantial portion of the human genome is devoted to host functions that affect host-microbe interaction: host barrier properties, antimicrobial defense, metabolic stress, and inflammation. Genes that are involved in such "gardening" of the microbiome are beginning to emerge. Indeed, global changes in the intestinal microbial community have been confirmed in mice with null mutations in molecules such as CD8 (T-cell coreceptor), perforin (lymphocyte cytotoxicity), interleukin-10 (immunoregulation), CD1d (innate-adaptive immune interactions), and NOD2 (microbial sensing). This would suggest that each person's allelic mosaic would shape the detailed composition of that individual's various microbial communities. However, because composition at the genus level is nearly equally similar between monozygotic or dizygotic twins, the genetic effect may be less important than maternal and other childhood environmental factors, notably diet and antibiotics.

Microbial communities at mucosal surfaces

Each mucosal surface has a unique physiologic role, with different cell types and different environmental stimuli, which result in a different microbiota. These communities of bacteria evolve with the host and are highly specialized to occupy different niches. The microbiota of the gut is the best-studied, but it is now also becoming clear that the microbiota of the airways and upper respiratory tract are also unique and diverse in health and disease.

19.1 Upper respiratory microbiome of nares is distinctive with similar phyla between individuals

The mucosa of the lower respiratory tract is minimally colonized by microbiota, although this assessment may change when it comes under study using culture-independent metagenomic analysis. In contrast, the microbiota of the nasal passage (nares) contains a rich, complex microbiome. The majority is derived from just two bacterial phyla, Actinobacteria and Firmicutes. A minor contribution is made by Proteobacteria, Bacteroidetes, and four other rare phyla. As expected, the nares microbiome is characteristic and is more similar between individuals than it is to other anatomic sites within the same individual. However, interindividual variation is also a feature, representing a relatively stable trait. Because aspects of the microbial fingerprint are shared among couples living together, the composition of this microbiome is likely to reflect a shared residential environment.

19.2 Oral microbiome is characterized by the formation of biofilms

It is currently estimated that over 750 different species of bacteria reside in the human mouth. In contrast to the skin, the mouth is a mucosal tissue with various subanatomical locations, including epithelial cells of the cheeks, gums, tongue, and palate, as well as enamel on teeth. Therefore, the surfaces that can be colonized differ substantially from the dry squamous epithelium of the skin. The major groups of microbes that inhabit the oral cavity include *Streptococcus, Actinomyces, Veillonella, Fusobacterium, Prevotella*, and *Treponema*, among others. It is believed that diverse consortia of microbial classes reside in close association in the mouth, and assemble into true polymicrobial communities where different species of bacteria network cooperatively in interactions that sustain the integrity of the microbiota. Although the surfaces of the mouth are varied, communities appear to form in aggregates on all anatomical niches due to the formation of biofilms. Biofilms are higher-order structures composed of bacteria and their secreted products (mostly polysaccharides), and they form intricate physical networks that sustain the microbial community. Within a biofilm, different microorganisms can exchange nutrients and metabolites that foster synergistic relationships that are required for the formation and maintenance of these microbial societies. Studies suggest that biofilm communities are highly structured, and that certain organisms (predominantly *Streptococcus*) are the primary colonizers that nucleate assembly and recruitment of other microbes into dental plaques. Within the biofilm, microbes appear to form organized partnerships, and specific bacterial species have evolved adhesions not only for host tissues but also for receptors on other bacteria. For example, *Fusobacterium nucleatum* can bind to epithelial cells as well as to *Streptococcus cristatus*, and this bacterial-bacterial interaction is required to suppress inflammatory responses against both microbes as a means to circumvent the host immune system during colonization. Specialized bacterial surface appendages have apparently evolved to mediate both bacterial and host associations in the oral cavity. Finally, the coordinated and regulated assembly of microbial communities suggests that perturbations in the "normal" composition of these microbes (a process termed *dysbiosis*) may affect oral health such as in periodontitis (see later).

Commensal microbes within biofilms may be composed of beneficial, symbiotic, or potentially pathogenic bacteria. In fact, inflammatory disease of the oral cavity is usually caused by members of the healthy microbiota. Periodontal diseases such as gingivitis and periodontitis result from immune responses to *Streptococcus, Actinobacteria*, and *Porphyromonas* species that are resident in the healthy mouth but form abnormal configurations in relation to the mucosal tissues in these diseases. Furthermore, there have been strong associations reported between poor oral hygiene and nonoral diseases such as endocarditis and lung disease. Although the central cause-and-effect relationships between the oral microbiota and disease are yet to be established, these studies show that the bacterial composition of the oral mucosa may have a considerable impact on human health. Further understanding of the dynamic microbial interactions within oral biofilms may provide clues into how the microbiota affects immunologic outcomes during health and disease in the mucosal surfaces that comprise the gastrointestinal tract.

19.3 Gut microbiome is most complex of mutualistic ecosystems of host

Of all anatomical locations, the lower gastrointestinal tract of mammals has the highest density and diversity of indigenous microorganisms and is a habitat permanently colonized by members of five of the six kingdoms of life. Bacteria predominate and reach astounding levels in the colon (approximately 100 trillion cells; Figure 19.3). Parenthetically, this means that on a per-cell basis, up to 90% of a human being is actually composed of the colonic microbiota. The aggregate human microbiome contains between 1000 and 1150 bacterial species (among all individuals sampled), with a given person harboring approximately 160 bacterial species. However, the species are largely different between individuals, and at the genus level, the overlap is less than 30%, and the overlap is only random. Thus, there does not yet appear to be a group of "keystone species" of the gut microbiota. In addition, the composition of the mucosal-surface microbiota is distinct from that recovered in the feces. Therefore, it is likely that there are a number of microbial compartments in the gut, including different anatomic regions of the intestinal mucosa, and the maturing fecal stream.

At a metagenomic level, the intestinal microbiome contains an enormous number of genes, more than 150 times the number of unique (nonredundant) genes found in the human genome. This metagenome spans the entire range of enzymatic functions and complements the human genome, particularly in genes contributing to nutrient uptake and digestion. In contrast to the interindividual diversity of microbial composition, the representation of genes by function is highly conserved. Although less studied, the intestinal

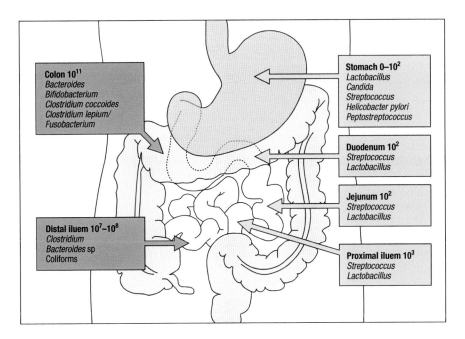

Figure 19.3 **Microbial colonization of the gastrointestinal tract.** The gut is home to a great density and diversity of commensal microbes. Although there is marked interindividual variation, particular organisms predominate in particular anatomical locations. For instance, organisms such as *Lactobacillus* and *Streptococcus* dominate the stomach, whereas *Bacteroides* and *Clostridium* are commonly found in the distal gut.

microbiome also includes diverse fungi, higher eukaryote organisms, and virus-like organisms. Unlike the bacterial microbiota, the intestinal "virome" seems to be a stable but highly individual trait, with minimal influence of either family or environment. It may also be of particular importance because its exceptional abundance and diversity may contribute an enormous genetic repertoire to intestinal function.

Host-microbe interactions

The host must sense, react to, and discriminate between pathogenic and commensal organisms at the barrier surfaces. The epithelium plays a critical role in this process by providing a physical barrier between the external environment, where microbes are abundant, and the underlying tissues, to which their access is restricted.

19.4 Host perceives and responds to microbiota

Organisms occupying the gastrointestinal tract are sensed by the intestinal epithelium and cells of the innate immune system. These cells are equipped with a diverse repertoire of receptors to recognize different conserved microbial components termed *pathogen-associated molecular patterns* (PAMPs) (see Chapter 18). Typical agonists of the pattern recognition receptors are microbial molecules such as the bacterial cell-wall components lipoteichoic acid, peptidoglycans, and lipopolysaccharide. The best known of these innate pattern recognition receptors are the toll-like receptors (TLRs), with 10 family members in humans and an additional 3 in mice. In addition to the TLRs, there are several other pattern recognition receptor families, including cytosolic receptors such as the NOD-like family of receptors. Together the pattern recognition receptor families allow the host to respond to a diverse range of different microbes. By controlling the repertoire, cellular distribution, and tissue localization of these receptors, the host can sense and respond to environmentally encountered microbes in different ways in different anatomical locations. For instance, TLRs may be differentially expressed on the apical surface of the mucosal epithelium where they interface with environmental bacteria, or alternatively on the basolateral aspect of the epithelium that in health is a "privileged" tissue compartment inaccessible to gastrointestinal commensals. However, in the event of loss of containment, as might occur secondarily to epithelial damage, this "hidden" pool of pattern recognition receptors becomes unmasked and exposed to bacteria richly endowed with PAMP signatures. A good example of this process is seen in acute gastrointestinal infection with invasive *Salmonella*. Infection with this flagellated pathogen results in colonic epithelial damage and *de novo* exposure of normally restricted basolateral TLR5 to the *Salmonella* flagellin. Flagellin ligates TLR5, triggering a powerful protective host inflammatory response. Humans with homozygous loss-of-function mutations in the TLR5 gene are more susceptible to Legionnaire's disease, a severe lung infection caused by the flagellated bacteria *Legionella pneumophila*.

Although one of the key roles of the pattern recognition receptors is to initiate host immune responses following exposure to PAMPs, it remains to be fully understood how the host manages to prevent igniting its immune response pathways in the face of exposure to the enormous antigenic load present at the epithelial barrier surfaces, particularly in the gut. It is often forgotten that mucosal surfaces are covered by a thick mucus layer, especially in the colon, so that the surface of the epithelium may in fact receive only minimal exposure to PAMPS derived from the microflora. Interestingly, defects in the mucus layer can lead to a spontaneous colitis in mice.

Despite their role in stimulating anti-pathogen immunity, it is also acknowledged that pattern recognition receptors interact with gut commensals to facilitate their relationship with the host. The critical importance of the cross talk between the innate immune system and the gut microbiota for the maintenance of intestinal homeostasis is readily demonstrated by experimental perturbation of either of these components. Mice reared in germ-free conditions have developmental defects in the mucosal immune system and impaired mucosal immunity to common infections. Disrupting innate immune recognition of luminal organisms by the introduction of genetic deficiencies in particular pattern recognition receptors or their signaling pathways also results in disordered homeostasis and altered susceptibility to inflammatory insult.

19.5 Life without microbiota results in profound changes in host

Germ-free mice have proven to be a particularly useful experimental tool for studying the interaction between intestinal microbiota and the host immune system. The animals are housed in sterile chambers completely free from the presence of all environmental microbes. Air, food, and water are sterilized to prevent microbial contamination, and individual bacterial species or defined bacterial communities are selectively introduced experimentally. Mice with defined flora (or no flora) are termed gnotobiotic (from *gnotos*, Greek for "to know," and *bios*, Greek for "life").

Since the 1960s it has been consistently noted that germ-free mice have numerous immunologic, physiologic, and anatomical differences from conventionally reared animals (Table 19.1). They require more calories because they lose the nutrients provided by microbial degradation of plant material in mouse chow. And mice are coprophagic (ingest feces of other cohoused individuals). The small intestine is longer and the villi in the

TABLE 19.1 ABNORMALITIES APPARENT IN GERM-FREE MICE

Anatomic/histologic	Immunologic	Functional
Enlarged cecum	Reduced secretory IgA production and low level of serum immunoglobulin	Increased susceptibility to infections: • *Salmonella* • *Listeria* • *Shigella* • *Leishmania* • *Bacillus anthracis*
Longer small intestine with taller villi in the duodenum and shorter villi in the ileum	Diminished numbers and activation of systemic T cells	
Poorly developed mesenteric lymph nodes	Reduced CD8+ T-cell cytotoxicity	Reduced susceptibility to autoimmune disease: • Experimental allergic encephalomyelitis • Arthritis • Some models of inflammatory bowel disease
Poorly developed Peyer's patches	Impaired lymphocyte homing to inflammatory sites	
Lower numbers of isolated lymphoid follicles	Reduced intestinal lamina propria lymphocytes and intraepithelial lymphocytes (IELs), especially $\alpha\beta$ T-cell receptor IEL; impaired helper T-cell T_H17 responses	
Small spleen	Reduced ability of granulocytes to kill bacteria	Reduced ability to induce oral tolerance

duodenum are longer than in normal mice but shorter in the ileum. The ceca of germ-free mice are grossly enlarged. There is poor development of gut-associated lymphoid tissue, including Peyer's patches which only contain primary follicles. Unlike conventionally colonized mice, few T cells can be found infiltrating the intestinal lamina propria, and there is an almost complete absence of secretory IgA. Intraepithelial lymphocyte numbers are very low, although there are more $\gamma\delta$ T cells than $\alpha\beta$ T cells. TLR and major histocompatibility class II expression on intestinal epithelial cells is also markedly reduced in germ-free animals, reflecting the reduced need for antigen detection and presentation.

Systemically germ-free mice have small spleens and very low levels of serum immunoglobulins. Granulocyte function in terms of their ability to kill bacteria is also reduced and, interestingly, is restored when mice are conventionalized. It is thought that systemic penetration of bacterial products into the bloodstream is responsible for this priming, and interestingly, priming is not seen in Nod1-deficient mice, implying that the effect is mediated via bacterial peptidoglycan.

Many studies have demonstrated that both the mucosal and systemic immune systems return to normal if germ-free mice are colonized with a complex flora, but in a fascinating observation made some 20 years ago, monoassociation with a single organism, namely, an unculturable segmented filamentous bacterium, restored systemic and mucosal immune populations to normality.

19.6 T-cell responses appear to be modulated by microbiota

Early studies demonstrated defective T-cell function in germ-free mice. For example germ-free mice show an impaired ability to generate T cells that mediate delayed-type hypersensitivity. Germ-free mice also appear to have deficiencies in their ability to generate the regulatory T cells that dampen responses to orally administered antigens. Recent data also show that the differentiation of effector T-cell lineages is also modulated by the community composition of the intestinal microbiota. Since the initial reports of the importance of T_H17 responses in the pathogeneses of different autoimmune diseases, different groups have observed that genetically identical mice housed in different facilities did not appear to have the same susceptibility to experimentally induced autoimmunity. One group of researchers also noticed that C57Bl/6 mice supplied by Jackson Laboratories, the largest vendor of research mice in the United States, were unable to generate IL-17 responses, whereas C57/B6 mice acquired from an alternative supplier, Taconic Farms, readily produced IL-17. Intestinal T_H17 responses could be generated in Jackson mice if cohoused with mice from Taconic Farms. In germ-free mice, T_H17 cells were markedly reduced in the intestinal lamina propria. However, transplantation of the intestinal microbiota from Taconic mice effectively restored T_H17 cells, whereas transplantation of the intestinal microbiota from Jackson mice failed to induce T_H17 cells. Thus, the capacity of genetically identical hosts to initiate T_H17 responses was dependent on the community profile of the intestinal microbiota. For the first time, these studies confirmed that intestinal microbes are involved in the differentiation of helper T-cell subsets, and that the microbial community interfacing with the host immune system may shape lineage commitment of helper T cells.

19.7 Specific bacteria specifically drive T_H17 versus T_{reg} responses

Sequencing of 16S ribosomal RNA genes confirmed differences in the intestinal microbiota community profile between Jackson Laboratories and Taconic Farms mice. In particular, a bacterium from the genus *Arthromitus*, termed *segmented filamentous bacterium* (SFB), was present in Taconic mice but not in Jackson mice (Figure 19.4). This is the same organism that was shown many years ago to restore systemic and mucosal immunity when monocolonized into germ-free mice. SFB could be transmitted to Jackson mice if they were cohoused in the same cages as Taconic mice, which also resulted in successful acquisition of mucosal T$_H$17 responses in Jackson mice. Critically, monoassociation of germ-free mice with SFB alone reinstated mucosal T$_H$17 cells. The T$_H$17-promoting property of SFB appears to be relatively selective because other T-cell lineages are less dramatically affected. SFB also reduced intestinal regulatory T cells, favoring a more inflammatory bias in gut-resident T cells.

The functional significance of SFB-permissive T$_H$17 responses has been demonstrated in experiments with the natural mouse pathogen *Citrobacter rodentium*. SFB colonizes the surface of epithelial cells causing distal colonic inflammation. In comparison with Taconic Farms mice, Jackson Laboratories mice lacking SFB and T$_H$17 cells were highly susceptible to *Citrobacter* infection. However, colonization of Jackson mice with SFB reduced their susceptibility to *Citrobacter* infection and restored T$_H$17 responses.

The mechanisms responsible for T$_H$17 induction by SFB are incompletely understood, although the close contact formed between long filaments from SFB and intestinal epithelial cells in the ileum is likely to be a key feature of their capacity to modulate T-cell immunity. SFBs appear to colonize the surface of Peyer's patches, where they may have a chance to orchestrate mucosal T-cell immunity at the inductive site. Other bacterial species that can also form attaching and effacing interactions with intestinal epithelial cells, including *Escherichia coli* O157 and *Citrobacter rodentium* are also capable of inducing intestinal IL-17 responses when monoassociated with germ-free mice, whereas selective mutant forms of these species that cannot form adhesions with intestinal epithelial cells lose the capacity to trigger T$_H$17 induction. Although SFBs do not colonize the human intestine, it is possible that bacteria with similar characteristics also exist in humans. Intriguingly, transfer of the cecal contents from humans with ulcerative colitis (which is characterized by expansion of intestinal T$_H$17 cells) results in induction of T$_H$17 responses in mice. Furthermore, a consortium of 20 anaerobic bacteria from these patients, which could induce intestinal IL-17 production, were observed to rapidly adhere to murine intestinal epithelial cells *in vivo* and induced a similar pattern of host genes to those induced by SFB colonization.

Little is known about the SFB-expressed molecular determinants responsible for modulating T$_H$17 responses. Analysis of the gene expression profile of the ileum following colonization of germ-free mice with SFB has been helpful in revealing which host genes are switched on or off in response to SFB colonization. SFB upregulates numerous immune response genes and genes involved in host defense, such as the antimicrobial peptide RegIIIγ and acute-phase proteins, such as serum amyloid A. The latter has also been shown to promote T$_H$17 differentiation *in vitro*; however, the *in vivo* relevance of this potential mechanism awaits verification. It needs to be emphasized, however, that the human gut does not contain bacteria that have the same intimate association with the epithelium as SFB has with mouse ileal epithelium. However, induction of the acute phase protein serum amyloid A (SAA) in the terminal ileum is believed to play an important role. SAA, typically induced during infection, tissue damage, and inflammation, is produced as three isoforms—*Saa1, 2,* and *3*. SAA1 and SAA2 are among the most highly upregulated genes in the intestinal epithelium after colonization of SFB. SAA stimulates lamina propria dendritic cells *in vitro* to induce T$_H$17 cell differentiation and amplifies T$_H$17 transcriptional programs *in vivo*. Importantly, it is likely that other epithelial cell–derived factors also contribute to the ability of SFB to induce T$_H$17 cells, as SAA1 and SAA2 deficiency is

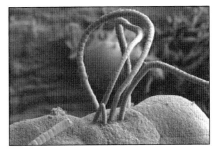

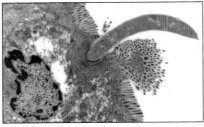

Figure 19.4 Segmented filamentous bacteria. Scanning and transmission electron micrographs of mouse terminal ileum reveal the presence of bacteria with long, segmental filaments in very close contact with the surface epithelium. These segmented filamentous bacteria (SFB) are observed in mice from Taconic Farms, mice that are good at generating mucosal T$_H$17 responses. SFB are not observed in mice from Jackson Laboratories that are poor inducers of T$_H$17 responses. Colonization of mice from Jackson Laboratories with SFB generates mucosal T cells that can make IL-17.

associated with only a partial inhibition of T_H17 cytokine production after SFB colonization.

Another extremely interesting precedent where a single molecule from a commensal bacterium can have profound effects on host immunity involves capsular polysaccharide A from the human commensal *Bacteroides fragilis*. This is a highly unusual molecule in that, despite being a polysaccharide, it can undergo processing by dendritic cells and be presented to T cells. In the transfer model of colitis used with lymphopenic mice and exacerbated by infection with *Helicobacter hepaticus*, cocolonization with *B. fragilis* ameliorates disease, but the effect is not seen with *B. fragilis* lacking polysaccharide A. Purified polysaccharide A given orally is also therapeutic in this model of colitis, in TNBS (trinitrobenzene sulfonic acid) colitis, and in experimental allergic encephalomyelitis (EAE), a mouse model for multiple sclerosis. Polysaccharide A promotes the generation of IL-10-secreting CD4+ Foxp3+ regulatory T (T_{reg}) cells that inhibit pro-inflammatory T_H1 and T_H17 responses.

A similar effect of promoting mucosal T_{reg} cells is conferred by a select consortium of 17 *Clostridia* strains isolated from the healthy human microbiome. Upon inoculation of germ-free mice, the *Clostridia* mixture induces anti-inflammatory molecules IL-10 and inducible T-cell costimulator (ICOS) from T_{reg} cells and promotes TGF-β to support differentiation of T_{reg} cells. Probiotic treatment with the *Clostridia* consortium attenuated symptoms in models of colitis and allergic diarrhea. In this case, the production of the short-chain fatty acid (SCFA) butyrate is thought to be important, because *Clostridia* are prominent producers of SCFAs, and butyrate, in particular, sufficiently induces the differentiation of T_{reg} cells *in vitro* and *in vivo*.

19.8 Intestinal microbiota influences autoimmunity

Although CD4 helper T-cell effector lineages are important in combating infections, inappropriate mobilization of these different T-cell subsets is also implicated in autoimmune and allergic disease. T_H17 cells, often in conjunction with T_H1 cells, are expanded in inflammatory lesions of patients with rheumatoid arthritis, psoriasis, type 1 diabetes mellitus, multiple sclerosis, and inflammatory bowel disease. Similarly, misdirected T_H2 cells, expressing IL-4, IL-5, and IL-13, support eosinophil and mast cell activation to drive allergic diseases such as asthma and rhinitis.

Although one might intuitively anticipate that intestinal microbes exert influence over helper T-cell polarization in the gut, it is more surprising that gut microbes can modulate extraintestinal T-cell responses. Germ-free mice have impaired immunity and increased susceptibility to particular infections. However, they are also less sensitive to T-cell–mediated autoimmune conditions. Experimental allergic encephalomyelitis (EAE) is a frequently studied animal model of autoimmunity that mirrors aspects of human multiple sclerosis. Disease is mediated by T cells responding to antigens present on the myelin sheath that coats nerves. EAE can be induced in rodents and primates by immunization with nerve tissue–derived proteins or peptides, such as myelin basic protein or myelin oligodendrocyte glycoprotein in adjuvant. About 2 weeks after immunization, mice develop relapsing and remitting symptoms of weakness and paralysis. EAE is characterized by expansion of T_H17 and T_H1 cells in inflamed nerve tissue. It has been shown that germ-free mice are much less sensitive to EAE induction than mice colonized with a conventional intestinal microflora. Many germ-free mice failed to develop any neurologic features. EAE resistance in germ-free mice is associated with an attenuated inflammatory response. These experiments strongly implied that gut microbes are necessary for at least some pathogenic T-cell responses outside the gut (**Figure 19.5**).

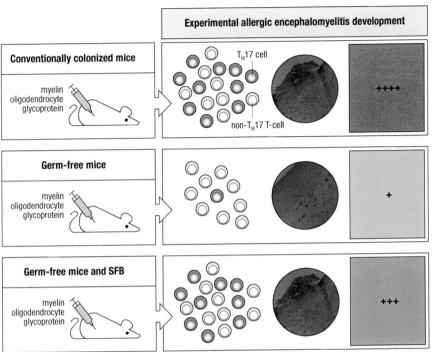

Figure 19.5 Segmented filamentous bacteria and autoimmunity. Immunization of conventionally colonized mice with myelin oligodendrocyte glycoprotein induces experimental allergic encephalomyelitis (EAE). Mice develop weakness and paralysis secondary to nerve damage characterized by leukocyte infiltration and erosion of the myelin sheath (hematoxylin and eosin histology panels of inflamed spinal cords at the onset of EAE are shown in the right panels). This is associated with increased initiation of inflammatory T-cell responses, including T_H17 cells (dark blue cells). Conversely, germ-free mice are largely resistant to EAE induction following myelin oligodendrocyte glycoprotein immunization with minor histologic changes with fewer inflammatory T_H17 cells. However, monoassociation of germ-free mice with SFB reinstates T_H17 responses and susceptibility to EAE induction.

Although the germ-free state conferred protection from EAE, subsequent monoassociation with SFB reinstated the host's capacity to initiate T_H17 responses and resulted in increased EAE development and severity. It is noteworthy that the recovery of T_H17 cells and EAE susceptibility in SFB monoassociated mice did not quite reach the levels observed in mice colonized with a complete repertoire of intestinal bacterial commensals, implying that alternative non-SFB bacterial species also contribute to the maturation of effective T-cell responses and disease development.

Similar results have been observed in an experimental model of autoimmune arthritis. K/BxN mice overexpress a T-cell receptor that recognizes a self peptide derived from glucose-6-phosphate isomerase presented by the major histocompatibility complex class II molecule Ag7. Mice expressing the transgenic receptor spontaneously develop autoimmune joint disease from about 4 weeks of age. However, in the same way as germ-free mice are very resistant to EAE, germ-free K/BxN mice are also protected from the development of joint disease. It was also observed that disease-resistant germ-free mice had marked reduction in the expression of T_H17-related genes. It was possible to recover the typical autoimmune disease manifestations characteristically observed in K/BxN mice following reconstitution of germ-free mice with monocultures of SFB. The triggering of arthritis in SFB monoassociated germ-free mice is also associated with the emergence of lamina propria and splenic T_H17 responses.

19.9 Microbiota appears to be involved in inflammatory bowel disease, particularly Crohn's disease

One of the most important examples of a disordered interaction between the host immune system and the intestinal microbiota is seen in inflammatory bowel disease (IBD). Crohn's disease and ulcerative colitis, the main forms, are characterized by intestinal inflammation and troublesome symptoms such as diarrhea, rectal bleeding, and abdominal pain, as well as potentially serious complications. Most investigators agree that IBD likely results from inappropriate immune activation in response to intestinal microbes. A

TABLE 19.2 EVIDENCE IMPLICATING THE MICROBIOTA IN INFLAMMATORY BOWEL DISEASE

	Crohn's disease	Ulcerative colitis
Antibiotics useful?	Effective in management of fistulae	Effective in pouchitis
	Effective in penetrating disease with abscess formation	
	Reduces postoperative recurrence	
Probiotics useful?	Not helpful	Effective in pouchitis
		Effective in maintaining remission
Prebiotics useful?	Not helpful	Yet to be established, likely small/no benefit
Diversion of fecal stream?	Helpful	Not routinely performed for ulcerative colitis
Disease location	Most common location is ileocecal disease (areas of high microbe exposure)	Colon only (occasional "backwash ileitis")
Adherent-invasive *Escherichia coli*	Common	Uncommon
Faecalibacterium prausnitzii	Protective effect	Unknown
T-cell reactivity to microbes	Present	Present
Genetics	Innate immune molecules handling the microbiota (e.g., NOD2)	No innate components implicated to date

number of lines of evidence have implicated the microbiota as key contributors to IBD pathogenesis (Table 19.2). In certain situations, antibiotics are therapeutically beneficial in patients with Crohn's disease, including patients with perianal disease or fistulae, and in the prevention of postoperative recurrence in patients with isolated ileocecal Crohn's disease. Antibiotics are also the first-line therapy for postoperative ulcerative colitis patients who develop pouchitis. Similarly, some probiotics can maintain remission in ulcerative colitis and can also be therapeutic in pouchitis. Probiotic bacteria appear to induce regulatory host immune responses favoring tolerance toward the intestinal microbiota. For instance, the probiotic mixture VSL3 induces production of IL-10 by dendritic cells and reduces pro-inflammatory cytokines such as IL-12. In severe IBD, some patients can be effectively treated by surgically diverting the fecal stream away from inflamed intestinal segments, reducing the exposure of inflamed tissue to luminal microbes. It is also well recognized that many commonly studied mouse models of intestinal inflammation rely on the intestinal microbiota. For instance, the spontaneous colitis that develops in mice deficient in the regulatory cytokine IL-10 is not seen when the mice are housed under germ-free conditions. The penetrance of colitis and ileitis in these mice varies according to the cleanliness of the animal facility.

It is unlikely that a single member of the indigenous microbiota drives IBD. This is well demonstrated in animal models. *Bacteroides vulgatus* and *Bacteroides thetaiotaomicron*, but not *Bacteroides distasonis* or *Escherichia coli*, induce colitis in germ-free HLA-B27 rats. However, in germ-free IL-10-null mice, colonization with any of these species fails to induce colitis. In contrast, *Enterococcus faecalis* is highly colitogenic in IL-10-null mice. Viral infections may also be involved in colitis. Murine norovirus is endemic in laboratory mice. When mice expressing the Crohn's disease mutated *ATG16L1* autophagy gene are infected with murine norovirus, there is marked abnormality of Paneth cells, and the animals are more susceptible to DSS (dextran sulfate sodium) colitis. Murine norovirus is related to several human norovirus species that cause acute gastroenteritis and have been associated with IBD.

Adherent and invasive *Escherichia coli*, which can invade intestinal epithelial cells and lamina propria macrophages (Chapter 25), where they can survive and replicate, may be involved in Crohn's disease (Chapter 23). However, these organisms are not present in all patients and are often present in the intestine of patients without Crohn's disease. Conversely, another bacterial component of the intestinal microflora, *Faecalibacterium prausnitzii*, may be important in patients with Crohn's disease. *F. prausnitzii* is markedly reduced in the gut of such patients. Patients with high levels of *F. prausnitzii* are less prone to disease recurrence following surgical resection of an inflamed segment of intestine. There is intense interest in the gut microbiome in IBD, but it is not certain whether any changes are causal or secondary to inflammation.

Host responses to pathogen-associated intestinal inflammation are also influenced by the microbiota. For example, *Helicobacter hepaticus* infection induces colitis and liver inflammation in conventionally colonized animals. However, germ-free mice monoassociated with *H. hepaticus* do not develop disease, implying that components of the conventional microbiota are not necessarily innocent bystanders in the company of pathogens. Interestingly, patients with IBD have detectable circulating antibodies and T-cell specificities directed against normal constituents of their intestinal microbiota.

19.10 Defective immune system handling of intestinal microbes leads to intestinal inflammation

Defects in the host immune system may result in disordered handling of the intestinal microbiota. The pattern recognition receptor families are crucial molecules involved in sensing and reacting to intestinal microbes, and selective genetic deficiencies in particular TLRs, such as TLR5, result in spontaneous intestinal inflammation. Similarly, deletion of MyD88, an adaptor molecule that is responsible for facilitating effective signal transduction of almost all of the TLRs, also results in markedly enhanced susceptibility to DSS colitis.

Recently, a novel mouse model of IBD was described in mice lacking the key immune transcription factor T-bet in the innate immune system. In addition to selective deletion of T-bet, these mice completely lack an adaptive immune system by additional knockout of the recombinase activating gene 2 (RAG2) that plays a critical role in the development of adaptive T-cell and B-cell receptors (**Figure 19.6**). These T-bet × RAG2 double-knockout mice develop spontaneous colitis that closely resembles many aspects of human ulcerative colitis, and they have been termed TRUC (T-bet × RAG2 ulcerative colitis) mice. The disease begins in the distal colon with relatively superficial inflammatory changes, and in the long term, colitis is often complicated by colon cancer, similarly to the situation in uncontrolled human ulcerative colitis. The initial phase of the disease is characterized

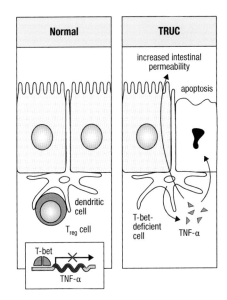

Figure 19.6 The TRUC (T-bet × RAG2 ulcerative colitis) model of colitis. In the healthy gut, intestinal dendritic cells sample luminal contents but are prevented from initiating inflammatory responses by regulatory T (T_reg) cells and a tightly regulated transcriptional program. For example, the transcription factor T-bet represses the pro-inflammatory cytokine tumor necrosis factor (TNF)-α. However, in the absence of T-bet, de-repressed expression of TNF-α leads to excess production of TNF-α, epithelial apoptosis, and spontaneous colitis. (Adapted from Hecht, G., *N. Engl. J. Med.* 2008, 358:528–530. With permission from Massachusetts Medical Society.)

by overproduction of the inflammatory cytokine tumor necrosis factor (TNF)-α by CD11c-expressing lamina propria myeloid cells. Colitis can be effectively treated by blocking TNF-α with monoclonal antibodies, again recapitulating human IBD. Like other IBD models, intestinal inflammation in TRUC mice is dependent on the intestinal microbiota. TRUC colitis can be successfully treated with antibiotics and probiotics, and fails to develop if mice are reared under germ-free conditions. Furthermore, isogenic colonies of T-bet × RAG2 double-knockout mice housed at different laboratories are not equally sensitive to colitis development, suggesting that a specific microbial component present in some colonies, but not others, is responsible for driving intestinal inflammation. T-bet represses the transcription of the TNF-α gene in myeloid cells, and consequently, in its absence, there is a tendency for deregulated TNF-α expression that drives epithelial apoptosis and intestinal inflammation.

TRUC and other spontaneous IBD models have provided some important lessons regarding the mechanisms underlying human IBD. It is becoming increasingly clear that interactions between host factors, such as genotype, and environmental factors, such as the microbiota, act in concert rather than isolation to determine disease occurrence, progression, and outcome. In recent years, considerable advances have been made in our understanding of the genetic basis of human IBD, and genome-wide association studies have identified many single nucleotide polymorphisms that are associated with IBD (Chapter 34). In Crohn's disease, genetic variations in genes encoding proteins involved in handling microbes have been identified, including NOD2, an intracellular pattern recognition receptor, and genes encoding autophagy pathway proteins, such as ATG16L1 and IRGM. Genetic variations in several components of the T_H17 pathway have also been singled out by genome-wide association studies, further underlining the critical importance of adaptive immunity in these complex diseases.

Influence of microbiota on host metabolism

Western societies are currently experiencing an obesity pandemic that is resulting in considerable economic and health costs. Obesity is associated with metabolic disturbances, including hyperglycemia, hyperlipidemia, and insulin resistance, collectively termed the *metabolic syndrome*. Patients with obesity and the metabolic syndrome have reduced life expectancy and are at high risk of developing diabetes, liver disease, hypertension, and cardiovascular disease. Although overeating and sedentary lifestyles play a central role in the development of obesity, recent insights indicate that colonic microbes also play a role in energy balance and contribute to the pathophysiology of obesity. Fat is also abundantly infiltrated with inflammatory macrophages secreting cytokines; thus, the obese state is also a pro-inflammatory state with, for example, raised levels of IL-6. Although the pathways are only beginning to be established, diet can influence the microbiota, which can then influence the host, perhaps via PAMPS, in addition to the direct effect of excess nutrients on host metabolism.

Experiments with germ-free mice provided clues that the microbiota exerts influence on host metabolism. As previously stated, germ-free mice eat more than genetically identical littermates with a conventional intestinal microflora because they cannot salvage nutrients from bacterial degradation of fiber in chow.

Colonization of germ-free mice with a conventional intestinal microbiota stimulates widespread changes in many processes involved in host metabolism. Intestinal glucose absorption is increased twofold, and systemic changes in the host such as increased hepatic lipogenesis, increased adipose triglyceride synthesis, and increased leptin production are also observed following *de novo* intestinal colonization. Gene expression profiling of epithelial cells also revealed that after immune response genes, the second largest group of genes induced by conventional intestinal microbes are genes involved in metabolism. High-resolution analysis of the plasma of germ-free and conventionally colonized mice also reveals dramatic differences in the profile of hundreds of metabolites, confirming the profound impact that the intestinal microbiota has on host metabolism.

In contrast to conventionally colonized mice that develop diet-induced obesity when they are fed a "Western"-style diet rich in fat, germ-free mice are relatively resistant to these effects, implying that the influence of the microbiota on host metabolism can adversely affect health. Critical insights into the role of the microbiota in obesity have come from studies evaluating the community structure of obese mice using 16S ribosomal RNA gene sequencing.

19.11 Alterations in community structure of colonic microbes are linked to obesity, growth, and metabolic syndrome

Two distinct divisions of bacteria overwhelmingly dominate in the gut: Firmicutes and Bacteroidetes. However, the relative abundance of these two phyla is consistently altered in obesity. The *ob/ob* mouse has a spontaneous mutation in the leptin gene, a hormone responsible for regulating appetite. Although these mice have normal birth weights, excess food consumption invariably leads to obesity and its attendant complications, including the metabolic syndrome. Leptin is also involved in regulating immune responses. The relative abundances of Bacteroidetes and Firmicutes in the colon of *ob/ob* mice are very different than that of wild-type mice, with a reduction in the abundance of Bacteroidetes and a reciprocal expansion of Firmicutes. Diet-induced obesity also prompted an increase in the ratio of Firmicutes to Bacteroidetes within the colonic microbiota, in keeping with observations in leptin-deficient *ob/ob* mice. Comparable shifts in the relative frequencies of Firmicutes and Bacteroidetes have also been observed in the colonic microflora of obese humans. Furthermore, obese volunteers who subsequently consumed a low-calorie diet and successfully lost weight restored the distorted ratio of the chief phyla back to normal.

Evidence directly implicating the colonic microbiota in a contributory role in obesity came from gnotobiotic studies. To directly demonstrate the role of the colonic microbiota in the development of obesity, an elegant series of experiments was conducted where the colonic microbiota was harvested from either obese or lean mice and then "transplanted" to germ-free mice, so that these mice with no colonic commensals of their own subsequently became colonized with the colonic microbial communities that prevail in obese or lean mice (Figure 19.7). Crucially, germ-free mice colonized with the colonic microbiota from obese mice developed a 50% increase in fat deposition compared with mice colonized with the microbiota from lean mice. These experiments demonstrated that the colonic microbial community present in obese mice contributed to the increased fat deposition and metabolic complications observed in obese mice. Remarkably, follow-up experiments that further transplanted human microbiome samples from twin pairs, wherein one twin was lean and the other obese, rendered the same results; germ-free mice that received the obesity-related human microbiome gained more weight than did those that received microbiomes from lean humans. These findings support the notion that specific disruptions to the microbiota could directly cause obesity-related weight gain. In a final twist to this story, young germ-free mice transplanted with microbiota from healthy or undernourished infants transferred the impaired growth phenotype to the mice. The growth failure could be reversed by adding two invasive bacterial species.

Additional studies have aimed to identify specific bacterial taxa in lean individuals that may be protective against metabolic syndrome. The indigenous gut bacterium *Akkermansia muciniphila* is of particular interest given its inverse correlation with body weight and diabetes in both mice and humans.

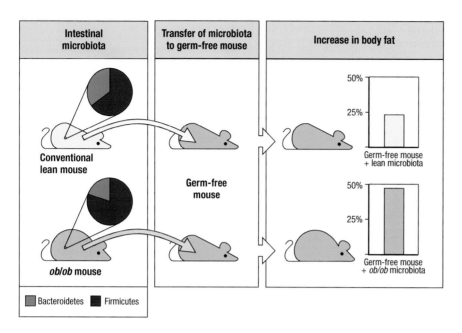

Figure 19.7 Obesity is associated with changes to the intestinal microbiota. It has been established that obese mice, and indeed humans, have an altered profile of the chief colonic bacterial phyla. Mice deficient in the appetite-controlling hormone leptin (*ob/ob* mice) develop marked obesity secondary to increased food consumption. In comparison with lean mice, these obese mice have an increase in the relative frequency of Firmicutes (red) and a reciprocal contraction of Bacteroidetes (blue). Fascinatingly, if the intestinal microbiota is harvested from *ob/ob* mice and "transplanted" to germ-free mice, these mice put on more body fat than germ-free mice transplanted with the microbiota harvested from lean mice. These changes are associated with increased energy harvest from the feces by the microbiota derived from obese mice due to more pronounced bacterial fermentation.

Aptly named from the Latin *mucin* for "mucus" and Greek *philos* for "friend or loving," *A. muciniphila* resides in the mucus layer of the intestinal epithelium and exhibits a unique ability to degrade mucins as an energy source. This close proximity to the host supports its important role in regulating intestinal barrier integrity and mucosal immunity. Treatment of mice fed a high-fat diet with *A. muciniphila* corrected symptoms related to metabolic syndrome, including fat-mass gain, endotoxemia, tissue inflammation, and glucose intolerance. These beneficial effects were seen only by treatment with live, and not heat-killed, bacteria, suggesting an active effect of *A. muciniphila* on homeostatic host functions.

19.12 Host genome interacts with microbiome to influence host phenotype

Recent data have also highlighted how defects in host perception of the microbiota can also impact host obesity. Pattern recognition receptors and especially the toll-like receptors are the major sensing apparatus of the innate immune system. Fascinatingly, mice lacking TLR5 receptors are highly susceptible to obesity and its attendant metabolic complications. In this situation, the structure of the major phyla appeared to be relatively preserved, hinting at the possibility that individual microbial species might also contribute to obesity and the metabolic syndrome.

   Given the association of particular community microbiota profiles with disease states, it follows that correction or manipulation of the microbiota might be a credible therapeutic option to reverse the deleterious effects on health. Unsophisticated manipulation of the intestinal microbiota through administration of broad-spectrum antibiotics has been shown to improve diet-induced obesity and its metabolic complications in *ob/ob* mice. Oral administration of probiotic mixtures, or even a single organism such as *E. coli* Nissle, has also found therapeutic success in ulcerative colitis and antibiotic-associated diarrhea. It is conceptually attractive to contemplate ways of manipulating the microbiota with the implicit intention of influencing host health. For example, it is conceivable that an artificial microbiota constructed from polysaccharide A–expressing *Bacteroides fragilis* that favors regulatory T-cell expansion at the expense of inflammatory $T_H 17$ cells could be harnessed to treat patients with IBD or other autoimmune diseases. In contrast, an artificial microbiota containing organisms that encourage inflammatory responses, such as SFB, might be employed to bolster protective immune responses against infectious agents. Alternatively, shaping the microbiota community profile by administering growth factors that encourage the growth of certain species at the expense of others might also allow us to shape inflammatory and regulatory immune responses *in vivo.*

SUMMARY

Prokaryotes are both our most enduring adversaries and our most constant allies. Present in phenomenal numbers in intimate contact with our barrier surfaces, they rarely threaten health, instead performing a diverse range of activities that benefit the host. Experiments in germ-free mice have demonstrated that these organisms influence host immune responses in qualitative and quantitative ways, and in their absence the immune system fails to develop properly, rendering the host vulnerable to infections. In some ways, the microbe-rich gastrointestinal tract might be considered a nursery or even *Ludus Magnus* for B cells and T cells, where they cut their teeth and subsequently mature into cells capable of mediating adaptive immunity. Although a number of these symbiotic activities can be carried out by many different intestinal commensals, it is now recognized that specific commensal species play a key role in shaping particular arms of adaptive immunity, including SFB for the initiation of T_H17 responses and polysaccharide

A–expressing *Bacteroides fragilis* in promoting regulatory T cells. In the coming years, it is likely that additional microbial factors, or unique species, will be identified that can also modulate specific aspects of host immunity. In recognition of the critical role played by the microbiota in influencing host health, an ambitious 5-year research program called the "Human Microbiome Project" was launched in 2008 by the National Institutes of Health costing in excess of $150 million. In 2016, the White House Office of Science and Technology announced the "National Microbiome Initiative" to continue support for advancing microbiome research by federal agencies and private-sector institutions. It is anticipated that these detailed analyses of the organisms that share our barrier surfaces will provide key insights into the inner workings of our own immune system and may uncover novel therapeutic strategies for combating immune-mediated diseases.

FURTHER READING

Blanton, L.V., Charbonneau, M.R., Salih, T. et al.: Gut bacteria that prevent growth impairments transmitted by microbiota from malnourished children. *Science* 2016, 351:aad3311-7.

Cadwell, K., Patel, K.K., Maloney, N.S. et al.: Virus-plus-susceptibility gene interaction determines Crohn's disease gene Atg16L1 phenotypes in intestine. *Cell* 2010, 141:1135–1145.

De Filippo, C., Cavalieri, D., Di Paola, M. et al.: Impact of diet in shaping gut microbiota revealed by a comparative study in children from Europe and rural Africa. *Proc. Natl Acad. Sci. USA.* 2010, 107:14691–14696.

Garrett, W.S., Lord, G.M., Punit, S. et al.: Communicable ulcerative colitis induced by T-bet deficiency in the innate immune system. *Cell* 2007, 131:33–45.

Goodman, A.L., and Gordon, J.I.: Our unindicted coconspirators: Human metabolism from a microbial perspective. *Cell Metab.* 2010, 12:111–116.

Hansen, J., Gulati, A., and Sartor, R.B.: The role of mucosal immunity and host genetics in defining intestinal commensal bacteria. *Curr. Opin. Gastroenterol.* 2010, 26:564–571.

Ivanov, I.I., Atarashi, K., Manel, N. et al.: Induction of intestinal Th17 cells by segmented filamentous bacteria. *Cell* 2009, 139:485–498.

Lee, Y.K., and Mazmanian, S.K.: Has the microbiota played a critical role in the evolution of the adaptive immune system? *Science* 2010, 330:1768–1773.

Ley, R.E., Turnbaugh, P.J., Klein, S. et al.: Microbial ecology: Human gut microbes associated with obesity. *Nature* 2006, 444:1022–1023.

Powell, N., Canavan, J.B., MacDonald, T.T. et al.: Transcriptional regulation of the mucosal immune system mediated by T-bet. *Mucosal Immunol.* 2010, 3:567–577.

Qin, J., Li, R., Raes, J., et al.: A human gut microbial gene catalogue established by metagenomic sequencing. *Nature* 2010, 464:59–65.

Sokol, H., Pigneur, B., Watterlot, L. et al.: *Faecalibacterium prausnitzii* is an anti-inflammatory commensal bacterium identified by gut microbiota analysis of Crohn disease patients. *Proc. Natl Acad. Sci. USA.* 2008, 105:16731–16736.

Thompson, G.R., and Trexler, P.C.: Gastrointestinal structure and function in germ-free or gnotobiotic animals. *Gut.* 1971, 12:230–235.

Turnbaugh, P.J., Ley, R.E., Mahowald, M.A. et al.: An obesity-associated gut microbiome with increased capacity for energy harvest. *Nature* 2006, 444:1027–1031.

Vijay-Kumar, M., Aitken, J.D., Carvalho, F.A. et al.: Metabolic syndrome and altered gut microbiota in mice lacking Toll-like receptor 5. *Science* 2010, 328:228–231.

Wu, H.J., Ivanov, I.I., Darce, J. et al.: Gut-residing segmented filamentous bacteria drive autoimmune arthritis via T helper 17 cells. *Immunity* 2010, 32:815–827.

GENITOURINARY TRACT

PART FOUR

The immune system of the genitourinary tract

20

DAVID A. MACINTYRE AND KENNETH W. BEAGLEY

The human genitourinary tract is characterized by a complex system of compartmentalized, highly integrated tissues, finely tuned hormonal regulation, and an immune system that has some similarities to, but also distinct differences from, the gastrointestinal immune system. Innate and adaptive immune defenses in the genitourinary tract influence, and are influenced by, the composition of commensal microbial communities, which provide a degree of protection against sexually transmitted diseases. The important role of sex hormones in regulating immune protection is a feature that distinguishes reproductive mucosal immunology from mucosal immune responses at other sites.

ANATOMY OF THE HUMAN MALE AND FEMALE REPRODUCTIVE TRACTS

The human male reproductive system includes the testis, epididymis, vas deferens, seminal vesicles, prostate, and the urethra (Figure 20.1, right). The testis of the male serves the dual function of producing both gametes (sperm) and sex hormones (testosterone). In the testis, androgen synthesis occurs in the Leydig cells, and spermatogenesis takes place in the seminiferous tubules. The two hormones (gonadotropins) produced by the anterior pituitary that regulate testicular function are follicle-stimulating hormone and luteinizing hormone (LH), each controlled by hormones produced in the hypothalamus. Spermatogenesis and androgen synthesis in Leydig cells are regulated by a negative feedback loop involving the hypothalamic-pituitary-gonadal

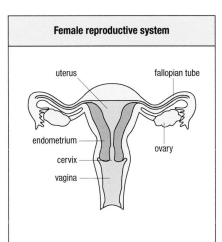
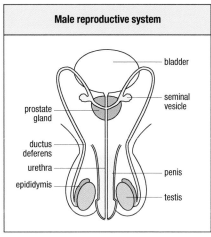

Figure 20.1 **Internal and external organs of the female and male human reproductive tracts.** Both the female and male reproductive systems are involved in the production of gametes and the secretion of hormones. In the female, the reproductive system produces ova or eggs and the hormones estradiol and progesterone. In the male, the reproductive system produces sperm as well as the hormone testosterone.

axis. Sperm formed in the seminiferous tubules enter the vas deferens, an androgen-dependent organ that transports sperm into the pelvis. The vas deferens joins the seminal vesicles to form the ejaculatory ducts that enter the prostatic urethra. Just prior to ejaculation, the testes are brought close to the abdomen, and fluid is rapidly transported through the vas deferens to the ejaculatory duct and subsequently into the prostatic urethra, the final pathway for both urine and semen.

The female reproductive tract has upper and lower components. The lower female reproductive tract structures (ectocervix, vagina, and vulva) enable sperm to enter the body and protect the internal genital organs from potential pathogens. From the vulva to the vagina, the epidermis transitions from keratinized to stratified, nonkeratinized squamous epithelium. The vaginal mucosa consists of three distinct layers; (1) the outer surface layer, which is lubricated by mucus produced primarily by the cervix; (2) the intermediate layer, which acts as an active site of glycogen production; and (3) the basal layer, made up of actively dividing cells. Internal reproductive structures include the endocervix, uterus, and fallopian tubes (see Figure 20.1, left). The ovary produces both gametes (ova) and sex hormones (estradiol and progesterone). The fallopian tubes attached to the upper part of the uterus provide passage for the ovum from the ovary to the uterus. Following fertilization of an egg by a sperm in the fallopian tubes, the fertilized egg then moves to the uterus, where it implants into the lining of the uterine wall. Each site in the female reproductive tract functions to insure passage of sperm to the site of fertilization, permit release of menstrual flow, as well as provide a route of passage of the baby at birth.

Hormonal regulation of female reproductive function

The menstrual cycle can be divided into proliferative, midcycle, and secretory phases (Figure 20.2). Follicle-stimulating hormone and LH made by the pituitary in response to hypothalamic signals are primarily responsible for the regulation of germ cell development and ovulation during the follicular phase. Follicle-stimulating hormone stimulates follicle growth and maturation by its action on granulosa cells, but ovulation does not take place. For follicular growth to occur, follicle-stimulating hormone acts in synergy with estradiol and LH. LH participates in the events leading to ovulation and acts on the preovulatory follicle to cause its rupture and the release of the ovum. Levels of these hormones in plasma are shown in Figure 20.2. The midcycle peak of LH is essential for ovulation, and occurs between 24 and 36 hours after estradiol levels peak in blood. Following ovulation, LH plays an important role in the formation and maintenance of the corpus luteum, the primary source of estradiol and progesterone during the secretory phase of the cycle. Human chorionic gonadotropin, with its intrinsic LH activity, rescues the corpus luteum during the cycle if fertilization occurs, and thereby maintains hormone production early in pregnancy. When fertilization does not occur, the corpus luteum involutes.

The mucosal immune system in the female reproductive tract contains an array of protective mechanisms that are under hormonal control. Resident epithelial cells and supportive stromal (lamina propria) cells, and the immune cells that migrate into the uterus, cervix, and vagina, confer immune protection to the female reproductive tract.

20.1 Regulation of cytokine, chemokine, and antimicrobial products contribute to protection of the female reproductive tract

Soluble factors, including cytokines, chemokines, and antimicrobial products, are secreted constitutively in the female reproductive tract and in response

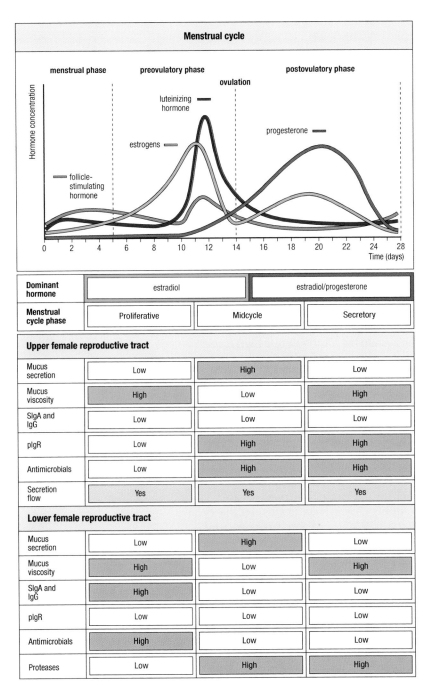

Figure 20.2 Variable levels of sex hormones secreted during the menstrual cycle regulate the secretion of protective factors in the reproductive tract of women. The menstrual cycle is divided into three phases: the proliferative phase during which estradiol is produced by the ovary; the midcycle phase when ovulation occurs; and the secretory phase during which estradiol and progesterone are the dominant hormones. These hormones, in turn, regulate the synthesis and secretion of mucus, antibodies, and antimicrobial products in the female reproductive tract. Soluble products optimize conditions for successful fertilization while protecting against potential pathogens. pIgR, polymeric immunoglobulin receptor; SIgA, secretory IgA.

to pathogenic challenge. Sensing of invading pathobionts in the reproductive tract occurs via pathogen recognition receptors such as toll-like receptors (TLRs) and nucleotide-binding oligomerization domain (NOD)-like receptors (NLRs), which act as bridging pathways between innate and adaptive immunity in the reproductive tract. At least 10 TLRs have been identified, each of which recognizes a limited number of highly conserved molecular structures known as "pathogen-associated molecular patterns." For example, TLR2 detects peptidoglycan found in gram-positive bacteria, whereas lipopolysaccharide in gram-negative bacteria is recognized by TLR4. TLR3 recognizes double-stranded viral RNA, while TLR3, TLR7–9, RIG-1, and MDA-5 sense nucleic acids. TLR signaling is mediated by four key adaptor proteins, MyD88, TRIF, MAL, and TRAM. All TLRs signal through MyD88, except TLR3, which signals solely through TRIF. TLR stimulation leads to activation and nuclear translocation of inflammatory transcription factors such as nuclear factor (NF)-κB and activator protein 1 (AP1), which subsequently upregulate expression of inflammatory cytokines such as interleukin (IL)-1β, IL-6, and tumor necrosis factor (TNF)-α, and chemokines such as IL-8 as well as a range of prostaglandins and extracellular matrix remodeling enzymes. Expression of TLRs throughout the female reproductive tract is highly tissue specific. For example, TLR7, 8, and 9 are constitutively expressed in the tissues of the upper component of the female reproductive tract, whereas TLR10, the ligands for which are unknown, is expressed exclusively in the fallopian tubes. Although TLR4 is expressed in high levels in the cervix and uterus, low expression in the vagina combined with the absence of MyD88 expression prevents lipopolysaccharide-induced activation of the innate immune response at this site. In contrast, NLRs, which are intracellular pattern recognition receptors, are expressed throughout the female reproductive tract. Of the 22 members to have been identified in the human genome to date, NOD1 and NOD2 are the best-characterized NLRs. These are activated by peptidoglycan-related molecules containing the γ-D-glutamyl-meso-diaminopimelic acid (iE-DAP for NOD1), and muramyl dipeptide (MDP for NOD2), respectively. Both NOD1 and NOD2 interact with RICK, a member of the receptor interacting protein kinase family, to induce NF-κB and downstream induction of cytokine and chemokine expression. Such high regulation of tissue-specific expression of pathogen recognition receptors is a likely reflection of an evolutionary driven process whereby the host innate immune system learns to tolerate high levels of commensal bacteria at specific sites such as the vagina, while maintaining the capacity to sense and respond to pathobiont invasion at other sites.

The cervix is a key organ involved in contributing to innate immune defense mechanisms that prevent infection in the female reproductive tract. Apart from providing a physical barrier to ascending infection, particularly during pregnancy, mucus produced by the cervix creates a high-viscosity barrier against microbial attachment to the underlying epithelium of the cervix and vagina. Further, the mucociliary escalator of the cervix actively propels microbes back into the vagina, with those that do attach to the epithelium being removed via desquamation. The cervix and vagina also secrete a wide variety of antimicrobial compounds including α- and β-defensins, secretory leukocyte protease inhibitor, elafin, lactoferrin, calgranulin, histones, and LL37, all of which inhibit bacterial, fungal, and viral pathogens, including human immunodeficiency virus-1 (HIV-1). MIP-3α and elafin are dual mediators in that they have antimicrobial activity as well as a chemoattractant function for the recruitment of T cells and immature dendritic cells to sites of pathogen invasion. Secreted factors therefore contribute to initial mucosal protection through their antimicrobial activity and recruit and activate a second line of cellular immune protection. Among those cells pivotal in conferring immune protection are mucosal epithelial cells. In addition to secreting antimicrobial factors that are both bactericidal and viricidal, epithelial cells provide barrier protection and transport immunoglobulins (IgA and IgG) into female reproductive tract secretions.

A spectrum of other factors such as immunoglobulin, antimicrobial peptides, and proteases are present in reproductive tract secretions. The secretion of these factors varies with the site (upper versus lower female reproductive tract) and the phase of the reproductive cycle (see **Figure 20.2**). For example, the level and viscosity of secreted mucus vary with the cycle phase throughout the female reproductive tract. Cervical and vaginal mucus production is low with high viscosity during both the proliferative and secretory phases of the cycle. In contrast, at midcycle, estradiol leads to increased secretion of mucus with decreased viscosity to facilitate sperm movement from the vagina to the upper female reproductive tract. IgA and IgG levels are reduced in the upper tract throughout the cycle. In the lower tract, levels are low at midcycle and during the secretory phase. The polymeric immunoglobulin receptor (pIgR), which transports IgA from tissue to lumen, is elevated at midcycle in the uterus. Antimicrobials, including secretory leukocyte protease inhibitor, which has antiprotease activity, and defensins, also vary with cycle stage in the female reproductive tract. In the upper tract, levels are elevated at midcycle and during the secretory phase. In contrast, they are lowest during this period in the lower female reproductive tract. Thus, sex hormones orchestrate a dynamic balance between humoral and innate immune components such that different components emerge as others are suppressed, optimizing chances for successful fertilization, while maintaining immune capacity.

20.2 Role of commensal microbiota in female reproductive tract immunity

The lower reproductive tract is host to rich and abundant microbial communities that vary throughout life. Emerging evidence suggests that the upper reproductive tract may also be host to a comparatively low biomass microbiota. The composition of these communities is strongly influenced by hormonal levels of the host, sexual activity, menses, contraceptive use, and hygiene habits. In reproductive-age women, the vagina is colonized primarily with *Lactobacillus* spp., but around 30% of women will harbor a *Lactobacillus* spp.-depleted, high-diversity bacterial community that can include commensals from the genera *Lactobacillus, Atopobium, Streptococcus*, and *Megasphaera*. Many of the species contained in these genera are lactic acid–producing bacteria, which suggests that alterations to the chemical environment of the mucosal layer of the lower reproductive tract may be as functionally important as the species themselves. These bacteria are by no means bystanders and instead actively modulate local innate immunity. For example, the production of lactate by commensal bacteria lowers the acidity in most women to a level (pH < 4.5) optimal for their own survival but that is inhibitory to other pathogenic species. For example, a low vaginal pH is associated with decreased risk of chlamydia, HIV, mycoplasma, and other urinary tract infection. *Lactobacillus* spp. also prevent infection and colonization of pathobionts through production of hydrogen peroxide, secretion of bacteriolytic enzymes, prevention of adhesion (e.g., lipoteichoic acids and glycoproteins), and bacteriocins (gassericins and acidocins). Viscosity of cervical mucus in the vagina is also influenced by bacterial community composition. For example, women with *Lactobacillus crispatus*-dominated communities have higher mucosal viscosity, which is associated with enhanced trapping of HIV particles.

20.3 Cellular immunity is also important in reproductive tract

Cellular components of the immune system in the female reproductive tract include dendritic cells, macrophages, natural killer (NK) cells, neutrophils, and CD4+, CD8+, and regulatory T cells (Figure 20.3). Leukocytes in the female reproductive tract make up approximately 12%–40% of the total number of cells, with a higher proportion in the upper tract (fallopian tubes and uterus) than the lower tract (cervix and vagina). T cells and NK cells are the major constituents of female reproductive tract leukocytes in all tissues. The fallopian tube also contains granulocytes, but these are much less common in the other tissues. All female reproductive tract tissues contain small numbers of B cells and macrophages. Immune cell numbers vary with the menstrual cycle, with T cells, neutrophils, and macrophages increasing in number as the menstrual cycle progresses from follicular to secretory phase (see Figure 20.3). Immune cell organization varies with tissue and stage of the menstrual cycle. Confocal scanning laser microscopy of vibratome-prepared uterine tissue sections shows aggregates of T cells exclusively in the uterus; these aggregates are oval in shape and located between glands in the

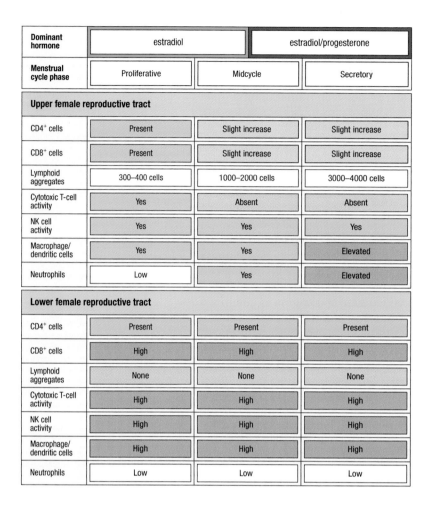

Dominant hormone	estradiol		estradiol/progesterone
Menstrual cycle phase	Proliferative	Midcycle	Secretory
Upper female reproductive tract			
CD4+ cells	Present	Slight increase	Slight increase
CD8+ cells	Present	Slight increase	Slight increase
Lymphoid aggregates	300–400 cells	1000–2000 cells	3000–4000 cells
Cytotoxic T-cell activity	Yes	Absent	Absent
NK cell activity	Yes	Yes	Yes
Macrophage/ dendritic cells	Yes	Yes	Elevated
Neutrophils	Low	Yes	Elevated
Lower female reproductive tract			
CD4+ cells	Present	Present	Present
CD8+ cells	High	High	High
Lymphoid aggregates	None	None	None
Cytotoxic T-cell activity	High	High	High
NK cell activity	High	High	High
Macrophage/ dendritic cells	High	High	High
Neutrophils	Low	Low	Low

Figure 20.3 Sex hormones released during the menstrual cycle regulate the recruitment and function of immune cells in the upper (fallopian tubes, uterus, and endocervix) and lower (ectocervix and vagina) female reproductive tissues. Estradiol and progesterone act directly and indirectly through growth factors, cytokines, and chemokines to regulate immune-cell trafficking and activation in a cyclic pattern of immune surveillance. In each tissue of the female reproductive tract, immune cells and their soluble products play a key role in protecting the mucosa from sexually transmitted pathogens.

functionalis region. The lymphoid aggregates contain a B-cell central region surrounded by T cells and an outer halo of macrophages. The B-cell core (CD19+) is most often identified in large aggregates in the late proliferative and secretory stages of the cycle. Phenotypic analysis indicates that the surrounding T cells are almost exclusively CD3+CD8+. CD3+CD4+ T cells are also present but usually are present outside the aggregates, mainly in the stroma. Macrophages (CD14+ cells) are present as a mantle around the T cells. Aggregates are smallest during the proliferative phase (300–400 cells per aggregate) and increase to 3000–4000 cells per aggregate during the secretory phase of the cycle. Similar menstrual cycle changes occur in the CD8+ cytotoxic T-lymphocyte (CTL) population. Although absolute numbers do not change in the uterus, CTL activity is demonstrable during the follicular phase but is suppressed during the secretory phase of the cycle, possibly to optimize conditions for successful implantation of the allogeneic fertilized egg. In contrast to CTL activity in the uterus, CTL activity in the cervix and vagina is sustained throughout the menstrual cycle and is not affected by hormonal balance.

Endocrine control of immune protection in reproductive tract

The reproductive tract in the female undergoes extensive remodeling during the menstrual cycle. This remodeling is regulated by an array of sex hormones. These hormones also regulate immune-cell function in tissues of the female reproductive tract as the immune system faces the unique challenge of protecting tissues that are changing shape, thickness, and vascularity.

20.4 Sex hormones directly and indirectly regulate immune-cell function in the reproductive tract

Sex hormones (estradiol and progesterone) directly and indirectly regulate immune-cell function in the female reproductive tract. Estradiol receptors are expressed by innate and adaptive immune cells, such as epithelial cells, stromal cells, macrophages, dendritic cells, natural killer cells, and lymphocytes.

Estradiol effects on stromal cells, which then influence other cell types in the female reproductive tract, were initially suggested by experiments in which vaginal epithelia from neonatal mice were grown in the presence of uterine stroma. Under such conditions, the epithelia developed into a uterine-like structure. In other studies, neonatal mice were implanted with a mixture of epithelial cells lacking the nuclear estrogen receptor-α (estradiol receptorα) and stromal fibroblasts that expressed the receptor. Treatment with estradiol stimulated uterine epithelial-cell proliferation, leading to the conclusion that estradiol stimulates mitogenesis of uterine (estradiol receptor α-positive) epithelial cells indirectly, via the underlying stroma.

Several growth factor–growth factor receptor systems, including epidermal growth factor, insulin-like growth factor-I, hepatocyte growth factor, keratinocyte growth factor, and transforming growth factor β (TGF-β), have been identified within the uterus and are involved in stromal fibroblast–epithelial cell–immune cell interactions. Within the local environment of the reproductive tract, interactions between estradiol, progesterone, and growth factors are complex and involve bidirectional cross talk between ligands and receptors, influencing both the reproductive cell function and immune cell migration, activation, and function.

Sex hormones also play a major role in shaping the composition of vaginal commensal bacteria. In particular, increasing concentrations of estrogen are associated with maturation of the vaginal epithelium leading to increased deposition of glycogen. Host α-amylase present in the vaginal mucosa then breaks glycogen down to complex sugar products (e.g., maltose, maltotriose, and maltotetraose) that can be preferentially utilized as carbon sources by *Lactobacillus* spp. Consistent with this, the onset of puberty and increased circulating estrogen concentration are associated with a reduction in the diversity of bacterial community composition seen in prepubescent females, and a shift toward *Lactobacillus* spp. (Figure 20.4) dominance, which continues for the majority of a woman's reproductive life. Expectedly, onset of menopause sees reduced *Lactobacillus* spp. dominance and increased

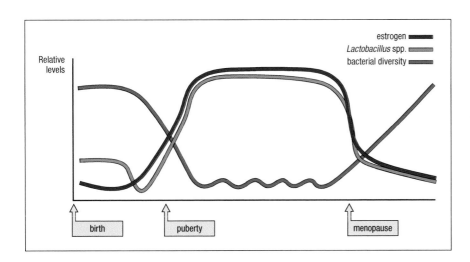

Figure 20.4 The relationship between vaginal microbiota diversity, *Lactobacillus* species abundance, and estrogen levels through a women's life cycle. The vagina is rapidly colonized after birth and maintains a microbiota structure characterized by high species diversity and low *Lactobacillus* spp. abundance until puberty. Increased estrogen levels preceding menarche promote reduced microbial diversity and increased *Lactobacillus* spp. dominance. This community structure is maintained throughout a woman's reproductive age but differs between ethnic groups. Fluctuation in composition is associated with cyclical secretion of estrogen and progesterone during the menstrual cycle. Following menopause, reduced estrogen and vaginal atrophy contribute to *Lactobacillus* spp. depletion and increased diversity.

diversity. Further evidence for estrogen-driven *Lactobacillus* spp. dominance of the vaginal niche is provided in pregnancy where a 1000-fold increase in circulating concentrations of estrogen is associated with increased *Lactobacillus* spp. dominance and a reduced richness and diversity of existing bacterial communities.

20.5 Immune protection is integrated in female reproductive tract during the menstrual cycle

Sex hormones downregulate many immunologic functions from midcycle and during most of the secretory phase of the menstrual cycle. Immunologic suppression occurs as an integral part of the physiologic processes that underlie successful reproduction. Selective immune suppression of certain components of the innate, humoral, and cell-mediated immune systems confers a window of vulnerability of approximately 7–10 days, raising the likelihood that viral, bacterial, and fungal pathogens can evade mucosal immune protection and infect the female reproductive tract. Immune suppression coincides with the recruitment of immune cells capable of hosting pathogens and the upregulation of receptors on target cells essential for pathogen infectivity.

20.6 Immune mechanisms contribute to protection of male reproductive tract

The male reproductive tract is less well studied than the female reproductive tract. There are, however, similarities in the restraints imposed on immune responses in both the male and female reproductive tracts. As the female reproductive tract must be regulated/suppressed to allow implantation and growth of a semi-allogeneic fetus, immunity to developing spermatozoa must be prevented in the male reproductive tract. Many genes expressed by developing male gametes are absent during late fetal and early neonatal life, the time when the developing immune system is "educated" to learn self and nonself. Consequently, many sperm antigens escape tolerance, making the developing gamete susceptible to autoimmune attack at puberty. The male reproductive tract, particularly the testis, is an "immune privileged" site. For example, survival of allografts in the testis is prolonged compared with survival in other tissues, and within the male reproductive tract multiple mechanisms prevent or suppress autoimmune responses to developing spermatozoa.

The male reproductive tract consists of the testis, the excurrent duct system (rete testis, efferent ducts, epididymis, and vas deferens), the accessory glands (seminal vesicles, prostate, and bulbourethral/Cowper's glands), the penile urethra, and the foreskin (**Figure 20.5**). The testis has a germinal compartment (the seminiferous tubules) containing Sertoli cells, immature germ cells and spermatozoa, and the interstitial tissues; the latter containing blood vessels, lymphatics, mast cells, and Leydig cells. To protect developing spermatozoa, the blood-testis barrier, formed by tight junctions between adjacent Sertoli cells, prevents entry of immunoglobulin, complement, and other macromolecules into the seminiferous tubules. Immune cells are rarely found in the germinal compartment, but macrophages and some T cells are common in the testicular interstitium. Cytokine secretion by interstitial macrophages probably contributes to both the suppression of immune responses in the testis and the regulation of Leydig cell function essential for spermatogenesis. Multiple immune-suppressive mechanisms contribute to testicular immune suppression. FasL expression on Sertoli cells may play a role in deleting activated T cells by apoptosis. Sertoli cells may also contribute to maintenance of immune suppression through the secretion of FasL, the tryptophan-degrading enzyme indoleamine 2,3 dioxygenase, TGF-β 1

	Innate				Adaptive				
	Macrophage/ dendritic cells	Neutrophils	NK cells	Anti-microbials	CD4+ T cells	CD8+ T cells	IgA B cells	IgG B cells	pIgR
Seminiferous tubules	++	–	undetermined	++	++	++	+/–	+/–	+
Epididymis	++	+	undetermined	++	++	++	+/–	+/–	+
Prostate	++	+	++	+	+	+	++	+	++
Ductus deferens	++	+	undetermined	undetermined	++	++	+/–	+/–	+
Seminal vesicles	–	–	–	++	–	–			+
Penis (urethra)	+++	+	+/–	undetermined	+++	+++	+++	++	+++
Prepuce	++	++	+++	undetermined	+	+	–	–	–

Figure 20.5 **Immune cells in the internal and external tissues of the male reproductive tract.** Immune cells are present throughout the tissues of the male reproductive tract. Located at sites adjacent to the site of sperm production (seminiferous tubules), immune cells protect the male reproductive tract from pathogens that invade the tract to cause ascending infection.

and 3, activin A, and inhibitors of complement activation and granzyme B. Maintenance of immune privilege may also depend on the continuous release of germ-cell autoantigens, perhaps important in maintaining immune tolerance. Multiple mechanisms may limit adaptive immunity to developing spermatozoa, but inflammatory responses to infection occur in the testis, although infections in the testis appear to be less common than in other regions of the male reproductive tract. Important questions that remain to be answered include the degree to which immune privilege exists in other regions of the male reproductive tract, including the epididymis, where sperm mature and are stored prior to ejaculation, and whether these immune mechanisms could adversely affect vaccine-induced immunity to sexually transmitted organisms.

20.7 Microbiota of male reproductive tract

The composition of the male reproductive tract microbiota is poorly characterized in comparison to the female reproductive tract. The coronal culcus (the base of the glans penis) and the urethra are colonized by different communities that are often composed of bacteria associated with bacterial vaginosis including *Prevotella*, *Atopobium*, *Megasphaera*, and *Gemella* species. However, much variability in the composition of these communities exists between individuals. The structures of these communities can be strongly influenced by sexual activity as well as by circumcision. Both culture-dependent and culture-independent approaches have shown that bacteria also colonize seminal fluid. Semen samples collected from healthy individuals are generally dominated by *Lactobacillus* taxa, whereas samples collected from donors with abnormal semen parameters are enriched for *Anaerococcus*, *Pseudomonas*, and *Prevotella* species. The role played by these bacteria in the male reproductive tract remains unclear; however, recent studies in mice indicate that seminal fluid microbiome composition is influenced by interactions between estrogen-regulated pathways and innate immunity. For example, mice that do not express the estrogen receptor α have reduced seminal fluid levels of *Propionibacterium acnes*, the causative agent of chronic prostatitis, and which is a risk factor for prostate cancer.

20.8 Innate immunity is an important component of male reproductive tract immune system

In the male reproductive tract, the suppression of adaptive immunity to maintain immune privilege complements enhanced innate immune protection against local infection. Seminal plasma has strong bacteriostatic and bactericidal activity due to a variety of mediators of innate immunity, such as zinc, lysozyme, and transferrin, and also large numbers of antimicrobial peptides. The antimicrobial peptides include β-defensins, human epididymis peptide 2 (HE2/EP2/SPAG11), hCAP-18, proteolysis-inducing factor (PIF, a dermacidin), epididymal protease inhibitor (EPPIN), semenogelins, and surfactant protein D. HE2/EP2/SPAG11 has antimicrobial activity against *Neisseria gonorrhoeae* and *Enterococcus faecalis*, whereas surfactant protein D inhibits infection of prostate epithelial cells by *Chlamydia trachomatis*. Antimicrobial peptide expression has been documented in most regions of the male reproductive tract, including the testis (Sertoli cells), seminal vesicles, epididymis, and prostate. Similar to estradiol regulation in the female reproductive tract, antimicrobial peptides may be regulated by androgens. NK cells have been identified in the testis and prostate (human and rat), but they are infrequently detected in the penile urethra.

20.9 Adaptive immunity in male reproductive tract is mediated predominantly by CD8+ T cells

Throughout the excurrent ducts, CD8+ T cells are present in the epithelium of the rete testis, the efferent ducts, and the epididymis, whereas CD4+ T cells are present in the lamina propria of these tissues. The majority of these T cells express the αβ T-cell receptor. Macrophages are abundant in most regions of the excurrent ducts, including the epithelium and lumen of the duct system. The pIgR, which transports polymeric IgA and IgM across mucosal epithelia, is expressed in the rete testis, efferent ducts, epididymis, and seminal vesicles. High levels of pIgR, as well as IgA and IgM plasma cells, have been identified in the prostate and the penile urethra. Most of these plasma cells contain J chain, suggesting that the secretory immune system is well developed at these sites. Large focal accumulations of lymphocytes, including CD8+ T cells, NK cells, and some B cells, are present in normal human prostate. Foxp3+CD4+ cells in the prostate may regulate responses to sperm or sexually transmitted pathogens. Antigen-presenting cells (dendritic cells and macrophages) and T cells are abundant in the human penile urethra. Most urethral intraepithelial T cells are CD8+, and many express the CD45RO marker characteristic of memory cells. A small population of CD4+ cells is present in the penile urethra. The presence of antigen-presenting cells and CD4+ cells in the penile urethra suggests that this tissue may be an inductive site, although this has not been demonstrated. The predominance of CD8+ cells at this site, however, is consistent with cell-mediated immunity to potential pathogens or, alternatively, immunosuppressive function. The function of T cells in the male reproductive tract has received little investigative attention, but adoptive transfer of immune CD4+ T cells from mice immunized intranasally with *Chlamydia* major outer membrane protein and cholera toxin into naive recipients provides partial protection against genital tract challenge with *Chlamydia*, raising the possibility that immune CD4+ cells, activated by intranasal immunization, can home to the male reproductive tract and mediate protective immunity.

IgG, not IgA, is the dominant isotype present in seminal plasma of healthy males. IgA in seminal plasma is predominantly IgA1, whereas IgA2 occurs in slight excess in female genital tract secretions. Seminal plasma IgA contains secretory IgA, polymeric IgA, and monomeric IgA, suggesting that local production and pIgR-mediated transport as well as serum contribute to the IgA present at this site. Both systemic and mucosal immunization can elicit antibody responses in seminal plasma. Systemic immunization of men with alum-adsorbed tetanus and diphtheria toxoids has been reported to elicit predominant IgG responses in seminal plasma (and serum), while oral immunization with a live attenuated *Salmonella enterica* serovar Typhi vaccine elicits secretory IgA (SIgA) in seminal plasma. In mouse models, intranasal immunization with chlamydial major outer membrane protein and cholera toxin as an adjuvant induces antigen-specific IgA plasma cells in the prostate and penile urethra, and *Chlamydia*-neutralizing IgA in prostatic secretions. Prostatic fluid from identically immunized pIgR knockout mice is unable to neutralize *in vitro* chlamydial infection, emphasizing the importance of SIgA in this model. Collectively, these findings indicate that both systemic and secretory antibodies contribute to humoral immunity in the male reproductive tract. IgG is predominantly serum derived, whereas IgA in seminal plasma is derived from both local mucosal production and from serum, suggesting that both systemic and mucosal immunization approaches should be considered as strategies for the induction of humoral immunity to sexually transmitted infections.

Infection and immune protection against sexually transmitted diseases in male and female reproductive tracts

Sex in mammals is invasive and offers a means by which pathogens can be transmitted into internal compartments of the body. Sexually transmitted diseases are a major and increasing health problem in humans and are also important in many domesticated and semidomesticated animals. The introduction of Gardasil and Cervarix, systemic vaccines against human papillomavirus that protect against some human papillomavirus serotypes, genital warts, and hopefully cervical cancer, represents a major breakthrough.

20.10 *Chlamydia trachomatis* is most common sexually transmitted disease worldwide

Approximately 90 million new *Chlamydia trachomatis* infections occur annually, mostly in the young (15–29 years). The target cells are the columnar epithelial cells of the endocervix in women, and the epithelial cells of the penile urethra in men. In women, more than 75% of infections are asymptomatic. Important sequelae of infection include pelvic inflammatory disease, infertility, and ectopic pregnancy, all of which are due to ascending infection, and chronic inflammation that occurs over a period of months or years. In men, up to 50% of infections are asymptomatic, and infection can result in prostatitis, epididymitis, and infertility.

Chlamydia are obligate intracellular gram-negative bacteria and have a unique biphasic life cycle. The infectious elementary body is metabolically inert and binds to unidentified receptors on target cells. Once internalized, the elementary body undergoes transformation to the replicative form, known as the reticulate body, which replicates by binary fission within a membrane-bound inclusion. After 24–72 hours, reticulate bodies convert back to elementary bodies, and the newly formed elementary bodies are released either by lysis or extrusion. A third form, the nonreplicating, persistent, or aberrant body, can be induced by nutritional stress, certain antibiotics, or immune pressure by cytokines such as interferon-γ (IFN-γ). Upon removal of the stress, *Chlamydia* differentiates back to the normal replicating forms. Natural immunity following infection is serovar-specific and short lived. Local host defense against *Chlamydia* involves the mucosal barrier and innate, humoral, and cell-mediated responses (Figure 20.6). *Chlamydia* possesses multiple immune evasion mechanisms that

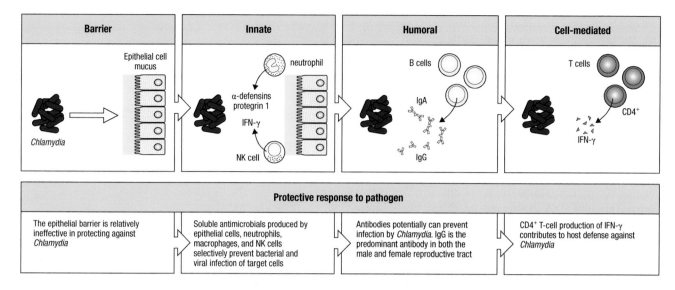

Figure 20.6 Levels of immune protection against *Chlamydia trachomatis* in the female and male reproductive tracts. Barrier protection and innate, humoral, and cell-mediated immunity provide integrated protection against genital pathogens such as *Chlamydia trachomatis*. Although *Chlamydia* has evolved to escape barrier protection, innate antimicrobial factors and adaptive responses restrict bacterial and viral infection of genital target cells. T cells also play a critical role in protecting against infection.

regulate host innate and adaptive immune responses. Resolution of infection coincides with recruitment of CD4+ T cells into the infected tissues, and studies in both humans and animal models have identified IFN-γ-producing CD4+ T cells as essential for resolution of natural infection and prevention of experimental infections in animal models. Interferon-mediated protection in humans involves tryptophan depletion mediated by activation of indoleamine 2,3-dioxygenase, which starves the replicating *Chlamydia* of this essential amino acid. Adoptive transfer of both IgG and IgA monoclonal antibodies against chlamydial surface antigens can protect mice against chlamydial challenge, and studies in animal models have shown that specific antibodies are essential for protection against reinfection. Activation of the immune system following chlamydial infection is a "double-edged sword," however, as both the innate and adaptive immune responses elicited by infection have been implicated in the development of inflammatory sequelae. Chlamydial infection of epithelial cells results in secretion of large amounts of pro-inflammatory chemokines and cytokines, including IL-8, GRO-α, GM-CSF, TNF-α, IL-1α, and IL-6. Unlike the short-lived innate immune response characteristic of most other bacterial infections, the secretion of inflammatory cytokines is prolonged in chlamydial infections, resulting in chemokine-mediated recruitment of inflammatory cells to the site of infection, and killing of fallopian tube ciliated epithelial cells by cytokines. Thus, the exaggerated innate immune response to chlamydial infection is a major contributor to the tissue scarring, fibrosis, and tubal occlusion associated with pelvic inflammatory disease, infertility, and ectopic pregnancy. Similarly, the adaptive immune response elicited by chlamydial infection, particularly antibody and T-cell–mediated immune responses against chlamydial heat shock protein 60 (cHSP60), has been implicated in tissue damage. cHSP60 and human HSP60 share significant sequence identity, such that cHSP60-specific antibody and T cells can recognize and potentially damage host tissues that express human HSP60 (molecular mimicry). Currently no vaccines are available to prevent genital chlamydial infections in humans. Important issues that need to be resolved are the identification of critical chlamydial antigens to elicit protective immunity, without causing tissue pathology, and the best adjuvant(s) and routes of immunization for inducing protection in the female and male reproductive tracts.

SUMMARY

The mucosal immune system in the male and female human reproductive tracts has evolved to meet the unique requirements of procreation and host defense against genital tract pathogens. Multiple levels of innate and adaptive immune protection in the female reproductive tract are regulated by sex hormones and optimize conditions for fertilization and pregnancy, while providing protection against potential pathogens (Figure 20.7). These ovary-derived hormones act directly and indirectly through soluble mediators on local immune cells,

modulating local immune function in a manner compatible with successful reproduction, yet influencing the response to sexually transmitted pathogens such as *Chlamydia*. Clearly, our understanding of the immunobiology of reproduction and genital mucosal infections has advanced substantially in recent years. Further elucidation of the cellular and molecular mechanisms of host defense in the female and male reproductive tracts should inform new approaches for therapy and vaccine prevention of infection in the human genitourinary tract.

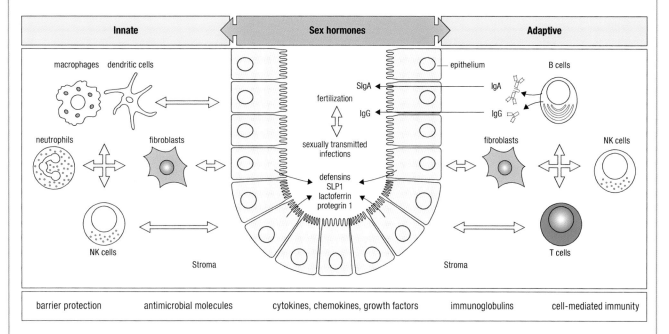

Figure 20.7 Integrated control of the mucosal immune system in the human female reproductive tract. The hormones estradiol and progesterone regulate immune cell and vascular cell receptor expression and immune cell function in tissues of the female reproductive tract. Communication between epithelial cells, local immune cells, and tissue cells, such as fibroblasts, is achieved through local release of cytokines, chemokines, and growth factors. Natural killer cell and dendritic cell activity is downregulated, whereas CD8$^+$ T-cell activity is maintained, except during the secretory phase of the cycle, in female genital tissue. Estradiol and progesterone also promote barrier protection, antimicrobial activity, local antibody secretion, and cell-mediated immunity. Acting in concert with cytokines, chemokines, and growth factors, sex hormones prepare the tract for fertilization, implantation, and pregnancy, while protecting the genital mucosa against bacterial and viral pathogens that may threaten reproductive health.

FURTHER READING

Brotman, R.M., Ravel, J., Bavoil, P.M. et al. Microbiome, sex hormones, and immune responses in the reproductive tract: Challenges for vaccine development against sexually transmitted infections. *Vaccine* 2014, 32:1543–1552.

Com, E., Bourgeon, F., Evrard, B. et al.: Expression of antimicrobial defensins in the male reproductive tract of rats, mice, and humans. *Biol. Reprod.* 2003, 68:95–104.

Griffin, J.E., and Wilson, J.D.: Disorders of the testes and the male reproductive tract, in Larsen, P.R., Foster, D.W., Kronenberg, H.M. et al. (eds): *Williams Textbook of Endocrinology*, 10th ed. Philadelphia, W. B. Saunders Company, 2003:709–770.

Hafner, L.M., Cunningham, K., and Beagley, K.W.: Ovarian steroid hormones: Effects on immune responses and *Chlamydia trachomatis* infections of the female genital tract. *Mucosal Immunol.* 2013, 6:859–875.

Hall, S.H., Hamil, K.G., and French, F.S.: Host defense proteins of the male reproductive tract. *J. Androl.* 2002, 23:585–597.

Hickey, D.L., Patel, M.V., Fahey, J.V. et al. Innate and adaptive immunity at mucosal surfaces of the female reproductive tract: Stratification and integration of immune protection against the transmission of sexually transmitted infections. *J. Reprod. Immunol.* 2011, 88:185–194.

Ma, B., Forney, L.J., and Ravel, J.: Vaginal microbiome: Rethinking health and disease. *Annu. Rev. Microbiol.* 2012, 66:371–389.

Meinhardt, A., and Hedger, M.P.: Immunological, paracrine and endocrine aspects of testicular immune privilege. *Mol. Cell. Endocrinol.* 2011, 15:60–68.

Moldoveanu, Z., Huang, W., Kulhavy, R. et al.: Human male genital tract secretions: Both mucosal and systemic immune compartments contribute to the humoral immunity. *J. Immunol.* 2005, 175:4127–4136.

Nasu, K., and Narahara, H.: Pattern recognition via the toll-like receptor system in the human female genital tract. *Mediators Inflamm.* 2010, 2010:976024.

Wira, C.R., Rodriguez-Garcia, M., and Patel M.V.: The role of sex hormones in immune protection of the female reproductive tract. *Nat. Rev. Immunol.* 2015, 15:217–230.

Mucosal immune responses to microbes in genital tract

21

AKIKO IWASAKI

Sexual reproduction provides an evolutionary advantage that generates a significant increase in the fitness of offspring. It is a fundamental activity for a species to procreate and survive in the ever-changing environment of its habitat. Not surprisingly, certain viruses, bacteria, fungi, and parasites have evolved to utilize sexual contact of the host organism for their own transmission. Sexually transmitted pathogens enter the human hosts through the male and female genital mucosae. The World Health Organization estimates that one million new cases of sexually transmitted diseases (STDs) occur daily. Yet, for the most part, vaccines that prevent STDs are not available.

Compared with the intestinal mucosa, female and male genital tracts are covered by distinct epithelial layer and mucus types, are inhabited by a unique set of microbial flora, and utilize distinct innate and adaptive effector mechanisms. In Chapter 20, we reviewed the anatomy and immunophysiology of the female and male genital mucosae. In this chapter, major human sexually transmitted pathogens are described with respect to our current understanding of their epidemiology, mode of entry, innate detection, initiation of adaptive immune responses, and clearance.

GLOBAL PREVALENCE OF SEXUALLY TRANSMITTED PATHOGENS

Certain bacteria, parasites, and viruses use genital mucosa as a portal of entry into human hosts. Clinically relevant STDs (with the estimated number of people infected globally indicated in brackets) include chlamydia (105.7 million), gonorrhea (106.1 million), syphilis (10.6 million), trichomonas (276.4 million), human immunodeficiency virus type 1 (HIV-1; 34 million), herpes simplex virus type 2 (HSV-2; 500 million), human papillomavirus (HPV; 630 million), and hepatitis B virus (HBV; 350 million). The prevalence of STDs in general is much higher in the developing countries. Since its discovery in the early 1980s, acquired immunodeficiency syndrome (AIDS) has already claimed the lives of more than 35 million people, the majority of whom lived in sub-Saharan Africa. In endemic countries, death and a reduced fertility rate caused by HIV-positive women have dramatically affected population demographics. AIDS has generated 14 million orphans in sub-Saharan Africa alone.

Diseases caused by sexually transmitted pathogens

Diseases caused by these sexually transmitted pathogens range from nuisance (warts) to lethal diseases (AIDS, cervical cancer). STDs caused by bacterial (*Chlamydia trachomatis, Neisseria gonorrhoeae, Treponema pallidum*) and parasitic (*Trichomonas vaginalis*) pathogens are curable by antibiotic treatment. In contrast, STDs caused by viral pathogens (HIV-1, HSV-2, HPV, HBV) are treated but not cured by antiviral therapy. Vaccines against HPV and HBV are highly effective in preventing infection and disease caused by the viruses.

21.1   Chlamydia is a disease caused by the bacteria *Chlamydia trachomatis*

Infection with *Chlamydia trachomatis* is the most common STD in the United States, with an estimated 2.3 million Americans aged 14–39 infected. Chlamydia can be cured with antibiotic treatment with tetracycline, azithromycin, or erythromycin. However, 80%–90% of infected men and women are asymptomatic and often are unaware of their carrier status. In infected women, untreated infection may lead to pelvic inflammatory disease, which can cause scarring of the fallopian tubes and result in infertility, and also increases the likelihood of an ectopic pregnancy. In addition, during pregnancy, the infection can cause premature labor and delivery. Moreover, the infant could develop chlamydial conjunctivitis and chlamydial pneumonia.

21.2   Gonorrhea is a sexually transmitted disease caused by a bacterium, *Neisseria gonorrhoeae*

Neisseria gonorrhoeae, a bacterium that replicates within the mucosae of the cervix, uterus, and fallopian tubes in women, and in the urethra in women and men, is the cause of gonorrhea. Gonorrhea can be cured by treatment with antibiotics including ceftriaxone and azithromycin. Untreated infection can lead to complications including pelvic inflammatory disease, fallopian tube damage leading to infertility, or increased chance of ectopic pregnancy. In men, gonorrhea can cause epididymitis that may lead to infertility. Gonococcal ophthalmia neonatorum, which could lead to blindness, is the most common manifestation in infants born to mothers with gonococcal genital tract infections.

21.3   Syphilis is caused by infection with the bacterium *Treponema pallidum*

Syphilis is transmitted through direct contact with a sore, which can occur on the external genitals, vagina, anus, mouth, or in the rectum. Penicillin is used to treat the causative agent of syphilis, the spirochete bacterium *Treponema pallidum*. However, if untreated, syphilis can result in three stages of disease progression. The primary stage manifests with a single or multiple sores (chancres). The chancre lasts 3–6 weeks, and it heals without treatment. However, if untreated, the infection progresses to the secondary stage. Skin rash (often on palms of the hands and bottoms of the feet) and mucous membrane lesions appear that may be accompanied by fever, fatigue, weight loss, and swollen lymph nodes during secondary syphilis. The symptoms of secondary syphilis will resolve without treatment, but the infection will progress to the latent and possibly late stages of disease. The latent stage of syphilis can last for years. Upon reactivation, a tertiary stage of syphilis ensues, in which the disease may damage the internal organs, including the brain, nerves, eyes, heart, blood vessels, liver, bones, and joints. This damage can lead to difficulty coordinating muscle movements, paralysis, numbness, gradual blindness, dementia, and even death.

21.4 *Trichomonas vaginalis*, the cause of trichomoniasis, is a protozoal parasite that infects the vagina

The vagina is the most common site of infection in women with *T. vaginalis*, and the urethra is the most common site of infection in men. While infected men rarely exhibit symptoms, infected women often experience vaginal discharge and discomfort during intercourse and urination. *Trichomonas* infection during pregnancy can cause premature birth or low birth weight. This infection can be cured by treatment with the antibiotics metronidazole or tinidazole.

21.5 Human immunodeficiency virus-1, the cause of acquired immunodeficiency syndrome, is an STD

HIV-1 is a member of the lentivirus family of retroviruses. HIV-1 is transmitted through sexual contact (vaginal, penile, rectal, or oral), via blood, or by vertical transmission from infected mother to child. There is no cure or vaccine against HIV-1. HIV-1 infection is treated with highly active antiretroviral therapy, but the treatment does not remove the latent viral pool and must be continued for the life of the infected individual. HIV-1 infection targets and kills $CD4^+$ T lymphocytes, whose function is essential in the adaptive immune system. Naive $CD4^+$ T cells differentiate into a variety of effector types and mediate protection against intracellular and extracellular pathogens. $CD4^+$ T-cell help is also often required for generating primary and memory $CD8^+$ T-cell and B-cell responses. As a consequence of $CD4^+$ T-cell depletion, the later stage of HIV-1 infection will manifest in AIDS, which is characterized by immunodeficiency leading to life-threatening opportunistic infections, cancers, and death.

21.6 Genital herpes is caused by infection with herpes simplex virus-2 and HSV-1

HSV is an enveloped virus containing a double-stranded DNA (dsDNA) genome (approximately 150 kb) within the capsid. HSV is transmitted through mucosal contact through vaginal, penile, and oral routes. Even though HSV-1 is traditionally associated with oral herpes, adults without immunity to HSV-1 who practice oral sex are at risk for genital HSV-1 infection. In the developed countries, HSV-1 has replaced HSV-2 as the leading cause of genital herpes. The World Health Organization estimates that, in addition to the 417 million people with genital HSV-2 infections, there are approximately 140 million people living with genital HSV-1 infections. HSV-1 and HSV-2 enter the host through genital epithelia, and upon replication, enter the innervating ganglia where the virus establishes latent infection. Reactivation of the virus results in infection of the genital epithelia and external skin around the genitals and anus, leading to blisters followed by ulcerative lesions. However, most people infected with HSV are asymptomatic and unaware of their carrier status. Vertical transmission of HSV-2 from an infected mother to newborn leads to neonatal herpes, which is often lethal. HSV infection can be treated with the antiviral drug acyclovir, but this is not always effective, does not remove the latent viral pool and does not cure disease. There is no approved vaccines for HSV.

21.7 Human papillomaviruses infect stratified squamous epithelium of skin and mucosa

HPV is a member of the *Papillomaviridae*, consisting of a capsid containing a circular dsDNA genome (approximately 8 kb). Upon sexual contact, HPV establishes productive infections only in the stratified epithelium of the skin or mucous membranes. While the majority of the nearly 200 known types of HPV cause no symptoms in most people, "high-risk" types of HPV are the causal agents of cervical cancer, the second most common cancer among women in the developing world. High-risk HPV types, such as HPV-16 and HPV-18, also cause oropharyngeal, anal, penile, and vulvar cancer; "low-risk" HPV types cause genital and oral warts. Genital warts are treated with a topical ointment containing imiquimod (a toll-like receptor 7 [TLR7] agonist) or surgery. HPV infection is also common, though usually asymptomatic, in men. HPV is the most common sexually transmitted disease in the world, causing 260,000 deaths annually, 80% of which occur in developing countries. The incidence of HPV infection and cervical cancer is declining in the developed countries, owing to Pap smear screening for early stages of cervical cancer. The incidence is expected to decline dramatically with the implementation of HPV vaccines.

21.8 Hepatitis B virus is a hepatotropic virus that is transmitted through sexual contact

Hepatitis B is caused by hepatitis B virus (HBV), which is a member of the hepadnavirus family. It has a circular genome composed of partially double-stranded DNA. The acute illness causes liver inflammation, vomiting, jaundice, and, in rare cases, death. Most infected individuals clear acute HBV infection. A small percentage of infected individuals develop chronic hepatitis B, which may cause liver cirrhosis and liver cancer over many years. Chronic HBV infection can be treated with antiviral drugs and immune modulators such as interferon-α (IFN-α). These agents cannot clear the infection, but they can stop the virus replication, thus minimizing liver damage.

Invasion mechanisms employed by sexually transmitted pathogens

All infectious pathogens must gain entry into the host by invasion of the skin or epithelium, and sexually transmitted pathogens are no exception. Each of the aforementioned organisms invades specific types of the genitourinary epithelia (simple columnar or stratified squamous) by distinct mechanistic pathways. As discussed later, although the strategies may vary, each microorganism seeks to either establish a niche within the epithelium or manage a means to traverse the epithelium in order to reach other desirable destinations, such as through the bloodstream into the target tissues. In some cases, the organism has developed phasic shifts in its mechanisms of growth (e.g., *Chlamydia* infection) to achieve these objectives.

21.9 *Chlamydia* exhibits distinct infectious (elementary body) and replicative forms (reticulate body)

C. trachomatis is an obligate intracellular gram-negative bacterium. *C. trachomatis* normally infects the single-cell columnar layer of the epithelium in the endocervix of women and the transitional epithelium of the urethra of men. Once inside the epithelial cell, *C. trachomatis* undergoes its entire life cycle within a vacuole, termed an *inclusion*. The chlamydial inclusion, which neither fuses with lysosomes nor acidifies, then traffics to the Golgi region. The acute infection consists of infectious and noninfectious stages. The infectious elementary body is small and metabolically inactive. Within the inclusion, the elementary body transforms into a larger metabolically active reticulate body, which divides by binary fission, giving rise to over a thousand progeny per host cell. The reticulate bodies transform back into infective elementary bodies, which are subsequently released from the inclusion vacuole to infect neighboring cells. *C. trachomatis* encode a type

III secretion system, which secretes proteins into the cytosol that bind to and transform the vacuole into a replication center.

21.10 Gonorrhea infects the apical surface of simple columnar epithelium and inhibits function of the immune system

Neisseria gonorrhoeae infects the cervix, uterus, and fallopian tubes in women, and the urethra in women and men. The bacterium can also be found in the mouth, throat, eyes, and anus. The initial attachment of the bacteria to the mucosal surface occurs via the hair-like pili, which recognize a putative receptor on mucosal epithelial cells. Subsequently, tight adherence to the apical surface of epithelial cells occurs and is mediated by opacity (Opa) outer membrane proteins. Adherence may also lead to the apoptosis and destruction of the epithelial layer, providing access to the basolateral compartment. Serum factors promote the infection of cells via the basolateral pole of the epithelium. Epithelial cell chemokines attract phagocytic cells, such as neutrophils and monocytes, which become readily activated and infected. The Opa proteins are also immunosuppressive by binding inhibitory proteins—such as carcinoembryonic antigen-related cell adhesion molecule 1 (CEACAM1), which contains immune tyrosine receptor-based inhibitory motifs as an immune-evasion mechanism—onto the cell surface of T cells. The release of inflammatory mediators, including reactive oxygen species, further increases inflammation and tissue destruction as well as upregulation of CEACAM1. *N. gonorrhoeae* uses its outer membrane porin (Por) molecules to bind the complement inhibitory proteins C4b-binding protein (C4BP) and factor H to evade killing by human complement. In addition, sialylation of gonococcal lipooligosaccharide also enables *N. gonorrhoeae* to bind factor H. Gonorrhea is a human-specific pathogen—rodent and primate species cannot be readily infected with this pathogen because gonorrhea does not bind to factor H or C4BP derived from these animals. *N. gonorrhoeae* also expresses IgA proteases that deter the function of this important immunoglobulin class.

21.11 *Treponema pallidum* invades humans at mucosal epithelia where it causes infection and gains access to blood and lymph systems

Treponema pallidum is a pathogenic spirochete bacterium that has a helical-shaped body with flagella on both ends. Laboratory investigation of *T. pallidum* is hampered by the fact that the bacteria cannot be cultured on artificial media and must be grown in rabbit testicles. *T. pallidum* is able to pass through the endothelial cell monolayer through intercellular junctions without altering tight junctions. Primary infection typically results in a single chancre at the site of inoculation. Within hours of inoculation, and during the evolution of the primary stage, *T. pallidum* disseminates widely, and organisms are deposited in a variety of tissues. Highly motile spirochetes may leave the circulation by invading the junctions between endothelial cells.

21.12 Trichomoniasis is initiated within either the female cervicovaginal epithelium or the male urethral epithelium

Trichomonas vaginalis is an anaerobic, flagellated protozoan. The only identified host for *T. vaginalis* is humans. This parasite encodes adherence factors that allow cervicovaginal epithelium colonization in women. In men, *T. vaginalis* can infect the urethra but is rarely symptomatic. The adherence to vaginal epithelial cells is regulated by pH and temperature. When the healthy acidic pH of the vagina is altered to a basic pH, *T. vaginalis* can multiply through binary fission. The life cycle of this parasite does not have a cyst stage, and it cannot survive in the external environment for long periods of time.

21.13 HIV-1 gains access to immune system by crossing epithelial cell barrier

HIV-1 uses the CD4 molecule as the viral receptor and CCR5 or CXCR4 as coreceptors to enter a target cell (CD4$^+$ T lymphocytes). The viral genome is reverse transcribed into DNA, which is integrated into the genome of the infected cells. Activation of CD4$^+$ T cells induces the transcription of the provirus, leading to viral synthesis and release. Several modes of entry into the host from the initial inoculums have been suggested including transcytosis through M cells (present over the mucosa-associated lymphoid tissue in type I mucosal epithelia), uptake by dendritic cells extending dendrites into the mucosal cavity, and entry through microabrasions in the epithelial layer. HIV-1 invasion is facilitated by physical abrasion, which occurs frequently even during consensual sex. Preexisting genital lesions caused by other STDs also contribute significantly to increasing risk of HIV-1 transmission. The majority of women acquire HIV-1 through heterosexual contact with infected men. Entry of the virus likely occurs near the transformation zone of the endocervix where the simple columnar epithelium of the endocervix transitions into the stratified squamous epithelium of the vagina. Here, the virus first replicates in local CD4$^+$ T lymphocytes and sets up founder cells. From there, the virus can disseminate through the lymph to local lymph nodes and further to the systemic circulation. However, the vaginal route of transmission is estimated to be only successful in 1/200 to 1/2000 encounters, owing to the protective stratified squamous epithelial layer present in this tissue. Risk for HIV-1 acquisition from females to males is even less, particularly in circumcised males. In contrast, HIV-1 can enter through the rectal mucosa much more efficiently, with transmission rates estimated to be as high as 1/10. This difference in transmission rate likely reflects the fact that rectal mucosa is covered by only a single layer of columnar epithelia, and the fact that the lamina propria hosts many effector/memory CD4$^+$ T cells that could serve as targets of HIV-1 infection.

21.14 Genital herpes infection initially targets stratified squamous epithelium

HSV-1 and HSV-2 enter the vaginal canal or the foreskin and first infect stratified squamous epithelial layers. HSV utilizes multiple cell-surface proteins for entry: heparan sulfate chains on cell-surface proteoglycans; a member of the tumor necrosis factor receptor family, herpesvirus entry mediator; and two members of the immunoglobulin superfamily related to the poliovirus receptor, nectin-1 and nectin-2. Primary replication in the keratinocytes results in the subsequent infection of sensory ganglia through nearby nerve endings. The de-enveloped virus migrates to the nerve cell body via retrograde axonal transport. Hiding in the ganglia, HSV establishes lifelong latent infection in the host. Immune suppression results in reactivation of the virus, causing lesions in the external genitalia. Notably, HSV virions are shed even in asymptomatic individuals, increasing the likelihood of transmission from carriers to uninfected partners.

21.15 HPV infection requires access to basement membrane for infection of basal keratinocytes of stratified epithelium

HPV is thought to infect external genital skin, the vagina, or the cervix in women, and the penile shaft, glans, and foreskin in men. HPV cannot infect an intact epithelial layer but requires disruption of the epithelial integrity, which allows access to the basement membrane. HPV utilizes an unconventional mechanism for infection, in that it must first bind to the basement membrane for entry into the basal keratinocytes. Upon binding to the basement membrane, the virus undergoes conformational changes that result in cleavage of the minor capsid protein L2, followed by the exposure of an *N*-terminal L2 epitope and transfer of the capsid to the epithelial cell surface. Once within the basal keratinocyte stem cells, the virus uses the differentiation program of the epithelial cells to complete its life cycle. HPV virions are shed from the superficial layer of the stratified squamous epithelium of the mucosa and transmitted to sexual partners. Most subclinical infection with HPV regresses spontaneously. Only a very small percentage of HPV-infected women, if untreated, go on to develop cervical cancer over many years.

21.16 About one-third of HBV transmission occurs through sexual contact

HBV is present in many mucosal secretions and in the blood of infected individuals and may reach 10^{10} copies per milliliter in serum. About 30% of HBV transmission occurs through sexual contact (the rest is related to perinatal, blood, or pericutaneous transmission). A liver-specific bile acid transporter named the sodium taurocholate cotransporting polypeptide (NTCP) has been recently identified as the cellular receptor for HBV. Once HBV enters the host circulation, the virus replicates in the hepatocytes expressing NTCP. In 95% of the infected adults and 10% of the infected neonates, HBV is cleared following acute infection. The remainder develop chronic viral infection. Hepatocellular carcinoma can develop in a small fraction of these patients over a period of 30–50 years.

Innate immune system of genital mucosa and its relation to infections

The genital mucosal surfaces are protected by a variety of mechanisms. Some of these mechanisms are described in Chapter 20. Here, we focus on and summarize those host factors that are specifically relevant to protecting against mucosal infections of the genitourinary tract. A pathogen entering the genital tract must pass through several hurdles in order to reach the epithelial layer and subsequently the distal target organs (**Figure 21.1**). The first hurdle is provided by the mucus layer, which contains a variety of antimicrobial factors and in the vagina has an inhospitable pH due to the *Lactobacillus*-rich microbiome composition. A pathogen must then invade the multilayered epithelial cells of stratified squamous epithelium, reach the lamina propria, and find the target cells before being engulfed by phagocytes or attacked by circulating antimicrobial substances such as complement.

21.17 Mucus is first-line epithelial defense

The front line of innate immune defense in the genital mucosa is provided by the physical barriers of the epithelium. In addition to the epithelial layer, mucus covers the internal surface of the vaginal tract, penis, and anal canal. Mucus is made up of mucins, which are complex high molecular weight glycoproteins that contain at least one and sometimes multiple protein domains that are sites of extensive *O*-glycan attachment. A major difference in the genital versus intestinal mucosae is the source of mucus-secreting cells. In the tissues that contain simple columnar epithelia (e.g., intestines, lung), goblet cells that are differentiated epithelial cells dedicated for mucus production and secretion are dispersed throughout the epithelial layer. Similarly, in simple columnar epithelium of the rectum and upper anal

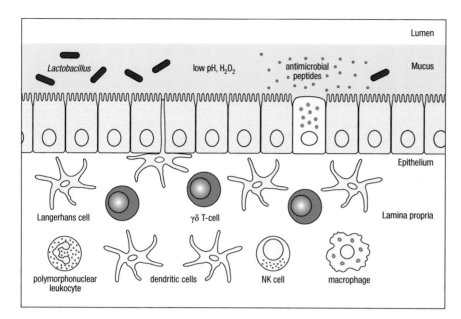

Figure 21.1 Innate immune system of the female genital mucosa. A mucus layer containing a variety of antimicrobial factors and endogenous microbiota, *Lactobacillus* spp., provides an inhospitable environment for an invading pathogen. The epithelial layer and the lamina propria contain phagocytes and other innate cells that recognize and eliminate incoming microbes.

canal, goblet cells secrete mucus (Table 21.1). In the genital mucosa, which contains stratified squamous epithelium (male urethra, vagina, lower half of anal canal, foreskin), and in the lower anorectal canal, mucus is secreted from glandular exocrine cells. During ovulation, crypts in the cervix secrete mucus that descends and covers the vaginal canal. In the male reproductive tract, the bulbourethral glands produce mucus, which is incorporated into semen. In the uncircumcised penis, the inner foreskin epithelium is kept moist by cells secreting mucin locally. In addition to mucins, mucus contains a variety of other defense molecules including immunoglobulins, complement, antimicrobial peptides, lysozyme, and lactoferrin (discussed in more detail later).

TABLE 21.1 DISTINCT FEATURES OF SIMPLE COLUMNAR AND STRATIFIED EPITHELIAL MUCOSAL SURFACES ASSOCIATED WITH GENITOURINARY TISSUES

	Type I mucosa	Type II mucosa
Epithelial type	Simple	Stratified
	Polarized	Noncornified
Presence of MALT	+	−
Immune priming	MALT, draining lymph nodes	Draining lymph nodes
Presence of pIgR	+	−
Presence of FcRn	+	+
Major Ig isotype	SIgA	IgG
Goblet cells	+	−
Langerhans cells	−	+

Abbreviations: FcRn, neonatal Fc receptor; Ig, immunoglobulin; MALT, mucosa-associated lymphoid tissue; pIgR, polymeric immunoglobulin receptor; SIgA, secretory IgA.

21.18 Female and male genitourinary secretions contain antimicrobial factors

Bacteria, fungi, parasites, and viruses within the mucus layer are confronted with a variety of antimicrobial factors in the genitourinary tract. Microbicidal molecules such as complement and antimicrobial peptides can directly bind and kill microbes before they can reach the host epithelial layer. The epithelial cells, cervical glands, and neutrophils contribute the majority of antimicrobial peptides to the vaginal fluid, including calprotectin, lysozyme, lactoferrin, secretory leukocyte protease inhibitor, human neutrophil peptides, and human β-defensins. The concentrations of these antimicrobial factors remain stable during different stages of the menstrual cycle. Calprotectin, highly abundant in vaginal and intestinal mucosae, is a calcium-binding protein that inhibits bacterial and fungal growth as well as bacterial adhesion. Lysozymes are a family of enzymes that catalyze hydrolysis of glycosidic bonds in peptidoglycans (present in the wall of gram-positive bacteria). Lactoferrin, a member of the transferrin family of iron-binding glycoproteins, binds to ferric iron (Fe^{3+}), which deprives microorganisms of this essential factor and has direct antimicrobial activities against a large panel of bacteria, viruses, fungi, and parasites. Secretory leukocyte protease inhibitor is a double-edged sword—its C-terminus protects tissues by inhibiting proteases such as elastase, cathepsin G, and trypsin, and its N-terminus exerts antibacterial and antifungal capabilities. Human neutrophil peptides (α-defensins secreted by neutrophils) and human β-defensins (secreted by epithelial cells) are small cationic proteins with significant and broad antimicrobial activities.

21.19 Female and male genitourinary systems possess endogenous (commensal) microbiota that provides colonization resistance against pathogenic infections

Commensal bacteria are essential in shaping intestinal immune responses in both health and disease. Recent advances in 16S ribosomal RNA sequencing and deep metagenomic sequencing have enabled a richer taxonomic identification of human microbial communities. Of more than 50 bacterial phyla on Earth, human intestine-associated communities are dominated by only four phyla: Firmicutes, Bacteroidetes, Actinobacteria, and Proteobacteria. More than 1000 species of bacteria are estimated to live in the intestines of humans. In addition to providing key metabolic, trophic, and defense functions to the mammalian host, the microbiota shapes the intestinal immune system in a variety of ways. This topic is discussed in further detail in Chapter 19.

In comparison with the intestinal tract, a normal vaginal microbiota is predominately composed of Firmicutes and especially *Lactobacillus* species, which perform key functions for the female host. First, vaginal H_2O_2-producing lactobacilli prevent the outgrowth of harmful bacteria that can cause bacterial vaginosis (see Figure. 21.1). Second, *Lactobacillus* spp. maintain the vaginal fluid acidity (pH 3.8–4.0) through lactic acid production. The acidic vaginal pH protects against STDs, including *Haemophilus ducreyi*, HSV-2, *T. vaginalis*, and *C. trachomatis* infection. By virtue of causing vaginitis, all of these infectious agents lead to a significant increase in the transmission of HIV-1. In addition to *Lactobacillus* (phylum Firmicutes), phyla Proteobacteria and Actinobacteria (mainly the genera *Pseudomonas* and *Gardnerella*, respectively) are also significant constituents of the commensal vaginal microbiota of healthy females. Within an individual, the vaginal microbiota is not homogeneous but differs significantly depending on the anatomical location and other factors such as hormonal state of the host. There is also significant interpersonal variation in the healthy vaginal microbiome depending on geographic locations. Mouse studies show that healthy microbiota composition is critical to promoting protective adaptive immune responses to HSV-2.

21.20 Innate immune cells provide defense against invading pathogens within female and male genitourinary tracts

Within the genital mucosa, a variety of innate immune cells provide defense against invading pathogens. At steady state, αβ T cells, γδ T cells, macrophages, Langerhans cells, and dendritic cells (DCs) survey the stratified squamous epithelia of vaginal and anal canals. In the human vagina, the numbers of neutrophils, macrophages, and T cells remain relatively constant throughout the menstrual cycle. Upon infection, a number of cell types are mobilized into the vaginal tissue, including polymorphonuclear cells (PMNs), monocytes, plasmacytoid DCs (pDCs), and natural killer (NK) cells. Subsequently, antigen-specific T cells and B cells enter the tissues to provide pathogen-specific immune defense. A large number of PMNs are recruited to HSV-2-infected vaginal mucosa in response to chemokine signals and are required for protection during primary and secondary challenge. NK cells sense infected cells by two different mechanisms: "missing" self (lack of major histocompatibility complex [MHC] class I molecule expression) and "altered" self (presence of NK-activating factors on target cells such as "stress" ligands including MHC class I chain-related gene A and B proteins in human). NK cells are rapidly recruited to the infected tissues where they are activated to lyse infected cells through perforin and Fas ligand–dependent mechanisms. NK cells are also a source of cytokines including interferon (IFN)-γ, which also has antiviral activity. Humans missing NK cells or having defective NK function often suffer from herpesvirus infections, indicating their role in defense against this family of viruses. The importance of NK cells in antiviral defense against HIV-1 is reflected by the evasion mechanism utilized by HIV-1 to specifically prevent NK recognition of infected cells. Viral Nef protein downregulates HLA-A and B to avoid recognition by cytotoxic T lymphocytes (CTLs) but spares HLA-C and E, which are the primary ligands for inhibitory NK receptors. Intraepithelial γδ T cells are present within the simple columnar and stratified squamous epithelia, and although the nature of antigens they recognize is unclear, they are thought to sense altered self upon viral infection. These cells contribute to immune protection against vaginal challenge with HSV-2. Langerhans cells within the epithelium and DCs in the submucosa patrol the genital mucosa for invading pathogens. At steady state, these cells are highly phagocytic and express a variety of pattern-recognition receptors (PRRs) that recognize a wide array of microbes (discussed further later). DCs and Langerhans cells depart the epithelium after capturing the relevant microbe and undergo a maturational program during migration to the draining lymph nodes to prime naive T cells and B cells. After becoming effector cells, αβ T cells can enter the vaginal and cervical mucosa and establish tissue residency. These tissue resident memory T cells (T$_{RM}$) provide long-term specific immune protection against subsequent challenges with the same pathogen.

Innate recognition of sexually transmitted pathogens

Pathogens are detected by multiple families of PRRs expressed by innate immune cells and the infected target cells. Stimulation of cells through PRRs results in the activation of genes encoding immediate defense responses (antimicrobial factors, chemokines, type I IFNs) as well as those required to elicit adaptive immune responses (cytokines, costimulatory molecules, chemokine receptors). Innate recognition systems utilized by the human host to detect sexually transmitted pathogens are described in this section.

21.21 *Chlamydia trachomatis* is recognized by multiple PRRs

While both TLR2 and TLR4 recognize *C. trachomatis*, TLR2 and MyD88 specifically colocalize with the intracellular chlamydial inclusion, suggesting that TLR2 is actively engaged in signaling from this intracellular location. TLR2 recognition of this organism is dependent on infection with live, replicating bacteria. In addition to TLRs, the NOD-like receptor, NOD1, is involved in sensing *Chlamydia* infection.

21.22 Host cells utilize immunoglobulin-related molecules such as CEACAM3 to initiate internalization and elimination of *Neisseria gonorrhoeae*

Carcinoembryonic antigen-related cell adhesion molecule 3 (CEACAM3) mediates the opsonin-independent recognition and elimination of a restricted set of human-specific gram-negative bacterial pathogens including *N. gonorrhoeae*. CEACAM3 binds to the Opa protein expressed by *N. gonorrhoeae* and internalizes the bacteria for degradation. *N. gonorrhoeae* expresses lipooligosaccharide, lacking the O antigen. This structure is recognized by TLR2, TLR4, and triggering receptor expressed on myeloid cells-2A (TREM-2A). In addition, porins and Lip lipoproteins of *N. gonorrhoeae* are recognized through TLR2.

21.23 *Treponema pallidum* lacks lipopolysaccharide but contains internal lipoproteins that can stimulate toll-like receptors

The outer membrane of *T. pallidum* does not contain lipopolysaccharide and has relatively few surface-exposed transmembrane proteins. Therefore, the extracellular bacterium is not recognized efficiently by TLRs. However, upon degradation in phagocytes, the lipoproteins present under the outer membrane can become accessible and stimulate TLR2.

21.24 Innate immune recognition of HIV-1 occurs after cellular infection

Innate recognition of HIV-1 or retroviruses in general remains unclear. pDCs can produce IFN-α in response to HIV-1 through a process that likely involves TLR7 (Figure 21.2). In infected cells, HIV-1 viral RNA is reverse transcribed into viral cDNA, which is recognized by a cytosolic DNA sensor, cGAS. The host factor TREX1, an exonuclease, inhibits innate immune detection of HIV DNA by metabolizing nonproductive reverse transcription products. Innate immune response to HIV-1 in DCs depends on interaction between newly synthesized HIV-1 capsid and cellular cyclophilin A (Figure 21.3a). In addition, HIV-1 induces global disruption of the innate signaling pathway within infected cells by degrading IRF3, making directly infected cells incapable of IFN production.

21.25 Toll-like receptors, RIG-I, cGAS, and IFI16 are involved in detecting HSV infection

HSV infection is sensed by multiple PRRs. In pDCs, HSV is recognized by TLR9 in the endosome, while in infected cell types, HSV genome is recognized by DNA sensors, cGAS and IFI16. In addition, in some infected cells, HSV-derived

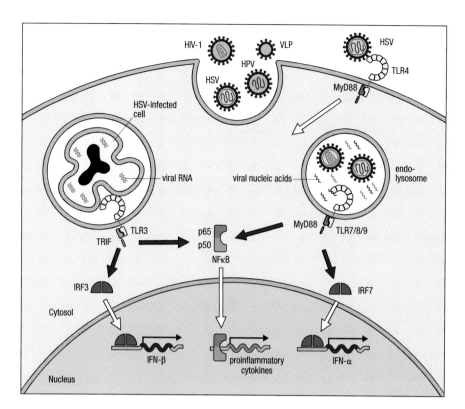

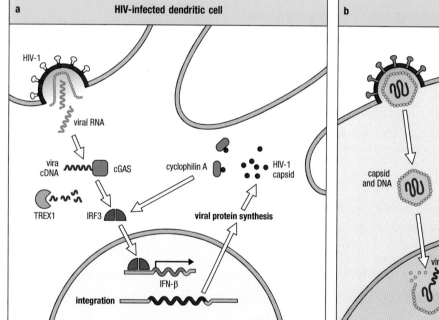

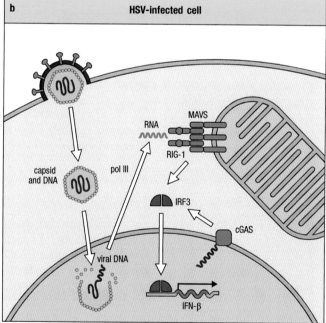

Figure 21.2 Cell-extrinsic recognition of sexually transmitted viruses by toll-like receptors (TLRs). Specialized viral sentinels such as the plasmacytoid dendritic cells (pDCs), B cells, and DCs recognize viral nucleic acids in the endolysosomes through TLR3, TLR7, TLR8, and TLR9. TLR3 recognizes higher-order viral RNA from infected cells (herpes simplex virus [HSV]); TLR7 and TLR8 recognize viral RNA (human immunodeficiency virus [HIV]-1); and TLR9 recognizes viral DNA (HSV). Human papillomavirus (HPV) virus-like particles (VLPs) also trigger MyD88-dependent signals. In B cells, HPV VLPs stimulate TLR4/MyD88 to induce an antibody response. Signals transduced from these TLRs lead to the induction of interferon and cytokine genes.

Figure 21.3 Cell-intrinsic recognition of sexually transmitted viruses. Virally infected cells recognize viral pathogen-associated molecular patterns in the cytosol. (**a**) Human immunodeficiency virus (HIV)-1 infection is recognized through multiple mechanisms, which are efficiently counteracted by the virus. HIV-1 viral DNA synthesized by reverse transcription is recognized through cGAS, activating IRF3. Exonuclease TREX1 degrades nonproductive reverse transcription products, inhibiting activation of IRF3. HIV-1 also degrades IRF3. In HIV-1-infected dendritic cells (DCs), cyclophilin A binds to newly synthesized capsid proteins and induces interferon (IFN) through the activation of IRF3. However, HIV-1 normally does not infect DCs and thus avoids activating this pathway altogether. In addition, both herpes simplex virus (HSV) and HIV-1 trigger inflammasomes, leading to secretion of interleukin (IL)-1β and IL-18 (not shown). (**b**) HSV viral DNA is sensed by DNA sensors such as cGas and IFI16. HSV viral DNA is transcribed by RNA polymerase III (pol III), and the resulting RNA serves as a substrate for RIG-I. Signals transduced from these pattern-recognition receptors lead to the induction of IFN and cytokine genes. MAVS, mitochondrial antiviral signaling protein.

RNA is recognized in the cytosol by a RIG-I-dependent pathway following transcription of HSV by DNA-dependent RNA polymerase III (see Figure 21.3b). In the cytosol, viral DNA is sensed by the DNA sensor cGAS and IFI16, and in the nucleus IFI16 can bind to nucleosome-free incoming genomic DNA to induce type I IFNs (see Figure 21.3b). Certain strains of HSV also trigger TLR2 on DCs and macrophages. In addition to these pathways, mutations in TLR3 and downstream signaling molecules are highly associated with uncontrolled HSV infection. TLR3 presumably detects higher-order viral RNA in infected cells within the endosomes (see Figure 21.2). In humans, a rare TLR3 P554S homozygous mutation and an L412F mutation, both resulting in impaired activity of TLR3, were found in patients with HSV-1 encephalitis and HSV-2-associated Mollaret meningitis, respectively.

21.26 HPV virus-like particles are sensed by innate and adaptive immune cells through toll-like receptors

HPV virus-like particles stimulate DCs via MyD88, and adaptive immunity to virus-like particles is compromised in the absence of MyD88. B cells recognize the HPV virus-like particles through TLR4 in a MyD88-dependent manner to induce class switching and costimulatory molecule expression (see Figure 21.2). The nature of the pathogen-associated molecular pattern associated with the intact HPV particles or within infected cells that stimulate the PRRs awaits further clarification.

21.27 HBV impairs innate immune responses as a means of immune evasion

The 3.2 kb genome of HBV consists of partially double-stranded, circular DNA. HBV infection impairs innate immune responses triggered both by specialized cells (such as pDCs) and in infected hepatocytes by downregulating functional expression of TLR. Remarkably, genetic analysis of HBV-infected cells showed no sign of innate gene activation. This "invisible" nature of the HBV to the innate immune system reflects its replication strategy, which sequesters the transcriptional template in the nucleus, generating viral transcripts indistinguishable from normal cellular mRNA, and the fact that the replicating viral genome is sheltered within viral capsid particles in the cytoplasm. In addition, evidence suggests that HBV viral proteins actively inhibit PRR and IFN signaling. In the absence of robust innate responses, how HBV infection results in robust activation of T- and B-cell responses remains unclear.

Adaptive immune responses against sexually transmitted pathogens

During the natural course of infection, sexually transmitted pathogens infect specific target cells and are taken up by local antigen-presenting cells. The type II mucosa is characterized by the absence of mucosa-associated lymphoid tissues. Priming occurs exclusively in the draining lymph nodes (see Table 21.1). Adaptive immunity that develops during the natural course of infection with sexually transmitted pathogens is required, but not always sufficient, to clear the primary infection. Effector mechanisms include antibody-mediated blockade of pathogen entry, antibody-dependent cellular cytotoxicity by NK cells, antibody-mediated phagocytosis, CD4+ T-cell–mediated suppression of intracellular pathogens, and CD8+ T-cell–mediated lysis of infected cells (Figure 21.4). Antibodies are transported across the epithelial layer by polymeric IgR in the type I mucosa or by neonatal FcR in type II mucosa (see Table 21.1).

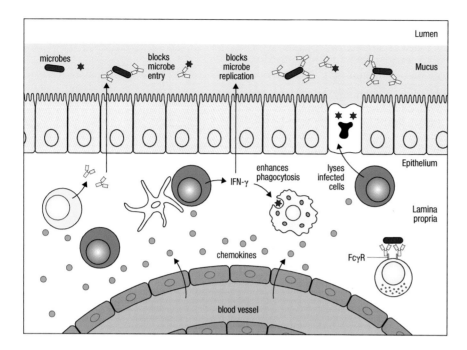

Figure 21.4 Adaptive immune system of the female genital mucosa. Once adaptive immune responses are initiated in the draining lymph nodes, B and T cells specific for the pathogen enter the infected tissue and eliminate the pathogen and the infected cells. B cells secrete antibodies to block the entry and further spread of pathogens. T_H1 CD4$^+$ T cells secrete IFN-γ, which suppresses viral replication and enhances phagocyte clearance of bacterial and parasitic pathogens. CD8$^+$ T cells lyse infected cells expressing MHC class I and limit the propagation of intracellular pathogens.

21.28 *Chlamydia* infection triggers the activation of local DCs to migrate to the draining lymph nodes and initiate T-cell activation

While both CD4$^+$ and CD8$^+$ T cells can contribute to controlling *Chlamydia* infection, T_H1 responses against chlamydia play a dominant role for resolving genital infection in an IFN-γ-dependent manner. IFN-γ can upregulate the phagocytic potential of macrophages, thereby promoting the engulfment and destruction of extracellular elementary bodies. IFN-γ can also limit the growth of intracellular bacteria by upregulating nitric oxide synthase. Recent work revealed that optimal *C. trachomatis* clearance requires both CD4$^+$ T_{RM} and infection-induced recruitment of circulating memory T cells. In contrast, the role of B cells and antibody response in defense against chlamydia is less clear. High titers of *C. trachomatis*-specific antibodies do not correlate with resolution of infection in humans.

21.29 T_H1 cells mediate immunity against gonorrhea

Protective immunity against *N. gonorrhoeae* is mediated by type 1 immunity, leading to the production of complement-fixing antibodies. Patients with defects in the complement pathways often suffer from recurrent and disseminated gonococcal infections. *N. gonorrhoeae* expresses lipooligosaccharide lacking the *O* antigen, which activates TLR2, TLR4, and TREM-2A. Human DCs in genital tissues also use different C-type lectins, MGL and DC-SIGN, to detect the *N. gonorrhoeae* types and induce CD4$^+$ T-cell activation.

21.30 Protective immunity to syphilis is undefined but likely exists

Clearance of *T. pallidum* during early infection is mediated by opsonizing antibodies. However, to date, no single antigen has been definitively identified in the outer membrane of *T. pallidum* that can elicit protective immunity.

This is owing to the ability of *T. pallidum* to undergo significant sequence change in the variable regions of its surface antigen, *tprK,* during infection. The proportion of variant bacteria increases as the adaptive immune response develops. This has led to the concept of the spirochete being a "stealth pathogen." This poses a challenge in the design of vaccines that can effectively bind to *T. pallidum* and block infection. However, patients who have been previously infected with *T. pallidum* but have not been treated with antibiotics appear to have at least some degree of immunity against repeated infection. In contrast, antibiotic-treated patients do not develop a similar degree of protective immunity, suggesting that either chronic infections or antigens exposed during the chronic stage of infections elicit a protective response against this pathogen.

21.31 Immunity to *Trichomonas* infection involves both B- and T-cell responses

Natural infection with *Trichomonas* results in the induction of both antibody and T-cell responses. However, immunity generated following natural infection is at best only partially protective. Using a mouse model of infection, mice that were vaginally infected, treated with metronidazole, and then reinfected vaginally did not develop protective immunity. In contrast, subcutaneous immunization with a whole *T. vaginalis* organism in complete Freund's adjuvant conferred protection against intravaginal challenge with *T. vaginalis*, protection that is not achieved as a result of prior vaginal infection. These results suggest that *T. vaginalis* has an efficient immune-evasion mechanism to prevent development of effective immunity following natural infection.

21.32 Adaptive immunity is mainstay of resistance to genital herpes infection

Following sexual transmission, HSV-2 infects stratified squamous epithelium of the vagina and ectocervix. Virus infection is detected by the infected epithelial cells and by submucosal dendritic cells in a MyD88-dependent manner. Recognition of infection by both hematopoietic and stromal compartments is required for successful induction of protective T_H1 immunity. However, directly infected cells are incapable of priming T cells because HSV-2 blocks MHC I and MHC II presentation and is highly lytic. Uninfected submucosal DCs pick up viral antigens and migrate to the draining lymph nodes, where they present antigenic peptides to cognate CD8$^+$ and CD4$^+$ T cells. Natural immunity that develops following genital HSV infection fails to clear the virus because the virus can invade and establish latent infection in the innervating ganglia prior to the onset of highly effective immunity. Latent virus cannot be cleared by T cells or antibody, although CD8$^+$ T cells provide important immune surveillance of the infected neurons through nonlytic mechanisms. Both CD4$^+$ and CD8$^+$ T_{RM} confer protection against disease following vaginal HSV-2 challenge in the mouse model, suggesting that vaccines that can induce a high frequency of tissue-resident T cells in the female reproductive tract are desirable. Multiple clinical trials of HSV-2 vaccines have failed to confer protection. The failure of vaccines may be due to the fact that conventional vaccines do not elicit T_{RM}s. In addition, antibody responses are rendered ineffective by the fact that the HSV-2 envelope is studded with glycoproteins that evade the effector function of antibodies. The gC binds complement component C3 and inhibits complement-mediated virus neutralization and lysis of infected cells. In addition, the gE/gI complex acts as an FcR decoy on viral envelope. These immuno-evasins collectively render Abs ineffective in control of HSV. Future vaccines against HSV-2 will need to overcome both of these hurdles. The ability of IgG antibodies to be transported into and out of luminal surfaces by FcRn, an IgG-transporting receptor (see Chapter 11) may also be important in protection against and responses to HSV-2.

21.33 HPV engenders limited immune responses leading to persistent, latent infection

Upon natural infection with HPV, only poor immune responses develop. This is owing to the ability of the virus to evade immunity by several mechanisms: (1) HPV E6 protein causes the depletion of Langerhans cells; (2) HPV E5 protein downregulates MHC I expression; and (3) HPV E7 protein blocks type I IFN signaling. Thus, activation of CD8$^+$ T-cell immunity to HPV requires cross-presentation of infected cells by noninfected DCs. Although the nature of the DCs that are responsible for this cross-priming is unknown, because HPV infection occurs only within the stratified squamous epithelial layer, Langerhans cells within the epithelial layer, and possibly submucosal DCs that extend their dendrites, are most likely the ones participating in this process. HPV infection results in the induction of antibody and T-cell responses. Since the majority of HPV infections are cleared within 2 years, and because HIV infection results in an increased prevalence of HPV infections, it is likely that naturally induced immune responses are responsible for controlling HPV infection in most cases. It is noteworthy that neutralizing antibodies induced by HPV vaccines are remarkably efficient in preventing infection by HPV, as discussed later.

21.34 Neutralizing antibodies to HIV-1 can block infection

During HIV-1 infection, the type of DCs that prime CD4$^+$ and CD8$^+$ T-cell responses is unknown. Langerhans cells can bind to HIV-1 gp120 through mannose C-type lectins. Using an *ex vivo* human organ culture system, Langerhans cells take up HIV-1 through endocytosis. As Langerhans cells exit the epithelium at the basal side and migrate to the lymph node, they transport intact virions, thereby enabling the infection to spread beyond the site of viral entry. All arms of the adaptive immune system are activated by HIV-1 infection. However, natural immunity to HIV-1 is not protective. HIV-1 superinfection can occur in individuals with a strong and broadly reactive virus-specific CD8$^+$ T-cell response. The immunologic correlates of protection against HIV-1 are currently unknown, making HIV/AIDS vaccine design difficult. However, recent studies have raised hope through the identification of broadly neutralizing IgG antibodies in long-term survivors. Some of these antibodies are able to bind to and neutralize more than 90% of infectious strains, thereby conferring broad coverage against circulating strains. How to elicit such antibodies by vaccines remains a key challenge in the field.

21.35 Adaptive immunity is critical to limiting and resolving HBV infection

Natural infection with HBV results in the generation of T- and B-cell responses. Because HBV is a noncytopathic virus, disease caused by HBV can be mediated by T cells specific to HBV antigens. Once primed, HBV-specific CTLs migrate into the liver and recognize and kill HBV-infected hepatocytes. In addition, T cells secrete IFN-γ, which leads to the expression of chemokines that enable further leukocyte recruitment. An inefficient T-cell response that is unable to completely clear HBV from the liver leads to a sustained continuous cycle of low-level hepatocyte destruction. The T-cell response involves both classical and nonclassical T cells such as CD1d-restricted NK

T cells. Over a long period of time, the chronic nature of these cycles may lead to recurrent immune-mediated liver damage, which contributes to the development of cirrhosis and hepatocellular carcinoma.

Challenges ahead

Sexually transmitted pathogens continue to propagate and cause significant morbidity and mortality throughout the world. Infection by sexually transmitted pathogens represents a complex problem, not only with respect to host-pathogen interaction, but also with respect to reproductive and emotional health. In addition, unlike for other infectious diseases, vaccines and interventions for STDs are put through religious and socioeconomic considerations. Having said this, prophylactic vaccines are the only option available to combat the incurable viral STDs. In this regard, two recent vaccines against HPV, Gardasil (Merck) and Cervarix (GlaxoSmithKline), have demonstrated spectacular efficacy, providing protection from infection in almost 100% of cases. Protection conferred by these HPV vaccines is mediated by neutralizing antibodies. The outcome of these trials highlights that successful vaccines against sexually transmitted viruses are possible. However, antibody-based vaccines are not effective against all STDs, and protective immunity to others such as HIV-1 and HSV-2 may require cellular immunity with the caveats described earlier. Future studies must focus on first identifying the protective immune mechanism tailored to each STD, and designing vaccines that elicit such immune responses at the relevant sites of exposure within the genital mucosae.

SUMMARY

STDs are caused by viral, bacterial, and parasitic pathogens. These pathogens have evolved to utilize the physiology of the female and male reproductive tracts for successful invasion, replication, and transmission. Currently, vaccines are not available for most of these pathogens and are desperately needed to combat the spread of incurable STDs. To this end, a clear understanding of the invasion mechanisms utilized by the pathogens, as well as innate and adaptive immune defense mechanisms for each sexually transmitted pathogen, would provide a rational basis for vaccine design. Innate recognition is well understood for some sexually transmitted pathogens but not for others. Natural infections with sexually transmitted pathogens often do not result in protective immunity, likely due to the evasion mechanisms employed by the pathogens. Protective immune responses to some STDs can be induced by vaccines, which are mediated by neutralizing antibodies, cytotoxic CD8+ T cells, and effector CD4+ T cells. Future challenges to the development of vaccines include identifying immune correlates of protection for each sexually transmitted pathogen and finding means of inducing such responses in humans.

FURTHER READING

Altfeld, M., Allen, T.M., Yu, X.G., et al.: HIV-1 superinfection despite broad CD8+ T-cell responses containing replication of the primary virus. *Nature* 2002, 420:434–439.

Dethlefsen, L., McFall-Ngai, M., and Relman, D.A.: An ecological and evolutionary perspective on human-microbe mutualism and disease. *Nature* 2007, 449:811–818.

Frazer, I.H.: Prevention of cervical cancer through papillomavirus vaccination. *Nat. Rev. Immunol.* 2004, 4:46–54.

Guidotti, L.G., and Chisari, F.V.: Immunobiology and pathogenesis of viral hepatitis. *Annu. Rev. Pathol.* 2006, 1:23–61.

Haase, A.T.: Targeting early infection to prevent HIV-1 mucosal transmission. *Nature* 2010, 464:217–223.

Hladik, F., and McElrath, M.J.: Setting the stage: Host invasion by HIV. *Nat. Rev. Immunol.* 2008, 8:447–457.

Iwasaki, A.: Mucosal dendritic cells. *Annu. Rev. Immunol.* 2007, 25:381–418.

Iwasaki, A.: Antiviral immune responses in the genital tract: Clues for vaccines. *Nat. Rev. Immunol.* 2010, 10:699–711.

Iwasaki, A.: Exploiting mucosal immunity for antiviral vaccines. *Annu. Rev. Immunol.* 2016, 34:575–608.

Kawai, T., and Akira, S.: The role of pattern-recognition receptors in innate immunity: Update on Toll-like receptors. *Nat. Immunol.* 2010, 11:373–384.

Medzhitov, R.: Recognition of microorganisms and activation of the immune response. *Nature* 2007, 449:819.

Peeling, R.W., and Hook, E.W. 3rd,: The pathogenesis of syphilis: The great mimicker, revisited. *J. Pathol.* 2006, 208:224–232.

Roan, N.R., and Starnbach, M.N.: Immune-mediated control of *Chlamydia* infection. *Cell. Microbiol.* 2008, 10:9–19.

Rottenberg, M.E., Gigliotti Rothfuchs, A.C., Gigliotti, D. et al.: Role of innate and adaptive immunity in the outcome of primary infection with *Chlamydia pneumoniae*, as analyzed in genetically modified mice. *J. Immunol.* 1999, 162: 2829–2836.

Stanberry, L.R.: Clinical trials of prophylactic and therapeutic herpes simplex virus vaccines. *Herpes* 2004, 11(Suppl 3):161A–169A.

Wu, J., and Chen, Z.J.: Innate immune sensing and signaling of cytosolic nucleic acids. *Annu. Rev. Immunol.* 2014, 32:461–488.

NOSE, AIRWAYS, ORAL CAVITY, AND EYES

PART FIVE

Nasopharyngeal and oral immune system

22

HIROSHI KIYONO AND KOHTARO FUJIHASHI

Mammals have evolved a complex immune system consisting of an integrated network of tissues, lymphoid and mucous membrane-associated cells, and effector antibody (Ab) molecules that serve to maintain homeostasis in exposed mucosal surfaces. This system is anatomically and functionally distinct from its blood-borne counterpart in the systemic (or peripheral) immune system and is strategically located at the portals through which most microorganisms enter the body. The development of this specific branch of the immune system may have been necessitated by the size of the mucosal surfaces, which cover an area of approximately 400 m^2 in the adult human, as well as the large number of exogenous antigens (Ags) to which these surfaces are exposed. Along with cytokines, chemokines, and their receptors, the effector Ab molecules, which are primarily of the IgA isotype, are key elements of mucosal immunity and appear to function in synergy with innate host factors. Thus, in order to induce Ag-specific immune responses at these mucosal barriers, one must consider the mucosal immune system, which is composed of functionally distinct mucosal IgA inductive and effector tissues (Figure 22.1). Mucosal inductive sites include the nasopharyngeal-associated lymphoid tissues (NALTs) and Peyer's patches in the gut-associated lymphoreticular tissues (GALTs) for the initiation of mucosal secretory IgA (SIgA) antibody (Ab) responses. The subsequent homing of memory and activated B and T lymphocytes from NALT to the nasal passage and upper respiratory tract effector regions or from GALT to the gastrointestinal lamina propria connects inductive and effector sites. This chapter focuses on the nasal-oral immune system in the induction of antigen (Ag)-specific SIgA Ab responses in the nasal and oral cavities, the portals of Ag entry into the aerodigestive tract.

NASOPHARYNGEAL-ORAL MUCOSAL IMMUNE SYSTEM

22.1 Mucosal inductive tissues

In the mammalian host, organized secondary lymphoid tissues have evolved in the upper respiratory and gastrointestinal tracts, which are collectively termed mucosa-associated lymphoreticular tissues (MALTs), to facilitate Ag uptake, processing, and presentation. Foreign Ags can be blocked by the barrier function of the epithelium that covers these mucosal surfaces or be selectively taken up into highly specialized inductive sites such as MALT for the initiation of an immune response. Peyer's patches are the largest component of GALT and NALT in humans and mice. The MALTs are covered by a lymphoepithelium containing microfold (M) cells, which

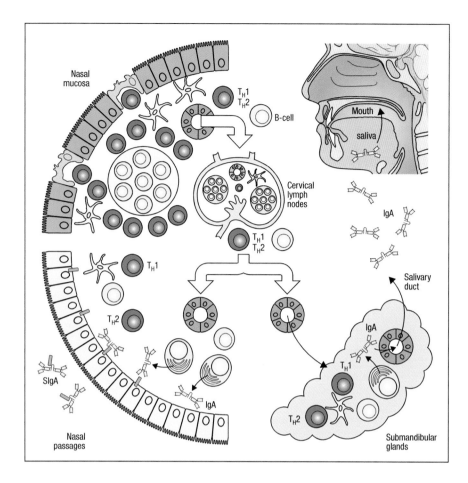

Figure 22.1 The nasopharyngeal-oral common mucosal immune system. Nasally administered antigens and vaccines are taken up mainly by the nasopharyngeal-associated lymphoreticular tissue (NALT) inductive site. Antigen-presenting cells, including dendritic cells, process antigens and subsequently stimulate T cells. B-cell follicles are formed, and IgA class-switch recombination is induced. Antigen-stimulated T cells and IgA-committed (sIgA+) B cells move into the systemic circulation via the cervical lymph nodes. These immune cells home to mucosal effector sites, including nasal passages and exocrine glands such as submandibular glands. Antigen-specific T- and B-cell interactions induce sIgA+ B-cell differentiation into antigen-specific IgA-producing plasma cells. The dimeric forms of IgA antibodies bind to polymeric Ig receptors on epithelial cells and ductal cells and transport the molecule as SIgA onto the nasal and oral mucosa.

are characterized by a pocket structure (M cell pocket) at the basolateral side that harbors dendritic cells (DCs), macrophages, and lymphocytes. M cells contribute to the transport of luminal antigens to underlying antigen-presenting cells (APCs) without Ag processing. In spite of their capacity to take up luminal Ags via pinocytosis and endocytosis, M cells contain very low numbers of lysosomes. In addition, IgA preferentially binds to the apical side of M cells, as mouse IgA and human IgA2, but not IgA1, bind to M cells via the Fc receptor for IgA, which differs from CD89.

MALT contains well-organized regions, including the follicle-associated epithelium (FAE) covering the follicle dome, a B-cell zone with germinal centers, adjacent T-cell areas with enriched APCs, and high endothelial venules (HEVs). Naïve, recirculating B and T lymphocytes enter MALT via the HEVs. The Ag-primed, activated, and/or memory B and T cells emigrate from the mucosal inductive sites via lymphatic drainage, circulate through the bloodstream, and home to mucosal effector sites. In the mouse, isolated lymphoid follicles (ILFs) in the small intestine are thought to be part of the GALT for the induction of mucosal immune responses. ILFs contain mainly B cells and DCs, as well as M cells, in the overlying epithelium. In addition to FAE-associated M cells in Peyer's patches and ILFs, M cell clusters were recently identified in non-FAE sites at the tips of the villi and have been termed *villous M cells* as alternative Ag sampling sites.

22.2   Nasopharyngeal-associated lymphoreticular tissue is a major IgA inductive site for nasal and oral mucosa

NALT is the major inductive tissue for nasal and inhaled Ags in humans, primates, mice, and rats (**Figure 22.2**). In rodents, NALT is present on both

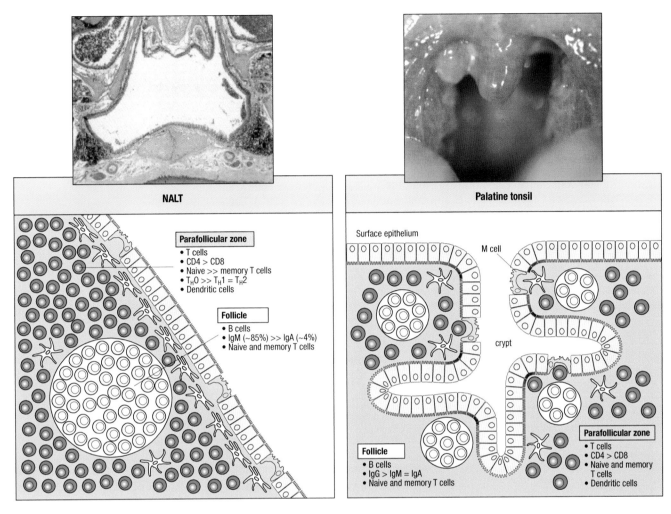

Figure 22.2 Nasopharyngeal-associated lymphoreticular tissue (NALT) in humans and mice is mucosal inductive tissue. Both human palatine tonsils and mouse NALT are covered by an epithelium that contains M cells. NALT consists of a follicle zone, which contains mainly B cells, and a parafollicular zone, which contains T cells and dendritic cells. Although sIgA$^+$ B cells are not the major subset of B cells in NALT, the immunopathobiological features, including spontaneous IgA class-switch recombination, are similar to the traditional mucosal inductive site, intestinal Peyer's patches.

sides of the nasopharyngeal duct dorsal to the cartilaginous soft palate. In humans, unpaired nasopharyngeal tonsils (adenoids) and paired palatine tonsils are present. The palatine tonsils constitute most of Waldeyer's ring in humans. Furthermore, a NALT-like lymphocyte aggregate with follicles has been identified in the human nasal mucosa, most notably in the middle concha in children less than 2 years old. In human tonsils, approximately 50% of tonsillar cells are B lymphocytes, present mainly in follicles containing germinal centers. Most human tonsillar B cells are surface IgG positive (sIgG$^+$), although significant numbers of sIgM$^+$ and sIgA$^+$ B cells are also present (see Figure 22.2). The human palatine tonsil also contains a distinct subepithelial B-cell population, similar to that beneath the FAE region of Peyer's patches, which differs from both germinal centers and follicular mantle B cells (see Figure 22.2). Subepithelial B cells located in NALT may play a crucial role in the production of antibodies to Ag taken up through M cells. Both human tonsils and adenoids are equipped with M cells for inhaled Ag uptake.

Murine NALT consists of bilateral strips of nonencapsulated lymphoid tissue under the epithelium on the ventral aspect of the posterior nasal tract and exhibits a bell-like shape in frontal cross sections (see Figure 22.2). Although dense aggregates of lymphocytes have been identified in the NALT of normal mice, germinal centers are absent but can be induced by nasal application of Ags. Thus, uncommitted B cells (sIgM$^+$) are present in high proportions (80%–85%), whereas low numbers of IgA (sIgA$^+$) and IgG (sIgG$^+$) committed B cells (3%–4% and 0%–1%, respectively) have been detected in mononuclear cells isolated from NALT. In contrast to Peyer's patches, which contain a high proportion (10%–15%) of sIgA$^+$ B cells, NALT contains fewer IgA committed B cells (see Figure 22.2). Nevertheless, nasal immunization induces higher numbers of IgA$^+$ (than IgG$^+$) B cells among NALT B cells in the memory compartment, showing the propensity of NALT for the induction of mucosal IgA Ab responses, the primary Ig involved in protecting mucosal surfaces in distant effector sites. Characterization of isolated NALT mononuclear cells has revealed that approximately 30%–40% of these cells are CD3$^+$ T cells with a CD4:CD8 ratio of approximately 3.0 (see Figure 22.2). The majority of NALT CD3$^+$ T cells coexpress CD45RB, suggestive of naïve, resting T cells. Since transcriptional single-cell analysis revealed the expression of mRNA for both helper T (T$_H$)1 and T$_H$2 cytokines, the majority of CD4$^+$ T cells are considered T$_H$0-type T cells (see Figure 22.2). Further, stimulation via the TCR-CD3 complex results in differentiation of both T$_H$1- and T$_H$2-type cells. These results support the notion that NALT exhibits characteristics of mucosal inductive sites.

NALT M cells mediate the entry of respiratory pathogens such as *Mycobacterium tuberculosis* and group A streptococcus, delivering antigenic substrate to underlying immune cells for the induction of Ag-specific immune responses. Reoviruses also initiate infection via M cells, an activity associated with the protein sigma one (pσ1). In this connection, a NALT M cell–targeting DNA vaccine complex consisting of plasmid DNA and the covalently attached reovirus pσ1 to poly-L-lysine induces high levels of Ag-specific mucosal IgA Ab responses in addition to systemic immunity following nasal immunization.

22.3 Nasal passage mucosa contains effector sites

After initial exposure to Ag in MALT, mucosal lymphocytes leave the inductive site and home to mucosal effector tissues, including lamina propria in the upper respiratory tract. The lamina propria is characterized by a diffuse collection of effector lymphoreticular cells, including IgA-producing plasma cells, B and T lymphocytes, as well as Ag-presenting DCs and macrophages. Nonclassical APCs such as epithelial cells, eosinophils, basophils, natural killer (NK) cells, and mast cells also may initiate innate mucosal responses. The interaction between these innate cells and acquired effector cells with exogenous antigenic stimulation promote mucosal (SIgA) and systemic (serum IgG) Ab and/or T-cell-mediated immune responses.

Nasal washes from humans, nonhuman primates, rats, and mice contain significant levels of IgA Abs. Nasal passage (NP) mucosal lamina propria contains a high frequency of IgA-producing plasma cells (Figure 22.3), similar to the intestinal lamina propria. At least two B-cell subsets, B-1 and B-2 B cells, are present in the mouse and human periphery. B-1 B cells, a minor subset comprising about 5% of the total B-cell population, arise during fetal development and have a restricted receptor repertoire. B-1 B cells differ from conventional B cells in cell surface protein CD5 expression, anatomical localization, and functional characteristics. These B-1 B cells display high levels of IgM, IgA, and IgG3 of low affinity and broad specificity for polysaccharides from gram-positive bacteria and lipopolysaccharides (LPS) from gram-negative bacteria. Although these Ig responses can be enhanced by T cells, they appear within 48 hours of exposure to Ag and are not dependent on T-cell help. Mononuclear cells in mouse NP lamina propria are approximately 20%–25% of B cells, consisting of conventional B-2 (∼15%) and B-1 B cells (see Figure 22.3). Although the precise immunobiological roles for

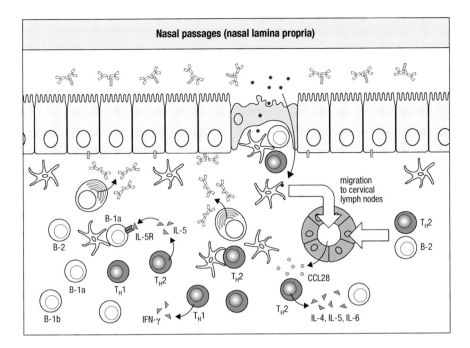

Figure 22.3 Nasal passage lamina propria is a mucosal effector tissue. Nasal immunization induces antigen-specific T- and B-cell migration into the nasal passages. Antigen-specific CD4+ T cells produce Th2 (IL-4, IL-5, and IL-6)-type cytokines, which are necessary for antigen-specific B-cell activation, proliferation, and differentiation of IgA antibodies. B-1 B cells expressing IL-5Rα in nasal passages play a key role in the induction of innate IgA antibodies. The nasal epithelium contains M cells that deliver the antigen to the M cell pocket. Nasal DCs likely take up the M cell–delivered antigen and migrate into the cervical lymph nodes, providing the inductive site for the generation of antigen-specific IgA antibody responses.

B-1 and B-2 cells in humans remain to be elucidated, mouse B-1 B cells have been further classified into B-1a (IgM^{high}, IgD^{low}, B220^{low}, CD5$^+$) cells and B-1b (IgM^{high}, IgD^{low}, B220^{low}, CD5$^-$) cells based on their expression of CD5. Thus, mouse NP mucosa contains both B-1b (~7%) and B-1a (~2%) cell subsets. Approximately 20% of NP lymphocytes are CD3$^+$ T cells, among which half express the CD4 and one-third CD8 (see Figure 22.3). The distribution of B and T cells is similar in NP lamina propria and intestinal lamina propria effector sites.

Langerhans-type (or DC-like) cells are present on the luminal side of the intestinal epithelium and may provide accessory function, including direct interaction with luminal Ags and epithelial cells. When confronted with microorganisms and even soluble proteins that transverse the tight junctions between epithelial cells, lamina propria DCs may take up and process the Ags and induce Ag-specific B- and T-cell responses. M cells also are present in the epithelium of the nasal mucosa (see Figure 22.3). M cells in NP mucosa take up and deliver soluble and particle-type Ags to the abundant DCs for immediate Ag processing (see Figure 22.3). These Ag-activated DCs may migrate into the cervical lymph nodes (CLNs) for the stimulation of naïve CD4$^+$ T cells and follicular B cells, leading to the induction of Ag-specific IgA Ab responses. Thus, the NP-CLN axis initiated by NP M cells may serve as an additional nasal inductive site for eliciting necessary protective immunity in the nasal mucosa.

22.4 Immune system of salivary glands

The oral cavity is protected by mucosal SIgA and systemic (blood plasma) IgG antibodies in saliva and gingival crevicular fluid, respectively (Figure 22.4). Saliva has been extensively used as an easily accessible external fluid for the measurement of Ag-specific SIgA Abs following mucosal immunization in both humans and experimental animals (see Figure 22.4). Based on the number of antibody-forming cells (AFCs) in submandibular glands (SMGs) such as salivary glands, the dominant isotype of Ig-producing cells in SMG is IgA, followed by small numbers of IgM and IgG AFCs. These findings indicate that the SMGs harbor the immunologic components to be an IgA effector tissue, including the dominant prevalence of IgA-producing cells (Figure 22.5).

Figure 22.4 Uniqueness of the oral mucosal immune system. The oral cavity is protected by both the mucosal and systemic immune systems since saliva contains large amounts of SIgA antibodies, whereas the gingival crevicular fluid contains mainly IgG antibodies derived from the blood plasma.

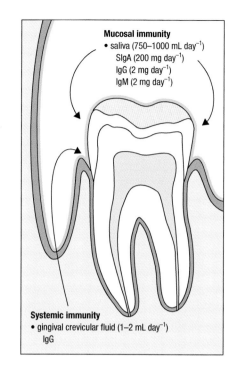

In addition to the dominance of IgA-producing cells in SMGs, the SMGs contain relatively high numbers of T cells, with approximately 50% of lymphocytes CD3$^+$ T cells (see **Figure 22.5**). When different subsets of CD3$^+$ T cells were assessed according to their expression of CD4 and CD8 molecules, three distinct subsets of T cells have been detected: (1) CD4$^+$, CD8$^-$ T cells; (2) CD4$^-$, CD8$^+$ T cells; and (3) CD4$^-$, CD8$^-$ T cells. Among CD4- and CD8-bearing T cells, approximately equal frequencies of CD4$^+$, CD8$^-$ and CD4$^-$, CD8$^+$ T cells are present in SMGs. Relatively high numbers of $\gamma\delta$ T cells (approximately 15%) are present in the SMGs. The increased number of $\gamma\delta$ T cells in SMGs is a characteristic of mucosa-associated tissues, as other mucosal effector sites such as intestinal lamina propria and epithelium also contain a relatively high frequency of $\gamma\delta$ T cells.

The T$_H$1 and T$_H$2 cytokine synthesis by SMG CD3$^+$ T cells is comparable to that of other IgA effector sites, such as the intestinal tract. When CD3$^+$ T cells are isolated from Con A- or anti-CD3-stimulated SMG lymphocyte cultures and then analyzed for cytokine synthesis, IL-2- and IL-4-producing cells are deleted, but the numbers of IFN-γ, IL-5, and IL-6 secreting cells are increased (see **Figure 22.5**). These findings demonstrate that although SMG CD3$^+$ T cells are capable of secreting an array of T$_H$1 and T$_H$2 cytokines, the cells are programmed to produce selected T$_H$1 (IFN-γ) and T$_H$2 (IL-5 and IL-6) cytokines, which provide a favorable molecular environment for IgA synthesis in mucosal effector tissues. Consistent with this view, SMG CD3$^+$ T cells support Peyer's patch B-cell differentiation into IgM-, IgG-, and IgA-producing cells. These findings further support the concept that the SMG CD3$^+$ T cells are programmed *in situ* to produce the IgA-enhancing cytokines IL-5 and IL-6, which support B-cell responses, including sIgA$^+$ B cells to become IgA-producing plasma cells in mucosal effector tissues.

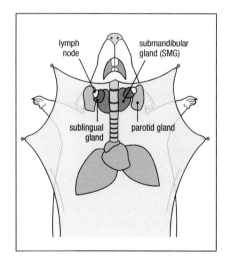

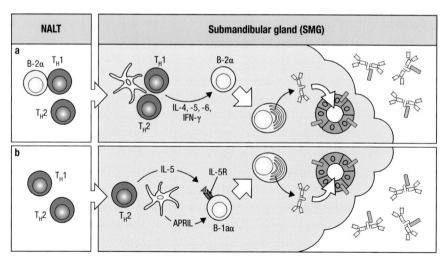

Figure 22.5 Submandibular glands as mucosal effector tissue for the oral immune system. (a) Antigen-specific CD4$^+$ T cells for antigen-specific salivary SIgA antibody responses in the submandibular glands. Upon nasal immunization, antigen-stimulated CD4$^+$ T cells, B-2 B cells, and dendritic cells (DCs) in nasopharyngeal-associated lymphoreticular tissue migrate into the submandibular glands. T$_H$1 and T$_H$2 cytokines produced by CD4$^+$ T cells elicit differentiation of conventional B-2 B cells to become antigen-specific IgA-producing plasma cells. **(b)** Potential roles for DCs in the induction of innate SIgA antibody responses. Upon T-independent antigen stimulation, DCs in submandibular glands may produce APRIL for the induction of IgA class-switch recombination in B-1a B cells. IL-5-producing CD4$^+$ T cells induce sIgA$^+$ B-1a B cells to differentiate into IgA-producing plasma cells.

Another important feature of the SMGs is the presence of a significant number of B-1 B cells (see Figure 22.5). In contrast, NALT and Peyer's patch inductive lymphoid tissues contain only conventional B-2 B cells, whereas mucosal effector lymphoid tissues (intestinal lamina propria, NPs, and SMGs) are enriched in B-1 B cells in addition to MALT-originating B-2 cells. A high frequency of B-1a B cells is more typical of the SMGs, whereas a predominance of B-1b B cells is characteristic of the intestinal lamina propria and NP lamina propria of wild-type mice. Interestingly, up to 40% of IgA plasma cells in the mouse intestinal lamina propria arise from a pool of B-1 precursors derived from the peritoneal cavity. Thus, B-1 B cells may represent a distinct lineage from that of conventional B-2 B cells originating in mucosal inductive lymphoid tissues. In addition, up to 50% of human intestinal B cells express the CD5 molecule, and a large number of these CD5$^+$ B cells secrete IgA Abs. Taken together, these findings suggest that B-1 B cells in SMGs are an alternative but important source for IgA-producing cells in addition to conventional B cells.

INDUCTION OF ACQUIRED IMMUNE RESPONSES VIA NASAL AND ORAL MUCOSAL IMMUNE SYSTEM

22.5 Nasal immunization with enterotoxin-based adjuvant induces antigen-specific immunity

The induction of effective immune responses in the mucosa generally requires mucosal adjuvants or attenuated live viral or bacterial vectors. Nasal delivery of purified Ag plus a mucosal adjuvant has emerged as an effective method to induce both systemic and mucosal immunity, including salivary SIgA Ab responses (see Figure 22.1). Most current protocols instill vaccine into each nostril (usually 5–10 μL per nostril), resulting in effective delivery of vaccine presumably into NALT and NPs. Two bacterial enterotoxins, native cholera toxin (nCT) and native *Escherichia coli* heat-labile toxin (nLT$_H$-1), and their derived artificial nontoxic mutants (mCTs and mLT$_H$-1s), are well-established mucosal adjuvants for the induction of both mucosal and systemic immunity to coadministered protein Ags. Mucosal administration of subunit or purified Ag vaccine together with nCT or nontoxic mCTs induces vaccine Ag-specific CD4$^+$ T$_H$2-type cells with characteristic plasma Ag-specific IgG1, IgG2b, IgE, and IgA, as well as mucosal SIgA Ab responses. Nasal immunization with the weak protein Ag ovalbumin (OVA) plus mCT elicits OVA-specific SIgA Ab responses in the SMGs and in other mucosal effector tissues.

Notable examples of microbe Ag plus adjuvant CT that induce Ag-specific responses include the following. Nasal vaccines containing tetanus toxoid and mCT induce protective immunity and the generation of tetanus toxin-specific neutralizing Abs. Mice immunized nasally with *Streptococcus pneumoniae* pneumococcal surface protein A (PspA) plus mCT induce antigen-specific mucosal IgA responses associated with effective protection against capsular serotype 3 *S. pneumoniae* A66. Nasal immunization with simian immunodeficiency virus p55gag plus CT results in selected T$_H$2- and T$_H$1-type cytokine profiles and Ag-specific IgA responses in mucosal secretions. Nasal vaccine containing outer membrane proteins of *Haemophilus influenzae* and adjuvant CT induces antigen-specific SIgA that blocks attachment of the bacterium to epithelial cells. Similarly, mice immunized with fimbrial protein of *Porphyromonas gingivalis* and CT generate antigen-specific IgA Abs in SMGs that inhibit *P. gingivalis* attachment and reduce subsequent inflammatory cytokine production by epithelial cells. Thus, nasal immunization induces both T$_H$1- and/or T$_H$2-type cytokine-producing Ag-specific CD4$^+$ T cells in nasal lamina propria and SMGs, as well as spleen, that support the generation of Ag-specific SIgA and plasma IgG responses, respectively.

The nasal administration of inactivated respiratory syncytial virus (RSV) with CT results in nasal IgA and serum IgG anti-RSV Ab responses. Influenza vaccine provides more effective protective immunity when administered nasally than by other routes. Nasal immunization with a trivalent influenza vaccine given with B subunit of CT (CT-B: a nontoxic portion of CT) containing a trace amount of nCT results in the induction of cross-protective SIgA antibodies against a broad range of viruses. Nasal immunization with influenza vaccine together with the B subunit of nLT_H-1 (with a trace amount of nLT_H-1) induces influenza-specific immune responses in humans. Together, these studies suggest that the nasal mucosal immune system is an attractive site for vaccine delivery to induce protective immunity in both the mucosal and systemic compartments.

22.6 Development of safe mucosal adjuvants and delivery systems for nasal vaccines

The olfactory neuroepithelium in the nasopharynx of mice constitutes approximately 50% of the nasal surface and has direct neuronal connections to the olfactory bulbs in the central nervous system (CNS). Consequently, a major safety concern in using mucosal adjuvants with nasal vaccination is that the adjuvant may enter and/or target olfactory neurons, bypassing the blood-brain barrier to gain access to olfactory bulbs and deeper structures in the brain parenchyma. Studies with enterotoxin adjuvants and more recently a recombinant adenovirus vector suggest that this adverse effect is in large part mediated by the ADP-ribosyl transferase activity via the targeted cellular receptors. Both nCT and nLT_H-1 bind to GM1 on epithelial cells and require endocytosis followed by transport across the epithelial cell to reach the basolateral membrane. GM1 gangliosides also are abundantly expressed by cells of the CNS, and their concentration on neuronal and microglial cells varies during the development of various cell types and different regions of the brain. GM1 also is thought to play a role in cell-to-cell and cell-matrix interactions and may act synergistically with several growth factors in development. GM1 ligation did not generate a calcium flux following exposure of microglial cells to CT-B; however, binding of CT-B to cerebellar granule cells *in vitro* improved survival and was associated with intracellular calcium increases through L-type voltage-sensitive channels during the first 7 days of culture with reduced calcium fluxes later in culture.

Nasally administered nCT and CT-B in mice enter the olfactory nerves and epithelium and the olfactory bulbs by mechanisms that are selectively dependent on GM1. As an adjuvant, nCT promotes the uptake of nasally coadministered, unrelated proteins into the olfactory nerves and epithelium. Nasal administration of mCT is reported to not induce pathologic changes in the CNS. However, nasal vaccination of humans with a subunit-type influenza with nLT_H-1 as mucosal adjuvant has resulted in Bell's palsy (facial paresis) in a very small number of subjects receiving a nonliving nasal influenza vaccine (Nasalflu), leading to withdrawal of the vaccine from the market.

To overcome nCT and CT-B toxicity, nontoxic mutants have been produced that sacrifice, or reduce, the mucosal adjuvanticity of nCT. mCT E112K, in which glutamic acid is replaced by lysine at position 112 in the CT-A subunit, has no detectable toxicity but retains adjuvant activity. Previous studies showed that nasal immunization with OVA, PspA of *S. pneumoniae* or diphtheria toxoid, plus mCT E112K, elicited both Ag-specific IgA and IgG Ab responses in mucosal and systemic immune compartments. Although these results indicate that mCT E112K did not elicit neuronal damage based on NGF-β1 production in the CNS, the studies did not provide direct evidence as to whether mCT E112K migrates into the CNS after nasal application. To further establish safety for enterotoxin-based nasal adjuvant, CT double mutants were developed by introducing two potent mutations in the ADP-ribosylation activity center of the A1 subunit and the ER retention signal tetrapeptide KDEL in the A2 subunit. Both of the double-mutant CTs exhibited low toxicity and induced enhanced OVA-specific immune responses in mucosal and systemic lymphoid tissues. A safe and effective nCT-based adjuvant that targets the Ig receptor on both naïve and memory B cells also has been developed to avoid GM1 ganglioside binding. The enzymatically active A1 subunit of nCT combined with a dimer of an Ig binding element from *Staphylococcus aureus* protein A (CT-A1-DD) is as strong an adjuvant as nCT without CNS toxicity following nasal immunization. Notably, CTA1-DD promoted a balanced T_H1/T_H2 response with little effect on IgE Ab production and did not induce inflammatory changes in the nasal mucosa or bind to or accumulate in the olfactory bulbs or the CNS. Thus, nonganglioside targeting adjuvants and delivery systems offer new tools for the development of safe and effective nasal vaccines.

The addition of a bioadhesive gel to an Ag increases the residence time, extends Ag release, and allows Ag retention by nasal epithelial cells. Such a bioadhesive gel has facilitated the induction of influenza-specific SIgA Ab responses in nasal fluids. Bioadhesive nanometer-sized (less than 100 nm) polymer hydrogels (nanogels) have attracted interest as nanocarriers, especially in drug-delivery systems and mucosal vaccine administration. Importantly, nanogels are a safe nasal delivery system since they do not induce CNS toxicity.

22.7 Innate molecule-based nasal adjuvants

Nasal coadministration of nonganglioside binding, regulatory cytokines, or chemokines also augment Ag-specific mucosal and systemic immune responses. For example, nasal immunization with tetanus toxoid or dinitrophenol-OVA plus IL-12 enhance Ag-specific Ab responses. Nasal application of IL-12 encoding a DNA vaccine for *Yersinia pestis* induced protective Ag-specific Ab responses against pneumonic plague. IL-1-α and IL-1-β also may serve as potent mucosal adjuvants. In addition, the human neutrophil peptides, lymphotactin and RANTES, possess nasal adjuvant activity for enhanced Ag-specific Ab responses.

Flt3 ligand (FL) also has been used as a mucosal adjuvant by targeting mucosal APCs, including DCs. FL binds to the *fms*-like tyrosine kinase receptor Flt3/Flk2, a growth factor that dramatically increases the numbers of DCs *in vivo* without inducing their activation. Treatment of mice by systemic FL injection induced marked increases in the number of DCs in both systemic (spleen) and mucosal lymphoid (intestinal lamina propria, Peyer's patches, and mesenteric lymph nodes) tissues. FL treatment favors the induction of immune responses after nasal and oral, systemic, or cutaneous vaccine delivery. Plasmid DNA encoding FL (pFL) also has been coadministered systemically with plasmids encoding protein Ags or linked to the Ag itself to induce Ag-specific immune responses. Further, nasal immunization with OVA plus pFL or adenovirus expressing FL as nasal adjuvant specifically targets nasal DCs for the induction of Ag-specific Ab responses in both the mucosal and systemic compartments.

Bacterial DNA and pathogen-associated molecular patterns contain a high proportion of unmethylated CpG motifs. These motifs are recognized by the innate immune system through toll-like receptor 9 (TLR9), which is expressed by B cells and pDCs. CpG DNA induces the maturation and stimulation of DCs, as well as Ag-specific Th1 and cytotoxic lymphocyte responses. In this regard, CpG oligodeoxynucleotides act as effective adjuvants for the induction of Ag-specific immunity, enhancing both Ab and cell-mediated immune responses to OVA in mice. An array of viral (influenza, hepatitis B, human immunodeficiency type-1) and toxoid (tetanus and diphtheria) vaccines administered with CpG oligodeoxynucleotides significantly increase levels of Ag-specific Ab and CTL responses.

Bacterial flagellin, the ligand for TLR5, also has been used as adjuvant and vaccine delivery systems. BALB/c mice subcutaneously immunized with a flagellin-enhanced green fluorescent protein (EGFP) fusion protein results in the effective generation of EGFP-specific T-cell responses. Further, *Salmonella typhimurium* flagellin enhances CD4$^+$ T-cell responses to cointravenously administered OVA peptides. Nasal immunization with purified flagellin from *Salmonella enterica* serovar *Enteritidis* alone or conjugated to starch microparticles induces mixed T_H1/T_H2-type response and high flagellin-specific SIgA Ab titers in fecal extracts. *Vibrio vulnificus* major flagellin coadministered with tetanus toxoid vaccine induces significant tetanus toxoid-specific IgA Ab responses in both mucosal and systemic compartments and IgG Ab responses in the systemic compartment. The flagella filament structural protein (FliC) of *S. enteritidis* stimulates human β defensin-2 mRNA in Caco-2 cells by mechanisms involving mitogen-activated protein kinase. In this regard, nasal administration of human neutrophil peptide defensins enhances systemic IgG and promotes B- and T-cell interactions to link innate immunity with the adaptive immune system. These findings underscore the importance of a full molecular understanding of the innate and acquired immune response initiated by microorganisms, which will facilitate the development of more appropriate, potent, and safe nasal adjuvants.

22.8   Oral cavity is an antigen-delivery site for the induction and modification of Ag-specific immune responses

The oral cavity is an effective inductive site for eliciting Ag-specific mucosal and systemic immune responses. The sublingual (SL) region of oral cavity in particular is an immunologically functional site for the initiation of Ag-specific immune responses. Since some nasal immunization strategies risk Ag tracking into the olfactory tissues and the CNS, SL immunization may be an alternative mucosal Ag delivery system without the safety concerns related to entry into the CNS. SL administration is a noninvasive route with the advantage of requiring lower inoculation doses of Ag than the oral route due to the reduced exposure to proteolytic enzymes and the lower pH of the stomach. The SL route may be a more efficient induction site due to the increased numbers of APCs in the SL region of the oral cavity.

Several recent studies have used the SL route for the delivery of vaccines against infectious agents. When plasmid DNA encoding hepatitis B surface Ag was administered via SL inoculation in mice, the humoral and CD8$^+$ CTL responses induced against hepatitis B were comparable to those elicited by intradermal injection. The SL delivery of dinitrophenol-bovine serum albumin in starch microparticles in combination with a penetration enhancer (α-lysophosphatidylcholine) enhances salivary IgA responses. SL immunization with influenza vaccine and mucosal non-toxin-based adjuvant mCT is reported to generate protective immunity. In addition, the SL delivery of lipopeptides induces elevated serum Abs and T-cell responses in the spleen and inguinal lymph nodes of mice. SL application of vaccine Ag preferentially induces IFN-γ-producing T cells and IgG2a Ab responses compared with subcutaneous injection, which elicit IL-4 and IgG1 Ab responses. More recently, SL vaccination with the outer membrane protein of *P. gingivalis* plus plasmid expressing FL cDNA has been shown to elicit increased numbers of DCs in submandibular lymph nodes and protective immunity against *P. gingivalis* infection in the oral cavity. The migration of CCR7-expressing DCs from SL mucosa to CLNs appears to play an essential role in the induction of Ag-specific immunity initiated by SL immunization. These findings suggest that by using the appropriate form of Ag, mucosal adjuvant, and delivery system, the SL route is an attractive and alternative vaccination route for the induction of mucosal, as well as systemic, responses.

22.9 Nasal immune system escapes mucosal aging

In contrast to the many studies of the impact of aging on gut immune responses, few studies have addressed the influence of advanced age on the upper respiratory tract immune responses. However, in an important study of 1-year-old (aging) mice, nasal administration of OVA with nCT was shown to induce impaired mucosal and systemic responses compared with the impaired responses following oral administration. Nasal administration of OVA equivalent induced levels of OVA-specific SIgA Ab in the nasal cavity of young adult and aging mice. In addition, administration of nasal tetanus toxoid and nCT resulted in equal protection from the tetanus toxin challenge in 1-year-old and young mice. When 2-year-old (aged) mice were immunized nasally with OVA and nCT adjuvant, the mice failed to undergo induction of Ag-specific SIgA Ab responses. However, the immunized mice displayed systemic OVA-specific Ab responses that were essentially identical to the responses in young adult mice. Similarly, OVA-specific $CD4^+$ T-cell proliferative responses as well as T_H1 and T_H2 cytokine responses in the spleens of aged mice were comparable to those of young adult mice when nCT was used as nasal adjuvant. These results suggest that mucosal immunosenescence occurs prior to systemic immune depression, even though NALT immunosenescence is less severe than GALT immunosenescence in aged mice. Thus, immune modulation of infectious diseases in the elderly will require consideration of mucosal immunosenescence in the development of nasal vaccines.

22.10 Induction of T-cell–independent mucosal IgA responses in nasal and oral cavities

Two distinct lineages of $sIgA^+$ B cells developed from B-1 and B-2 B cells are involved in the induction of SIgA Abs for mucosal immunity. As discussed earlier (in Sections 22.3 and 22.4), the oral-nasopharyngeal mucosa, including NPs and SMGs, contain high frequencies of B-1 B cells. NP lamina propria contains a higher proportion of B-1b B cells, similar to the proportion of B-1 B cells in the intestinal lamina propria, whereas the SMGs display a more dominant B-1a B-cell repertoire. However, the majority of $sIgA^+$ B cells in both SMGs and NP lamina propria are B-1a type, suggesting a similarity to oral-nasopharyngeal mucosal effector tissues. Mouse B-1 B cells are capable of isotype switching as well as proliferation and differentiation in the presence of IL-5, even in the absence of CD40-CD40L signaling, which are distinctive characteristics compared with B-2 cells. Further, SMGs contain significant numbers of $sIgM^+$ IgA^- B-1a B cells. Studies showing that DCs induce CD40-independent Ig class switching support their consideration in the T-cell–independent induction of B-1 cells.

Respiratory tract DCs play an important role in the induction of IgA Ab responses. DCs interact directly with B cells through the B-cell activation factor of the TNF family (BAFF), also called lymphocyte stimulator protein (BLyS), and a proliferation-inducing ligand (APRIL) in order to induce $sIgA^+$ B cells. Since B-1 B cells express receptors for these ligands, mucosal DCs in the NPs and SMGs may play a key role in the induction of Ag-specific B-1 B-cell IgA class switching in a T-cell–independent manner (see **Figure 22.5**), despite the fact that continuous IgA isotype class switching occurs effectively for the B-2 cells in the organized mucosa-associated tissues such as NALT, Peyer's patches, and ILFs. In this regard, nCT as a nasal adjuvant is reported to elicit increased levels of LPS-specific SIgA Ab responses through IL-5–IL-5 receptor interaction between $CD4^+$ T cells and IgA^+ B-1 B cells in murine SMGs and NPs. Thus, DCs in the SMGs and NPs stimulated by T-cell–independent Ag plus nCT may play a role in the induction of B-1 B-cell IgA class-switch recombination for the enhancement of T-cell–independent mucosal SIgA Ab responses. These findings suggest that the B-1a B cells and DCs in SMGs and NPs play key roles in the induction of T-cell–independent Ag-specific mucosal IgA Ab responses in the oral-nasopharyngeal mucosa.

Microbiota at oral and nasal mucosa

The oral cavity contains a complex consortium of approximately 2×10^{10} bacteria from six phyla, including Firmicutes, Bacteroidetes, Proteobacteria, Actinobacteria, Spirochaetes, and Fusobacteria. The oral microbiome contributes to local mucosal homeostasis and certain metabolic functions, including deglycosylation of complex carbohydrate, sulfate reduction, and amino acid absorption. The overgrowth of a pathogenic bacteria or an alteration in the normal microbiome, termed *microbial dysbiosis*, may contribute to dental caries or periodontal diseases. For example, periodontal disease is associated with a significant alteration in the oral microbiome. Bacteria such as *P. gingivalis, Tannerella forsythensis,* and *Treponema denticola* are closely associated with severe periodontitis and are termed as "red-complex" bacteria, but they do not induce periodontal disease in the absence of the microbiota, reflected in observations that germ-free mice infected with *P. gingivalis* do not develop inflamed gingival tissues and the associated alveolar bone loss. In addition, specific pathogen-free mice do not develop gingival inflammation. These findings indicate that dysbiosis of the oral microbiome with increased numbers of red-complex bacteria likely contributes to the induction of chronic severe periodontitis. The role of the oral microbiota in systemic diseases such as gastrointestinal and head and neck cancers is under investigation. Thus, the oral cavity microbiota impacts stomatological and possibly systemic diseases.

The nasopharyngeal microbiota, which is less diverse than oropharyngeal microbiota, contains bacteria in the Actinobacteria, Firmicutes, and Proteobacteria phyla. Virulent strains of bacteria such as methicillin-resistant *S. aureus* are associated with nasal dysbiosis. The nasal microbiota also changes with the different seasons, with the dominant strains of bacteria in the nasal cavity during fall and winter being different from those present in spring. Interestingly, the oropharyngeal microbiota appears to be influenced by aging and lifestyle, including such factors as alcoholism and smoking.

SUMMARY

The nasal and oral cavities are intimately associated with the entrance to the aerodigestive tract and are consequently exposed to Ags, pathogens, and allergens that enter the host through inhalation and ingestion. The mucosal surfaces of the nasal and oral cavities contain inductive and effector cellular elements and are protected by mucosal SIgA and systemic IgG Abs in nasal fluid and saliva, distinct from antibodies in the gastrointestinal and lower respiratory tracts. Organogenesis and lymphocyte trafficking in the nasopharyngeal-oral mucosa are distinctly regulated, contributing to the compartmentalization of the nasopharyngeal-oral cavity. The nasopharyngeal-oral mucosa exhibits a different phenotype of B-1 B cells that express SIgA as innate Ig-producing cells. Resident microbiota in the nasopharyngeal and oral cavities likely contribute to local homeostasis and, in the setting of dysbiosis, to local disease and possibly pathologic processes beyond the cavities. Understanding the common and distinct characteristics of the nasopharyngeal-oral immune system will enhance the development of new strategies for mucosal vaccines for an array of infectious and inflammatory diseases.

FURTHER READING

Brandtzaeg, P.: Do salivary antibodies reliably reflect both mucosal and systemic immunity? *Ann. N.Y. Acad. Sci.* 2007, 1098:288–311.

Czerkinsky, C., and Holmgren, J.: Mucosal delivery routes for optimal immunization: Targeting immunity to the right tissues. *Curr. Top. Microbiol. Immunol.* 2012, 354:1–18.

Fujihashi, K., and Kiyono, H.: Mucosal immunosenescence: New developments and vaccines to control infectious diseases. *Trends Immunol.* 2009, 30:334–343.

Hajishengallis, G.: Periodontitis: From microbial immune subversion to systemic inflammation. *Nat. Rev. Immunol.* 2015, 15:30–44.

Kiyono, H., and Fukuyama, S.: NALT- versus Peyer's-patch-mediated mucosal immunity. *Nat. Rev. Immunol.* 2004, 4:699–710.

Kunisawa, J., Nochi, T., and Kiyono, H.: Immunological commonalities and distinctions between airway and digestive immunity. *Trends Immunol.* 2008, 29:505–513.

McGhee, J.R., Kunisawa, J., and Kiyono, H.: Gut lymphocyte migration: We are halfway "home." *Trends Immunol.* 2007, 28:150–153.

Mora, J.R., Iwata, M., and von Andrian, U.H.: Vitamin effects on the immune system: Vitamins A and D take centre stage. *Nat. Rev. Immunol.* 2008, 8:685–698.

Moutsopoulos, N.M., Greenwell-Wild, T., and Wahl, S.M.: Differential mucosal susceptibility in HIV-1 transmission and infection. *Adv. Dent. Res.* 2006, 19:52–56.

Pascual, D.W., Riccardi, C., and Csencsits-Smith, K.: Distal IgA immunity can be sustained by $\alpha_E\beta_7{}^+$ B cells in L-selectin$^{-/-}$ mice following oral immunization. *Mucosal Immunol.* 2008, 1:68–77.

Schneider, P.: The role of APRIL and BAFF in lymphocyte activation. *Curr. Opin. Immunol.* 2005, 17:282–289.

Takatsu, K., Kouro, T., and Nagai, Y.: Interleukin 5 in the link between the innate and acquired immune response. *Adv. Immunol.* 2009, 101:191–236.

Taubman, M.A., and Nash, D.A.: The scientific and public-health imperative for a vaccine against dental caries. *Nat. Rev. Immunol.* 2006, 6:555–563.

Williams, I.R.: Chemokine receptors and leukocyte trafficking in the mucosal immune system. *Immunol. Res.* 2004, 29:283–292.

Yu, X., Tsibane, T., McGraw, P.A. et al.: Neutralizing antibodies derived from the B cells of 1918 influenza pandemic survivors. *Nature* 2008, 455:532–536.

Bronchus-associated lymphoid tissue and immune-mediated respiratory diseases

23

DALE T. UMETSU AND BART LAMBRECHT

The major function of the lung is gas exchange, to allow uptake of oxygen and disposal of carbon dioxide. Given the amount of air inhaled on a daily basis to perform this function, it comes as no surprise that the lung is a major portal of entry of pathogens or other toxic substances contained in air. Evolution has created a unique and rapid defense mechanism to shield off these pathogens and potential insults, while at the same time preserving the highly specialized microanatomy of the gas exchange apparatus that is so crucial for survival. A failure of these homeostatic mechanisms can lead to chronic respiratory inflammatory diseases like asthma and chronic obstructive pulmonary disease (COPD).

GENERAL ANATOMY AND PHYSIOLOGY OF THE CENTRAL AND LOWER AIRWAYS

The respiratory tract can be divided into three compartments. The first is the upper airways, which include the nasal passages, the sinuses, and the pharynx; these were discussed in Chapter 22. The other two immunologic compartments of the respiratory tract are the central airways, which include the trachea, bronchi, and bronchioles, and the lower airways, which include the respiratory bronchioles, alveolar ducts, alveolar sacs, and alveoli. The lower airways all perform gas exchange functions, while the central airways are seen more as conductive tubes allowing flow of high volumes of air. The immune inductive sites within these two compartments are the bronchus-associated lymphoid tissue (BALT) and draining nodes of the lungs that are situated around the hili (the roots of the large vessels and bronchi near the heart) and mediastinum (in the vicinity of branching points of vessels and left and right bronchi). In humans, BALT exists during childhood as isolated lymphoid follicles in close contact with the surface epithelium. However, such structures are not regularly identified in healthy adults. In the absence of BALT, mucosal-draining lymph nodes take over the important role of induction of mucosal immunity in the airways. Just as the superficially and deeply located cervical lymph nodes receive lymphatic vessels from both nasopharynx-associated lymphoid tissue (NALT) and effector sites in the upper respiratory tree (see Chapter 22), the parabronchial, hilar, and paratracheal lymph nodes drain the airway mucosa distal to the pharynx, including the lung parenchyma (Figure 23.1). Once immune responses in T and B cells are induced in lymphatic tissue, primed lymphocytes return to the lung effector sites. The lamina propria of the conducting airways, and the lung parenchyma (also called lung interstitium) recruit many effector cells in steady state and during the course of an immune response. A specific lung compartment is the alveolar space, which contains highly specialized and long-lived alveolar

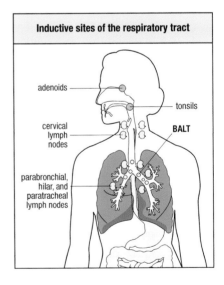

Figure 23.1 Inductive sites of the respiratory tract. Nasopharynx-associated lymphoid tissue (NALT), consisting mainly of the palatine tonsils and adenoids, is strategically located in the aerodigestive tract, whereas bronchus-associated lymphoid tissue (BALT) is located in the central airways. The cervical lymph nodes drain the upper airways, and the lymph nodes located adjacent to the bronchi drain the lower part of the airway mucosa as well as the lung parenchyma.

macrophages and recruits many inflammatory and immune cells that are available to clinicians via bronchoscopic lavage of the compartment. A prime function of alveolar macrophages is to clear inhaled dust and particles, but they have an equally important role in immunoregulation to prevent overt immune-mediated damage to the delicate gas exchange apparatus, i.e., the alveolo-capillary membrane.

23.1 Bronchus-associated lymphoid tissue and lung draining nodes as immune inductive site

Normal mice do not have a clearly identified BALT under homeostatic conditions, but a distinguishable BALT can be rapidly induced in the lungs and around the bronchi and vessels in response to antigen or microbial/inflammatory challenge, like viral infection or endotoxin exposure. In both humans and mice, this induction of BALT structures is more efficient in very young compared to older individuals. After inducible BALT is established, lymphoid aggregates with B-cell areas containing germinal centers and T-cell areas around high endothelial venules efficiently support the induction of primary immune responses to unrelated antigens. Importantly, mice with inducible BALT that lack peripheral lymphoid organs survive higher viral challenge doses than normal mice, suggesting that BALT functions to initiate rapid mucosal responses to airway challenges. Like for other MALT tissues, BALT structures contain resident memory T cells (T_{rm}), dendritic cells (DCs), and plasma cells producing local IgA. Because of the inducible nature of BALT, prior environmental exposures (e.g., chronic endotoxin, cigarette exposure, or repeated viral insults that can induce BALT structures in mice) likely have important effects on the development of human lung diseases such as asthma and COPD.

In comparison with the well-established role for NALT as an inductive site of mucosal immune responses (see Chapter 22), the role of BALT as an inductive site in humans is more controversial. Similarly, whether BALT structures have an overlying specialized epithelium containing microfold (M) cells, as do Peyer's patches, is controversial. The bronchial mucosa of infants and children contains large numbers of isolated lymphoid follicles in close contact with the surface epithelium (**Figure 23.2**). These lymphoid aggregates contain many small B cells surrounded by naive T cells, regulatory T (T_{reg}) cells, and high endothelial venules, as well as many DCs. Under steady-state conditions, only a minority of the aggregates show germinal centers. Thus, BALT may be an important inductive site for immune responses, especially during childhood or in patients with lung disease.

In addition to NALT and BALT, regional lymph nodes that drain the airways are also important inductive sites for the respiratory tract. These lymph nodes receive afferent lymphatic vessels from NALT and BALT, as well as effector sites, including the airway mucosa and the lung parenchyma (see **Figure 23.1**). T and B cells differentiated in NALT and BALT, and also immune cells within the effector sites of the airways, pass with free antigens through the lymphatics to the draining lymph nodes. Of particular interest for the induction of local immune responses is the dynamic population of DCs that are strategically positioned in the airway mucosa, lung parenchyma, and alveolar space. These cells capture luminal antigens for transfer to the draining nodes. Such migrating DCs typically trigger T-regulatory (T_{reg})

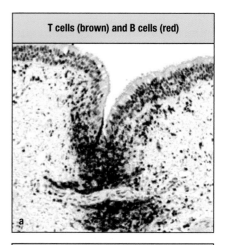

T cells (brown) and B cells (red)

a

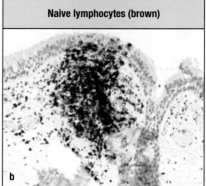

Naive lymphocytes (brown)

b

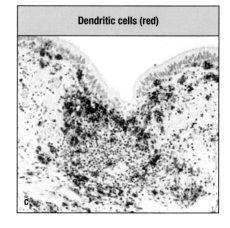

Dendritic cells (red)

c

Figure 23.2 Micrographs of lymphoid tissue in the airway mucosa.
Immunostaining for CD3 indicating T cells (brown) and CD20 indicating B cells (red) (**a**), CD45RA indicating naive lymphocytes (**b**), and CD11c indicating dendritic cells (**c**) that visualizes isolated lymphoid follicles in sections of bronchial mucosa of a 2-year-old child.

responses and helper T (T_H) cell T_H2-dominated responses under homeostatic conditions, particularly in very young mice and children. However, through pathogen-associated molecular pattern (PAMP) recognition, cytokines, and other innate signals, DCs become activated and acquire the capacity to induce pathogen-specific T_H1 or T_H17 responses, and as such are also crucial in initiating pulmonary adaptive immunity. As in many other mucosal sites, the steady-state lung also contains many innate cell types like tissue resident eosinophils, mast cells, basophils, and innate lymphoid cells (ILC2) that also help in the surveillance of incoming pathogens and instruct DCs to initiate a tailored adaptive immune response. These cells are typical of type 2 immune responses yet seem to be present in baseline, particularly in childhood, explaining why the lung is a "T_H2 biased" organ.

23.2 Lung contains effector sites similar to those in other mucosal tissues

From the outer part of the nose to the respiratory bronchioles immediately distal to the trachea and bronchi, the airway mucosa is covered by a pseudostratified ciliated epithelium that is interspersed with goblet cells producing mucus that covers the cilia. Large numbers of exocrine glands are present in the underlying mucosa, which together with goblet cells produce mucus that covers the epithelial surface. The glands are surrounded by plasma cells that produce predominantly dimeric IgA, which is translocated from the basolateral to the luminal side of the epithelium by the polymeric immunoglobulin receptor. Secretory IgA, present in the mucus layer, participates in immune defense by neutralizing microbial toxins and pathogens and by preventing commensal bacteria from breaching the mucosal surface. The mucosal secretions of NALT and conducting systems of the upper airway also contain significant quantities of IgG that are transported by the neonatal Fc receptor (FcRn) expressed by the epithelium. The mucus and its content are constantly translocated upward by the ciliated cells, which provide mechanisms for mucociliary clearance of inhaled antigens, including antigens complexed with immunoglobulins. The adequate functioning of these cilia and the mucociliary escalator is critical in lung defense, and people who have primary ciliary dyskinesia (PCD) or cystic fibrosis frequently have infections with *Pseudomonas* or S*taphylococci*.

The airway mucosa also contains DCs, macrophages, T cells, mast cells, eosinophils, and other minor populations of cells. In the human airway mucosa, both DCs and macrophages are strategically positioned within and beneath the surface epithelium. Airway mucosal DCs are sentinel cells that link the innate and the adaptive immune system by transporting antigenic material from the mucosal effector sites to the draining lymph nodes, where they interact with T and B cells. DCs located within the airway epithelium extend cellular projections into the airway lumen as they do within the gut. These extensions allow the DCs to sample antigens directly from the airway lumen through an intact epithelium, which may have important implications for the induction of immune responses under homeostatic situations and for the early detection of infectious threats. Subepithelial resident macrophages participate in the phagocytosis and elimination of incoming pathogens. Along the airway mucosa, large numbers of T_{rm} are present within and beneath the epithelium. T cells in the epithelial compartment are mainly CD8, similar to intraepithelial lymphocytes in the intestinal mucosa, whereas most T cells in the lamina propria express CD4. Both T-cell subsets express CD45RO, consistent with a memory/effector phenotype, and many T_{rm} cells have specificity for viral antigens, although the precise T-cell receptor repertoire has not been carefully studied. T cells with a regulatory phenotype (Foxp3$^+$) are also present in the lamina propria of the airways.

Although the respiratory epithelium has been traditionally thought to be primarily a physical barrier and a base for ciliary activity, it is now clear that respiratory epithelial epithelium has a major role in sensing the environment, maintaining homeostasis, and repairing injury. Airway epithelial cells respond to microbial challenge and environmental insults by rapidly producing an array of cytokines and growth factors that initiate innate immunity, prime adaptive immunity, and establish homeostasis. For example, airway epithelial cells respond to PAMPs associated with antigens or microbes that enter the airways, through toll-like receptors (TLRs), nucleotide-binding oligomerization domain (NOD)-like receptors, and C-type lectin receptors. Signals generated by these receptors or due to direct injury of epithelial cells induce lung epithelial cells to produce cytokines such as interleukin (IL)-1, IL-25, IL-33, thymic stromal lymphopoietin (TSLP), granulocyte-macrophage colony-stimulating factor (GM-CSF), and transforming growth factor (TGF)-β, as well as chemokines (CCL20 [MIP-3α; ligand of CCR6], CCL17 [TARC; ligand of CCR4], and CCL22 [MDC; ligand of CCR4]) and antimicrobial peptides such as defensins. Epithelial cells are also a copious source of endogenous danger signals like adenosine triphosphate (ATP), uric acid, and high mobility group box 1 (HMGB1) that can alert immune cells. Thus, epithelial cells can initiate innate immunity that can later affect adaptive mucosal immunity.

RESPONSE OF THE LUNG TO ENVIRONMENTAL CHALLENGES

The lung, like the gut, interfaces with the environment. This interface of the lung with the environment is enormous: about 160 m^2 in an adult, which is equivalent to the floor area of a medium-sized house. At this interface, the lung encounters a multitude of environmental factors, including pathogens, particulate matter, allergens, and inert material as well as oxidizing agents, related to the absorption of oxygen. Therefore, the lung, like the intestinal tract, has developed several mechanisms to deal with these encounters, but these mechanisms sometimes result in inflammation and disease.

23.3 Two major chronic inflammatory diseases of the lung are chronic obstructive pulmonary disease and asthma

Asthma and chronic obstructive pulmonary disease (COPD) are two major chronic inflammatory diseases of the lung. Asthma is a major public health problem, affecting 300 million persons worldwide, and has increased markedly in prevalence in Westernized countries over the past three decades. Asthma is a complex trait caused by multiple environmental factors in combination with more than 100 major and minor susceptibility genes. Asthma is characterized by reversible obstruction of the airways, with symptoms of shortness of breath, coughing, wheezing, and chest tightness, and is a major cause of emergency room visits, hospitalization, and school absences. The estimated health-related annual costs for asthma amount to $80 billion in the United States alone. COPD is also a major public health problem; worldwide it is the fourth leading cause of death (shared with human immunodeficiency virus/acquired immunodeficiency syndrome). Both asthma and COPD are characterized by airway obstruction, which in asthma is variable and reversible, but in COPD is progressive and irreversible, and associated with symptoms of shortness of breath, which is progressive, and sputum overproduction. Both chronic diseases are characterized by acute exacerbations, with a great increase in symptoms and inflammation; these exacerbations are frequently caused by infections or air pollution. In addition, the medications for these two diseases overlap, and patients with long-standing asthma may evolve into a picture that mimics COPD, particularly when there is a history of smoking in asthmatics.

Although many differences exist between asthma and COPD, the similarities between asthma and COPD suggest that asthma and COPD may be the ends of a single spectrum of inflammatory lung diseases modulated by environmental and genetic factors (the Dutch hypothesis, more recently called Asthma COPD Overlap Syndrome [ACOS]). An alternative hypothesis (the British hypothesis), suggests that asthma and COPD are fundamentally different diseases, with asthma caused by allergy and COPD by cigarette smoking.

23.4 Inflammation in COPD is characterized as either chronic bronchitis or emphysema

COPD, which is most commonly associated with a prior history of cigarette smoking or with in-house cooking on wood fires, includes chronic obstructive bronchitis, emphysema, and small-airways disease, resulting in progressive, poorly reversible disease. Chronic bronchitis is diagnosed when there is chronic coughing, with sputum production due to inflammation in the large airways and increased airway mucus production resulting in airway narrowing. Emphysema is characterized by enlargement and destruction of the air spaces (alveoli), with reduced elasticity of the lung tissue. The inflammation in COPD is primarily associated with smoking, other environmental irritants, oxidative stress, and aging and is characterized by the presence of neutrophils, T_H1 cells, and CD8 T cells, with fibrosis around small airways and with alveolar disruption, although subsets of patients, e.g., those with ACOS, have type 2 airway inflammation associated with eosinophils and type 2 cytokines. Telomere length shortening is consistently found in COPD, as is dysregulation in the clearance of apoptotic epithelial cells, suggesting that cellular senescence is a characteristic of COPD. In end-stage COPD, there is also formation of BALT structures that contain class-switched B cells and plasma cells. These B cells might be autoreactive or produce antibodies to bacteria or viruses that colonize the distorted airways.

23.5 Allergy plays a central mechanism in pathogenesis of classical form of asthma

Classically, asthma has been considered a disease caused by adaptive immunity and allergy, associated with a predominantly allergic, eosinophilic inflammation in the airways. Allergy to aeroallergens is present in 70%–80% of patients with asthma, and allergic asthma is the most common form of asthma. In addition, the risk of developing asthma is directly related to the serum concentration of total IgE. Moreover, sensitization to aeroallergens has been shown to be a major risk factor for persistent wheezing in children, which is consistent with the idea that allergy is common in patients with asthma. Further, mast-cell infiltration of airway smooth muscle has been shown to be an important component of asthma. Importantly, allergen-specific T_H2 cells and type 2 cytokines are thought to be present in the lungs of most patients with asthma, particularly in the lungs of patients with allergic asthma. T_H2 cells as well as type 2 innate lymphoid cells (ILC2s) produce an array of cytokines that greatly enhance the allergic inflammatory response. These cytokines include interleukin-4 (IL-4), which is essential for the differentiation of T_H2 cells; in the absence of IL-4, T_H2 cell differentiation is severely impaired. IL-4 is also required for the induction of IgE synthesis and for the upregulation of the low-affinity IgE receptor (CD23) on B cells and the high-affinity IgE receptor (FcεRI) on basophils. IL-4 also induces eotaxin expression by lung cells, induces the production of mucus by airway epithelial cells, and causes the upregulation of cell adhesion molecules VCAM-1 and ICAM-1 on lung vessels, allowing extravasation of eosinophils.

IL-5 is another important type 2 cytokine, usually produced in coordination with IL-4 and/or IL-13. Together with IL-3 and GM-CSF, IL-5 promotes the

Figure 23.3 Depiction of airway hyperreactivity (AHR). AHR is a cardinal feature of asthma because it is present in all forms of asthma and correlates with disease severity. AHR is characterized by hyperresponsiveness of the airways to nonspecific irritants, such as smoke, air pollutants, and cold air, and results in symptoms of asthma, such as wheezing and coughing. AHR can be measured by responsiveness to methacholine, a parasympathomimetic, which causes bronchospasm. A normal individual has only a small decrease in pulmonary function (measured as FEV_1, forced expiratory volume in 1 second) after challenge with increasing concentrations of methacholine, whereas an individual with asthma will have a significant decline in pulmonary function after challenge with even small concentrations of methacholine. An individual with mild asthma will have a smaller decrease in pulmonary function on challenge with methacholine. The level of AHR can change over time, worsening with viral infection or with chronic exposure to allergen.

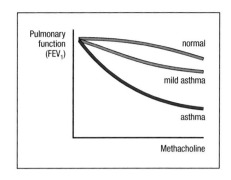

differentiation and maturation of eosinophils, and the release of eosinophils from the bone marrow. Eosinophils are strongly associated with allergic responses and asthma, although their precise role is not fully understood. IL-13 is another important type 2 cytokine that has homology with IL-4 and shares one receptor chain with it—the IL-4Rα chain—that signals through STAT6. Thus, IL-13 shares some functions with IL-4, including the induction of mucus secretion by epithelial cells in the lungs and intestines. However, the IL-13 receptor is absent on T and B cells, and IL-13 therefore has minimal effect on T and B cells. In contrast, because the IL-13 receptor is expressed by airway epithelial cells, smooth muscle cells, and fibroblasts (and macrophages), and although IL-4 can bind to the IL-13 receptor, IL-13 but not IL-4 is thought to increase collagen deposition and airway fibrosis, as well as airway hyperreactivity (AHR), a cardinal feature of asthma. AHR is defined as an exaggerated bronchoconstrictor response to nonspecific stimuli and irritants, such as cold air, smoke, and odors. AHR correlates with asthma severity and can be induced directly in mice by the administration of IL-13, or the adoptive transfer of T_H2 cells, ILC2, or natural killer T (NKT) cells producing IL-13. The degree of AHR in a given patient can change over time, worsening with environmental exposures such as viral infection and allergic reactions. Airway remodeling, smooth muscle hyperplasia, collagen deposition, and fibrosis also contribute to the development of airway obstruction. Airway obstruction and AHR can be quantitated by measuring decreases in pulmonary function in response to graded doses of bronchoconstrictors such as methacholine or histamine, or by measuring the volume of air that can be exhaled from the lung during the first second of a maximally forced expiratory maneuver (FEV_1) (**Figure 23.3**). In asthmatics, this FEV_1 value typically increases sharply after inhalation of a bronchodilator drug (like the β_2-adrenergic agonist formoterol), whereas in COPD patients, airway obstruction is fixed and unresponsive to this bronchodilator. All these observations indicate that type 2 cytokines have a major role in asthma and bronchial hyperreactivity.

23.6 Complexity of asthma and asthma phenotypes

Although allergy and the T_H2 paradigm can explain many features of asthma, clinical and experimental observations over the past 10 years suggest that asthma is much more heterogeneous and complex than suggested by the T_H2 paradigm. These observations include heterogeneity in the response to corticosteroids, the mainstay of asthma treatment, with a sizable fraction of patients failing to respond to corticosteroids; diversity with regard to the underlying genetics; and importantly, heterogeneity in clinical characteristics, including response to environmental exposures and triggers (allergens, exercise, aspirin, cigarette smoke, etc.), age of onset, sex, body

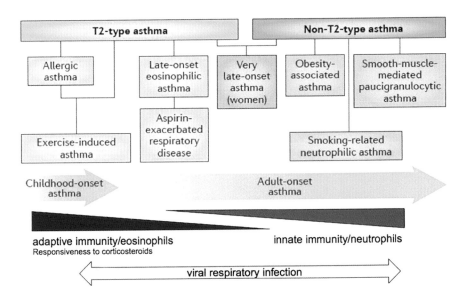

Figure 23.4 **Asthma phenotypes.** Asthma is heterogeneous, and several different phenotypes have been described. In the majority of patients with asthma, type 2 inflammation is present, which is characterized by the presence of eosinophils. The most common form begins in childhood, is associated with allergies to environmental factors, and responds to corticosteroids, often administered by inhalation. Aspirin-exacerbated respiratory disease (associated with aspirin intolerance and nasal polyps) and late-onset eosinophilic asthma are generally not associated with allergies and are less responsive to corticosteroids. In a minority of patients, non–type 2 asthma occurs, sometimes associated with obesity or with smoking, the presence of neutrophils in the airways, later onset in life, and poor response to corticosteroids. (Modified from, Holgate et al. *Nat. Rev. Dis. Prim.* 2015, 1:Article number: 15025.)

mass index (BMI), degree of severity, and response to some other asthma medications. In some patients, eosinophilic asthma is also associated with chronic rhinosinusitis with nasal polyps, and these patients are notoriously resistant to treatment with corticosteroids.

These differences in clinical patient characteristics have led to the idea that asthma is not a single disease but rather a syndrome with multiple distinct phenotypes, also called endotypes (short for endophenotypes) because they are caused by specific disease mechanisms. Some phenotypes are associated with airway eosinophils and type 2 cytokines (type 2 asthma), but others are associated with airway neutrophils, IL-17 or IFN-γ (non–type 2 asthma) (Figure 23.4). While the distinct phenotypes help to segregate patients into groups that help predict prognosis and optimal response to treatment, the various phenotypes may not be mutually exclusive. For example, early onset, allergic asthmatics may become obese, developing obesity-associated asthma, and therefore some individuals might have complex phenotypes, with complex overlapping mechanistic pathways, and with characteristics of several differing phenotypes that complicate assessment and treatment. Importantly, however, a feature that appears to be shared by virtually all asthma phenotypes is exacerbation of asthma symptoms following viral respiratory infection, through innate pathways that are not fully understood.

GENETIC BASIS OF ASTHMA, A COMPLEX DISEASE

Asthma is heterogeneous in part because it is a complex genetic trait, in which asthma develops as a result of specific environmental exposures in genetically susceptible hosts. The heritability of asthma is estimated to range from 35% to 70%, and more than 100 genes have been implicated.

23.7 Common variants in some genes have a role in asthma

Common variants of the genes implicated in asthma function in a variety of ways and can be grouped into those that affect an atopic diathesis versus those that affect supporting structures in the lung, such as epithelial cells, smooth muscle cells, or mucus-secreting cells. For example, the genes *IL4*, *IL13*, *TNF*, *HLA-DR*, *FCER1B*, *TIM1* (*HAVCR1*), and *IL4RA* are expressed by immune cells and presumably predispose toward type 2–biased immune responses and allergy, which underlie the most common form of asthma,

TABLE 23.1 GENES ASSOCIATED WITH ASTHMA

Gene	Mechanism or function of gene
IL4	Enhanced production of IL-4
IL13	Enhanced production of IL-13
IL4RA	Increased signaling in response to IL-4 and IL-13
TNF	Increased production of TNF-α
HLA-DR	Enhanced presentation of allergens, enhanced sensitization
FCER1B	Enhanced activation of mast cells and dendritic cells
TIM1 (HAVCR1)	Altered regulation of T_H1/T_H2 balance; receptor for hepatitis A virus
ADAM33	Altered response of airway smooth muscle
GPRA	Altered response of airway
β_2-Adrenergic receptor (ADRB2)	Reduced lung responsiveness to β-adrenergic medications
5-Lipoxygenase (ALOX5)	Variation in leukotriene production and response to medications
FLG	Reduced barrier function in skin; enhanced sensitization
TSLP	Innate cytokine produced by epithelial cells
IL33	Innate cytokine produced by epithelial cells
ORMDL3	Associated with childhood nonallergic asthma, and dysregulated sphingolipid synthesis
GSDMB	Associated with childhood nonallergic asthma; involved in terminal differentiation of epithelial cells
CDHR3	Receptor for rhinovirus type C
IL1RL1	Receptor for IL-33
IL6R	Receptor for IL-6
RORA	Transcription factor, important for type 2 innate lymphoid cells

allergic asthma (Table 23.1). In contrast, *ADAM33* and *GPRA* are expressed in the lung, whereas *FLG* is expressed primarily in the skin and predisposes to atopic dermatitis, which is a risk factor for asthma. Genome-wide association studies (GWAS) have repeatedly identified additional genes, such as *ORMDL3, IL1RL1, IL33, TSLP, TNFAIP3*, and several others. These genes affect innate immunity (discussed in Section 23.8), and emphasize that innate pathways, distinct from, and/or in parallel with, adaptive T_H2 immune responses, play important roles in asthma. The large number of genes involved in the predisposition to asthma suggests that the pathways to asthma are complex and may involve multiple routes, consistent with the idea that asthma is heterogeneous with several distinct phenotypes.

Importance of environmental effects on asthma

As with most complex traits, the environment plays a critical role in the expression of asthma. This idea is underscored by the fact that the prevalence of asthma has increased enormously over the past several decades. This increase in prevalence is not due to changes in the genetic composition of the population, but rather to environmental changes that have occurred in

Westernized countries. These environmental changes include increases in air pollution (sulfur dioxide, nitrogen dioxide, ozone, and particulate matter like diesel exhaust), increased exposure to indoor and outdoor allergens (due in part to increased time spent indoors, and to global warming, which increases pollen output), increased prevalence of obesity, and an increase in the prevalence of vitamin D deficiency. However, the environmental effect that has received the most attention recently is changes in microbial exposures, related to the use of antibiotics, improved public health measures such as improved hygiene and indoor plumbing, and greater use of cesarean section (decreasing exposure to vaginal bacteria) for delivery.

The notion that decreased microbial exposures in young children might increase the expression of asthma was first proposed in 1989 by David P. Strachan, a theory that is now known as the Hygiene Hypothesis. This idea has been strengthened by multiple studies demonstrating that chronic exposure of young children to farming environments or to households with multiple pets reduces the risk of developing allergic asthma, although the specific microbial agents and the mechanisms by which these agents might reduce atopy are poorly defined. These farming environments and households with multiple pets have been shown to contain high levels of endotoxin as well as diverse microbiota, suggesting that the diversity of microbial exposure, particularly in early childhood, may provide the important immune-modifying effects, resulting in the induction of protection against the development of asthma and allergy. Early life exposures to specific constellations of microbiota may be important in determining the later life susceptibility to asthma by proper microbial-induced programming of the immune response.

The understanding of the role of microbiota in the development of immunity in general has progressed rapidly after the introduction of methods to identify microbes through sequencing of bacterial DNA rather than by culture (a method that greatly increases the sensitivity of detection) and methods to raise experimental mice under germ-free conditions. These methods together have shown that multiple components of the immune system fail to develop normally under germ-free conditions. For example, germ-free mice have reduced production of IgA, IgG, and IgE, reduced CD4 and CD8 T cells, and reduced CD4 T_{reg} cells. In addition, germ-free conditions result in profound effects on innate immunity, with reductions in the secretion of antimicrobial peptides. Furthermore, germ-free mice have an increased proclivity to develop allergic airway and intestinal diseases that are ameliorated by the reconstitution with normal diverse microbiota in early life.

Exposure of mice to diverse microbiota affects the microbiome not only of the intestines but also of the lung, which is no longer thought to be sterile, at least as assessed by nonculture methodologies, i.e., sequencing for bacterial DNA. Moreover, exposure of the lung to microbiota or extracts from nondomesticated sites, i.e., from farming environments, results in a great increase in the cellularity of the lungs, with 10- to 20-fold increases in the number of mature T_H1, T_H2, T_H17, T_{reg}, innate lymphoid, and CD8 T cells present in the lung. Whether these cells accumulate in BALT structures is not yet clear; nevertheless, it appears that exposure to diverse microbiota results in a more mature immune system that is more capable of preventing the development of type 2 inflammation and the development of allergic asthma. Moreover, clinical studies suggest that exposure to diverse microbiota and environments with diverse microbiota may also improve immunity against viral respiratory disease that often triggers asthma exacerbations. Farm dust extracts, which contain endotoxin, also heavily influence the barrier function of the lung epithelium. The threshold of signaling by TLR and IL-1R family cytokines in epithelial cells can be increased by farm dust inhalation, explaining how farm dust might suppress responses to allergens that are relatively harmless compared with bacteria or viruses.

INNATE IMMUNE CELL TYPES IN ASTHMA

The pathogenesis of asthma depends heavily on innate immune cells, including DCs, alveolar macrophages, mast cells, basophils, eosinophils, natural killer T (NKT) cells, and recently described innate lymphoid cells.

23.8 Dendritic cells and alveolar macrophages are the major antigen-presenting cells in asthma

The induction of T_H2-biased adaptive immunity to inhaled allergens requires APCs such as DCs, which are the most potent type of APC (**Figure 23.5**). In the lung, several different subsets of DCs can be found throughout the conducting airways, interstitium, vasculature, and pleura, and in bronchial lymph nodes. Lung DCs also express numerous receptors such as TLRs, NOD-like receptors, and C-type lectin receptors, as well as CD11c, and can upregulate the expression of several costimulatory molecules (such as CD80 and CD86) and the production of chemokines (such as CCL17 and CCL22) and pro-inflammatory cytokines that attract T cells, eosinophils, and basophils into the lungs. DCs take up antigen that might accumulate in the lungs; they then migrate to the draining lymph nodes, where they encounter and activate naive antigen-specific T cells. Like in other mucosal sites, we now recognize the existence of several subsets of DCs that have distinct functions. cDCs all express CD11c and CD26 and derive from a circulating common DC progenitor that depends on the cytokine receptor Flt3. cDC1s rely on the transcription factors Batf3, Id2, and IRF8 for further differentiation and express XCR-1 and in many cases the integrin CD103 in addition to CD26 and CD11c. cDC2s rely on IRF4 and ZEB2 for further differentiation and express CD172 in addition to CD26 and CD11c. In many cases, these cDC2s also express CD11b. Circulating monocytes can also give rise to "monocyte-derived DCs" (mo-DCs or inflammatory DCs) that also express CD11c and major histocompatibility complex (MHC) class II and CD11b like cDC2s. However, gene expression data

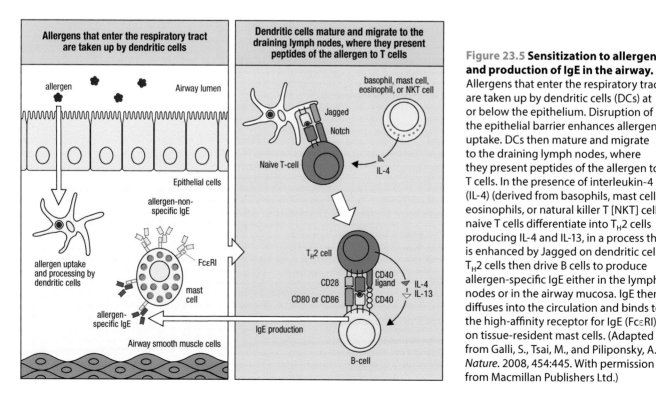

Figure 23.5 Sensitization to allergens and production of IgE in the airway. Allergens that enter the respiratory tract are taken up by dendritic cells (DCs) at or below the epithelium. Disruption of the epithelial barrier enhances allergen uptake. DCs then mature and migrate to the draining lymph nodes, where they present peptides of the allergen to T cells. In the presence of interleukin-4 (IL-4) (derived from basophils, mast cells, eosinophils, or natural killer T [NKT] cells), naive T cells differentiate into T_H2 cells producing IL-4 and IL-13, in a process that is enhanced by Jagged on dendritic cells. T_H2 cells then drive B cells to produce allergen-specific IgE either in the lymph nodes or in the airway mucosa. IgE then diffuses into the circulation and binds to the high-affinity receptor for IgE (FcεRI) on tissue-resident mast cells. (Adapted from Galli, S., Tsai, M., and Piliponsky, A. *Nature.* 2008, 454:445. With permission from Macmillan Publishers Ltd.)

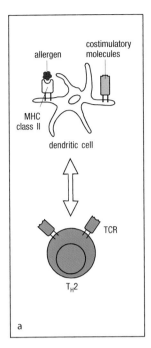

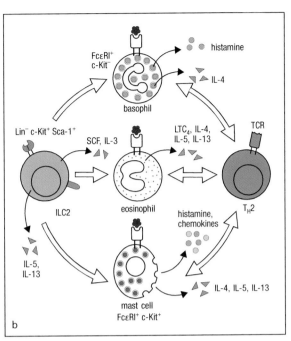

Figure 23.6 Dendritic cells (DCs) are key antigen-presenting cells (APCs) in the lung. After antigen challenge, lung DCs process antigen and induce antigen-specific T_H2 cell responses (**a**). Other cells can also function as APCs to activate T_H2 cells (**b**). Basophils, eosinophils, mast cells, and natural helper cells express MHC class II and costimulatory molecules. These cells of the innate immune system can therefore be potential APCs in the lung and can also be the initial sources of T_H2 cytokines. Lin, lineage; LTC_4, leukotriene C_4; SCF, stem cell factor; TCR, T-cell receptor. (Adapted from Kim, H., DeKruyff, R., and Umetsu, D. *Nat. Immunol.* 2010, 11:577. With permission from Macmillan Publishers Ltd.)

suggest that these cells should best be regarded as specialized macrophages as they also express the macrophage markers CD64 and MerTK and fail to migrate to draining lymph nodes.

In mouse models, depletion of CD11c$^+$ DCs and macrophages from the airways using CD11c-diphtheria toxin receptor transgenic mice or thymidine kinase-transgenic mice treated with the antiviral drug ganciclovir abolished the characteristic features of asthma, including eosinophilic inflammation and T_H2-cytokine production. Furthermore, in the absence of CD11c$^+$ DCs, T_H2 cells cannot produce IL-4, IL-5, and IL-13. In humans, monocyte-derived DCs promote T_H2-cell responses by secreting pro-inflammatory cytokines and upregulating the expression of costimulatory molecules after stimulation with antigen (Figure 23.6a). Taken together, these findings indicate that lung DCs are the primary APCs and are necessary for T_H2-cell stimulation during airway inflammation. More recently, studies in mice in which subsets of cDCs were depleted via CD11cCre mediated deletion of either IRF4 or IRF8 have demonstrated that particularly IRF4-dpendent cDC2s are crucial in initiating T_H2 immunity to allergens (i.e., during the early sensitization period early in life), whereas both moDCs and cDC2s can maintain ongoing effector T_H2 responses in the lung without migrating to the lymph nodes. Irrespective of the precise type of DCs performing these functions, it is now clear that DCs communicate very closely with the lung epithelial cells and innate immune cells to perform these functions. In early life, for example, lung epithelial cells secrete significant quantities of IL-33, and the IL-33 produced suppresses the potential of cDC2s and moDCs to produce IL-12, while at the same time upregulating the pro-T_H2 costimulatory molecule OX40L. Not surprisingly, immune responses in young children and newborn mice are T_H2 biased. In adult mice, allergen exposure also induces IL-33, TSLP, IL-25, and GM-CSF production in lung epithelial cells, and these cytokines all induce OX40L and suppress IL-12 in adult DCs. Innate lymphoid cells and basophils also closely collaborate with cDC2s to promote T_H2 immunity by providing IL-13 that promotes the migration of cDCs.

In the lung, another DC subset called plasmacytoid DCs (pDCs) is also present, although these pDCs are primarily found in the peripheral lung tissues rather than in the conducting airways. pDCs have a very important role in respiratory tolerance and in the induction of T_{reg} cells, which can suppress T_H2-biased adaptive immunity (discussed in Section 23.23).

Depletion of pDCs abolishes respiratory tolerance and exacerbates airway inflammation, whereas adoptive transfer of feline McDonough sarcoma-related tyrosine kinase 3 (Flt3L)-induced bone marrow–derived pDCs, thought to be precursors of pDCs, enhances respiratory tolerance. Thus, lung DCs can not only initiate pulmonary inflammation but also downmodulate it.

Another important cell type in the lung that expresses CD11c and that participates in lung immune homeostasis is the alveolar macrophage, which must be distinguished from DCs. Alveolar macrophages, which develop shortly after birth and derive from fetal monocytes that self-renew in the lungs after birth, are large, highly phagocytic cells that clear particulate matter or pathogens inhaled into the lungs. Until recently, the role of alveolar macrophages in asthma was not well characterized. It is now known that under homeostatic conditions, alveolar macrophages play an anti-inflammatory role by removing antigens and particles and limiting antigen presentation by DCs. However, when activated by infection or interferon (IFN)-γ (classical activation of alveolar macrophages, resulting in an M1 cell type), alveolar macrophages produce large quantities of chemokines and cytokines, such as type I interferons, and have an increased capacity to kill intracellular organisms. M1 cells can also produce type 1 (IFN-γ) and/or T_H17 cytokines and can modulate immune responses and inhibit the development of allergic asthma. Alveolar macrophages do not migrate to the draining lymph nodes, but because they express class II MHC antigen, they may be able to activate memory antigen-specific T cells, and T_{reg} cells as well.

23.9 Alternatively activated macrophages promote T_H2 inflammation in asthma

The phenotype and function of alveolar macrophages can change in an opposite direction, for example, with activation by IL-4, IL-13 or IL-25, and IL-33 into what is known as alternatively activated macrophages (AAM) or M2 cells. Many of these cells also adopt a more DC-like phenotype termed moDC (see Section 23.8). M2 cells express arginase 1, chitinase-like molecules, resistin-like molecule-α (also known as FIZZ1), and IL-17RB (IL-25 receptor) and can actively participate and promote allergic asthma by producing type 2 cytokines, including IL-4, IL-13, and IL-33. However, the precise role of AAM/M2 macrophages in type 2 asthma remains to be clarified.

23.10 Mast cells, basophils, and eosinophils are innate immune cells that are recruited to the airways during asthma

When the lungs are challenged with allergen, several inflammatory cell types, including basophils, mast cells, eosinophils, and innate lymphoid cells are recruited to airways. Recent studies also suggest that these cells are also recruited to the lung shortly after birth under the influence of epithelial IL-33 and play an important role in lung development and growth during the first years of life when the lung is still forming new alveoli. Mast cells express FcϵRI and c-Kit and reside in tissues near mucosal surfaces and blood vessels (see **Figure 23.6b**). Classically, mast cells can be activated through cross-linking of antigen-specific IgE bound to FcϵRI, by degranulating and releasing preformed mediators such as histamine and proteases, as well as newly formed mediators such as cysteinyl leukotrienes, cytokines (IL-1, IL-3, IL-4, IL-5, IL-6, IL-8, IL-10, IL-13, IL-16, tumor necrosis factor α [TNF-α], and transforming growth factor-β [TGF-β]), and chemokines (e.g., IL-8, lymphotactin, TCA3, RANTES, MCP-1, and MIP-1α), and thus initiate and greatly amplify allergic reactions. In addition, mast cells can rapidly degranulate in response to innate immune signals, such as TLR agonists, or to cytokines such as IL-33 (discussed in Section 23.15). In allergic disease, widespread, systemic mast cell (and basophil) degranulation can result in anaphylaxis, a severe, life-threatening reaction, particularly when involving the respiratory tract. In some but not all mouse models of allergic asthma, the presence of mast cells can enhance the development of asthma, and mast cell–deficient strains (e.g., W/W_v, FcεRI$^{-/-}$, and Kit$^{W\text{-}sh/W\text{-}sh}$) have decreased AHR, decreased airway inflammation, decreased goblet cell hyperplasia, and lowered concentrations of IgE. Like basophils and eosinophils, mast cells can also act as APCs, but this function is poorly understood. IL-3 (which is essential for mast-cell growth), IL-4, and GM-CSF increase the expression of MHC class II in mast cells, and allow mast cells to induce T-cell proliferation.

Basophils, like mast cells, express FcεRI (but not c-Kit) and can amplify immediate hypersensitivity responses by rapidly degranulating after IgE-FcεRI cross-linking, releasing preformed mediators (histamine-containing granules) and newly synthesized mediators (cysteinyl leukotrienes and cytokines, including large quantities of IL-4). Basophils are granulocytes that normally circulate but are highly enriched in lung tissue of patients with asthma. Although the precise role of basophils versus mast cells has not been fully elucidated, the fact that asthma improves after treatment with anti-IgE mAb suggests that IgE-mediated effects through both mast cells and basophils play an important role in at least some forms of asthma. Recent studies also suggest that in chronic human asthma, basophils can be a copious source of T_H2 cytokines in the airways, secreting cytokines to the same extent as effector T_H2 cells.

Another circulating granulocyte that is prominent at sites of allergic inflammation is the eosinophil. Upon stimulation, eosinophils have an important pro-inflammatory role by producing cysteinyl leukotrienes, as well as T_H1 (IFN-γ and IL-2) and T_H2 (IL-4, IL-5, IL-10, IL-13, and TNF-α) cytokines. In some but not all experimental models of asthma, eosinophils are required for AHR, possibly as a result of the activity of IL-13 and cysteinyl leukotrienes and owing to the toxicity of eosinophil granule proteins, such as major basic protein, eosinophil peroxidase, and eosinophil derived neurotoxin, which could contribute to airway inflammation and epithelial damage. Improved symptoms in patients with asthma treated with anti-IL-5 mAb, which reduces the number of circulating and tissue eosinophils in asthma phenotypes characterized by increased circulating and tissue eosinophils (e.g., allergic asthma), suggests that eosinophils play an important role in at least some forms of asthma. More and more, eosinophils are seen as immunoregulatory cells that do more than just damage the airway structures. Eosinophils, like basophils, can also function as APCs. GM-CSF induces MHC class II and costimulatory molecule expression on eosinophils. Class II–expressing eosinophils, when loaded with antigen, promote the production of IL-4, IL-5, and IL-13 from antigen-specific CD4 T cells in a dose-dependent manner. Although this APC function of eosinophils has been contested, some of the discrepancies might be explained by the methods used for isolating eosinophils that can decrease antigen processing. Therefore, these studies together indicate that eosinophils have important effector cell functions and might also modulate adaptive T_H2 immunity. Finally, in steady-state mouse and human lungs, eosinophils expressing MHC class II and the adhesion molecule CD101 can be found, which appear to have an important immunoregulatory role.

23.11 Lung epithelial cells have a central role in induction and maintenance of asthma

As mentioned in Section 23.2, airway epithelial cells play a key role beyond their barrier function, in sensing and responding to the environment and in maintaining homeostasis. Airway epithelial cells express TLR, NOD-like receptors, and C-type lectin receptors; produce mucus and antimicrobial

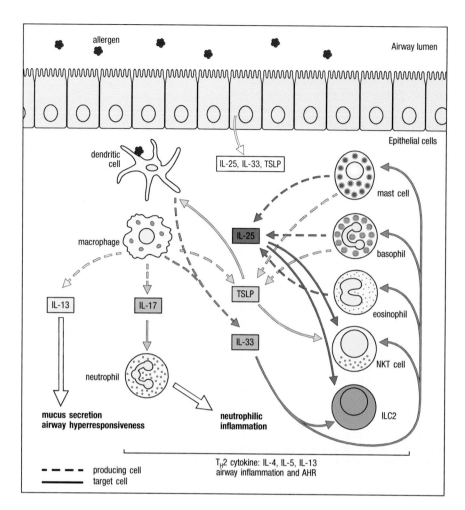

Figure 23.7 Innate immune mechanisms in the lung. Although adaptive immunity is critical for asthma pathogenesis, asthma also involves innate, cognate antigen-independent immune responses. IL-25 induces cytokines such as IL-5 and IL-13 from ILC2s in the absence of T$_H$2 cells and stimulates invariant NKT cells to produce IL-13, thereby promoting AHR and airway remodeling. IL-33 acts on multiple targets; it stimulates mast cells, eosinophils, basophils, and NKT cells to elicit cytokines. TSLP activates mast cells and NKT cells to secrete type 2 cytokines. The finding of these cytokines, IL-25, IL-33, and TSLP, and cells of the innate immune system greatly extends the understanding of the pathogenesis of asthma. (Adapted from Kim, H., DeKruyff, R., and Umetsu, D. *Nat. Immunol.* 2010, 11:577. With permission from Macmillan Publishers Ltd.)

peptides that facilitate clearance of inhaled particles, toxins, and microbes; and produce a large array of cytokines and chemokines that regulate adaptive immunity. Under homeostatic conditions, airway epithelial cells express CD200 and PD-L1, which interact with CD200R-expressing alveolar macrophages and PD-1 expressing T cells, respectively. These interactions dampen the activity of effector cells and help to maintain the anti-inflammatory, tolerance-inducing environment in the lung. In contrast, epithelial cells activated by TLR engagement or cell injury (e.g., due to protease activity, toxic damage, or direct physical injury) release IL-25, IL-33, and TSLP (**Figure 23.7**). These three innate cytokines are critical in allergic responses and are discussed in Sections 23.15, 23.16 and 23.17. All three innate cytokines can activate mast cells, basophils, eosinophils, NKT cells, and type 2 innate lymphoid cells (ILC2), producing IL-5, IL-13, and IL-9, as described in Section 23.12.

23.12 Innate lymphoid cells have a role in lung inflammation

In 2011, a novel innate lymphocyte was identified that responded to IL-25, IL-33, TSLP, and IL-2 and produced large amounts of IL-5, IL-9, and IL-13, but not IL-4. These cells, now called type 2 innate lymphoid cells (ILC2s), as reviewed in Chapter 8, do not express lineage (Lin) markers but express c-Kit, Sca-1, IL-7R, IL-33R, and IL-17RB and the transcription factor GATA-3 and Rorα, a gene that is associated with asthma susceptibility. After stimulation with IL-25 or IL-33, ILC2s rapidly produce IL-5 and IL-13. These cytokines can drive the development of type 2 adaptive immunity but can also mediate effector function such as the production of mucus from epithelial cells, enhance eosinophil activation, and AHR, all in the absence of T_H2 cells. ILC2s have been found in the lungs of patients with asthma as well as in nasal polyp tissue along with eosinophils. Since nasal polyps are often found in patients with severe eosinophilic asthma associated with aspirin sensitivity (often in the absence of allergen sensitization), ILC2s may play a role in type 2 asthma phenotypes that are nonallergic, i.e., innate and distinct from allergic asthma, in which adaptive immunity plays an important role. Importantly, activated ILC2s, in contrast to T_H2 cells, appear to be corticosteroid resistant, which may explain some asthma phenotypes that are eosinophilic and yet corticosteroid resistant.

23.13 IL-17 and IFN-γ may enhance neutrophilic responses in asthma

ILC2s are related to type 1 innate lymphoid cells (ILC1s) producing IFN-γ and type 3 innate lymphoid cells (ILC3s) producing IL-17 and IL-22. ILC3s express RORγt and respond to IL-1β and IL-23. IL-17 (also known as IL-17A) is a member of the six-member IL-17 family of cytokines, IL-17A–F. IL-17 increases the influx of neutrophils into inflammatory sites, possibly by increasing GM-CSF and G-CSF production, suggesting an important role for ILC3s and IL-17 in asthma phenotypes associated with the presence of airway neutrophils. IL-17 has been shown to directly induce the development of AHR, independently of IL-13, but can also synergize with IL-13 to induce mixed eosinophilic and neutrophilic airway inflammation and AHR. Endotoxin stimulation of TLR4 can induce strong IL-17 responses, promoting airway neutrophilia and enhancing IL-13-induced AHR. The ability of IL-17 to enhance neutrophil inflammation and induce AHR emphasizes the role of innate immune mechanisms in the development of asthma, particularly in asthma phenotypes associated with neutrophils.

In this regard, asthma associated with obesity is characterized by measureable increase in airway neutrophils, as well as by severe disease, refractoriness to corticosteroid therapy, and adult onset with reduced serum IgE levels with few allergies. IL-17 and IL-17-producing ILC3s may play a role in this phenotype, particularly since obesity is associated with chronic inflammation with metabolic dysfunction and an increase in IL-1β and IL-6 production, which can enhance innate immunity and ILC3 expansion.

IFN-γ production from ILC1s or T_H1 cells can either protect or exacerbate allergic asthma. IFN-γ can inhibit T_H2 cell development and reduce eosinophil accumulation in the lungs. But IFN-γ can increase ongoing inflammatory responses, by promoting airway neutrophil accumulation, reducing eosinophil accumulation, and preventing resolution of airway inflammation.

INNATE CYTOKINES IMPORTANT IN LUNG INFLAMMATION

23.14 Thymic stromal lymphopoietin is a novel IL-17-like cytokine that promotes type 2 immune responses

TSLP, encoded by an important asthma susceptibility gene, is released by epithelial cells in response to certain microbial products such as peptidoglycans, lipoteichoic acid, and double-stranded RNA, or to physical injury or inflammatory cytokines (such as IL-1β and TNF-α). TSLP expression is detected in the airways of patients with asthma, and TSLP mRNA expression levels correlate with disease severity. Moreover, lung-specific expression of TSLP induces airway inflammation and AHR, by activating ILC2s and NKT cells. TSLP can also activate DCs by increasing their OX40L expression and can therefore enhance T_H2 inflammatory responses via OX40 and OX40L interaction. These findings suggest that TSLP has an important role in the initiation of allergic and adaptive airway inflammation through innate pathways that initially activate airway epithelium.

23.15 IL-25, an IL-17 cytokine family member that amplifies type 2 immunity

IL-25, also known as IL-17E, is a member of the IL-17 cytokine family. Airway epithelial cells release IL-25 after exposure to allergens and helminth infection. IL-25 is also produced by activated eosinophils, basophils after FcεRI cross-linking, and mast cells. Increased IL-25 has also been detected in the lungs of patients with asthma. IL-25 amplifies type 2 cytokine production and eosinophilia and enhances AHR. Furthermore, IL-25 induces the production of IL-5 and IL-13 by T_H2 cells, ILC2s, and subsets of NKT cells. Thus, IL-25 acts on both the innate and adaptive immune systems to amplify type 2 responses. In the intestines, IL-25 is produced primarily by tuft cells (also called brush cells), which are taste-chemosensory epithelial cells, based on studies in intestines. These cells are also in the epithelial layer of the trachea and salivary glands and greatly expand with IL-13 signaling. The precise comparative role of epithelial cells versus tuft cells in the secretion of IL-25 in lung immunity is not yet clear.

23.16 IL-33 is an IL-1 family member that enhances the activity of mast cells, basophils, and eosinophils

IL-33 is a member of the IL-1 cytokine family, and IL-33 transcripts can be detected in many diverse cell types, including airway epithelial cells, mast cells, NKT cells, and alveolar macrophages. IL-33 synergizes with SCF and IgE receptor to activate mast cells. IL-33 activates DCs and induces their expression of OX40L while suppressing IL-12 production. IL-33 also enhances the survival of eosinophils and promotes eosinophil degranulation in humans. This cytokine is made spontaneously in the developing lung up until the weaning period in mice, explaining the T_H2 bias of this early period of life. In mice, viral infection or exposure to pathogen-derived products, irritants, or allergens can enhance IL-33 production from airway epithelial cells. When IL-33 is administered with ovalbumin, IL-33 enhances airway inflammation in an IL-4-independent manner. Consistent with this, administration of neutralizing antibodies against IL-33 or against ST2, the receptor for IL-33, attenuates eosinophilic pulmonary inflammation and AHR. Furthermore, administration of IL-33 alone can induce AHR and even enhance airway inflammation in *Rag1*-deficient mice. The gene for IL-33 and its receptor, *IL1RL1*, are both very important asthma susceptibility genes, consistent with the idea that IL-33 has an important role in the development of asthma, possibly by enhancing both innate and adaptive pathways. There is also a splice variant of IL-33 that seems to be very actively transcribed in chronic asthmatics and drives the activation of basophils. Also the receptor for IL-33 has a soluble variant that acts as an antagonist (sST2). Some polymorphisms in the *IL1RL1* gene mainly affect the relative abundance of this soluble antagonist over the membrane expressed proinflammatory receptor.

23.17 IL-25, IL-33, and TSLP synergize to induce T_H2 and ILC2 effector cell differentiation

Type 2 responses appear to be initiated by the triggering of the release of IL-25, IL-33, and TSLP from epithelial cells, mast cells, NKT cells, and some macrophages (see Figure 23.7). While each of these tissue-released innate cytokines alone can activate ILC2s and T_H2 cells, the three cytokines when present together are most effective, causing maximal differentiation and expansion of T_H2 cells and ILC2s. ILC2s are primarily found in mucosal tissues, whereas T_H2 cells present in lymph nodes and in mucosal tissues. However, because IL-25, IL-33, and TSLP are produced primarily in mucosal tissues and not in lymph nodes, the expansion and differentiation of ILC2s and T_H2 cells to produce high levels of IL-5 and IL-13 occurs primarily at mucosal sites, such as the lungs.

23.18 IL-22 promotes epithelial cell responses associated with the asthma phenotype

IL-22, which is produced by T_H17 cells, $\gamma\delta$ T cells, NKT cells, and by ILC3s, is a member of the IL-10 family of cytokines and has a very important role in mucosal immunology. Triggering of the IL-22 receptor, expressed by airway epithelial cells and also by intestinal epithelial cells, keratinocytes, and hepatocytes, activates STAT3, causing epithelial and stromal cell proliferation, improvement of barrier function, and production of antimicrobial peptides such as S100A7, RegIIIγ, β-defensins, and psoriasin. IL-22 therefore has an important role in host defense against extracellular bacteria, including *Klebsiella pneumonia* and *Streptococcus pneumonia*. IL-22 also protects hepatocytes against apoptosis but may play a pathogenic role in psoriasis and colon cancer. IL-22 production is triggered by IL-1β and IL-23, produced by alveolar macrophages and DCs, particularly after TLR and other pattern-recognition receptors (PRR) activation.

ADAPTIVE IMMUNITY IN PATHOGENESIS OF ASTHMA

23.19 Allergen-specific IgE and T_H2 cells

Approximately 70%–80% of patients with asthma have allergies to environmental agents, including plant pollens, dust mites, pet dander, and fungal allergens. Allergy is a risk factor for the development of asthma in children, and high serum total IgE levels correlate with a higher risk for developing asthma in children and in adults. Allergen exposure leads to sensitization with the production of allergen-specific IgE and allergen-specific T_H2 cells. Allergen reexposure of sensitized individuals causes mast cell and basophil degranulation, release of histamine, leukotrienes, and type 2 cytokines. This by itself, a manifestation of adaptive immunity, can cause an asthma attack, but when coupled with T_H2 cell activation, it can also lead to the development of more robust inflammation, characterized by the infiltration of eosinophils, basophils, and lymphocytes into the lungs and respiratory mucosa. Leukotrienes, released by mast cells, basophils, eosinophils, and macrophages, also enhance ILC2 activation, particularly in coordination with IL-33. T_H2 cells, by releasing IL-4, IL-5, and IL-13 are thought to coordinate the allergic inflammatory response that leads to allergic asthma. Allergic asthma tends to be responsive to corticosteroids, due to the steroid sensitivity of T_H2 cells and eosinophils. However, although allergic asthma is common and can explain many features of asthma, other forms of asthma exist that are independent of allergy and T_H2 cells and are often not fully responsive to corticosteroids.

23.20 Phases of allergic asthma

The T_H2-biased inflammatory response that underlies allergic asthma includes both an early phase and a late-phase reaction (Figure 23.8). The early phase reaction occurs within minutes of allergen challenge, as a result of the binding of allergen to allergen-specific IgE bound to FcϵRI on the surface

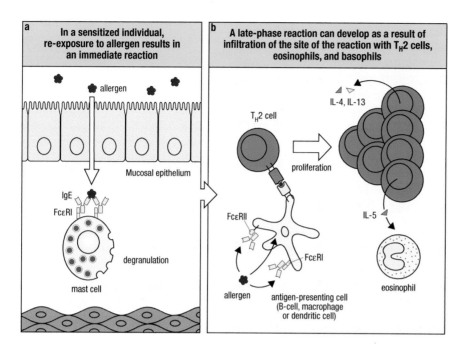

a In a sensitized individual, re-exposure to allergen results in an immediate reaction

allergen

Mucosal epithelium

IgE
FcεRI

degranulation

mast cell

b A late-phase reaction can develop as a result of infiltration of the site of the reaction with T$_H$2 cells, eosinophils, and basophils

T$_H$2 cell

proliferation

IL-4, IL-13

IL-5

FcεRII

FcεRI

allergen

antigen-presenting cell (B-cell, macrophage or dendritic cell)

eosinophil

Figure 23.8 Allergic reactions. In a sensitized individual, reexposure to allergen results in an immediate reaction (**a**) (early phase reaction) that can occur within minutes of allergen exposure and is due to the cross-linking of allergen-specific IgE on the surface of a mast cell. This results in the release of preformed and newly synthesized mediators from mast cells, leading to localized reactions (acute asthma, allergic rhinitis, or gastrointestinal reactions in food allergies) or systemic reactions (anaphylaxis). Mast-cell mediators include histamine, proteases, prostaglandins, leukotrienes, platelet-activating factor, and cytokines, and cause vasodilation, edema, bronchoconstriction, mucus secretion, hives, and diarrhea. At 6–12 hours after the early phase reaction, a late-phase reaction (**b**) can develop as a result of infiltration of the site of the allergic reaction with T$_H$2 cells, eosinophils, and basophils. This inflammatory reaction is characterized by swelling, itching, warmth, and erythema (redness), and in the lungs by airway narrowing, mucus hypersecretion, and the development of airway hyperreactivity. (Adapted from Valenta, R. *Nat. Rev. Immunol.* 2002, 2:446. With permission from Macmillan Publishers Ltd.)

of mast cells at mucosal sites, resulting in mast-cell degranulation and the release of preformed mediators, such as histamine, resulting in the symptoms that can include hives, sneezing, wheezing, vomiting, and anaphylaxis. The late-phase reaction occurs in some individuals 6–12 hours after the early phase reaction and is caused by the infiltration of eosinophils, basophils, and lymphocytes into the tissues at sites of allergy. In the lung, the late-phase reaction is associated with significant bronchial obstruction that is resistant to bronchodilator therapy, as well as with the development of AHR.

23.21 Characteristics of allergens and their effects on innate immunity

Whereas viral, bacterial, and fungal infections induce T$_H$1 and T$_H$17 responses by activating the immune system through PRRs such as TLRs on antigen-presenting cells, type 2 immunity, e.g., in response to most allergens (and parasitic worms), develop for the most part independently of PRRs triggering on APCs. Rather, allergens (and parasitic worms) appear to activate type 2 immunity by affecting or causing injury to epithelial cells, resulting in the release of IL-25, IL-33, and TSLP, which together cause T$_H$2 and ILC2 effector cell differentiation and expansion. In some cases, TLR triggering on epithelial cells is required to elicit these epithelial responses, such as is the case with house dust mite or cat allergens triggering TLR4 on lung epithelial cells. Allergens are thought to be small molecules that have high aqueous solubility and can rapidly diffuse across mucosal tissue and stimulate or injure cells that then produce IL-25, IL-33, and TSLP, such as epithelial cells, mast cells, NKT cells, and some macrophages. Allergens can also bind to IgE and trigger mast cell degranulation. Many allergens, such as dust mites, fungi, and pollens, have protease activity that injures and/or activates epithelial cells. Bee venom allergen contains phospholipase A2, PLA2, and enzymatic activity that hydrolyzes membrane phospholipids, damaging epithelial cell membranes and releasing IL-33 that then activates ILC2s. Parasitic worms also appear to activate tuft cells in the intestines (taste-chemosensory epithelial cells, see IL-25 in Section 23.15), possibly through sensing changes in nutrient and metabolite levels and resulting in the production of IL-25, which can then activate ILC2s. In addition, cockroach and dust mite fecal particles have associated endotoxin activity, and dust mite allergen Der p 2 has structural homology with MD-2, the lipopolysaccharide-binding component of the TLR4 signaling complex, that can enhance DC function and immunity (and induce IL-17 production). Allergen-specific IgE is primarily produced against allergenic components that are proteins, which can be presented by DCs to induce the development of T$_H$2 cells and T-follicular-helper (T$_{FH}$) cells in the context of MHC class II molecules. T$_{FH}$ cells producing IL-4 induce the production of allergen-specific IgE from B cells.

23.22 T$_{reg}$ cells are also recruited or induced locally during asthma and provide restraint on inflammatory response

There are several mechanisms that downregulate allergic inflammation. For example, allergen-specific T$_{reg}$ cells can downregulate the function of T$_H$2 cells and other airway effector cells, such as ILC2s. These T$_{reg}$ cells are thought to be very important in regulating allergic diseases including asthma, and atopic individuals are thought to have a relative deficiency in these cells, either in terms of numbers or function of these cells. There are several subsets of T$_{reg}$ cells, including natural T$_{reg}$ cells, which develop in the thymus and are important in recognizing self-antigens and preventing autoimmunity, and T$_{reg}$ cells that develop in the periphery in response to nominal antigens and allergens including adaptive T$_{reg}$ cells and T$_R$1 cells. T$_H$1 cells have also been thought to inhibit the development of T$_H$2 cells and may play a role in preventing the development of allergy. This may occur, for example, in nonallergic individuals, where allergen-specific T$_H$2 cells may not develop or may be deleted, possibly due to suppression from IFN-γ secreted by T$_H$1 cells.

The induction of adaptive allergen-specific T$_{reg}$ cells is thought to be related to the level of allergen exposure: lower doses of allergen induce greater numbers of allergen-specific adaptive T$_{reg}$ cells, whereas higher doses induce greater deletion of antigen-specific cells. The precise mechanism by which T$_{reg}$ cells control inflammation is not fully understood, but it includes the direct inhibition of T$_H$2 cells, or the inhibition of antigen presentation by DCs to T$_H$2 cells. TGF-β and/or the production of IL-10 by T$_{reg}$ cells may also inhibit the development of airway inflammation and AHR. Subsets of adaptive T$_{reg}$ cells may also develop under different conditions, involving inducible costimulatory molecule (ICOS)–ICOS ligand (ICOSL) pathways, IL-10, and CTLA-4, and in the presence of IL-4 or IFN-γ, resulting in some T$_{reg}$ cells expressing T-bet and others expressing GATA-3. Allergen immunotherapy is thought to boost the number of allergen-specific T$_{reg}$ cells and decrease the number of allergen-specific T$_H$2 cells. Both natural and adaptive T$_{reg}$ cells express the transcription factor Forkhead box p3 (Foxp3), which is essential for their development and function. Mutations in Foxp3 result in multiorgan inflammatory disease, including autoimmune diseases, as well as food allergy with high levels of IgE, and in older children with IPEX (immune dysregulation, polyendocrinopathy, and enteropathy, X-linked), asthma. Allergen-specific T$_R$1 cells do not express Foxp3, develop on exposure to IL-10, and produce high levels of IL-10.

Recent studies of allergic individuals suggest an alternative view of the role of T$_{reg}$ cells in allergy. These studies suggest that the number of T$_{reg}$ cells or function for aeroallergens is not reduced but rather directed against antigens that are different than those to which T$_H$2 cells respond. Human aeroallergen-specific T$_{reg}$ cells, the specificity of which has in the past been largely unknown, appear to respond to antigens that are distinct from the major allergens that have been identified for effector T$_H$2 cells. Thus, in allergic individuals, T$_H$2 cells respond and expand to small, water-soluble protein allergens, whereas T$_{reg}$ cells appear to respond to antigens that are associated more strongly with pollen particles, which possibly have altered antigen-processing steps that lead to T$_{reg}$ cell and not to T$_H$2 cell development.

Allergic inflammation and resistance to viral infection

Allergic inflammation involving excess production of type 2 cytokines (IL-4, IL-5, and IL-13) and reduced production of type 1 cytokines (IFN-γ) alters the baseline state of macrophages, epithelial cells, and other tissue resident cells. For example, macrophages develop an altered activations state, called alternatively activated macrophages or M2 macrophages, which allows them to produce IL-4, IL-13, and IL-33 and less IFN-γ. DCs and epithelial cells may also react to a type 2 cytokine environment with diminished type I and type III interferon responses. Since IFNs play a critical role in host defense against viral respiratory infection (e.g., with rhinovirus and influenza) and in limiting the development of type 2 responses, allergic inflammation is thought to predispose toward more severe virus-induced exacerbations that occur in allergic asthma patients. Whether this also requires concomitant allergen exposure or genetic predisposition toward severe asthma exacerbations (e.g., polymorphisms in *ORMDL3*, *GSDMB*, or *CDHR3*) is not clear. Moreover, the relationship between viral respiratory disease and asthma is further strengthened by observations strongly suggesting that early viral respiratory illness in childhood, e.g., with rhinovirus or respiratory syncytial virus, predisposes toward the development of asthma.

T$_H$17 cells and asthma

In some models of asthma, IL-17 produced by T$_H$17 cells can induce AHR and airway inflammation. The combination of IL-6 and TGF-β skews T-cell development toward T$_H$17 cell differentiation by suppressing Foxp3 and inducing the expression of RORγt, the lineage-specific transcription factor for T$_H$17 cells. In various mouse models, IL-17 is required for AHR, e.g., triggered by the repeated exposure of mice to ozone, can induce AHR independently of IL-13, can greatly enhance allergic asthma, or can be responsible for corticosteroid resistant asthma. These studies together suggest that IL-17-secreting cells (including T$_H$17 cells, macrophages, $\gamma\delta$ T cells, NKT cells, or ILC3s) are important regulators of asthma, although the importance of IL-17 in human asthma has not been established.

NEW BIOLOGICS FOR TREATMENT OF ASTHMA

As our understanding of the pathogenesis of asthma has grown, particularly with regard to cytokines, new biopharmaceutical products that target the immune system, called biologics, have become available for the treatment of asthma. For the most part, these include anticytokine monoclonal antibodies (mAbs), such as anti-IL-5, anti-IL-5 receptor, anti-IL-4 receptor α, and anti-IL-13, as well as anti-IgE mAb. Understanding the specific patients who respond to each of these mAbs has provided new insight into the role of the target molecules in disease pathogenesis. Some of the newer therapies are listed in Sections 23.23–23.29.

23.23 Anti-IgE mAb

A humanized anti-IgE mAb called omalizumab was approved by the U.S. Food and Drug Administration (FDA), initially in 2003, for the treatment of patients with moderate to severe allergic asthma, age 6 years or older, and in 2014 for a disease called *chronic idiopathic urticaria*. Omalizumab reduces immediate and late-phase pulmonary responses to allergen challenge but is also effective in reducing asthma exacerbations caused by viral infections. It is thought that omalizumab prevents asthma exacerbations not only by affecting mast cells and basophils but also by increasing type 1 interferon production by plasmacytoid dendritic cells (pDCs). pDCs express FcϵR1, which reduces interferon production when cross-linked by allergen.

23.24 Anti-IL-5 mAb

Humanized anti-IL-5 mAbs were approved by the FDA for the treatment of severe asthma (mepolizumab, GSK, in 2015 for patients aged 12 years or older, and reslizumab, Teva, in 2016 for adults with severe asthma). An anti-IL-5 receptor (CD125) mAb (benralizumab, AstraZeneca) was approved for severe eosinophilic asthma. These antibodies significantly reduce the number of eosinophils in the periphery and in the lung, and target patients with an eosinophilic asthma phenotype, which can be identified by elevated levels of eosinophils in peripheral blood.

23.25 Anti-IL-4 Rα mAb

An antibody that targets IL-4Rα (dupilumab, Regeneron) has been approved for the treatment of severe asthma, atopic dermatitis, and chronic rhinosinusitis with nasal polyps. IL-4Rα is utilized by both IL-4 and IL-13, and therefore, this mAb reduces signaling of both IL-4 and IL-13.

23.26 Anti-IL-13 mAb

Several monoclonal antibodies against IL-13 (lebrikizumab, Dermira and tralokinumab, AstraZeneca) are being studied for the treatment of atopic diseases, including asthma and atopic dermatitis. IL-13 has been shown to be an important effector molecule in asthma and allergic disease.

23.27 Anti-IL-17RA mAb

Brodalumab (Amgen/Valeant), by targeting IL-17RA, which is utilized by both IL-17 and IL-25, blocks the effects of IL-17 and IL-25. Brodalumab has been approved by the FDA for plaque psoriasis. It has also been studied for asthma, but early studies suggest minimal effect in patients with moderate to severe asthma.

23.28 Anti-TSLP mAb

Tezepelumab/AMG157 (Amgen) has been shown to reduce airway changes following allergen challenge in patients with mild to moderate asthma, suggesting an important role for TSLP in asthma. Tezepelumab is currently being studied in patients with asthma and with atopic dermatitis.

23.29 Anti-IL-33R/ST2 mAb and anti-IL-33 mAb

Anti-IL-33R/ST2 (Roche/GSK) and anti-IL-33 (etokimab, AnaptysBio) mAbs are in early stages of development for the treatment of asthma, food allergy, and/or atopic dermatitis.

SUMMARY

Asthma and COPD are extremely important respiratory diseases, affecting very large numbers of individuals. Although it is clear that type 2–biased immunity has a very important role in asthma, it has become increasingly apparent that asthma is a heterogeneous disease, and that many clinical forms or phenotypes exist, with distinct pathogenic mechanisms. These mechanisms include allergic (adaptive immunity) and nonallergic (innate) pathways, induced by allergen exposure, viral infection, and oxidative stress and involving a large number of cell types belonging to both the innate and adaptive immune systems, including eosinophils, basophils, T_H2 cells, ILC2s, DCs, neutrophils, NKT cells, epithelial cells, T_H17 cells, ILC3s, and M1 and M2 macrophages. In addition, asthma involves not only cytokines and chemokines produced by innate and adaptive lymphoid and myeloid cells, but also cytokines such as IL-25, IL-33, and TSLP produced by epithelial cells. Although asthma can develop through several independent pathways, these pathways can also coexist and interact. However, the recent recognition of multiple molecular and cellular pathways and with diverse underlying disease mechanisms and genetic effects leading to different forms of asthma provides us with a greater understanding of distinct asthma phenotypes and endotypes. Such an understanding is likely to lead to new and more effective therapies for asthma, which must be individualized and personalized, taking into account the distinct pathogenic mechanisms that occur in each patient.

FURTHER READING

Anderson, G.P.: Endotyping asthma: New insights into key pathogenic mechanisms in a complex, heterogeneous disease. *Lancet* 2008, 372:1107–1119.

Bacher, P., Heinrich, F., Stervbo, U. et al.: Regulatory T cell specificity directs tolerance versus allergy against aeroantigens in humans. *Cell* 2016, 167:1067–1078.

Buonocore, S., Ahern, P.P., Uhlig, H.H. et al.: Innate lymphoid cells drive interleukin-23-dependent innate intestinal pathology. *Nature* 2010, 464:1371–1375.

Gaboriau-Routhiau, V., Rakotobe, S., Lecuyer, E. et al.: The key role of segmented filamentous bacteria in the coordinated maturation of gut helper T cell responses. *Immunity* 2009, 31:677–689.

Gordon, S.: Alternative activation of macrophages. *Nat. Rev. Immunol.* 2003, 3:23–35.

Holgate, S.T., and Polosa, R.: Treatment strategies for allergy and asthma. *Nat. Rev. Immunol.* 2008, 8:218–230.

Holgate, S.T., Wenzel, S., Postma, D.S., Weiss, S.T., Renz, H., Sly, P.D. Asthma. *Nat. Rev. Dis. Prim.* 2015, 1:Article number: 15025.

Ivanov, I.I., Atarashi, K., Manel, N. et al.: Induction of intestinal Th17 cells by segmented filamentous bacteria. *Cell* 2009, 139:485–498.

Iwasaki, A., Foxman, E.F., and Molony, R.D.: Early local immune defences in the respiratory tract. *Nat. Rev. Immunol.* 2017, 17:7–20.

Kim, E.Y., Battaile, J.T., Patel, A.C. et al.: Persistent activation of an innate immune response translates respiratory viral infection into chronic lung disease. *Nat. Med.* 2008, 14:633–640.

Kim, H.Y., DeKruyff, R.H., and Umetsu, D.T.: The many paths to asthma: Phenotype shaped by innate and adaptive immunity. *Nat. Immunol.* 2010, 11:577–584.

Lambrecht, B.N., and Hammad, H.: Biology of lung dendritic cells at the origin of asthma. *Immunity* 2009, 31:412–424.

Matangkasombut, P., Pichavant, M., DeKruyff, R.H. et al.: Natural killer T cells and the regulation of asthma. *Mucosal Immunol.* 2009, 2:383–392.

Moro, K., Yamada, T., Tanabe, M. et al.: Innate production of T_H2 cytokines by adipose tissue-associated c-Kit$^+$Sca-1$^+$ lymphoid cells. *Nature* 2010, 463:540–544.

Neill, D.R., Wong, S.H., Bellosi, A. et al.: Nuocytes represent a new innate effector leukocyte that mediates type-2 immunity. *Nature* 2010, 464:1367–1370.

Palm, N.W., Rosenstein, R.K., Yu, S. et al.: Venom phospholipase A2 induces a primary type 2 response that is dependent on the receptor ST2 and confers protective immunity. *Immunity* 2013, 39:976–985.

Perrigoue, J.G., Saenz, S.A., Siracusa, M.C. et al.: MHC class II-dependent basophil-CD4$^+$ T cell interactions promote T_H2 cytokine-dependent immunity. *Nat. Immunol.* 2009, 10:697–705.

Saenz, S.A., Siracusa, M.C., Perrigoue, J.G. et al.: IL25 elicits a multipotent progenitor cell population that promotes T_H2 cytokine responses. *Nature* 2010, 464:1362–1366.

Sokol, C.L., Chu, N.Q., Yu, S. et al.: Basophils function as antigen-presenting cells for an allergen-induced T helper type 2 response. *Nat. Immunol.* 2009, 10:713–720.

Tait Wojno, E.D., and Artis, D.: Emerging concepts and future challenges in innate lymphoid cell biology. *J. Exp. Med.* 2016, 213:2229–2248.

Van Dyken, S.J., Nussbaum, J.C., Lee, J. et al.: A tissue checkpoint regulates type 2 immunity. *Nat. Immunol.* 2016, 17:1381–1387.

von Mutius, E.: Gene–environment interactions in asthma. *J. Allergy Clin. Immunol.* 2009, 123:3–11; quiz 12–13.

Xystrakis, E., Kusumakar, S., Boswell, S. et al.: Reversing the defective induction of IL-10-secreting regulatory T cells in glucocorticoid-resistant asthma patients. *J. Clin. Invest.* 2006, 116:146–155.

Yoshimoto, T., Yasuda, K., Tanaka, H. et al.: Basophils contribute to T_H2-IgE responses *in vivo* via IL-4 production and presentation of peptide–MHC class II complexes to CD4$^+$ T cells. *Nat. Immunol.* 2009, 10:706–712.

Ocular surface as mucosal immune site

24

RACHEL R. CASPI AND ANTHONY ST. LEGER

Good vision is a very strong evolutionary selective pressure. Therefore, the eye and its associated structures (known as adnexa and including the eyelids, conjunctiva, and the tear-producing and draining structures) have evolved not only anatomically but also immunologically to protect vision. The process of inflammation, while important to eradicate infectious agents, is itself harmful to the eye. Even small perturbations of the light refracting, transmitting, and gathering ocular structures (Figure 24.1) can have very deleterious consequences to vision. The eye, perhaps more than any other tissue, depends on immune protection mechanisms that preserve tissue integrity and minimize collateral damage. The interior of the eye resists inflammatory processes, termed *immune privilege*. The inside of the globe is devoid of immune cells and has no lymphatic drainage. The neural retina, where the light-sensitive photoreceptor cells are located, is protected by a tight blood-retinal barrier that is largely impermeable to the passage of cells and even of large molecules. As long as the eye is physically intact, this separation from the immune system is largely sufficient to protect the internal ocular structures from the environment. If immunocompetent cells enter the eye from a disrupted blood vessel, for example, they encounter a profoundly

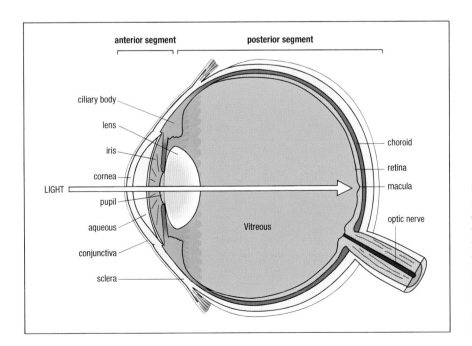

Figure 24.1 **Scheme of an eye showing major anatomical structures and organization.** Light passes through the ocular media and concentrates on the macula (area of retina responsible for sharp color vision), whereupon the photoreceptor cells sense the signal and transmit it to the brain via the optic nerve. Any damage to the ocular structures along the visual axis would likely result in a visual deficit.

inhibitory microenvironment composed of cell-bound and soluble inhibitory molecules. If these mechanisms prove insufficient to control inflammation, systemic eye-dependent regulatory processes are invoked to minimize tissue damage.

The exterior of the eye is constantly exposed to the environment and therefore contains a complex system of lymphoid structures whose primary role is to protect the health of the ocular surface and preserve the transparency of the cornea. Working in concert with the tear film that lubricates the ocular surface, efficient mucosal immune mechanisms resist microbial colonization of the warm and moist area, while tolerating self-antigens and commensal microorganisms. Activation of immunity must be tightly controlled and requires precise cooperation of the innate and adaptive arms of the immune system. Although the mechanisms that are used to maintain homeostasis at the ocular surface and inside the globe are different, the purpose is the same: to prevent destructive inflammation that would compromise visual function. Thus, the notion of privilege should perhaps be extended to include also the ocular surface and its associated mucosal lymphoid tissue.

ORGANIZATION OF EYE-ASSOCIATED LYMPHOID TISSUE

The eye-associated lymphoid tissue (EALT), which maintains homeostasis at the ocular surface, is a functionally important but relatively little-studied part of the mucosal immune system. It can be anatomically divided into conjunctiva-associated lymphoid tissue (CALT) and the tear (lacrimal)-duct-associated lymphoid tissue (variably known as TALT or LDALT), which lines the tear-draining system (Figure 24.2). Also included as part of EALT (but having no accepted acronym to its name) are the lymphoid cells found in the tear gland itself, which contain a rich population of B cells and plasma cells. Despite the nomenclature, which implies separateness, the lacrimal gland, CALT, and TALT systems are anatomically continuous with each other,

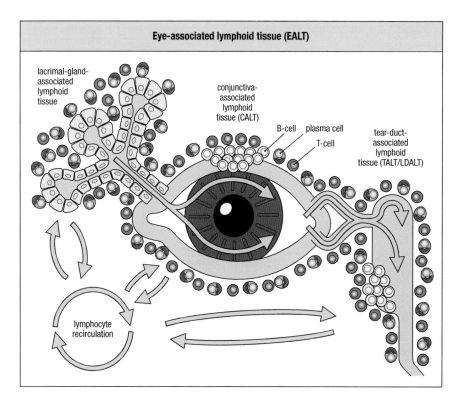

Figure 24.2 Eye-associated lymphoid tissue (EALT). EALT consists of, in order, lacrimal-gland-associated lymphoid tissue, conjunctiva-associated lymphoid tissue (CALT), and TALT (also known as LDALT) along the tear-draining ducts. The arrows show the direction of movement of tears and of antigens encountered at the ocular surface. Diffuse lymphoid tissue found throughout the EALT contains numerous T lymphocytes and plasma cells. Additionally, the organized lymphoid tissue contains lymphoid follicles with defined T- and B-cell zones. Cells primed within these follicles recirculate back to the tissue where they were primed, as well as to other mucosal lymphoid sites. (Adapted from Knop, E., and Knop, N. *Chem. Immunol. Allergy* 2007, 92:36–49. With permission from Karger.)

and are, moreover, connected by the flow of tears. Thus, they necessarily all contribute to and share effector molecules and together form a functional unit for antimicrobial defense of the ocular surface.

24.1 Eye-associated lymphoid tissue contains organized and diffuse populations of lymphocytes

As in other mucosa-associated lymphoid tissues (MALTs), components of EALT reside within two tissue layers divided by a basement membrane: the epithelium and the underlying loose connective tissue of the lamina propria. The fine structure and cellular composition of these layers change along the ocular tract, depending on location and degree of interaction with the external environment.

Again recapitulating MALT in other sites, EALT occurs in two forms: "organized" and "diffuse." The diffuse EALT is mainly populated by differentiated effector cells capable of dealing directly with antigens and is therefore considered an effector site. This diffuse organization is typical of the tear gland. The organized EALT contains lymphoid follicles and serves as an inductive site where antigens from the ocular surface are sampled and acquired, immune responses initiated, and germinal centers formed. CALT and TALT, which come in direct contact with the environment, contain both diffuse and organized lymphoid tissue (Figure 24.3).

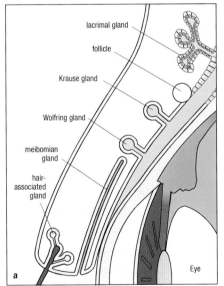
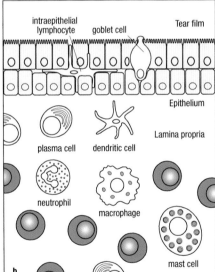
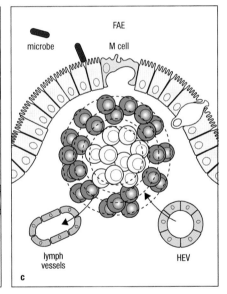

Figure 24.3 Organization of human eye-associated lymphoid tissue (EALT) and associated structures (adnexa). As shown in (**a**), mucosal immune tissue in the lacrimal gland and accessory lacrimal glands of Krause and Wolfring is continuous with the mucosal surface of the conjunctiva into the periacinar tissue of the lacrimal gland and the conjunctival accessory lacrimal gland. Organized accumulations of lymphocytes into lymphoid follicles occur in the conjunctiva. (**b**) Components of the diffuse lymphoid tissue that occurs in the lamina propria and epithelium throughout the EALT. Intraepithelial lymphocytes (IELs) are mainly CD8 cells. In the lamina propria, there are CD4 and CD8 and plasma cells as well as monocyte-macrophages, mast cells, occasional neutrophilic granulocytes, and stromal fibroblastic cells. The epithelium is covered by the tear film and contains mucus-producing goblet cells. (**c**) Organized lymphoid tissue in conjunctiva-associated lymphoid tissue and tear-duct-associated lymphoid tissue. Antigens are sampled at the ocular surface and translocated into the underlying lymphoid follicles, composed of discrete T- and B-cell regions, where effector cells generated from lymphocytes enter lymphatic tissues through high endothelial venules (HEVs) and leave via lymphatic vessels, from which they eventually reenter the circulation. FAE, follicle-associated epithelium.

24.2 Lacrimal gland and lacrimal gland-associated lymphoid tissue are anatomically located within upper eyelid

The lacrimal gland, located within the upper eyelid, contains acini whose secretory epithelial cells produce the tear fluid that continuously drains through 10–12 secretory ducts into the upper fold of the eyelid or fornix. Tear production is tightly controlled by neural input from the ocular surface tissues: the conjunctiva and cornea. In fact, the cornea is the most highly innervated surface tissue of the body, and even very minor irritating stimuli activate copious production of tear fluid. A basement membrane separates the epithelium from the layer of loose connective tissue constituting the lamina propria, which is similar to and anatomically contiguous with the connective tissue of the conjunctiva, to which it connects through the secretory tear ducts. The connective tissue contains accumulations of lymphocytes as well as many plasma cells. The vast majority of lacrimal gland plasma cells in mice are dedicated to the secretion of IgA and are primarily of the B-1 lineage. The acinar epithelium of the lacrimal gland produces the polymeric immunoglobulin receptor (pIgR). As discussed in Chapter 10, pIgR binds to the J chain in the dimeric IgA (and also pentameric IgM) and transports the antibody molecules that have been produced in the lamina propria through the epithelial cell layer into the lumen of the acini in covalent association with the secretory component of pIgR. The secretory IgA mixes with the tear fluid where it provides protection to the eye surface. Lymphocytes other than plasma cells and B cells include intraepithelial lymphocytes (IELs) which are mostly CD8 cells, and some CD4 cells as well as innate immune cells of various types. Of note are the unusually high numbers of natural killer T (NKT) cells and $\gamma\delta$ T cells in the lacrimal gland, which contribute to innate-like immune defense of the eye.

The lymphoid tissue in the lacrimal gland is diffuse and is not organized into well-defined follicular structures where priming takes place, which implies its effector nature. Lymphoid follicles are rarely observed and may not be physiologically relevant. The tear gland comprises a normally sterile environment, and studies have shown that there seems to be no retrograde transport of antigens from the ocular surface into the tear gland. Thus, the antibody-producing plasma cells most likely have reached the lacrimal gland by a process of recirculation, after having been primed and matured in lymphoid follicles elsewhere (see **Figure 24.2**). While $\alpha_4\beta_7$ integrin is known to have a major role in directing homing to the gut, the homing molecules that direct the cells to the lacrimal gland are not well defined. That said, the lacrimal gland, CALT, and TALT express a complement of integrins, addressins, and chemokines that differs from those expressed in gut-associated lymphoid tissue (GALT) and that shows similarities to the nasopharynx-associated lymphoid tissue (NALT). The tear fluid leaves the ocular surface through the tear-draining ducts at the inner corner of the lower and upper eyelids, which connect to the nasolacrimal duct leading into the nasal cavity (see **Figure 24.2**). The NALT thus receives the antigens from the ocular surface and is the logical place, in addition to CALT and TALT, where priming of cells that end up in the tear gland could occur. Cells primed in the CALT, TALT, and NALT may then recirculate to the tear gland to secrete their effector molecules there. In addition to IgA, tears also contain many constitutively produced and inducible antimicrobial peptides such as lactoferrin and defensins; cytokines such as interleukin-1 (IL-1), IL-4, IL-6, and transforming growth factor-β (TGF-β); surfactant proteins; and mucins, including MUCs 4, 5AC, 5B, and 7 that are secreted into the tear film.

24.3 Conjunctiva-associated lymphoid tissue covers external surface of eye with three anatomical regions (palpebral, bulbar, and fornix)

The conjunctiva is a continuous structure lining the inner surface of the eyelids and continuing over the front part of the eye to terminate around the cornea in a region known as the limbus. Its functions are protection, lubrication of the ocular surface by secretions from goblet cells and small accessory tear glands, and immunity. We recognize three principal conjunctival regions: palpebral, lining the inner surface of the eyelids; bulbar, covering the scleral surface; and the fornix, at the angle between the eyelids and the eyeball, where the secretory tear ducts discharge the tear fluid. The epithelium is thinnest in the palpebral part and gradually thickens toward the limbus of the eye. The lamina propria in all three regions contains organized as well as diffuse lymphoid tissue.

In humans, the conjunctiva has a well-developed lymphoid tissue, which mostly develops after birth, at around 3–4 months of age. Organized primary and typical secondary follicles with T cells, B cells, and high endothelial venules (HEVs) are present in CALT, and the follicle-associated epithelium contains M cells that sample environmental antigens and transport them into the follicle. Also present is diffuse lymphoid tissue containing intraepithelial lymphocytes (IELs) and abundant plasma cells in the lamina propria. Many of these plasma cells secrete IgA, and the conjunctival epithelial cells, as well as epithelium in the conjunctiva-associated crypts, express the pIgR that is responsible for the transcytosis of IgA. Thus, in keeping with its immunologic context of direct contact with the environment, the CALT is both an inductive and an effector site.

In contrast to the IELs, which contain a large proportion of CD8$^+$ T cells as in other mucosal surfaces, the lymphocytes in the lamina propria of the conjunctiva are mostly CD4$^+$. About 20% of conjunctival leukocytes are plasma cells actively producing antibodies, primarily IgA. The conjunctiva contains about two-thirds as many plasma cells as the lacrimal gland. Therefore, the conjunctiva is able to contribute considerably to its own antimicrobial defense and does not appear to be solely dependent on a passive supply of IgA from the lacrimal gland.

Interestingly, in contrast to humans, the mouse and rat conjunctiva has a virtual absence of diffuse lymphoid cells including plasma cells, although some follicular structures and cells resembling M cells are present in the nictitating membrane of the eye. The nictitating membrane is part of the conjunctiva, is typically transparent, and slides across the eye like a third eyelid in some species that include rodents and other mammals as well as birds and reptiles. The CALT of larger animals, including rabbits, dogs, and cats, resembles the organization in humans more closely. It is likely that the tear fluid compensates for this lack of diffuse CALT and contains sufficient IgA and other effector molecules for the ecological situation of the rodent. The conjunctiva (including that of laboratory rodents) also contains many mast cells, whose mediators are released after binding allergen-IgE complexes to receptors expressed on their membrane, which are responsible for the unpleasant consequences of hay fever and allergy affecting the ocular surface.

24.4 Tear duct–associated lymphoid tissue is present in rodents and humans and is subject to a pathway of organogenesis that is distinct from GALT

Tear duct–associated lymphoid tissue is present and well developed in laboratory rodents. Therefore, its developmental requirements have been well studied at the molecular level. Interestingly, the developmental requirements of TALT are very different from those of other mucosal lymphoid tissues.

Organogenesis of secondary lymphoid tissues such as Peyer's patches (PP) and peripheral lymph nodes is driven by CD3⁻CD4⁺CD45⁺ lymphoid tissue inducer (LTi) cells. The development of LTi cells is dependent on the inhibitor of DNA binding/differentiation 2 (Id2) and the retinoic acid–related orphan receptor (ROR) γt. Mice deficient in either one of these molecules lack LTi cells and do not develop lymph nodes and PP. LTi cells migrate to sites of lymphoid tissue genesis in response to chemokine signals including CXCL13, CCL19, and CCL21 and are triggered by IL-7 to produce lymphotoxin $\alpha_1\beta_2$ ($LT\alpha_1\beta_2$) that signals through the lymphotoxin β receptor (LTβR) to activate the mesenchyme in peripheral lymph nodes and PP anlagen. Production of CXCL13, CCL19, and CCL21 as well as adhesion molecules vascular cell adhesion molecule-1 (VCAM-1) and peripheral lymph node addressin (PNAd) then promotes migration and recruitment of leukocytes and permits formation of organized lymphoid tissue. Mice deficient in component(s) of these pathways largely or completely lack development of PP and peripheral lymph nodes, although they do develop nasopharynx-associated lymphoid tissue (NALT; Table 24.1).

Unlike PP and peripheral lymph nodes, which develop in embryonic life, TALT develops postnatally in mice. Despite its postnatal development, the appearance of TALT does not require environmental microbial stimuli, as TALT develops normally even in germ-free mice. TALT genesis is also normal in both Id2-deficient and RORγt (*Rorc*)-deficient mice, which lack LTi cells. Furthermore, TALT formation is present, though reduced in size, in mice deficient in molecules downstream of LTβR signaling. This indicates that TALT formation utilizes different pathways than formation of classical secondary lymphoid tissues, and even than formation of other mucosal lymphoid tissues. Interestingly, LTi-like CD3⁻CD4⁺CD45⁺ cells are observed in TALT anlagen

TABLE 24.1 PRESENCE OF DIFFERENT MUCOSA-ASSOCIATED LYMPHOID TISSUES IN MICE DEFICIENT IN GENES CONTROLLING LYMPHOID ORGANOGENESIS PATHWAYS

Genetic deficiency	Lymphoid tissue development						
	TALT	NALT	PP	CLN	MLN	ILF	Cryptopatch
Id2 ⁻/⁻	+++	−	−	−	−	ND	ND
Rorc ⁻/⁻	+++	+++	−	−	−	−	+/−
Ltα ⁻/⁻	+	+	−	−	+/−	−	+/−
aly/aly	+	+	−	−	−	−	++
Il7rα ⁻/⁻	++	++	−	+/−	++	++	−
Cxcl13 ⁻/⁻	++	+	+/−	+/−	++	ND	ND
plt/plt	+++	++	++	++	++	ND	ND
Cxcl13 ⁻/⁻ *plt/plt*	+	+	+/−	−	++	ND	ND

Source: Adapted from Nagatake, T. et al., *J. Exp. Med.* 2009, 206:2351–2364.
Note: *aly/aly* mice carry a null mutation of NFκB-inducing kinase (NIK), resulting in a failure to signal through LTβR, and *plt/plt* mice have a null mutation of Ccl19 and Ccl21 genes.
Abbreviations: CLN, cervical lymph node; ILF, isolated lymphoid follicle; MLN, mesenteric lymph node.

of the Id2- and RORγt-deficient mice, suggesting that the inducer cells that drive TALT formation may be a population that is distinct from the classical LTi that control organogenesis of secondary lymphoid tissues such as PP and peripheral lymph nodes.

24.5 Cornea constitutes an atypical but immunologically active mucosal tissue

Maintaining the architecture of collagen within the cornea is critical for an optically clear cornea and pristine vision. Disruption of collagen at the ocular surface by excessive inflammation can be a cause of impaired vision or blindness. Under steady-state conditions, the cornea is considered immune privileged due to the absence of vascular and lymphatic vessels that is believed to be maintained by constitutive production and release of vascular endothelial growth factor receptors (VEGFRs), which neutralize VEGF and actively inhibit the in-growth of vessels. In addition, the cornea expresses Fas ligand, complement regulatory proteins, programmed death-ligand 1 (PD-L1), and other immunosuppressive factors like IL-1 receptor antagonist (IL-1RA) and TGFβ, which all contribute to the reduced immunogenicity of self and nonself antigens.

More recently, several groups have begun to identify immunologic roles of corneal nerves in ocular mucosal immunity. Specifically, corneal nerves, similar to sensory nerves of the gut, secrete the immune suppressive molecules, vasoactive intestinal peptide (VIP) and substance P (SP), which limit ocular inflammation. Further, severing corneal nerves during corneal transplant and potentially other corneal procedures abolishes the production of active VIP and SP, increasing the propensity for ocular inflammation, corneal graft rejection, and potential blindness.

Despite the absence of defined mucosal immunologic structures, the cornea possesses macrophages and dendritic cells, which have the ability to uptake antigen, migrate to the draining lymph nodes, and induce the innate and adaptive immune responses. In many instances, antigen-presenting cells (APCs) from the ocular surface induce antigen-specific T_{reg} (iT_{reg}) cells to limit the inflammation. However, after the introduction of inflammatory stimuli during an infectious or traumatic insult, corneal DCs are primary drivers of immune responses within the draining lymph nodes and have been shown to be critical in generating immune responses during allergy and infection.

Induction and expression of immunity at ocular surface: The good and the bad

Because the CALT and TALT contain lymphoid follicles into which antigens sampled by M cells and dendritic cells at the ocular surface are transported, processed, and presented, the ocular surface is a rich environment for inducing immune responses. Antigens from the ocular surface also drain with the tears through the lacrimal drainage system, that is, the nasolacrimal duct and its associated structures, where there is also an associated lymphoid tissue (see Figure 24.2). Thus, mucosal vaccination through the ocular surface inevitably also stimulates the nasal immune system. Furthermore, whether induction of immunity takes place locally in the mucosal lymphoid follicles or occurs in the draining lymph nodes, the responses also become systemic such that antibodies to ocularly encountered antigens are present in the serum, and immune cells can migrate not only back to the ocular surface, but also to other mucosal and lymphoid tissues. Conjunctival epithelial cells in culture can induce the expression of human mucosal lymphocyte antigen $\alpha_E\beta_7$ (CD103) on CD8$^+$ (and to a lesser extent also on CD4$^+$) lymphocytes. This may on the one hand allow the retention of CD8$^+$ and CD4$^+$ lymphocytes within the epithelial compartment of the conjunctiva and on the other hand play a part in homing of lymphocytes to other mucosal tissues. Immune responses induced in other mucosal tissues can have protective effects in the eye. For example, nasal vaccination of mice with herpes simplex virus type 1 (HSV-1) can ameliorate the ocular manifestations of HSV-1 and prevent herpetic stromal keratitis.

That said, immune responses occurring at the ocular surface must strike a balance between protection from a pathogen and direct bystander damage to the ocular surface as a result of the inflammatory process leading to pathogen eradication. Ocular herpes is a prime example where an exuberant local response to eradicate infection causes serious harm to the corneal surface, known as herpes keratitis, which can culminate in opacification of the cornea and blindness.

24.6   Regulatory mechanisms exist in the EALT that restrict an inflammatory response

To reduce inflammation-induced pathology, a number of mechanisms control the type and the intensity of the immune responses in the lacrimal gland and the ocular surface, thereby limiting inflammatory immunity in favor of less destructive effector mechanisms. Much of the ocular protective response is through secretory IgA (SIgA), which does not activate complement but is effective at neutralizing viruses and at promoting antibody-dependent cellular cytotoxicity by eosinophils bearing $Fc\alpha R$. The numerous $CD8^+$ T cells in the epithelial layer and lamina propria contain cytotoxic T lymphocytes, which kill in a targeted fashion through perforin and granzyme B rather than through effector cytokines such as tumor necrosis factor-α (TNF-α), interferon-γ (IFN-γ), and IL-17 that are harmful to many cells, promote recruitment of phagocytes, and result in considerable bystander tissue damage. Thus, the activity of T_H1 and T_H17 effector T cells that exert effector function through pro-inflammatory cytokines is fortunately dampened by "natural" $Foxp3^+CD4^+CD25^+$ regulatory T (nT_{reg}) cells, which are found in conjunctival tissues and are believed to dampen or inhibit the inflammatory, autoimmune, and pathogen-directed immune responses on the ocular surface. Research in rabbits has revealed that nT_{reg} cells present in the conjunctiva suppress virus-specific $CD4^+$ and $CD8^+$ effector T cells during experimental ocular HSV infection. More recently, it has been shown that thrombospondin-1 (TSP-1), which is produced by cells at the intestinal and ocular surface, can activate TGF-β_2, aiding the induction of nT_{reg} cells and thus peripheral tolerance. Other T cells in the epithelium and lamina propria of the ocular surface that have been shown to be regulatory in other mucosal sites and control responses to commensal microbiota include $CD8^+$ T cells, $\gamma\delta$ T cells, and NKT cells; however, functional information about their relative importance in the eye is lacking.

24.7   Immunization through ocular mucosal immune system can induce an immune response

Despite efficient immunoregulatory mechanisms that operate at the ocular surface, it is possible to take advantage of the mucosal system of the eye for active induction of immunity by vaccination. Experiments in rabbits (which, unlike mice and rats, do have well-developed organized CALT) show that antigen-sampling M cells, present in the follicle-associated epithelium above the lymphoid follicles of the conjunctiva, bind and translocate SIgA from the tear film. Topical conjunctival immunization leads to generation of antibody-secreting plasma cells not only in the local but also in distant mucosae together with detectable antibody titers in serum. Natural antibodies, present in the tear film before immunization, may contribute to antigen uptake, potentiating immune responses induced through the eye.

Mucosal adjuvants such as cholera toxin (CT) can potentiate these responses, which they do at least in part by enhancing antigen uptake. CT has been well studied as an oral and nasal adjuvant. In mice, despite their paucity of organized CALT, eye drop administration of a model antigen (ovalbumin) or microbial protein antigens plus CT induces immune responses in both mucosal and systemic tissues and affords effective protection against subsequent infection with the immunizing microbial agent. It is notable that the responses are induced even after occlusion of tear drainage from the eye to the nose and that they are CCR6 but not CCR7 dependent, thereby excluding the NALT and the draining regional lymph nodes (which require CCR7 for homing), respectively, as necessary priming sites. A local priming mechanism is further supported by the finding that the eye drops containing protein antigen plus CT induce organogenesis of CALT and increase the numbers of M-like cells on the nictitating membrane, which is part of the conjunctiva, though not in the conjunctiva proper. Moreover, efficient induction of CALT in the nictitating membrane also follows administration of a live microbial agent (*Chlamydia trachomatis*) without CT, supporting the notion that immunization through the ocular surface can be an effective route of mucosal vaccination.

24.8 Modulation of immunity at ocular surface by microbiota

The mucosal surfaces of the body harbor diverse strains of bacteria, and it is now clear that the commensal/mucosal immune system axis is necessary to prevent pathogenic infection, maintain barrier function, and limit inflammation. In models of Sjögren's syndrome and corneal *Pseudomonas aeruginosa* infection, dysbiosis of the gut microbiome resulted in more severe disease, leading to the conclusion that composition of the intestinal microbiome has direct effects on diseases at the ocular surface.

Due to the huge difference in the biomass associated with bacteria found in the gut ($\sim 10^{12}$ bacteria) compared to the ocular surface (0–10^3 bacteria), many have discounted a role for ocular commensal bacteria in modulation of local immunity. More recently, however, focus has shifted toward understanding if a resident ocular microbiome truly exists or detectable bacteria have simply arrived via the environment and are promptly eliminated (which would not necessarily preclude their ability to affect local immune responses). In agreement with past hypotheses, the central cornea, under normal conditions, is likely sterile and devoid of most if not all organisms, probably due to the profoundly antimicrobial nature of the tears and continual mechanical cleaning action of the eyelids. However, the conjunctiva, which attaches to the peripheral cornea and extends as a liner underneath the eyelid, appears to be a slightly less inhospitable environment, as conjunctival swabs routinely yield culturable bacteria, though very limited in diversity and numbers. In addition to studies using 16S rRNA gene sequencing analyses to reveal bacterial "signatures" between healthy and diseased conjunctival tissue, recent studies provide definitive proofs of concept that bacteria can colonize the ocular mucosa and induce IL-17 from γδ T cells in part through interaction with CD11b$^+$ dendritic cells. In this case, the host immune response toward the bacterium, *Corynebacterium mastitidis*, was critical for the prevention of infectious disease at the ocular surface, which could be mostly attributed to the effects IL-17 had on neutrophil recruitment and the production and release of antimicrobial compounds within the tears. This bacterium was passed from mother to her pups but did not appear to be passed between adult animals sharing the same cage, supporting the notion that this bacterium, and potentially others like it, are true residents of the ocular surface. Together, these data confirmed that an ocular microbiome can, indeed, exist and plays a critical role in mucosal immune homeostasis and host defense at the ocular surface.

It is important to point out that under some circumstances, commensal microbes may have deleterious effects. In the case of autoimmune uveitis, a retinal disease, antigens from gut commensals can mimic ocular peptides, which activate self-reactive T cells and trigger disease. A similar situation may occur at the ocular surface. It is conceivable that microbes that elicit protective immune responses in normal subjects may become pathobionts in subjects who are immunodeficient or who respond too strongly. These data suggest that specific microbes may be able to elicit protective or destructive immune responses, depending on the context. Exactly which bacterial products stimulate immunity remains unresolved and is currently being investigated. By illuminating systemic and local microbe-dependent immunologic mechanisms, better therapeutics may be developed to help promote inflammatory or regulatory immunity to alleviate infectious and autoimmune disease, respectively.

Diseases associated with dysfunction of ocular mucosal immune system

Ocular surface inflammation can occur due to a wide range of causes: infection, chemical or mechanical injury, allergy, and autoimmune responses. This can manifest as conjunctivitis (i.e., inflammation of the conjunctiva), keratitis (inflammation of the cornea), or keratoconjunctivitis if both are involved. In many cases, dysfunction of the ocular mucosal immune system and ocular surface disease can be something of a vicious cycle, so that it becomes difficult to distinguish which came first.

24.9 Loss of tear fluid results in dry eye syndrome

Dry eye affects up to 34% of all people globally and is one of the most frequent ocular surface problems, whose primary manifestation is a dysfunction of the tear film, resulting in a gritty feeling, dryness, irritation, and inflammation of the ocular surface. Dry eye can stem from a number of causes: from neurogenic, due to dysfunction of parasympathetic nerves that control tear production, through autoimmune damage to the tear gland, to infections that can cause physical blockage of tear ducts or inhibition of tear production by inflammatory mediators. The classification of dry eye diseases, established in a 1995 National Eye Institute/Industry workshop, provides a compartmentalized classification system for these syndromes (**Figure 24.4**), but in practice, combined disorders are common, and it is not always possible to determine a single cause for the disease. As an example, chronic irritation of the ocular surface due to evaporative dry eye can lead to secondary autoimmunization to autologous tissue components of the ocular surface and lacrimal gland components, shifting from "simple" dry eye to Sjögren's syndrome. As a therapy, lipids have garnered more attention recently due to their ability to increase tear stability and modulate excessive inflammatory responses. Specifically, studies are investigating ways to increase natural production of lipids or more effectively deliver synthetic lipids to the ocular surface.

24.10 Sjögren's syndrome is an autoimmune disease of lacrimal and salivary glands

Sjögren's syndrome, an inflammatory condition of the lacrimal and salivary glands that appears to be autoimmune, affects predominantly women (9 out of 10 Sjögren's syndrome patients are women) usually during or past middle age. There is extensive evidence that Sjögren's syndrome has an autoimmune basis. A major antigen has been reported to be α-fodrin, but there are antibodies to other autoantigens such as Ro, La, antinuclear antibodies, and rheumatoid factor, resembling other systemic autoimmune diseases like systemic lupus erythematosus. Association of disease susceptibility

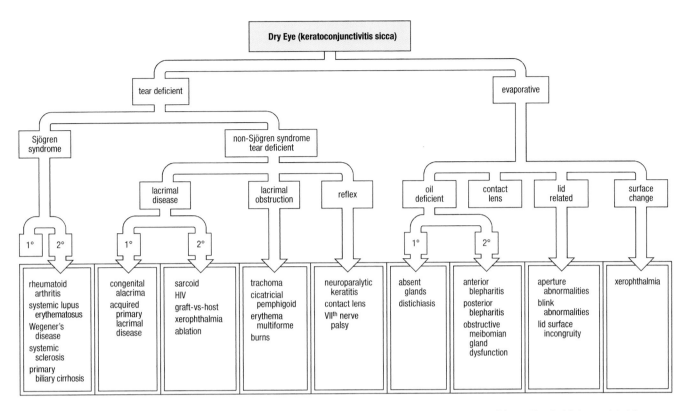

Figure 24.4 NEI/Industry workshop—classification of dry eye diseases. (Reprinted from *The CLAO Journal,* 21(4), 1995.)

with HLA-B8, HLA-Dw3, HLA-DR3, and with the DQA1*0501 allele supports the notion of autoimmunity, as these are antigen-presenting molecules. A recently developed mouse model, in which mechanical desiccating stress to the ocular surface results in induction of self-reactive lymphocytes (antigen currently unidentified) that can transfer the condition from a diseased host to healthy recipients, further supports the notion that autoimmune mechanisms can become induced by, and can perpetuate, dry eye disease. Similarly, cytokines like IFN-γ and IL-4 have shown to be important for the development of disease. It has been found that sex and sex steroid hormones are critical factors in the regulation of ocular surface tissues, as well as in the pathogenesis of dry eye syndromes. Androgens reduce the inflammation in, and enhance the functional activity of, lacrimal glands in mouse models of Sjögren's syndrome through unknown mechanisms that involve androgen binding to receptors in lacrimal gland epithelial cells, whereas estrogens have the opposite effect, providing a potential explanation for the predominance of the disease in women.

24.11 Ocular cicatricial pemphigoid is an autoimmune disease directed against components of the basement membrane

Ocular cicatricial pemphigoid (OCP) is part of a group of systemic autoimmune diseases known as mucous membrane pemphigoid, characterized by circulating autoantibodies directed against a variety of epithelial basement membrane components, the major ones being bullous pemphigoid antigens 1 and 2 (BPAG1 and BPAG2), laminin 5 (epiligrin), and in some cases also laminin 6, collagen 7, and β_4 integrin subunit. As a result of antibodies binding to the basement membrane, the connection between the epidermis and dermis is weakened such that the two layers separate easily, forming blisters.

The antibodies are mainly of the IgG4 subclass but also IgM (complement binding) and sometimes IgA. Intravenous immunoglobulin (IVIg) and anti-B-cell treatment with rituximab (anti-CD20) have been reported to have beneficial effects, supporting a contribution of anti–basement membrane autoantibodies and, thus, B cells to OCP. Despite the presence of complement C3 components at the basement membrane of human conjunctival biopsies, complement-deficient neonatal mice still develop blistering after infusion of anti-laminin-5 antibodies. This would suggest the circulating autoantibodies can induce dermal–epidermal cleavage directly, without the need for complement activation. Demonstration of circulating autoantibodies in serum of OCP patients often requires sensitive detection methods, suggesting that most of them are bound to the tissue. Usually a patient will demonstrate only one antibody reactivity, suggesting that there are subsets of the disease, but clinical differences between patients with different reactivities have not been observed.

In cases where OCP patients do not respond to immunosuppressants or B-cell inhibition therapy, anti-tumor necrosis α factor therapy has proven efficacious. Specifically, this therapy has improved conjunctival fibrosis and reduced inflammatory infiltrates. These data suggest that OCP progression may be similarly linked to diseases such as psoriasis and inflammatory bowel disease (IBD), as the effects of both are alleviated with similar therapies.

The morbidity in OCP is due to the progressive fibrosis and contraction of the conjunctiva that results from scarring when the blisters heal. This consequently results in shortening of the conjunctival fornices and blockage of the lacrimal ducts causing a secondary severe dry eye disease that leads to more keratinization and fibrosis, often ultimately causing blindness. Mechanistic studies have now shown a critical involvement of aldehyde dehydrogenase family 1 (ALDH1) produced by classical dendritic cells (DCs) in the conjunctiva. Through enzymatic activity, ALDH1 can cleave retinol into active retinoic acid, which directly contributes to fibrosis within the conjunctiva during OCP as well as allergy.

Data from conjunctival biopsies reveal an increase in CD4$^+$ T cells and DCs in the epithelium and activated IL-2R-bearing T cells (CD4$^+$ and CD8$^+$), macrophages, and B cells in the stroma. An increase in IL-17-producing cells has also been reported. T_H2 cells are present and may amplify the fibrotic process through production of cytokines, such as IL-13, that stimulate resident fibroblasts to produce profibrotic cytokines, such as TGF-β, platelet-derived growth factor, basic fibroblast growth factor, and connective tissue growth factor, all of which are increased in the conjunctiva. All of these contribute to indirect fibrogenesis, which can then lead to fibrosis and disease.

Several reports have estimated the incidence of OCP to be 0.87 cases per million of the population per year in Germany and 1.16 per million in France, with a 2:1 female predominance. Associations with several HLA antigens have also been reported, depending on the study and ethnic group, in keeping with the autoimmune nature of the disease.

24.12 Allergy is a common clinical manifestation in the eye due to its exposed surfaces

The conjunctiva is a highly exposed environmental surface and contains many mast cells. In allergic individuals who have IgE antibodies directed against environmental allergens such as pollen, which bind to the FcεR of mast cells and basophils, contact with the allergen through the ocular surface precipitates release of mast cell granules whose mediators cause enhancement of vascular permeability and conjunctival swelling (chemosis), itching, and tearing (see Chapter 14). While it is clear that allergy can be expressed when allergen encounters the ocular surface, evidence that allergy can be induced through the ocular surface is circumstantial, as it is difficult to exclude contact through additional mucosal surfaces, including the NALT and even the lung. In the classical animal models of ocular allergy, for example to ragweed pollen, induction is systemic by immunization with the allergen in alum, and instillation of eye drops containing allergen brings about allergic manifestations in the eye. However, as was previously stated, it is possible to induce pathogen-related immune responses through the eye experimentally even when the ducts draining the tears into the nasal cavity are blocked, excluding significant participation of NALT. Furthermore, it has been suggested that local mucosal epithelial cells can help initiate ocular allergic inflammation by producing a novel pro-allergic cytokine, thymic stromal lymphopoietin (TSLP), which activates dendritic cells to prime T_H2 differentiation and allergic inflammation through the TSLP-TSLPR and OX40L-OX40 signaling pathway. Microbial products such as LPS and flagellin can induce DC production of TSLP, suggesting a link between systemic or local microbes and allergic conjunctivitis. NKT cells, probably through their production of IL-4, are needed for maximal expression of allergic conjunctivitis. IL-33, a cytokine within the IL-1 family that binds to ST2 together with IL-1 receptor accessory protein to form the IL-33 receptor complex on eosinophils and $CD4^+$ T cells and promotes T_H2 responses, has also been implicated in the induction and augmentation of allergic conjunctivitis. More recently, focus has transitioned to the DCs in allergy, which act to prime allergic T cells that cause disease. Specifically, DCs can suppress allergy through the production of thrombospondin-1 (TSP-1), which inhibits allergic T-cell responses. Conversely, DCs can also directly induce mucosal fibrosis in the conjunctiva during allergy via retinoic acid synthesis, suggesting that proper maintenance of DC function is crucial to prevent disease.

Most types of allergic conjunctivitis are fairly benign, if unpleasant. However, there are exceptions. Vernal keratoconjunctivitis is a relatively rare but serious allergic response that can leave permanent scarring and visual deficit. It occurs most often in spring (vernal) in young boys around 8 years of age and usually resolves spontaneously at puberty, indicating a hormonal influence. It is characterized by conjunctival giant papillae, hyperemia, and frequent involvement of the cornea. Although allergic mechanisms, including presence of typical T_H2-type immunity and IgE, are well documented in vernal keratoconjunctivitis, up to 50% of patients do not have a family or medical history of allergic diseases and do not show IgE sensitization, suggesting that this disease is not solely IgE mediated. Recent studies have also pointed out the role of resident conjunctival nonlymphoid cells in the pathogenetic processes of vernal keratoconjunctivitis. Thus, the pathogenesis of the condition is likely to be multifactorial and complex, involving not only the ocular mucosal immune system but also, similarly to Sjögren's syndrome, the nervous and endocrine systems.

24.13 Herpetic stromal keratitis is a pathologic response against the cornea after herpes virus infection has been cleared

Infection of the ocular surface with herpesvirus (HSV) causes keratitis, which in about one-fifth of the cases induces an inflammatory reaction in the corneal stroma and permanent scarring of the cornea known as herpetic stromal keratitis (HSK). This is a classic example in which an exuberant and insufficiently controlled immune response to infection does as much harm as good, by causing serious bystander tissue damage. The pathology continues even after antiviral immunity has contained viral replication, suggesting a secondary autoimmune component. Mouse models of HSK have shown a strong participation of T_H1 cells and their mediators in the immune-mediated damage, as well as of innate immune cells recruited by them. The site of priming of these T cells may be local within the CALT as well as in the lymph nodes that drain the ocular surface, but the nature of the priming antigen(s) remains unknown. At least four different effector mechanisms of T-cell activation have been proposed. These include specific activation by viral antigens, secondary responses to cornea-derived self-antigens, nonspecific activation by CpG motifs in viral DNA, and bystander activation by inflammatory mediators or cytokines against corneal-derived autoantigens. These mechanisms are not mutually exclusive, and in fact may well occur together and synergize to cause immune-mediated tissue damage. Experimental results in the mouse model of HSK have shown that vascular endothelial growth factor and matrix metalloproteinase 9 (MMP9), which are produced by the immune cells invading the cornea, promote pathologic formation of new blood vessels that encroach on the corneal surface (corneal neovascularization) and are an important part of HSK pathogenesis.

Corneal HSV infections are thought to be a leading cause for the development of neurotrophic keratitis (NTK), which is a degenerative disease that leads to the loss of corneal epithelial cells, tear production, and sensation. In addition, this effect is also thought to reduce the amount of immune suppressive molecules released by nerves, which increases the propensity of immune cell priming and subsequent autoimmunity. Recently, in a mouse model, neutralization of IL-6 or the excision of sympathetic nerves promoted the survival of sensory nerves after infection, which substantially reduced the prevalence of HSK.

24.14 Interior of eye is an immune-privileged site due to powerful regulatory mechanisms that resist immune activation

The internal cavity of the eye is not a mucosal system in the classical sense, nor does it contain lymphoid tissue resembling the mucosal lymphoid tissue. Nevertheless, there are some notable similarities in the biological requirement to control inflammatory immunity within the eye to protect the physical integrity and preserve the visual function of the neural retina, and in the molecular mediators that fulfill this requirement.

Ocular immune privilege is a term that has been coined to define the special immune environment within the eye. The healthy retina is protected by a tight blood-retinal barrier, composed of tight junctions between adjacent vascular endothelial cells, which prevents cells from freely entering the eye. The ocular fluids contain TGF-β, inhibitory neuropeptides (VIP, α-MSH, somatostatin), and a high concentration of retinoic acid; the ocular tissues constitutively express FasL, PDL-1, thrombospondin, galectins, and other inhibitory cell-surface molecules. These conditions help maintain resident ocular APCs in a nonactivated state with low major histocompatibility complex class II expression and create an inhibitory environment against immune cells from the circulation that may enter the eye if the barrier breaks down due to physical trauma, inflammation, activation of uveitogenic T cells by commensal bacteria of the gut. After injection of an immunogenic protein into the anterior chamber, tolerogenic APCs are also "exported" from the eye into the spleen, where they induce CD4$^+$ and CD8$^+$ T$_{reg}$ cells; this phenomenon is known as anterior chamber-associated immune deviation (ACAID) and may involve CD1d-restricted NKT cells. Another mechanism involves the direct conversion, within the ocular microenvironment, of conventional T cells specific to retinal antigens associated with the eye into Foxp3-expressing T$_{reg}$ cells. This latter process requires local antigen presentation in the presence of retinoic acid (which is abundant in the eye due to its function in the chemistry of the visual process) and TGF-β, and resembles T$_{reg}$ induction in mucosal tissues by presentation of antigen on retinoic acid–producing CD103$^+$ DCs.

Thus, the mucosal tissues and "classical" immune-privileged sites such as the interior of the eye share the need and the ability to actively control immune responses taking place in their "territory" in order to maximize protection of the tissue. It therefore seems appropriate to extend the concept of immune privilege to the ocular surface and to its associated mucosal tissues, as well as to other mucosal sites.

SUMMARY

The ocular mucosal immune system has to fulfill the function of protection from microbial infection while also providing protection from the deleterious consequences of inflammation on vision. To do this, it has adapted to use immune mechanisms at both the humoral and the cellular levels that are effective in antimicrobial defense but minimize inflammation and the associated collateral damage to the surrounding tissue. Nevertheless, the system is not perfect, and an overactive effector response or defects in regulatory mechanisms can precipitate a vicious circle of immune, mechanical, and finally autoimmune or allergic processes that irritate and inflame the ocular surface. At best, they can be difficult to control and require chronic treatment, and at worst they bring about permanent ocular surface damage with disastrous consequences to vision.

FURTHER READING

Knop, E., and Knop, N.: The role of eye-associated lymphoid tissue in corneal immune protection. *J. Anat.* 2005, 206:271–285.

Knop, E., and Knop, N.: Anatomy and immunology of the ocular surface. *Chem. Immunol. Allergy.* 2007, 92:36–49.

Knop, N., and Knop, E.: Regulation of the inflammatory component in chronic dry eye disease by the eye-associated lymphoid tissue (EALT). *Dev. Ophthalmol.* 2010, 45:23–39.

Kugadas, A., and Gadjeva M.: Impact of microbiome on ocular health. *Ocul. Surf.* 2016, 3:342–349.

Mebius, R.E.: Organogenesis of lymphoid tissues. *Nat. Rev. Immunol.* 2003, 3:292–303.

Nagatake, T., Fukuyama, S., Kim, D.Y. et al.: Id2-, RORγt-, and LTβR-independent initiation of lymphoid organogenesis in ocular immunity. *J. Exp. Med.* 2009, 206:2351–2364.

Paulsen, F.: Functional anatomy and immunological interactions of ocular surface and adnexa. *Dev. Ophthalmol.* 2008, 41:21–35.

Seo, K.Y., Han, S.J., Cha, H.R. et al.: Eye mucosa: An efficient vaccine delivery route for inducing protective immunity. *J. Immunol.* 2010, 185:3610–3619.

Stern, M., Schaumburg, C., Dana, R. et al.: Autoimmunity at the ocular surface: Pathogenesis and regulation. *Mucosal Immunol.* 2010, 3:425–442.

Streilein, J.W.: Ocular immune privilege: The eye takes a dim but practical view of immunity and inflammation. *J. Leukoc. Biol.* 2003, 74:179–185.

INFECTIOUS DISEASES OF MUCOSAL SURFACES

PART SIX

Mucosal interactions with enteropathogenic bacteria

25

NADINE CERF-BENSUSSAN, PAMELA SCHNUPF,
VALÉRIE GABORIAU-ROUTHIAU, AND PHILIPPE J. SANSONETTI

The gastrointestinal epithelium is the main body interface with the microbial world. The density of bacteria increases along the gastrointestinal tract from 10^3 and 10^5 resident bacteria per milliliter luminal content in the stomach and duodenum, respectively, to 10^8 in the ileum, and 10^{12} in the colon. The colon thereby is one of the most densely populated microbial ecosystems known. In humans, the resident (or commensal) microbiota is composed of approxihundreds of species. The complexity is enormous and contains 10 divisions of bacteria as well as viruses, eukarya, and archaea. Despite this complexity, recent studies suggest that a surprising quantity of these bacteria can be cultured using specialized microbiologic approaches. Among the 10 dominant divisions (phyla), two represent more than 90% of the bacteria. These are the strictly anaerobic bacteria of the Firmicutes (gram-positive bacteria) and the Bacteroidetes (gram-negative bacteria) divisions. A few other phyla, such as Proteobacteria (which includes the Enterobacteriaceae order), Actinobacteria, and Verrucomicrobia, are minor components of the normal microbiota. In healthy adults, Proteobacteria represent less than 1% of the enteric microbiota. Yet, these gram-negative aerotolerant microorganisms have a particular importance as they include a vast spectrum of pathogenic and opportunistic bacteria such as *Salmonella*, *Shigella*, *Helicobacter*, and *Escherichia coli* that are major causes of intestinal and extraintestinal diseases. Intestinal bacteria can thus be schematically divided into commensal and pathogenic bacteria even though the frontiers between these two categories have been blurred by genomic studies. Owing to their long coevolution, resident bacteria or symbionts (from the Greek for "living with") and their hosts have developed mutualistic interactions that promote each other's fitness. Notably, resident bacteria stimulate a "state-of-alert" within the innate and adaptive immune components in mucosal tissues of their hosts and form a competitive barrier that opposes mucosal colonization by pathogens ("colonization resistance"). In contrast, pathogens behave as intruders that occasionally colonize and/or invade their eukaryotic hosts to promote their own life cycle. These bacteria possess virulence factors that encode a wide range of effectors that are able to subvert and control eukaryotic cellular functions in order to aid their adherence, replication, and/or dissemination and often eliminate resident bacteria. The continuous presence of resident bacteria, as well as the repeated intrusions of pathogens that are a major threat to their hosts, has pressured the mucosa-associated lymphoid tissues to undergo evolutionary adaptations to build a finely tuned and tightly regulated immune system that balances homeostasis with rapid responses to pathogens. The simple epithelium (see Chapter 5) of the gastrointestinal tract, which is the most well-studied mucosal interface, is the first line of defense against pathogenic intrusions and a central component of host innate immune defenses. This chapter compares interactions between resident bacteria and pathogens with the gastrointestinal epithelium, which is a model for understanding microbial interactions with other mucosal surfaces, and discusses the consequences for the host and for bacteria.

GASTROINTESTINAL EPITHELIUM: PHYSICAL AND CHEMICAL BARRIER AGAINST BACTERIA

Besides its central function in digestion and absorption, the monolayer of polarized (simple columnar) epithelial cells that lines the luminal surface of the gastrointestinal tract forms a robust physicochemical barrier that efficiently opposes the entry of luminal bacteria. Tight junctions, which maintain the cohesion between epithelial cells, are impermeable to bacteria. Hydrochloric acid produced by gastric parietal cells, together with bile salts released into the duodenal lumen, exert potent, nonimmunologic bactericidal effects that reduce the density of bacteria in the upper part of the gastrointestinal tract; this density is further reduced by the very active peristalsis of the small intestine. Mucus produced by gastric mucous cells and intestinal goblet cells, and microbicidal peptides produced by enterocytes, goblet cells, and ileal Paneth cells, play key and complementary roles to contain bacteria (and other microbes) within the lumen and reduce their contact with epithelial cells. Thus, in healthy individuals, most resident bacteria are contained within a central and relatively fluid mucus layer that is largely separated from the apical cell surface of the polarized epithelium. The mucus layer varies significantly, however, in depth and composition along the length of the gastrointestinal tract, which emphasizes the importance of the redundant pathways of microbial containment that exist within mucosal tissues. In the colon, where the bacterial load increases massively, goblet cells are extremely dense and secrete an additional compact sterile layer of mucus that limits bacterial contact with the epithelial surface. In mice lacking Muc2, a major component of the colonic mucus, resident bacteria come into close contact with the epithelial surface and induce spontaneous colonic inflammation, attesting to the protective function of mucus. In addition, the physicochemical gradient of pH, oxygen, and bile salts not only limits the number of bacteria but also selects for bacteria that can more or less withstand these factors, thereby effectively leading to a shift in species diversity along the gastrointestinal tract. Production of microbicidal peptides by enterocytes and Paneth cells is enhanced on stimulation of their pattern recognition receptors (PRRs) by bacteria-derived signals (see Chapter 18 and later in this chapter).

25.1 Gastrointestinal epithelium acts as a sensor of luminal microbes and functions as an innate immune barrier

Similarly to professional immune cells, epithelial cells express a vast array of PRRs including membrane-associated toll-like receptors (TLRs), intracellular nucleotide-binding oligomerization domain-containing protein (NOD)-like receptors (NLRs), and retinoic acid-inducible gene-related receptors (RIG) that recognize common structures derived from microorganisms (microbe-associated molecular patterns [MAMPs]). Six TLRs (TLR1–5 and TLR9) are able to recognize bacteria-derived motifs although their exact distribution along the intestine and within epithelial cells *in situ* is not fully delineated. Among the four intracellular PRRs known to interact with bacterial motifs, two of them, NOD1/CARD4 and NOD2/CARD15, have been described in epithelial cells. These two molecules act as cytosolic sensors for peptidoglycan-derived peptides derived from the bacterial cell walls of both gram-positive and gram-negative bacteria. NOD2 can also sense mycobacteria and viruses. NOD1 is constitutively expressed in most cell types including epithelial cells; NOD2 is primarily present in hematopoietic cells but is also expressed by Paneth cells and may be induced in enterocytes during inflammation. In addition, recognition of endogenous danger-associated molecular patterns (DAMPs), such as extracellular adenosine triphosphate (ATP) or extracellular actin, may be used to alert the host of breaches of the integrity of the intestinal monolayer.

Interactions of epithelial TLR and NLR with MAMPs trigger two intracellular cascades important for host defense against bacteria: the canonical NF-κB and the mitogen-activated protein (MAP) kinase cascades. These synergize and activate several pathways that are useful to reinforce the epithelial antimicrobial barrier and maintain intestinal homeostasis. MAMP and PRR interactions in epithelial cells induce the synthesis of microbicidal peptides, such as β-defensins by enterocytes and RegIII proteins by Paneth cells. They also stimulate the production of chemokines and cytokines that recruit and/ or activate immune cells of hematopoietic origin along the basolateral surface of the epithelium which, in turn, complement or sustain epithelial barrier functions. TLR activation of epithelial cells can also reinforce tight junctions (an effect described for TLR2) and promote humoral protection of epithelial cells by stimulating the synthesis of the polymeric immunoglobulin receptor (pIgR) and that of B-cell activation factor (BAFF) and a proliferation-inducing ligand (APRIL), two cytokines that stimulate IgA production by B cells potentially independently of T cells ("T-cell independent B-cell switching"). Finally, NLR and TLR stimulation can promote the production of reactive oxygen species. These properties underscore the role of epithelial PRRs in host protection against either resident or pathogenic bacteria. How resident bacteria and pathogens can respectively interact with epithelial PRRs and downstream intracellular cascades is discussed later in this chapter.

25.2 Epithelium is regulated gatekeeper for commensal bacteria

Despite being an efficient and inducible physicochemical barrier, the intestinal epithelium is not entirely impermeable even to commensal bacteria. Thus, small numbers of resident bacteria can be detected in mesenteric lymph nodes (MLNs) of immune-competent mice. The best-established mechanism of bacterial entry involves transcytosis across M cells in the specialized epithelium of Peyer's patches (PPs) or of isolated follicles, both structures that predominate in the lower part of the intestine and notably in the ileum (Figure 25.1). In the follicle-associated epithelium (FAE), reduced numbers of goblet cells and decreased expression of defensins and pIgR facilitate bacterial contact with the apical surface of the epithelium. Moreover, the lack of a brush border and the intense endocytic activity of M cells allow for rapid transcytosis of intact bacteria and their delivery to underlying dendritic cells and macrophages (see Chapter 15). M cells may also express specific receptors. Thus, a microarray approach of M cell-specific molecules identified glycoprotein 2 as a receptor for FimH, a component of type I pili expressed by many commensal and pathogenic *E. coli* strains (see later). Glycoprotein 2 binds FimH in a mannose-dependent fashion and seems necessary for the transcytosis of type I–piliated commensal *E. coli* and pathogenic *Salmonella* strains. Interestingly, expression of glycoprotein 2 by FAE and of FimH by bacteria seems necessary to observe efficient *Salmonella*-specific responses in PPs and MLNs. Thus, under physiological conditions, M cell sampling of commensal microbiota is important for the maintenance of tolerance pathways and induction of IgA secretion on the one hand and the initiation of anti-infective immune responses to invasive pathogens on the other hand.

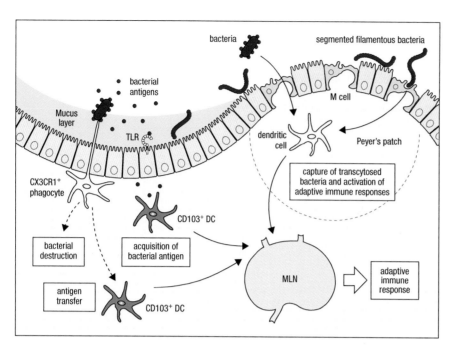

Figure 25.1 Interactions of resident bacteria with host epithelial cells. Most commensal bacteria reside in the mucus and have little direct contact with epithelial cells. One exception is segmented filamentous bacteria, which can strongly adhere to epithelial cells in ileum and Peyer's patches (PPs). Small numbers of resident bacteria enter the host by transcytosis via PP M cells and initiate adaptive immune responses. A second potential pathway involves their capture by lamina propria CX3CR1+ phagocytes that extend dendrites in between epithelial cells. Bacteria-derived products released in the lumen, such as lipopolysaccharide and peptidoglycan, can interact with TLR on the epithelial cells. Some can be translocated across the epithelium and reach the lamina propria immune cells and bloodstream. In healthy individuals, this passage is very limited and its mechanism is not fully elucidated. It likely involves a transcellular pathway because tight junctions are impermeable to molecules with a molecular weight over 600 daltons. DC, dendritic cell. (Reprinted from Cerf-Bensussan, N., and Gaboriau-Routhiau, V. *Nat. Immunol.* 2010, 10:735–744. With permission from Macmillan Publishers Ltd.)

On the basis of confocal and two-photon microscopy studies, a second mechanism of bacterial transepithelial passage has been described. This involves subsets of lamina propria dendritic cells (DCs) and potentially macrophages. Several studies, mainly performed in *in vivo* mouse models and *in vitro* cellular models with human epithelial cells, have shown that CX3CR1+ cells can extend protrusions between the tight junctions of epithelial cells and capture commensal and pathogenic bacteria within the intestinal lumen. This process depends on the synthesis of CX3CR1 ligand by the epithelium and is dependent on MyD88-associated signaling. In contrast to CD103+CD11c+ DCs, which mainly confer regulatory function on naive T cells within MLNs, CX3CR1^hi cells do not migrate to MLNs, are poor antigen-presenting cells, and might be closer to macrophages than to DCs. The role of CX3CR1+ cells and of their transepithelial dendrites may be to serve multiple purposes. They could perhaps pass captured bacteria to lamina propria CD103+ DCs, which subsequently migrate to MLNs and present bacteria-derived antigens to naive T and B cells and buttress regulatory pathways to commensal microbiota. Alternatively, CX3CR1+ cells may serve as a first line of defense by phagocytosing and killing bacteria adjacent to the epithelial surface or in the lamina propria. Accordingly, it has been demonstrated that CX3CR1+ cells can migrate into the lumen in response to a pathogenic *Salmonella* strain but not a nonpathogenic strain, via a MyD88/TLR5-dependent mechanism, suggesting a role in cell-mediated immune exclusion.

In an immune-competent host, the entry of small quantities of resident bacteria typically has no deleterious consequences, and as such bacteria can be readily eliminated by phagocytes present in the intestinal mucosa and associated lymphoid tissues. Accordingly, such translocated commensal microbiota cannot be detected beyond the MLNs, and notably cannot be cultured from the spleen. Such steady-state, low-grade levels of bacterial translocation may be viewed then as a component of the pathways that generate tolerance to the commensal microbiota as well as the maintenance of commensalism. This physiologic entry of resident bacteria may be mandatory for the postnatal maturation of the gut-associated lymphoid system and stimulation of IgA-secreting plasma cell and activated T-cell migration into the intestinal mucosa, where they reinforce the epithelial barrier and the composition of the intestinal microbiota through the pIgR-mediated transcytosis of IgA and thereby help maintain local homeostasis.

One species of bacteria in mice, segmented filamentous bacterium (SFB), appears to play a special role in the postnatal maturation of innate and adaptive host responses. This unusual symbiont, which only recently was cultured *in vitro* for the first time, is related to gram-positive clostridia. SFB has been identified in many species from arthropods to mammals (most notably rodents and some evidence exists for their presence in humans) on the basis of its unusual morphology and/or by its 16S rRNA composition. In contrast to most resident species, which remain entrapped in mucus within the intestinal lumen, SFB can attach to the surface of intestinal epithelium within the ileum and of PPs via an as yet unknown receptor. This behavior mimics that of a pathogen (see later), although SFB is not known to cause pathology due to its intimate attachment to the host. SFB settles in the mouse intestine at the time of weaning and induces strong innate and adaptive immune responses, including the production of RegIII, defensins, and the activation and migration of IgA plasma cells, CD8$^+$ intraepithelial lymphocytes, and CD4$^+$ lamina propria T cells that have been immune-mediated to either a T_H1, T_H17, T_H2, or regulatory function. SFB seems indispensable for the induction of intestinal T_H17-secreting CD4$^+$ T cells. The mechanisms underlying its strong stimulatory effect are not yet elucidated, but recent work has shown an absolute requirement for attachment. In addition, actin rearrangements below the point of contact as well as epithelial production of reactive oxygen species and serum amyloid A protein, which can activate interleukin-23 (IL-23) production by dendritic cells, are involved in the SFB-mediated T_H17 induction. In immune-competent hosts, colonization by SFB reinforces the mucosal barrier and protects against enteropathogens and has been a very useful model for understanding host and commensal microbial interactions.

Epithelium: A gateway for enteropathogens

In order for enteropathogens and other systemic pathogens to gain access to the host, they must first invade or traverse the intestinal (or mucosal) epithelium. For this purpose, they have evolved a wide variety of strategies based on modifications of their genome, notably through the loss and gain of mobile genetic elements and horizontal gene transfer. Bacterial pathogenicity thus largely depends on large clusters of virulence genes called *pathogenicity islands* that are either found on plasmids or integrated into the chromosome and are usually flanked by mobile genetic elements. Some virulence traits and particularly toxin-encoding genes are carried by prophages. Although each pathogen has developed distinctive ways to colonize the intestine, intestinal pathovars share many virulence strategies (Table 25.1). They use fimbriae and/or adhesins to harness host membrane molecules as receptors and promote their adherence to epithelial cells and/or their internalization. As for resident bacteria, the initial steps of adhesion and entrance preferentially take place in PPs across M cells. Many pathogens also use needle-shaped protein assemblages, called type III, IV, or VI secretion systems, to inject a wide spectrum of specialized proteins, called effectors, into the cytosol of host cells. These bacterial effectors can sequentially target multiple host cellular pathways in order to promote bacterial adhesion and/or entry into epithelial cells, evade phagocytic degradation and immune cell responses, eliminate resident microbiota from their intestinal niches, and ultimately facilitate their replication and spread.

Adherence to the epithelium by a pathogen is necessary to avoid mechanical clearance and is the first step of colonization by enteropathogens. Bacterial pathogens exhibit a large variety of cell surface adhesins, including fimbriae (a term used to define pili with a function in adhesion, but not in conjugation) and afimbrial adhesins, some of which participate in a subsequent step of internalization in epithelial cells. We now focus on these various types of pathogenic mechanisms.

TABLE 25.1   MAIN HUMAN ENTEROPATHOGENS

Enteropathogens	Site of colonization	Clinical expression	Virulence genes	Intestinal histology	Adhesion to epithelial cells		Replication and invasion	Toxins	Secretion system
					Pili/fimbriae	Bacterial adhesin/receptor			
Gram-negative proteobacteria									
Enteropathogenic *E. coli* (EPEC)[a]	Small intestine	Diarrhea	35 kb LEE pathogenicity islands EAF plasmid	Attaching/effacing epithelial lesions forming pedestals	Bundle-forming pili	Intimin/bacterium encoded Tir	No replication at epithelial cell surface		T3SS
Enterohemorrhagic *E. coli* (EHEC) Common serotype: O157:H7	Ileum, colon	Bloody diarrhea Hemolytic uremic syndrome	92 kb LEE pathogenicity islands Shiga toxin encoding phage pO157 plasmid	Attaching/effacing epithelial lesions forming pedestals	Type IV pili?	Intimin/bacterium encoded Tir	No replication at epithelial cell surface	Shiga toxin	T3SS T6SS
Enterotoxigenic *E. coli* (ETEC)	Small intestine	Diarrhea			Multiple fimbriae, including type IV pili	Autotransporter TibA/epithelial cell receptor Tia adhesin/host heparan sulfate proteoglycan	Extracellular replication Invasion possible in epithelial cells	Heat-stable toxins (Sta, STb, AEST1) Heat-labile toxins (AB_5 toxin) Pore-forming cytotoxin	T3SS T6SS
Enteroinvasive *E. coli* (EIEC) *Shigella*	Colon	Bloody diarrhea (dysentery)	220 kb plasmid	Epithelial destruction/crypt abscesses	No dedicated structure	T3SS insertion into epithelial cell membrane IpaB effector/host CD44	M cells and PP macrophages Dissemination via laterobasal membranes of epithelial cells		T3SS T6SS (*S. sonnei*)
Enteroaggregative *E. coli* (EAEC)	Small intestine, colon	Diarrhea	100 kb plasmid	Aggregative adhesion/biofilms Exfoliating lesions (some isolates)	Dr-related aggregative fimbriae			Exfoliating Plasmid-encoded toxin	
Diffusely adhesive *E. coli*	Small intestine	Diarrhea	Undetermined	Diffuse adherence	Dr fimbriae Afa adhesins	Afa-Dr/host CD55 and CEACAM6	Extracellular replication Replication possible in epithelial cells	Secreted autotransporter toxin	
Adherent-invasive *E. coli* (AIEC) Prototype strain: LF82	Ileum	Associated with Crohn's disease (ileal lesions)	Undetermined		Type I pili	Type I pili (mannose residues)/CEACAM6 (induced by IFN-γ and TNF-α)	Epithelial cells Macrophages		

Intestinal E. coli

(Continued)

TABLE 25.1 (CONTINUED) MAIN HUMAN ENTEROPATHOGENS

Enteropathogens	Site of colonization	Clinical expression	Virulence genes	Intestinal histology	Adhesion to epithelial cells		Replication and invasion	Toxins	Secretion system
					Pili/fimbriae	Bacterial adhesin/receptor			
Extraintestinal E. coli									
Uropathogenic E. coli	Bladder, kidney	Urinary infections	Multiple pathogenicity islands (depending on strains)	None	Type I pili	FimH/host β_1, α_3 integrins	Translocation into bloodstream		
Neonatal meningitis E. coli (NMEC)	Brain	Neonatal meningitis		None	Type I pili	FimH/host CD48	Translocation into bloodstream but extracellular replication	Cytotoxic necrotizing factor	
Vibrio cholerae	Small intestine	Watery diarrhea	41.2 kb phage-derived pathogenicity islands Cholera toxin (CT)-encoding prophage		Type IV pili			Cholera toxin (AB_5)	TSS6
Yersinia enterocolitica Yersinia pseudotuberculosis	Ileum	Diarrhea Ileitis Mesenteric lymphadenitis Septicemia Autoimmunity	Chromosomal pathogenicity islands 70 kb virulence plasmid (pYV)	Peyer's patch abscess		Invasin/host β_1 integrins YadA/host extracellular matrix proteins	M cells and PP macrophages followed by extracellular replication in PP and MLN		T3SS
Salmonella Typhoid: Salmonella Typhi/Paratyphi Non-typhoid: Salmonella Enteritidis, Salmonella Typhimurium	Small intestine	Diarrhea, enteric fever Gastroenteritis	7–40 kb chromosomal pathogenicity islands (SPI-1 to 5, SPI-7) Salmonella Typhimurium: 90 kb virulence plasmid ± prophages	Variable epithelial destruction	Type IV pili	T3SS insertion into epithelial cell membrane	M cells, macrophages, epithelial cells		T3SS SPI- for intestinal colonization T3SS SPI- for survival in macrophages T6SS
Helicobacter pylori	Stomach	Gastritis Gastric carcinomas, lymphomas	Cag chromosomal pathogenicity islands	Chronic gastric inflammation		BabA/host Lewis[b] (MuC5AC) SabA/host Sialyl Lewis[x] (FUT4) CagL/host α_5 β_1	Extracellular replication	H. pylori vacuolating cytotoxin	T4SS
Gram-positive bacteria									
Listeria monocytogenes	Brain/placenta	Abortion Meningoencephalitis in immunocompromised hosts	Chromosomal LIP1-1/LIP1/2	None		Internalin A/host E-cadherin	Epithelial cells, macrophages	Listeriolysin 0	

Abbreviations: LEE, locus of enterocyte effacement; EAF, EPEC adherence factor; MLN, mesenteric lymph nodes; PP, Peyer's patches; T3SS, type III secretion system; T4SS, type IV secretion system.
[a] Related strains: rabbit diarrheagenic *E. coli* (RDEC), murine pathogen *Citrobacter rodentium*, and recently identified human pathogen *Escherichia albertii* (formerly known as *Hafnia alvei*).

25.3 Some bacteria express fimbriae that allow for adhesion to epithelium

Fimbriae are a type of pili with adhesive properties and are hairlike organelles that are exported by complex machineries to the bacterial surface. They comprise a scaffold-like rod anchored to the bacterial outer membrane and a bacterial adherence factor, or adhesin, located at the external tip. Type I pili expressed by many commensal and pathogenic *E. coli* strains possess FimH adhesins, the different variants of which bind to host mannose-containing glycoproteins. They notably bind to glycoprotein 2 expressed by M cells (see earlier). FimH from uropathogenic *E. coli* can also bind β_1 and α_3 integrins and thereby promote bacterial internalization, a process, however, only described in urinary epithelial cells. Type IV pili are another category of polymeric adhesive surface structures expressed by many gram-negative bacteria including pathogens such as *Salmonella* spp., enteropathogenic *E. coli* (EPEC), enterohemorrhagic *E. coli* (EHEC), and *Vibrio cholerae*. In the latter species, type IV pili can aggregate laterally and form bundles. These bundles are required for the initial adhesion of EPEC and EHEC to epithelial brush borders as a prelude to the development of the typical "attaching/ effacing" lesions. A striking feature of type IV pili is their ability to retract through the bacterial wall while the pilus tip remains firmly attached to the surface of target cells. Another heterogeneous family of adherent structures is the Afa/Dr adhesin family identified in uropathogenic *E. coli* and in diffusely adherent *E. coli* strains, some of which were shown to assemble into fimbriae. Afa/Dr adhesion fimbriae bind brush border-associated complement decay-accelerating factor (also called CD55).

Still others can interact with carcinoembryonic antigen-related adhesion molecules (CEACAM), a family of receptors that may be preferentially associated with lipid rafts. These interactions trigger intracellular signals that can disrupt the brush border and/or promote bacterial internalization. An illustrative example is an *E. coli* variant that was originally isolated from human subjects with inflammatory bowel disease (IBD)—adherent-invasive *E. coli* (AIEC).

25.4 Adherent-invasive *Escherichia coli* utilize fimbriae to adhere to CEACAM6 and invade epithelia

It had long been known that patients with IBD possessed increased circulating IgG antibodies with specificities for *E. coli* and exhibited increased numbers of mucosa-associated *E. coli* with invasive properties or the presence of intramucosal *E. coli*. These pathogenic *E. coli*, compared with commensal *E. coli*, have acquired specific virulence factors that increase their ability to adapt to new niches and allow them to cause disease. *E. coli* strains associated with the intestinal mucosa from IBD, and in particular Crohn's disease, are highly adherent to intestinal epithelial cells and are also invasive. On the basis of the pathogenic traits of Crohn's disease–associated *E. coli*, a new potentially pathogenic group of *E. coli* was designated AIEC. The interaction between AIEC and intestinal epithelial cultured cells induces inflammatory responses such as upregulated expression of IL-8 and CCL20, leading to transmigration of polymorphonuclear leukocytes and dendritic cells in coculture models. AIEC can also disrupt the integrity of the polarized epithelial cell monolayer, allowing bacteria to breach the intestinal barrier and to penetrate into the gut mucosa. AIEC are also able to survive and to replicate extensively within large vacuoles in macrophages. In contrast to many pathogens that escape from the normal endocytic pathway, Crohn's disease–associated invasive *E. coli* are taken up by macrophages within phagosomes, which mature without diverting from the classical endocytic pathway, and which share features with phagolysosomes. To survive and replicate in the harsh environment encountered inside these compartments, including acid pH and the proteolytic activity of cathepsin D, such bacteria have developed adaptation mechanisms in which acidity constitutes a key signal for activating the expression of virulence genes. Macrophages infected with AIEC release large amounts of tumor necrosis factor-α (TNF-α), and in an *in vitro* model of human granuloma formation, AIEC-infected macrophages aggregate, fuse to form multinucleated giant cells, and subsequently recruit lymphocytes. These characteristics of AIEC have many similarities to pathogenic *Shigella* spp. as summarized in Figure 25.2.

In clinical samples, AIEC can be isolated from ileal specimens in nearly 40% of patients with Crohn's disease in comparison with less than 10% of controls. It is likely that the higher prevalence of AIEC bacteria in patients with Crohn's disease might arise from a variety of factors. These include the abnormal expression of host factor(s) involved in the epithelial cell gut colonization, an inability of the intestinal mucosa to control AIEC infection, for example, because of defects in Paneth cell function and subsequent decreased secretion of antimicrobial peptides, and a loss of control of intracellular AIEC replication related to autophagy deficiencies (see Chapter 34).

Among the host factors that facilitate AIEC colonization is the abnormal ileal expression of CEACAM6, which has been observed in patients with ileal Crohn's disease. In contrast to CEACAM1, which is constitutively expressed on the apical cell surface of intestinal epithelial cells, CEACAM6 is an induced molecule. AIEC adhere to the brush border of primary ileal enterocytes isolated from patients with Crohn's disease, but not those from individuals without IBD, due to the abnormal expression of CEACAM6. Most AIEC strains associated with Crohn's disease ileal mucosa express type I pili variants that increase the interaction between AIEC and ileal epithelial cells. Increased expression of CEACAM6 can result from interferon-γ (IFN-γ) or TNF-α stimulation and also from infection with AIEC bacteria, indicating that AIEC can promote their own colonization in patients with Crohn's disease. The presence of AIEC bacteria and their ability to induce the secretion of pro-inflammatory cytokines by infected macrophages could lead to an amplification loop of colonization and inflammation. This has been confirmed *in vivo* with the LF82 strain of AIEC, but not nonpathogenic *E. coli* K-12, wherein LF82 persists in the gut of mice that transgenically express human CEACAMs and induces severe colitis.

Another host factor that participates in AIEC colonization of the ileal mucosa is the increased expression of the endoplasmic reticulum (ER) stress response protein Gp96 on the apical surface of ileal epithelial cells in patients with Crohn's disease. This glycoprotein is able to bind to *E. coli* outer membrane protein A present on the surface of AIEC outer membrane vesicles and serve as a means for the AIEC bacteria to deliver bacterial effectors into host cells and

Figure 25.2 Adherent-invasive *Escherichia coli* (AIEC) exhibit a similar mechanism of pathogenesis as *Shigella* infection. In the first panel, *S. flexneri* invades the epithelium from the intestinal lumen through M cells and is further phagocytosed by resident macrophages. *Shigella* escapes the phagosome by rapidly disrupting the phagosomal membrane, thereby avoiding phagosome–lysosome fusion and degradation. It induces macrophage cell death by triggering both caspase 1–dependent (pyroptosis) and caspase 1–independent cell death pathways. The dying macrophages release the pro-inflammatory cytokines IL-1β and IL-18, critical mediators of the acute and massive inflammatory responses. Together with IL-8 secreted from the invaded epithelial cells, they induce transmigration of polymorphonuclear leukocytes (PMNs) to the site of infection. Infiltrating PMNs destroy the integrity of the epithelial lining, thus enabling more luminal bacteria to reach the submucosa. In the second panel, similar to *Shigella*, AIEC have the ability to target M cells via binding to CEACAM6 to cross the intestinal barrier, invade intestinal epithelial cells through their basolateral cell surface, and induce the secretion of IL-8. The major difference between *Shigella flexneri* and AIEC is the ability of AIEC to replicate within macrophages without inducing cell death and to continuously activate infected macrophages to secrete large amounts of TNF-α.

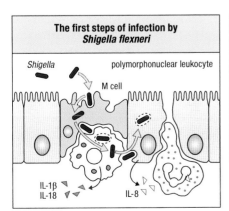

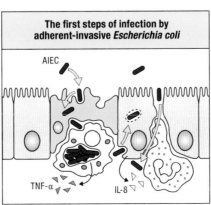

promote bacterial invasion. As such, AIEC bacteria can take advantage of the ER stress response occurring within the intestinal epithelium of patients with IBD. Because microbiota have an important role in the establishment of the ER stress response, AIEC, as opportunistic pathogens, might take advantage of such changes in the host innate immune response.

AIEC have another interesting effect on the intestinal epithelium. Increased numbers of membranous, microfold cells (M cells) have been observed in Nod2-deficient mice. AIEC express long polar fimbriae that allow the bacteria to increase their interaction with mouse and human Peyer's patches and their translocation across M cell monolayers. Thus, in the presence of NOD2 mutations as observed in Crohn's disease, higher numbers of AIEC bacteria are able to interact with Peyer's patches. This observation is particularly relevant because recurrent ileal Crohn's disease originates from small erosions in the FAE that lies over PPs.

Finally, the susceptibility of patients with Crohn's disease to epithelial infection with AIEC is dependent on a variety of genetic risk factors as discussed in Chapter 34. Moreover, AIEC bacteria can take advantage of defects in autophagy to replicate within host epithelial cells. Interestingly, the behavior of related *E. coli* strains, including nonpathogenic, environmental, commensal, or pathogenic enterotoxigenic *E. coli*, EPEC, diffusely adherent *E. coli*, and enteroinvasive *E. coli* bacteria with regard to autophagy, is different from that of AIEC bacteria. Defects in autophagy mainly affect the ability of the host to handle AIEC but not other pathogenic *E. coli* variants. This has been specifically shown for alterations due to hypomorphic function of both ATG16L1 and IRGM in epithelial cells, autophagy pathway proteins that are genetic risk factors for Crohn's disease. Hypomorphic function of these leads to increased numbers of AIEC LF82 intracellular bacteria, indicating that autophagy has a key role in controlling intracellular AIEC replication.

25.5 Bacteria use a large number of afimbrial surface structures to adhere to and/or invade host cells

Afimbrial adhesion occurs by many different mechanisms. Autotransporters, for example, possess C-terminal domains that form pores in the outer membrane, allowing the exposition of a passenger domain that can participate in adhesion to host cells. Thus, AIDA-A is expressed by diffusely adherent *E. coli* strains and is responsible for their diffusely adhesive property to epithelial cells. *Helicobacter pylori* BabA binds to the Lewis[b] blood group antigen (also called MuC5AC) in the gastric mucosa. *Yersinia* YadA mediates adhesion to collagen, laminin, and fibronectin.

A distinct original adhesion pathway has been developed by EPEC and EHEC strains that induce the characteristic "attaching and effacing" epithelial lesions. After intimate attachment to intestinal epithelial cells via their adhesins and bundle-forming pili, these bacteria use their type III secretion system (T3SS) to rapidly translocate a protein called Tir (for translocated intimin receptor) into the cytosol of host cells (see later for T3SS description). Tir is displayed at the host-cell surface and acts as a receptor for the bacterial outer membrane protein intimin, resulting in Tir clustering and initiation of the cytoskeleton rearrangements that lead to attaching and effacing lesions (Figure 25.3).

Other pathogens have evolved molecules that can harness host transmembrane cell-adhesion receptors and trigger intracellular cascades that lead to epithelial cell invasion. This "zipper" strategy is used by *Yersinia* spp. and *Listeria* spp. to invade PPs (Figure 25.4). Depending on temperature, pH, and growth phase, two strains of enteropathogenic *Yersinia*, *Y. enterocolitica* and *Y. pseudotuberculosis*, express invasin (Inv), a 101 kDa outer membrane protein encoded by a 70 kb virulence plasmid called pYV. Bacterial

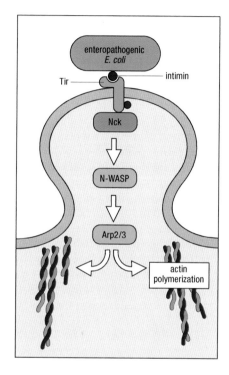

Figure 25.3 Tir/intimin-mediated interaction of enteropathogenic *Escherichia coli* (EPEC) with epithelial cells. Adhesion to epithelial cells is initiated by EPEC bundle-forming pili that bind host *N*-acetyl lactosamine-containing host receptors. Intimate attachment is then triggered by injection of Tir protein into the epithelial cell via EPEC T3SS apparatus. Tir is displayed at the epithelial cell surface and acts as a receptor for bacterial intimin. Tir/intimin interaction induces clustering of Tir, which is phosphorylated by various host kinases. Phosphorylated Tir recruits host adaptor Nck, which activates N-WASP and the actin nucleator Arp2/3 complex, resulting in actin polymerization and pedestal formation at the site of attachment. While stabilizing bacteria and host cell interactions, pedestal formation promotes T3SS-mediated injection into epithelial cells of additional effector proteins able to subvert host cell pathways. (Adapted from Pizarro-Cerda, J., and Cossart, P. *Cell*. 2006, 124:715–727. With permission from Nature Publishing Group.)

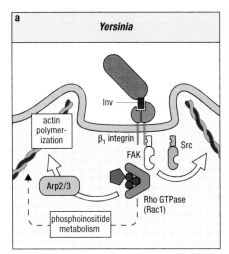

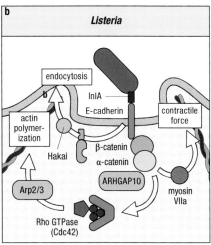

Figure 25.4 Invasion of epithelial cells by *Yersinia* and *Listeria* via hijacking of host adhesion receptors. (**a**) *Yersinia* binds to β₁ integrin at the surface of PP M cells resulting in the activation of Rho GTPases, recruitment of the Arp2/3 complex, and actin polymerization. This process leads to the formation of a phagocytic vacuole, which engulfs the bacterium. β₁ integrin–associated host kinases such as FAK or Src kinases participate in the process. (**b**) *Listeria* Internalin A (InlA) interacts with host E-cadherin and activates the "adherens junction machinery" leading to the successive recruitment of β-catenin, the Rho GTPase-activating protein ARHGAP10, α-catenin, and myosin VIIa. ARHGAP10 promotes the activation of Cdc42 and of the Arp2/3 complex, resulting in actin polymerization and cup formation around the bacterium. Myosin VIIa may generate the contractile forces required for internalization of engulfed bacteria. InlA-mediated *Listeria* internalization also activates the Src tyrosine kinase, inducing the recruitment of the ubiquitin ligase Hakai and the ubiquitination of E-cadherin. Clathrin-mediated endocytosis of ubiquitinylated E-cadherin further promotes bacterial internalization. (Adapted from Marlovits, T., and Stebbins, C. *Curr. Opin. Microbiol.* 2010, 13:47–52. With permission from Elsevier Ltd.)

Inv binds β₁ integrins on the luminal surface of M cells. Inv binding results in the clustering of β₁ integrins and subsequent activation of a signaling cascade that induces cytoskeleton rearrangements and bacterial internalization. Studies using *inv* mutants suggest that Inv is critical for dissemination from the intestinal lumen (even in animals lacking PPs) and for abscess formation in mesenteric lymph nodes but dispensable for overall virulence. Besides *Yersinia*, other bacteria may target host β₁ integrins to adhere to the mucosa. For example, *H. pylori* uses its type IV secretion system (T4SS) to target β₁ integrins and subsequently translocate its effector protein CagA, a protein with carcinogenic properties. Thus, an adhesin called CagL is present at the T4SS pilus surface. This adhesin was shown to bridge and activate the α₅β₁ integrin on the basolateral membrane of gastric epithelial cells.

Listeria monocytogenes is a food-borne gram-positive bacterium that makes use of two surface proteins, Internalin A and B, to engage in a species-specific manner host adhesion molecules E-cadherin and hepatocyte growth factor receptor Met, respectively, and to induce its internalization. After entry within epithelial cells, *L. monocytogenes* can spread from cell to cell and disseminate to its target organs, the blood-brain and placental barriers in humans. Analysis of *inl* mutants indicates that only Internalin A is critical for crossing the intestinal epithelial barrier. Internalin A and B both participate in a minor way to bacterial crossing of the placenta, while another recently described internalin, InlP, is critical. Internalin A is covalently anchored to the bacterial surface and possesses a leucine-rich repeat that interacts with the first ectodomain of human E-cadherin, a transmembrane glycoprotein that mediates homophilic interactions below tight junctions of polarized epithelial cells and is inaccessible to luminal bacteria. However, *L. monocytogenes* is able to use E-cadherin that is luminally accessible around mucus-expelling goblet cells to hijack the endogenous transcytosis pathway in goblet cells in order to reach the basolateral side of the intestinal epithelial monolayer. The Internalin A and E-cadherin interaction and molecular details of the signaling pathways involved in *L. monocytogenes* epithelial cell entry have been well characterized (see Figure 25.4).

Beyond adhesion: Bacterial injection systems

Interestingly, some pathogens use adhesion to interact with the cellular machinery directly by injecting effectors that regulate actin skeleton dynamics. *Shigella* and *Salmonella* use this "trigger" mechanism of bacterial uptake into nonphagocytic cells. For *Shigella*, contact with epithelial cells depends on the interaction of the T3SS effector OspB with the host CD44 receptor and of the adhesin IcsA with a yet unidentified receptor. IcsA has dual functions:

when *Shigella* is extracellular, it is upregulated in response to the bile salt deoxycholate and mediates adhesion to epithelial cells; during cytosolic growth of *Shigella*, IcsA functions to mediate actin-based motility. Conversely, *Salmonella enterica* uses an array of pili, fimbriae or autotransporters, or large secreted repetitive adhesins SiiE and BapA, depending on the serotype to adhere to epithelial cells. Adhesion for *Shigella* and *Salmonella* serves to facilitate the contact of the molecular syringe, the T3SS, which injects effector proteins into host cells causing massive cytoskeletal changes that result in the formation of a macropinocytic pocket, loosely bound to the bacterial body to initiate the internalization of the bacteria.

25.6 At least six different (type I–VI) bacterial "injection" systems have been defined

Among the six secretion systems currently described in bacteria, three systems (type III, IV, and VI) allow penetration of host-cell membranes and intracytosol injection of effectors. The best characterized are type III systems (T3SS), which are evolutionarily related to the flagellar system and are well conserved across a variety of taxa (in contrast to the effector proteins that are generally specific to each pathogen and govern their adaptation to their host). T3SS are formed of 20–30 proteins, the assembly of which is tightly regulated. Following assembly of a basal body into the inner and outer bacterial membranes, early substrates including the needle subunit and the needle-length control protein are targeted across this apparatus and assemble to form the extracellular needle (Figure 25.5). Tip proteins are then secreted and assembled at the tip of the needle, but further secretion remains blocked until host-cell contact. Upon host-bacterium contact, a second switch triggers

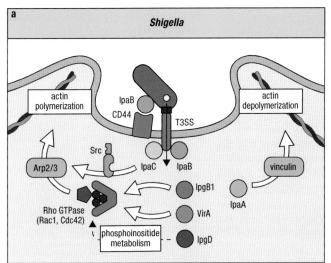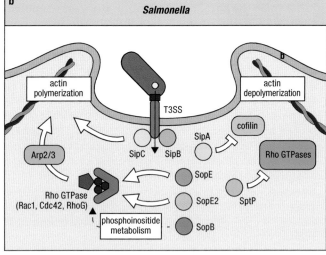

Figure 25.5 **T3SS-mediated invasion of epithelial cells by *Shigella* and *Salmonella*.** Upon contact with epithelial cells, *Shigella* and *Salmonella* T3SS are activated and deliver bacterial effectors via the IpaB/C and SipB/C pores, respectively, thereby triggering internalization. (**a**) Shigella key effector IpaC activates Src kinases at the site of bacterial contact and ultimately recruits Arp2/3, causing actin polymerization and ruffle formation. This process is favored by effectors IpgB1 and VirA through the activation of Rac1. IpgD activates the phosphoinositide pathway and membrane ruffling, while IpaA recruits vinculin and induces actin depolymerization, which allows closing of the phagocytic vacuole. IpaB (probably expressed at the tip of T3SS) can allow bacterial adhesion to the CD44 hyaluronic receptor at the basolateral membrane of the epithelial cell and further promote invasion. (**b**) Translocated Salmonella effectors SopE, SopE2 (both acting as guanine nucleotide exchange factors), and inositol phosphatase SopB activate Rho GTPases Rac1, Cdc42, and RhoG, resulting in actin polymerization, which is further promoted by SipC and SipA. SipC can nucleate actin while SipA antagonizes the function of cofilin, an actin depolymerizing factor. Subsequently, effectors such as SptP inactivate Rac1 and Cdc42, inducing actin depolymerization and closure of the phagocytic cup. (Adapted from Pizarro-Cerda, J., and Cossart, P. *Cell*. 2006, 124:715–727. With permission from Elsevier Ltd.)

the secretion of translocators that form a pore into the host-cell membrane. Bacterial effector proteins are then translocated into the host-cell cytosol. The mechanism activating this second switch has been deciphered for *Shigella flexneri*. In the anaerobic intestinal environment, *Shigella* is primed for invasion and expresses extended needles, which may help establish contact with the intestinal epithelium. These processes are tightly regulated. For example, a transcriptional regulator of anaerobic metabolism (called fumarate and nitrate reduction) represses effector secretion. This repressor is inactivated in the presence of oxygen. Because of the numerous capillaries underlying the epithelium, the luminal zone immediately adjacent to epithelial cells contains sufficient oxygen to inactivate fumarate and nitrate reduction. Thus, when *Shigella* comes close to the epithelial surface, the anaerobic block of secretion is reverted, allowing secretion of Ipa effectors into epithelial cells and host invasion. Whether this mechanism can be generalized to other T3SS-expressing enteropathogens remains to be defined, but the presence of fumarate and nitrate reduction boxes upstream of genes required for T3SS functions in *Yersinia* spp. and serovars of *Salmonella enteritidis* suggests that indeed this is likely to be the case.

Type IV secretion systems (T4SS) are large assemblies of variable lengths of at least 12 distinct proteins used by gram-positive and gram-negative bacteria for one-step transport not only of virulent proteins but also of DNA. In *H. pylori*, the T4SS are long and large needle-like pili that are encoded by the *c*ag pathogenicity islands and not only transport the CagA effector but also participate in adhesion to the gastric epithelium (see earlier).

Type VI secretion systems (T6SS) are expressed by many gram-negative bacteria and function in interbacterial competition and pathogenesis (Figure 25.6). T6SS are contractile secretion systems that resemble phage-tail complexes and are made up of multiprotein needle-like apparatuses that inject toxin or effector proteins into either eukaryotic or prokaryotic target cells. The T6SS is a contact-dependent weapon that bacteria use to kill off other bacterial competitors through the translocation of proteinaceous toxins. The mechanisms of toxicity for the target bacterium are diverse, and only bacteria that have the appropriate antitoxin immunity protein are resistant to killing. This interbacterial warfare is exploited by pathogens to cause disease. For example, for *Salmonella enterica* serovar *typhimurium*, the T6SS is activated by bile salts and functions to kill off specific commensal bacteria to establish infection in the gut (see Figure 25.6). Notably, the T6SS and T3SS systems of *S. typhimurium* are inversely regulated to first lower colonization resistance encountered in the gut before engagement of the T3SS to invade the eukaryotic host. In other pathogens such as *Vibrio cholera* and *Pseudomonas aeruginosa*, T6SS effectors are translocated into host cells to prevent phagocytosis and facilitate invasion, respectively. The T6SS are also involved in modulating the host innate immune system. In the fish pathogen *Edwardsiella tarda*, the T6SS effector EvpP was shown to suppress NLRP3-inflammasome activation through a Ca^{+2}-mediated suppression of c-Jun *N*-terminal kinase activity, which was important for intracellular replication and virulence. Conversely, the T6SS effector TecA of *Burkholderia cenocepacia*, an opportunistic pathogen of the cystic fibrosis lung, facilitates survival

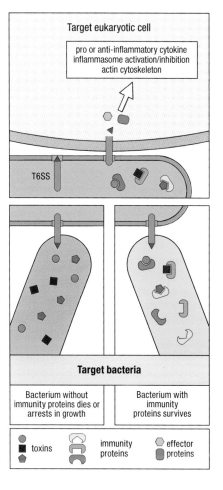

Figure 25.6 The T6SS functions in intrabacterial competition and pathogenesis. The T6SS is produced by many gram-negative bacteria, including commensals and pathogens. The secretion system is a contractile phage-related nanosystem that can infect toxins and effectors into either bacteria or eukaryotic cell. Toxins injected into bacteria can be neutralized if appropriate immunity proteins are present. In the absence of immunity, proteins toxins function in diverse ways to kill or arrest growth of the target bacterium. Injection of toxins or effectors into eukaryotic cells can affect processes such as the host innate immune response and its actin cytoskeleton.

of the pathogen in macrophages, but TecA ultimately activates the pyrin inflammasome leading to pathogen clearance. For *H. hepaticus*, a bacterium often present in the murine microbiota, T6SS might even promote symbiotic relationships via the release of as yet unknown effectors downregulating host pro-inflammatory responses.

25.7 Bacterial secretion systems inject specific bacterial effectors into epithelial cells to hijack the host cytoskeleton

The most commonly described cellular target of pathogens is the cytoskeleton. Each pathogen has its own repertoire of effectors and therefore of cellular targets. However, there are several common functions for such bacterial effector systems: to enable further firm adhesion, cell entry, and dissemination between cells. In this section, we discuss each of these.

Extracellular pathogens such as EPEC and EHEC induce the formation of actin-rich pedestal structures on the host epithelial surface where they can reside and grow. Via T3SS, they deliver the effector protein Tir, which embeds itself into the host membrane and binds the bacterial outer membrane protein intimin. Tir intracellular domain, either directly (EPEC) or indirectly via another bacterial effector (EHEC), stimulates an intracellular cascade that activates the host protein N-WASP (neural Wiskott–Aldrich syndrome protein). In turn, N-WASP recruits the seven-protein Arp2/3 complex that mediates actin polymerization and formation of the pedestal beneath the extracellular bacterium (see Figure 25.3).

Invasive pathogens such as *Salmonella* and *Shigella* use their T3SS to inject effectors that activate host Cdc42 and Rac1 GTPases, albeit via different intracellular relays. Activation of these G proteins leads to the successive recruitment of N-WASP or related proteins, and of the Arp2/3 complex that stimulates actin polymerization and thereby the formation of a macropinocytic pocket. Another set of bacterial effectors, acting either directly or indirectly, can then mediate actin depolymerization, closing the macropinocytic pocket into an intracellular vacuole containing the bacterium (see Figure 25.5).

To invade host cells, *Yersinia* and *Listeria* use a different strategy. Following their binding to host adhesion receptors, they hijack downstream cellular machinery. Thus, binding of *Yersinia* invasin to β_1 integrin activates the focal adhesion tyrosine kinase and triggers a complex cascade implicating a Rac1-Arp2/3 pathway but also phosphoinositide-3-kinase (PI3K). These events lead to formation and then closure of a phagocytic cup (see Figure 25.4). In the case of *Listeria*, engagement of E-cadherin by Internalin A initiates activation of the adherens junction machinery and induces the recruitment of β-catenin, Rho GAP protein ARHGAP10, α-catenins, and unconventional myosin VIIa to the site of entry. Internalization is further mediated by Rac- and Arp2/3-dependent actin polymerization. Internalin A and E-cadherin interaction can also activate the kinase Src. This kinase promotes the recruitment of the ubiquitin ligase Hakai, allowing clustering and clathrin- or caveolin-dependent internalization of ubiquitinylated E-cadherin bound, or not, to *Listeria* (see Figure 25.4).

Finally, following their internalization, intracellular pathogens such as *Listeria* and *Shigella* can escape from the phagocytic vacuole, replicate in the cytosol, and use actin polymerization as the force to propel the bacterium within the cytosol. When bacteria reach the plasma membrane, they form protrusions into neighboring cells, which lead to secondary vacuoles, whereby the bacterium is surrounded by a double plasma membrane. These vacuoles can be lysed and the cycle repeated, allowing cell-to-cell spread and tissue dissemination. *Listeria* uses a pore-forming toxin called listeriolysin O to escape from the vacuole; its surface protein ActA can then bind and activate the Arp2/3 complex. Similarly, the effector IpaB permits *Shigella* to escape from the phagocytic vacuole. *Shigella* then expresses an outer membrane protein IcsA (also called VirG) that recruits N-WASP, which subsequently binds and activates the Arp2/3 complex.

Mechanisms of bacterially induced epithelial cell dysfunction

Many enteric pathogens can inhibit intestinal absorption and/or stimulate secretion resulting in diarrhea, a mechanism thought to ensure their dissemination outside their hosts. Bacteria accomplish this by the intracellular injection of bacterial effectors as discussed earlier or via the intraluminal release of toxins that can bind epithelial cell-surface receptors. These various mechanisms are discussed later.

25.8 Bacteria secrete toxins that affect epithelial cell function into intestinal lumen

Bacteria secrete many types of toxins that are capable of affecting epithelial cell function. The heat-stable toxin produced by some strains of enterotoxigenic *E. coli* (ETEC), for example, binds a brush border guanyl cyclase, stimulating the production of cyclic GMP that activates the cystic fibrosis transmembrane conductance regulator (CFTR) and Cl^- secretion (Figure 25.7). A more complex mechanism is used by *Vibrio cholerae* toxin and by the homolog heat-labile enterotoxin secreted by ETEC strains. Via their B subunits, these AB_5 toxins interact with monoganglioside GM1 at the surface of epithelial cells, allowing internalization of the A subunit via lipid rafts. After a complex intracellular trafficking, the A subunit is released into the cytosol and ADP-ribosylates adenylate cyclase, resulting in increased cyclic AMP (cAMP) production. cAMP in turn stimulates CFTR phosphorylation and activation. cAMP can also decrease the activity of Na^+/H^+ exchangers and inhibit Na^+ reabsorption. This effect combined with the activation of CFTR results in enhanced NaCl at epithelial cell surfaces.

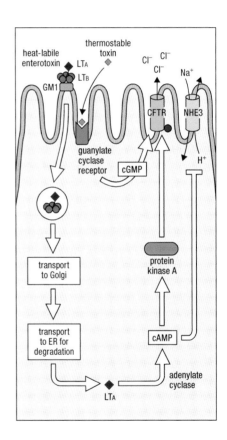

Figure 25.7 Alteration of epithelial transport by ETEC-derived toxins. Thermostable toxin binds to a guanylate cyclase receptor on the epithelial cell brush border, inducing production of cyclic GMP, which activates CFTR. Heat-labile AB_5 enterotoxin (LT) is a homolog of *Vibrio cholerae* toxin. Via its B subunits, heat-labile enterotoxin binds to epithelial cell surface GM1 ganglioside, inducing toxin internalization at lipid rafts. After retrograde shuttling through Golgi and endoplasmic reticulum (ER), the heat-labile enterotoxin A subunit is delivered into the cytosol where it activates adenylate cyclase and increases the level of cyclic AMP. cAMP activates protein kinase A, which in turn phosphorylates and activates CFTR, resulting in increased Cl^- secretion. cAMP can also inhibit Na^+/H^+ exchanger 3 (NHE3), resulting in decreased Na^+ absorption.

25.9 Bacteria use their secretion machinery to directly inject toxic effector molecules into epithelial cells to co-opt their function

A second mechanism to induce diarrhea through functional effects on the epithelium or effects on epithelial cell survival does not involve extracellular toxins but involves bacterial effectors that are injected into or secreted within host cells. Thus, EPEC strains use their T3SS to inject two effectors (EspG, EspG2) that induce tubulin degradation and disruption of the host microtubule network, and subsequently the dramatic internalization of the apical Cl^-/OH^- exchanger DRA (downregulated in adenoma).

In other cases, intestinal pathogens have devised many ways to manipulate host survival mechanisms. As an example, apoptosis induction is used by *Shigella* and *Salmonella* to promote their dissemination. Thus, following their translocation across M cells, both bacteria can be phagocytosed by underlying macrophages. Using their T3SS, the bacteria secrete the effector proteins IpaB and SipB, respectively, which both activate the cysteine protease caspase 1 from the phagosome into the cytosol. This enzyme induces the apoptotic death of infected macrophages and the maturation of pro-inflammatory IL-1β and IL-18. Bacteria escaping dying macrophages can invade the basolateral side of enterocytes with the aid of their T3SS and subsequently propagate between epithelial cells, a process thought to be enhanced by the inflammatory reaction initiated by IL-1β and IL-18. In contrast, once free in the enterocyte cytosol, *Shigella* releases several bacterial effectors that prevent cell death and cell sloughing, thereby protecting its replication niche to maintain infection. Thus, IpaB targets the anaphase inhibitor Mad2L2 and prevents intestinal epithelial cell renewal; OspE interacts with the integrin-linked kinase ILK and prevents epithelial cell detachment; and IpgD promotes cell survival via the activation of PI3K and Akt proteins. Beyond these examples, an ever-growing number of mechanisms have been unraveled through which pathogens and their effectors can manipulate host-cell survival.

Regulation of host epithelial innate immune responses by bacteria

Bacteria have evolved multiple mechanisms to escape host innate and adaptive immune responses. This section focuses on their manipulation of innate epithelial responses. A complex cross talk exists between bacteria and epithelial PRRs. For the host, PRR activation is necessary to regulate homeostasis through effects on epithelial barrier function on the one hand, and recruitment of inflammatory immune cells on the other hand, for the ultimate elimination of an invading pathogen. However, inflammation can lead to severe epithelial damage that emphasizes the need for tight control mechanisms in order to avoid excessive stimulation by resident bacteria. For bacteria, host inflammation is dangerous and must be dampened to avoid their own destruction. Yet, several pathogens have been shown to use the host inflammatory response during the early phase of infection to eliminate competitors within the resident microbiota, colonize the emptied intestinal niches, and invade tissues. We now consider the mechanisms by which pathogens stimulate PRRs to elicit these responses.

25.10 Pathogenicity, in comparison with commensalism, is often determined by the structure-function relationships between microbe-associated molecular patterns and host pattern-recognition receptors

MAMPs derived from commensal and pathogenic bacteria are similar. It has been suggested that MAMPs from commensals have undergone biochemical modifications that render them "stealthy" to pathogenic PRR activation. This is notably the case for lipopolysaccharide (LPS) because its agonist effect on TLR4 depends on the number of acyl chains within its lipid A component. However, modification of lipid A acylation lessens or even abrogates LPS recognition by TLR4. Thus, *Yersinia pestis* fully acylates its lipid A at 27°C but exhibits partial acylation at 37°C, leading to significant loss of its agonistic effect on TLR4 during its life within the mammalian host. A mutant strain expressing a fully acylated lipid A at 37°C becomes avirulent, indicating that avoiding TLR4 recognition facilitates invasion at body temperatures. Along the same line, *H. pylori* and *Campylobacter jejuni* produce flagellin molecules that are not recognized by TLR5, and this property is thought to promote chronic gastric colonization. In contrast, *Shigella* expresses an extra gene on its virulence plasmid that achieves full acylation of lipid A, thus significantly enhancing host release of IL-8 and recruitment of polymorphonuclear leukocytes in response to TLR4 stimulation, an event thought to promote *Shigella* dissemination within the epithelial layer, its main replication niche.

25.11 Commensal and pathogenic microbiota engage PRRs in a distinctive manner

While further work is needed to evaluate the extent to which MAMPs from resident bacteria have evolved to avoid recognition by host PRRs, there is good evidence that hosts have evolved mechanisms that help discriminate signals from resident and pathogenic bacteria. There are many examples of mechanisms by which this may occur. One such mechanism involves the distribution and regulation of PRRs within the intestinal epithelium that may limit their activation by resident bacteria while preserving their activation by pathogens. At steady state, TLR4 is preferentially expressed by epithelial crypt cells and is localized in endosomes. Its coreceptor MD-2 is not, or only weakly, expressed, and brush border alkaline phosphatase can dephosphorylate LPS lipid A, thereby reducing its stimulatory activity on TLR4. TLR5, a receptor for bacterial flagellin expressed by pathogens, but also by many commensal strains, is preferentially localized at the basolateral epithelial membrane at least in the colon away from its ligand. TLR9 ligands (CpG derived from bacterial DNA) can activate NF-κB when applied at the epithelial basolateral membrane but repress this cascade when applied at the apical membrane. Through spatial recognition of TLR engagement, the host cells can thereby tune their response toward either inflammation if the epithelial barrier has been breached, or tolerance, when signals are sensed from the apical side, constituting normal gut colonization.

In contrast to the resident bacteria, which mainly remain within the intestinal lumen, enteropathogens can gain access to the basolateral membrane of epithelial cells where TLRs can be more readily activated. Furthermore, pathogens can inject peptidoglycan moieties from the bacterial cell wall and thereby activate cytosolic NOD proteins and their downstream pro-inflammatory signaling cascades. Epithelial PRR activation by pathogens can thus increase the production of microbicidal products and reactive oxygen species, and stimulate the recruitment and activation of host immune cells, notably of phagocytes and DCs, which can in turn initiate adaptive immune responses; all of which are mechanisms that participate in clearing pathogens. As already alluded to, some pathogens can exploit host inflammatory responses to their benefit. Thus, host inflammation induced by *Salmonella typhimurium* infection results in the elimination of resident bacteria and promotes colonization. During *Shigella* infection, IL-8 induced via NF-κB and MAP kinase (MAPK) activation of epithelial cells stimulates the recruitment and transmigration of polymorphonuclear leukocytes, which enhances access of *Shigella* to the basolateral membrane of the epithelium and thereby promotes bacterial dissemination. Or pathogens can subvert the host to respond in an inappropriate manner and to activate the wrong arm of the immune system. Thus, *L. monocytogenes* infection leads to a potent type 1 interferon response, which functions to protect the host from viral intruders. Listeria does this by releasing a bacterial-specific second messenger, cyclic-di-AMP, during cytosolic growth, which can be sensed by the cytosolic DNA sensor STING (stimulator of interferon genes) (Figure 25.8). STING activation in macrophages leads to type I interferon production, whereby epithelial cells respond by IL-1β and IL-18 production. In the absence of interferon signaling, *Listeria* virulence is attenuated, indicating the benefit of activating this

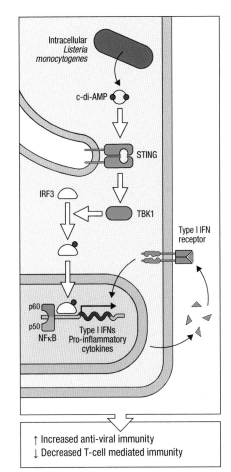

↑ Increased anti-viral immunity
↓ Decreased T-cell mediated immunity

Figure 25.8 Activation of the cytosolic DNA sensor STING by bacterial second messengers. *Listeria monocytogenes* replicates in the host cell cytosol and releases the bacterial-specific second messenger cyclic di-AMP. cAMP, and cGMP from other bacteria, are recognized by the DNA sensor STING. During viral infection, the host protein cGAS converts viral DNA into a cyclic GMP-AMP (cGAMP), which is the endogenous ligand for STING. Activated STING interacts with the kinase TBK1, which in turn phosphorylates IRF3. IRF3 and NF-κB act together to induce type 1 interferon genes that in turn induce interferon-regulated genes to induce an antiviral state. *L. monocytogenes* benefits from this host response through an inhibition of T-cell-mediated immunity, which is required for host defense against *L. monocytogenes*.

antiviral pathway for *Listeria* pathogenesis. It is thought that *Listeria* benefits from STING activation through a subsequent inhibition of T-cell-mediated immunity (see **Figure 25.8**).

In another set of examples, TLR activation by the microbiota can exert negative feedback on signaling pathways. One such pathway is that associated with NF-κB (**Figure 25.9**). Thus, immediately after birth in mice, microbiota-derived LPS can downregulate expression of the interleukin-1 receptor-associated kinase-1 (IRAK1), the proximal activator of the NF-κB cascade downstream of MyD88, resulting in desensitization of the TLR pathway. Via the induction of reactive oxygen species, commensal bacteria can also inhibit the common ubiquitin ligase E3-$^{SCF\beta-TrCP}$ and thereby prevent polyubiquitination and degradation of IκB, a key step in NF-κB activation. Furthermore, TLR4 stimulation by LPS stimulates the expression of peroxisome proliferation-activated receptor-γ (PPARγ), which in turn can divert NF-κB from the nucleus. Since PPARγ positively controls the expression of the colonic microbicidal peptide, human β-defensin-1, PPARγ can simultaneously maintain the gut barrier and prevent excessive activation of the NF-κB cascade (see **Figure 25.9**). Altogether these mechanisms may explain the overall beneficial effect of epithelial TLR activation and the corollary observation that selective inactivation of two key elements of the NF-κB pathway (IKKβ or IKKγ) within

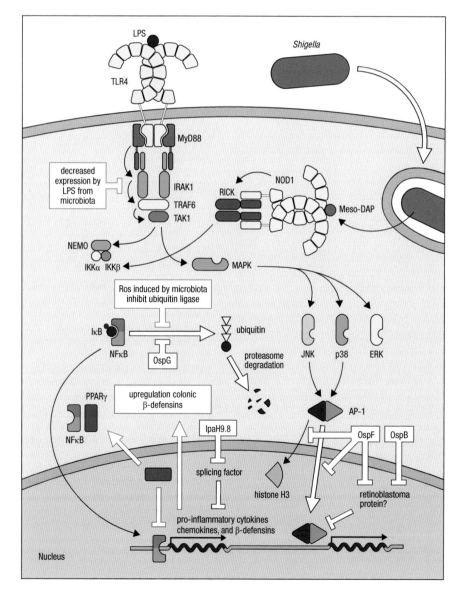

Figure 25.9 Modulation of epithelial cell pro-inflammatory signaling pathways by intestinal bacteria. NF-κB and MAP kinases can be activated by bacteria-derived products via membrane-associated TLR and intracytosolic NOD1 leading to the production of microbicidal peptides and pro-inflammatory cytokines and chemokines. These pathways can be inhibited by commensal bacteria and enteropathogens. Steps inhibited by resident bacteria are indicated in blue. Steps inhibited by *Shigella*, a paradigm of an invasive enteropathogen, are shown in red.

the epithelium is highly detrimental. These observations further indicate that activation of epithelial PRRs by the microbiota limits epithelial damage and/ or accelerates epithelial repair in several mouse models of colitis. In contrast, epithelial PRRs do not seem to participate in the induction of inflammation, but rather the presence of PRRs in the hematopoietic compartment is necessary for the induction of intestinal inflammation within the colon. The beneficial effect of epithelial TLRs has been ascribed to the recruitment of immune cells producing cytokines that stimulate epithelial repair (e.g., IL-11, IL-22) or to the induction of antimicrobial peptides (e.g., defensins) that limit bacterial adhesion to the epithelium. Another beneficial effect of epithelial TLR stimulation by the microbiota may result from the induction of BAFF and APRIL, and of pIgR, which respectively promote the production of secretory IgA (SIgA) and its translocation into the intestinal lumen, wherein SIgA can complex bacteria and promote their entrapment within the mucus. This mechanism is important to limit bacterial translocation and also to reduce epithelial cell production of reactive oxygen species in response to the microbiota.

25.12 Bacterial effectors from enteropathogens manipulate epithelial intracellular signaling cascades

While host inflammatory responses can benefit some pathogens, at least at the early phase of infection, they are clearly a threat to the pathogen, and many pathogens have evolved strategies to dampen host inflammatory responses in order to promote their persistence or dissemination. Thus, many enteropathogens use their T3SS to inject effectors able to interfere with intracellular signaling cascades threatening their survival.

Many inhibitory effects of bacterial effectors have been described in hematopoietic cells and notably in macrophages. This is the case for the *Salmonella* secreted factor L, which impairs IκBα ubiquitination and NF-κB activation in infected macrophages, and for *Yersinia* YopJ/P effectors, multifunctional proteins that inhibit NF-κB, p38, ERK, and IRF3 pathways and thereby induce the apoptosis of infected macrophages.

Yet other effectors act on epithelial cells. A close homolog of YopJ, the *Salmonella* effector AvrA, is also a multifunctional protein with transacetylase and deubiquitinase activities. This protein is injected into epithelial cells during *Salmonella typhimurium* infection and potently inhibits c-Jun *N*-terminal kinase and NF-κB signaling pathways. *In vivo* analysis of mutants indicates that AvrA is a strong inhibitor of intestinal inflammation and epithelial apoptosis that promotes persistent infection. Two other effectors (SptP and SspH1) injected by *Salmonella typhimurium* strains into epithelial cells can also inhibit the NF-κB cascade.

In the case of *Shigella*, several effectors injected via T3SS into epithelial cells also cooperate to inhibit NF-κB activation (see Figure 25.9). This mechanism is thought to limit IL-8 production and delay *Shigella* destruction by neutrophils. For example, OspG impairs the ubiquitination and thus the degradation of IκBα, preventing nuclear translocation of NF-κB. OspF is a phosphatase targeted to the nucleus where it impairs MAPK-dependent phosphorylation of histone H3. In addition, OspF can irreversibly dephosphorylate ERK. OspF can thus prevent the activation of numerous NF-κB-dependent genes. IpaH9.8, another effector targeted to the nucleus, can interact with a splicing factor involved in the maturation of inflammatory cytokines mRNA (see Figure 25.9). Besides interfering with NF-κB, other effectors injected via T3SS into epithelial cells can block the production of the β3-defensin to which *Shigella* is very sensitive, perhaps by interfering with the JAK/STAT pathway. Finally, *Shigella* can inhibit autophagy, a cytoplasmic mechanism by which cells entrap internalized microorganisms into autophagosomes and degrade these as well as damaged organelles and misfolded protein aggregates.

Upon *Shigella* infection of epithelial cells, autophagy is initiated via NOD1-dependent recruitment of the autophagy protein ATG16L1 to the plasma membrane. Delivery of *Shigella* to the autophagosome further depends on the interaction between the autophagy protein ATG5 and IscA, a bacterial effector that accumulates at one pole of the bacterium and induces actin-based motility. Wild-type *S. flexneri* can partially escape autophagy using the IcsB effector, another protein translocated by the T3SS that competitively inhibits IscA binding with ATG5. Analysis of *Shigella icsB* mutant indicates that this camouflage role of IcsB is important for efficient intraepithelial dissemination of the bacterium.

These observations underscore the multiple mechanisms evolved by pathogens to manipulate host epithelial cells in order to promote their local survival and replication or their access to other host cells and tissues appropriate for their replication. For some enteropathogens, such as *Shigella* and many enteropathogenic *E. coli*, the infection remains limited to the intestine. If the host survives epithelial damage and diarrhea, the infection can be controlled by host innate immune cells, particularly by neutrophils, and by adaptive immune responses that include the production of specific secretory IgA and T-cell responses, notably helper T_H17 cells that recruit neutrophils. In contrast, other pathogens such as *Salmonella*, and *Yersinia* have evolved multiple additional mechanisms to escape phagocytosis by macrophages, downmodulate host adaptive immune response, and disseminate toward other tissues, resulting in systemic infections. On the contrary, *Listeria* co-opts macrophages and uses them as a replicative niche and a Trojan horse to escape the host immune system and reach deeper tissues.

SUMMARY

The gastrointestinal epithelium is the main body interface between the host and the microbial world. The epithelium in this location is exposed daily to 10^{13} resident bacteria and can be occasionally colonized by a wide spectrum of enteropathogens, most of which are gram-negative Proteobacteria. The gastrointestinal epithelium has evolved as a central component of host innate immune defenses and forms a potent and inducible physicochemical and antibacterial barrier that can be finely tuned via the recognition of bacteria-derived motifs through pathogen-recognition receptors.

The gastrointestinal epithelium is, however, also a gateway for bacteria. A small fraction of resident bacteria can cross the epithelium, mainly via PP M cells. These bacteria are rapidly eliminated by host phagocytes and elicit adaptive immune responses that reinforce the epithelial barrier and promote their confinement into the intestinal lumen.

In contrast with resident bacteria, which mainly replicate and thrive within the mucus layer covering the intestinal epithelium, enteropathogens have the ability to adhere to and/or invade the gastrointestinal epithelium in order to promote their life cycles. For this purpose, they have evolved a wide variety of strategies depending on large clusters of virulence genes called *pathogenicity islands*.

In order to adhere to epithelium, enteropathogens express a large variety of cell surface adhesins, including polymeric fimbriae and afimbrial adhesins. Some pathogens such as *Yersinia* spp. and *Listeria* spp. can harness host transmembrane cell-adhesion receptors and trigger intracellular cascades that lead to epithelial cell invasion. Other invasive bacteria, such as *Shigella* and *Salmonella*, use their T3SS to inject bacterial effectors and induce the formation of a macropinocytic pocket, which enables phagocytic uptake.

Following adhesion, enteropathogens use their T3SS to inject a wide spectrum of bacterial effectors that interfere with epithelial cell functions. Some effectors provoke massive cytoskeleton changes that reinforce adherence to the epithelial surface, or promote bacterial entry or cell-to-cell dissemination. Other T3SS-injected effectors influence host-cell survival or, together with toxins released in the intestinal lumen, alter epithelial transport functions, inducing diarrhea, which allows bacteria dissemination.

Finally bacteria, both resident and pathogenic, can regulate intracellular signaling cascades that control inflammation and autophagy. The distribution and regulation of PRRs in epithelial cells avoid their excessive activation by resident bacteria, so that their activation by the commensal bacteria results predominantly in the production of microbicidal peptides and/or of repair factors that preserve intestinal homeostasis. Enteropathogens are more easily recognized by epithelial PRRs and can induce strong inflammatory reactions that facilitate their elimination. Yet, they have evolved multiple mechanisms to downmodulate inflammatory cascades and/or autophagy, helping their escape from host defense mechanisms.

FURTHER READING

Abreu, M.T.: Toll-like receptor signalling in the intestinal epithelium: How bacterial recognition shapes intestinal function. *Nat. Rev. Immunol.* 2010, 10:131–144.

Carvalho, F.A., Barnich, N., Sivignon, A. et al.: Crohn's disease adherent-invasive *Escherichia coli* colonize and induce strong gut inflammation in transgenic mice expressing human CEACAM. *J. Exp. Med.* 2009, 206:2179–2189.

Cerf-Bensussan, N., and Gaboriau-Routhiau, V.: The immune system and the gut microbiota: Friends or foes. *Nat. Rev. Immunol.* 2010, 10:735–744.

Chassaing, B., and Darfeuille-Michaud, A.: The commensal microbiota and enteropathogens in the pathogenesis of inflammatory bowel diseases. *Gastroenterology* 2011, 140:1720–1728.

Cianfanelli, F.R., Monlezun, L., and Coulthurst, S.J.: Aim, load, fire: The type VI secretion system, a bacterial nanoweapon. *Trends Microbiol.* 2016, 2:51–62.

Cossart, P., and Sansonetti, P.J.: Bacterial invasion: The paradigms of enteroinvasive pathogens. *Science.* 2004, 304:242–248.

Croxen, M.A., and Finlay, B.B.: Molecular mechanisms of *Escherichia coli* pathogenicity. *Nat. Rev. Microbiol.* 2010, 8:26–38.

Goodman, A.L., Kallstrom, G., Faith, J.J. et al.: Extensive personal human gut microbiota culture collections characterized and manipulated in gnotobiotic mice. *Proc. Natl Acad. Sci. USA.* 2011, 108:6252–6257.

Kim, M., Ashida, H., Ogawa, M. et al.: Bacterial interactions with the host epithelium. *Cell Host Microbe.* 2010, 8:20–35.

Pizarro-Cerda, J., and Cossart, P.: Bacterial adhesion and entry into host cells. *Cell* 2006, 124:715–727.

Radoshevich, L., and Dussurget, O.: Cytosolic innate immune sensing and signaling upon infection. *Front. Microbiol.* 2016, 7:313.

Sansonetti, P.J., and Di Santo, J.P.: Debugging how bacteria manipulate the immune response. *Immunity* 2007, 26:149–161.

Shames, S.R., Auweter, S.D., and Finlay, B.B.: Co-evolution and exploitation of host cell signaling pathways by bacterial pathogens. *Int. J. Biochem. Cell Biol.* 2009, 41:380–389.

Sun, J.: Pathogenic bacterial proteins and their anti-inflammatory effects in the eukaryotic host. *Antiinflamm. Antiallergy Agents Med. Chem.* 2009, 8:214–227.

Travassos, L.H., Carneiro, L.A., Girardin, S. et al.: NOD proteins link bacterial sensing and autophagy. *Autophagy* 2010, 6:409–411.

Viswanathan, V.K., Hodges, K., and Hecht, G.: Enteric infection meets intestinal function: How bacterial pathogens cause diarrhoea. *Nat. Rev. Microbiol.* 2009, 7:110–119.

Wells, J.M., Loonen, L.M., and Karczewski, J.M.: The role of innate signaling in the homeostasis of tolerance and immunity in the intestine. *Int. J. Med. Microbiol.* 2010, 300:41–48.

Helicobacter pylori infection

26

DIANE BIMCZOK, ANNE MÜLLER, AND PHILLIP D. SMITH

Helicobacter pylori is a gram-negative bacterium acquired usually during early childhood through fecal-oral or oral-oral transmission that colonizes the gastric mucosa in approximately 50% of the world's population. Signature features of *H. pylori* infection are chronic inflammation in virtually all colonized subjects and persistence for decades due to highly effective immune evasion strategies. Among infected persons, about 10%–15% develop one or more inflammatory sequelae, which include gastroduodenal ulceration, atrophic gastritis, intestinal metaplasia, distal gastric adenocarcinoma, and mucosa-associated lymphoid tissue (MALT) lymphoma. Importantly, *H. pylori* colonization with its attendant inflammation is the strongest known risk factor for gastroduodenal ulceration and gastric adenocarcinoma, the latter causing over 700,000 deaths per year worldwide. *H. pylori* induces two alternative patterns of mucosal infection and inflammation. Antral predominant gastritis leads to overproduction of gastrin that stimulates acid production by parietal cells in the gastric body and increases the risk for duodenal ulceration. In contrast, corpus predominant gastritis or pan-gastritis causes a reduction in acid production, likely due to destruction of parietal cells, leading to atrophic gastritis, a key premalignant lesion. In this chapter, we discuss the bacterial virulence factors and the host immune mechanisms, influenced by environmental and host genetic factors, that modulate *H. pylori*-induced mucosal inflammation and disease.

HELICOBACTER PYLORI IS BOTH A COMMENSAL AND A PATHOGEN

H. pylori has colonized humans for approximately 100,000 years and was dispersed by modern hominids during prehistoric migrations out of East Africa. Similar to human genetic diversity, the greater the distance from East Africa, the less the genetic diversity in the *H. pylori*, supporting a common geographic origin and strong association between humans and *H. pylori*. Acquisition of the bacteria early in life and persistent habitation in the stomach through an array of immune evasive strategies summarized in Table 26.1 support *H. pylori*'s status as a commensal. *H. pylori* is predominantly a commensal in early life and becomes a pathogen in only a subset of subjects later in life. *H. pylori* proteins that detoxify reactive oxygen and nitrogen species released by innate cells, proteins that repair bacterial DNA damage, adhesins that bind glycoproteins on gastric epithelial cells, and genomic recombination and mispairing that promote diversity, enhance bacterial persistence. At the epidemiologic level, however, improved hygienic living conditions in many countries, widespread use of antibiotics, and modern diet (e.g., reduction in salted and smoked foods) have led to a reduction in the

TABLE 26.1 *HELICOBACTER PYLORI* ADAPTATIONS THAT PROMOTE SURVIVAL

H. pylori adaptation	Host function evaded
Anergic lipopolysaccharide and flagella	Innate immune activation
Residence in the mucus layer	Microbe clearance and host recognition
Urease hydrolysis of urea to ammonia	Gastric acid production
Genomic plasticity	Innate and adaptive immunity
Antioxidant enzymes	Phagocyte reactive oxygen species and reactive nitrogen species production
Induction of T_{reg} cells	T_H1 and T_H17 T-cell responses
VacA-induced T-cell suppression	Adaptive immunity
Membrane engraftment of cholesterol	Phagocytosis

prevalence of *H. pylori* (the "disappearing microbiota") and its inflammation-associated sequelae in developed regions of the world. In contrast, the high prevalence of corpus predominant *H. pylori* infection in persons residing in resource-poor regions cause *H. pylori* to remain one of the world's leading causes of cancer and cancer-related mortality. The ability of *H. pylori* to cause gastric cancer, as well as gastric and duodenal ulceration, clearly classifies the bacterium as a pathogen, despite some of the commensal properties listed earlier.

26.1 *H. pylori* express an array of virulence factors

Isolates of *H. pylori* display a remarkably high level of genetic variability. Despite such variability, common bacterial virulence factors have evolved to promote *H. pylori* survival in the hostile habitat of the stomach. The major virulence factors include urease, flagellin, VacA, CagA, and outer membrane proteins (Table 26.2). These factors play a key role in *H. pylori* persistence, the host immune response to the bacteria, and the associated inflammatory and oncogenic potential in the gastric mucosa.

26.2 Urease enables *H. pylori* survival and epithelial cell adhesion

Expressed by all strains, *H. pylori* urease is essential for establishing and maintaining bacterial colonization in the acidic environment of the stomach by hydrolyzing urea into carbon dioxide and ammonia (NH_3), which neutralizes gastric acid. The pH change induced by urease activity also reduces the viscosity of gastric mucus, promoting bacterial motility. At the epithelial surface, the released ammonia may disrupt tight junctions, thereby damaging the epithelial barrier. In addition to its enzymatic function, urease mediates bacterial adhesion to the gastric epithelium and serves as a chemotactic factor for monocytes and neutrophils. Urease also can directly damage host cells, especially antigen-presenting and epithelial cells, by triggering apoptosis through a major histocompatibility complex (MHC)-II-dependent mechanism. Urease is an immunodominant antigen, with 25% of T-cell clones in infected subjects having specificity for urease epitopes. Consequently, many vaccine constructs currently being developed include urease as a major vaccine antigen.

TABLE 26.2   *HELICOBACTER PYLORI* VIRULENCE FACTORS AND THEIR ACTIONS

Virulence factor	Expression	Role in bacterial pathogenesis
Urease	Constitutive; secreted, surface expressed, cytoplasmic	Essential for colonization, increases local pH to allow bacterial survival and motility, disrupts tight junctions
Flagellin	Constitutive; surface expressed	Bacterial motility and colonization
Cag T4SS	Strain-specific, surface expressed	Delivery of CagA and peptidoglycan into host epithelial cells; peptidoglycan activates innate signaling pathways through NOD1 activation
CagA	Strain specific, released through T4SS	Activation of SHP2- and NF-κB-dependent signaling pathways; disruption of cell morphology, tight junctions, cell polarity
VacA	Sequence variation, secreted	Induction of cell death and inhibition of T-cell proliferation
γ-Glutamyl-transferase	Highly conserved, constitutive	Causes glutathione consumption and oxidative damage; induces tolerance by dendritic cell skewing, anti-proliferative effects on T cells
BabA	Variable, surface expressed	Adhesion to host cells
SabA	Variable, surface expressed	Adhesion to host cells
OipA	Variable, surface-expressed	Adhesion, β-catenin activation, pro-inflammatory effects
HP-NAP	Highly conserved, secreted	Activates neutrophil respiratory burst, chemotactic activity, protects bacteria from oxidative damage, various immunomodulatory effects

26.3 Flagellin mediates *H. pylori* motility in gastric mucus

H. pylori possess several unipolar, sheathed flagella containing FlaA and FlaB proteins that mediate *H. pylori* motility. This motility is necessary for bacterial colonization and persistence in the gastric mucus, reflected in the inability of aflagellated mutant bacteria to colonize the gastric mucosa. Interestingly, in contrast to flagellin expressed by intestinal bacterial pathogens such as *Salmonella, H. pylori* flagellin does not bind to, or activate immune cells through, toll-like receptor (TLR)5. In addition, *H. pylori* flagellin delivered to the cytosol does not trigger NLRC4 inflammasome activation, whereas flagellins from other gram-negative bacteria activate this pathway. Thus, *H. pylori* flagellin is a highly conserved virulence factor but has limited activity in activating innate immune responses as a pathogen-associated molecular pattern (PAMP).

26.4 *H. pylori cag* pathogenicity island enhances risk for disease

The *H. pylori* cytotoxin-associated gene (*cag*) pathogenicity island is an important determinant of disease severity. A 40 kb segment of *H. pylori* DNA, the *cag* pathogenicity island encodes more than 25 proteins, including Cag3, CagM, and CagH, that form a type IV secretion system (T4SS), which mediates the translocation of CagA into host epithelial cells. Notably, recent studies show that iron depletion enhances the assembly and function of the *cag* T4SS and the severity of *H. pylori*-induced premalignant lesions in humans, consistent with epidemiological evidence that links iron deficiency and accelerated *H. pylori*-induced carcinogenesis.

H. pylori utilizes host integrins, especially $\alpha_5\beta_1$ integrin, to bind to target cells via CagL. The proteins CagA, CagI, and CagY also bind β_1 integrin and induce conformational changes in integrin heterodimers, which permit CagA translocation. CagY may act as a regulator of T4SS activity, since *cagY* contains many DNA repeat regions, making it especially susceptible to spontaneous homologous recombination that results in loss of T4SS function during chronic infection. Attachment of *H. pylori* to β_1 integrin also initiates activation of host kinases Src and focal adhesion kinase, ensuring that translocated CagA is phosphorylated at its site of injection. Phosphorylated CagA activates a cellular phosphatase (SHP-2), leading to the aberrant activation of multiple intracellular signaling cascades, including Wnt/β-catenin signaling, which is associated with carcinogenesis, and pro-inflammatory nuclear factor (NF)-κB signaling. Nonphosphorylated CagA also exerts effects within the cell. By binding to the epithelial tight-junction scaffolding protein ZO-1 and the transmembrane protein junctional adhesion molecule-A (JAM-A), nonphosphorylated CagA causes ineffective assembly of tight junctions at sites of bacterial attachment. By interacting with PAR1b, a central regulator of cell polarity, nonphosphorylated CagA induces the loss of epithelial cell polarity. As a consequence of these changes, *cag*+ strains are more strongly associated with gastric cancer than *cag*− strains.

The T4SS and outer membrane vesicles translocate *H. pylori* peptidoglycan into host epithelial cells. *In vitro* studies showed that *H. pylori* peptidoglycan delivered into the host cell cytoplasm was sensed by NOD1, a cytoplasmic pathogen-recognition molecule that triggers pro-inflammatory signaling. The contribution of *H. pylori* peptidoglycan sensing by NOD1 to the inflammatory response *in vivo*, however, has not been fully explored. Underscoring the complex array of cellular factors involved in *H. pylori* activation of NF-κB in innate cells, the *H. pylori* lipopolysaccharide (LPS) metabolite heptose 1,7-bisphosphate (βHBP) recently was shown to trigger NF-κB through α kinase 1 (ALPK1) and TRAF-interacting protein with FHA domain (TIFA) in a T4SS-dependent manner (**Figure 26.1**). The activation of NF-κB in gastric epithelial cells leads to the production of inflammatory chemokines such as interleukin (IL)-8 and cytokines such as type I and II interferons (IFNs), which, among other activities, promote Th1 responses. Thus, *cag*+ strains activate multiple signaling pathways involved in the regulation of inflammation and oncogenic cellular responses that may increase the risk for cell transformation during prolonged *H. pylori* colonization.

26.5 *VacA* encodes a cytotoxin that mediates long-term persistence

The *H. pylori* gene *vacA* encodes a secreted protein (VacA) that induces vacuolation in cultured epithelial cells, which is also linked to the induction of gastric malignancy. All *H. pylori* strains possess *vacA*, but *vacA* sequences vary markedly among strains. VacA interacts with host cells by binding to multiple epithelial cell surface components, including fibronectin, epidermal growth factor receptor (EGFR), sphingomyelin, and CD18 (integrin β_2) expressed on T cells. A notable consequence of prolonged exposure of epithelial cells to VacA is disrupted autophagy, which normally contributes to protection against the infection and tumor suppression.

VacA is a pore-forming toxin best known for its ability to cause vacuolation in host cells, which leads to cell death through programmed necrosis. In addition to inducing vacuolation, VacA specifically interferes with the induction of T-cell responses. Purified VacA inhibits processing and presentation of antigenic peptides to human CD4+ T cells and can specifically block antigen-dependent T-cell proliferation by interfering with IL-2-mediated signaling. *H. pylori* co-opts CD18 as a VacA receptor on human T cells. After entering the cell, VacA inhibits activation of the transcription factor nuclear factor of

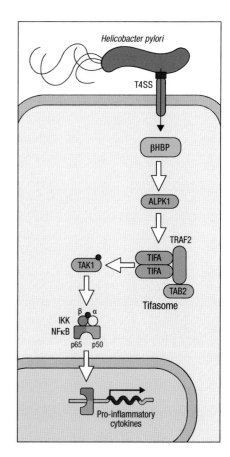

Figure 26.1 *Helicobacter pylori* **delivery of heptose 1,7-bisphosphate (βHBP) leads to the activation of NF-κB.** After adherence, *H. pylori* can translocate the lipopolysaccharide metabolite βHBP into the host cell by the T4SS, which activates the α kinase 1 (ALPK1) and TRAF-interacting protein with FHA domain (TIFA). The ALPK1-TIFA axis is an innate immune pathway, which leads to the activation of the transcription factor NF-κB that induces pro-inflammatory cytokine genes. (Adapted from Zimmermann, S. et al. *Cell Rep.* 2017, 20:2384.)

activated T cells (NFAT) and thus expression of NFAT target genes such as IL-2 and IL-2 receptor. Additionally, VacA can suppress IL-2-induced proliferation of primary T cells in an NFAT-independent manner. Thus, VacA can inhibit the clonal expansion of *H. pylori*-specific T cells, thereby promoting *H. pylori* evasion of the adaptive immune response and long-term persistence.

26.6 Additional *H. pylori* determinants influence disease pathogenesis

Other virulence factors that contribute to *H. pylori* infection and pathogenesis include several outer membrane proteins that act as adhesion factors and the neutrophil activating protein HP-NAP. The key *H. pylori* virulence factors are summarized in Table 26.1.

H. pylori and the innate immune system

Innate immune cells in the gastric mucosa play a central role in recognizing *H. pylori* and initiating the host response to the bacteria. Understanding the interaction between *H. pylori* and these cells, which include epithelial cells, neutrophils, macrophages, and dendritic cells, and the molecules they utilize to recognize *H. pylori* is critical for understanding the pathogenesis of the infection.

26.7 *H. pylori* is recognized by innate immune cell receptors

H. pylori expresses an array of PAMPs that induce innate immune cell signaling pathways through multiple pattern recognition receptors (PRRs). Among the PRRs expressed by gastric epithelial cells, dendritic cells (DCs) and macrophages, TLRs have been implicated in *H. pylori* recognition. Early studies of *H. pylori* recognition focused on TLR4, which is involved in innate cell detection of bacterial LPS. However, *H. pylori* LPS contains modifications that render it an ineffective activator of the TLR4/CD14/MD-2 receptor-induced signal pathway. Similarly, *H. pylori* flagellin contains point mutations in the TLR5 recognition domain that prevent flagellin from activating TLR5, as discussed later.

In contrast, *H. pylori* recognition involves TLR2 activation. TLR2 is an important surface-expressed PRR that collaborates with TLR1, TLR6, TLR10, and other receptors to recognize bacterial and fungal products. An array of *H. pylori*-derived molecules, including LPS, heat shock protein 60 (HP-HSP60), and HP-NAP, induce TLR2 activation. Activation of TLR2 leads to NF-κB signaling and induction of cytokine expression by epithelial cells, as well as macrophages, DCs, neutrophils, and B cells. In addition, TLR2 engagement by *H. pylori* leads to activation of the inflammasome pathway through the Nod-like receptor family member NLRP3, which in turn causes activation of IL-1β, a key pro-inflammatory mediator. *H. pylori* also signals through TLR9, an intracellular PRR activated by bacterial DNA that can activate both pro-inflammatory and anti-inflammatory pathways. Interestingly, *H. pylori* DNA is delivered to the intracellular TLR9 through the T4SS encoded by the *cag* pathogenicity island, through a similar pathway that delivers *H. pylori* peptidoglycan to NOD1.

26.8 Gastric epithelial cells contribute to the inflammatory response to *H. pylori*

Gastric epithelial cells are the first line of defense against *H. pylori*. Although some reports have shown *H. pylori* invasion into the gastric lamina propria or individual cells, the bacteria are generally considered noninvasive. Thus, many *H. pylori*-induced inflammatory effects are initiated by the interaction between the bacteria or bacterial products and gastric epithelial cells, which respond by expressing mediators that communicate with innate and adaptive immune cells. Multiple gastric epithelial receptors enable this interaction. *H. pylori* SabA and BabA bind to epithelial Lewis antigens, the T4SS binds to epithelial $\alpha5\beta1$ integrin, and urease binds to epithelial MHC-II and CD74. Bacterial virulence factors, specifically CagA and VacA, are predominantly delivered into epithelial cells, leading to multiple epithelial cell responses. Gastric epithelial cells also express innate PRRs such as TLRs that trigger innate signaling pathways, as discussed earlier. *H. pylori* recognition by the gastric epithelium results in the activation of NF-κB and other pro-inflammatory signaling cascades that lead to the secretion of cytokines and chemokines. This response includes the secretion of IL-8, IL-6, TNF-α, IL-1β, IL-1α, granulocyte-macrophage colony-stimulating factor (GM-CSF), MIF, and TGF-β, which activate, regulate, or recruit innate and adaptive immune cells. IL-8 (CXCL8) is a signature chemokine released by the gastric epithelium upon *H. pylori* infection that plays a key role in neutrophil recruitment. *H. pylori*-induced epithelial cell release of CCL2 may contribute to DC recruitment through CCR2 engagement. *H. pylori* also induces epithelial cell death and disrupts epithelial barrier function through an array of mechanisms, with downstream effects on immune cell activation and recruitment. Overall, the epithelium plays a central role in initiating the host immune response to *H. pylori*.

26.9 Neutrophil accumulation is a characteristic feature of *H. pylori*-infected mucosa

A signature feature of *H. pylori* infection is the migration of neutrophils into the gastric mucosa. Even after decades of infection, neutrophils are still present in the *H. pylori*-infected gastric mucosa; however, neutrophils are unable to control the infection, resulting in chronic active gastritis. Neutrophils are highly responsive to *H. pylori*-induced chemokines such as IL-8 secreted by epithelial cells, and IL-17 secreted by Th17 cells further enhances the release of neutrophil-attracting chemokines. In addition to chemokines secreted by the gastric epithelium, HP-NAP, a virulence factor released by *H. pylori*, promotes neutrophil recruitment. HP-NAP also induces production of reactive oxygen species (ROS) by the neutrophils through NADPH oxidase 2 (NOX2) activation. However, *H. pylori* inhibits NOX2 assembly at the phagosome to evade oxidative killing, resulting in NOX2 assembly on the neutrophil cell membrane. NOX2 expression on the cell membrane allows ROS release into the extracellular space, which may cause neutrophil-derived oxidative damage to neighboring cells. In addition to altering phagosome function, experimental studies suggest that *H. pylori* may limit or delay neutrophil phagocytosis through several molecular pathways. Thus, *H. pylori* actively recruits neutrophils to the gastric mucosa but renders the neutrophil response ineffective, another potential mechanism by which the bacterium promotes its own survival.

26.10 Gastric macrophages contribute to inflammatory and carcinogenic responses to *H. pylori*

Macrophages reside in the gastric lamina propria in close proximity to the epithelium and thus are positioned to interact with *H. pylori*-derived products translocated across the epithelium. Upon *H. pylori* infection, bacterial products and chemokines recruit macrophages and DCs to the gastric mucosa. Gastric macrophages contribute to the gastric inflammatory response and to Th1 polarization by secreting cytokines, including IL-12, IL-1, TNF-α, and IL-6. In *H. pylori*-infected mice, gene expression profiles implicate an M1 macrophage response, with ornithine decarboxylase a critical regulator of M1 activation during *H. pylori* infection. Macrophages also may be a major source of IL-10, which promotes a tolerizing environment that favors *H. pylori* colonization. This anti-inflammatory gastric macrophage phenotype is driven by the regulatory transcription factor PPAR-γ. *H. pylori* limits phagocytic killing by macrophages through the ability of VacA toxin to prevent fusion of the phagosome with the lysosomes required for killing. However, macrophages can kill *H. pylori* in the absence of phagocytosis, likely through a nitric oxide (NO)–dependent mechanism. Overall, macrophages are a major source of local ROS and reactive nitrogen species (RNS), which promote oxidative tissue damage, inflammation, and carcinogenesis.

Elegant molecular studies have identified multiple strategies by which *H. pylori* limits the production of ROS and RNS and resists oxidative damage. *H. pylori* arginase can compete efficiently with macrophages for the inducible nitric oxide synthase (iNOS) substrate L-arginine (L-Arg), impairing host NO production and enhancing bacterial survival. Bacterial arginase converts L-Arg to urea, which is then hydrolyzed to NH_3 to neutralize gastric acid. Moreover, *H. pylori* induces macrophage arginase expression, consistent with an M2 phenotype, leading to inhibition of iNOS and reduced RNS production. Increased arginase activity also leads to the production of polyamines such as spermine, which may induce macrophage apoptosis. To escape the detrimental effects of ROS, *H. pylori* expresses superoxide dismutase, catalase and other enzymes to degrade ROS. Overall, *H. pylori* elicits dual pro-inflammatory and anti-inflammatory effects through its interaction with macrophages, contributing to both persistent colonization and tissue damage.

26.11 Dendritic cells initiate the immune response to *H. pylori*

Gastric DCs are key initiator cells in the host adaptive response to *H. pylori,* and their cytokine secretion patterns determine the type of T-cell response that is induced. Primary DCs isolated from human gastric mucosa phagocytose and process *H. pylori*, undergo maturation, and induce T-cell secretion of IFN-γ, but not IL-4, and only low levels of IL-10, consistent with a Th1 response. Outer membrane proteins Omp18 and HpaA in particular have been shown to stimulate DC maturation and antigen presentation capacity. DCs recognize *H. pylori* and its products through multiple receptors, including DC-specific ICAM-3-grabbing non-integrin (DC-SIGN), TLR2, and TLR9. *H. pylori* induces signaling through both MyD88-dependent and -independent pathways. Notably, gastric stromal and epithelial-derived mediators such as retinoic acid render gastric DCs less susceptible to bacterial activation, resulting in a dampened pro-inflammatory Th1 response and reduced induction of proliferative T-cell responses. This reduced responsiveness of gastric DCs to *H. pylori* infection may contribute to the maintenance of persistent *H. pylori* infection.

DCs in the mucosa of the human stomach are distributed throughout the lamina propria, frequently adjacent to epithelial cells. Additionally, the gastric mucosa contains a subset of intraepithelial DCs that likely are involved in luminal antigen sampling (Figure 26.2). Gastric DCs are more prevalent, mature, and activated in the inflamed lamina propria of *H. pylori*-infected subjects and have been reported within the T-cell areas of tertiary lymphoid follicles that are formed during chronic gastritis. Supporting these findings, DCs have been shown to recruit to the gastric mucosa in experimental *H. pylori* infection in mice. Recent studies in mice indicate that the gastric mucosa contains multiple subsets of DCs that differ in CD11c,

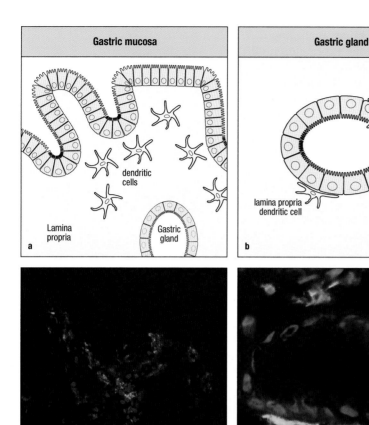

Figure 26.2 Distribution of dendritic cells (DCs) in non-*Helicobacter pylori*-infected human gastric mucosa. Gastric mucosa contains intraepithelial and lamina propria HLA-DR⁺ CD11c⁺ DCs, which are shown by schematic representation (**a,b**) and immunofluorescence (**c,d**) in longitudinal (bar = 50 μm) (**a,c**) and cross (bar = 20 μm) (**b,d**) sections of gastric mucosa. HLA-DR-expressing cells are green, CD11c-expressing cells red, and cell nuclei blue; HLA-DR/CD11c double-positive cells are yellow. (Adapted from Bimczok, D. et al. *Mucosal Immunol.* 2010, 3:60–69.)

CD11b, and CD103 expression, with a predominant CD11b⁺ population that samples *H. pylori* bacteria. Interestingly, in contrast to small intestinal DCs, CD103 expression by gastric DCs is generally low in both humans and mice. Mouse studies also have shown that the gastric response to *Helicobacter* can be modulated through interactions with DCs at extragastric sites. Thus, DCs in Peyer's patches may phagocytose coccoid forms of *H. pylori*, leading to the priming of CD4⁺ T cells that migrate to the stomach and initiate gastric inflammation. Determining which DC subset is the most efficient at inducing a protective immune response to *H. pylori* is a subject of ongoing investigations and has large implications for mucosal vaccine design.

Adaptive immune responses to *H. pylori*

In addition to the innate mucosal cell response, *H. pylori* induces prominent adaptive immune responses involving T and B cells, which are discussed next. The interdependent innate and adaptive immune responses to *H. pylori* are depicted in **Figure 26.3**.

26.12 *H. pylori* induces T$_H$1 and T$_H$17 cell responses

Gastric DCs located in the epithelium and subepithelial space respond to *H. pylori* antigens, releasing IL-12 and IL-23 that induce T$_H$1 and T$_H$17 cells, respectively. T$_H$1 cells, which release IFN-γ and TNF-α, appear to be the most prominent and earliest T-cell population that accumulates in the inflammatory lesion. IFN-γ, which also is released by gastric CD8 T cells and NK cells, activates phagocytes that in turn release pro-inflammatory cytokines. Increasing numbers of local IFN-γ-producing cells correlate

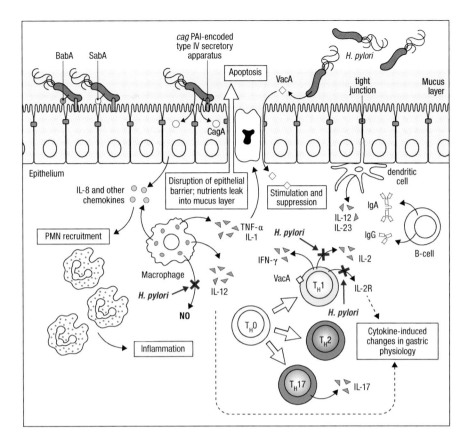

Figure 26.3 Gastric immune cell response to *Helicobacter pylori*. *H. pylori* binds to gastric epithelial cells through adhesins such as BabA. Utilizing a type IV secretory apparatus, the bacterium then inoculates CagA into the epithelial cell, leading to the release of the neutrophil chemokine IL-8 and other cytokines and alterations in epithelial cell integrity. The *H. pylori* product VacA induces epithelial apoptosis, and together with an array of other products induce T-cell, polymorphonuclear neutrophil (PMN), and macrophage pro-inflammatory responses, leading to gastric inflammation and cytokine-induced changes in gastric physiology. (Adapted from Monack, DM. et al. *Nat. Rev. Microbiol.* 2004, 2:747.)

with the severity of the gastritis. Studies in *H. pylori*-infected mice show that neutralization of IFN-γ and *Ifng* gene deletion cause marked reduction in the gastric inflammation. Conversely, high levels of IFN-γ are associated with reduced *H. pylori* colonization levels. Thus, T_H1 cells may protect against infection, while contributing to inflammation. Importantly, IFN-γ also induces gastrin, which stimulates the secretion of gastric acid in early infection. Over time, however, *H. pylori* gastritis causes gastric mucosal atrophy, a reduction in the parietal cell mass and consequently reduced acid secretion. Eventually, the inflammatory response leads to the progressive loss of *H. pylori's* ecological niche and diminishing *H. pylori* colonization, but the inflammation persists.

In addition to the induction of Th1 activity, *H. pylori* infection is associated with a prominent T_H17 response, reflected in the IL-23-directed accumulation of T_H17 cells and increased tissue levels of IL-17. IL-17 is a dominant cytokine in the gastric mucosa of infected humans and one of the earliest cytokines detected in the gastric mucosa of mice after *H. pylori* challenge. IL-17 mediates the recruitment and activation of neutrophils and likely contributes to maintenance of the barrier function of the gastric mucosal epithelium. A marked reduction in neutrophil infiltration into the gastric mucosa of IL-17$^{-/-}$ mice infected with *H. pylori* supports the role of IL-17 in regulating the neutrophil contribution to the inflammatory lesion.

26.13 *H. pylori* induces a prominent regulatory T-cell response

H. pylori induces regulatory T (T_{reg}) cells of the CD4$^+$CD25$^+$Foxp3$^+$ phenotype in the gastric mucosa of infected subjects. Both naturally occurring and inducible T_{reg} cells have been identified in the stomach. In mouse studies, the depletion of T_{reg} cells is associated with increased *H. pylori* gastritis, decreased colonization, and enhanced T-effector-cell responses. In adult humans, an attenuated T_{reg} response is associated with an increased risk for peptic ulcer disease. Compared with *H. pylori*-infected adults, infected children have less gastric inflammation and ulcer disease and reduced gastric $T_H 1$ and $T_H 17$ responses in parallel with a reciprocal increase in gastric T_{reg} responses. Studies in infant mice have confirmed that early life infection with *H. pylori* induces T_{reg} cells that suppress inflammatory T-effector-cell responses, which in turn limits the development of precancerous lesions. Overall, increased gastric T_{reg} responses correlate with increased *H. pylori* and prolonged *H. pylori* survival but also reduced destructive pathology.

Interestingly, the T_{reg} response to *H. pylori* infection in young hosts may have beneficial effects at other mucosal sites. Humans and mice infected with *H. pylori* at a young age appear to be protected against or have a delayed response to allergen-induced asthma due to increased T_{reg} activity in the lungs. In this connection, the reduced prevalence of *H. pylori* infection in some age groups coincides with an increasing prevalence of asthma.

26.14 *H. pylori* infection is associated with an altered gastric microbiota

The impact of *H. pylori* infection on the host microbiota has been the subject of recent studies in both mice and humans. Mice persistently infected with *H. pylori* display altered structures of microbial composition in the stomach, as well as in the intestine. Notably, *H. pylori* has greater impact on the gastric microbial composition in younger mice than older mice. In humans, *H. pylori* has been shown to induce a greater restructuring of commensal microbiota in the stomachs of children than adults residing in Chile, where *H. pylori* is endemic. In parallel, the gastric mucosa of infected children displayed a stronger T_{reg} response, raising the possibility that commensal microbes are involved in the induction of the more prominent gastric T_{reg} response in infected children. Notably, clearance of *H. pylori* in children restores the microbiota altered by the presence of *H. pylori* to that of uninfected children. Studies also have begun to address the potential role of *H. pylori*-induced microbial changes on gastric cancer. Transgenic FVB/N insulin-gastrin (INS-GAS) mice, which develop spontaneous atrophic gastritis and intraepithelial neoplasia, develop the lesions much sooner in the presence of *H. pylori* plus commensal microbiota than *H. pylori* alone. Gastric microbial composition differs in certain human populations with different levels of risk for gastric cancer, and differences in gastric microbial richness and diversity accompany the different histological stages of gastric carcinogenesis. Thus, in the presence of *H. pylori*, changes in the gastric microbiota may influence the progression to gastric cancer.

26.15 Nonprotective local and systemic antibody responses

H. pylori infection initiates the recruitment of IgA- and IgM-producing plasma cells and induces strong local and systemic antibody responses. In chronic infection, lymphoid follicles that contain activated B cells and antibody-producing plasma cells form within the gastric mucosa. Serum IgG anti-*H. pylori* antibodies can be detected within 21 days of initial infection. The IgG1 to IgG4 ratio of greater than 1 is consistent with a $T_H 1$ dominant response. Remarkably, more than 300 antigenic *H. pylori* proteins that induce specific antibody responses have been identified, including urease, flagellin, CagA, HpaA, and various membrane proteins. However, the anti-*H. pylori* antibodies do not clear the bacteria. In chronically infected hosts, *H. pylori* antibody responses remain stable. After eradication, however, the antibody titer typically declines 20%–50% within the first 9 months but may remain elevated for several years. Antibody titers do not differ between asymptomatic patients and those with ulcer disease. Importantly, in 20%–30% of infected persons, *H. pylori* induces autoantibodies, particularly to the gastric proton pump, which contribute to achlorhydria or lead to autoimmune gastritis. Thus, antibodies to *H. pylori* do not provide protection but may elicit harmful effects.

26.16 *H. pylori* utilizes multiple mechanisms for immune evasion

Despite innate and adaptive immune responses to *H. pylori*, the bacteria persist in all infected persons who have not been treated with antibiotics, indicating that *H. pylori* has highly effective immune evasion mechanisms (see Table 26.1). After inoculation into the stomach, *H. pylori* utilize their helical shape, flagellar motility, and urease-mediated hydrolysis of urea to NH_3 to penetrate the mucus layer and establish colonization. *H. pylori* can detoxify the reactive oxygen and nitrogen products of innate inflammatory cells through production of antioxidant enzymes, including superoxide dismutase, catalase, and arginase, as discussed earlier. *H. pylori* also has developed multiple mechanisms to evade phagocytosis by neutrophils and macrophages, also discussed earlier. Importantly, the tetra-acetylated lipid A with fatty acids of 16 and 18 carbons and the reduced phosphorylation of the LPS lipid A backbone contribute to rendering *H. pylori* LPS 1000-fold less biologically active than the LPS of other gram-negative bacteria. In addition, modification of the flagellin *N*-terminal TLR5 recognition site, with compensatory changes in the flagellin molecules that preserve motility, allows *H. pylori* to evade TLR5 binding. Also, the remarkable genomic plasticity of *H. pylori*, reflected in gene loss, gain, recombination, and mutation, has promoted bacterial strain divergence and altered outer membrane proteins such as BabA and BabB, which mediate attachment to Lewis B blood group antigens on gastric epithelial cells. In addition to evasion tactics and genomic plasticity, *H. pylori* activates anti-inflammatory responses through several mechanisms, including the activation of DCs, resulting in MyD88-dependent production of anti-inflammatory cytokines such as IL-10. *H. pylori* fucose, in contrast to mannose-expressing organisms, dissociates DC-SIGN signalosome proteins, thereby enhancing IL-10 expression in addition to decreasing secretion of pro-inflammatory cytokines IL-6 and IL-12. The induction of gastric T_{reg} cells, which downregulate effector T-cell responses, is another prominent mechanism by which *H. pylori* modulates host inflammatory responses, as discussed earlier. Thus, *H. pylori* evades recognition, modifies attachment molecules, and downregulates the ensuing inflammatory response, thereby promoting persistent infection.

26.17 Complex immunobiology of *H. pylori* infection has limited development of an effective vaccine

The high prevalence of *H. pylori*, the potential sequelae of chronic inflammation, and the increasing rates of antibiotic-resistant strains underscore the need for an effective *H. pylori* vaccine. Vaccines, including vector vaccines, epitope vaccines, and fusion protein vaccines, using both mucosal and systemic application routes and several adjuvants, have been tested in mouse models. A significant reduction in bacterial load after *H. pylori* challenge has been achieved due to the induction of T_H1 immunity, but none of the vaccines have consistently led to bacterial clearance. Few human trials have been conducted, and no human vaccine is commercially available. However, encouraging results from a large-scale clinical trial of an *H. pylori* urease–*Escherichia coli* heat-labile toxin fusion protein vaccine in more than 4000 healthy, *H. pylori*-negative children showed an efficacy of 72% in the first year and 55% in the second year after vaccination. The partial success of this human vaccine may stimulate further research and industry investment into a much-needed preventive vaccine against *H. pylori*.

SUMMARY

H. pylori has coevolved with humans for approximately 100,000 years, developing unique mechanisms for colonization, persistence, and immune evasion. After the stomach has been colonized by *H. pylori*, bacterial, host, and local environmental factors promote chronic mucosal inflammation, the predisposing lesion for atrophic gastritis, leading in some subjects to intestinal metaplasia, dysplasia, and ultimately adenocarcinoma. The gastric commensal microbiota, which is restructured in the presence of *H. pylori*, may in turn influence the mucosal response to *H. pylori*. Tolerance to the bacteria, rather than immunity, provides protection against the preneoplastic lesions induced by chronic *H. pylori* infection. The prevalence of *H. pylori* infection is declining in the United States and Europe, largely due to improved hygiene and frequent antibiotic usage, but the infection is still the leading cause of gastroduodenal inflammatory disease in developed countries and the third leading cause of cancer-related mortality in many resource-poor nations. Further elucidation of the immunobiology of *H. pylori* infection should provide new approaches for limiting the sequelae of *H. pylori* infection, including gastric cancer, and for developing an effective vaccine.

FURTHER READING

Amieva, M., and Peek, R.M., Jr.: Pathobiology of *Helicobacter pylori*-induced gastric cancer. *Gastroenterology* 2016, 150:64–78.

Arnold, I.C., Dehzad, N., Reuter, S. et al.: *Helicobacter pylori* infection prevents allergic asthma in mouse models through the induction of regulatory T cells. *J. Clin. Invest.* 2011, 121:3088–3093.

Arnold, I.C., Lee, J.Y., Amieva, M.R. et al.: Tolerance rather than immunity protects from *Helicobacter pylori*-induced gastric preneoplasia. *Gastroenterology* 2011, 140:199–209.

Bimczok, D., Grams, J.M., Stahl, R.D. et al.: Stromal regulation of human gastric dendritic cells restricts the Th1 response to *Helicobacter pylori*. *Gastroenterology* 2011, 141:929–938.

Bimczok, D., Kao, J.Y., Zhang, M. et al.: Human gastric epithelial cells contribute to gastric immune regulation by providing retinoic acid to dendritic cells. *Mucosal Immunol.* 2014, 8:533–544.

Brawner, K.M., Kumar, R., Serrano, C.A. et al.: *Helicobacter pylori* infection is associated with an altered gastric microbiota in children. *Mucosal Immunol.* 2017, 10:1169–1177.

Chaturvedi, R., Asim, M., Hoge, S. et al.: Polyamines impair immunity to *Helicobacter pylori* by inhibiting L-arginine uptake required for nitric oxide production. *Gastroenterology* 2010, 139:1686–1698.

Chen, Y., and Blaser, M.J.: Inverse associations of *Helicobacter pylori* with asthma and allergy. *Arch. Intern. Med.* 2007, 167:821–827.

Coker, O.O., Dai, Z., Nie, Y. et al.: Mucosal microbiome dysbiosis in gastric carcinogenesis. *Gut* 2018, 67:1024–1032.

Falush, D., Wirth, T., Linz, B. et al.: Traces of human migrations in *Helicobacter pylori* populations. *Science* 2003, 299:1582–1585.

Hardbower, D.M., Asim, M., Luis, P.B. et al.: Ornithine decarboxylase regulates M1 macrophage activation and mucosal inflammation via histone modifications. *Proc. Natl. Acad. Sci. USA.* 2017, 114: E751–E760.

Harris, P.R., Wright, S.W., Serrano, C. et al.: *Helicobacter pylori* gastritis in children is associated with a regulatory T-cell response. *Gastroenterology* 2008, 134: 491–499.

Noto, J.M., Gaddy, J.A., Lee, J.Y. et al.: Iron deficiency accelerates *Helicobacter pylori*-induced carcinogenesis in rodents and humans. *J. Clin. Invest.* 2013, 123:479–492.

Salama, N.R., Hartung, M.L., and Muller, A.: Life in the human stomach: Persistence strategies of the bacterial pathogen *Helicobacter pylori*. *Nat. Rev. Microbiol.* 2013, 11:385–399.

Serrano, C., Wright, S.W., Bimczok, D. et al.: Downregulated Th17 responses are associated with reduced gastritis in *Helicobacter pylori*-infected children. *Mucosal Immunol.* 2013, 6:950–959.

Mucosal responses to helminth infections

27

WILLIAM GAUSE AND RICHARD GRENCIS

Helminths, parasitic worms, are elongated (8–400 mm), multicellular, invertebrate organisms, most of which require a host for their maturation and reproduction. These parasites belong to three major animal groups, the nematodes or roundworms, the trematodes or flatworms, and the cestodes or ribbon worms. They have coevolved with vertebrates for hundreds of millions of years, remain ubiquitous in wild vertebrate populations, and are a significant concern in agriculture. Of these, roundworms are the most prevalent, with approximately 2 billion people worldwide infected by soil-transmitted roundworms including *Ascaris lumbricoides*, *Necator americanus* (hookworms), and *Trichuris trichiura* (whipworms). These roundworm infections are long lasting and can develop into chronic disease states. They are transmitted to the human host by either ingestion through contaminated food or skin penetration. Most nematodes cannot replicate in the host; therefore, the severity of infection is based on the frequency of the host's exposure to the infective stages. Opportunity for exposure is associated with the condition of the host's living environment. Roundworm infections persist in regions of poor hygiene and sanitation since the larvae survive using the organic material provided in feces and waste. Additionally, these areas coincide with areas of inadequate medical care, limiting preventive measures, and treatments that exacerbate the spread of disease. Many of these areas are found in subtropical and tropical regions, providing ideal temperature for survival in soil, and range from sub-Saharan Africa and South America to India and East Asia. Helminths can live for many years within a host, yet in many individuals they remain asymptomatic. In others, helminths can cause adult morbidity and impaired childhood development. Given this selection pressure, it is not surprising that a specific host immune response, the type 2 response, has evolved that can specifically target these harmful pathogens. Together, components of this response can contribute to the control of parasite burden and, just as importantly, the mitigation of tissue damage that occurs as the parasites migrate through vital organs and tissues. A hallmark of the type 2 immune response is that many of the same innate and adaptive immune cell populations important in protection against microbial pathogens also play a significant role in the response to helminths. However, these immune cells are differentially activated, expressing distinct cell surface and secreted molecules that together mediate the host protective response to this group of parasites. As helminths are too large to be phagocytosed, the immune response uses quite different mechanisms to destroy or expulse invading helminths. Control of tissue damage involves release of cytokines and other factors that both downregulate harmful type 1 inflammation and directly promote wound healing. A potential downside of the helminth-induced type 2 immune response is that in some cases it may increase susceptibility to microbial pathogens. This is due to the type 2 response producing factors that can downmodulate type 1 immunity important in resistance to many viruses and bacteria. At the same time, factors associated with type 2 immunity may also control harmful type 1 immune responses, which can contribute to inflammatory diseases. As such, components of the type 2 immune response may have the potential to be used for the development of new treatments for these conditions. The next few sections discuss characteristics of roundworm infection that inhabit or transverse mucosal tissues, and of the type 2 immune response elicited by these important pathogens.

INTRODUCTION TO HELMINTHS AND HELMINTH INFECTIONS

Infection by multicellular parasites such as soil-transmitted helminths (STHs) presents a considerable antigenic challenge to the host immune system. Not only do they contain multiple antigens, but they change the antigens they express as they move across mucosal surfaces and other tissues traversing the body to complete their life cycle in the definitive host. A good example of this is the human hookworm, with recent estimates indicating 439 million people infected and 3.23 million years lived with disability (YLDs) attributable to these parasites. In humans, hookworm infection constitutes two species, *Ancylostoma duodenale* and *Necator americanus*, that have similar but not identical life cycles.

27.1   Life cycle of a common human helminth

Humans become infected when infective larvae (known as the third larval stage, L3) that live in the soil, penetrate through the skin of the host (**Figure 27.1**). These then enter the vasculature and move through afferent circulation and the pulmonary circulation, eventually breaking through the lung capillaries into the lung parenchyma where they actively move up the

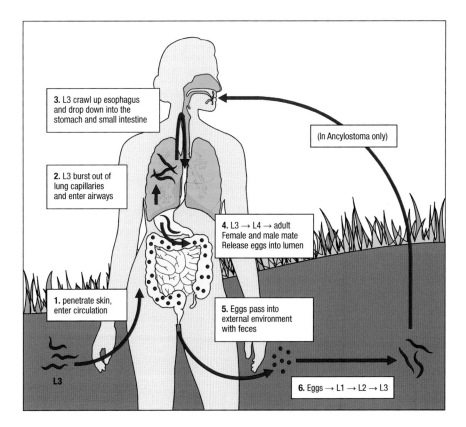

3. L3 crawl up esophagus and drop down into the stomach and small intestine

(In Ancylostoma only)

2. L3 burst out of lung capillaries and enter airways

4. L3 → L4 → adult
Female and male mate
Release eggs into lumen

1. penetrate skin, enter circulation

5. Eggs pass into external environment with feces

L3

6. Eggs → L1 → L2 → L3

Figure 27.1 Life cycle of human hookworm. (1) Infective L3 larvae are found in the soil and vegetation. On contact with the host, they can penetrate through the skin and enter the circulation (*Necator americanus* and *Ancylostoma duodenale*) or be ingested (*A. duodenale* only). (2) For skin-penetrating L3, the larvae migrate through the circulation into the lung capillaries, burst through into the lung spaces, and (3) crawl up esophagus/are coughed up and are swallowed, passing into the small intestine where (4) the L3 molt twice to become male and female adult worms, mate and female parasites release fertilized eggs into the intestinal lumen, which (5) pass out into the feces. (6) Under suitable conditions, the eggs hatch and molt twice to become infective L3.

airways and are then swallowed into the gastrointestinal (GI) tract, passing through the stomach. This is the only mode of infection for *N. americanus*, but infection by *A. duodenale* can also occur via oral ingestion of the L3 larvae. Once in the small intestine, the L3 parasites molt twice to become adult worms (approximately 1 cm in length) of separate sexes, mate, and the female worm then begins to release eggs into the intestinal lumen approximately 5–8 weeks after skin invasion. Female hookworms are capable of producing thousands of eggs per day over the course of their lives, which is generally 1–3 years for *A. duodenale* and 3–10 years for *N. americanus* (although can be for much longer). The eggs pass into the external environment via the feces where, if suitable conditions of warmth and moisture are present, they hatch and give rise to first-stage larvae (L1). These feed, undergo another two molts, and become nonfeeding L3 larvae that are about 600 μm in length. The L3 are very active and exhibit behavior that facilitates their chances of contacting a host; they can survive for several weeks before losing their infective potential.

Transit of larvae through the mucosal tissue of the lungs causes damage and hemorrhage, which is associated with cough, sore throat, and fever. Once adults are present in the small intestine, they attach to the mucosa via their mouthparts that have the characteristic "hooks" or cutting plates. They are able to browse the mucosa and as they feed, rupturing intestinal blood vessels. The parasites secrete anticoagulants and digestive enzymes that result in blood loss from the host that can lead to anemia, which is a particular concern in pregnancy due to iron loss. *A. duodenale* is associated with greater loss of blood from the intestine than *N. americanus*, but the latter has a wider distribution worldwide.

Individuals can become infected as early as 6 months of age. These individuals are likely to be exposed to L3 larvae continually throughout their lives, and epidemiological studies would suggest that host protective immunity is slow to develop and incomplete. The slow acquisition of acquired immunity is common with the other major STH of humans, *Trichuris trichiura* (whipworm) and *Ascaris lumbricoides* (roundworm). The latter two species have distinct life cycles, i.e., whipworm is contracted by ingestion of eggs found in the soil and has no migratory phases as an adult (2–3 cm in length), instead infecting the large intestine. Roundworm is contracted by ingestion of eggs from the soil but has a migratory phase through the lung with intestinal worms being up to 30 cm in length. For all the STH infections, children suffer the most severe symptoms, including anemia, malnutrition, intestinal obstruction, chronic dysentery, and rectal prolapse, particularly in heavily infected individuals. This may lead to impaired physical and cognitive development. At a population level in endemic regions, most people will be infected with relatively low parasite numbers and are likely to be infected for most of their lives, often with more than one species of STH.

27.2 Experimental models of helminth infections

Our understanding of the mechanisms of resistance and susceptibility to infection by helminth parasites has largely been determined through the use of infections in laboratory rodents, notably the mouse (see later sections). These have been extremely informative at defining the roles of various immune cells and molecules involved in protective immunity. By controlling conditions and infection levels (that are impossible to do in the field), these models have generated clear insight into how protective immunity can operate against these large multicellular parasites. Extensive use has been made of *Nippostrongylus brasiliensis*, which is a natural roundworm parasite of the rat but that also readily infects mice and exhibits many of the features of human hookworm in that it infects via skin penetration by the L3 larval stage and migrates through the lungs prior to residence in the small intestine. In mice, this infection is readily expelled during the intestinal phase of infection and animals become resistant to subsequent infection.

Another species used to model hookworm infection is *Heligmosomoides polygyrus bakeri*, a natural parasite of rodents, including mice, which resides in the small intestine. In this case, infection occurs by ingestion of the L3 larval stage, and no systemic migration occurs. A single infection usually results in a long-lived infection in most inbred strains of mouse, although strong resistance to infection can be generated experimentally by clearing a primary infection with anthelmintic drugs (see Section 27.8) prior to challenge with a secondary infection. Helminth models are therefore available that allow investigation of mucosal responses to parasites that live in different niches within the intestine.

Trichinella spiralis is a cosmopolitan and unusual parasite of mammals that infects the host as an L1 larval stage found in striated muscle (infected meat normally) and invades the epithelium of the small intestine. The value of this model in mice is that it generates a very robust immune response in the intestine with rapid clearance of the parasites. *Trichuris muris* is a natural parasite of the mouse, lives embedded within the epithelium of the large intestine, and is an excellent model for human whipworm infection (the second most prevalent human STH, *T. trichiura*), which occupies the same intestinal niche. This model can also be manipulated in the laboratory to induce strong host protective immunity or chronic infection. The use of this array of model systems has generated the current view of host immunity to this group of mucosal pathogens in addition to identifying novel aspects of mucosal immune responses.

HELMINTHS AND THE TYPE 2 IMMUNE RESPONSE

The type 2 immune response triggered by invading multicellular helminths is quite different from the type 1 immune response induced by much smaller microbial pathogens, such as viruses and bacteria. Generally, although many of the same immune cell lineages are involved in both types of responses, their activation states are quite different, resulting in the production of distinct arrays of cytokines and other factors mediating regulatory and effector functions for each response (see Table 27.1). As such, the immune response is tailored to the specific group of pathogens invading the host.

27.3 Danger-associated molecular patterns and alarmins released by helminths initiate the type 2 immune response

Microbe-associated molecular patterns (MAMPS), produced by microbes and recognized by various innate immune cells, are critical in triggering type 1 responses. In contrast, few helminth-associated MAMPs have been identified. Instead, danger-associated molecular patterns (DAMPS) appear to be preferentially important in triggering the helminth response. These endogenous danger signals appear to be released by host epithelial cells that are damaged by the large multicellular helminths as they invade barrier surfaces, such as the skin or intestinal epithelial cells. Not only is their large size an important factor, but these eukaryotic pathogens also secrete proteolytic enzymes that help create passageways as they traffic through organs and tissues. ATP, trefoil factor 2, and other DAMPS are quickly released by damaged cells at early stages of the helminth-induced immune response. They in turn stimulate release of alarmins, such as IL-33, IL-25, and thymic stromal lymphopoietin (TSLP) (Figure 27.2). These molecules play essential roles in driving initiation of the type 2 immune response. Although not yet well characterized, helminths also produce excretory/secretory (ES) products that can modulate the host immune response. These products may include factors that both trigger type 2 immunity and downregulate type 1 and type 2 inflammatory responses. For example, the helminth *Schistosoma mansoni*

TABLE 27.1 GENERAL OVERVIEW OF CYTOKINE EXPRESSION AND ACTIONS DURING THE TYPE 2 IMMUNE RESPONSE

Overall function	Cytokine	Produced by	Cells affected	Main function
Induce/promote type 2 immune response	TSLP IL-33 IL-25	Epithelial cells	Dendritic cells Eosinophils Mast cells Basophils Neutrophils ILC2 T_H2 cells	• Alarmins • Suppress IL-12 production in dendritic cells • Initiate T-cell differentiation by activating dendritic cells • Initiate type 2 response
	IL-4 IL-13	ILC2s Basophils Neutrophils Eosinophils T_H2 cells	Macrophages Eosinophils Dendritic cells Neutrophils T_H2 cells	• Suppress IL-12 production in dendritic cells • Initiate T-cell differentiation by activating dendritic cells • Initiate type 2 response • Induce AAM differentiation and proliferation • Induce increased mucus production by goblet cells • Increase intestinal fluid flow and epithelial turnover
	IL-5		Eosinophils	• Activate eosinophils
Resistance	IgG/IgE	B cells (activated by T_{FH} cells)	Basophils AAM Mast cells	• Enhance memory response when bound to FcεR1 receptors • Activate degranulation of mast cells • Activate basophils and AAM • Attach to surface of helminth to enhance immune cell recognition
	IL-15	Dendritic cells	Goblet cells	• Induce mucus production by goblet cells
	IL-22	ILC2s T_H17 cells		
	IL-13 IL-4	ILC2s Basophils Neutrophils Eosinophils T_H2 cells	Macrophages Eosinophils Goblet cells Epithelial cells	• Bind to goblet cells to increase mucus production • Increase intestinal fluid flow and epithelial turnover to expel helminths
Tissue repair	IGF-1	AAM	Damaged epithelial cells	• Induce fibroblast proliferation
	RELMα	AAM Eosinophils		• Promote collagen cross-linking
	TGF-α and -β	Eosinophils AAM		• Promotes matrix production and tissue repair
	VEGF	AAM		
	MMP	Eosinophils		
	Amphiregulin	ILC2s T_{reg}		• Promote tissue repair • Downregulate type 1 and type 17 inflammation
	Arginase 1	AAM ILC2s Mast cells		• Cell proliferation • Collagen synthesis
Immunosuppression	IL-10 TGF-β	T_{reg}	T_H1 and T_H17 cells	• Downregulate type 1 and type 17 inflammation

Note: Specific responses vary depending on pathogen species.

Abbreviations: AAM, alternatively activated macrophages; IGF-1, insulin-like growth factor 1; ILC2, innate lymphoid cells 2; MMP, matrix metalloproteinase; T_{FH}, T-follicular-helper cell; TGF, transforming growth factor; TSLP, thymic stromal lymphopoietin; VEGF, vascular endothelial growth factor.

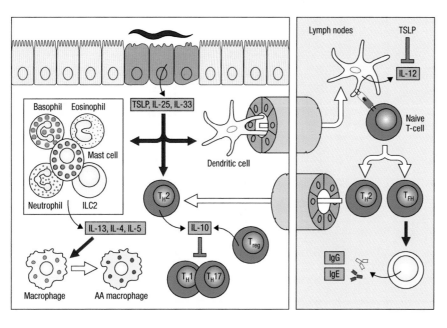

Figure 27.2 Alarmins initiate type 2 immune response through cytokine production and IL-12 suppression. Alarmins, secreted by damaged cells, are released into the intestine. Cells activated by the alarmins produce large quantities of IL-13, IL-4, and IL-5, which enhance the type 2 immune response by stimulating T_H2 cells and inducing alternatively activated macrophage differentiation. Alarmins additionally suppress IL-17 expression in dendritic cells which migrate to the lymph node to promote T_H2 cells differentiation from naïve T cells. T-follicular-helper cell (T_{FH}) activation induces IgE and IgG production through B cells.

releases the glycoprotein ω1, which attaches to lectin exposed on dendritic cells and conditions the dendritic cells to promote T_H2 cell differentiation by naïve T cells. Another ES molecule, in this case produced by *Heligmosomoides polygyrus*, has been identified as a TGFβ mimic. Although it shows no homology to the natural ligand, it can still bind the host TGFβ receptor and trigger its downstream signaling pathways. The contribution of these and other ES molecules to the initiation and development of the helminth-induced immune response is currently a significant area of research.

27.4 Innate immune cells initiate the type 2 immune response

Within hours after alarmin release, upregulated expression of type 2 cytokine genes, such as IL-4, IL-5, IL-9, and IL-13, can be detected in a number of innate immune cells, including innate lymphoid cells (ILCs) and granulocyte populations, such as basophils, neutrophils, and eosinophils. Additionally, alarmins suppress the production of IL-12 by dendritic cells, in this way further promoting a polarized type 2 immune response (see **Figure 27.2**). Each of these distinct innate immune cell lineages can differentiate to produce type 2 cytokines and other specific factors in the context of helminth invasion, with some innate immune cell types being preferentially important in specific tissues or at specific stages of the immune response. The capability of these quite different immune cell lineages to respond with such plasticity allows tailoring of each of the components of the immune response to the type of pathogen invading the host. The ILC2 subset has gained particular prominence in response to intestinal helminth infection by the production of the canonical type 2 cytokines. Of particular note is the recent demonstration that intestinal tuft cells, a chemosensory cell located in the gut epithelium, are a potent source of IL-25, that drives expansion of IL-5, IL-9, and IL-13 producing ILC2 at mucosal surfaces, particularly following intestinal helminth infection. Also, neuropeptides released from intestinal cholinergic neurons can activate ILC2s to proliferate, providing support for a coordinated response network following infection. These activated innate immune cells, through cytokine production, and potentially through other mechanisms, may then drive hematopoietic cells such as macrophages and dendritic cells and nonhematopoietic cells such as epithelial cells to differentiate along specific activation pathways. For example, recent studies have shown that

IL-4 producing basophils help drive the differentiation of antihelminth effector macrophages in the skin, while IL-13 producing neutrophils similarly promote effector macrophages in the lung. Communication between different innate immune cell lineages at the site of infection is still little understood.

27.5   T and B cells are initially activated in the lymph nodes

Activated dendritic cells migrate from the site of helminth invasion to the draining lymph nodes, where they interact with naïve T cells stimulating them through T-cell receptor (TCR) recognition of Ag complexed with MHCII. Innate immune cells are also significant sources of type 2 cytokines in the lymph node, and consequentially, CD4 T_H2 cells producing IL-4, IL-5, and IL-13 develop within several days (see Figure 27.2). T_H2 cells migrate to the periphery and are there further stimulated by ILC2 cells and other innate immune cells. CD4$^+$ T-cell production of T_H2 cytokines helps to further promote and amplify the type 2 immune response. In addition, follicular T-helper cells also develop in the lymph node and stimulate B-cell activation and Ig class switching to IgE and specific IgG isotypes. IgE and IgG secretion by B cells in the lymph nodes and spleen results in marked elevations of these antibodies in serum (see Figure 27.2). Although IgA is also elevated in response to intestinal helminth infection, surprisingly there are little data supporting a protective role for secretory antibody against helminths.

27.6   Controlling tissue injury is an important component of the antihelminth immune response

Several days after an initial helminth infection, a strong, polarized type 2 immune response has developed that includes both innate and adaptive components. This immune response can have potent host protective effects, which can be generally divided into tolerance and resistance mechanisms. Here tolerance refers to the ability to reduce the impact of the parasite on the host without affecting the parasite burden. The helminth-induced type 2 immune response has important tissue repair components that contribute to enhanced tolerance. Previously, macrophages were primarily associated with resistance to microbes via phagocytosis and destroyed and intracellular degradation. This class of macrophages is known as M1 macrophages. However, increased levels of IL-4 and IL-13 induce the differentiation and proliferation of alternatively activated macrophages (AAMs). This specific macrophage phenotype aids in antihelminth immunity; however, it can also contribute to tissue repair and regeneration (Figure 27.3). For example, AAMs produce insulin-like growth factor 1 (IGF-1), RELMα, and arginase 1, each of which is important in directly promoting tissue repair. IGF-1 induces fibroblast proliferation and extracellular matrix production, while arginase 1 induces cell proliferation and collagen synthesis, and RELMα promotes collagen cross-linking. Additionally, injury induced by helminths recruits eosinophils, whose granules contain RELMα, TGF-α and -β, fibroblast growth factors, collagen regulating enzymes, and many other molecules that play a role in wound healing. ILC2s can also help with tissue repair by producing amphiregulin (AREG), which can promote muscle and epithelial cell repair (see Figure 27.3).

For an effective tissue repair response, inflammation must be suppressed. Many of the cells and molecules involved in tissue repair also have the ability to suppress the pro-inflammatory response. Immune regulatory components are activated including FOXP3+ T-regulatory cells that can produce IL-10 and TGF-β (see Figure 27.3). These cells can downregulate harmful type 1– and type 17–mediated inflammation as parasites migrate through tissues,

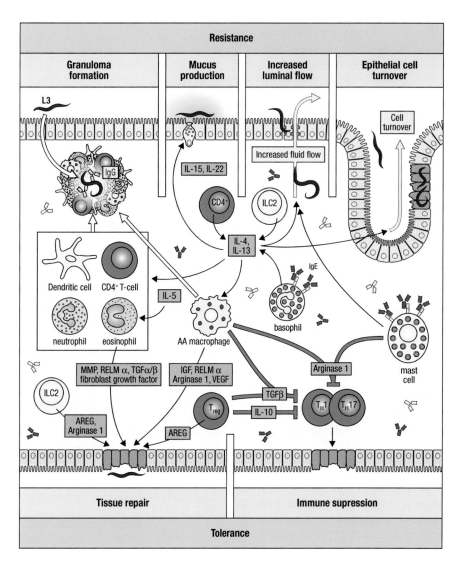

Figure 27.3 Overview of resistance and tolerance mechanisms in the intestine during a type 2 immune response. The general overview of the overlapping cytokine and cell-signaling pathways involved in both tolerance and resistance mechanisms. Cellular response and cytokine activity mediate different responses depending on the invading pathogen species. Granuloma formation generally responds to *Heligmosomoides polygyrus* yet can result from a number of other different pathogens. Mucus production and increased luminal flow are commonly activated by hookworm species and *Trichinella spiralis*. Epithelial turnover is induced by parasites that reside in the epithelial layer (i.e., *Trichuris muris* which attaches to the epithelial layer through a pocket of dead cells).

and which would otherwise contribute to tissue damage. T_{reg} cells may also produce AREG, thereby directly promoting tissue repair. As such, the type 2 immune response includes various immune cells producing cytokines and other factors that both downregulate harmful inflammation and directly promote tissue repair (see **Figure 27.3**). However, as with type 1 inflammation, the tissue repair mechanisms of the type 2 response must also be regulated. Chronic type 2 inflammation can result in excessive collagen production and increased deposition of extracellular matrix, which, if untreated, can lead to defective organ function and ultimately death of the individual.

27.7   Multiple mechanisms contribute to resistance

Many of the same cell types and even molecules that mediate tolerance also contribute to resistance against helminths (see **Figure 27.3**). For example, the AAMs activated by helminths have been shown to contribute to parasite damage at several barrier sites including the skin, lung, and intestine. This alternative activation of macrophages unleashes cellular signaling pathways and factors, which appear tailored to mediate multicellular parasite damage. These include adhesion of macrophages to helminths and upregulation of arginase which, through as yet undetermined mechanisms, can promote parasite killing. Other cell types that may contribute to parasite damage include eosinophils, basophils, and mast cells, although their mechanism of action, besides amplifying the type 2 response through production of cytokines, is still unclear. Resistance mechanisms may also depend on the site of helminth infection. In the intestine, increased goblet cell mucus production, triggered by IL-13, may coat parasites, impeding their capacity to damage intestinal epithelial cells and gain essential nutrients. Also, increased luminal fluid flow and increased muscle contractility, both of which can be elicited through IL-4 receptor (IL-4R) signaling, may contribute to expulsion of luminal parasites. With parasites, such as the whipworm (*Trichuris* sp.), that reside in the epithelial layer, IL-4R signaling can enhance epithelial cell turnover, essentially creating an epithelial cell escalator that propels the parasite back into the lumen. In the case of *Trichinella spiralis*, the larvae reside in the intestinal epithelial cells and here recruited mast cells may be particularly important in releasing mediators that degrade the tight junctions between the epithelial cells of the paracellular channels. Dissolving these junctions increases the fluid flow into the gut, driving worms into the lumen.

27.8 Memory response to helminth infection

The memory response to helminth parasites is essentially an accelerated, more potent type 2 immune response. In some cases, such as the immune response to the mouse parasite, *Heligmosomoides polygyrus bakeri*, memory is dependent on CD4$^+$ T cells, while in other cases, including the immune response to *Nippostrongylus brasiliensis*, acquired resistance does not require T cells at the time of the memory response. Instead, a persistent AAM phenotype develops that is capable of mediating accelerated resistance even in the absence of CD4$^+$ T cells. Memory B cells, through their production of antibodies IgG and IgE, also play an important role in resistance during the memory response. IgG may play an important role, in part through its contribution to parasite trapping and alternative macrophage activation, through binding Fc receptors and triggering effector-cell signaling pathways. IgE may be important in arming mast cells and basophils, through binding FcεR1 and thereby potentiating more rapid responses.

27.9 Type 2 response varies among helminth species

The preceding subchapters cover the type 2 immune response observed predominantly in response to nematodes. However, different helminth species can induce various immune responses in the host. This variation is dependent on the structure of the helminth in addition to their life cycle and migration pattern throughout the host. For example, while nematodes almost immediately induce a strong immunomodulatory and immunosuppressive type 2 response, other helminth species such as tapeworms and schistosome vary in the response they initiate. Tapeworms that most commonly infect livestock initially induce an equal balance of both an inflammatory type 1 and immunosuppressive type 2 response. Unlike the previously described type 2 tolerance/resistance response produced by nematodes to control the infection, the type 2 response during tapeworm infection increases the host's susceptibility to infection, while the early type 1 response promotes resistance and clearance. During trematode infections, the parasite itself does not induce a strong type 2 response. Instead, this response is initiated later in the infectious life cycle in response to egg production. With limited detrimental effects to the parasite and continued egg production, type 2 immune responses during schistosome infections can develop into chronic schistosomiasis leading to increased inflammation and fibrosis.

27.10 Helminth infections may impact effective immune responses against microbial pathogens

Geographic locations of high endemic helminth infections tend to overlap with high prevalence of microbial diseases, such as tuberculosis and AIDS. As discussed, type 2–mediated suppression of the type 1 inflammatory response is beneficial for both tolerance and resistance mechanisms during a helminth infection. However, type 1 immunity is essential in control of microbial infections and for effective vaccines. It is very clear from a variety of experimental systems that helminth infection can negatively influence the protective responses to a number of viral, bacterial, and protozoan infections including respiratory syncytial virus, norovirus, influenza, salmonella, mycobacteria, *Plasmodium* species, and *Leishmania* spp. Recent mechanistic studies suggest that helminth infections can impede clearance of microbial pathogens by increased levels of type 2 and other immunosuppressing cytokines, including TGF-β and IL-10. However, mechanisms are complex and vary between different interacting species, and influences of altered intestinal microbiome are also likely to be involved. In humans, effects of helminths on coinfection are much more difficult to define, and different studies can often generate conflicting data. Nevertheless, there is good evidence to suggest that intestinal helminth infection can influence the immune response to a number of coinfecting microbial pathogens such as *Mycobacterium tuberculosis*, *Plasmodium* spp., and *Leishmania* spp., although this does not always translate to a clear alteration in clinical outcome of the coinfection.

CURRENT TREATMENTS

The fact that helminth infections by STH are often associated with iron deficiency, malnutrition, mucosal pathology, dysentery, and rectal prolapse indicates that removal of the parasites would have considerable health benefits, particularly for children. The most cost-effective strategy implemented so far is the periodic administration of anthelmintic drugs (usually one dose, annually if possible), particularly to preschool and school-age children. This is a goal of the World Health Organization; however, there are many practical and economic challenges to achieving this in parts of the world endemic for intestinal dwelling helminths.

27.11 Common anthelmintic drugs and increasing resistance

There are various anthelmintic drugs used to clear STH in man. Currently the most commonly used are the benzimidazoles, albendazole and mebendazole. Both drugs work relatively well against *A. lumbricoides*, while only albendazole shows good efficacy against hookworm. For the whipworm, *T. trichiura*, both drugs show relatively poor cure rates. This group of drugs works by disrupting the tubulin-microtubule system in cells, particularly β-tubulin, which results in parasites being unable to survive effectively in their chosen intestinal niche. Unfortunately, as demonstrated extensively from studies in animals, particularly domestic stock, the extensive use of these drugs drives resistance. It is generally agreed that changes in parasite β-tubulin are the basis for the resistance observed in gut-dwelling nematodes. Changes to the structure of the molecule driven through mutations to the β-tubulin gene result in structural alterations resulting in the failure of benzimidazoles to bind their target sites or have decreased binding efficiency. Other anthelmintic groups such as the nicotinic agonists (e.g., pyrantel) that target nicotinic acetylcholine (Ach)–gated cation channels found at the parasite's neuromuscular junctions are also used but generally have no better efficacy and again generate drug resistance.

Immunologically, it is important to remember that clearance of the parasites by drugs does not mean that reinfection will not occur. Indeed, it does, repeatedly, and often to pretreatment levels within 12–18 months. Thus, because of the strategies evolved by the helminths to maximize their efficiency of infection (i.e., large numbers of resistant/persistent infectious stages in the environment), reducing levels that will lead to a break in transmission are very difficult to achieve. When coupled with the fact that the helminths are able to modulate or evade host protective immunity leading to lifelong infection, it can be appreciated that long-term effective intervention is challenging.

27.12 Vaccine development

There are currently no vaccines against STH of man. Vaccines against helminths *per se* are rare even for veterinary use. Many issues contribute to this situation. Genomic information on the major STH is now available for all the major species that infect man, although genetic manipulation of the parasites remains problematic. Choosing the most effective parasite stage to target (e.g., intestinal dwelling, lung or skin invading) is critical. Identification of host-protective antigens of the chosen stage as potential vaccine candidates is also difficult, as antibody recognition profiles in infected individuals do not clearly associate with host protection, a stronger association being cellular immune responses and type 2 cytokine profiles. Thus, screening potential antigens with antibody is not informative. IgE may be the exception here for certain helminth infections, although the use of IgE-inducing antigens as helminth vaccines in endemic regions is unlikely to progress into widespread use based on recent work with a potential hookworm vaccine candidate. In this case, vaccinating individuals previously infected with hookworm using an antigen from the skin-invading L3 stage caused unacceptable dermal immediate hypersensitivity reactions. As a consequence, efforts are currently directed toward other major targets such as the digestive proteases used by the adult stages of the worm in the intestine.

Another complication is that STH infections often begin around 6 months of age and continue at some level for life. Thus, vaccines are most likely to have to be given to individuals already infected, or possibly those that have had recent anthelmintic treatment. Thus, vaccinees will be individuals who are not mounting very effective host-protective immune responses to the parasites and are also subject to the parasite's immunomodulatory mechanisms (see later). Presently, very few vaccine development studies take these multiple confounding factors into account.

HELMINTHS AND INFLAMMATORY DISORDERS

In 1989 Strachan noted a correlation between increased incidences of seasonal rhinitis with decreased family size in developed countries. This insightful observation led to the hypothesis that allergic responses may be related to lower sibling-related childhood infections and decreased exposure to microbes—the "hygiene hypothesis."

27.13 Hygiene hypothesis

The hygiene hypothesis postulates that the coevolution of infectious agents and microbes with humans and the consequent adaptive immunomodulatory and immunoregulatory strategies has led to a balanced but dynamic relationship. In the absence of high microbial diversity and infectious challenge, immunoregulatory processes are disrupted and lead to a higher frequency of allergies and inflammatory disorders, including autoimmune diseases, particularly in industrialized countries. The microbes that underpin this remain to be defined but will probably include those associated with rural living, i.e., components of the "archaic" microbiome. Helminth parasites fit well into this scenario with nematodes believed to have first invaded primitive tetrapods around the time of the colonization of land by animals. Thus, helminth infection has been a constant pressure throughout evolution and influenced the development of our protective mechanisms that operate at mucosal barriers.

The fact that STH infection is associated with immunoregulation and modulates the incidence and severity of inflammatory disease raises the possibility of identifying new therapeutic approaches to these conditions. Numerous studies in the laboratory have demonstrated that experimental GI nematode infections in mice modulate a variety of inflammatory insults such as experimentally induced colitis, allergic challenge in the lung and skin, and models of autoimmune disease.

Remarkably, this approach has been rapidly translated to experimental studies in humans. These involve infecting individuals with inflammatory conditions with live, fully pathogenic helminths. This has included both whipworm and hookworm in studies designed to modulate conditions such as inflammatory bowel disease of various types, autoimmune conditions such as multiple sclerosis, and seasonal allergies. Success has been limited so far, but the notion clearly holds promise.

27.14 Excretion/secretion molecules and their therapeutic potential

At the other end of the spectrum, great efforts are aimed at defining the parasite-derived molecules that initiate and maintain immunomodulatory responses, with most of this work using parasites in mice. A starting point for helminth-derived immunomodulatory molecules is a detailed analysis of the excretions/secretions (ESs) from various parasite stages. It is reasoned that these will most readily gain access to cells and tissues of the immune system. The ESs usually comprise hundreds of different molecules of all types including proteins, glycoproteins, carbohydrates, and lipids. Once immunomodulatory function of the complex mixture can be demonstrated either *in vitro* or *in vivo*, the hunt for specific molecules begins. Relatively few have been defined to date, but there are a number of promising candidates from several helminths including gut-dwelling nematodes, e.g., a mimic of host transforming growth factor (TGF-β) from *H. polygyrus bakeri* that can drive the induction of T_{reg} cells *in vitro*.

It is also known that GI helminth infection can be associated with marked changes in the intestinal microbiome of the host in man and animals. Experimental laboratory studies have shown that the modulation of allergic responses in the lung-accompanying infection can be due to the altered microflora and the products they secrete (e.g., fatty acids), rather than directly by parasite-derived molecules. This emphasizes the complex and dynamic interplay between intestinal-dwelling helminths, the intestinal microflora, and the mucosal immune system, and the importance of each to overall health.

SUMMARY

Helminth parasites can seriously compromise host fitness in man, vertebrates in the wild, and animals used in agriculture. In humans, they remain a major global health problem contributing to morbidity and impaired childhood development. The type 2 immune response is uniquely adapted to mediate protective immunity against these relatively large multicellular eukaryotic pathogens. Although many of the same immune cell types are involved in helminth and microbial infections, their activation states can be quite different. Host-protective immune responses against helminths can mediate enhanced resistance and mitigation of tissue injury, which is particularly important as these large parasites migrate through vital organs such as the lung and intestine. However, in some cases, chronic type 2 immune responses can lead to harmful type 2 inflammation contributing to fibrosis and components of the type 2 immune response can cause immediate hypersensitivity reactions.

Epidemiological studies have identified a correlation between helminth infections and reduced inflammatory and autoimmune diseases in developing countries. Helminth excretory/secretory products, an outcome of the long coevolutionary relationship with vertebrates, have numerous immunomodulatory effects. More research is needed to develop effective vaccines against helminths and to potentially harness immunomodulatory effects of these helminths to treat inflammatory diseases and enhance wound healing.

ACKNOWLEDGMENT

Ms. Katherine Lothstein is a doctoral student in the laboratory of Dr. William C. Gause, where she is engaged in the study of how helminth excretory/secretory molecules modulate host immune function. She made significant contributions to the text and figures of this chapter.

FURTHER READING

Babu, S., and Nutman, T.B.: Helminth-tuberculosis co-infection: An immunologic perspective. *Trends Immunol.* 2016. 37(9):597–607.

Bancroft, A.J., McKenzie, A.N., and Grencis, R.K.: A critical role for IL-13 in resistance to intestinal nematode infection. *J. Immunol.* 1998, 160(7):3453–3461.

Chen, F., Wu, W., Millman, A., Craft, J.F., Chen, E., Patel, N., Boucher, J.L., Urban, J.F. Jr., Kim, C.C., and Gause, W.C.: Neutrophils prime a long-lived effector macrophage phenotype that mediates accelerated helminth expulsion. *Nat. Immunol.* 2014, 15(10):938–946.

Gause, W.C., Wynn, T.A., and Allen, J.E.: Type 2 immunity and wound healing: Evolutionary refinement of adaptive immunity by helminths. *Nat. Rev. Immunol.* 2013, 13(8):607–614.

Grainger, J.R., Smith, K.A., Hewitson, J.P. et al.: Helminth secretions induce *de novo* T cell Foxp3 expression and regulatory function through the TGF-β pathway. *J. Exp. Med.* 2010, 207(11):2331–2341.

Grencis, R.K.: Immunity to helminths: Resistance, regulation, and susceptibility to gastrointestinal nematodes. *Annu. Rev. Immunol.* 2015, 33:201–225.

Howitt, M.R., Lavoie, S., Michaud, M. et al.: Tuft cells, taste-chemosensory cells, orchestrate parasite type 2 immunity in the gut. *Science.* 2016, 351(6279):1329–1333.

Loukas, A., Hotez, P.J., Diemert, D., Yazdanbakhsh, M., McCarthy, J.S., Correa-Oliveira, R., Croese, J., and Bethony, J.M.: Hookworm infection. *Nat. Rev. Dis. Primers.* 2016, 2:16088.

Neill, D.R., Wong, S.H., Bellosi, A. et al.: Nuocytes represent a new innate effector leukocyte that mediates type-2 immunity. *Nature* 2010, 464(7293):1367–1370.

Osbourn, M., Soares, D.C., Vacca, F. et al.: HpARI protein secreted by a Helminth parasite suppresses interleukin-33. *Immunity.* 2017, 47(4):739–751.

Ramanan, D., Bowcutt, R., Lee, S.C. et al.: Helminth infection promotes colonization resistance via type 2 immunity. *Science* 2016, 352(6285):608–612.

Rivera, A., Siracusa, M.C., Yap, G.S., and Gause, W.C.: Innate cell communication kick-starts pathogen-specific immunity. *Nat. Immunol.* 2016, 17(4):356–363.

Vannella, K.M., Ramalingam, T.R., Borthwick, L.A. et al.: Combinatorial targeting of TSLP, IL-25, and IL-33 in type 2 cytokine-driven inflammation and fibrosis. *Sci. Transl. Med.* 2016, 8(337):337.

Webb, L.M., and Tait Wojno, E.D.: The role of rare innate immune cells in type 2 immune activation against parasitic helminths. *Parasitology.* 2017, 144(10):1288–1301.

Wynn, T.A., and Vannella, K.M.: Macrophages in tissue repair, regeneration, and fibrosis. *Immunity.* 2016, 44(3):450–462.

28

Viral infections

SARAH ELIZABETH BLUTT, MARY K. ESTES, SATYA DANDEKAR,
AND PHILLIP D. SMITH

Infection of mucosal surfaces by an array of viral pathogens is an important cause of morbidity, and occasionally mortality, in residents of both developed and developing countries. Among such viruses, certain RNA viruses have evolved unique properties by which they exploit mucosal surfaces to achieve or induce entry, survival, disease, and transmission to naive hosts. In this chapter, we discuss two RNA viruses: rotavirus, which infects predominantly infants and children, and human immunodeficiency virus-1 (HIV-1), which infects people of all ages but especially adults. We emphasize the unique bidirectional interactions between each virus and the mucosal immune system. Understanding the biology of rotavirus and interaction between the virus and the mucosal immune system has led to the development of highly effective live-attenuated oral vaccines, making life-threatening rotavirus diarrhea a vaccine-preventable disease. However, current vaccines have suboptimal effectiveness in low-income countries stimulating new efforts to understand the causes and improve vaccine effectiveness in those settings. In contrast, the biology of HIV-1 and the interactions between HIV-1 and cells of the mucosal immune system are more complex and not fully elucidated, preventing the successful development of an effective vaccine.

ROTAVIRUS INFECTION

Sarah E. Blutt and Mary K. Estes

Rotavirus infects nearly every child in the world and is the leading cause of life-threatening diarrheal disease among infants and young children in many countries (Table 28.1). Unlike many bacterial pathogens, rotavirus occurs with high frequency in both temperate and tropical climates and in both developed and less-developed settings. Worldwide, severe rotavirus-induced gastroenteritis accounts for almost 40% of hospital admissions for diarrhea and over 200,000 deaths annually, the latter occurring primarily in developing countries, in children younger than 5 years of age. These grim statistics have stimulated efforts to develop rotavirus vaccines, several of which are licensed and being used now in young children.

Rotavirus causes an acute diarrheal disease, usually lasting about 5 days in immunocompetent individuals (see Table 28.1). The clinical symptoms in young children include vomiting and diarrhea. The diarrheal stools do not contain blood, mucus, or inflammatory cells, hallmarks of inflammation that help distinguish viral from bacterial enteritis. Rotavirus, which is environmentally highly stable, is shed in large numbers (approximately 10^{10} particles per gram of diarrheal stool). This explains why rotaviruses can be identified in the stools of ill children by electron microscopy, which requires

TABLE 28.1   CLINICAL AND EPIDEMIOLOGICAL FEATURES OF ROTAVIRUS INFECTION

Variable	Feature
Age predisposition	Primarily affects young children 6 months to 3 years old in developed countries; younger children affected in developing countries. Adults infrequently infected, mostly asymptomatic.
Seasonality	Seasonal infection in developed countries with epidemic peaks in cooler months of the year. No seasonality in tropical climates.
Settings	Households, day care centers, hospitals, schools, favoring person-to-person spread.
Asymptomatic infections	Common in adults and newborns. Can occur in all age groups.
Incubation period	Generally 24–48 hours.
Symptoms	Sudden onset of vomiting and diarrhea. Diarrheal stools do not contain blood, mucus, or leukocytes. Fever common.
Severity of illness	Generally more severe than many other infectious diarrheal illnesses, leading to dehydration and hospitalization. Malnutrition increases disease severity.
Duration of illness	Typically 3–5 days. Longer illness in immunocompromised.
Virus shedding	Peaks 1–3 days after onset of illness. Shedding can be prolonged in immunocompromised individuals. Antigenemia and viremia detectable early after infection.
Mode of transmission and vehicles	Fecal-oral; aerosol/vomitus; contact with fomites. Food, water, and environmental contamination may transmit infection.
Immunity	Repeated infections with or without illness can occur with the same or different strains. Long-term immunity occurs with repeated exposure.
Treatment	Supportive rehydration therapy to prevent dehydration. Live attenuated vaccines are available and highly effective in developed countries; vaccine effectiveness is less for children in developing countries.
Reservoir	Humans and animal reservoirs.

approximately one million particles per gram of stool for visualization, and why the virus is so infectious and easily spread by the fecal-oral route: as few as 10 particles can cause infection in a susceptible host. Asymptomatic infections frequently occur, and multiple infections can lead to immunity. Treatment is supportive to alleviate the associated dehydration. Rotavirus disease has been significantly reduced by use of new, live attenuated vaccines.

28.1   Rotavirus structure impacts pathogenicity in the intestinal mucosa

Rotaviruses are members of the *Rotavirus* genus of the *Reoviridae* family of viruses, which contains viruses with segmented double-stranded RNA

Figure 28.1 **Rotavirus structure and proteins.** Rotavirus particles are visualized in (**a**) by negative stain electron microscopy. (**b**) The rotavirus genome of 11 segments of double-stranded RNA analyzed by polyacrylamide gel electrophoresis. Each gene encodes at least one protein with at least one major function. (**c**) A cartoon of a rotavirus particle shows the designated proteins that make up each concentric protein layer. (Adapted from Greenberg, H., and Estes, M. *Gastroenterology*. 2009, 136:1939–1951. With permission from Elsevier Ltd.)

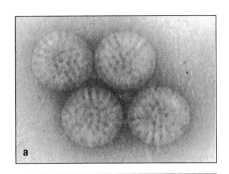

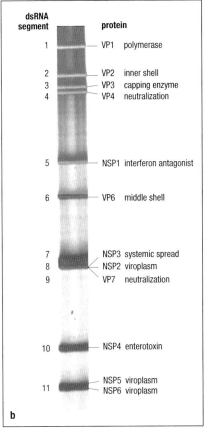

genomes. Rotavirus particles are large (1000 Å; Figure 28.1a) and complex with three concentric protein layers that surround the viral genome of 11 segments of double-stranded RNA (see Figure 28.1b). The rotavirus genome segments code for six structural proteins that make up virus particles (viral protein [VP]) and six nonstructural proteins (NSPs; see Figure 28.1b). The NSPs are synthesized in infected cells and function in some aspect of the viral replication cycle or interact with host proteins to influence pathogenesis and/or the immune response. Two rotavirus proteins, VP7, which makes up the outer capsid protein shell, and VP4, which forms spikes that emanate through the outer capsid shell, induce neutralizing antibody and are the basis of a binary classification system for viral serotypes (see Figure 28.1c). Thus, VP7 (a glycoprotein or G-type antigen) and VP4 (a protease-sensitive protein or P-type antigen) are used to classify rotaviruses. To date, investigators have identified 32 G and 28 P types of rotavirus.

Rotaviruses exhibit unusual structural complexity and properties that are relevant to their success as gastrointestinal pathogens. Entry into, and replication in, intestinal epithelial cells is critical for this success (Figure 28.2). The stable outer proteins first bind to receptors on differentiated epithelial cells near the tips of the small intestinal villi. These mature intestinal epithelial cells express factors required for efficient infection and replication. The 60 spikes that protrude from the surface of the viral capsid act as initial attachment proteins to bind host receptors. The VP4 spike protein is susceptible to proteolytic cleavage, a common feature of attachment proteins of many viruses that infect mucosal surfaces. Proteolytic cleavage by trypsin at the mucosal surface induces a remarkable conformational change in the structure of the spike, exposing additional attachment sites on the spike and the surface glycoprotein of rotavirus for interaction with a series of host cell coreceptors. The multistep attachment and entry process into intestinal epithelial cells results in virus delivery across the plasma membrane by endocytosis and into the cell where the outer capsid shell is removed and double-layered particles are delivered into the cytoplasm. These particles function as molecular machines and produce capped viral mRNAs that are extruded from transcribing particles into the cytoplasm to be translated to produce new proteins and replicated to produce new genomic RNA. The proteins in the core of the incoming particles possess all the enzymatic activities required to produce the capped viral transcripts from the viral genome dsRNA, because eukaryotic cells do not express RNA polymerases capable of transcribing mRNA from dsRNA templates.

28.2 Rotavirus replicates in intestinal epithelial cells

Unique aspects of rotavirus replication include the incomplete uncoating of the virus, resulting in the absence of free dsRNA within the cell. Viral replication is restricted to the cell cytoplasm and occurs within specialized electron-dense structures called viroplasms localized adjacent to the cell nucleus and near the endoplasmic reticulum (see Figure 28.2). Viroplasms are composed of nascent viral proteins, and their size and shape change during

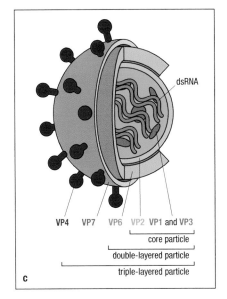

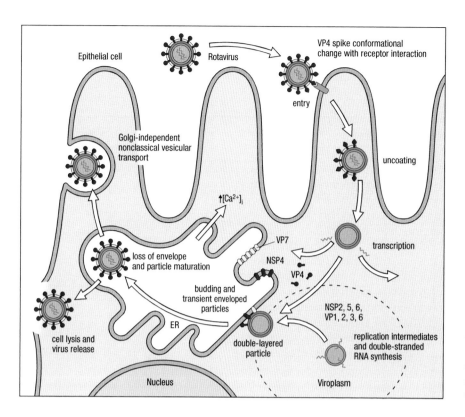

Figure 28.2 Stages of the rotavirus life cycle. Interaction of rotavirus with epithelial cell receptors leads to rearrangement of the virus spike protein and virus entry and uncoating. Removal of the outer capsid proteins produces double-layered particles and activates the endogenous RNA-dependent RNA polymerase that produces mRNA transcripts from each dsRNA segment. The mRNAs are translated on ribosomes that are free or associated with the endoplasmic reticulum (ER). Some proteins associate to form viroplasms in which viral RNA is replicated and encapsidated into newly made particles. The nascent particles bind to the C-terminus of the NSP4, which acts as an intracellular receptor, and particles acquire the outer capsid proteins VP4 and VP7 while budding into the lumen of the ER. The transient membrane is lost and particles undergo maturation. Particles are released by cell lysis or from the apical surface of polarized cells by a Golgi-independent nonclassical vesicular transport mechanism.

the replication cycle. Newly produced double-layered particles containing replicated dsRNA bud from viroplasms into the endoplasmic reticulum (ER) in an unusual process during which particles become transiently enveloped. The budding of particles into the ER is initiated by the binding of newly made double-layered particles to an intracellular viral receptor. This receptor consists of the cytoplasmic tail of a rotavirus nonstructural protein (NSP4) that is a transmembrane ER glycoprotein. The outer capsid proteins are incorporated into new particles during the budding process through protein rearrangements that occur as the transient envelope is lost. Mature viral particles are released from epithelial cells either by cell lysis or by delivery of particles to the apical plasma membrane of polarized cells by a nonclassical trafficking pathway. The epithelial cells die and are extruded from the villus surface, causing a potential, albeit transient, disruption in the epithelial barrier.

28.3   Rotavirus infection is calcium (Ca^{2+}) dependent

Rotaviruses enhance and exploit Ca^{2+} signaling as a key cellular target to regulate replication, morphogenesis, and pathogenesis (**Figure 28.3**). Thus, rotavirus infection results in at least threefold increases in intracellular calcium $[Ca^{2+}]i$ and up to 10-fold increases in uptake of Ca^{2+} into cells. Ca^{2+} also plays an important role in virion assembly and disassembly processes. Ca^{2+} maintains the integrity of the rotavirus outer capsid layer; VP7 is a Ca^{2+}-binding protein; and Ca^{2+} chelation is a mechanism that activates the endogenous RNA polymerase. NSP5 also is a Ca^{2+}-binding protein, and viroplasm formation requires Ca^{2+}. Rotavirus morphogenesis is dependent on the presence of sufficient levels of $[Ca^{2+}]i$. In the absence of Ca^{2+}, virus morphogenesis is terminated at the double-layered particle step, and VP7 is excluded from hetero-oligomeric complexes made of NSP4 and VP4 that participate in the budding of double-layered particles into the ER. Furthermore, Ca^{2+} depletion of the ER by the sarco-/endoplasmic reticulum Ca^{2+}-ATPase

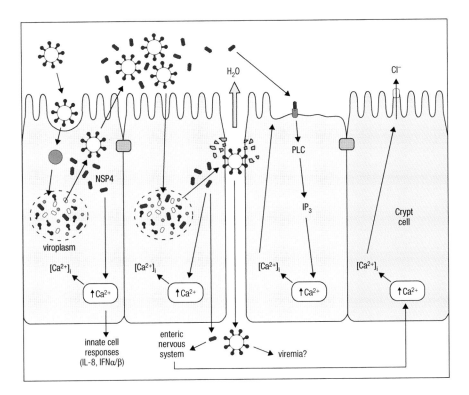

Figure 28.3 Mechanisms by which rotavirus causes diarrhea. Cellular and viral events that occur after rotavirus infection of enterocytes are shown in order from left to right. Rotavirus infection produces virus particles and viral proteins, such as NSP4, which are released by a nonclassical secretory pathway. Intracellular NSP4 also induces the release of Ca^{2+} from internal stores, primarily the endoplasmic reticulum, leading to increased levels of intracellular calcium, $[Ca^{2+}]i$. Innate intestinal epithelial cell responses to infection result in the release of pro-inflammatory cytokines. A cell is secondarily infected after rotavirus is released from the adjacent cell. Intracellular expression of NSP4 results in elevated levels of $[Ca^{2+}]i$ through a PLC-independent mechanism. The increase in $[Ca^{2+}]i$ also disrupts the microvillar cytoskeleton. Infected villus enterocytes may stimulate the enteric nervous system by the basolateral release of NSP4 or other effector molecules. NSP4 produced by the infection disrupts tight junctions, allowing paracellular flow of water and electrolytes (blue arrow). Viral particles may also gain access to the blood by this route. NSP4 released from previously infected cells binds to a specific receptor and triggers a signaling cascade through phospholipase C (PLC) and inositol 1,4,5-trisphosphate (IP_3) that results in release of Ca^{2+} and an increase in $[Ca^{2+}]$i. One receptor for NSP4 is $\alpha_2\beta_1$, and NSP4 binding to this receptor can elicit diarrhea in neonatal mice. A crypt cell can be acted on directly by NSP4, or NSP4 can stimulate the enteric nervous system, which in turn signals an increase in $[Ca^{2+}]i$ that induces Cl^- secretion. IFN, interferon; IL-8, interleukin-8. (Adapted from Ramig, R. *J. Virol.* 2004, 78:10213–10222. With permission from American Society for Microbiology.)

pump inhibitor thapsigargin inhibits VP7 and NSP4 glycosylation and virus maturation. NSP4 is the only rotavirus protein that mobilizes $[Ca^{2+}]i$ in cells by functioning as a viroporin in the ER membrane. Release of $[Ca^{2+}]i$ from the ER alters plasma membrane permeability and compensatory entry of extracellular Ca^{2+} into cells. Changes in calcium homeostasis also alter other cellular functions that affect pathogenesis.

Rotavirus pathogenesis

Rotavirus infects and replicates in the mature nondividing enterocytes and enteroendocrine cells in the middle and tip of small intestinal villi, sparing the less-differentiated crypt cells. Infection typically resolves within 7 days. The outcome of infection is affected by both viral and host factors. The major host factor that affects the clinical outcome of infection is the age of the host. Thus, neonates may be infected with rotavirus, but they rarely have symptomatic disease, likely due to the protection provided by transplacentally acquired maternal antibody. The waning of such antibody coincides with the age of maximum susceptibility of infants to severe rotavirus-induced disease (3 months to 2 years of age). As a consequence of waning acquired immunity, adults can experience severe symptomatic disease, and such illness may result from infections with an unusual virus strain or infection with high doses of virus.

28.4 Rotavirus virulence is mediated by five genes

Rotavirus virulence is related to properties of the proteins encoded by a subset of the 11 viral genes. Virulence is multigenic and associated with the products of genes 3, 4, 5, 9, and 10 (see **Figure 28.1**). The involvement of these genes in virulence is only partially understood. Gene 3 encodes the capping enzyme that affects levels of viral RNA replication and can antagonize innate immunity, and genes 4 and 9 encode the outer capsid proteins required to initiate infection. Gene 5 encodes a protein (NSP1) that functions as a major interferon antagonist and is discussed later (see "Rotavirus immunity").

Gene 10 codes for the nonstructural protein NSP4, which regulates calcium homeostasis and virus replication and functions as an enterotoxin.

28.5 Diarrheal illness caused by rotavirus is multifactorial

Diarrhea and vomiting are the main clinical manifestations of rotavirus infection in infants and young children. Disease pathogenesis is multifactorial, with malabsorption and diarrhea the consequences of virus-mediated destruction of absorptive enterocytes, downregulation of the expression of absorptive enzymes, and functional changes in tight junctions between enterocytes that lead to paracellular leakage (see Figure 28.3). Our understanding of disease pathogenesis is based primarily on studies in animal models. A secretory component of rotavirus diarrhea is also recognized and is thought to be mediated by activation of the enteric nervous system and the effects of NSP4, the first identified virus-encoded enterotoxin. Studies of virus and NSP4 in cultured cells and animal models indicate that rotavirus-induced diarrhea results, in part, from activation of cellular Cl⁻ channels, inducing an increase in Cl, and consequently, water secretion (see Figure 28.3). Cl⁻ secretion is induced by a mechanism that does not involve the cystic fibrosis transmembrane regulator, as rotavirus and NSP4 induce diarrhea in mouse pups lacking this channel and in children with cystic fibrosis. Villus ischemia and alterations in intestinal motility are recognized in some animal models, but their role in disease in children is poorly documented.

28.6 Mucosal rotavirus infection is associated with extraintestinal rotavirus

Extraintestinal rotavirus can be detected in the blood for at least a short period after infection in immunocompetent hosts. Virus may enter the circulation by uptake through Peyer's patches or through breaches in the intestinal epithelium, accounting for the detection of virus in several tissues. The clinical consequences of systemic infection remain unclear, but one consequence may be enhanced innate responses, including the release of cytokines that induce fever. Although many case reports associate rotavirus with systemic illnesses, currently there is no proof of causation from extraintestinal spread of rotavirus.

Rotavirus immunity

The first level of immunity to human rotavirus is an innate susceptibility or resistance to infection based partially on an individuals' fucosyltransferase 2 (FUT2) genotype. FUT2 controls fucosylation and the expression of ABH histoblood group antigens (HBGAs) at the intestinal mucosa. These carbohydrates function as binding ligands and are thought to be initial receptors required for rotavirus binding to host cells. Susceptibility or resistance to a particular rotavirus strain depends on whether the cell surface in the gut expresses the appropriate binding glycan. Human rotaviruses bind HBGAs, while most animal virus strains bind sialoglycans. The distribution of HBGA types in different human populations affects the prevalence of common rotavirus genotypes across the globe and susceptibility to some strains of animal origin. Innate and adaptive immune cells direct the host response to rotavirus in both primary infection and reinfection (Figure 28.4). Innate immunity that regulates interferon and the induction of interferon (IFN)-stimulated genes also has an important role in modulating rotavirus infection *in vitro* and in animal models, but its role in humans is not fully understood. Protection from reinfection and disease generally correlates positively with markers of mucosal immunity, including levels of anti-rotavirus intestinal IgA, enteric rotavirus-reactive antibody-secreting cells, and IgA memory cells.

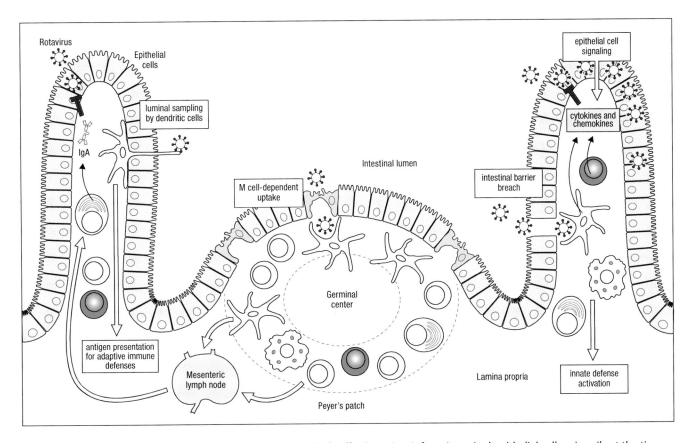

Figure 28.4 **Rotavirus interaction with intestinal cells.** Rotavirus infects intestinal epithelial cells primarily at the tips of the villi. The virus enters the mucosa through several routes, including lamina propria dendritic cells, M cells located on Peyer's patches, and possibly breaches in the epithelium. The ensuing innate immune response helps control primary infection, facilitate rotavirus clearance, and regulate the adaptive immune response to the virus. During rotavirus infection, B cells are activated and proliferate. Protection from reinfection is due at least in part to rotavirus-specific antibody produced by IgA antibody-secreting cells that migrate back into the lamina propria. (Adapted from Niess, J., and Reinecker, H. *Cell. Microbiol.* 2006, 8:558–564. With permission from John Wiley & Sons.)

28.7 Multiple immune cells contribute to host responses to rotavirus

CD8[+] T-cell responses contribute to rotavirus clearance and shortening the course of primary infection, whereas B-cell responses are the primary determinant of protection against rotavirus reinfection, as shown in mouse models. CD4[+] T cells provide help to CD8[+] T-cell and B-cell responses to rotavirus. After immunization with recombinant VP6, CD4[+] T cells appear to contribute to protection in an IFN-γ-dependent manner. Regulatory T cells appear not to affect the level of rotavirus replication and shedding or the severity of disease during primary infection in an animal model. Lymphocyte homing also plays a prominent role in the regulation of immune responses to rotavirus in the intestinal mucosa, and B-cell trafficking to the intestine appears critical for the induction of optimal protection in a mouse model of chronic infection.

Passive transfer studies of monoclonal antibodies in the mouse demonstrate that neutralizing antibody to either VP4 or VP7 can transfer either homotypic or heterotypic protection against rotavirus. Nonneutralizing IgA antibodies to the antigenically conserved VP6 protein contribute to protection, apparently via intracellular antiviral activity during viral transcytosis through intestinal epithelial cells. Rotavirus-induced diarrhea is also significantly reduced in mouse pups born to dams immunized with several forms of the enterotoxin NSP4. The role, if any, of these protective mechanisms in human immunity is unclear.

Innate immune responses may also participate in the control of rotavirus infection and the regulation of immune responses to the virus. Bacterial flagellin, which activates NF-κB-mediated pro-inflammatory cytokine gene expression, provides temporary protection against rotavirus infection in mice. In a gnotobiotic porcine model, probiotic treatment with *Lactobacillus acidophilus* significantly enhances both B- and T-cell responses to attenuated live virus infection. However, the role of innate responses in the protection of the human small intestine remains to be determined.

28.8 Type I and type III interferon responses contribute to rotavirus clearance

The role of IFN-induced antiviral activity in rotavirus infection has been examined in *in vivo* and *in vitro* studies. Increasing evidence suggests that type I IFN responses contribute to innate immune-mediated clearance of rotavirus *in vitro* and in mice. For example, elevated expression of a panel of genes related to type I IFN responses, as well as other pro-inflammatory genes, including interleukin (IL)-8, are detected in gene-profiling studies of rotavirus-infected intestinal epithelial cells. Pretreatment of cultured cells with type I IFNs limits rotavirus infections. Levels of type I and II IFNs are elevated in rotavirus-infected children and animals, and administration of exogenous type I IFN reduces viral replication in mice and disease sequelae in cattle and pigs. Moreover, rotavirus NSP1 suppresses IFN signaling, and mutations in NSP1 that ablate rotavirus's ability to interfere with IFN-related signaling attenuate rotavirus's spread to uninfected cells, further supporting a role for IFN signaling as a potential hindrance to rotavirus infection. While mice deficient in type I and II IFN receptors are able to clear rotavirus, loss of the transcription factor STAT1, which mediates much of the gene expression induced by type I IFN, severely impairs the ability of the host to contain rotavirus.

Type III IFN is the predominant response to human rotavirus infection in a nontransformed human intestinal enteroid culture; type III IFN also is induced in mice. Type III IFN is being studied as the primary response to several mucosal infections. However, current evidence suggests that type III IFN appears to be less effective in clearing rotavirus infection in mice and in cultured enteroids, and it may have another function in the intestine that remains to be fully understood.

Emerging evidence suggests a role for dendritic cells (DCs), particularly mucosal DCs, in the redundant host defense mechanisms against rotavirus. Human primary plasmacytoid DCs (pDCs), DCs in Peyer's patches in adult mice, as well as intestinal pDCs and conventional DCs from piglets are activated after exposure to rotavirus *in vitro* to produce both pro-inflammatory and in some cases anti-inflammatory cytokines. These responses are thought to play a critical role in controlling the infection and in limiting an excessive inflammatory response in the mucosa. DC responses to rotavirus likely regulate the subsequent adaptive immune response.

Rotavirus vaccines

Remarkable progress has been made in the reduction of rotavirus-associated disease and deaths by the development and implementation of oral, live attenuated rotavirus vaccines. Two vaccines licensed for global use in 2006 are broadly used worldwide, two newly vaccines licensed initially in India are beginning to be used, and two additional oral rotavirus vaccines are being used in specific countries (Vietnam and China). A number of other vaccine candidates are in different stages of clinical development. These vaccines are the result of over 30 years of vaccine research and development that had initial success followed by unexpected complications.

28.9 Currently licensed vaccines are live attenuated rotavirus strains

RotaShield vaccine was the first live attenuated rotavirus vaccine to be licensed. This vaccine was withdrawn by its commercial manufacturer owing to an increased relative risk (1 excess case per 5000–10,000 vaccinated children) of intussusception within 10 days of vaccination. The mechanism underlying this association remains unknown, but intussusception has been observed at low rates (approximately 1 excess case per 50,000–100,000 vaccinated children) with subsequent licensed live attenuated vaccines.

Since 2006, three live-attenuated rotavirus vaccines with similar efficacy and safety have been licensed and are used throughout the world. The RV5 vaccine (RotaTeq, Merck, West Point, Pennsylvania) is a pentavalent bovine-human reassortant rotavirus and contains four human G-type reassortants and a single P-type reassortant on a bovine rotavirus backbone. This vaccine is given in three doses. The RV1 vaccine (Rotarix, GlaxoSmithKline biologicals, Risensart, Belgium) is a monovalent human vaccine given in two doses. A third rotavirus vaccine (ROTAVAC, Bharat Biotech International, Hyderabad, India), derived from a naturally occurring bovine-human reassortant neonatal strain received prequalification from the World Health Organization (WHO) for global use in 2017. This vaccine is given in three doses and has been implemented in routine immunization in nine Indian states. Other rotavirus vaccines have been used regionally or are in the process of receiving WHO prequalification. A pentavalent bovine-human reassortant vaccine (ROTASIIL, Serum Institute of India, Pune, India) given in three doses has been administered out of the cold chain in clinical trials, is approved for universal immunization in India, and is being considered by WHO for global recommendation. A monovalent attenuated human strain (Rotavin-M1, PolyVac, Hanoi, Vietnam) given in two doses is used in Vietnam. A monovalent lamb rotavirus vaccine (Lanzhou Institute of Biological Products, Landhou, China) given in multiple doses is used China.

28.10 Severe rotavirus diarrhea is a vaccine-preventable enteric infection

As of April 2018, routine vaccination of infants against rotavirus has been implemented in 95 countries globally. After vaccine implementation, substantial declines in the incidence of severe rotavirus gastroenteritis have been documented, reductions in hospitalization for diarrhea are seen, and childhood deaths from diarrhea are significantly reduced. In several countries, reductions in severe gastroenteritis have been documented in children who are unvaccinated, and in older adults, suggesting that vaccination of young infants indirectly protects these groups by reducing rotavirus transmission in the community, indicating that the vaccines provide some level of herd as well as individual immunity.

At present, the efficacy of both live-attenuated vaccines is high in developed countries, with no reliable data to suggest substantial differences in efficacy between the RV5 and RV1 vaccines. Thus, severe rotavirus diarrhea is now a vaccine-preventable disease based on successful outcomes in vaccinated children in developed countries. However, rotavirus vaccines show lower vaccine efficacy in developing countries than in developed countries (50%–64% versus 85%–98%), a phenomenon seen previously with live poliovirus vaccines. Although the factors that determine the diminished effectiveness in less-developed countries are not fully understood, both vaccines save more lives in low-income countries because of the higher rotavirus disease burden and mortality in these countries. Several factors that may reduce vaccine effectiveness in some low-income countries are being studied including other coinfections with other enteric pathogens, concurrent illnesses, environmental enteropathy, malnutrition, interference with coadministered oral poliovirus vaccine, maternal antibodies, the intestinal microbiome, and host genetic differences; it is predicted that understanding these factors will help develop methods to enhance the public health value of rotavirus vaccines.

28.11 New rotavirus vaccines are under development

Additional vaccines are important to meet the need to vaccinate the global birth cohort and reduce cost. A monovalent neonatal human vaccine strain (RV3-BB) given in neonatal or infant schedules, respectively, with the first dose being a birth dose to neonates, has shown good efficacy in a trial in Indonesia. This strategy of administering a neonatal rotavirus vaccine at birth to target early prevention of rotavirus gastroenteritis may address some of the barriers to global rotavirus vaccination. Several other live-attenuated vaccines are being developed by emerging-market manufacturers.

In addition, several third-generation rotavirus vaccines are currently being developed based on the following rationale. First, concern lingers about possible safety issues associated with the three licensed live virus vaccines, which are associated with intussusception that occurs at a rate of about one to five excess cases per 100,000 vaccinated infants. This risk is low compared to the numerous health benefits of vaccination. Second, global immunization for poliovirus has changed from live attenuated vaccines to nonreplicating vaccines in the United States, and the WHO is working to achieve this transition to inactivated poliovirus vaccination globally. When this occurs, it could be beneficial to have nonreplicating rotavirus vaccines available to be given with the inactivated poliovirus vaccines. Therefore, several groups are pursuing nonreplicating vaccines including a trivalent recombinant VP8 subunit protein, inactivated rotavirus, and recombinant virus-like particle vaccine approaches. Parenteral immunization with such candidate vaccines has proven effective in animal models, but no proof of principle for efficacy of these approaches exists for humans. The VP8 subunit vaccine was safe and immunogenic in phase I testing in toddlers and infants in South Africa, and the other candidate nonreplicating vaccines are ready for phase I testing in humans.

HIV-1 INFECTION

Phillip D. Smith and Satya Dandekar

Mucosal surfaces of the genital and gastrointestinal tracts play fundamental roles in the pathogenesis of HIV-1 infection. First, these surfaces are the major route through which HIV-1 enters the body. Second, the intestinal mucosa is the site of the earliest and most intense viral replication, leading to profound local and systemic CD4$^+$ T-cell depletion. Third, as a consequence of compromised immunity, the mucosa is a site of disabling infections due to opportunistic viral, parasitic, bacterial, and fungal pathogens. Consequently, HIV-1 disease may be viewed as a mucosal infection with systemic manifestations. Accordingly, in this section of the chapter, we discuss the role of the mucosal surfaces of the genital and gastrointestinal tracts in HIV-1 pathogenesis.

28.12 Mucosal surfaces mediate HIV-1 entry

Excluding infections acquired parenterally, virtually all HIV-1 infections are transmitted through a mucosal surface. In heterosexual transmission, HIV-1 enters the body through the genital tract mucosa, in male-to-male transmission through the rectal or upper gastrointestinal tract mucosa, and in mother-to-child (vertical) transmission through the upper gastrointestinal tract mucosa

TABLE 28.2 FACTORS THAT INCREASE HIV-1 TRANSMISSION ACROSS THE MUCOSA

High viral load during primary and end-stage HIV-1 infection in the donor

Mucosal infection in the donor

Mucosal infection, trauma, inflammation, erosion, ulceration, and altered microbiota in the recipient

Increased frequency of sexual contacts

Unprotected sexual contact

Receptive anal intercourse

Absence of circumcision in a male index partner

and placenta. Vertical transmission through the mucosa occurs when the fetus or infant ingests HIV-1-infected amniotic fluid *in utero*, infected blood or cervical secretions during delivery, or infected breast milk postpartum.

Worldwide, 36.7 million people were living with HIV-1 infection in 2016, and nearly 1.8 million new infections occurred the same year. Approximately 80% of these infections are assumed to have been acquired through heterosexual contact. Transmission by this route occurs after cell-free and/or cell-associated HIV-1 in semen or secretions is inoculated onto recipient genital mucosa. Several factors contribute to HIV-1 transmission across genital, as well as gut, mucosal surfaces (Table 28.2). The donor is most infectious during primary HIV-1 infection and late-stage disease, when the level of virus in serum, and consequently semen and other body fluids, is highest. The level of virus in semen may be increased during mucosal inflammation or ulceration caused by sexually transmitted disease in the donor genital tract. Similarly, the recipient is more susceptible to HIV-1 when genital infection or inflammation is present, conditions that may increase the number and activation of local target mononuclear cells and their expression of HIV-1 receptors. Also, microtrauma to the vaginal mucosa may provide inoculated virus direct access to the microcirculation and lymphatic vessels of the subepithelial tissue. Microtrauma may be more injurious to single-cell columnar epithelium than multilayer squamous epithelium and thus contribute to the higher HIV-1 transmission rate for rectal inoculation compared with genital inoculation. By extension, in simian immunodeficiency virus (SIV) infection of macaques, the clustering of SIV-infected cells in the endocervix (lined by columnar epithelium) and the transformation zone between the endocervix and the ectocervix (lined by squamous epithelium) may be due in part to the greater susceptibility of columnar epithelium to disruption. Recent studies indicate that changes in the cervicovaginal microbiota composition may influence reproductive health in women and increase the incidence of HIV transmission. Highly diverse cervicovaginal bacteria with low abundance of *Lactobacillus* strains and high abundance of anaerobic bacteria correlate with impaired vaginal epithelial barriers, decreased micobicide efficacy, and increased susceptibility to HIV acquisition.

28.13 Cervicovaginal mucosa provides both barrier function and pathways for HIV-1 entry

Following inoculation onto the vaginal mucosa, HIV-1 first encounters cervicovaginal mucus, which can trap both cell-associated and cell-free virus. Leukocytes that may carry HIV-1 are immobilized and then killed by normally acidic mucus; cell-free virus is trapped but not neutralized by acidic mucus, potentially limiting contact between inoculated virus and the underlying epithelium. In addition, cervicovaginal secretions contain an array of antimicrobial factors with variable antiviral activity. Produced by local mucosal cells, these products include α- and β-defensins, lactoferrin, secretory leukocyte protease inhibitor, and macrophage inhibitory protein-3α. The next physical barrier encountered by HIV-1 is the nonkeratinized squamous epithelium of the vagina and ectocervix in the lower female reproductive tract and the columnar epithelium of the uterus and endocervix in the upper tract.

The multiple layers of cells in the squamous epithelium of the vagina and ectocervix present a formidable barrier to HIV-1. However, the absence of tight junctions in squamous epithelium may permit diffusion of virions through gaps between cells in the loosely stratified upper epithelium. Vaginal and ectocervical epithelial cells express glycosphingolipids and heparan sulfate proteoglycans, adhesion molecules that can bind HIV-1, but, like columnar epithelial cells, squamous epithelial cells do not support viral replication. Importantly, in the absence of polarity, squamous epithelial cells also do not support transcytosis, the transcellular transport process characteristic of columnar epithelial cells. However, after diffusion through gaps between upper layer squamous cells, HIV-1 may encounter Langerhans cells and CD4$^+$ T cells, which support HIV-1 capture and productive infection, respectively.

28.14 Vaginal Langerhans cells capture and internalize HIV-1

The ability of Langerhans cells in the vaginal epithelium to capture and internalize HIV-1 has been conclusively demonstrated by electron microscopy. Langerhans cells are present in the upper layers of the epithelium and may extend processes between epithelial cells into the apical surface of the epithelium, positioning the cells to capture and internalize HIV-1 inoculated onto the mucosal surface (Figure 28.5). Langerhans cells express CD207 (langerin, a C-type lectin) and CD4 and CC-chemokine 5 (CCR5), the HIV-1 primary receptor and coreceptor, but blocking studies suggest that only CD4 and CCR5 are involved in HIV-1 endocytosis. Once captured, HIV-1-containing Langerhans cells can exit the upper epithelium in an *ex vivo* tissue system in which the underlying stroma has been removed. However, Langerhans cells do not migrate through explanted vaginal mucosa in which the subjacent

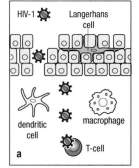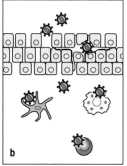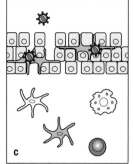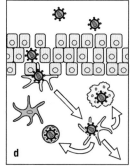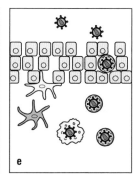

Figure 28.5 Pathways of HIV-1 entry in cervicovaginal mucosa. HIV-1 can enter the ectocervix and vaginal tissue by multiple routes. (**a**) Microtrauma provides cell-free (and cell-associated) virus direct access to target cells in the stratified squamous epithelium and, depending on the depth of the tear or disruption, access to the lamina propria and the microcirculation and lymphatic vessels. (**b**) The upper layer of loosely stratified epithelial cells allows cell-free virus to pass through gaps between the cells and penetrate into the epithelium, where cells such as dendritic cells (DCs) or Langerhans cells may take up the virus. (**c**) Langerhans cells in the epithelium bind and internalize HIV-1 via langerin, leading to virus degradation in Birbeck granules; dermal Langerhans cells migrate to lymph nodes, but cervicovaginal Langerhans cell migration is controversial. (**d**) DCs in the epithelium capture virus, migrate into the lamina propria, and *trans* infect target mononuclear cells, or continue to the draining lymph nodes, where the DCs also *trans* infect target cells. (**e**) HIV-1 infects epithelial CD4$^+$ T cells in the epithelium through receptor-mediated fusion, and the T cells become productively infected.

stroma remains attached to the epithelium, and Langerhans cells have not been identified in the draining lymph nodes. Moreover, langerin, at least on skin Langerhans cells, mediates the binding and internalization of HIV-1 into Birbeck granules, which degrade the virus. Thus, dermal Langerhans cells inhibit T-cell infection by viral clearance through langerin-mediated internalization, but whether vaginal Langerhans cells also serve as a barrier to HIV-1 infection is unclear.

28.15 Vaginal CD4$^+$ T cells bind, take up, and support HIV-1 replication

CD4$^+$ T cells in the vaginal epithelium also can take up HIV-1, likely by CD4/CCR5 receptor-mediated fusion, and support viral replication (see Figure 28.5). In contrast, Langerhans cells do not support HIV-1 replication. Studies using epithelial sheets and fluorescence microscopy to detect green fluorescent protein-encoding pseudotyped HIV-1 indicate that HIV-1 can productively infect vaginal intraepithelial CD4$^+$ T cells independent of Langerhans cells. The HIV-1-infected T cells can emigrate from the outer layer of the epithelial sheets and likely can migrate and disseminate virus systemically. Thus, in the outer vaginal epithelium, CD4$^+$ T cells and Langerhans cells can take up or capture HIV-1, but of these two cell types, only the CD4$^+$ T cells support productive infection.

28.16 Cervicovaginal dendritic cells (DCs) capture, disseminate, and *trans* infect mononuclear cells

In contrast to Langerhans cells, few DCs are present in the outer layers of the cervicovaginal epithelium (see Figure 28.5). Rather, CD11c$^+$ DCs, which do not express langerin, are prominently distributed in the papillae, at the epithelial-stromal border and throughout the stroma. These myeloid DCs express high levels of human leukocyte antigen (HLA)-DR and moderate levels of C-type lectins, including intercellular adhesion molecule 3–grabbing non-integrin (DC-SIGN) and mannose receptor (CD206). The C-type lectins mediate DC binding to carbohydrates on immune cells and certain pathogens, critical functions by which DCs mediate HIV-1 *trans* infection of mononuclear cells. Vaginal DCs also express variable levels of CD4 and the coreceptors CCR5 and CXC chemokine receptor 4 (CXCR4). In studies with explanted human cervix and vagina, DCs appear capable of transmitting HIV-1 to mononuclear cells by *trans* infection. Thus, in the vaginal mucosa, DCs in the deeper epithelial layers and lamina propria likely play a critical role in the uptake, dissemination, and possibly *trans* infection of HIV-1. Targets of DC *trans* infection in vaginal mucosa include resident lamina propria CD4$^+$ T cells and macrophages, whereas in the intestinal mucosa, predominantly CD4$^+$ T cells support viral replication (see later). Nevertheless, much remains to be elucidated regarding the immunobiology of vaginal DCs in HIV-1 transmission, including whether DCs participate in the preferential selection of the R5 virus and genetically unique transmitted/founder virus that characterizes acute HIV-1 infection.

28.17 Cell-associated HIV-1 is transmitted across inner foreskin mucosa

Female-to-male and male-to-male transmission of HIV-1 may occur after the inoculation of virus in cervicovaginal and rectal secretions onto the male foreskin. Novel models of explanted foreskin epithelium and foreskin epithelium reconstructed from isolated primary epithelial cells have been used to elucidate the early events in this transmission pathway (Figure 28.6).

Cell-associated virus is significantly more efficiently transmitted across the inner foreskin through the formation of viral synapses between the infected donor cell and apical keratinocytes. Synapse formation induces polarized budding of virus from the donor mononuclear cell. After virus capture by the Langerhans cells, conjugates of Langerhans cells and T cells promote HIV-1 infection of the T-cell. In contrast to the less keratinized inner foreskin, heavily keratinized outer foreskin traps virus within the keratin layers, preventing viral transmission. These findings provide a potential explanation for the reduced transmission of HIV-1 associated with circumcision.

28.18 HIV-1 entry into gut mucosa is accomplished by translocation of virus across the mucosal epithelium

The small intestine, colon, and rectum are lined by columnar epithelium (**Figure 28.7**). The integrity of this single layer of cells may be altered in the small intestine by enteric pathogen infection or in the rectum by sexually transmitted disease or trauma associated with receptive anal intercourse. Tight junctions between epithelial cells promote polarity in which the cell is separated into apical and basal domains, the hallmark of columnar epithelium. In contrast to stratified squamous epithelium in the ectocervix and vagina, gut columnar epithelium is capable of transcytosis, a rapid process in which cargo endocytosed at the apical surface is transported in vesicles in a nondegradative manner to the basolateral surface for release into the subepithelial space. In studies utilizing epithelial cell line monolayers, the transcytosis of cell-associated HIV-1 is 100- to 1000-fold more efficient than the transcytosis of cell-free virus, likely due to the formation of viral synapses between the infected donor mononuclear cell and the epithelium, concentrating virus in the resultant pocket or cleft (see **Figure 28.7**). The heparan sulfate proteoglycan agrin on epithelial cells participates in the formation of the synaptic cleft and also serves as an attachment molecule for a conserved region in the HIV-1 envelope glycoprotein gp41; attachment then promotes galactosylceramide-mediated endocytosis of the virus by the epithelial cells (see later). HIV-1 transcytosis across columnar epithelium in the intestine is relevant to vertical and genital-oral transmission and in the rectum to genital-rectal transmission.

Epithelial cell transcytosis of HIV-1 was first explored in epithelial cell lines. Subsequently, *in vitro* studies with primary human jejunal epithelial cells showed that intestinal epithelial cells express galactosylceramide, an alternative primary receptor for HIV-1, and CCR5, but not CXCR4. Intestinal epithelial cells can efficiently transcytose infectious R5, but not X4, viruses for subsequent infection of CCR5$^+$ cells. Importantly, the epithelial cells did not support productive HIV-1 infection. These findings suggest that human small intestinal epithelial cells may preferentially select R5 viruses for entry. In addition, *ex vivo* studies suggest that human rectal mucosa also transcytoses HIV-1 across the epithelium and that antibodies to the conserved membrane proximal external region of gp41 in the HIV-1 envelope can block transcytosis. Thus, antibodies to the virus, or perhaps blockade of the transcytosis machinery itself, could be developed to prevent this important entry step in HIV-1 transmission across gut mucosa.

In addition to the likely role of intestinal epithelial cells in HIV-1 entry, intestinal DCs have been shown in *ex vivo* studies to participate in HIV-1 entry (see **Figure 28.7**). Intestinal CD11c$^+$ DCs are distributed throughout the lamina propria and, as shown in immunohistochemical studies of normal human small intestine, extend processes between epithelial cells, a process that in mice is dependent on the activation of the myeloid differentiation primary-response gene 88 (MyD88) signal pathway. CD11c$^+$ intestinal

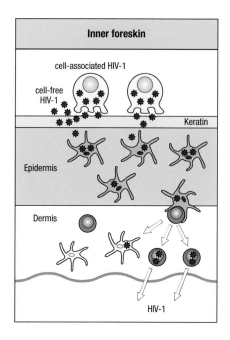

Figure 28.6 Cell-associated HIV-1 enters the inner, but not the outer, foreskin. The inner foreskin is the site of HIV-1 entry in female-to-male and some male-to-male transmissions. HIV-1-infected donor cells form viral synapses, resulting in viral budding and the concentration of virus in the synaptic cleft. Released virus can be captured, internalized, and degraded by Langerhans cells, or high levels of virus can induce Langerhans cell accumulation, HIV-1 capture, and migration to the epidermal-dermal junction, where the Langerhans cells can form conjugates with T cells and then transmit virus to the T-cell. The conjugates may emigrate from the dermis or via productive T-cell infection transmit virus to DCs, which migrate to the draining lymph nodes. (Adapted from Ganor, Y. et al. *Mucosal Immunol.* 2010, 3:506–521. With permission from Macmillan Publishers Ltd.)

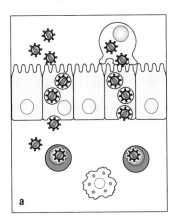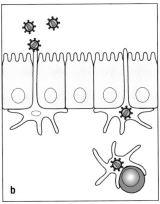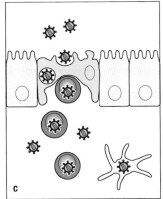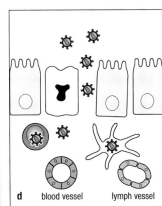

Figure 28.7 **Pathways of HIV-1 entry through gut epithelium.** Cell-associated and cell-free HIV-1 can cross gut columnar epithelium by four possible routes. (**a**) Intestinal epithelial cells expressing galactosylceramide and CCR5 endocytose cell-associated or cell-free R5 HIV-1 and transcytose virus in vesicles to the basolateral surface, where fusion with the cell membrane leads to release of virus into the lamina propria. (**b**) Lamina propria dendritic cells with processes that extend between epithelial cells capture virus and migrate into the lamina propria or to the mesenteric lymph node where they *trans* infect mucosal or systemic CD4+ T cells, respectively. (**c**) M cells translocate virus *in vitro* from the apical surface in vesicles by a nondegradative process, but this pathway has not been confirmed *in vivo*. (**d**) Disrupted epithelium, in the small intestine due to enteric infection or in the rectum due to trauma or sexually transmitted disease, could provide HIV-1 direct access to lamina propria target cells and to the microcirculation and/or lymphatic vessels.

DCs, like vaginal DCs, express the C-type lectins DC-SIGN and moderate levels of CD206, potential attachment receptors for HIV-1. Importantly, approximately 30%–50% of intestinal DCs are double positive for CD4/CCR5 and CD4/CXCR4, the receptors utilized by R5 and X4 viruses, respectively, for mononuclear cell entry.

Primary human intestinal DCs rapidly take up HIV-1 in suspension, and, after inoculation onto explanted human jejunum, lamina propria DCs transport virus through the mucosa and *trans* infect intestinal and blood CD4+ T cells. However, whether DC processes capture the virus above the apical surface of the epithelium or capture virus transported by epithelial cells into the lamina propria has not been conclusively determined. Nevertheless, intestinal DCs are clearly capable of transporting HIV-1 through the intestinal mucosa and transmitting infectious virus to mucosal and blood target cells. In the intestinal mucosa, lamina propria CD4+ T cells are the predominant target cell; lamina propria macrophages are nonpermissive to HIV-1 due to stromal downregulation of macrophage CD4/CCR5 expression and NF-κB inactivation, requirements for cell entry and viral transcription, respectively.

Another potential entry route for HIV-1 is through M (microfold) cells, which are present in the follicle-associated epithelium overlying Peyer's patches in the small intestine and isolated follicles in the rectum and tonsils (see Figure 28.7). In contrast, genital mucosa does not contain inductive sites, and, consequently, M cells are not present in ectocervical or vaginal mucosa. M cells endocytose macromolecules and certain microorganisms at the apical membrane and then translocate the cargo in membrane-derived vesicles by a nondegradative, transcellular process to the basolateral cell surface for delivery to interdigitating DCs and lymphocytes (see Chapter 15). Thus, M cell-mediated HIV-1 entry in human gut mucosa is theoretically possible but has not been established due to the inherent difficulty in tracking M cell transcytosis of HIV-1 in *in vivo* transmission studies. Recently, mast cells in human colorectal tissue were shown to express HIV-1 attachment molecules, including DC-SIGN, heparan sulfate proteoglycan, and α4β7 integrin, and to capture HIV-1 and *trans*-infect CD4+ T cells through mast-cell and T-cell conjugates.

28.19 Few HIV-1 virions are transmitted in acute HIV-1 infection

Nearly all new HIV-1 infections are caused by CCR5-tropic (R5) viruses. In addition, approximately 80% of such infections are caused by a single virus (or virus-infected cell) and 20% by two to five viruses, which investigators have elegantly shown using single-genome amplification of plasma virion RNA in combination with viral sequencing. Irrespective of the transmission route (male-to-male, male-to-female, female-to-male), transmitted/founder viruses productively infect CD4$^+$ T cells but not monocyte-derived macrophages for as yet unclear reasons. The genotype- and R5 phenotype-restriction of *in vitro* studies shows that transmitted/founder virus is strikingly different from that of genetically heterogeneous, CCR5-tropic and CXCR4-tropic virus populations that characterize chronic HIV-1 infection. A mixture of such viruses is present in the body fluids of chronically infected people, and consequently, a mixture of such viruses is typically inoculated onto the mucosal surface of a recipient. However, the transmission of only one or a few R5 viruses suggests that a "bottleneck," possibly at the level of the mucosae, restricts entry to the founder virus. The mechanism for such restriction has not been determined but may be linked to the level and composition of *N*-linked glycosylation on the V1–V3 variable loops of the envelope gp120.

CD4$^+$ T-cell depletion in early HIV-1 infection

Within days of crossing the genital or gut epithelium, the single or few viruses that are transmitted infect target CD4$^+$ T cells in the small intestinal lamina propria, and viral replication is initiated (**Figure 28.8**). Because the gut mucosa contains the largest number of memory CD4$^+$CCR5$^+$ cells in the body, the intestinal lamina propria is the major site of viral replication and CD4$^+$ T-cell depletion, regardless of whether the virus was acquired through genital or gut mucosa. The loss of this population of cells has profound immunologic consequences for the host, as discussed next, and sets the stage for the subsequent acquisition of opportunistic infections, as discussed in the conclusion.

28.20 HIV-1 causes rapid, profound, and prolonged CD4$^+$ T-cell depletion in the intestinal mucosa

HIV-1-induced depletion of gut CD4$^+$ T cells is severe and persists through all stages of infection in the absence of antiretroviral therapy. The majority of activated memory CD4$^+$CCR5$^+$ T cells in the body are located in the intestinal lamina propria (less than 20% in blood and lymph nodes), and HIV-1 targets these cells, resulting in the profound depletion of CD4$^+$ T cells. Thus, the gut mucosa plays a critical role in establishing HIV-1 infection and in the depletion of CD4$^+$ T cells in humans and in SIV infection of rhesus macaques.

The rapid and severe CD4$^+$ T-cell depletion in the gastrointestinal mucosa is due mainly to the HIV-1 cytopathic effect and partly to virus-induced bystander apoptosis. The loss of CD4$^+$ T cells in gut mucosa coincides with a reciprocal increase in CD8$^+$ T cells. Although major histocompatibility complex (MHC) class I–restricted cytotoxic T lymphocytes (CTLs) are present in the mucosa, they emerge too late to eradicate the HIV-1-infected cells and prevent the resultant massive CD4$^+$ T-cell loss. Natural killer (NK) cells and γλT cells at mucosal surfaces may contribute to anti-HIV-1 defense, but with limited effect. Thus, intense viral replication and severe CD4$^+$ T-cell loss progress relentlessly despite mucosal innate and subsequent adaptive antiviral responses.

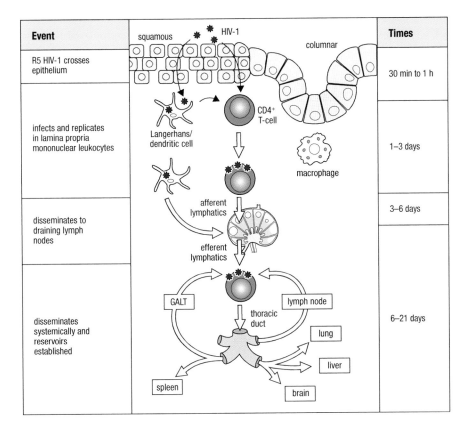

Event		Times
R5 HIV-1 crosses epithelium	squamous HIV-1 columnar	30 min to 1 h
infects and replicates in lamina propria mononuclear leukocytes	Langerhans/ dendritic cell CD4⁺ T-cell macrophage	1–3 days
disseminates to draining lymph nodes	afferent lymphatics efferent lymphatics	3–6 days
disseminates systemically and reservoirs established	GALT lymph node thoracic duct lung liver spleen brain	6–21 days

Figure 28.8 HIV-1 dissemination after mucosal entry. HIV-1 captured by dendritic cells (DCs) or Langerhans cells *trans* infects mucosal CD4⁺ T cells and possibly macrophages. The DCs also traffic via afferent lymphatic vessels to the draining lymph nodes to *trans* infect systemic target cells. Infected CD4⁺ T cells may support local viral replication or also enter the lymphatics and traffic to the lymph nodes, where massive HIV-1 replication occurs. Infected cells emigrate from the nodes, migrating through the efferent lymphatic vessels and thoracic duct to enter the circulation and disseminate HIV-1 to multiple tissues. GALT, gut-associated lymphoid tissue. (Adapted from Pope, M., and Haase, A. *Nat. Med.* 2003, 9:847–852. With permission from Macmillan Publishers Ltd.)

The recently discovered T_H17 CD4⁺ T-cell subset plays a critical role in host defense against bacterial and fungal pathogens at mucosal sites (see Chapter 7). T_H17 CD4⁺ T cells are preferentially infected by HIV-1 (and SIV in macaques) and rapidly depleted due to the cytopathic effect of the virus. Linking gut HIV-1 infection with systemic manifestations of the disease, investigators have shown that the loss of mucosal T_H17 cells in SIV-infected macaques results in systemic dissemination of *Salmonella typhimurium* from the gut in animals infected with this enteric pathogen. The T_H17 cells secrete IL-17, IL-6, TNF-α, IL-22, and granulocyte-macrophage colony-stimulating factor (GM-CSF) and promote the renewal of intestinal epithelial barrier by inducing extracellular signal-regulated kinase/mitogen-activated protein kinase (ERK/MAPK)-dependent synthesis of tight junction proteins and claudins. In addition, IL-17 in normal mucosa induces expression of antimicrobial defensins and chemokines in epithelial cells. Thus, loss of T_H17 cells likely contributes to an ineffective epithelial barrier and increased microbial (and microbial product) translocation, stimulating systemic immune activation, a key feature of HIV-1 disease progression.

In contrast to the loss of memory CD4⁺CCR5⁺ and T_H17 T cells, the prevalence of T regulatory (T_{reg}) cells is increased in HIV-1 and SIV infections. T_{reg} cells mediate suppressive function through tolerance to self and foreign antigens. Consequently, the increase in the number of T_{reg} cells in gut mucosa may dampen adaptive immune responses and together with the decrease in the number of T_H17 cells contribute to local and systemic immune activation.

Highly active antiretroviral therapy (HAART) suppresses viral replication, resulting in restoration of the CD4⁺ T-cell population in HIV-1-infected people. However, restoration of the CD4⁺ T cells in gut mucosa is delayed compared with that of the peripheral blood compartment. This delay is likely due to incomplete suppression of viral replication in the mucosa despite suppressive activity of HAART. Residual HIV-1 in the mucosa limits local immune restoration and the eradication of viral reservoirs. In the SIV model, however, early initiation of antiretroviral therapy causes near complete restoration of gut mucosal CD4$^+$ T cells. Recent studies of HIV-1-infected humans have shown that the early initiation of HAART results in better clinical outcome. Thus, early therapeutic intervention in HIV-1-infected subjects is necessary to restore effective immune function in the mucosal and systemic immune systems. Although HAART causes viral suppression, it fails to eradicate latent HIV reservoirs, which pose a major obstacle for HIV cure. Multiple molecular mechanisms contribute to the establishment of HIV latency and long-term silencing of the viral genomes. New strategies are being explored to activate these silent viral reservoirs and target them for immune clearance.

In contrast to the relentless progression of disease in the majority of subjects with HIV-1 infection, a small subset of HIV-1-infected people called "long-term nonprogressors" remain clinically asymptomatic for many years in the absence of therapy. Their plasma viral loads are usually undetectable or very low, and CD4$^+$ T-cell numbers in peripheral blood are in the normal range. Long-term nonprogressors retain normal levels of CD4$^+$ T cells and high levels of virus-specific cellular responses in the gastrointestinal mucosa, consistent with the critical role of mucosal immunity in limiting HIV-1 disease. In SIV-infected sooty mangabeys and African green monkeys, important nonhuman primate models of HIV-1 infection, maintenance of the integrity of the gut mucosal immune system correlates with nonpathogenic SIV infection. Thus, maintenance of the integrity of the gut mucosal immune system may become an important measure of the efficacy of HIV-1 vaccines and therapeutic agents.

28.21 HIV-1 alters intestinal epithelial permeability, promoting microbial translocation

Early in the acquired immunodeficiency syndrome (AIDS) epidemic, investigators showed that HIV-1 frequently causes low-grade atrophy of the intestinal mucosa with a maturational defect in enterocytes and reduced brush border enzyme activity. Studies of gut mucosa in SIV-infected macaques indicate that epithelial cell turnover is accompanied by increased expression of genes that regulate cell repair and regeneration. Epithelial cell apoptosis also occurs in the intestinal and colonic mucosa during early SIV infection. Together, villus atrophy, epithelial cell maturational defect, and epithelial cell apoptosis promote reduced absorptive capacity, manifesting as diarrhea and malabsorption, and can occur in HIV-1-infected subjects. Termed AIDS enteropathy, manifestations of this syndrome improve or resolve with HAART-induced restoration of immune function.

People chronically infected with HIV-1 may have increased levels of circulating lipopolysaccharide (LPS), a measure of microbial translocation, which correlates with innate and adaptive immune activation. Translocation of gut bacteria and bacterial components into the circulation may help maintain a pool of activated immune cells as targets for HIV-1 infection. A possible mechanism for the microbial translocation is suggested by *in vitro* studies in which HIV-1 envelope glycoprotein (gp120) induced primary genital and intestinal epithelial cell lines to secrete TNF-α, which in turn dysregulated the tight junction proteins claudin, occludin, and ZO-1, causing a reduction in transepithelial resistance and enhanced permeability. The increased permeability permitted bacteria placed on the apical surface to leak into the subepithelial space. However, circulating levels of LPS did not increase during disease progression in a study of HIV-1-infected subjects in Africa. Nevertheless, the microbial translocation studies underscore the connection between compromised gut mucosa and systemic disease in HIV-1 infection.

Mucosal infections associated with HIV-1 infection

The profound CD4$^+$ T-cell depletion initiated in the intestinal lamina propria leads to severe local and systemic immunosuppression. HIV-1 also dysregulates the host defense activities of monocyte-derived macrophages, although the effect of virus on the defense capabilities of mucosal macrophages has not been reported. As a consequence of HIV-1-induced immunosuppression, the mucosa is a site of infection by an array of viral, parasitic, bacterial, and fungal pathogens.

Several principles govern the interaction between opportunistic pathogens and HIV-1 infection (Table 28.3). First, susceptibility to mucosal, as well as systemic, opportunistic infections is inversely related to the level of immunosuppression, and clearance of the infection usually occurs after restoration of immune function with antiretroviral therapy. For example, the presence of active cytomegalovirus infection is inversely proportional to the number of circulating CD4$^+$ T cells, with mucosal cytomegalovirus-induced inflammation and ulceration usually occurring after the number of CD4$^+$ T cells declines to less than 100 cells mm^{-3}, indicating severe immunosuppression. Conversely, cytomegalovirus disease resolves with HAART-induced restoration of immune function. Second, replication of some opportunistic pathogens may enhance HIV-1 transcription and vice versa in dual-infected cells and in lymphoid tissue. For example, the bidirectional upregulation of *Mycobacterium avium* complex replication and HIV-1 transcription has been detected in dual-infected macrophages in lymph nodes. Bidirectional upregulation likely occurs in dual-infected immune cells in the mucosa. Third, the route of HIV-1 entry does not determine the site of mucosal infections. Thus, HIV-1 transmitted by the genital route can lead to opportunistic infections in the gastrointestinal mucosa and/or the genital mucosa. Fourth, the specific mucosal pathogen acquired reflects in part the pathogens endemic to a specific geographical area and the sexual practices of the HIV-1-infected host. For example, among HIV-1-infected people, *Cystoisospora belli* is more common in Africa and the Caribbean region than in the United States, and *Histoplasma capsulatum* is more common in the Mississippi and Ohio River valleys than other regions of the United States. Regarding the influence of sexual practices on mucosal infection, *Chlamydia trachomatis* proctitis, for example, occurs only in men who practice receptive anal intercourse.

TABLE 28.3 PRINCIPLES OF MUCOSAL PATHOGEN DISEASE IN PEOPLE INFECTED WITH HIV-1

Acquisition of mucosal opportunistic pathogens is directly related to the severity of the immunosuppression

Infection by certain opportunistic pathogens resolves with highly active antiretroviral therapy (HAART)-induced restoration of immune function

Opportunistic pathogen replication and HIV-1 transcription may be enhanced in a bidirectional manner

The route of HIV-1 entry does not determine the mucosal site of opportunistic infection

Acquisition of some mucosal infections reflects the pathogens endemic to a particular geographic area

Infection by certain mucosal pathogens reflects the sexual practices of the HIV-1-infected person

28.22 HIV-1-induced immunosuppression leads to opportunistic infection of gut mucosa by viral, parasitic, bacterial, and fungal pathogens

The parasitic, viral, bacterial, and fungal microorganisms that cause enteric disease in HIV-1-infected people include opportunistic pathogens and common pathogens that acquire features of opportunistic pathogens in immunosuppressed HIV-1-infected people (Table 28.4). Features of mucosal opportunistic pathogens include (1) the ability to cause severe mucosal disease during immunosuppression but no disease in the presence of normal immune function, (2) resistance to common antimicrobial drugs, (3) prolonged infection, and (4) systemic manifestations. These features are illustrated by the enteric bacteria *Campylobacter jejuni*, *Shigella flexneri*, and *Salmonella* species, which normally cause easily treated diarrhea but in immunosuppressed HIV-1-infected subjects are associated with incapacitating and recurrent diarrheal illness, antimicrobial drug resistance, and bacteremia. Some opportunistic infections resolve after HAART restoration of immune function. Similarly, the AIDS enteropathy syndrome, which was common in the pre-HAART era of the AIDS epidemic, usually resolves after immune function has been restored by HAART.

TABLE 28.4   MUCOSAL PATHOGENS IN HIV-1 INFECTION

Viral	Parasitic	Bacterial	Fungal
Cytomegalovirus	*Cryptosporidium* species	*Salmonella* species	*Candida albicans*
Herpes simplex virus	*Microsporidia* species	*Shigella flexneri*	*Histoplasma capsulatum*
Adenovirus	*Cystoisospora belli*	*Campylobacter jejuni*	
Epstein-Barr virus	*Cyclospora*	*Mycobacterium avium* complex	
	Trichomonas vaginalis	*Mycobacterium tuberculosis*	
		Chlamydia trachomatis	

SUMMARY

Nearly 50 years of basic and clinical research have provided fundamental insights into the molecular structure and biology of rotavirus and the host immune response to intestinal rotavirus infection. As a result, rotavirus diarrhea is now a vaccine-preventable disease, reflected in the successful outcomes of vaccinated children in developed countries. Despite this impressive progress, the mechanisms of virus-induced immunity and vaccine-induced protection, as well as the molecular basis for species restriction and cell tropism, have not been elucidated, but the development of a tractable reverse genetics system may help resolve these issues. In even less time, namely, since the first report almost 40 years ago of the syndrome eventually defined as AIDS, the pathogenesis, molecular virology, and immunobiology of HIV-1 have been substantially elucidated. Understanding the life cycle of HIV-1 has led to the development of efficient antiretroviral drugs that have helped control the infection and reduce the rate of transmission. Although the mucosa is recognized as playing critical roles in HIV-1 transmission, virus expression, CD4$^+$ T-cell depletion, and secondary opportunistic infections, a successful vaccine

has eluded investigators. However, the accelerating pace of discoveries that continue to elucidate the biology of HIV-1 and the pathogenesis of mucosal HIV-1 infection should provide the scientific basis for a mucosal vaccine to halt the relentless HIV-1/AIDS epidemic.

FURTHER READING

ROTAVIRUS INFECTION

Aoki, S.T., Settembre, E.C., Trask, S.D. et al.: Structure of rotavirus outer-layer protein VP7 bound with a neutralizing Fab. *Science* 2009, 324:1444–1447.

Banyai, K., Estes, M.K., Martella, V. et al.: Viral gastroenteritis. *Lancet.* 2018, 392:175–186.

Bines, J.E., At Thobari, J., Satria, C.D. et al.: Human neonatal rotavirus vaccine (RV3-BB) to target rotavirus from birth. *N. Engl. J. Med.* 2018, 378:719–730.

Blutt, S.E., Fenaux, M., Warfield, K.L. et al.: Active viremia in rotavirus-infected mice. *J. Virol.* 2006, 80:6702–6705.

Blutt, S.E., Warfield, K.L., Estes, M.K. et al.: Differential requirements for T cells in virus-like particle- and rotavirus-induced protective immunity. *J. Virol.* 2008, 82:3135–3138.

Deal, E.M., Jaimes, M.C., Crawford, S.E. et al.: Rotavirus structural proteins and dsRNA are rewired for the human primary plasmacytoid dendritic cell IFNα response. *PLOS Pathog.* 2010, 6:e1000931.

Greenberg, H.B., and Estes, M.K.: Rotaviruses: From pathogenesis to vaccination. *Gastroenterology* 2009, 136:1939–1951.

Groome, M.J., Koen, A., Fix, A. et al.: Safety and immunogenicity of a parenteral P2-VP8-P[8] subunit rotavirus vaccine in toddlers and infants in South Africa: A randomised, double-blind, placebo-controlled trial. *Lancet. Infect. Dis.* 2017, 17:843–853.

Hu, L., Crawford, S.E., Czako, R. et al.: Cell attachment protein VP8* of a human rotavirus specifically interacts with A-type histo-blood group antigen. *Nature* 2012, 485:256–259.

Kirkwood, C.D., Ma, L.F., Carey, M.E. et al.: The rotavirus vaccine development pipeline. *Vaccine* 2017, S0264-410X:30410–30413.

Lin, J.D., Feng, N., Sen, A. et al.: Distinct roles of type I and type III interferons in intestinal immunity to homologous and heterologous rotavirus infections. *PLOS Pathog.* 2016, 12:e1005600.

Nordgren, J., Sharma, S., Bucardo, F. et al.: Both Lewis and secretor status mediate susceptibility to rotavirus infections in a rotavirus genotype–dependent manner. *Clin. Infect. Dis.* 2014, 59:1567–1573.

Parker, E.P., Ramani, S., Lopman, B.A. et al.: Causes of impaired oral vaccine efficacy in developing countries. *Future Microbiol.* 2017, 13:97–118.

Patel, N.C., Hertel, P.M., Estes, M.K. et al.: Vaccine-acquired rotavirus in infants with severe combined immunodeficiency. *N. Engl. J. Med.* 2010, 362:314–319.

Saxena, K., Simon, L.M., Zeng, X.L. et al.: A paradox of transcriptional and functional innate interferon responses of human intestinal enteroids to enteric virus infection. *Proc. Natl Acad. Sci. USA.* 2017, 114:E570–E579.

HIV-1 INFECTION

Alfsen, A., Yu, H., Magérus-Chatinet, A. et al.: HIV-1-infected blood mononuclear cells form an integrin- and agrin-dependent viral synapse to induce efficient HIV-1 transcytosis across epithelial cell monolayer. *Mol. Biol. Cell.* 2005, 16:4267–4279.

Brenchley, J.M., Price, D.A., Schacker, T.W. et al.: Microbial translocation is a cause of systemic immune activation in chronic HIV infection. *Nat. Med.* 2006, 12:1365–1371.

Farcasanu, M. and Kwon, D.S.: The influence of cervicovaginal microbiota on mucosal immunity and prophylaxis in the battle against HIV. *Cur. HIV/AIDS Rep.* 2018, 15:30–38.

Ganor, Y., Zhou, Z., Tudor, D. et al.: Within 1 h, HIV-1 uses viral synapses to enter efficiently the inner, but not outer, foreskin mucosa and engages Langerhans-T cell conjugates. *Mucosal Immunol.* 2010, 3:506–521.

Hladik, F., and McElrath, M.J.: Setting the stage: Host invasion by HIV. *Nat. Rev. Immunol.* 2008, 8:447–457.

Keele, B.F., Giorgi, E.E., Salazar-Gonzalez, J.F. et al.: Identification and characterization of transmitted and early founder virus envelopes in primary HIV-1 infection. *Proc. Natl. Acad. Sci. USA.* 2008, 105:7552–7557.

Kim, Y., Anderson, J.L., and Lewin, S.R.: Getting the "kill" into "shock and kill" strategies to eliminate latent HIV. *Cell Host Microbe.* 2018, 23:14–26.

Lusic, M., and Siliciano, R.F.: Nuclear landscape of HIV-1 infection and integration. *Nat. Rev. Microbiol.* 2017, 15:69–82.

Meng, G., Wei, X., Wu, X. et al.: Primary intestinal epithelial cells selectively transfer R5 HIV-1 to CCR5+ cells. *Nat. Med.* 2002, 8:150–156.

Nazli, A., Chan, O., Dobson-Belaire, W.N. et al.: Exposure to HIV-1 directly impairs mucosal epithelial barrier integrity allowing microbial translocation. *PLOS Pathog.* 2010, 6:e1000852.

Raffatellu, M., Santos, R.L., Verhoeven, D.E. et al.: Simian immunodeficiency virus-induced mucosal interleukin-17 deficiency promotes *Salmonella* dissemination from the gut. *Nat. Med.* 2008, 14:421–428.

Sankaran, S., Guadalupe, M., Reay, E. et al.: Gut mucosal T cell responses and gene expression correlate with protection against disease in long-term HIV-1-infected nonprogressors. *Proc. Natl. Acad. Sci. USA.* 2005, 102:9860–9865.

Shen, R., Meng, G., Ochsenbauer, C. et al.: Stromal down-regulation of macrophage CD4/CCR5 expression and NF-κB activation mediates HIV-1 non-permissiveness in intestinal macrophages. *PLOS Pathog.* 2011, 7:e1002060.

Shen, R., Richter, H.E., Clements, R.H. et al.: Macrophages in vaginal but not intestinal mucosa are monocyte-like and permissive to human immunodeficiency virus type 1 infection. *J. Virol.* 2009, 83:3258–3267.

Smith, P.D., and Shen, R.: Target cells for HIV-1/SIV infection in mucosal tissues. *Curr. Immunol. Rev.* 2019, 15:28–35.

Infection-driven periodontal disease

THOMAS E. VAN DYKE AND MIKE CURTIS

Periodontal diseases are defined as a series of inflammatory diseases of one or more of the periodontal tissues, including the alveolar bone, the periodontal ligament, the cementum on the tooth root surface, and the gingiva. There are many different diseases affecting the supporting structure of the teeth; however, bacterial biofilm (plaque)-induced inflammatory lesions comprise the vast majority of periodontal diseases. There are two major categories of plaque-induced periodontal diseases: gingivitis and periodontitis. The distinction between the two forms is based on the reversibility of the lesion; tissue inflammation and connective tissue loss are reversible in gingivitis with removal of bacterial deposits on the teeth. In contrast, the tissue destruction associated with periodontitis is irreversible. Gingivitis may never progress to periodontitis; gingivitis always precedes periodontitis.

Periodontitis is a complex of multiple diseases that share common clinical manifestations. Within the plaque-induced periodontitis category, there is a chronic, slowly progressing form of disease that usually affects adults, termed chronic periodontitis; a rapidly progressive form that usually has early onset and exhibits an unusual distribution of affected teeth termed aggressive periodontitis; and necrotizing disease with marginal destruction of soft tissues. These are considered distinct clinical entities, but it remains unclear whether they differ in etiology and pathogenesis.

The oral cavity and particularly the periodontium is the only organ system in the body where hard tissue traverses soft tissue into a contaminated external environment. The teeth are constantly bathed in saliva and colonized by microorganisms exposing the teeth and gingiva to their products. In health, there is a continous neutrophil infiltrate of the gingiva and the epithelium lining the gingival sulcus around the teeth. The *initial lesion* of periodontitis, as originally defined in 1976 by Page and Schroeder based on histological characterization, is a more robust neutrophil infiltration in response to the bacteria colonizing the teeth with a possible shift in the phenotype of the neutrophil to a more aggressive/destructive form. Loss of perivascular collagen is characteristic, and an infiltration of macrophages and lymphocytes begins.

The early lesion of periodontitis is a more mature leukocyte infiltrate in which neutrophils no longer dominate. The absolute number of neutrophils does not decline, but the main infiltrating cell types are lymphocytes with increasing numbers of lymphoblasts and a few peripheral plasma cells. Mononuclear phagocytes and macrophages also accumulate and maintain a proinflammatory phenotype contributing to the chronicity of the lesion. Collagen loss may reach 80%, and there is loss of fibroblasts and matrix. An exudate forms and flows through the gingival sulcus.

The *established lesion* of periodontitis can also be termed the chronic lesion of gingivitis. It is characterized by plasma cells as the dominant inflammatory cell type, but there is no bone loss. There are extravascular immunoglobulins directed against plaque bacteria in the connective tissue and junctional epithelium.

The *advanced lesion* does not differ significantly from the established lesion in terms of cell composition, but it extends into the alveolar bone and periodontal ligament and loss of alveolar bone is seen. Loss of collagen continues, and clinically there is a detachment of the gingiva from the tooth (pocket formation).

ETIOLOGY OF PERIODONTAL DISEASE

In the mid-1960s, classical studies in "experimental gingivitis" showed that healthy volunteers who refrained from toothbrushing for 21 days developed gingivitis. The findings resulted in a major shift in defining the etiology of periodontal diseases. Gingivitis developed in all volunteers, and reinstatement of daily dental plaque removal resulted in a return to gingival health. The accumulation of bacteria on the teeth caused gingivitis, and removal of the bacteria returned the gingival condition to health.

Likewise in classic studies in the 1960s in dogs, dental plaque accumulation led to the development of periodontal disease. Plaque removal every other day for 18 months maintained clinically healthy periodontal tissues; however, teeth in the same mouth that were not cleaned developed gingival inflammation and attachment loss. Experimental gingivitis in most dogs progressed to periodontitis over a 4-year period if dental plaque was not removed on a daily basis. Control animals with daily plaque removal during the same period did not develop gingivitis or periodontitis.

The bacterial etiology of periodontal diseases was further established in the 1960s when it was shown that periodontitis was transmissible from diseased Syrian hamsters to healthy animals with *Actinomyces viscosus* isolated from the diseased animals. At this stage, a specific microbial etiology for the periodontal diseases in humans was not established, and it was considered plausible that the amount of dental plaque accumulation rather than the microbial composition of the biofilm was the critical factor. However, advances in microbiological techniques, notably anaerobic culture in the 1970s and DNA hybridization techniques in the following two decades resulted in the identification of complexes of indigenous bacteria in the subgingival microbiota that are increased in abundance in the presence and progression of periodontitis compared to health. Specifically, *Porphyromonas gingivalis*, *Tannerella forsythia*, *Treponema denticola*, *Campylobacter rectus*, *Micromonas micros*, *Streptococcus intermedius*, *Eubacterium nodatum*, *Aggregatibacter* (formerly *Actinobacillus*) *actinomycetemcomitans*, and *Prevotella intermedia* were associated with periodontitis, although causation has yet to be established.

The major conclusions of these early classic studies were that periodontal infections are polymicrobial; the causative agents are part of the indigenous (normal) microbiota; the amount of dental plaque is of etiologic importance; and some bacteria in dental plaque are potentially more pathogenic than others.

29.1 Biofilms and environmental conditions are important for the development of periodontitis

Dental plaques are highly organized bacterial biofilms. Biofilms are complex polymicrobial communities that are resistant to externally applied antimicrobial agents and antibacterial host mechanisms. Other than mechanical removal, the disruption of biofilms remains a clinical problem.

Until very recently, only 50% or less of the oral microbiota was cultivable in the laboratory. To further characterize the oral microflora, genetic techniques for bacterial detection are now routinely used. In particular, high-throughput sequence analysis of variable regions of 16S ribosomal RNA facilitates identification of virtually all the bacterial cells in a given sample, including those of very low abundance. The findings reveal that the oral flora in health and disease is extraordinarily diverse and that the level of variability appears to be the individual, not the disease, although there are marked differences in the composition of the flora in health and disease in the same individual. Comparison of the microbial composition in health versus disease using these more recent tools reveals whole scale shifts in the microbial community composition. The change from health to disease is therefore viewed as a disturbance to the normally well-tolerated microbial plaque composition in *symbiosis* with the host tissues to a *dysbiotic* community structure where potentially pathogenic organisms or complexes of organisms normally present at low abundance become predominant. These include those bacterial species identified in earlier cultural investigations as well as a significant number of previously unrecognized organisms including *Filifactor alocis*, *Peptostreptococcus stomatis*, and species drawn from the *Desulfolobus*, *Dialister*, and *Synergistetes* genera.

The mechanisms of dysbiosis of the periodontal microbiome remain to be established. However, the influence of local environmental conditions, particularly those factors associated with elevated inflammation, may be of paramount importance: the varying abilities of different bacteria to grow and proliferate under different environmental conditions will dictate the balance of microbial communities at any given site on the tooth surface. For example, in a periodontal pocket where the pH can rise to well over pH 7, those bacteria most adapted to grow at alkaline pH will be able to out-compete those bacteria most suited to more acidic conditions. Similarly, organisms able to withstand the antimicrobial properties of the host's inflammatory response will be more predominant at inflamed sites in the periodontium than those bacteria ill equipped for this injurious environment. Hence, the composition of microbial communities in disease will be intimately linked to the environmental conditions prevalent at a diseased site.

Shifts in the environmental conditions due to, for example, the introduction of different nutrients due to the arrival of a plasma exudate in the form of gingival crevicular fluid, will lead to concomitant shifts in the microbial community. Organisms previously limited in their growth due to, for example, only very low concentrations of the iron source, hemin, will have the nutritional capacity to increase in number and potentially out-compete

Figure 29.1 Evasion of the host response by periodontal bacteria. Gingivitis is the natural inflammatory response to accumulation of bacteria on the surfaces of the teeth. Persistence of inflammation leads to tissue breakdown and an inflammatory exudate that contains collagen peptides and heme-containing compounds that change the nutrient environment of the gingival sulcus and lead to shifts in the microbiome and dysbiosis, possibly mediated by increases in specific organisms such as *Porphyromonas gingivalis*. Overgrowth of bacteria leads to excess inflammation, which, coupled with mechanisms of immune subversion employed by certain bacteria, leads to the formation of a chronic immune lesion. The combination of a persistent proinflammatory dysbiotic dental plaque and a proinflammatory immune lesion has the potential to develop into periodontitis involving the irreversible destruction of hard and soft tissues.

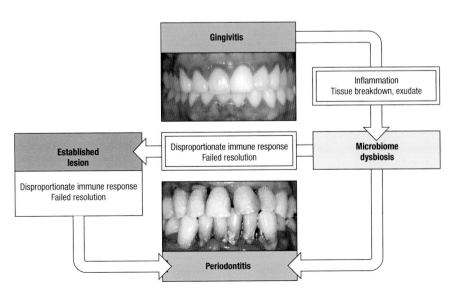

those bacteria most frequently found in health where there is little plasma exudation from the tissues. Those bacteria able to withstand the killing effects of migratory phagocytic cells will be able to increase in number at the expense of those organisms susceptible to these killing mechanisms. In so doing, the newly selected microbial community will present a different and potentially more injurious challenge to the periodontal tissues; hence, the escalation of increasing inflammation coupled to frustrated bacterial clearance will continue.

Importantly this *ecological plaque hypothesis* allows for the fact that potential periodontal pathogens may be present in health, albeit in relatively low numbers, but with the capacity to become more dominant members of the community when the environmental conditions favor their competitiveness over the other, more health-associated, members of the microbiota (**Figure 29.1**).

Bacterial and host mucosal interactions in periodontal disease

The current consensus is that periodontal disease is an inflammatory disease initiated by bacteria that causes the destruction of the supporting tissues of the teeth in a susceptible host. There are several bacteria associated with the disease, but which organisms actually initiate disease remains unknown. Organized biofilms enhance bacterial survival. Bacterial virulence factors further enhance survival. Molecules such as toxins, proteases, and glycosidases are classified as virulence factors that hide the bacteria from host detection and provide nutrients.

The relationship of specific periodontal bacteria and the host response is complex, and how these bacteria evade clearance by the inflammatory/immune response is an area of significant interest. Much of the focus has been given to the putative pathogen *Porphyromonas gingivalis* because of its frequent association with periodontal disease lesions and its capacity as a keystone pathogen, as described earlier. Evasion of the host response takes many forms and, from a natural selection standpoint, benefits the oral organisms by promoting chronic inflammation that produces abundant sources of collagen peptides and heme-containing compounds. Proposed mechanisms that have been demonstrated *in vitro* or in animal models include inhibition of leukocyte recruitment by *Porphyromonas gingivalis* lipopolysaccharide (LPS) binding to host adhesion molecules (ICAM-1, E-selectin), inhibition of IL-8 production by intracellular dephosphorylation of serine 536 on the p65 subunit of nuclear factor (NF)-κB, degradation of FMLP and C5a peptides by gingipains that directly disrupt a chemotactic gradient, and PI3K activation and inhibition of RhoA GTPase. LPS and lipid A have been shown to delay neutrophil apoptosis through toll-like receptor (TLR)2 signaling. LPS of *Porphyromonas gingivalis* displays significant structural heterogeneity that is mediated by lipid A phosphatases produced by the organism. It is postulated that the lipid A remodeling by the organism is necessary for colonization and contributes to the initiation of commensal organism dysbiosis.

However, the *in vitro* findings are not always consistent with observations *in vivo*. For instance, injection of *Porphyromas gingivalis* into a dorsal air pouch or dorsal chamber in a mouse induces a robust neutrophil influx suggesting that it is greater than with other gram-negative organisms, suggesting that disruption of chemotactic gradients observed *in vitro* does not play a role *in vivo*. Interestingly, pharmacologic reduction of inflammation improves bacterial clearance and reduces tissue damage suggesting that excess inflammation contributes to bacterial persistence.

Along these same lines, as outlined earlier, there is a growing body of evidence that the host inflammatory response has a major impact on the composition of the resident biofilm by altering the growth environment locally. The potential mechanisms by which this may occur are examined in the following sections.

29.2 Pathogen-associated molecular patterns are important in periodontal disease

To combat the onslaught of microorganisms at mucosal surfaces and to maintain homeostasis in contaminated environments such as the oral cavity, the host has evolved mechanisms to detect bacteria: the recognition of structural components of the bacterial surface. Lipopolysaccharides (LPSs), peptidoglycan (PGN), and other cell-surface components such as fimbriae are essential structural components of bacteria. Structural variation of these bacterial components between various species, or even between different strains of the same species, creates incredible structural diversity in prokaryotes. Despite structural heterogeneity, there are conserved motifs known as pathogen-associated molecular patterns (PAMPs). PAMPs and their receptors are considered in more detail in Chapter 16. Host cells have PAMP receptors termed *pattern-recognition receptors* (PRRs). These innate immune receptors are highly conserved and presumably evolved to detect invading bacteria. Binding of PAMPs by PRRs initiates an inflammatory response. On mucosal surfaces colonized by commensal organisms, the balance between constant inflammatory stimulus and maintenance of homeostasis is often lost, resulting in local inflammation of colonized tissues, as in gingivitis.

The best-characterized PAMPs include LPS, PGN, lipoteichoic acids, fimbriae, proteases, heat shock proteins, formylated bacterial peptides, and toxins. The host's PRRs include TLRs, nucleotide-binding oligomerization domain (NOD)-like receptors and C-type lectin receptors.

The oral cavity is the entrance to the gastrointestinal tract and is naturally colonized by a wide variety of bacteria. The colonization patterns and symbiotic function of the bacteria are physiologic. The tooth and gingival interface is the site of a variety of natural, innate host defenses. The regular shedding of epithelial cells, saliva, and the gingival crevicular fluid, and the continuous migration of neutrophils across the junctional epithelium into the gingival sulcus, maintains homeostasis, the sometimes delicate balance between the host tissues and the potentially pathogenic bacteria. Gingival inflammation results when equilibrium is lost and bacteria overgrow. Overt gingival inflammation may never proceed to destructive periodontitis, but the parameters that determine the transition remain unknown. In periodontitis, the overgrowth of pathogenic bacteria including *P. gingivalis*, *T. forsythia*, and *T. denticola* is common. Periodontal bacteria attach to epithelial cells via fimbriae; PRR recognition of PAMPs induces epithelial cell secretion of pro-inflammatory cytokines (tumor necrosis factor-α [TNF-α], interleukin [IL]-1β, IL-6) and the chemokine IL-8 in the connective tissue. The inflammation changes the growth environment of the biofilm leading to the overgrowth of pathogens. Several periodontopathogens (i.e., *P. gingivalis*, *A. actinomycetemcomitans*) can invade and transverse epithelial cells to the connective tissue *in vitro,* and PAMPs that are either shed or secreted further promote the development of the inflammatory lesion.

PAMP-mediated amplification of inflammation impacts leukocytes, fibroblasts, mast cells, endothelial cells, dendritic cells, and lymphocytes. Release of pro-inflammatory cytokines (TNF-α, IL-1β, IL-6, and IL-12), chemoattractants (CXCL8, CCL3, CXCL2, CCL2, and CCL12), and prostaglandin E$_2$ (PGE$_2$) by neutrophils, macrophages, fibroblasts, and mast cells in the connective tissue promotes collagen degradation. Mast-cell degranulation results in the secretion of histamine and leukotrienes.

Endothelial cells activated by the cytokines are active participants in the pathogenesis of periodontal inflammation. Secretion of endothelial cell chemokines (CXCL8, CCL2) and expression of adhesion molecules (P- and

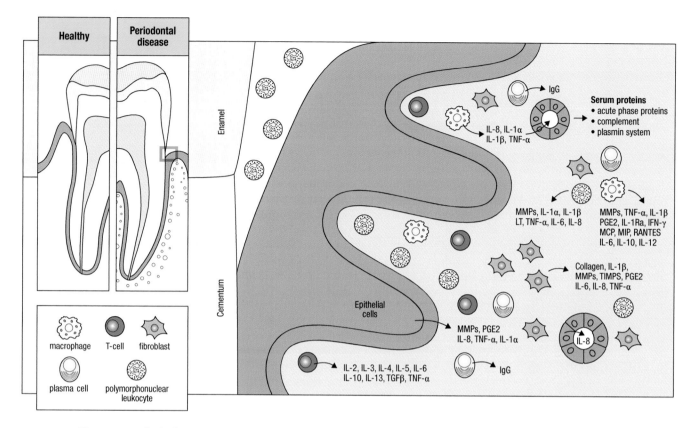

Figure 29.2 The inflammatory response to the bacterial biofilm. Healthy gingiva, by definition, does not contain inflammatory cells. In response to the buildup of a complex biofilm on the surfaces of teeth, an inflammatory lesion is established with the accumulation of inflammatory cells (T cells, plasma cells, polymorphonuclear leukocytes, and macrophages) resulting in the secretion of cytokines, chemokines, and matrix metalloproteinases (MMPs). The increased cytokine concentration in the environment causes fibroblasts to secrete collagen, MMPs, cytokines, and tissue inhibitors of metalloproteinases (TIMPs).

E-selectin and intercellular adhesion molecule (ICAM)-1 and ICAM-2) on the surface of endothelial cells promote leukocyte extravasation. P- and E-selectins bind glycoproteins on leukocytes and promote rolling of cells along the vessel wall. Firm binding of leukocytes to endothelium is mediated by integrins (leukocyte function-associated antigen-1 [LFA-1]), and leukocytes attach firmly to ICAM-1 expressed on endothelial cells; TNFα, PGE$_2$, and histamine increase vascular permeability and permit leukocyte diapedesis. Chemokines, such as IL-8, that are produced at the site of infection, along with bacterial chemoattractants (the tripeptide fMLP [*N*-formyl-methionine-leucine-phenylalanine]), form a concentration gradient for leukocytes to migrate to the focus of infection. Circulating pro-inflammatory cytokines from the site of inflammation activate hepatocytes in the liver to release acute-phase proteins, including LPS-binding protein and sCD14 that are important for the recognition of LPS. Complement proteins and C-reactive protein (CRP) are also released to opsonize bacteria for phagocytosis. **Figure 29.2** illustrates the initiation sequence of inflammation at the gingiva.

Acquired immunity and the chronic lesion

Failure to clear the bacterial insult and persistence of the lesion leads to activation of the acquired immune response. Gingival dendritic cells take bacterial antigens to the draining lymph nodes in the neck. Antigenic peptide binding to a class II major histocompatibility complex (MHC) receptor is required for antigen presentation to antigen-specific effector T cells, and antibody-secreting B cells are generated by clonal expansion and differentiation over the course of several days. Antigen-specific T cells are released into the blood and home to the site of infection. Likewise, macrophages that have engulfed bacteria present bacterial antigens to T cells. Antigen-specific T cells respond and activate macrophages. Secreted antibodies protect the host by (1) inhibiting toxicity or infectivity of pathogens (neutralization), (2) opsonizing pathogens and promoting phagocytosis, and (3) activating the complement system. Failure to clear the infection leads to further tissue damage. Activated macrophages produce oxygen radicals, nitric oxide, and proteases that are toxic to host cells. Moreover, recent work in mice reveals that the induction of an adaptive immune response to colonizing pathogens results in receptor activation of NF-κB ligand (RANKL)-dependent periodontal bone loss.

The trigger that causes the shift from tissue homeostasis to pathology remains unclear. Studies of the etiology of the periodontal lesion are cross-sectional, and definitive cause/effect relationships have not been demonstrated. One recent longitudinal study of periodontal disease progression failed to implicate any single organism or group of organisms; longitudinal animal studies suggest that inflammation significantly impacts the composition of the biofilm. In clinical studies, inflammation is a stronger predictor of future periodontal attachment loss than the quality or quantity of the biofilm.

29.3 T-cell and B-cell immune responses are involved in periodontitis

The induction of an innate immune response following the interaction between the periodontopathogens and host cell via PRR results in the production of cytokines and chemokines mediating an inflammatory response, as well as the induction of an acquired immune response. The TLRs that recognize ligands on periodontopathogens include TLR4 and TLR2 that heterodimerize with TLR1 or TLR6 to recognize a triacylated or diacylated form of lipoproteins, respectively. Once the ligand interacts with its appropriate TLR, a sequence of cellular events follows including the recruitment of adaptor molecules, the phosphorylation of signaling molecules, and the activation and translocation of transcription factors, all of which result in the expression of gene products that are essential for activation of cells. In dendritic cells, this process leads to the production of cytokines and chemokines, as well as an upregulation in the expression of costimulatory molecules, which are important for the induction of an adaptive immune response because they mediate the priming and differentiation of naive T cells that ultimately leads to the activation of B cells and antibody production. Furthermore, the nature of the cytokines produced by the activated dendritic cell influences the differentiation of the naive T-cell into a helper T (T_H) T_H1, T_H2, or T_H17 cell, or regulatory T (T_{reg}) cell, depending on the cytokine profile they produce and their regulatory functions, which in turn affect the nature of the acquired immune response induced. In the case of chronic periodontitis, the early lesion is characterized by a cellular infiltrate of macrophages and T cells, then shifts to increased numbers of B cells and plasma cells in the more advanced lesion (*vide supra*). The histology of the early lesion is more reflective of a T_H1 response, whereas the advanced lesion is more reflective of a T_H2 response. Analyses of gingival tissues from periodontitis patients for cytokine mRNA have often revealed a predominance of T_H1 or T_H2 cytokines or a mixed T_H1 and T_H2 cytokine profile. This is not totally unexpected considering the complexity of the subgingival plaque microflora and their ability to activate an inflammatory response via TLRs. However, the interplay between the microflora and TLRs for the response has not been fully elucidated.

Systemic and local (gingival crevicular fluid) antibodies to periodontopathogens are present in periodontitis patients, and the nature of the antibody response, including the magnitude, specificity, and IgG subclass, has been extensively investigated in order to determine correlates with disease severity and local cytokine responses. Although some investigations have shown a relationship between the induction of antibodies to periodontopathogens resulting from infection or from immunization and a reduction in the microflora and/or progression of disease, other studies have revealed extremely variable results. The variability observed in these studies reflects the complexity of this disease, questions the functionality of the antibodies induced, and challenges future studies in this area.

29.4 Cytokines are involved in bone resorption in periodontitis

A major consequence of the inflammatory response associated with periodontal disease is resorption of alveolar bone. Under normal conditions, bone is periodically resorbed while new bone forms. In the case of periodontal disease, there is a shift to bone resorption due to an increase in inflammatory cells and cytokines, and an increase in osteoclasts in the local tissue. (Osteoclasts are large, multinucleated cells derived from the monocyte-macrophage lineage, whose major function is to resorb bone.) Differentiation of osteoclasts is regulated by macrophage colony-stimulating factor (M-CSF) and receptor activator of NF-κB ligand (RANKL) found on several cell types including T cells and osteoblasts. M-CSF promotes proliferation and survival of osteoclast progenitors, whereas RANKL promotes cells to differentiate along the osteoclast lineage and acts as an activating and survival factor for mature osteoclasts. RANKL binds to RANK present on osteoclast precursors and mature osteoclasts resulting in the recruitment of TNF receptor-associated factor (TRAF) family proteins, especially TRAF6, which leads to activation of various signaling pathways and transcription factors, and the secretion of bone-degrading factors.

Osteoclastogenesis is positively regulated by RANK–RANKL signaling and negatively regulated by osteoprotegerin (OPG), a soluble decoy receptor. OPG is produced by various cells, including dendritic cells and osteoblasts, and binds to RANKL, thus preventing RANKL from interactions with its biologically active receptor RANK. A main mechanism regulating bone resorption and formation is the relative concentration of RANKL and OPG. When RANKL expression is high relative to OPG, RANKL can bind RANK on osteoclast precursors, which tips the balance to favor activation of osteoclast formation and bone resorption; whereas when OPG levels are higher than RANKL expression, OPG binds free RANKL inhibiting it from binding to RANK, which leads to a reduction in osteoclast formation and apoptosis of osteoclasts. *In vivo* studies have shown that overexpression of OPG in transgenic mice resulted in severe osteopetrosis, while osteoporosis was seen in OPG-deficient mice due to an increase in the number of osteoclasts. Furthermore, analysis of RANKL and OPG levels in gingival tissue extracts or from gingival crevicular fluid has revealed a higher RANKL/OPG ratio from diseased sites than from healthy sites. During an inflammatory response, pro-inflammatory cytokines such as IL-1 or TNF-α can induce osteoclastogenesis by increasing the expression of RANKL by T cells and B cells as well, whereas the anti-inflammatory cytokine IL-10 inhibits RANKL but enhances OPG production.

Because osteoclast precursors express TLRs, it is likely that microbial components in dental plaque can affect tissue homeostasis leading to a more destructive lesion; however, the evidence is conflicting. Some studies have shown that bacteria or their products, for example, LPS, induce osteoclast differentiation, whereas others have shown that TLR ligands, including LPS, inhibit osteoclast differentiation, although LPS was able to stimulate osteoclastogenesis in RANKL-committed cells. The contributions of TLRs and their microbial ligands in shifting bone homeostasis toward a destructive host inflammatory response that contributes to excessive bone resorption as seen in cases of chronic periodontitis have yet to be determined.

Inflammation drives pathogenesis of periodontitis

Inflammation has been demonstrated to play a central role in the progression of periodontal disease and a number of systemic diseases not traditionally considered inflammatory diseases, such as cardiovascular disease. While the etiology is bacterial, periodontal destruction is mediated primarily by the inflammatory response. Genetic and epigenetic factors may also be responsible for a hyperinflammatory phenotype that impacts susceptibility of the host to periodontal disease and tissue destruction.

29.5 Arachidonic acid pathway is important in periodontal disease

Initiation of periodontal inflammation stimulates the accumulation of leukocytes in the affected area. Early events in neutrophil stimulation include activation of neutrophil lysosomal phospholipase to release free arachidonic acid from membrane phospholipids (Figure 29.3). Free arachidonic acid is the substrate for two different pathways: (1) the cyclooxygenase (COX) pathway that leads to the production of prostaglandins (e.g., PGE_2) and (2) the lipoxygenase pathways that lead to the production of a series of hydroxyl acids depending on the cell type. 5-Lipoxygenase (LO) products, including the leukotrienes (i.e., LTB_4), are derived from myeloid cells: the 12-LO is expressed in platelets and the 15-LO in epithelial and endothelial cells. The end products of 12- and 15-LO are 12- and 15-hydroxytetraenoic acids, respectively. PGE_2 is a potent activator of osteoclast-mediated bone resorption and has been shown to mediate inflammation and periodontal tissue destruction. LTB_4 is a neutrophil chemoattractant that stimulates granule enzyme release from neutrophils and contributes to further tissue damage.

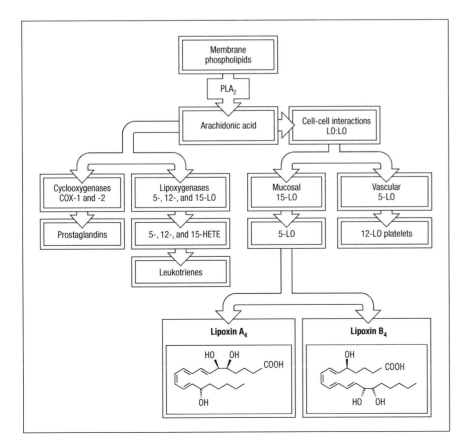

Figure 29.3 Metabolism of arachidonic acid. Arachidonic acid is cleaved from the sn-2 position of membrane phospholipids and metabolized by cyclooxygenases (COX) or lipoxygenases (LO). The pro-inflammatory products are well known and include the prostaglandins from COX-1 and COX-2 and the leukotrienes and other hydroxyl acids (hydroxytetraenoic acid, HETE) from 5-, 12-, and 15-LO. Later in the inflammatory lesion, lipoxygenase interactions take place; a single arachidonic acid molecule interacts with two different LOs yielding a new class of molecule called lipoxins. PLA_2, phospholipase A_2.

29.6 Regulation of inflammation is major determinant of bacterial colonization and disease

Various aspects of innate and acquired immunity are the determinants of the pathogenesis of periodontitis; however, the role of each aspect in susceptibility to disease remains unclear. There are few studies investigating alteration of the suggested susceptibility determinant and the resultant impact on the outcome of disease.

Throughout the modern era of medicine, periodontal inflammation has been investigated in the context of pro-inflammatory mediators such as prostaglandins, leukotrienes, and interleukins. The resolution of inflammation and return to homeostasis was thought to result from the passive decay of pro-inflammatory molecules over time and a deterioration of the response. It is now understood that late in the inflammatory molecular cascade, a new class of lipid mediators derived from arachidonic acid is produced, the lipoxins (LXA_4, LXB_4). The lipoxins bind to the FPRL1 receptor on inflammatory cells resulting in a change of cell phenotype that alters cell function to drive the lesion to resolution and healing. Extensive structure/function relationship elucidation and characterization of this new genus of lipid mediators established that resolution is an *active* process, rather than a passive decay of pro-inflammatory signals as was once widely believed.

Lipoxins derived from arachidonic acid comprise the endogenous fatty acid resolution pathways. There is also a broad array of dietary fatty acid pathways exemplified by the omega-3 polyunsaturated fatty acids (PUFAs) that also participate in active resolution of inflammation. Simply stated, PUFAs are an alternate substrate for endogenous lipoxygenases to produce other classes of resolving molecules with properties similar to lipoxins termed *resolvins* that act through unique receptor mechanisms at the cellular level (**Figure 29.4**). This new genus of lipid compounds has been termed

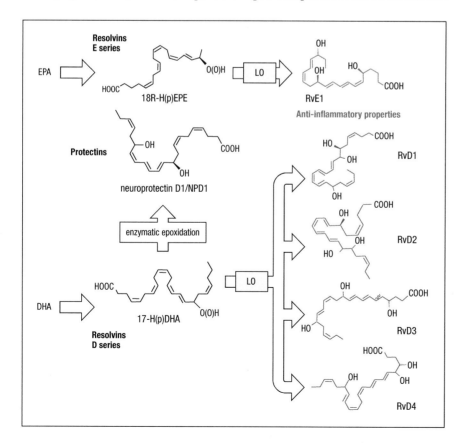

Figure 29.4 Metabolism of omega-3 polyunsaturated fatty acid (PUFA). The same LO enzyme system that generates lipoxins from arachidonic acid metabolizes omega-3 PUFA into resolvins and protectins. The two major omega-3 PUFA are eicosapentaenoic acid (EPA), which is metabolized into resolvins of the E series, and docosahexaenoic acid (DHA), which is metabolized into resolvins of the D series and the protectins. The role of aspirin-modified COX-2 is also indicated. In the presence of aspirin, COX-2 is acetylated and becomes a new active enzyme, a 15R-lipoxygenase. Aspirin-modified COX-2 is responsible for the production of aspirin-triggered lipoxins, the 18R-resolvins of the E series, and 17R-resolvins of the D series.

specialized proresolving mediators (SPMs). The term "active" in this context cannot be overstated. SPMs bind to specific receptors exerting their actions through feed-forward mechanisms, not inhibition. The potential use of agonists to resolve inflammation, as opposed to enzyme inhibitors or receptor antagonists to block pro-inflammation, has led to the emergence of the new field of *Resolution Pharmacology.*

29.7 Endogenous generation of SPM is important in periodontal disease

Lipoxins are the product of lipoxygenase and lipoxygenase interactions through transcellular biosynthetic pathways. Experimental evidence suggests that as an inflammatory lesion matures, the lipoxygenase product of one cell type (the product of the platelet 12-LO: 12-hydroxytetraenoic acid) is further metabolized by a myeloid cell 5-LO to produce lipoxin A_4. The known lipoxin synthetic pathways are illustrated in **Figure 29.3**. LXA_4 binds to different inflammatory cells to produce both anti-inflammatory and proresolution actions driving resolution of inflammation. LXA_4 binding to FPRL receptors on inflammatory neutrophils inhibits chemotaxis, superoxide generation, and secretion of pro-inflammatory and proteolytic molecules, including PGE_2, and induces neutrophil apoptosis. Concurrently, LXA_4 is chemotactic for nonphlogistic recruitment of mononuclear phagocytes; the resolving monocyte/macrophage exhibits enhanced phagocytosis of bacteria and apoptotic neutrophils but does not secrete pro-inflammatory mediators. In the context of periodontal disease, there is detectable LXA_4 in the gingival crevicular fluid of periodontitis patients, albeit at insufficient concentrations to exert clinical actions. These observations have suggested that periodontal diseases, and other inflammatory diseases, are characterized by a failure of resolution pathways in addition to, or instead of, overproduction of pro-inflammatory signals.

The main endogenous fatty acid substrate for the production of resolution agonists is arachidonic acid. However, dietary fatty acids are also substrates for the same metabolic pathways. The principal dietary fatty acids are omega-3 polyunsaturated fatty acids (omega-3 PUFA). Eicosapentaenoic acid, C20:5, and docosahexaenoic acid, C22:6, are the substrates for lipoxygenases generating a new genus of molecules termed the *resolvins* of the E and D series and docosatrienes. E- and D-series resolvins bind to distinct receptors on inflammatory cells to elicit anti-inflammatory and proresolution actions that are similar but not exactly the same as the lipoxins. For instance, the receptors for resolvin E1 (RvE1) are ERV1 (a.k.a. chemR23) on macrophages and BLT1 on neutrophils. It is also interesting to note that multiple ligands can bind to the same receptor: ERV1 is also a receptor for chemerin, a peptide that is chemotactic for macrophages. Recent work has also identified new mediators produced from DHA by macrophages called maresins. Notably, receptor distribution of different cell types is also more widespread than formerly known. ERV1, for instance, is also expressed on osteoclasts, osteoblasts, and other stromal cells. It is hypothesized that these omega-3 PUFA pathways contribute, in large part, to the systemic anti-inflammatory activity of omega-3 fatty acids providing protection from cardiovascular disease and other inflammatory conditions. The resolvins derived from eicosapentaenoic acid (RvE1–4), and the D-series (RvD1–6), maresins, and protectins derived from docosahexaenoic acid, are defined by their potent bioactivity and novel chemical structures. First identified in resolving exudates, the biological activity of these compounds includes protective functions in neural tissues, retina, and corneal epithelial injury in the eye. The known actions of resolvins include reduction of neutrophil trafficking, regulation of cytokines and reactive oxygen species production, and a general dampening of the inflammatory response and enhancement of bacterial clearance. The protectins were so named because of their anti-inflammatory and protective actions in stroke, human Alzheimer's disease, and other neural systems.

Aspirin plays an interesting role in both the arachidonic acid and omega-3 PUFA pathways. Among the known actions of aspirin is inactivation of cyclooxygenase-2 (COX-2). Unlike other nonsteroidal drugs, aspirin modification of COX-2 through acetylation converts COX-2 into another active enzyme, a 15R-lipoxygenase. The stereochemical change in the resulting lipoxin and resolvin products yields resolving molecules that are more stable and longer acting. The potent anti-inflammatory properties of aspirin were recognized when it was discovered in the early 1900s; decades later, the inhibition of prostaglandin synthesis by aspirin was recognized as a mode of action. Discoveries further clarifying the mode of action of aspirin suggest that the action of aspirin is due in part to production of "aspirin-triggered" lipoxins and resolvins. The previously unappreciated aspects of the action of aspirin have interesting implications for drug design: the ligands are receptor agonists, not inhibitors or antagonists, precluding the requirement for maintaining an effective blood level, and the R-epimers of the natural molecules are resistant to degradation by specific endogenous enzymes.

29.8 Resolution phase of acute inflammation is as important as onset phase

The goal of the acute inflammatory response is the elimination of the injury or invader and return to homeostasis (Figure 29.5). The process is well known, involving leukocyte (neutrophils) recruitment, phagocytosis of invaders, recruitment and differentiation of mononuclear phagocytes

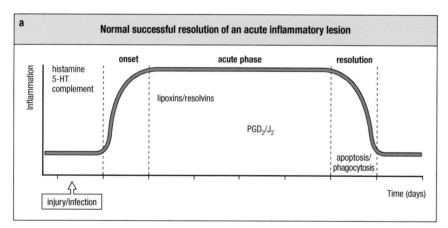

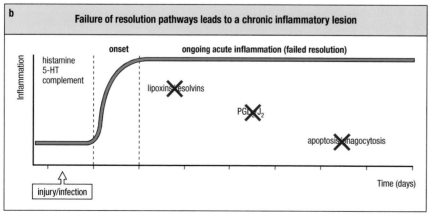

Figure 29.5 Failure of resolution pathways leads to chronic inflammation. (a) The normal successful resolution of an acute inflammatory lesion with the temporal emergence of resolution agonists that drive return to homeostasis. **(b)** A failure of resolution pathways leading to a chronic inflammatory lesion. 5-HT, 5-hydroxytryptamine.

into macrophages, and clearance of the lesion. Return to homeostasis is not complete until all inflammatory cells, including neutrophils, are removed from the lesion. Failure to resolve because of persistent infection or destruction of the cellular matrix may result in scarring and fibrosis with incomplete healing or chronic inflammation. The early events in conditions such as periodontitis, for example, controlling neutrophil biology, can mediate conversion of acute gingivitis to chronic periodontitis. Scarring and fibrosis in periodontitis (repair rather than regeneration) prevent the return to homeostasis.

It is now known that resolution of inflammation is an active process and that anti-inflammation is not the same as resolution. Anti-inflammation is the downregulation, or pharmacologic blocking or inhibition, of pro-inflammatory mediators such as PGE_2 and LTB_4. Resolution is mediated by the active recruitment of nonphlogistic macrophages to clear the lesion of apoptotic neutrophils and remaining microorganisms. The distinction is important because the current pharmacologic approach to the control of inflammation only addresses half of the process. Cyclooxygenase inhibitors such as ibuprofen or flurbiprofen block the production of PGE_2 and the cardinal signs of inflammation, but they do not promote resolution. In fact, this class of drug is resolution toxic, prolonging the time to resolve the lesion and extending tissue inflammation temporally. Resolution pharmacology is an approach that focuses on receptor agonists that drive resolution. This approach is not limited to SPM; other molecules, e.g., melanocortins, also fulfill the role of receptor mediators of active resolution of inflammation.

Endogenous resolving molecules as therapeutic agents in periodontal disease

A new hypothesis for the pathogenesis of periodontitis and other inflammatory conditions is that a failure of resolution contributes as much to inflammatory disease as overproduction of pro-inflammatory mediators. Proof-of-principle experiments for periodontitis were performed in a series of intervention studies in an animal model: *P. gingivalis*–induced periodontitis in the New Zealand white rabbit. Periodontitis was induced by tying 3/0 silk ligatures around the second premolars in the mandibular quadrants, and 10^9 colony-forming units (cfu) *P. gingivalis* in a methyl cellulose slurry was applied three times per week for 6 weeks. Ligature alone is not sufficient to induce disease in this model, and the disease progression induced by *P. gingivalis* was inhibitable by systemically administered metronidazole, confirming the bacterial etiology.

In separate experiments, prevention of disease and treatment of established disease with a resolution agonist as a monotherapy were assessed. In the prevention protocol, a parallel design compared application of resolving molecules three times weekly (applied at the same time as the *P. gingivalis*) with application of a placebo (ethanol vehicle) throughout the experimental period. In the treatment protocol, disease was induced for 6 weeks with topical *P. gingivalis*, which established a stable *P. gingivalis* infection; *P. gingivalis* was not applied during the treatment phase.

29.9   Resolvins can help prevent periodontitis

The preventive regimen in the rabbit model was the application of 5 μL of a 1 mg mL^{-1} solution of RvE1 in ethanol topically three times per week; the control was ethanol alone (placebo). At the end of the 6 weeks, animals were sacrificed and periodontal disease progression was quantified morphologically and histologically. Placebo treatment had no impact on the progression of disease, which resulted in significant bone and attachment loss. RvE1 application prevented the onset and progression of periodontal destruction including all clinical signs of inflammation (**Figure 29.6**).

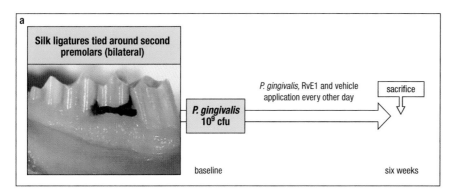

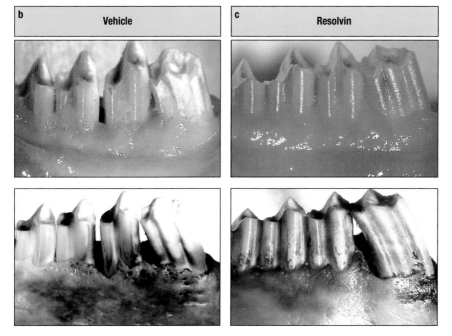

Figure 29.6 **Prevention of periodontitis with exogenous topical application of a resolution agonist.** (**a**) The experimental design of inducing rabbit periodontitis as described in the text. Two groups of animals were compared, with and without RvE1 application. The animals given the vehicle (**b**) show bone loss and loss of soft tissue around the ligated tooth. In contrast, RvE1 treatment (**c**) prevents bone loss and the loss of soft tissue.

29.10 Resolvins can treat periodontitis

Periodontal lesions induced by *P. gingivalis* were established over 6 weeks. After 6 weeks, application of *P. gingivalis* was discontinued. During a second 6-week interval, the established periodontal lesions were treated with topical RvE1 or placebo. At the end of the second 6 weeks, animals were sacrificed and periodontal destruction quantified. Periodontal disease characterized by soft tissue and bone loss was observed at 6 weeks and after an additional 6 weeks of placebo treatment. Progression of disease was significant with deepening of pocket depth and additional loss of crestal bone with infrabony pocketing. In the placebo group, disease progressed in the second 6 weeks although *P. gingivalis* application was discontinued; microbial analyses demonstrated the persistence of the *P. gingivalis* infection. In the RvE1-treated group, no clinical inflammation was observed; soft tissues returned to preligature architecture. Bone regeneration to predisease architecture with restoration of crestal height, elimination of infrabony defects, and regeneration of new cementum and bone with an organized periodontal ligament was also observed (**Figure 29.7**).

Biofilms and inflammatory periodontitis

The relationship of the biofilm to the inflammatory response has been well documented as cause and effect based on intervention studies that clearly demonstrate that the biofilm induces inflammation. However, the initiating

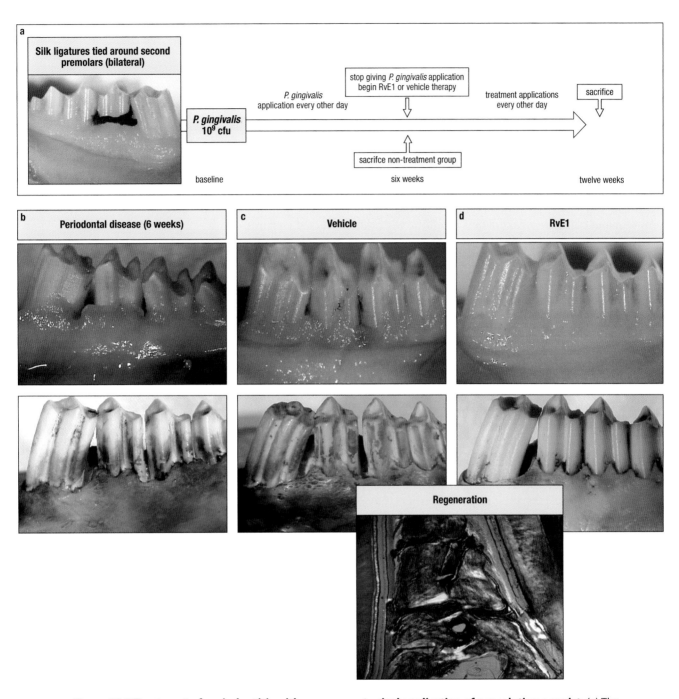

Figure 29.7 **Treatment of periodontitis with exogenous topical application of a resolution agonist.** (**a**) The experimental design for the induction and treatment of rabbit periodontitis as described in the text. After induction of periodontitis, one group of animals was sacrificed at 6 weeks to determine the extent of periodontal destruction. The other two groups were treated for an additional 6 weeks with RvE1 or placebo and sacrificed at 12 weeks. In the lower panels, (**b**) illustrates the soft and hard tissue periodontal destruction after 6 weeks of disease progression. (**c**) illustrates persistent soft tissue inflammation and bone destruction after 6 weeks of treatment with vehicle and (**d**) after 6 weeks of RvE1 treatment. Note the appearance of healthy, pink soft tissue and the regrowth of bone after RvE1 treatment. The inset marked "Regeneration" is a polarized light photomicrograph of an undecalcified section of regenerated tissues showing reestablishment of the periodontal ligament with new bone and new connective tissue fibers uniting the new bone and new cementum on the tooth root surface.

pathogens in the biofilm remain unknown. There is a large literature associating specific organisms in cross-sectional studies, but cause and effect have not been established. Based on the available data, a second explanation of the cross-sectional observations is that *P. gingivalis* is present in high numbers in periodontitis *because* of the deep pockets.

In the previously discussed rabbit experiments, longitudinal assessment of microbial changes with RvE1 treatment using DNA checkerboard analysis revealed the presence of *P. gingivalis* and an increased complexity of the biofilm in placebo-treated animals. RvE1 treatment eradicated *P. gingivalis* from the biofilm, and the resident flora returned to its normal composition and numbers despite no antibacterial activity of the resolvins. The spontaneous elimination of the pathogen and the return of a "normal" biofilm suggest that the composition of the biofilm is mediated by the inflammatory milieu providing nutrients to support the pathogenic organisms. For instance, *P. gingivalis* is an asaccharolytic organism that depends on essential amino acids as a food source. It is well appreciated that *P. gingivalis* has an armamentarium of proteolytic enzymes, the gingipains, that break down peptides into essential amino acids. In the absence of inflammation, tissue protein and collagen breakdown is minimal, which eliminates the source of peptides that support the organism.

This principle was further investigated in a rat periodontitis model. In these experiments, exogenous addition of human pathogens was not used to induce periodontitis. Instead, ligatures applied around teeth were used to induce dysbiosis and resulting periodontitis. In these experiments, next-generation sequencing of 16S rDNA was used to identify the entire microbiome associated with rat periodontal health and periodontitis. Two separate experiments were performed. In the first, ligature-induced periodontitis was prevented with topical application of RvE1; in the second, established periodontitis was treated with topical application of RvE1. Results revealed that ligature-induced periodontitis caused a significant shift in the microbiome that was prevented by RvE1. Treatment of established periodontitis with RvE1 reversed the microbiome shift back toward a microbiome more associated with health and also led to reversal of disease. The implication is that the pathogenesis of periodontitis is not a linear progression of colonization by a pathogen followed by an inflammatory lesion. The inflammatory lesion may in fact alter the microbiome leading to disease. The dynamic interaction of the biofilm and the host response requires further study. Moreover, implications for therapeutic strategies that target etiology may also not be linear. Future therapeutics for periodontitis will likely target the inflammatory response to effect changes in the pathogenic biofilm.

SUMMARY

Periodontitis is an inflammatory disease initiated by bacteria that initiates the host-mediated destruction of the supporting tissues of the teeth in a susceptible host. The periodontal microbiota can bind and activate gingival epithelial cells via TLRs and other PAMPs leading to the production of pro-inflammatory cytokines and chemokines. The inflammatory process is amplified following the activation of the underlying cells including macrophages, dendritic cells, lymphocytes, neutrophils, and fibroblasts via TLRs. This, in turn, also leads to the induction of an adaptive immune response, all of which shifts the local environment leading to overgrowth of the colonizing pathogens. If left untreated, a major consequence of the host response is resorption of alveolar bone, and a main mechanism regulating bone resorption and formation is the relative concentration of RANKL and OPG. When RANKL expression is high compared with OPG, RANKL binds RANK on osteoclast precursors, which results in osteoclast formation and bone resorption, whereas when OPG levels are higher than RANKL expression, OPG acts as a decoy receptor for RANKL leading to a reduction in osteoclast formation and apoptosis of preexisting osteoclasts. The lipoxins, lipid mediators derived from arachidonic acid, as well as other fatty acid products termed *resolvins*, have now been shown to play an important role in the resolution of inflammation by binding to specific receptors on inflammatory cells to elicit active resolution of inflammation, such as a reduction of neutrophil trafficking, regulation of cytokine production, and general dampening of the inflammatory response and enhancement of bacterial clearance. This active process differs from that seen with pharmacologic anti-inflammatory agents, such as ibuprofen, which block the production of PGE_2 but do not promote resolution. Our growing knowledge of the relationship between the etiologic agents of periodontitis and the inflammatory pathogenesis will have a major impact on treatment strategies and pharmacotherapeutics for the treatment of inflammatory diseases.

FURTHER READING

Colombo, A.P., Boches, S.K., Cotton, S.L. et al.: Comparisons of subgingival microbial profiles of refractory periodontitis, severe periodontitis, and periodontal health using the human oral microbe identification microarray. *J. Periodontol.* 2009, 80:1421–1432.

Darveau, R.P.: Periodontitis; a polymicrobial disruption of host homeostasis. *Nat. Rev. Microbiol.* 2010, 8:481–490.

Darveau, R. P., Hajishengallis, G., and Curtis, M. A.: *Porphyromonas gingivalis* as a potential community activist for disease. *J. Dent. Res.* 2012, 91:816–820.

Dewhirst, F.E., Chen, T., Izard, J. et al.: The human oral microbiome. *J. Bacteriol.* 2010, 192:5002–5017.

El-Sharkawy, H., Aboelsaad, N., Eliwa, M. et al.: Adjunctive treatment of chronic periodontitis with daily dietary supplementation with omega-3 fatty acids and low-dose aspirin. *J. Periodontol.* 2010, 81:1635–1643.

Kolenbrander, P.E., Palmer, R.J., Jr., Periasamy, S. et al.: Oral multispecies biofilm development and the key role of cell-cell distance. *Nat. Rev. Microbiol.* 2010, 8:471–480.

Lorenzo, J., Horowitz, M., and Choi, Y.: Osteoimmunology: Interactions of the bone and immune system. *Endocrine. Rev.* 2008, 29:403–440.

Manicassamy, S., and Pulendran, B.: Modulation of adaptive immunity with Toll-like receptors. *Semin. Immunol.* 2009, 21:185–193.

Marsh, P.D., and Devine, D.A.: How is the development of dental biofilms influenced by the host? *J. Clin. Periodontol.* 2011, 38(Suppl 11):28–35.

Olsen, I., and Hajishengallis, G.: Major neutrophil functions subverted by *Porphyromonas gingivalis. J. Oral Microbiol.* 2016, 8:30936.

Schenkein, H.A.: Host responses in maintaining periodontal health and determining periodontal disease. *Periodontology 2000.* 2006, 40:77–93.

Serhan, C.N., Chiang, N., and Van Dyke, T.E.: Resolving inflammation: Dual anti-inflammatory and pro-resolving lipid mediators. *Nat. Rev. Immunol.* 2008, 8:349–361.

Takeuchi, O., and Akira, S.: Pattern recognition receptors and inflammation. *Cell* 2010, 140:805–820.

Van Dyke, T.E.: Proresolving lipid mediators: Potential for prevention and treatment of periodontitis. *J. Clin. Periodontol.* 2011, 38(Suppl 11):119–125.

Zenobia, C., Hasturk, H., Nguyen, D. et al.: Porphyromonas gingivalis lipid A phosphatase activity is critical for colonization and increasing commensal load in the rabbit ligature model. *Infect. Immun.* 2014, 82:650–659.

Mucosal vaccine strategies

30

NILS LYCKE, JAN HOLMGREN, AND HARRY B. GREENBERG

Mucosal immunization was introduced in the 1960s with the Sabin live attenuated oral polio vaccine (OPV). The success of this vaccine, attributable not only to its needle-free character, but foremost to its ability to stimulate protective immunity, has not been followed by the series of efficient mucosal vaccines one might have anticipated. Rather, only a handful of internationally licensed, commercially available mucosal vaccines exist today for human use (Table 30.1). In fact, these vaccines, which aside from the OPV are directed against cholera, typhoid fever, rotavirus, and influenza virus, have little in common, except that a majority are live attenuated vaccines, which could in part explain their efficacy. The exception is the predominating inactivated whole-cell oral cholera vaccines (OCV) which contain large numbers of killed *Vibrio cholerae* bacteria ± cholera toxin B subunit, and hence, even though it is inactive, it hosts inbuilt immunomodulating properties. The great problem is that mucosal vaccine candidates that in animal models show strong protection often fail in clinical trials. For example, trials in humans with different types of *Helicobacter pylori* vaccines, including live attenuated vaccines, have failed despite animal studies showing great promise. We believe there are several explanations as to why this is the case, the most important being the lack of safe and effective clinically acceptable vaccine adjuvants for mucosal administration. At the same time, it should be emphasized that highly protective vaccines against a number of mucosal infections are administered by a parenteral rather than a mucosal route and still convey strong protection, especially those that are directed against infections in or from "less tight" mucosal membranes and stimulate strong serum IgG antibodies that cross the mucosal epithelium by neonatal Fc receptor (FcRn)–mediated transport and, perhaps, transudation (see Chapter 5).

PRINCIPLES OF MUCOSAL VACCINATION

Today we have fairly precise information about signals necessary for the stimulation of adaptive immune responses. This involves the stepwise stimulation of naive T cells through recognition by the T-cell receptor of major histocompatibility complex (MHC) class I or II together with peptide, and the more complicated second signals that are provided by cytokines and costimulatory molecules on the membranes of antigen-presenting cells (APCs). Costimulatory molecules such as CD40, CD80, CD86, and OX40L have been found critical for driving T-cell–dependent immune responses. The linking of innate responses to the induction of adaptive immunity is the key to understanding how to construct effective mucosal vaccines. It is noteworthy that while we have a wealth of data relevant to the induction and preservation of an immune response (both systemic and mucosal) in animals, there is a

TABLE 30.1   CURRENTLY INTERNATIONALLY LICENSED MUCOSAL VACCINES FOR HUMAN USE[a]

Disease/ Pathogens	Trade names	Composition	Dosage	Immunological mechanism	Efficacy
Cholera *Vibrio cholerae*	Dukoral Shanchol Euvichol	Inactivated *V. cholerae* O1 bacteria + Cholera toxin B subunit (Dukoral), or + inactivated *V. cholerae* O139 bacteria (Shanchol; Euvichol)	Oral, 2–3 doses	Antibacterial LPS-specific and toxin-specific IgA (& antibacterial and antitoxic IgA memory)	70%–85% (Over 90% effectiveness due to added herd protection)
Influenza Influenza virus A + B	FluMist	Cold-adapted live attenuated influenza viruses	Nasal, 1 dose	Mucosal IgA and systemic neutralizing IgG	Variable, ca 60%–70% (similar as for injected vaccine)
Polio virus	Orimune OPV Poliomyelitis vaccine	Live attenuated trivalent, bivalent and monovalent polioviruses	Oral, 3 doses	Mucosal IgA and systemic IgG	Over 90% in most of the world (less effective in poor countries)
Rotavirus diarrhea	Rotarix; RotaTeq; Rotavac	Live attenuated, monovalent or pentavalent rotaviruses	Oral, 3 doses	Mucosal IgA and systemic neutralizing IgG	Over 70%–90% against severe disease in industrialized countries (less effective in poor countries)
Typhoid fever *Salmonella Typhi*	Vivotif	Live attenuated *S. typhi* bacteria	Oral, 3–4 doses	Mucosal IgA, systemic IgG and cell-mediated responses	Variable, ca 50%–70%

[a] Additional only nationally licensed or otherwise restricted mucosal vaccines include, e.g., additional inactivated oral cholera vaccines (OCVs) mOrcVax (Vietnam; same composition as Shanchol/Euvichol) and Oravac (China; similar to Dukoral), a live attenuated OCV VaxChora (USA) and an oral adenovirus vaccine restricted to use in US military staff.

significant lack of mechanistic data as to how this information pertains to the performance of licensed vaccines in humans (Table 30.1).

The mucosa-associated lymphoid tissues (MALTs) are composed of anatomically defined lymphoid subcompartments, which serve as the principal mucosal inductive sites where immune responses are initiated. The Peyer's patches (PPs) in the small intestine, and the tonsils and adenoids at the entrance of the aerodigestive tract, are examples of such inductive sites. Also viewed primarily as inductive sites are the colon patches, the appendix, and the abundant small isolated lymphoid follicles in the small and large intestines and in the upper respiratory tract. The mesenteric lymph nodes (MLNs), draining the intestinal lymph, also belong to the inductive sites, but for induction of immune tolerance rather than for IgA B-cell immunity, although this is still incompletely investigated. By contrast, immune effector sites are localized to the lamina propria and intraepithelial compartments of the mucosal membranes. At mucosal sites, IgA B-cell responses can be directed against T-cell–independent antigens, commonly provided by the gut microbiota, but the vast majority of IgA responses are T-cell–dependent. A few studies have indicated that IgA B-cell responses, primarily against the microbiota, can occur in the nonorganized lamina propria, but most studies have identified organized lymphoid tissues, and PP in particular, as the sites for class-switch recombination (CSR) from IgM to IgA and for somatic hypermutations (SHMs), which increase antibody affinity. This is because B-cell proliferation is critically required for IgA CSR and for SHM, both of which are prominent in PPs but not well documented in the gut lamina propria. In addition, mRNA for the AID enzyme, responsible for IgA CSR and SHM, has not been detected in nonorganized lamina propria. Nevertheless, all T-cell–dependent IgA B-cell responses require organized lymphoid tissue, such as that found in the PPs of the gut-associated lymphoid tissues (GALT).

Today we know that mucosal administration of vaccines, as opposed to parenteral immunization, provides two essential components to generating strong local immune protection. The first is the establishment of tissue-specific resident memory T cells, which have attracted recent interest for their critical role in resistance against, for example, lung or genital tract viral infections. The second component is the stimulation of locally produced secretory IgA (SIgA) antibodies. In fact, SIgA is the classical hallmark of mucosal immune responses and important for the barrier function of the mucosal membrane. In particular, the neutralizing function of SIgA for antitoxin and antiviral protection has been well documented, while mechanism(s) of SIgA-mediated protection against pathogenic bacterial colonization are less well defined. However, bacterial clumping has been found a major function of SIgA for eliminating bacteria from the intestine, as in the case of nontyphoidal *Salmonella*, *Escherichia coli*, and *Vibrio cholerae*. In fact, there is good correlation between SIgA levels and resistance against infection, for example, with rotavirus or *Vibrio cholerae*. SIgA also controls translocation of the microbiota to the draining MLN. In fact, in recent years it has become increasingly clear that the composition of the microbiota shapes the IgA production, and conversely, the SIgA produced influences the composition of the microbiota. Noteworthy is that in germ-free mice, little IgA is produced, but levels rapidly increase upon colonization. Monocolonization experiments have indicated that bacterial species differ in their ability to stimulate IgA production. Intestinal segmented filamentous bacteria (SFBs) are particularly effective at triggering SIgA production, while other bacterial strains exert a weaker effect on IgA production. Thus, under homeostatic conditions, a large fraction of gut SIgA is constitutively produced against the microbiota. Whether these SIgAs are dominated by polyreactive or cross-reactive SIgA of low affinity, as has been observed, or strain- or taxa-specific SIgA of higher affinity that coat and contain the commensal microbiota, is debated. In this context, it can be foreseen that an expanded knowledge about the glycobiology of the microbiota would contribute to a better understanding of SIgA production and function in a steady state. The phenomenon of cross-species reactive SIgA may well turn out to reflect the recognition of specific glycan structures rather than a promiscuous binding to multiple bacterial species. Nevertheless, the function of nonaggregating SIgAs still awaits to be further explored, and their role for eliminating or retaining certain bacterial species in the gut intestine warrants further investigation.

Of critical importance to the induction of mucosal immunity is the presence of a specialized epithelium over the organized lymphoid follicles. This follicle-associated epithelium (FAE) in the gut, for example, can sample antigen from the lumen and allow APCs in the PPs to activate specific $CD4^+$ or $CD8^+$ T cells, under strict MHC I and II control. As more extensively discussed in Chapters 5 and 13, antigens may either penetrate or be taken up at mucosal inductive sites through a variety of mechanisms. The microfold (M) cells are most prominent over the PPs and have special properties for engulfment and transcellular transport of antigens across the epithelial barrier. In the subepithelial dome of the PPs, APCs can take up antigen and activate T cells present in the interfollicular region. Activated B cells in the subepithelial dome (SED) can also acquire antigen from the M cells, and in addition, evidence has shown that they can undergo IgA CSR at this site as well. An alternative mechanism for antigen uptake has been proposed, namely that antigens can be captured by macrophages that express CX3CR1, which protrude antigen-sampling dendrites across the tight junctions of the intestinal epithelium. Because macrophages do not leave the tissues, the sampled antigen has to be loaded onto dendritic cells (DCs) in the lamina propria. These DCs undergo maturation and subsequently migrate to the draining lymph node where T cells are primed. This pathway for antigen uptake appears to be more important for tolerance-induction rather than for SIgA responses, which rely heavily on the M cells overlaying the PPs (see Chapter 13). A third pathway for antigen uptake across the mucosal membrane is through the goblet cells, which can sample antigen and deliver these to the lamina propria in a retrograde process. Also this pathway is used more for tolerance induction than for SIgA responses. In both cases, DCs transport antigen to the draining MLNs, which is the prime site for tolerance induction of CD4$^+$ T cells and less involved in IgA B-cell responses. Taken together, most observations favor uptake via M cells and specialized FAE, as this is most important for IgA B-cell responses. Because M cells are in close contact with activated B cells in the SED, it has been proposed that these B cells can migrate from the SED to the germinal center in the follicle and, in this way, an ongoing immune response in PPs can secure the presence of antigen in the germinal center to maintain the response. This highly selective M cell–B-cell antigen-delivery pathway was found to be independent of DCs in the SED region. On the other hand, an exception to the M cell dependence for antigen uptake is the genital tract mucosa, which seems to lack FAE, and may, therefore, rely on local DCs in the lamina propria for antigen sampling, but this has as of yet not been completely investigated.

30.1 Problem for mucosal vaccines is to avoid tolerance

In most animal model systems, the vast majority of foreign antigens encountered at mucosal membranes generate unresponsiveness; that is, tolerance. Classically, oral administration of soluble proteins results in "oral tolerance" unresponsiveness to a subsequent systemic challenge with the same protein. Such tolerance is not only seen in the gastrointestinal mucosa, but in fact, most mucosal membranes have tolerance as their default response pathway. Although oral administration of antigen has been reported to concomitantly induce both mucosal IgA and systemic tolerance, in most studies a strong specific mucosal IgA response requires breaking the tolerogenic pathway. This is achieved when pathogens colonize or invade the mucosal membrane, because they then elicit pro-inflammatory "danger signals," which are recognized by pathogen recognition receptors (PRRs), leading to IgA B-cell responses and CD4$^+$ T-cell help, while avoiding tolerance. It can be assumed that live attenuated vaccines enhance their potency by expressing various pathogen-associated molecular patterns (PAMPs), a property probably shared by the inactivated whole-cell vaccines that stimulate via PRRs and induce mucosal immunity (Table 30.1). Therefore, one strategy is to incorporate one or more PAMPs into nonliving mucosal vaccines that mimic "danger signals" observed during natural infection and in this way overcome tolerance and increase immunogenicity of the vaccine.

The dominant mechanism behind mucosal tolerance is the induction of regulatory CD4$^+$ T (T$_{reg}$) cells following mucosal antigen exposure. These cells have a dampening effect on the effector phase of an immune response following systemic, and probably also local, mucosal, exposure to antigen. For example, T$_{reg}$ cells can reduce priming of T$_H$1 or T$_H$17 effector CD4$^+$ T cells and limit interferon-γ and interleukin (IL)-17 production, thereby preventing tissue damage. Inducible T$_{reg}$ cells express the transcription factor Foxp3 and produce transforming growth factor (TGF)-β and/or IL-10, which inhibit T-cell priming and effector functions. The T$_{reg}$ populations in MALT that exert dampening effects on effector-T-cell responses have been incompletely studied. Nevertheless, modulation of the function of T$_{reg}$ cells may prove to be a promising avenue in the search for future vaccine adjuvant strategies. For example, blocking Foxp3 T-cell development could be an effective strategy to enhance T-cell priming and antibody production, as has been explored with the checkpoint strategy used in anticancer vaccines, where PD-1 or PD-L1 are blocked by specific antibodies and in this way inhibit T_{reg} suppression of a cancer-specific immune response. Of note, though, at present, few, if any, vaccine strategies have included targeted downmodulation of T_{reg} cells as a means to achieve enhanced mucosal immunity against a pathogen. However, immune checkpoint inhibitors have been shown to enhance *ex vivo* effector T-cell responses in chronic viral, bacterial, and parasitic infections, including human immunodeficiency virus (HIV), tuberculosis, and malaria. Hence, it is likely that these inhibitors have great potential also for treating chronic infections when combined with therapeutic vaccines.

30.2 Unique compartmentalization and cell-migration pathways in mucosal immune responses

In the early days of mucosal immunology, it was assumed that immune responses initiated at one mucosal site would be disseminated widely to multiple mucosal tissues. Had this common mucosal immune system existed, it could have meant that immunization by any mucosal route, such as oral immunization, could be used for inducing effective immune responses, not only in the gastrointestinal tract, but also in the airways and the urogenital tract. However, further work showed that mucosal immune responses are generally compartmentalized, not only between separate mucosal organs but also between regions from the same mucosal organ, such as the gut proximal duodenum and rectum. Irrespective of sampling mechanism, antigens taken up at the mucosal surface are transported to draining lymph nodes by DCs or are directly captured by FAE and delivered to professional APCs in the PPs and presented to CD4$^+$ and CD8$^+$ $\alpha\beta$ T cells. Also, it appears that certain antigens may be processed and presented directly by epithelial cells to T cells located close to mucosal membranes, although the significance of this induction pathway is poorly understood.

Antigen-triggered B and T cells in the inductive site, such as the nasopharynx-associated lymphoid tissue (NALT) or the PPs, leave the site of initial antigen encounter, transit through the lymph, enter the blood circulation, and then seed the nonorganized lamina propria of the mucosal membrane at selected sites (Figure 30.1). Hence, there is a distinct compartmentalization for antigen-activated T and B cells, which home preferentially to the mucosa of origin where they differentiate into effector T cells and plasma cells, respectively. The anatomical affinity of such cells is determined through site-specific integrins—"homing receptors" ($\alpha_4\beta_7$) and chemokine receptors (CCR9, CCR10) on mucosal lymphoid cells—and the complementary tissue-specific endothelial cell adhesion molecules ("addressins") and chemokines that are expressed differentially in the various mucosal tissues. For instance, gut-homing PP IgA B cells and plasmablasts, as well as mucosal CD4$^+$ and CD8$^+$ T cells, express $\alpha_4\beta_7$ integrin that can specifically attach to vascular mucosal addressin cell adhesion molecule-1, a tissue-specific addressin expressed selectively on the high endothelial venules in the gut intestinal mucosa. As mentioned previously, the homing properties are imprinted on the antigen-activated lymphocytes by mucosal DCs acting at the inductive sites, more specifically in the PPs and MLN. Interestingly, this property is the function of CD103$^+$ DCs and is directly dependent on their ability to metabolize vitamin A because the concentration of retinoic acid appears to be responsible for the induction of *Aldh1a2* expression and aldehyde dehydrogenase activity, which have been linked to the imprinting properties of these DCs. During this process, DCs can also be influenced by epithelial cells that produce cofactors (such as IFNλ) in response to pathogens, which may then indirectly affect homing and differentiation of specific B and T cells. An example of

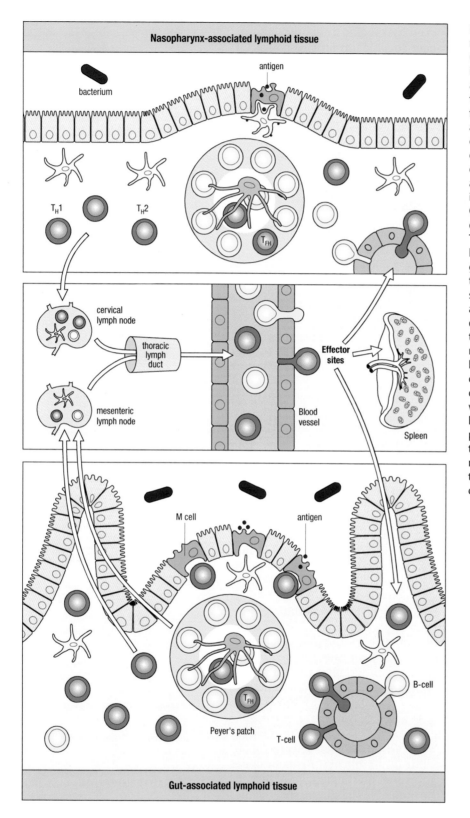

Figure 30.1 The organization of the mucosa-associated lymphoid tissue (MALT) relevant to vaccination. The inductive sites at the Peyer's patches and nasopharynx-associated lymphoid tissue take up antigen through the follicle-associated epithelium, which is a specialized type of epithelium with M cells channeling antigen to underlying dendritic cells in the subepithelial dome. These cells process and present antigen to naive CD4+ T cells in the interfollicular spaces. The activated CD4+ T cells can differentiate into different types of effector cells: T_H1, T_H2, and some will become follicular helper T (T_{FH}) cells, which enter the germinal center to support and direct the IgA class-switch recombination and somatic hypermutations of antigen-activated B cells in the follicle. Following differentiation in the germinal center, the antigen-triggered B cells and T cells migrate to the mesenteric or cervical lymph nodes and then enter the circulation via the lymphatics. From the circulation, these cells can distribute widely to mucosal effector sites and home back to the lamina propria, preferentially to the tissue from where the cells originated but also to other mucosal sites. Some activated B cells also traffic to the spleen, where memory B cells may reside after oral immunizations.

such a cofactor-induced molecule is thymic stromal lymphopoietin (TSLP), which appears to promote DC-priming of IL 4-producing follicular helper T (Tfh) cells. Moreover, chemokines produced by epithelial cells promote chemotaxis of immune cells with cognate chemokine receptors. For instance, colon CCL28, also known as mucosa-associated epithelial chemokine, selectively attracts IgA B cells and plasmablasts expressing the chemokine receptor CCR10; whereas CCL25 (TECK), which is abundantly produced in the small intestine, can selectively attract B and T cells expressing the CCR9 receptor to preferentially disseminate from the blood into the small intestinal mucosa. Thus, homing and chemokine receptors selectively expressed on antigen-activated lymphocytes (e.g., $\alpha_4\beta_7$, CCR9, and CCR10) and addressins and chemokines produced in the mucosal membrane help explain both the segregation of secretory immune responses from systemic immune responses and the preferential dissemination of T- and B-cell responses to uniquely compartmentalized mucosal sites.

30.3 Preferential dissemination of mucosal immune responses after different routes of vaccination

Compartmentalization within the mucosal immune system places constraints on the choice of vaccination route for inducing effective immune responses at the desired sites. Administration of antigens by rectal, vaginal, and more recently sublingual routes has been explored, but only for experimental purposes, and in humans mainly for studying SIgA antibody responses. In general, the strongest immune response is usually obtained at the site of initial vaccine exposure and in anatomically adjacent mucosal sites. However, a few notable exceptions have been found, such as intranasal immunizations being effective for eliciting genital tract antibody responses, which allows for more practical vaccine administrations than would otherwise be required. This has obvious practical implications for the development and deployment of mucosal vaccines against sexually transmitted diseases.

Traditional routes of mucosal immunization in humans include the oral and nasal routes. If appropriate antigens with inherent immunogenicity are used, either alone or coadministered with an effective adjuvant, oral immunization may induce a substantial SIgA antibody response in the proximal part of the small intestine, in the ascending colon, in the stomach, and in the mammary and salivary glands (Figure 30.2). Oral immunization, however, is relatively inefficient at evoking an SIgA antibody response in the distal segments of the large intestine, tonsils, lower airway mucosa, or reproductive tract mucosa. Conversely, rectal immunization evokes a strong local SIgA antibody response in the rectum and sigmoid colon, a weaker response in the descending colon, and little, if any, response in the proximal colon or small intestine. In contrast, nasal or tonsillar immunizations in humans result in antibody responses in the upper airway mucosa and regional secretions (saliva, nasal secretions) without evoking an immune response in the gut. Nasal immunization, however, has been found to give rise to substantial IgA and IgG antibody responses in the human cervicovaginal mucosae. The magnitude of the response achieved in the genital mucosa of women after intranasal immunization appears to be fully comparable to that

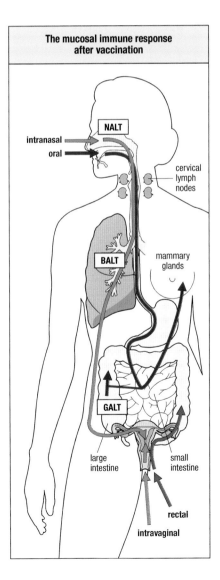

The mucosal immune response after vaccination

Figure 30.2 **The compartmentalization of the mucosal immune response following oral and intranasal vaccination.** This figure depicts the different routes used for mucosal immunizations. In humans, the most common routes are the oral and the nasal routes. A clear difference in the distribution of specific B-cell responses is elicited by different routes, such that oral immunization (red line) leads to preferential expression of IgA plasma cells in the proximal part of the small intestine, in the ascending colon, in the stomach, and in the mammary glands. Intranasal vaccination (blue line) results in IgA responses in the upper airway mucosa, regional secretions (saliva, nasal secretions), and cervicovaginal mucosa without evoking an immune response in the gut. Rectal vaccination (purple line) gives strong IgA immunity in the rectum and sigmoid colon. Intravaginal vaccination (orange line) effectively stimulates B- and T-cell responses in the genital tract.

seen when the vaccine is given by topical vaginal application. Whether this is also the case for priming of CD4$^+$ and CD8$^+$ T cells in the genital tract has not been completely investigated, but studies in mice support vaginal rather than intranasal immunization for strong genital tract T-cell immunity.

Apart from the anatomical differences in the dissemination of SIgA antibody responses induced by orogastric and intranasal immunization, respectively, the kinetics of the responses also appear to be markedly different. Several studies have shown that the intestinal immune response after oral immunization is rapid and relatively short-lasting, but it is associated with long-lasting immunologic memory. Data from field trials have shown that after oral cholera vaccination, protection from the acute intestinal IgA response appears to wane after 6–9 months, while overall protection lasts for several years. It has been proposed that the longer duration of protection reflects the ability of IgA memory B cells to elicit a rapid recall response to a renewed exposure to the pathogen and subsequently the capacity to shut off an infection before it causes disease. In both mice and humans, the kinetics of the antibody-secreting cell recall response to a mucosal challenge is rapid; within days, as documented after oral booster vaccinations in previously cholera-vaccinated subjects. These individuals exhibited strong duodenal-specific SIgA antibody-secreting cell responses, peaking after 1 week and decreasing over a 5-month period. Hence, observations like this are consistent with the notion that immunologic memory following oral mucosal vaccination in humans can last for at least 10 years.

Because of the relative limitations of primarily oral vaccination for specific SIgA immunity elicited at distant nonintestinal locations, other alternative routes have been explored. In this context, the potential of the sublingual route for administering vaccines is gaining interest. This is based on recent studies indicating that the sublingual route may promote induction of broad and exceptionally disseminated mucosal and systemic immune responses. Several studies have shown that sublingual administration of a variety of soluble and particulate antigens, including live and killed bacteria and viruses, can evoke a wide spectrum of immune responses in mucosal and extramucosal tissues, ranging from SIgA and systemic IgG antibody production to systemic cytotoxic T-lymphocyte (CTL) responses. Although mainly studied in murine models, sublingual administration appears to require 10- to 50-fold lower antigen doses than oral immunizations. Sublingual immunization induces strong immune responses and protection in the lungs of mice, as seen with both killed and live attenuated influenza vaccines. Similarly to nasal immunization, sublingual vaccination appears to effectively stimulate specific SIgA and IgG immune responses and CTL in the genital tract. Also, sublingual immunization with virus-like particles from human papillomavirus provided protection against a genital human papillomavirus challenge infection. Moreover, sublingual immunization with an experimental *Helicobacter pylori* vaccine given together with cholera toxin (CT) adjuvant effectively induced B- and T-cell responses in the stomach and intestine, respectively, resulting in significant protection against challenge infections.

Mucosal vaccine formulations and delivery systems

The underlying strategies behind the still relatively few licensed mucosal vaccines for human use are largely empirical and draw attention to the need for novel approaches to mucosal vaccine development that are currently being actively pursued by researchers across the globe. Significant problems have been encountered in translating successful vaccines in animal models into clinical practice. It remains a challenge to carry out clinical studies on new vaccines in target populations, usually in developing countries with poor infrastructures.

30.4 Ability to replicate, present native antigens, and stimulate innate immunity are key features of live attenuated vaccines

Live attenuated bacteria and viruses have several attractive features for use as mucosal vaccines. Being based on the target pathogen, they usually have properties allowing them to colonize and multiply at the appropriate mucosal site. Their ability to self-replicate also provides sufficient antigen concentrations to overcome the requirement for relatively large amounts of antigen to stimulate mucosal SIgA immunity. These amounts are more readily reached using a self-replicating live vaccine, which, as an extra advantage, continuously produces antigen, thereby prolonging stimulation of the mucosal immune system in comparison with inactivated or subunit vaccines. In addition, live vaccines, by definition, express all or most of the antigenic targets of protective immunity present in the parental pathogen. Finally, an important advantage compared with mucosally administered subunit vaccines in particular is that the live attenuated vaccines are generally "self-adjuvanting" by providing an appropriate selection of PAMPs. Hence, one strategy to explore in nonliving mucosal vaccines would be to mimic the "self-adjuvanting" potency of the few live attenuated mucosal vaccines that are listed in Table 30.1. Nevertheless, the preparation of an attenuated live mucosal vaccine is generally a multistep process that requires obtaining sufficient attenuation to render the vaccine safe and genetically stable while retaining sufficient replicative capacity to permit immunization.

The development of the licensed rotavirus vaccines available on the market today serves as an example where considerations were brought into place to launch a successful live attenuated mucosal vaccine. It was already known that a natural rotavirus infection conveyed strong protective immunity against subsequent severe disease, irrespective of serotype. Hence, a live attenuated rotavirus vaccine was predicted to do the same. Therefore, attenuation by passage of the candidate bovine or human virus strains through culture with host cells in multiple steps was undertaken. The final result was a stable attenuation, although the molecular basis for the attenuation is currently unknown. A more refined way to achieve attenuation is the introduction of selected gene deletions as used in the *Salmonella typhimurium*–based experimental live vectors for expression of recombinant vaccine antigens. It should be noted, though, that even when trying to mimic already licensed attenuated vaccines, such as the oral Ty21A typhoid vaccine, selected gene deletions can fail to generate a safer vaccine. Hence, as with all live attenuated vaccines, including the licensed rotavirus vaccines, there are always concerns about stability and safety. Another difficult problem to address, especially for live attenuated vaccines, is the extent of attenuation without loss of immunogenicity. This problem has been especially prominent for live attenuated bacterial vaccines against enteric infections intended for use in both industrialized and developing countries: marked differences have been noted in adverse reactions and poor vaccine "take," relating to nutritional status, gut microflora, and preexisting natural or maternal antibodies. To date, these problems have prevented development of effective safe live vaccines against a number of enteric bacterial pathogens, such as *Shigella*.

30.5 Inactivated whole-cell bacterial oral cholera vaccines are effective

There is only one type of inactivated mucosal vaccine licensed for human use, namely, the oral whole-cell cholera vaccine with or without added cholera toxin B subunit (CTB); it is therefore difficult to define with certainty which particular features of this vaccine can explain its protective properties. Thus, at the moment, we cannot define which properties of this vaccine could be further explored to successfully launch other oral mucosal vaccines. The following facts are, however, well established: (1) This vaccine induces substantial SIgA antibacterial immunity, mainly anti-lipopolysaccharide (LPS) antibodies in the intestine and, when CTB is included, an SIgA antitoxin together with long-lasting memory IgA responses to both these antigens. (2) As shown in experimental challenge models, and supported by observations in breastfed children in Bangladesh, both anti-LPS and anti-CTB antibodies can independently protect against cholera disease by inhibiting bacterial colonization and toxin binding, respectively. These antibody specificities have been found to act synergistically, strongly improving protection. (3) As observed in a large field trial, CTB-specific immunity added significantly to protection stimulated by the whole-cell-only vaccine for the first 9 months of follow-up, which closely corresponded to the SIgA antibody response. By contrast, protection at later times—approximately 60% protection for the next 2–3 years in adults—was similar, irrespective of the CTB component, arguing for a predominant antibacterial SIgA long-term memory function, perhaps reinforced by repeated bacterial exposure. In fact, mucosal memory induced by the oral cholera vaccine has proved to be of very long duration, being readily demonstrable upon single-dose revaccination in Swedish volunteers who received their primary immunization more than 10 years ago. Finally, (4) the antitoxin response induced by the CTB component also conferred significant short-term protection against diarrheal disease caused by enterotoxigenic *E. coli* (ETEC) bacteria. Based on these observations, an analogous mucosal vaccine against ETEC diarrhea has been developed, which is composed of a cocktail of inactivated *E. coli* bacteria engineered to express large amounts of four different colonization antigens together with a heat-labile enterotoxin B-like B subunit. This vaccine, when given orally to mice, induces strong intestinal SIgA antibody responses against each of the different components in the vaccine, especially when given together with an enterotoxin-based adjuvant, and recent studies in humans, both adults and children, support the safety and effective mucosal immunogenicity of the vaccine. Other oral mucosal vaccines that might be developed along similar principles include vaccines against, for example, *Campylobacter*, *Salmonella*, and *Shigella* infections.

30.6 Plant-based "edible" vaccines may represent a strategy for mucosal vaccination

For more than a decade, there has been work exploring plant expression of relevant antigens for use in edible vaccines. However, despite considerable efforts and substantial investments, such vaccines are still at an early stage for clinical use. Successful production of protective vaccine antigens has been achieved in many types of plants including tobacco, tomatoes, lettuce, potatoes, and rice. An attractive aspect of recombinant protein expression in plants is that large amounts of protein can be produced at a very low cost. Clinical trials initially reported very convincing immunogenicity data, but optimism has waned because of difficulties with regulatory authorities and the fact that most successful trials targeted norovirus or enterotoxin-induced diarrheal diseases, while trials of other vaccine candidates were not successful. Therefore, it appears likely that these vaccines boosted already existing immunological memory while being poor in priming naïve lymphocytes. A fundamental problem has been dosage of vaccine and quality control of the vaccine production. Therefore, the early ideas that inexpensive and effective vaccination could be instituted simply by eating, for example, a vaccine-containing banana, have long since been abandoned. A difficult technical problem has also been that glycosylation sites differ between proteins expressed in mammalian and plant cells, resulting in immune responses directed against "incorrect" polysaccharides rather than against the target pathogen proteins. Nevertheless, clinical trials are underway against several pathogens, but the issue of tolerance remains controversial in the edible vaccine field. The experimental MucoRice-CTB edible vaccine against cholera and ETEC holds promise and is planned to go into clinical trial. It appears to provide a stable and cold-chain-free alternative for mucosal vaccination.

30.7 Mucosal vaccines can be nano- or microparticles and use lectin-targeting of the follicle-associated epithelium

Exposure of antigens to an often harsh mucosal microenvironment, hosting many proteases and other degrading substances, can dramatically reduce vaccine immunogenicity. Therefore, mucosal vaccine formulation strategies have extensively explored various ways to protect the antigenic structures of the vaccine. An attractive approach has been to formulate the vaccine in nano- or microparticles to achieve this protection. This has proven productive in many ways, but few particles are effectively taken up by the M cells, which is required for a strong SIgA response. Therefore, specifically targeting the vaccine to the M cells has been found effective. Three principal strategies have been tried. The first has explored substances like β-glucan, a polysaccharide derived from barley, or lectins, which are proteins that bind to M cells. Of these, *Ulex europaeus* agglutinin-1 has attracted much interest. Experimental studies have shown that lectin-coated beads can be effectively taken up by M cells and in this way are rendered highly immunogenic. Unfortunately, *Ulex europaeus* agglutinin-1 also binds to human enterocytes and is not selective for M cells in a human vaccine. No specific lectin has so far been found suitable for targeting of human M cells. Perhaps more promising is the second strategy based on the generation of monoclonal antibodies that specifically bind to M cells through recognition of surface molecules such as $\alpha(1,2)$-fucose moieties. However, no clinical testing of this strategy has yet been done. The nano- or microparticle technology that targets the vaccine to the inductive site was first demonstrated in the 1990s with synthetic biodegradable particles, which were taken up by M cells, a function that was shown to be dependent on size, surface charge, and hydrophobicity. Since then, many nanoparticle-based vaccines have been tested in experimental models, and some have reached clinical trials. Despite technical difficulties and other problems, a growing interest for nanoparticle delivery of nasal and oral vaccines can be seen in recent years. However, realistically, this approach will likely entail the combination of particles, lectins, targeting antibodies, or substances, such as CTB or invasins, that make the nanoparticle specifically bind to M cells and in this way improve the effectiveness of this vaccine strategy. However, we are still far from clinical testing of this potentially exciting approach to a mucosal nanoparticle-based vaccine with M cell targeting.

Limitations of mucosal vaccines due to unique conditions in tropics and age of vaccinees

Although initially noted for oral polio vaccine, several other live oral vaccines, including vaccines against rotavirus and cholera, have underperformed in developing countries compared with industrialized countries. This feature, sometimes referred to as the "tropical barrier" to oral vaccines, is attributed mainly to chronic environmental enteropathy, also called tropical enteropathy. This is characterized by malabsorption, which is associated with mucosal inflammation and blunting of small intestinal villi. Factors that may contribute to chronic environmental enteropathy include poor sanitation and intestinal flora overgrowth, as seen in children living under conditions of extreme poverty. Other factors that might reduce the efficacy of oral vaccines in developing countries include deficiencies in micronutrients such as vitamin A (retinoic acid) and zinc, which can influence the response to oral adjuvants and vaccines by affecting discrete subpopulations of intestinal DC and T cells. Also, persistent activation of the gut immune system by infectious agents, such as helminths, or concomitant immunosuppressive viral and bacterial infections could simply exhaust the immune system. Likewise, in breastfed infants and children, in developing countries, antibacterial antibodies in the breast milk could interfere with vaccine immunogenicity or "take." Possible strategies for overcoming the reduced vaccine efficacy in developing countries could include giving the vaccines together with agents that improve gut integrity, such as zinc, vitamin A, and possibly probiotics. Of course, withdrawing breast milk shortly before oral vaccination and providing antihelminth treatment prior to oral immunization could be other means of improving mucosal vaccination. Recently, much attention has been placed on investigating the role of the microbiota in defining the responsiveness to oral vaccines with the hope that a combination of vaccines with a microbiota-normalizing probiotic treatment might improve the vaccine performance; to date, however, there have been no clear results that convey optimism. Finally, it may be that immunization routes such as transcutaneous or sublingual are not affected by tropical and age barriers and may, therefore, be more suitable routes for vaccination against mucosal disease in developing countries.

Mucosal adjuvants and their function

Over the last 10 years, our knowledge about different substances that exert adjuvant effects has grown immensely. This has, however, not helped in the design of more effective mucosal vaccine adjuvants. CT and ETEC heat-labile enterotoxins are among the most effective mucosal adjuvants, and their adjuvant function was first described in the 1980s. Unfortunately, these holotoxins are too toxic to be used in the clinic. Despite this, it is believed that the mechanism for the adjuvanticity of CT and heat-labile enterotoxin can be modified and exploited so that better and safer nontoxic constructs can be produced.

Because of the requirement for adjuvants in most nonliving vaccines, there is a growing interest in immunomodulating substances. All types of immune responses can be augmented: antibody formation as well as cell-mediated immunity, including CTL activity. Historically, adjuvants, such as toll-like receptor (TLR) agonists, are mostly of microbial origin, many with known modes of action. However, we must conclude that our understanding of how adjuvants work *in vivo* is still incompletely known, but many will trigger a pro-inflammatory response from the APCs. In fact, the adjuvant efficacy could in many ways be linked directly to the level of local inflammation. Thus, in this regard, adjuvants can also be responsible for tissue destruction, pain, and distress, which are key components to consider in any vaccine development. Hence, the selection of adjuvant is often as critical as which antigen or combination of antigens to include in a vaccine. Adjuvants profoundly affect the vaccine properties and function, such as quality and longevity of the vaccine response. Recent experimental and human data have documented that the choice of adjuvant can dramatically affect the long-term protective effect of a vaccine. For example, the formation of germinal centers (GCs) appears to directly influence the size of the long-lived plasma cell pool in the bone marrow as well as the presence of memory B cells following vaccination. Because of the recent progress in identifying surface markers for human (CD27) and mouse memory B cells (CD73, CD80, and PD-L2), one can predict that many new studies will identify strategies for generating functionally effective memory B cells, and principles will be defined for how adjuvant selections can be made that convey long-term protective immunity in the context of a particular vaccine.

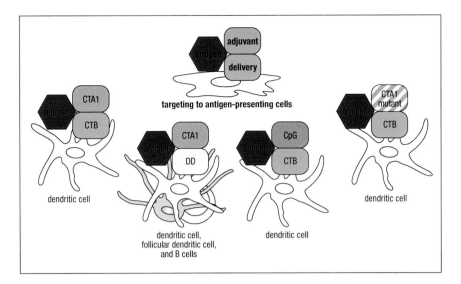

Figure 30.3 An ideal system for mucosal adjuvant function. The system outlined schematically in the figure is the hypothetical ideal construct for an effective optimally adjuvanted mucosal vaccine based on the cholera toxin molecule. It capitalizes on the specific targeting of the antigen and adjuvant linked together in a complex that can bind to and be taken up by antigen-presenting cells (APCs) in the mucosal immune system. The complex provides sufficient shielding to protect the antigen and adjuvant against degradation as well as facilitating the specific targeting to the APC across the mucosal barrier. The CTA1 adjuvant component may be exchanged for a TLR agonist such as CpG or mutated to reduce its ADP-ribosylating function. On the other hand the targeting element could be either the CTB moiety of the holotoxin or an alternative targeting unit, such as the DD dimer derived from *Staphylococcus aureus* protein A, which by virtue of its ability to bind to surface immunoglobulin, targets vaccines to B cells and follicular dendritic cells.

In principle, adjuvants may exploit three types of modulating effects on the innate immune system that will impact the adaptive immune response and promote improved immunogenicity (Figure 30.3). The first type of effect occurs with adjuvants that contain TLR- or NOD-like receptor (NLR) agonists. These are dominated by the former, such as MPL, a chemically modified derivative of LPS and a TLR4 agonist, flagellin (TLR-5 agonist), or CpG-oligodeoxynucleotide (TLR9 agonist), which have been found to be the most effective. However, relative to the bacterial holotoxins CT and heat-labile enterotoxin, they are poor mucosal adjuvants, especially for mucosal SIgA responses (see Figure 30.3). The holotoxins, on the other hand, do not use TLRs or NLRs to activate the innate immune system. A third category of mucosal adjuvants are functionally less distinct. These are oil-in-water emulsions, immune-stimulating complexes containing cytokines, such as IL-1, or mucoadhesive substances, such as chitosan. This third category exploits many different mechanisms of action; eventually it may be shown that some act through TLR binding, but in most cases, their function is poorly known, and hence, we cannot explain how they work in any greater detail when part of a mucosal vaccine.

30.8   ADP-ribosylating toxins and their derivatives are mucosal adjuvants

The bacterial enterotoxins or derivatives thereof, typified by CT and *E. coli* heat-labile enterotoxin, are AB_5 complexes and carry an A1 subunit that is an ADP-ribosylating enzyme and five B subunits that bind distinct ganglioside receptors present on most nucleated mammalian cells. Because gangliosides reside in the cell membrane of all nucleated cells, the binding is promiscuous; hence, the enzyme can affect virtually all cells in the human body. Following binding to the receptor, the holotoxins ADP-ribosylate the $Gs\alpha$ membrane protein and stimulate adenylate cyclase, leading to an increase in intracellular cyclic adenosine monophosphate (cAMP). This property makes them highly toxic but also contributes to their excellent adjuvant function *in vivo*. Thus, simply admixing or coupling of antigens to CT and repeated oral immunizations are highly effective ways to generate mucosal immune responses to soluble antigens and haptens (Figure 30.4). The B subunits of CT and heat-labile enterotoxin are responsible for binding to the ganglioside receptors on the cell membrane. Following mucosal administration, CT and heat-labile enterotoxin have been reported to host variable effects on priming of T_H1, T_H2, T_H17, and T_{reg} cells. Whereas CT has mostly been associated with T_H2 responses in the past and somewhat weaker T_H1 immunity, heat-labile

Figure 30.4 Specific gut IgA antibody responses require multiple oral immunizations. Oral mucosal immunizations require multiple administrations of the vaccine in order to stimulate strong antigen-specific IgA plasma cell responses in the gut lamina propria. The first chart in (**a**) depicts the specific IgA response against the hapten NP (4-hydroxy-3-nitrophenyl) conjugated to CT. The IgA anti-NP antibody-secreting cell numbers (SFC, spot forming cells) in the small intestinal lamina propria increase dramatically after a second or third immunization. CT on its own elicits no response. NP-specific IgA B cells in Peyer's patches undergo significantly more hypermutation after a third immunization (3×), as shown in the second chart. Using the 107 G to T mutation as a marker of high-affinity maturation, the third chart shows that repeated immunization increases the affinity of the antigen-specific IgA response to NP. (**b**) The presence of NP-specific IgA plasma cells in the lamina propria following three oral immunizations with NP-CT. A PP labeled with green fluorescent antibody to GL7, a sialic acid glycan moiety on the surface of germinal center cells, is clearly visible at the bottom of the image. In the lamina propria, roughly 10% of all IgA plasma cells are NP-specific (red). These cells exhibit strong clonal relationships with a relatively low number of dominant clones.

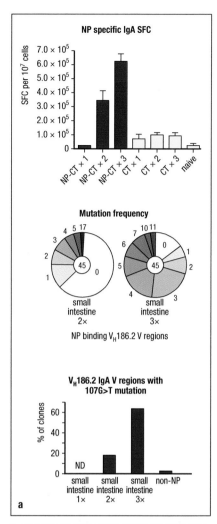

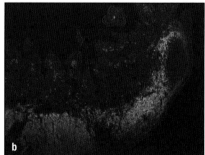

enterotoxin has been found to have a more balanced effect on priming of T_H1 and T_H2 types of immunity. However, with a growing knowledge about the immunobiology of T_H17 cells, it has been found that both holotoxins are excellent inducers of these cells. Nonetheless, recent investigations identify classical DCs to be the adjuvant targets for CT and have found that priming into T_H1, T_H2, and T_H17 as well as follicular helper T cells (T_{FH}) are all well enhanced by CT adjuvant. Whereas CT has been found to depend on IL-17 for some adjuvant functions, this has been linked to an effective T_H17 priming ability, but it is still unclear to what extent its adjuvant effects are directly associated with T_H17 cell functions. Rather, its immunomodulating effects on the DCs in various tissues would be expected to be the mechanism of action, and the antigen- and tissue-specific factors influence impacts on the adjuvant outcome. Furthermore, a few studies have pointed out that T_{reg} cells may be enhanced by CT adjuvant, but this needs to be further investigated. In fact, it is more likely that CT reduces Treg-development and favors Tfh differentiation and this could be the key role of CT acting on DCs in the PP. Finally, other studies have indicated that CD8$^+$ T cells are selectively lost following CT administration, as are CD8α^+ DCs. Taken together, it is still poorly understood what the adjuvant mechanism of these holotoxins is at the molecular level, and in particular, whether their adjuvant function is dependent on cAMP induction, while it has recently been demonstrated that ADP-ribosylation of Gsα in DCs is critically required for the adjuvant effect.

A major limitation of the holotoxins is the promiscuous binding to the GM1 ganglioside receptors present on all nucleated cells, including epithelial cells and nerve cells. This binding has rendered the holotoxins unattractive for clinical use because when used for intranasal immunizations, it was found that an intranasal flu vaccine with heat-labile enterotoxin as an adjuvant caused facial nerve paralysis (Bell's palsy) in some patients; moreover, when used orally, some cases of diarrhea were observed in vaccinated subjects. The problem with toxicity has prompted a search to find alternative strategies to lower toxicity but retain adjuvant function. A strategy to introduce single-amino-acid point mutations, most frequently in the A1 subunit, dramatically lowers toxicity with retained adjuvant function. Unfortunately, mutants, such as the LTK63 adjuvant, are also associated with cases of Bell's palsy as observed in clinical trials after intranasal administration. More promising, though, are holotoxin derivatives that are mutated in the A1 binding region that connects to the A2 region, leaving the nicotinamide adenine dinucleotide-binding and enzymatically active site intact. In clinical trials, the LTR192G mutant and, recently, the LTR192G/L211A double mutant were effective mucosal adjuvants, and a similarly mutated nontoxic cholera toxin adjuvant (mmCT) has also been developed. Studies are underway to dissect their adjuvant function in humans, and preliminary findings convey optimism.

A second alternative strategy has capitalized on the fact that the CTA1 enzyme is the most critical component for the adjuvant function. Therefore, a novel molecule was constructed by combining the enzymatic activity of CTA1 with a synthetic dimer of the D fragment of *Staphylococcus aureus* protein A in a gene fusion protein. The resulting CTA1-DD adjuvant was completely devoid of ganglioside binding because it lacked the CTB element of CT, while it had retained the full enzymatic activity of the CTA1 enzyme. CTA1-DD was found to be nontoxic and safe in mice and monkeys, and importantly, it was not found to interact with the nervous system following intranasal administration. No inflammatory response was seen at the site of administration, and hence, CTA1-DD represented a truly noninflammatory vaccine adjuvant. Numerous studies have testified to its potency as a mucosal adjuvant, comparable to that of CT and heat-labile enterotoxin, but it is completely nontoxic and safe. Thus, CTA1-DD represents an exciting new generation of targeted vaccine adjuvants, which holds promise for further exploitation in human vaccines. Following intranasal, rectal, or parenteral immunizations in mice, it effectively augmented a wide range of immune responses: from SIgA antibodies at mucosal sites to IgG antibodies in the serum, or $CD4^+$ T-cell and CTL immunity in different tissues. The adjuvant effect is achieved through two different and independent mechanisms, the first greatly augmenting the GC reaction, by direct stimulation of the follicular dendritic cells (FDCs) that subsequently enhance secretion of CXCL13, the main chemokine that recruits activated B and Tfh cells to the FDC and the GC. The second mechanism of action is the direct effect on classical DCs leading to augmented priming of T_H1, T_{FH}, and, in particular, T_H17 cells. The adjuvant effect of CTA1-DD has been shown to depend on an intact CTA1 enzyme. In fact, disruption of this activity through single point mutations in the CTA1 completely blocks the adjuvant function. Unexpectedly, though, this inactivation facilitates the induction of tolerance against incorporated peptides when such constructs are given intranasally. Thus, the CTA1R7K-OVA-DD, hosting a peptide from ovalbumin, has been shown to specifically stimulate tolerance mediated by IL-10-producing T_{reg} cells, and in this regard, the mutant molecule represents a novel vector for antigen-specific tolerance induction, which is potentially useful therapeutically against autoimmune disease, as discussed next.

30.9 Autoimmune disease may be treated by mucosal vaccination

Theoretically, targeting of DC subsets via, for example, the CD103 surface molecule may be a means to stimulate mucosal immunity by parenteral vaccination, because DCs will imprint mucosal homing of primed T cells. Immature DCs residing in mucosal tissues are known to take up antigen and, if maturation occurs, migrate to regional lymph nodes. In the secondary lymphoid tissues, DC immigrants expressing costimulatory molecules are highly effective at priming T cells. However, whether poorly activated DC immigrants or resident DCs that take up antigen via the blood promote tolerance in the draining lymph nodes by stimulating T_{reg} differentiation has been incompletely investigated. It has been found, though, that migratory $CD103^+$ DCs that travel from mucosal tissues under steady-state conditions support T_{reg} development and tolerance. Hence, exploiting this DC function for treatment of autoimmune conditions has attracted interest from researchers developing tolerance-inducing vaccines.

Mucosal administration, mostly oral or intranasal, has been found effective at ameliorating or preventing autoimmune disease in a large number of animal, primarily murine, models. For example, oral feeding of collagen or insulin can prevent collagen-induced arthritis, a model for rheumatoid arthritis, and type 1 diabetes, respectively, in mice and rats. Using mucosal vaccination for reinstating tolerance in patients suffering from various autoimmune diseases has attracted much attention lately. The last decade has seen many clinical trials exploring mucosal tolerization, but it has consistently been difficult to achieve the same promising results as seen in the rodent models. Notwithstanding these negative findings, the strategies to vaccinate against autoimmune diseases in humans using mucosal pathways are still attempted. Except for administering large doses of antigen, few tolerogenic vectors have been described, although a number of possible tolerogenic pathways have been presented in recent years. In particular, those influenced by IL-10 and mTOR/rapamycin are known to turn DCs into tolerogenic APCs. Type II collagen, for example, has been the antigen of choice in several clinical trials principally because it is the structural component of cartilage against which both B- and T-cell immunity is directed in rheumatoid arthritis patients. Therefore, one strategy to improve the efficacy of treatment is to specifically target the antigen to tolerogenic pathways in DCs and in this way enhance unresponsiveness and curb effector T-cell activity and tissue destruction. This can be achieved by exploiting certain targeting and immunomodulating proteins, among which the CTB and the mutant CTA1-DD molecules have been found to promote tolerance induction. Using the CTB or heat-labile enterotoxin B-subunit molecules as carriers of whole protein or peptides to stimulate mucosal tolerance has revealed promising results. Several studies have documented the efficacy of this strategy not only in mice but also in a human trial, where CTB linked to a Behçet's disease–associated peptide was given orally to patients to treat uveitis. As mentioned previously, the enzymatically inactive mutant of CTA1-DD, when carrying immunodominant peptides incorporated in the molecule, as exemplified with a collagen-specific peptide in CTA1R7K-COL-DD given nasally, was a treatment that greatly reduced arthritis in a mouse model for rheumatoid arthritis. Several other peptide constructs have been tried, and today experimental data in mouse models for type-1 diabetes, multiple sclerosis (EAE), and myasthenia gravis (EAMG) also exist. These studies all demonstrate therapeutically convincing results following vaccination with the CTA1R7K-X-DD platform. Collectively, CTB/heat-labile enterotoxin B and CTA1R7K-X-DD could be termed *tolerance-inducing* vectors. It is expected that these molecules will contribute to a shift of paradigm in the treatment of autoimmune diseases by reinstating tolerance using mucosal vaccines.

SUMMARY

Safe and effective mucosal vaccines currently in the market are few (Table 30.1), but many more are being developed, and some are ready for clinical trials. However, it is fair to say that despite early success with the live attenuated oral polio vaccine, only limited success has been seen in the nearly 50 years that have followed. This is partly due to a lack of safe mucosal adjuvants that would make nonliving vaccines more effective and the fact that many live attenuated vaccines are either genetically unstable or insufficiently or overly attenuated. Therefore, intense research is ongoing to identify better and more effective mucosal adjuvants, and studies in animals and in humans are dissecting routes of administration, antigen dose requirements, and effective formulations. However, in recent years, live attenuated mucosal vaccines against influenza virus, rotavirus, and typhoid fever have been launched, and a parenteral vaccine against human papillomavirus has also been highly successful. An inactivated oral cholera vaccine has provided proof-of-concept that mucosal vaccines based on nonliving components can be effective. But why this vaccine is effective while other nonliving vaccines are not is far from clear. Presently, we also suffer from a knowledge gap as to why so many candidate vaccines in experimental small animal models have been highly promising, while few have proven effective in human trials. Notwithstanding this, the overall prospects for mucosal vaccine administration now look brighter, as the mechanisms responsible for priming of mucosal SIgA immune responses are better understood. Based on the newly acquired knowledge about adjuvant functions and innate immunity at mucosal membranes and the buildup of IgA B-cell responses, including long-term memory development, we should expect many new and effective mucosal vaccines in the future.

FURTHER READING

Angel, J., Franco, M.A., and Greenberg, H.B.: Rotavirus immune responses and correlates of protection. *Curr. Opin. Virol.* 2012, 2:419–425.

Chiu, C., and Openshaw, P.J.: Antiviral B cell and T cell immunity in the lungs. *Nat. Immunol.* 2015, 16:18–26.

Czerkinsky, C., and Holmgren, J.: Vaccines against enteric infections for the developing world. *Philos. Trans. R. Soc. Lond. B. Biol. Sci.* 2015, 19:370.

Greenberg, H.B., and Estes, M.K.: Rotaviruses: From pathogenesis to vaccination. *Gastroenterology* 2009, 136:1939–1951.

Holmgren, J., Parashar, U.D., Plotkin, S., Louis, J., Ng, S.P., Desauziers, E., Picot, V., and Saadatian-Elahi, M.: Correlates of protection for enteric vaccines. *Vaccine* 2017, 35:3355–3363.

Holmgren, J., and Svennerholm, A.M.: Vaccines against mucosal infections. *Curr. Opin. Immunol.* 2012, 24:343–353.

Iwasaki, A.: Exploiting mucosal immunity for antiviral vaccines. *Annu. Rev. Immunol.* 2016, 34:575–608.

Lambrecht, B.N., Kool, M., Willart, M.A. et al.: Mechanism of action of clinically approved adjuvants. *Curr. Opin. Immunol.* 2009, 21:23–29.

Lycke, N.: Recent progress in mucosal vaccine development: Potential and limitations. *Nat. Rev. Immunol.* 2012, 12:592–605.

Lycke, N., and Bemark, M.: The regulation of gut mucosal IgA B-cell responses: Recent developments. *Mucosal Immunol.* 2017, 10:1361–1374.

Komban, R.J., Strömberg, A., Biram, A., Cervin, J., Lebrero-Fernandez, C., Mabbott, N., Yrlid, U., Shulman, Z., Bemark, M., and Lycke, N.Y.: Activated Peyer's patch B cells sample antigen directly from M cells in the subepithelial dome. *Nat. Commun.* 2019, 10:2423–2432.

Mattsson, J., Schön, K., Ekman, L., Fahlén-Yrlid, L., Yrlid, U., and Lycke, N.Y.: Cholera toxin adjuvant promotes a balanced Th1/Th2/Th17 response independently of IL-12 and IL-17 by acting on Gsα in CD11b⁺ DCs. *Mucosal Immunol.* 2015, 8:815–827.

Nakaya, H.I., and Pulendran, B.: Vaccinology in the era of high-throughput biology. *Philos. Trans. R. Soc. Lond. B. Biol. Sci.* 2015, 370(1671).

Pabst, O. and Slack, E.: IgA and the intestinal microbiota; The importance of being specific. *Mucosal Immunol.* 2019, 13(1):12–21.

Pasetti, M.F., Simon, J.K., Sztein, M.B. et al.: Immunology of gut mucosal vaccines. *Immunol. Rev.* 2011, 239:125–148.

Rappuoli, R., Bottomley, M.J., D'Oro, U., Finco, O., and De Gregorio, E.: Reverse vaccinology 2.0: Human immunology instructs vaccine antigen design. *J. Exp. Med.* 2016, 213:469–481.

Sohrab, S.S., Suhail, M., Kamal, M.A., Husen, A., and Azhar, E.I.: Edible vaccine: Current status and future perspectives. *Curr. Drug Metab.* 2017, 18:831–841.

Yoshida, T., Mei, H., Dorner, T. et al.: Memory B and memory plasma cells. *Immunol. Rev.* 2010, 237:117–139.

Celiac disease

31

BANA JABRI AND LUDVIG M. SOLLID

Celiac disease is a chronic, T-cell–mediated, inflammatory disorder of the small intestine resulting from an inappropriate immune response to gluten. Gluten is a complex of proteins present in wheat, barley, and rye. Wheat gluten proteins consist of α-, γ-, and ω-gliadins as well as high and low molecular weight glutenin subcomponents. Both gliadin and glutenin proteins can induce the disease, which is treated with a lifelong gluten exclusion diet. The mucosal pathology associated with celiac disease is localized to the proximal small intestine and is characterized by villus atrophy and prominent infiltration of leukocytes into the epithelium and lamina propria. The mucosal absorptive surface is reduced as a consequence of villus atrophy, leading to malabsorption that may result in anemia and steatorrhea. Additional symptoms, including fatigue, infertility, neurologic manifestations, and enamel defects, may not indicate an intestinal disease, and some patients have few or even no symptoms. The eclectic nature of the symptoms is a challenge to clinicians, and consequently, many patients are not diagnosed or experience long delays before diagnosis.

THE CELIAC LESION

Even though celiac disease is an intestinal inflammatory disorder induced by a food antigen, the disease and its associated histology are more similar to organ-specific autoimmune disorders than to food allergies or intestinal inflammatory bowel disease. Celiac disease, like type 1 diabetes, is mediated by CD4$^+$ and CD8$^+$ T cells and is associated with the presence of autoantibodies that can precede the disease and are not thought to be directly pathogenic. The multifaceted aspect of celiac disease makes the disorder especially fascinating and highly relevant to mucosal immunology in particular. We begin with a discussion of the inflammatory lesion associated with celiac disease.

31.1 Villous blunting and crypt cell hyperplasia are characteristic histological features

The intestinal lesion in celiac disease is characterized by loss of the finger-like villous structures that normally increase the absorptive surface of the gut and by increased cell division of epithelial cells in the crypts (crypt cell hyperplasia). The development from the normal state to an overt lesion follows a uniform pattern, often described as Marsh stages (Figure 31.1). Marsh 0 is the normal state. The Marsh 1 stage is characterized by an increased number of intraepithelial lymphocytes (IELs) but without crypt cell hyperplasia or villous blunting. In Marsh 2, crypt cell hyperplasia occurs without villous blunting. Marsh 3 is often divided into Marsh 3a, Marsh 3b, and Marsh 3c.

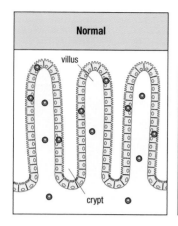
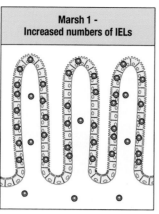
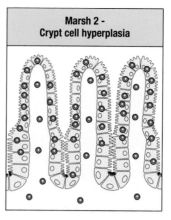
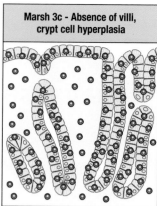

Figure 31.1 Histologic alterations of the small intestinal mucosa in celiac disease. The normal mucosa with slender intestinal villi and few leukocytes is shown on the left, and the Marsh stages of development of the celiac lesion are shown on the right 3 panels. Marsh 1 is characterized by increased numbers of intraepithelial lymphocytes (IELs) but no crypt cell hyperplasia; Marsh 2 by crypt cell hyperplasia but no villous blunting; and Marsh 3c by absence of villi, crypt cell hyperplasia, and massive infiltration of leukocytes into the epithelium and lamina propria. The degree of villous blunting increases from Marsh 3a to Marsh 3c, with total loss of villi in Marsh 3c.

The degree of villous shortening increases from Marsh 3a to Marsh 3c, with total loss of villi in Marsh 3c. Different biopsies from the same patient may reveal varying degrees of mucosal alteration because the celiac lesion is not uniformly expressed throughout the intestine.

31.2 Intraepithelial lymphocytes are prominent

The histological lesion in active celiac disease is strikingly different from that of other T-cell–mediated enteropathies, such as graft-versus-host disease and autoimmune enteropathy. In celiac disease, the crypts are intact and regenerative, but the surface epithelium is profoundly altered. Furthermore, a major infiltration of lymphocytes in the epithelium can develop into enteropathy-associated T-cell lymphoma in rare cases. In contrast, crypt destruction is common in other enteropathies, and intraepithelial lymphocytosis is minor or appears late in the disease. Enteropathy-associated T-cell lymphoma has not been reported in intestinal inflammatory disorders other than celiac disease. These histological differences suggest that the surface epithelium is preferentially targeted, and IELs are abnormally activated in celiac disease.

In the normal human small intestine, three major lineages of IELs participate in the defense of the mucosal surface. The most prominent IEL is the cytolytic $CD8\alpha\beta^+CD4^-$ T-cell receptor (TCR) $\alpha\beta$ population (IE-CTLs); $CD4^+ TCR\alpha\beta^+$ and $TCR\gamma\delta^+$ populations are also present. The IE-CTLs in humans simultaneously encompass characteristics of conventional adaptive T cells (expression of heterodimeric $CD8\alpha\beta$ coreceptor and antigen-driven variable T-cell receptors) and innate T cells (expression of natural killer receptors recognizing inducible nonclassical major histocompatibility complex [MHC] class I molecules). Interestingly, unlike in the peripheral blood and the liver, human IELs selectively express natural killer (NK) receptors of the C-type lectin family. The most represented natural killer receptors are NKRP1-A, receptors of the CD94/NKG2 family, and the NKG2D receptor. Under physiologic conditions, IELs express high levels of inhibitory CD94/NKG2A receptors and low levels of activating NKG2D receptors. As IELs are effector cells, they can exert TCR-mediated cytolysis *ex vivo*. However, the level of cytolysis is low in the absence of additional stimuli. Furthermore, the IELs cannot kill NK cell targets.

IEL, in addition to T cells, comprise also intraepithelial innate-like lymphocytes (IE-ILCs). These are mature effector lymphocytes that express CD103 and lack expression of antigen-specific receptors. Two main subsets of

CD103$^+$ IE-ILCs have been identified. An IE-ILC1 subset that lacks intracellular CD3 and does not display signs of TCR engagement, and an ILC1-like subset named iCD3$^+$ innate IEL that expresses intracellular CD3 and shows signs of TCR rearrangements, especially TCRγδ T-cell rearrangements. Both subsets express IL-15Rβ, T-box transcription factor TBX21 (T-bet), and granzyme B; they produce interferon-γ; and display NK-like properties.

The numbers of IE-CTLs and TCRγδ$^+$ cells are increased in celiac disease. By contrast, the frequency of intraepithelial CD4$^+$ T cells is unchanged. In addition to increased frequency, IE-CTLs acquire strong cytolytic properties (see later). When patients adopt a gluten-free diet, the number of IE-CTLs returns to normal within weeks or months, whereas the number of TCRγδ$^+$ IELs remains permanently elevated TCRγδ$^+$ IELs in healthy individuals take on an innate-like phenotype characterized by the unique expression of the activating natural cytotoxicity receptors (NCRs) NKp44/NKp46 and a semi-invariant TCR expressing Vδ1Vγ4 gene segments. The heathy subset is displaced in patients with active celiac disease in favor of newly recruited Vδ1$^+$ IELs that produce IFN-γ in a gluten dependent manner and whose TCRs show evidence of adaptively driven expansions with the emergence of a non-germline encoded TCR motif shared across individuals. Importantly, although a gluten-free diet restores the heathy architecture of the intestine, the healthy subset of TCRγδ$^+$ IELs does not recover, thus leaving the compartment permanently devoid of these unique innate-like cells. Furthermore, the production of IFN-γ by Vδ1$^+$ IELs only when patients eat gluten is compatible with them playing a pathogenic role in celiac disease. The identification of the specificity of expanded TCRγδ$^+$ IELs in celiac disease patients will further increase our understanding of their role in the disease. Finally, while IE-CTL and TCRγδ$^+$ IELs are significantly increased in active celiac disease, the number of IE-ILC is decreased. However, in a subset of celiac disease patients who have developed refractory celiac disease, iCD3$^+$ innate IELs represent more than 20% of total IELs and can comprise more than 90% of IELs in some patients. In refractory celiac disease patients, iCD3$^+$ innate IELs often display mutation in signal transducer and activator of transcription 3 (STAT3) and Janus kinase 1 (JAK1) and can undergo malignant transformation.

31.3 Immune cells infiltrate lamina propria

An array of immunocompetent cells is present in the small intestinal mucosa in celiac disease. In the celiac lesion, large numbers of leukocytes infiltrate into the epithelium and lamina propria. The increase of plasma cells in the lamina propria is particularly striking, but the number of T cells also increases (Figure 31.2). The number of plasma cells in the lesion more than doubles,

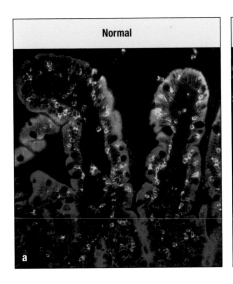
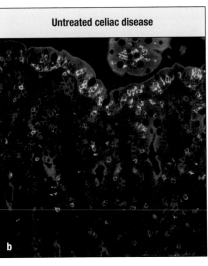

Figure 31.2 **Distribution of T cells and plasma cells in the duodenal mucosa.** The numbers of CD3-positive T cells (green) and CD138-positive plasma cells (red) in healthy mucosa (**a**) increase in active celiac disease (**b**). The epithelial cells stained in blue also express some CD138. In celiac disease, the number of plasma cells increases substantially in the lamina propria. (Photographs courtesy of A-C.R. Beitnes.)

and, as in normal mucosa, IgA is the dominant isotype. Some of the lamina propria CD4$^+$ T cells are specific for gluten antigens and, as described later, are master regulators of the immune response that initiate formation of the celiac lesion. Few memory B cells are present in the lamina propria of celiac patients or normal subjects. The celiac lesion contains dendritic cells and macrophages that express high levels of human leukocyte antigen (HLA)-DQ molecules and thus should be able to present gluten epitopes to CD4$^+$ T cells, in contrast to epithelial cells, which do not express HLA-DQ molecules.

Celiac disease genetics

Like most chronic inflammatory diseases, celiac disease has a multifactorial etiology, as both genes and environmental factors contribute to disease pathogenesis. In this regard, exclusion of gluten from the diet results in resolution of the disease, and the absence of the HLA predisposing genes is associated with reduced likelihood of developing celiac disease (99% of patients have the HLA-DQ2 or HLA-DQ8 haplotypes, see later). A high degree of familial clustering (10% of first-degree relatives affected compared to a population prevalence of about 1%) and a large difference in the concordance rate between monozygotic and dizygotic twins (75% versus 10%) suggest a genetic contribution. Both HLA genes and non-HLA genes are implicated in the disease. Most people who express HLA risk genes do not develop the disease, and the concordance rate among HLA-identical dizygotic twins who share their HLA genes in addition to half of their non-HLA genes is much lower than that of monozygotic twins.

31.4 Human leukocyte antigen genes are associated with celiac disease

Celiac disease has one of the strongest HLA associations. Although several genes within the HLA gene complex on chromosome 6 may contribute, the primary association is with the DQA1*05 and DQB1*02 alleles. This pair of genes is part of the autoimmune-associated HLA haplotype DR3-DQ2 and encodes the heterodimeric HLA-DQ2 molecule often termed DQ2.5 (**Figure 31.3**). The DQ2 variant DQ2.2, encoded by the DQA1*02:01 and DQB1*02 alleles of the DR7-DQ2 haplotype, is weakly associated with celiac disease on its own. However, people heterozygous for DR7-DQ2.2 and DR5-DQ7 are at high risk for disease, and in some populations, a third of celiac disease patients express this genotype. The DR5-DQ7 haplotype carries the DQA1*05 allele, and the DR7-DQ2.2 haplotype carries the DQB1*02 allele, indicating that these individuals also encode the DQ2.5 molecule, but by genes located on opposite chromosomes. The few patients lacking this HLA-DQ2.5 heterodimer are either DQ2.2 or DQ8. DQ8 is encoded by the DQA1*03 and DQB1*03:02 alleles being part of the DR4DQ8 haplotype. Interestingly, expression of HLA-DQ2.5 and HLA-DQ8 molecules predisposes to both celiac disease and type 1 diabetes; HLA-DQ2.5 molecules are predominant in celiac disease, whereas HLA-DQ8 molecules are predominant in type 1 diabetes.

Almost all celiac disease patients, as well as many healthy subjects, express HLA-DQ2.5 or HLA-DQ8 molecules. Therefore, certain HLA genes are necessary but not sufficient factors for disease development (**Figure 31.4a**).

31.5 Non-HLA genes also contribute to celiac disease predisposition

Whereas HLA is the single most important locus that predisposes to celiac disease, the combined effect of the non-HLA genes is larger than that of the HLA locus. Thus far, 42 non-HLA loci have been identified by genome-wide association studies. The phenomenon of linkage disequilibrium makes it hard to pinpoint the causal polymorphisms of these loci. However, it seems

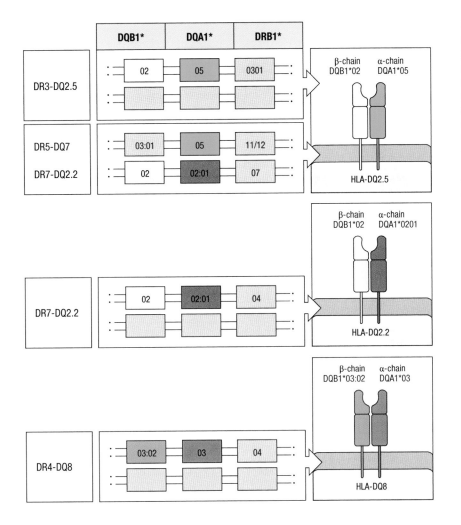

Figure 31.3 Human leukocyte antigen (HLA) association in celiac disease. The great majority of celiac disease patients express the HLA-DQ2.5 heterodimer encoded by the DQA1*05 and DQB1*02 alleles. These two alleles are carried on either the DR3-DQ2.5 haplotype or on opposite chromosomes in individuals who are DR5-DQ7 and DR7-DQ2.2 heterozygous. HLA-DQ2.2 alone, encoded by the DQA1*02:01 and DQB1*02 alleles, confers a low risk to develop celiac disease. The patients who are not DQ2.5 or DQ2.2 are almost always HLA-DQ8. The HLA-DQ8 molecule is encoded by the DQA1*03 and DQB1*03:02 alleles.

clear that a great majority of the single nucleotide polymorphisms do not localize to exons where they affect protein sequence, but rather localize to the noncoding genome where they exert an effect on gene regulation and gene expression. The size of the effect of each of the identified non-HLA loci is small; the identified non-HLA risk loci altogether account for about 15% of the genetic variance (see **Figure 31.4b**). In contrast, HLA contributes about 40% of the genetic variance, suggesting that many risk factors have not yet been discovered.

Most of the established loci harbor candidate genes with immune functions. These genes can be classified into different pathways. Many of the genes belong to pathways of T- and B-cell costimulation. Such genes include *CTLA4, CD80, SH2B3, PTPN2, TAGAP, ICOSLG,* and *CD247*. Also frequently represented are cytokine and cytokine receptor genes, including *IL2/IL21,*

Figure 31.4 Key features of celiac disease genetics. (a) Human leukocyte antigen (HLA) is a necessary but not sufficient factor for the development of celiac disease. Almost all people with celiac disease express HLA-DQ2.5, HLA-DQ2.2, or HLA-DQ8, but only a fraction of subjects who express these HLA-DQ molecules develop the disease. People who do not express HLA-DQ2.5, HLA-DQ2.2, or HLA-DQ8 rarely develop celiac disease. **(b)** The relative contributions to the total genetic variance of celiac disease by loci. HLA is the single most important locus contributing about 40% of the genetic variance. A total of 39 non-HLA loci that harbor celiac disease genes have been identified and collectively explain about 15% of the genetic variance. Most of the genetic variance has not been explained, suggesting the presence of additional non-HLA celiac disease genes.

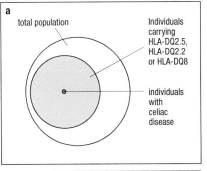

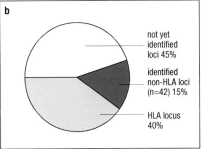

IL12A, and *IL18RAP*; genes involved in migration of immune cells such as the chemokine receptors *CCR3/CCR5/CCR1* and *CCR4*; and the integrin gene *ITGA4*. Another interesting pathway involves molecules important for T-cell development in the thymus, such as THEMIS, which plays a role in both positive and negative T-cell selection during late thymocyte development, and RUNX3, which is involved in CD8+ T-cell differentiation. One network includes genes involved in nuclear factor (NF)-κB signaling, such as *REL*, which encodes a component of the NF-κB complex, and *TNFAIP3*, which encodes a molecule that inhibits NF-κB activity. Finally, a pathway involving molecules implicated in innate immune detection such as toll-like receptor 7 (TLR7), TLR8, and IRF4 has been identified. Both TLR7 and TLR8 recognize viral RNA, whereas IRF4 is a transcriptional activator that is part of the TLR7 pathway. This latter finding suggests involvement of viruses in the pathogenesis of celiac disease, likely initial events important for inducing the antigluten CD4+ T-cell response.

Interestingly, many of the regions identified in the genome-wide association studies of celiac disease have also been identified in the genome-wide association studies of autoimmune diseases and other chronic inflammatory traits.

Adaptive immune response to gluten in celiac disease

The adaptive immune response plays a fundamental role in celiac disease pathogenesis. Central to this response are gluten-reactive CD4+ T cells, which are present in the intestinal mucosa. These T cells recognize gluten epitopes presented by HLA-DQ2.5, HLA-DQ2.2, or HLA-DQ8, but not other HLA molecules, suggesting that the mechanism underlying the HLA association in celiac disease is preferential antigen presentation.

31.6 Gluten-reactive CD4+ T cells in the intestinal mucosa recognize specific gluten residues

Consistent with the complexity of the gluten proteins are the many similar, but distinct, T-cell epitopes (**Figure 31.5**). Strikingly, most gluten-reactive

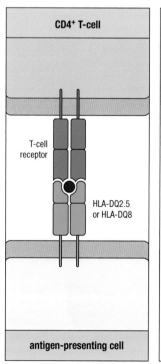

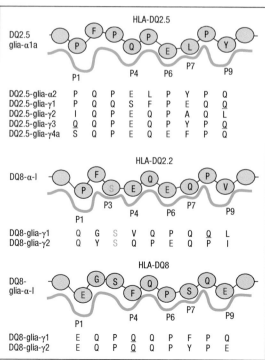

Figure 31.5 T-cell recognition of deamidated gluten epitopes in the context of the human leukocyte antigen (HLA)-DQ2.5, HLA-DQ2.2, and HLA-DQ8 molecules. The sequence of the DQ2.5-glia-α1a, DQ2.2-glia-α1, and DQ8-glia-α1 epitopes, as well as some other epitopes, which bind to HLA-DQ2.5, HLA-DQ2.2, or HLA-DQ8, are shown. The HLA molecules have pockets at the relative positions P1, P4, P6, P7, and P9 that accommodate amino acid side chains of the peptides. Glutamate residues (E) induced by transglutaminase 2 (TG2) deamidation of glutamine residues are important for peptide binding and/or recognition by T cells and are marked in red. Glutamine residues targeted by TG2 but not important for T-cell recognition are underlined. Note the many proline residues (P) in the epitopes. In the DQ2.5-restricted epitopes, the critical glutamate residues are at the P4, P6, or P7 positions, whereas in the DQ8-restricted epitopes, the critical glutamate residues are at the P1 or P9 positions. Serine (S) at position P3 serves as a key anchor residue for binding of peptides to HLA-DQ2.2.

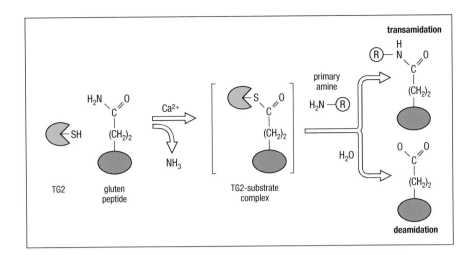

Figure 31.6 Transamidation and deamidation reactions catalyzed by the enzyme transglutaminase 2 (TG2). TG2 targets certain glutamine residues of polypeptides and either cross-links them to primary amines (transamidation) or converts them to glutamate by hydrolysis (deamidation). If the primary amine of the transamidation reaction is the lysine residue of another polypeptide, a covalent linkage of polypeptides is formed.

T cells preferentially recognize epitopes that have become posttranslationally modified by the enzyme transglutaminase 2 (TG2). This enzyme selects specific glutamine residues in polypeptides and either cross-links the glutamine to a primary amine, such as lysine, in another polypeptide or converts it to glutamate via a process called *deamidation* (Figure 31.6). Deamidation generates gluten peptides with negatively charged glutamate residues that bind much better to the positively charged binding pockets of DQ2.5, DQ2.2, and DQ8 than their native (nondeamidated) counterparts. Even though DQ2.5, DQ2.2, and DQ8 molecules share physicochemical properties, for example, all of them lack an aspartate at position β 57 and show a preference for negatively charged peptides; a smaller repertoire of gluten peptides appear to be presented by DQ8 and DQ2.2 than by DQ2.5. Furthermore, the epitopes presented by the three HLA-DQ molecules are distinct for each HLA molecule, as the preference for negatively charged anchor residues is different between the HLA molecules; DQ2.5 and DQ2.2 prefer a negative charge in positions P4, P6, or P7, whereas DQ8 prefers a negative charge in P1 or P9 (see Figure 31.5). Although DQ2.5 and DQ2.2 share many features of peptide binding, the gluten epitopes they present are different. Gluten epitopes presented by DQ2.2, in contrast to those presented by DQ2.5, typically have a serine residue at position P3 (see Figure 31.5). Serine at P3 is thus an additional anchor residue for peptides binding to DQ2.2. DQ2.5-restricted epitopes, when binding to DQ2.2 molecules *in vitro*, make unstable peptide: MHC complexes, and this is likely the reason why no T-cell responses are generated to these peptides in DQ2.2-expressing patients. This difference in peptide binding between DQ2.5 and DQ2.2 is controlled by residue 22 of the DQα chain, where DQ2.5 carries a tyrosine whereas DQ2.2 carries phenylalanine.

Notably, HLA gene dosage influences risk for celiac disease. Homozygosity for both DQ2.5 and DQ8 is associated with higher risk. DQ2.5-expressing individuals are at a higher risk of developing disease even when the second copy of the DQB1*02 allele derives from the DR7-DQ2 haplotype (i.e., DR3-DQ2.5/DR7-DQ2.2 individuals). For DQ2, the gene dose effect is related to the magnitude and breadth of gluten-specific T-cell responses. On the basis of these observations, threshold effects have been suggested for disease development, with HLA-DQ expression and the available number of T-cell stimulatory gluten peptides being critical limiting factors.

Gluten proteins of rye and barley also contain epitopes that are recognized by CD4+ T cells of patients with celiac disease. This is the reason why barley and rye are also not tolerated by celiac patients.

X-ray crystal structures of T-cell receptors recognizing deamidated gluten peptides in the context of HLA-DQ2.5 and HLA-DQ8 have been resolved. While the T-cell receptors are specific for the deamidated peptides, in no case were the T-cell receptors shown to make direct contact with the glutamate side chain of the deamidated peptide. The glutamate residues typically face the HLA-DQ molecules as anchor residues, and it appears likely that the T-cell receptor specificity for deamidated peptide is caused by indirect effects on peptide structure when the glutamate residue makes tight interaction with HLA.

31.7 Transglutaminase 2 is involved in celiac disease pathogenesis

The discoveries that TG2 was the target of autoantibodies in celiac disease and increased the binding affinity of gluten peptides to HLA-DQ2.5 and HLA-DQ8 molecules placed the enzyme centrally in celiac disease pathogenesis. Gluten peptides are excellent substrates for TG2, and T-cell epitopes are among the preferred substrates among the peptides in a digest of gluten, suggesting that TG2 plays a key role in T-cell epitope selection. The finding that gluten is a good substrate for TG2 also provides interesting insights into the generation of gluten-dependent anti-TG2 antibodies in celiac disease (see later).

Despite the recognition that TG2-mediated deamidation plays an important role in celiac disease pathogenesis, whether it plays a role in the initiation and/or the amplification of the antigluten immune response is unclear. Observations in children and the humanized HLA-DQ8 mouse model suggest that deamidation is not required for the initiation of the immune response but may be involved in the amplification of the immune response (see earlier). Furthermore, TG2 is inactive in the intestinal environment, posing the question of how its activation occurs and whether it precedes or follows the initiation of the antigluten T-cell response. In one scenario, the inflammatory antigluten immune response is responsible for the initial activation of TG2. This scenario is compatible with a study in the humanized HLA-DQ8 mouse, which showed that deamidation is not required for the initiation of an HLA-DQ8-restricted antigluten CD4$^+$ T-cell response and that HLA-DQ8 can amplify the antigluten immune response by allowing the recruitment of cross-reactive TCR that recognizes native and deamidated peptides. In the other scenario, an environmental trigger, such as an intestinal viral infection, activates TG2 and initiates the adaptive antigluten immune response. This scenario is compatible with reports suggesting that repeated rotavirus infections are associated with an increased incidence of celiac disease and that double-stranded RNA (as during rotavirus infection) can trigger TG2 activation in mice either directly or indirectly by causing tissue damage. It is important to point out that these different scenarios are *not* mutually exclusive, and that one scenario can predominate, depending on the genetic background.

The role of posttranslational modifications in generating immunogenic peptides is also reported for other autoimmune diseases. For example, in the case of rheumatoid arthritis, the posttranslational modification and the MHC molecules associated with the disease are citrullination and DR4, respectively. The overall underlying theme involves a favorable combination of MHC molecule, tissue enzyme, and antigen that optimizes binding of the causative antigen to the disease-associated MHC molecule and promotes T-cell responses of sufficient amplitude to induce a pathogenic immune response associated with tissue damage. Because posttranslational modifications of antigen can promote loss of immune tolerance to that antigen, enzymes promoting posttranslational modifications must be under tight control. Understanding the regulation of TG2 and its role in celiac disease pathogenesis will also enhance more general understanding of the role of posttranslational modifications by tissue enzymes in autoimmunity.

Figure 31.7 **Model depicting how gluten-specific CD4⁺ T cells can provide help to TG2-specific B cells.** Production of antibodies to TG2 is dependent on ingestion of gluten and only occurs in subjects who carry HLA-DQ2.5 or HLA-DQ8. The formation of the TG2 antibodies is likely dependent on T-cell help, probably provided by gluten-specific T cells. Gluten peptides and TG2 readily form complexes, which when bound by antibodies on the surface of a TG2-specific B-cell are internalized and processed, and the peptide is presented in the context of HLA-DQ2.5 or HLA-DQ8 to a gluten-specific CD4⁺ T-cell. B cells expressing low-affinity anti-TG2 immunoglobulins can receive T-cell help and undergo maturation to plasma cells, producing high-affinity IgA and IgG anti-TG2 antibodies.

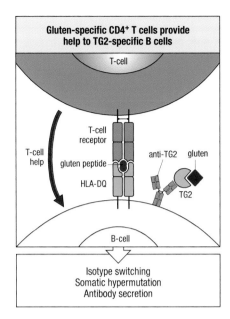

31.8 Autoantibodies are present in celiac disease

IgA and IgG antibodies to TG2, which are dependent on gluten exposure, are a hallmark of celiac disease. Detection of these antibodies is increasingly important for the diagnosis of celiac disease due to the sensitivity and specificity of the antibodies. The formation of anti-TG2 antibodies is also dependent on the presence of HLA-DQ2 and HLA-DQ8, as people who do not carry either of these HLA molecules do not make the antibodies. The strict HLA and gluten dependence suggest that gluten-reactive CD4⁺ T cells are involved in the generation of the anti-TG2 antibodies. This could be explained by a mechanism in which complexes of TG2 and gluten, which readily form when incubated together, act as hapten-carrier complexes so that gluten-reactive T cells can provide help to TG2-reactive B cells (Figure 31.7). B cells as antigen-presenting cells, both of which are specific for TG2 and deamidated gliadin peptides, can thus play an important role for amplifying the T-cell response to gluten.

IgA and IgM plasma cells express surface immunoglobulin, and this allows for staining of antigen-specific cells. Strikingly, on average, 10% of the plasma cells in duodenal mucosa of untreated celiac disease patients are specific for TG2, while a lower number of plasma cells are specific for deamidated gliadin. Both the plasma cells specific for TG2 and deamidated gliadin peptides have few somatic mutations in the immunoglobulin genes compared to other plasma cells in the gut mucosa. The reason for this is unknown.

31.9 Oral antigen induces an inflammatory response

Resistance of gluten to proteolytic digestion and posttranslational modifications lead to the generation of peptides of higher affinity for HLA-DQ2.5, HLA-DQ2.2, and HLA-DQ8, which then are presented to T cells. However, why inflammatory antigluten CD4⁺ T-cell responses occur in an environment where oral antigens normally lead to tolerogenic T-cell responses is less clear. Celiac disease patients display an upregulation of IL-15 not only in the epithelium but also in the lamina propria where IL-15 can signal in dendritic cells in conjunction with retinoic acid that acquires adjuvant properties in the presence of IL-15. In a mouse model overexpressing IL-15 in the lamina propria, intestinal dendritic cells acquire the ability to produce IL-12p70 and promote T-helper-1 immunity to dietary gluten. In accordance with a role for IL-15 in inducing a proinflammatory phenotype in dendritic cells in celiac disease patients, IL-15 upregulation was correlated with increased IL-12p70 protein expression in the lamina propria. Studies have also shown the presence of high levels of interferon-α expression in the intestinal mucosa of celiac disease patients, and genome-wide association studies point to genes involved in viral responses, suggesting that viral infections may also be involved in the initiation of pathogenic antigluten T-cell immunity; however, the role of viral infections in celiac disease pathogenesis remains to be determined. Alternatively, gluten may have intrinsic innate properties that lead to activation of dendritic cells and production of pro-inflammatory mediators.

Innate immunity and stress response to gluten

Whether gluten induces an innate immune response has been debated. Gluten-mediated innate effects could explain some elements of celiac disease pathogenesis. However, to date, no reproducible gluten-induced innate signaling pathway has been reported. Numerous gluten-mediated innate effects have been reported, but these effects are eclectic and difficult to integrate in a single biological pathway. Nevertheless, numerous studies point toward a role for innate immunity in the cytolytic intraepithelial T-cell response that is likely involved in epithelial cell destruction.

31.10 Innate immune cell responses may be induced by gluten

Interestingly, TCR$\gamma\delta$ IELs increase in number after rectal gluten challenge in HLA-DQ2.5$^-$ and HLA-DQ8$^-$ siblings of celiac disease patients, suggesting that celiac disease can be induced in the absence of HLA-DQ2.5 or HLA-DQ8 molecules. Another striking finding is the upregulation of mucosal interferon-α and IL-15 in subjects with celiac disease while on a gluten-containing diet. Organ culture studies have suggested that IL-15 can be induced in response to the α-gliadin 31–43 peptide (LGQQQPFPPQQPY), which does not induce an adaptive antigluten T-cell response. Interestingly, this peptide also was implicated in mitogen-activated protein kinase activation, inhibition of epidermal growth factor receptor degradation, and upregulation of the nonclassical stress-inducible MHC class I molecule MIC. Studies in mice suggest that the effect of α-gliadin 31–43 peptide is dependent on MyD88 and type I interferons, but not TLR4. In addition to the α-gliadin 31–43 peptide, other gluten peptides with the ability to induce maturation of antigen-presenting cells, chemokine receptor activation, and upregulation of the nonclassical MHC class I molecule HLA-E have been described. It has also been suggested that the innate immunity stimulatory capacity of wheat is mediated by α-amylase/trypsin inhibitors (ATIs), which are found as contaminator or gluten antigen preparations and can drive intestinal inflammation via activation of TLR4. What remains unexplained from available studies is why gluten mediates innate immunity effects only in some individuals. Furthermore, the involvement of multiple, unrelated innate mechanisms by which gluten was suggested to drive activation of antigen-presenting cells and epithelial cells, and the fact that they have not been confirmed by independent investigators, suggest that additional studies are required to confirm the pathogenic innate role of gluten.

31.11 Intraepithelial cytotoxic T lymphocytes with natural killer cell receptors, epithelial cells with nonclassical major histocompatibility complex class I molecules, and IL-15 contribute to disease pathogenesis

The induction of an inflammatory antigluten immune response may not be sufficient to induce epithelial cell destruction, tissue remodeling, and ultimately, villous atrophy. In this connection, another histological hallmark of celiac disease is the dramatic increase in IE-CTLs (CD8$\alpha\beta^+$TCR$\alpha\beta^+$) that reverts under a gluten-free diet and correlates with recovery of villous structures. The role of IELs was disregarded before gluten-specific IELs could be identified. In contrast, gluten-specific CD8$\alpha\beta^+$TCR$\alpha\beta^+$ T cells were identified in the lamina propria. Strong evidence now indicates that IELs do not mediate destruction by direct recognition of gluten peptides presented by MHC class I molecules, but by the killing of epithelial cells in a noncognate

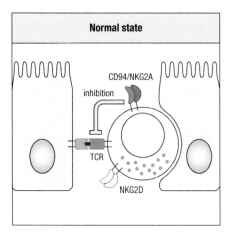

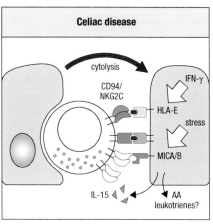

Figure 31.8 **Natural killer receptors and their ligands in intestinal mucosa.** Under physiologic conditions, epithelial cells lack expression of nonclassical major histocompatibility complex (MHC) class I molecules and release low levels of IL-15. IELs express inhibitory CD94/NKG2A receptors and low levels of activating NKG2D receptors. In active celiac disease, IELs lose expression of the inhibitory CD94/NKG2A receptors, acquire activating CD94/NKG2C receptors, and upregulate NKG2D receptors. These natural killer receptors recognize the nonclassical MHC class I molecules HLA-E and MIC induced on epithelial cells and inflammatory and stress signals respectively. Furthermore, epithelial cells express high levels of IL-15, which license intraepithelial lymphocytes to kill by costimulating the NKG2D cytolytic pathway and decreasing the activation threshold of the T-cell receptor. AA, arachidonic acid; IFN-γ, interferon-γ.

manner through NK receptors (Figure 31.8). This noncognate destruction of epithelial cells could be mediated in a TCR-dependent and TCR-independent manner. Regarding the former, NK receptors would lower the TCR activation threshold in such a way that IELs would recognize low-affinity self or microbial peptides. Regarding the latter, killing would be mediated only via NK receptors that recognize their ligands on epithelial cells. Two NK receptor-ligand pairs are involved in celiac disease pathogenesis, but other yet-to-be-identified NK receptors also may be involved.

NKG2D is a C-type lectin receptor that associates selectively with the adaptor molecule DAP10 in humans. This adaptor molecule has a phosphoinositol-3-kinase binding motif but lacks an immunoreceptor tyrosine-based activation motif. However, NKG2D can mediate direct cytolysis in T cells via a cytolytic signaling pathway that involves cytosolic phospholipase A2 and arachidonic acid. In order for NKG2D to mediate direct cytolysis, IE-CTLs must be activated by IL-15 expressed on intestinal epithelial cells. IL-15 acts as a costimulatory molecule for the NKG2D cytolytic pathway. In addition, IL-15 upregulates NKG2D and DAP10 expression.

The second well-described NK receptor involved in celiac disease pathogenesis is CD94/NKG2C. Interestingly, normal IELs predominantly express the inhibitory CD94/NKG2A receptor that can block TCR and NKG2D signaling. In IE-CTLs of celiac disease patients, CD94/NKG2A receptors are conspicuously decreased, and activating CD94/NKG2C receptors are expressed. Importantly, CD94/NKG2C receptors are associated with the adaptor molecule DAP12 that contains an immunoreceptor tyrosine-based activation motif and consequently can induce cytokine secretion and T-cell proliferation in addition to cytolysis. Furthermore, analysis of IE-CTLs expressing NKG2C showed that IE-CTLs had actually undergone a global natural killer reprogramming, resulting in the expression of a panel of natural killer receptors and molecules normally only found in natural killer cells. The ligand for CD94/NKG2C receptors is the nonclassical MHC class I molecule HLA-E that is induced in the presence of interferon-γ. Celiac disease epithelial cells upregulate HLA-E, MIC, and IL-15, indicating that they have received inflammatory and stress signals. Together, these observations suggest that ingested gluten interacts with epithelial cells, leading to the expression of IL-15 and nonclassical MHC class I molecules in genetically susceptible individuals. IL-15 and nonclassical MHC class I molecules then activate NK receptors that induce IE-CTl activation and epithelial cell destruction. The role of tissue cells in licensing cytolytic T cells to mediate tissue damage may constitute a "checkpoint" to limit tissue destruction in the presence of harmless antigens.

While villous atrophy in active celiac disease is mediated by IE-CTLs, iCD3+ innate IELs that display major cytolytic functions are the main effector lymphocyte subset mediating tissue destruction in patients with refractory

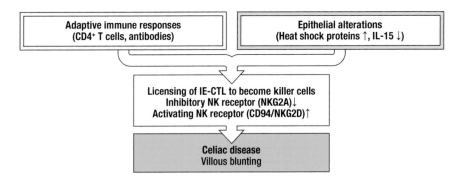

```
┌─────────────────────────────────┐   ┌─────────────────────────────────┐
│   Adaptive immune responses      │   │     Epithelial alterations       │
│   (CD4⁺ T cells, antibodies)     │   │  (Heat shock proteins ↑, IL-15 ↓)│
└─────────────────────────────────┘   └─────────────────────────────────┘

            ┌────────────────────────────────────────────┐
            │  Licensing of IE-CTL to become killer cells  │
            │     Inhibitory NK receptor (NKG2A)↓          │
            │   Activating NK receptor (CD94/NKG2D)↑       │
            └────────────────────────────────────────────┘

                     ┌──────────────────────┐
                     │    Celiac disease     │
                     │    Villous blunting   │
                     └──────────────────────┘
```

Figure 31.9 Epithelial stress and adaptive antigluten immunity synergize to license cytotoxic T cells to become killer cells in celiac disease. Potential celiac disease patients who display adaptive antigluten CD4⁺ T-cell immunity have a normal epithelium with low levels of IL-15 expression and IE-CTLs with a similar phenotype to the one observed in control patients; (i.e., they express predominantly inhibitory CD94/NKG2A receptors and have low levels of activating natural killer [NK] receptors). Conversely, intestinal epithelial cells in a subset of apparently healthy family members of celiac disease patients have ultrastructural anomalies associated with upregulation of IL-15 and heat shock proteins Hsp70 and Hsp25. In these individuals, IE-CTLs conserve expression of inhibitory NK receptors but have increased expression of activating NK receptor, albeit at lower levels than in active celiac disease patients. IE-CTLs acquire a complete killer phenotype, and villous atrophy develops only in individuals who combine epithelial distress with high levels of IL-15 expression and expression of nonclassical major histocompatibility complex class I molecules and inflammatory adaptive antigluten immunity.

celiac disease type 2. As for IE-CTLs in active celiac disease, IL-15 plays a critical role in the activation and expansion of iCD3⁺ innate IELs and their ability to mediate NK-like cytolytic functions.

Taken together, in both active celiac disease and refractory celiac disease, IL-15 and NK-like functions exerted by IEL play a critical role in epithelial cell destruction and development of villous atrophy. This explains how IE-CTLs and innate IELs can mediate tissue destruction based on recognition of stress signals without being gluten specific. Because epithelial cell destruction involves recognition of nonclassical class I molecules by NK receptors and requires IL-15 that is only upregulated in epithelial cells in response to gluten ingestion, the killing of epithelial cells remains specific and dependent on gluten despite not involving direct recognition of gluten by the T-cell and NK receptors. The mechanisms underlying the upregulation of IL-15 and nonclassical class I molecules in intestinal epithelial cells remains to be determined. Analysis of potential celiac disease patients who display dysregulated immunity to gluten, as assessed by the presence of anti-TG2 and antigluten antibodies, but lack of villous atrophy, suggested that activation of gluten-specific CD4⁺ T cells is not sufficient to induce an NK-like phenotype in IELs and epithelia cell distress. Interestingly, in around 30% of apparently healthy family members of celiac disease patients, epithelial cells showed ultrastructural anomalies associated with IL-15 upregulation in the absence of anti-TG2 and antigluten antibodies. These findings suggest that acquisition of a complete activating killer phenotype by IE-CTLs as defined by loss of inhibitory NKG2A receptors and induction of activating NK receptors requires activation of gluten-specific CD4⁺ T cells as well as epithelial distress associated with IL-15 upregulation in the epithelium (Figure 31.9). Future studies will characterize the interplay between epithelial cells, gluten-specific CD4⁺ T cells and IELs, and their role in tissue destruction, and they will define the mechanisms underlying epithelial alterations in celiac disease and refractory celiac disease.

Immunologic tests in the diagnostic workup of celiac disease

The diagnosis of celiac disease is based on the histopathologic assessment of small intestinal biopsies, supplemented by other newly available tests. These tests include serum antibodies to TG2 and gluten and HLA genotyping. In children, the diagnosis can be made on the basis of highly elevated levels of anti-TG2 antibodies and the presence of HLA-DQ2 or HLA-DQ8. In adults, altered histology of gut biopsies is still considered mandatory for diagnosis.

31.12 Antibodies to transglutaminase 2 and gluten are used to diagnose celiac disease

Detection of serum IgA and IgG antibodies to TG2 and gluten is widely used in the diagnosis of celiac disease. Measurements of IgA antibodies to TG2 and IgG antibodies to deamidated gliadin peptides are the most sensitive and specific tests, depending on the study population. In selected populations with high risk for the disease, the specificity and sensitivity are usually higher than 99%, but these values become lower when testing general populations. Patients with IgA deficiency are overrepresented among patients with celiac disease, and for such patients IgG antibodies to TG2 and deamidated gliadin represent reliable markers for disease. A point-of-care test based on the detection of IgA serum antibodies to TG2 has also been developed.

The IgA and IgM produced in the gut are transported across the epithelium to the gut lumen, and it is likely that the serum IgA-TG2 and IgG-deamidated gliadin antibodies are produced by plasma cells located outside of the gut, like in the bone marrow and the spleen. Antibodies produced in the mucosa can react with the TG2-antigen expressed locally, and thus, IgA anti-TG2 can be detected in the mucosa without elevated levels of TG2 antibodies in serum. The presence of IgA in a typical subepithelial distribution is typical of celiac disease and is an early marker of disease activity.

31.13 Other tests have a complementary role in diagnosis of celiac disease

HLA is a necessary but not sufficient factor for celiac disease development. HLA typing can help exclude the diagnosis of celiac disease. Specifically, the absence of DQB1*02 is a strong negative predictor of the disease. Importantly, detection of either DQB1*02 or DQB1*03:02 has very little diagnostic value, as both these DQB1* alleles are frequent in the normal population.

An elevated level of $\gamma\delta$ T cells in the epithelium is a typical feature of celiac disease but may accompany other inflammatory conditions of the upper small bowel. As this is not a specific test, its use in clinical practice is diminishing.

The diagnosis of celiac disease in a person on a gluten-free diet is problematic, because in the absence of gluten ingestion, the clinical signs, serology, and morphology return to normal. Thus, several months of gluten ingestion are required for a proper diagnosis. Many patients are reluctant to engage in a prolonged gluten challenge, but such a challenge may not be necessary in the future. After oral gluten consumption for 3 days, most celiac disease patients have T cells specific to deamidated gluten peptides in their peripheral blood. Such T cells can be detected by ELISPOT (enzyme-linked immunospot assay) or by staining with HLA-DQ2–gliadin peptide tetramers. Anti-TG2 antibodies can be detected in patients with extraintestinal diseases, such as dermatitis herpetiformis, but biopsy is required when the levels of anti-TG2 antibodies are low.

31.14 Immune-based therapy may have a future in celiac disease

Because a gluten-free diet is difficult to follow and more than half of patients on such a diet have incomplete recovery, three immune-based therapeutic strategies have been considered (Figure 31.10). The first strategy seeks to prevent recognition of gluten peptides by CD4+ T cells by increasing gluten digestion using bacterial enzymes that can cleave at proline and glutamine residues by sequestering gluten proteins in the gut lumen, by decreasing intestinal permeability with pharmacologic reagents that act on tight junctions, by inhibiting TG2 activity, or by blocking presentation of gluten peptides by HLA-DQ molecules using an inhibitor of cathepsin S to prevent degradation of the invariant chain. The second strategy aims to delete antigluten CD4+ T cells and/or divert inflammatory, antigluten immune responses into regulatory, antigluten immune responses in which T cells produce anti-inflammatory cytokines such as IL-10 or transforming growth factor-β. This strategy employs agents that induce a tolerogenic phenotype in

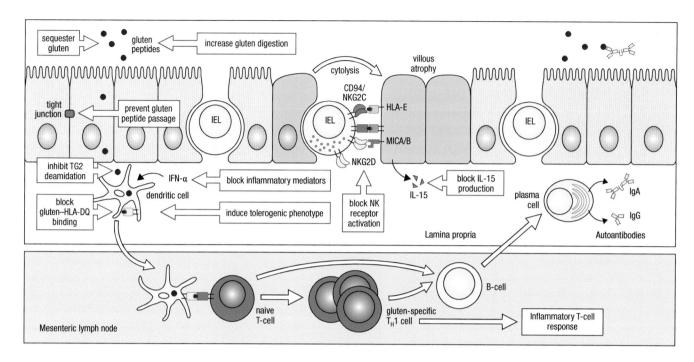

Figure 31.10 Pathogenesis of celiac disease with targets for therapy. Three main therapeutic strategies are currently envisioned. In the first strategy, decreased gluten peptide presentation can be achieved by increasing digestion of gluten via introduction of bacterial enzymes, by sequestering gluten proteins in the gut lumen, or by preventing passage of gluten peptides into the lamina propria by targeting tight junctions, inhibiting deamidation by TG2, or blocking the binding of gluten peptides to HLA-DQ2.5 or HLA-DQ8 molecules using competitive HLA ligands. The second strategy aims to prevent the differentiation of pathogenic inflammatory antigluten T cells by blocking inflammatory mediators such as interferon-α, educating dendritic cells to acquire a tolerogenic phenotype, or injecting low concentrations of immunodominant peptides multiple times in the absence of adjuvant. A third strategy seeks to prevent lysis of epithelial cells by blocking IL-15, which licenses intraepithelial lymphocytes to kill, or by using antibodies or molecules that block activating natural killer receptors.

dendritic cells or repeatedly present low concentrations of immunodominant gluten peptides. The third strategy blocks the activation of intraepithelial lymphocytes by inhibiting IL-15 or NK receptor activation, such as NKG2D, with either blocking antibodies or molecules that interfere with their signaling pathway. Although still in their infancy, immune intervention therapies offer hope to patients with celiac disease that therapy other than gluten restriction may someday be possible.

SUMMARY

Celiac disease is caused by hypersensitivity to the food protein gluten and resembles features of T-cell–mediated autoimmune disease. Manifestation of the disease is multifactorial, involving environmental factors such as gluten and possibly viral infections, as well as genetic factors. HLA is the single most important genetic factor, reflected in a strong association between the disease and the presence of HLA-DQ2.5, HLA-DQ2.2, and HLA-DQ8. The total number of non-HLA genes is estimated to represent 60% of the total genetic risk. To date, only a fraction of these genes, which mainly encode molecules with important immune functions, has been identified. HLA-DQ2.5, HLA-DQ2.2, and

HLA-DQ8 molecules present gluten epitopes to CD4$^+$ T cells in the lamina propria. The epitopes are harbored in proline- and glutamine-rich gluten peptides that are resistant to intestinal proteolysis and that have become posttranslationally modified by conversion of certain residues from glutamine to glutamate. This process, termed *deamidation*, is mediated by the enzyme TG2, the presence of which is dependent on carriage of HLA-DQ2.5 or HLA-DQ8 and dietary exposure to gluten. The TG2 enzyme is also the target of disease-specific IgA and IgG autoantibodies. Gluten-specific CD4$^+$ T cells are necessary but not sufficient to induce the celiac lesion, which is characterized by villous blunting, crypt cell hyperplasia, and leukocyte infiltration in both epithelium and the lamina propria. Several effector mechanisms appear to be involved in the lesion formation, but CD8[+] T cells with NK cell receptors infiltrating the epithelium appear to be particularly critical. These cells lose inhibitory and acquire activating NK cell receptors under the influence of IL-15, licensing the cells to kill enterocytes expressing stress-induced ligands of the NK cell receptors. Immunologic assays play an important role in the diagnostic workup of celiac disease, and novel therapies to interdict the key immune processes are under development.

FURTHER READING

Arentz-Hansen, H., Körner, R., Molberg, O. et al.: The intestinal T cell response to α-gliadin in adult celiac disease is focused on a single deamidated glutamine targeted by tissue transglutaminase. *J. Exp. Med.* 2000, 191:603–612.

DePaolo, R.W., Abadie, V., Tang, F. et al.: Co-adjuvant effects of retinoic acid and IL-15 induce inflammatory immunity to dietary antigens. *Nature* 2011, 471:220–224.

Dieterich, W., Ehnis, T., Bauer, M. et al.: Identification of tissue transglutaminase as the autoantigen of celiac disease. *Nat. Med.* 1997, 3:797–801.

Di Niro, R., Mesin, L., Zheng, N.Y. et al.: High abundance of plasma cells secreting transglutaminase 2-specific IgA autoantibodies with limited somatic hypermutation in celiac disease intestinal lesions. *Nat. Med.* 2012, 18:441–445.

Ettersperger, J., Montcuquet, N., Malamut, G. et al.: Interleukin-15-dependent T-cell-like innate intraepithelial lymphocytes develop in the intestine and transform into lymphomas in celiac disease. *Immunity.* 2016, 45:610–625.

Hüe, S., Mention, J.J., Monteiro, R.C. et al.: A direct role for NKG2D/MICA interaction in villous atrophy during celiac disease. *Immunity* 2004, 21:367–377.

Jabri, B., and Sollid, L.M.: Tissue-mediated control of immunopathology in celiac disease. *Nat. Rev. Immunol.* 2009, 9:858–870.

Junker, Y., Zeissig, S., Kim, S.J. et al.: Wheat amylase trypsin inhibitors drive intestinal inflammation via activation of toll-like receptor 4. *J. Exp. Med.* 2012, 209: 2395–2408.

Mayassi, T., Ladell, K., Gudjonson, H. et al.: Chronic inflammation permanently reshapes tissue-resident immunity in celiac disease. *Cell.* 2019, 176:967–983.

Meresse, B., Chen, Z., Ciszewski, C. et al.: Coordinated induction by IL-15 of a TCR-independent, NKG2D signaling pathway converts CTLs into natural killer-like, lymphokine activated killer (LAK) cells in celiac disease. *Immunity* 2004, 21:357–366.

Molberg, O., McAdam, S.N., Körner, R. et al.: Tissue transglutaminase selectively modifies gliadin peptides that are recognized by gut-derived T cells in celiac disease. *Nat. Med.* 1998, 4:713–717.

Petersen, J., Montserrat, V., Mujico, J.R. et al.: T-cell receptor recognition of HLA-DQ2-gliadin complexes associated with celiac disease. *Nat. Struct. Mol. Biol.* 2014, 21:480–488.

Setty, M., Discepolo, V., Murray, J. A. et al.: Epithelial stress and adaptive anti-gluten immunity synergize to license cytotoxic T cells to become killer cells in celiac disease. *Gastroenterology* 2015, 149:681–691.

Trynka, G., Hunt, K.A., Bockett, N.A. et al.: Dense genotyping identifies and localizes multiple common and rare variant association signals in celiac disease. *Nat. Genet.* 2011, 43:1193–1201.

Vader, L.W., de Ru, A., van der Wal, Y. et al.: Specificity of tissue transglutaminase explains cereal toxicity in celiac disease. *J. Exp. Med.* 2002, 195:643–649.

Withoff, S., Li, Y., Jonkers, I. et al.: Understanding celiac disease by genomics. *Trends Genet.* 2016, 32:295–308.

IgA nephropathy

32

JAN NOVAK, BRUCE A. JULIAN, AND JIRI MESTECKY

IgA nephropathy is the most common form of primary glomerulonephritis and an important cause of renal failure. A mesangioproliferative glomerulonephritis, IgA nephropathy is characterized by mesangial deposits of IgA1-containing immune complexes. IgA nephropathy was initially described in 1968 by Berger and Hinglais on the basis of its unique renal immunohistologic features as an IgA:IgG immune-complex renal disease.

CLINICAL PRESENTATION OF IgA NEPHROPATHY

IgA nephropathy is a relatively common renal disease that accounts for 5%–20% of renal diseases diagnosed by native-kidney biopsy in Europe and the United States, and up to 30%–45% in China and Japan. In contrast, IgA nephropathy is rare in central Africa. Some individuals probably have subclinical disease for years. A study in Japan demonstrated mesangial IgA deposits in 82 of 510 renal allografts (16%) at implantation, of which 19 showed mesangioproliferative glomerulonephritis by light microscopy.

32.1 IgA nephropathy is characterized by sporadic and familial forms

The initial clinical features of IgA nephropathy manifest most frequently in adolescents and young adults, often with a 2:1 predominance of males over females, although the sexes are commonly affected equally in Asia. Asymptomatic proteinuria and hematuria are typical clinical presentations. Painless macroscopic hematuria is common in children and often coincides with mucosal infections, including infections of the upper respiratory and gastrointestinal tracts.

The two forms of IgA nephropathy are familial and sporadic. The familial form is defined by the presence of at least two biopsy-proven cases in a family. Clinically, the two forms are indistinguishable. Regardless of whether IgA nephropathy is sporadic or familial, 20%–40% of patients with IgA nephropathy progress to end-stage renal disease within 20 years of diagnosis; such patients require functional renal replacement, dialysis, or transplantation. Unfortunately, the disease recurs in 50%–60% of patients 5 years after transplantation.

Secondary forms of IgA nephropathy occur in association with diseases of other organ systems, including hepatobiliary diseases (cirrhosis due to alcoholic liver disease or viral hepatitis), gastrointestinal diseases (ulcerative colitis), and infectious diseases (infection with human immunodeficiency virus).

Disease-specific treatment of IgA nephropathy is not currently available. Instead, therapy focuses on minimizing proteinuria and maintaining normal blood pressure, with emphasis on the use of pharmacologic agents to suppress the effects of angiotensin II. In this chapter, we focus on primary IgA nephropathy.

32.2 Definitive diagnosis of IgA nephropathy requires renal biopsy

Evaluation of a renal biopsy specimen is required for a definitive diagnosis of IgA nephropathy and the assessment of severity of the tissue injury. Typically, samples of the kidney cortex obtained by percutaneous needle biopsy are divided into three portions and processed for immunofluorescence, light, and electron microscopy.

Dominant or codominant IgA deposits are detected in the mesangium by immunofluorescence (Figure 32.1) with variable IgG and/or IgM codeposits. The IgA is exclusively of the IgA1 subclass. Complement component C3 is almost always present whereas C1q is usually absent or minimally present. Analysis of immunofluorescence staining for IgA and C3 by confocal microscopy has revealed co-localization of these components. In approximately 50% of renal biopsy specimens, IgG co-deposits are detected by routine immunofluorescence and immunohistochemistry. However, using new approaches and reagents and high-resolution imaging, it has recently been shown that all patients with IgA nephropathy contain IgG in the glomerular immunodeposits.

Mesangial cellular proliferation and expansion of the extracellular matrix, detected by light microscopy, are characteristic for IgA nephropathy (Figure 32.2). Glomerular sclerosis and interstitial fibrosis are associated with progressive disease that leads to renal insufficiency. An "Oxford classification" for IgA nephropathy was developed by an international group of nephrologists and pathologists on the basis of the identification of specific light microscopy features that predict the risk of progression of renal disease in adults and children, independent of clinical or laboratory parameters. Four pathologic features, mesangial hypercellularity, segmental glomerulosclerosis, endocapillary hypercellularity, and tubular atrophy/interstitial fibrosis, are each independently associated with a worse outcome. The value of this classification system has been validated in additional cohorts, and the scoring has become the standard for pathology reports of IgA nephropathy.

Electron microscopy typically shows electron-dense deposits in the mesangial or paramesangial areas of glomeruli, confirming the presence of immune complexes detected by immunofluorescence (Figure 32.3). Electron microscopy also may reveal ultrastructural changes such as the effacement of podocytes in association with proteinuria.

Immunopathogenesis of IgA nephropathy

Considerable evidence indicates that the mesangial deposits originate from circulating IgA1-containing immune complexes. This evidence includes (1) the frequent recurrence of the disease in patients who receive a renal allograft; (2) in the few cases in which a kidney was transplanted from a donor with subclinical IgA nephropathy into a patient with non-IgA nephropathy renal disease, the immune deposits cleared from the engrafted kidney within weeks; (3) patients with IgA nephropathy have elevated levels of IgA1-containing immune complexes in the circulation; (4) idiotypic determinants are shared between the circulating complexes and the mesangial deposits, although a disease-specific idiotype has not been identified; (5) IgA1 in the circulating immune complexes and in the renal deposits has an unusual chemical composition, with galactose (Gal) deficiency of some of the O-glycans of the

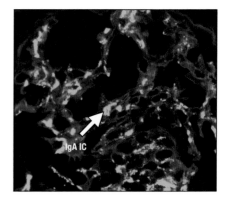

Figure 32.1 Immunofluorescence confocal microscopy image of a glomerulus from a patient with IgA nephropathy stained for IgA. Example of a granular IgA-containing immunodeposit (IgA IC, arrow). (Staining and photograph courtesy of L. Novak.)

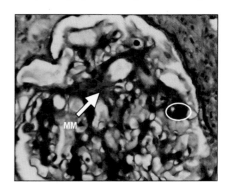

Figure 32.2 Light microscopy image of a glomerulus from a patient with IgA nephropathy. MM, expanded mesangial matrix; oval marks a cluster of proliferating mesangial cells. (Photograph courtesy of L. Novak.)

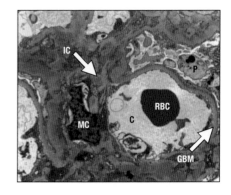

Figure 32.3 Electron microscopy image of an area of a glomerulus from a patient with IgA nephropathy. C, capillary; IC, electron-dense immunodeposits; GBM, glomerular basement membrane; MC, mesangial cell; P, podocyte; RBC, red blood cell. (Photograph courtesy of W. J. Cook.)

hinge region unique to IgA1 (Gal-deficient IgA1); and (6) IgG in renal deposits is enriched for autoantibody specific for Gal-deficient IgA1.

32.3 IgA, key component of IgA-containing immune complexes, is the major immunoglobulin isotype produced in the body

The daily production of IgA (about 66 mg kg^{-1} body weight per day) far exceeds the biosynthesis of immunoglobulins of all other isotypes combined. About two-thirds of IgA is produced in a polymeric form in mucosal tissues, particularly in the intestinal tract, and then selectively transported by a receptor-mediated pathway into external secretions (Chapters 10 and 11). The IgA subclass distribution of IgA1- and IgA2-producing cells and corresponding IgA subclasses in secretions displays a characteristic pattern: IgA1 is more abundant than IgA2 in tears, saliva, respiratory-tract secretion, and secretions from the small intestine. In contrast, secretions of the large intestine and of the female genital tract contain a slight excess of IgA2. Most IgA in the circulation is represented by monomeric (m) IgA with a dominance of IgA1 (about 85% of total IgA). In contrast to IgG, circulatory IgA has a short half-life—about 5 days—and only trace amounts of mIgA1 from the circulatory pool enter external secretions. In humans, IgA-producing cells in the bone marrow and, to a lesser degree, in lymph nodes and spleen are the most important sources of the circulatory IgA pool. Experiments performed in mice and in primates clearly indicate that the liver is the major organ involved in the catabolism of IgA in both subclasses, as well as of other glycoproteins. The asialoglycoprotein receptor (ASGP-R) expressed on cell surfaces of hepatocytes binds glycoproteins through their terminal Gal- and *N*-acetylgalactosamine (GalNAc)-containing saccharides. The glycoproteins are subsequently internalized and degraded. Human hepatoma cell lines also have been examined, with identical results. The Gal deficiency of *O*-glycans within the hinge region of IgA1 in patients with IgA nephropathy, as described earlier, should not profoundly decrease such binding to ASGP-R due to the presence of terminal GalNAc. The apparent basis for the impaired clearance of nephritogenic IgA1 has been elucidated by detailed analyses of circulating immune complexes. These studies demonstrated that in circulating immune complexes, GalNAc residues on IgA1 are recognized by Gal-deficient-IgA1-specific autoantibodies (predominantly of IgG isotype); thus, this terminal GalNAc is inaccessible for interactions with the hepatic ASGP-R. Furthermore, the size of these circulating immune complexes prevents their penetration through the fenestrae in endothelial cells of hepatic sinuses to enter the space of Disse and reach the ASGP-R for subsequent binding and catabolism. Because the fenestrae of the endothelial cells of the glomerular capillaries are significantly larger, circulating immune complexes are diverted from the normal hepatic catabolic pathway for IgA to deposit in the mesangium and induce glomerular injury. The validity of this concept has been reinforced in recent studies in mice. High-molecular-mass immune complexes composed of Gal-deficient IgA1 and the corresponding IgG autoantibodies specifically displayed this deposition pattern, with ensuing pathogenic consequences.

32.4 Human IgA1 has a unique hinge region that contains sites amenable to *O*-linked glycosylation

The glycosylation of IgA1 is unique and provides insight into the mechanism underlying immune-complex formation and deposition in the glomerular mesangium of patients with IgA nephropathy. Heavy (α) chains of human IgA1

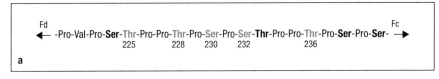

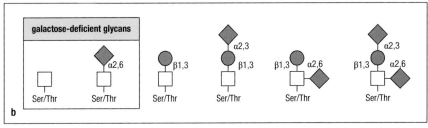

a

b

Figure 32.4 Hinge region of serum IgA1 and its *O*-glycans. (a) IgA1 has *O*-glycans attached in the hinge region between the first (C$_\alpha$1) and second (C$_\alpha$2) constant domains of the heavy chain. The hinge-region amino acid sequence is composed of two octapeptide repeats with as many as six of the nine potentially *O*-glycosylated sites occupied. Serine and threonine amino acids that can be theoretically glycosylated are in bold, and the five residues that have been shown to be most frequently glycosylated on circulatory IgA1 are numbered and shown in blue. **(b)** Variants of *O*-glycans on circulatory IgA1. Gal-deficient glycans present in elevated amounts on IgA1 in the circulation of patients with IgA nephropathy are in a large rectangle. The largest glycan on circulatory IgA1 is a GalNAc-Gal with two sialic acids, a tetrasaccharide. Symbols: open rectangle, GalNAc; filled circle, Gal; filled diamond, sialic acid.

possess a unique hinge region, located between the C$_\alpha$1 and C$_\alpha$2 domains, that contains a sequence rich in proline, serine, and threonine residues. Nine serine and threonine amino acids, the potential sites of *O*-glycan attachment, are in each hinge region (see Chapter 11 for details). However, up to only six of these sites are usually glycosylated. Hinge-region glycoforms with four and five glycans are the most common in the circulatory IgA1.

The *O*-glycans of serum IgA1 are core 1 glycans in that they consist of GalNAc with a β1,3-linked Gal that are *O*-linked to either serine or threonine; both residues of the core 1 glycan may be sialylated. The carbohydrate composition of the *O*-linked glycans on normal serum IgA1 is variable; the prevailing forms include the Gal-GalNAc disaccharide and its mono- and di-sialylated forms (**Figure 32.4**).

O-glycosylation is initiated by the attachment of GalNAc to serine or threonine residues by UDP-GalNAc:polypeptide *N*-acetylgalactosaminyltransferases (GalNAc-Ts). The sites to be *O*-glycosylated and the order of the process are determined by the specific set of GalNAc-Ts expressed in a particular cell type. The IgA1 *O*-glycans are attached predominantly by GalNAc-T2, although other GalNAc-Ts (such as GalNAc-T1 and -T11, which can efficiently catalyze the same reaction) can use the IgA1 hinge region as an acceptor (**Figure 32.5**). GalNAc then is modified by the addition of β1,3-linked Gal in a reaction catalyzed by UDP-Gal:GalNAc-α-Ser/Thr β1,3-galactosyltransferase 1 (C1GalT1). This reaction results in a core 1 structure. Formation of the active C1GalT1 depends on core 1 β1,3-galactosyltransferase-specific chaperone (Cosmc).

The GalNAc-Gal structure in the hinge region of circulatory IgA1 can be further modified by attaching the sialic acid from cytidine monophosphate

Figure 32.5 Biosynthesis of IgA1 *O*-glycans. *O*-linked glycans in the IgA1 hinge region are synthesized in a stepwise manner. The initial step is the attachment of GalNAc to the oxygen atom of the hydroxyl group of some of serine or threonine amino acids. The *O*-glycan chain then is extended by the sequential attachment of Gal and/ or sialic acid residues to the GalNAc. The addition of Gal is mediated by core 1 β1,3-galactosyltransferase (C1GalT1), which transfers Gal from UDP-galactose to the GalNAc residue. The stability of this enzyme during biosynthesis depends on interaction with its chaperone protein, Cosmc, which assists in protein folding. In the absence of Cosmc, the C1GalT1 protein is rapidly degraded. The glycan structure is completed by sialyltransferases that attach sialic acid to the Gal and/or GalNAc residues. If sialic acid is linked to GalNAc before the attachment of Gal, this "premature" sialylation precludes the subsequent attachment of Gal. An imbalance in the activities or expression of specific glycosyltransferases accounts for the increased production of Gal-deficient *O*-linked glycans in the IgA1 hinge region with increased sialic acid residues: decreased C1GalT1 and Cosmc and increased ST6GalNAcII (marked by red arrows). Gal-deficient glycans present in elevated amounts on the IgA1 in the circulation of patients with IgA nephropathy are shown in red boxes. Symbols are as in **Figure 32.4.**

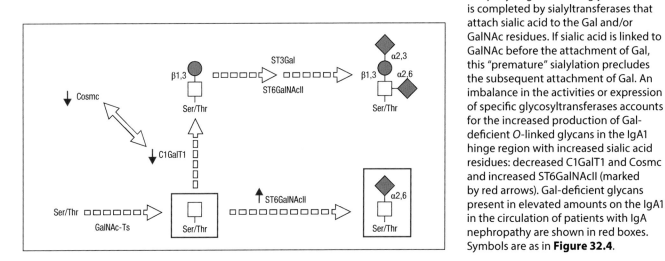

(CMP)-*N*-acetylneuraminic acid to the Gal residues in a reaction catalyzed by a Galβ1,3GalNAc α2,3-sialyltransferase (ST3Gal) and/or to the GalNAc residues catalyzed by a α2,6-sialyltransferase (ST6GalNAc). A ST3Gal, probably ST3GalI, is responsible for the sialylation of Gal on IgA1 *O*-glycans. Among the α2,3-sialyltransferases, IgA1-producing cells do not express ST6GalNAcI, but express ST6GalNAcII, which is apparently responsible for the sialylation of GalNAc in both the terminal and Gal-containing GalNAc on IgA1.

32.5 Circulatory IgA1 in IgA nephropathy contains aberrant *O*-linked glycans

The blood of patients with IgA nephropathy contains elevated levels of IgA1 molecules with some aberrantly glycosylated hinge-region *O*-glycans that are deficient in Gal. These aberrant molecules are present in circulating immune complexes. Furthermore, IgA1 in the mesangial deposits is enriched for this aberrantly glycosylated form of IgA1. IgA1-secreting cell lines derived from Epstein-Barr-virus-immortalized B cells from the blood of patients with IgA nephropathy (and from healthy-control subjects) have been used to characterize the glycosylation of the secreted immunoglobulins. The cells from patients with IgA nephropathy secrete IgA1 that is mostly polymeric and contains Gal-deficient *O*-linked glycans, with terminal or sialylated GalNAc. These glycans also exhibit an increased content of sialic acid, as shown by an increased binding to *Sambucus nigra* agglutinin, a lectin specific for α2,6-bound sialic acid. In contrast, the IgA1 that is secreted by cells from healthy control subjects has a normal content of Gal. Thus, the cells from patients with IgA nephropathy secrete IgA1 that is Gal-deficient and contains terminal and sialylated GalNAc. These aberrancies are due to the altered expression of specific glycosyltransferases: increased expression and activity of ST6GalNAcII and decreased expression and activity of C1GalT1, as well as decreased expression of *Cosmc*. Decreased amount of Cosmc protein further reduces the amount C1GalT1 enzyme as a result of its increased degradation in the absence of Cosmc. The changes in enzyme activities mean that there will be fewer Gal but more sialic acid residues added to the GalNAc in the IgA1 hinge region. Overexpression of ST6GalNAcII may result in premature sialylation of GalNAc that would block other modifications, including galactosylation (see Figure 32.5).

32.6 IgG autoantibodies develop against aberrantly glycosylated IgA1 to form immune complexes in IgA nephropathy

Autoantibodies specific for terminal GalNAc-containing glycoproteins have been described in patients with IgA nephropathy, IgA vasculitis with nephritis (Henoch-Schönlein purpura nephritis), and the rare Tn syndrome, also called *permanent mixed-field polyagglutinability*. Tn syndrome is an autoimmune disease in which subpopulations of blood cells in all lineages carry incompletely *O*-glycosylated cell-surface glycoproteins. GalNAc-specific antibodies also are present in the sera of healthy persons, although at lower levels, and in sera of a diverse array of vertebrate species. The origin of such ubiquitous antibodies is unclear. However, many microorganisms, including bacteria and viruses, express glycoconjugates with GalNAc residues on their surfaces that may induce the synthesis of antibodies cross-reactive with analogous structures in the hinge region of Gal-deficient IgA1 and other cell-associated glycoproteins.

In IgA nephropathy, Gal-deficient hinge-region glycans with terminal GalNAc on IgA1 represent neoepitopes that are recognized by the previously mentioned autoantibodies, resulting in the formation of circulating immune

Figure 32.6 **CDR3 of heavy chain of IgG auto antibodies specific for Gal-deficient IgA1.** Sequence analyses of the cloned heavy- and light-chain antigen-binding domains of the IgG autoantibodies produced by immortalized lymphocytes from patients with IgA nephropathy identified unique features in CDR3 of the IgG heavy chain. Specifically, the third position in CDR3 in the heavy-chain variable region was serine in six of seven patients with IgA nephropathy (the first three amino acids were Tyr-Cys-Ser); in contrast, all six healthy controls had Ala in that position (the first three amino acids were Tyr-Cys-Ala). Mutagenesis experiments determined a critical role of serine for effective binding to Gal-deficient IgA1. Parentheses show the range of lengths of CDR3 regions. Note, they are longer in the CDR3 regions of patients with IgA nephropathy. (Data from Suzuki, H. et al. *J. Clin. Invest.* 119:1668–1677, 2009. With permission from American Society for Clinical Investigation.)

	CDR3 amino acid sequence	Binding to Gal-deficient IgA1
IgAN	Tyr-Cys-**Ser** (X)$_n$ n = 10–24	High
Controls	Tyr-Cys-**Ala** (X)$_n$ n = 6–9	Low

complexes. These complexes are relatively large, are not efficiently cleared from the circulation, and tend to deposit in the renal mesangium thereby binding and activating mesangial cells to induce glomerular injury.

As the circulating lymphocytes from patients with IgA nephropathy produce IgG that is capable of binding to Gal-deficient IgA1, an approach has been developed to select Epstein-Barr-virus-immortalized lymphocytes that secrete IgG specific for Gal-deficient IgA1. Using this method, the IgG from the immortalized lymphocytes from patients with IgA nephropathy has been shown to bind Gal-deficient IgA1 in a glycan-dependent manner, as does the patients' polyclonal serum IgG. Moreover, the binding of these autoantibodies to Gal-deficient IgA1 is blocked by preincubation with the GalNAc-specific lectin (*Helix aspersa* agglutinin), suggesting that the binding to IgA1 hinge region may be glycan-mediated. Sequence analyses of the cloned heavy- and light-chain antigen-binding domains of the Gal-deficient IgA1-specific autoantibodies produced by the immortalized lymphocytes from patients with IgA nephropathy have identified unique features in complementarity-determining region 3 (CDR3) of the IgG heavy chain. Specifically, the third position in CDR3 in the heavy-chain variable region is typically serine in IgA nephropathy (in which the first three amino acids are Tyr-Cys-Ser) as opposed to alanine in that position (in which the first three amino acids are Tyr-Cys-Ala) (Figure 32.6). Site-directed mutagenesis has proven that serine in the third position is necessary for efficient binding of IgG to Gal-deficient IgA1. The Ser in CDR3 in the heavy chains of IgG autoantibodies in patients with IgA nephropathy originates from somatic mutations rather than from rare variants of *VH* genes. Thus, the IgG anti-Gal-deficient IgA1 autoantibodies in patients with IgA nephropathy display an array of unique features.

Elucidation of the mechanism of immune-complex formation in which Gal-deficient IgA1 acts as an antigen reacting with ubiquitous, naturally occurring, GalNAc-specific antibodies provides a potential opportunity to interfere with the formation of pathogenic immune complexes. Non-cross-linking, monovalent reagents (such as single-chain antibodies) with high affinity for GalNAc would theoretically prevent the binding of IgG autoantibodies and, thus, the formation of pathogenic complexes. This approach may provide a new strategy for the development of a disease-specific treatment of IgA nephropathy.

32.7 Gal-deficient IgA1-containing immune complexes in IgA nephropathy are pathogenic

Cultured primary human mesangial cells provide a convenient model for evaluating the biological activities of IgA1-containing complexes. Immune complexes from patients with IgA nephropathy with Gal-deficient IgA1-bind to the cells more efficiently than do complex-free IgA1 or immune complexes from healthy control subjects. Moreover, the high-molecular-mass complexes associated with IgA nephropathy stimulate cellular proliferation and the production of cytokines such as interleukin-6 (IL-6) and transforming growth factor-β (TGF-β). IgA1-depleted fractions are devoid of such stimulatory activities. Consistent with this finding, when small quantities of desialylated polymeric Gal-deficient IgA1 are added to the sera of patients with IgA nephropathy, new immune complexes are formed, resulting in an increase in the amount of stimulatory complexes with molecular mass of 800–900 kDa. These complexes contain IgG and IgA1. In contrast, complex-free IgA1 does not affect cellular proliferation in this model. Complexes in the native sera of patients with IgA nephropathy enhance cellular proliferation more than do complexes of similar molecular mass from sera of healthy volunteers. In addition, immune complexes from sera of patients with IgA nephropathy collected during episodes of macroscopic hematuria stimulate more cellular proliferation than do complexes obtained during a later quiescent phase of disease. IgA1 complexes with high levels of Gal-deficient IgA1 induce proliferation of cultured human mesangial cells to a higher degree than do complexes with a lower content of Gal-deficient IgA1.

Although the pathogenic role of IgA1-containing immune complexes in IgA nephropathy is well accepted the cellular mechanisms and receptors involved in the binding of IgA1 complexes and activation of mesangial cells are much less well understood. Several lines of evidence point to a specific IgA receptor expressed by mesangial cells that bind and at least partly internalize and catabolize IgA. IgA receptors reported to be present on human cells include ASGP-R on hepatocytes; polymeric immunoglobulin receptor specific for J-chain-containing polymeric IgA and IgM on epithelial cells; FcαR (CD89) on monocytes, neutrophils, and eosinophils; and Fcα/μ receptor expressed by B cells and macrophages. However, none of these receptors is expressed on human mesangial cells. In addition, a surface-bound Gal-transferase can also bind immunoglobulins *via* the glycans, but a role for this protein in the binding of IgA1 to mesangial cells has not been proven. A transferrin receptor (CD71) is another mesangial-cell receptor that has been shown to bind polymeric IgA1. Moreover, CD71 on human mesangial cells effectively binds polymeric Gal-deficient IgA1 and immune complexes containing Gal-deficient IgA1, which further enhances the expression of CD71. Binding of immune complexes to CD71 creates a positive-feedback loop that leads to overexpression of CD71 on proliferating mesangial cells in glomeruli. Transglutaminase is needed for the activation of mesangial cells by these IgA1-containing immune complexes. However, whether CD71 is the only receptor involved in the binding of IgA1-containing immune complexes and whether it has a direct pathogenic role in IgA nephropathy are not known.

Mesangial cells activated by immune complexes containing Gal-deficient IgA1 proliferate and overproduce extracellular matrix proteins, cytokines, and chemokines. These processes, if they continue unchecked for substantial periods, may lead to expansion of the glomerular mesangium and, ultimately, glomerular sclerosis with the loss of glomerular filtration function. Furthermore, humoral factors (such as tumor necrosis factor and TGF-β) are released from mesangial cells activated by IgA1-containing immune complexes alter podocyte gene expression and may thus alter glomerular permeability. This mesangio-podocyte communication may explain the proteinuria and tubulointerstitial injury in IgA nephropathy.

32.8 Immunoglobulins are potential biomarkers of IgA nephropathy

Currently, a definitive diagnosis of IgA nephropathy requires renal biopsy because there is no valid noninvasive test that can be reliably used as a diagnostic alternative. Unfortunately, renal biopsy entails a risk for serious bleeding complications. Consequently, early detection of the disease is frequently not possible, and monitoring disease activity is compromised. Some patients with IgA nephropathy will progress to end-stage renal disease, even when the initial histology display shows relatively minor glomerular abnormalities. Consequently, a reliable noninvasive diagnostic test will be very useful for detecting subclinical IgA nephropathy, assessing activity, monitoring the progression or reduction in renal damage, and evaluating the response to treatment.

Efforts to identify disease-specific markers have focused on urine since it is easy to obtain and collection is noninvasive. The urinary concentrations of several cytokines involved in cellular proliferation and repair have been evaluated as potential markers of histopathologic glomerular and tubulointerstitial changes and predictors of long-term prognosis. For example, urinary levels of IL-6 are elevated in patients with glomerulonephritis, but elevated levels do not differentiate between various forms of the disease. However, urinary excretion of IL-6 predicts long-term renal outcome in patients with IgA nephropathy, and excretion of renal IL-6 and epidermal growth factor (EGF) has been shown to correlate with the degree of tubulointerstitial damage. Consequently, the ratio of urinary IL-6/EGF has been proposed as a prognostic marker for the progression of renal damage in IgA nephropathy. Among other potential urinary markers for renal disease, the excretion of monocyte chemotactic peptide-1 (MCP-1) and IL-8 correlates with tubulointerstitial damage but does not distinguish between the specific types of glomerulonephritis. Excretion of the membrane-attack complex may be elevated in patients with IgA nephropathy.

Urinary excretion of immunoglobulins has offered a promising new possibility in the search for surrogate markers of ongoing renal injury of IgA nephropathy. Urinary IgA and IgG concentrations are higher in patients with IgA nephropathy than in normal healthy volunteers and patients with other renal diseases. Moreover, the amounts excreted correlate with the serum creatinine concentration and with proteinuria. Furthermore, a recent study has shown a correlation between glomerular filtration rate, urinary IgG excretion, and pathologic grading of renal biopsies in patients with nephritis. These findings demonstrate the presence of variable levels of proteins and protein complexes, including those of high molecular mass, in the urine of patients with IgA nephropathy, but clinically useful tests relying on these levels have not been developed.

The urine of patients with IgA nephropathy contains many polypeptides and/or their proteolytic fragments that are disease-specific. The characterization of these peptides by modern proteomic techniques eventually may result in their use as biomarkers of the disease. These proteins include high-molecular-mass immune complexes and their components, immunoglobulin molecules, and smaller polypeptides. Enzyme-linked immunosorbent assay (ELISA) and Western blots are adequate for the analysis of large polypeptides and their complexes. Using ELISA, urinary levels of IgA:IgG-containing immune complexes in patients with IgA nephropathy are elevated compared with levels in patients with non-IgA nephropathy glomerulonephritis or in healthy control subjects. For the analysis of smaller polypeptides, modern separation techniques can be combined with mass-spectrometry-based identification and quantitation of the separated component.

Mass-spectrometry-based technologies, including capillary electrophoresis–mass spectrometry (CE-MS), have been used for proteomic analyses of clinical samples. CE-MS has helped establish markers for different renal diseases, including focal segmental glomerulosclerosis and diabetic nephropathy. CE-MS has been used to identify naturally occurring peptide markers of IgA nephropathy. Although urinary proteins are potential biomarkers, more studies are required for their clinical application.

As the role of IgA1 and IgA1-containing immune complexes in the pathogenesis of IgA nephropathy has been identified, serum also has become a promising source of potential biomarkers. Levels of Gal-deficient IgA1 and the corresponding autoantibodies have been targeted as possible diagnostic markers of IgA nephropathy, as well as markers of disease progression. An ELISA using *Helix aspersa* agglutinin, a lectin specific for GalNAc of the neoantigen of the nephritogenic IgA1, has shown that 74% of patients with IgA nephropathy have a serum level of Gal-deficient IgA1 that is above the 90th percentile for healthy controls. Furthermore, serum levels of Gal-deficient IgA1 show significant heritability, suggesting a basis for the familial pattern of disease for some patients. A dot-blot test has been developed to semiquantitatively measure the amount of circulating IgG specific for Gal-deficient IgA1. As these methods are refined to improve their sensitivity and specificity, they may become lead candidates for better clinical assays.

SUMMARY

In patients with IgA nephropathy, Gal-deficient IgA1 is produced at elevated levels and is recognized by unique IgG autoantibodies in a glycan-dependent manner. Immune complexes form, and on deposition in the mesangium, become capable of inducing mesangioproliferative glomerular injury (Figure 32.7). Thus, formation of the Gal-deficient IgA1-containing circulating immune complexes is a critical factor in the pathogenesis of IgA nephropathy. The autoantibodies that bind to Gal-deficient IgA1 and the resultant immune complexes are key factors contributing to the development of IgA nephropathy. On the basis of the understanding of the disease pathogenesis, biomarkers specific for IgA nephropathy can be identified and developed into clinical assays to aid in the diagnosis and assessment of disease progression. Interference with specific pathogenetic pathways may become useful for disease-targeted therapy in IgA nephropathy.

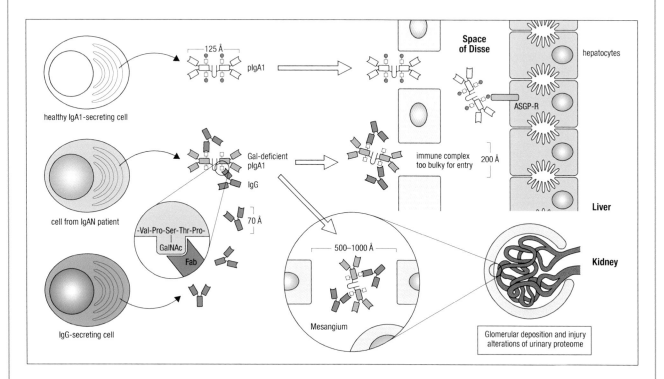

Figure 32.7 Model of IgA nephropathy pathogenesis. A fraction of polymeric IgA1 (pIgA1) produced by B cells and plasma cells in patients with IgA nephropathy (IgAN) is Gal-deficient and is recognized in the circulation by autoantibodies. The resultant immune complexes are too bulky to enter the space of Disse in the liver to reach the asialoglycoprotein receptor (ASGP-R) on hepatocytes, a major site of catabolism for normally glycosylated circulating IgA1, but the complexes are able to pass through the larger fenestrae in glomerular capillaries overlying the mesangium. These complexes bind to and activate mesangial cells, thereby initiating glomerular injury to induce alterations in the urinary proteome.

FURTHER READING

Berger, J., and Hinglais, N.: Les dépôts intercapillaires d'IgA-IgG. [Intercapillary deposits of IgA-IgG.]. *J. Urol. Nephrol.* 1968, 74:694–695.

Coppo, R., Troyanov, S., Camilla, R. et al.: The Oxford IgA nephropathy clinicopathological classification is valid for children as well as adults. *Kidney Int.* 2010, 77:921–927.

Gharavi, A.G., Moldoveanu, Z., Wyatt, R.J. et al.: Aberrant IgA1 glycosylation is inherited in familial and sporadic IgA nephropathy. *J. Am. Soc. Nephrol.* 2008, 19:1008–1014.

Huang, Z.Q., Raska, M., Stewart, T. et al.: Somatic mutations modulate autoantibodies against galactose-deficient IgA1 in IgA nephropathy. *J. Am. Soc. Nephrol.* 2016, 27:3278–3284.

Kiryluk, K., and Novak, J.: The genetics and immunobiology of IgA nephropathy. *J. Clin. Invest.* 2014, 124:2325–2332.

Knoppova, B., Reily, C., Maillard, N. et al.: The origin and activities of IgA1-containing immune complexes in IgA nephropathy. *Front. Immunol.* 2016, 7:117. doi: 10.3389/fimmu.2016.00117.

Lai, K.N., Tang, S.C., Schena, F.P. et al.: IgA nephropathy. *Nat. Rev. Dis. Primers.* 2016, 2:16001 doi:10.1038/nrdp.2016.1.

Maillard, N., Wyatt, R.J., Julian, B.A. et al.: Current understanding of the role of complement in IgA nephropathy. *J. Am. Soc Nephrol.* 2015, 26:1503–1512.

Mestecky, J., Raska, M., Julian, B.A. et al.: IgA nephropathy: Molecular mechanisms of the disease. *Annu. Rev. Pathol.* 2013, 8:217–240.

Moura, I.C., Arcos-Fajardo, M., Sadaka, C. et al.: Glycosylation and size of IgA1 are essential for interaction with mesangial transferrin receptor in IgA nephropathy. *J. Am. Soc. Nephrol.* 2004, 15:622–634.

Novak, J., Barratt, J., Julian, B.A. et al.: Aberrant glycosylation of the IgA1 molecule in IgA nephropathy. *Semin. Nephrol.* 2018, 38:461–476. doi: https://doi.org/10.1016/j.semnephrol.2018.05.016.

Onda, K., Ohsawa, I., Ohi, H., Tamano, M. et al.: Excretion of complement proteins and its activation marker C5b-9 in IgA nephropathy in relation to renal function. *BMC Nephrol.* 2011, 12:64. doi: 10.1186/1471-2369-12-64.

Rizk, D.V., Saha, M.K., Hall, S. et al.: Glomerular immunodeposits of patients with IgA nephropathy are enriched for IgG autoantibodies specific for galactose-deficient IgA1. *J. Am. Soc. Nephrol.* 2019, 30:2017–2026.

Suzuki, H., Fun, R., Zhang, Z. et al.: Aberrantly glycosylated IgA1 in IgA nephropathy patients is recognized by IgG antibodies with restricted heterogeneity. *J. Clin. Invest.* 2009, 119:1668–1677.

Suzuki, H., Moldoveanu, Z., Hall, S. et al.: IgA1-secreting cell lines from patients with IgA nephropathy produce aberrantly glycosylated IgA1. *J. Clin. Invest.* 2008, 118:629–639.

Suzuki, H., Raska, M., Yamada, K. et al.: Cytokines alter IgA1 *O*-glycosylation by dysregulating C1GalT1 and ST6GalNAc-II enzymes. *J. Biol. Chem.* 2014, 289:5330–5339.

Woof, J.M., and Mestecky, J.: Mucosal immunoglobulins, in Mestecky, J., Strober, W., Russell, M.W. et al. (eds): *Mucosal Immunology*, 4th ed. Amsterdam, Elsevier Academic Press, 2015:287–324.

Wyatt, R.J., and Julian, B.A.: IgA nephropathy. *N. Engl. J. Med.* 2013, 368:2402–2414.

Mucosal manifestations of immunodeficiencies

33

SCOTT SNAPPER, JODIE OUAHED, AND LUIGI D. NOTARANGELO

For many years it has been considered almost axiomatic that the innate and acquired immune system must collectively defend against ingested and/or inhaled toxic foreign substances, and the ever-present potential assault of microbial pathogens, while simultaneously coexisting with the luminal commensal microflora. A critical role(s) for the innate and adaptive immune system in maintaining mucosal homeostasis has become evident through the study of humans with naturally occurring primary immunodeficiencies that result from dominant mutations in immune-related genes, and of murine strains with similar genetically defined mutations. Gastrointestinal pathology is a common clinical feature of many human primary immunodeficiencies that are inherited as simple Mendelian traits, with attendant implications for common polygenic disorders such as inflammatory bowel disease (IBD), celiac disease, and asthma. In order to highlight the specific roles of unique aspects of the mucosal immune system in maintaining mucosal homeostasis, this chapter focuses on those human immunodeficiencies where gastrointestinal manifestations usually occur. For this discussion, immunodeficiencies are categorized based on those that present primarily as (1) antibody deficiencies; (2) combined (B-cell and T-cell) immunodeficiencies; (3) immunodeficiencies with immune dysregulation; and (4) immunodeficiencies with defects in innate immune cells (Table 33.1). Since recent large genome-wide association studies and other mapping approaches have implicated causative polymorphisms in several immune-related genes in the etiology of IBD, celiac disease, and asthma, an understanding of the specific role of the immune system in regulating mucosal immune function has acquired a much broader significance.

ANTIBODY DEFICIENCIES

As discussed in Chapter 10, the presence and secretion of immunoglobulins are important to the maintenance of mucosal homeostasis and the proper architecture of microbial commensalism and defense against invading pathogens. It is not surprising, therefore, that the dysfunction in the ability to properly regulate the production of immunoglobulins represents an important class of immunodeficiencies. Such immunodeficiencies result from either primary defects in B cells or defects in T-cell regulation of B-cell function.

33.1 Congenital agammaglobulinemia, or X-linked agammaglobulinemia, is the prototypic disorder of genetically determined antibody deficiency

X-linked agammaglobulinemia (XLA) results from mutations of the Bruton's tyrosine kinase (*BTK*) gene. *BTK* is located on the X chromosome and

TABLE 33.1 PRIMARY IMMUNODEFICIENCIES THAT MOST COMMONLY PRESENT WITH MUCOSAL MANIFESTATIONS

Antibody deficiencies	Combined or primary T-cell deficiencies	Immunodeficiency with immune dysregulation[a]	Immunodeficiencies with defects in innate immune cells
X-linked agammaglobulinemia (XLA)	Severe combined immunodeficiency (SCID)	Wiskott-Aldrich syndrome (WAS)	Chronic granulomatous disease (CGD)
Common variable immunodeficiency[b] (CVID)	CD40 ligand (CD40L) or CD40 deficiency	Immune dysregulation, polyendocrinopathy, enteropathy, X-linked (IPEX) syndrome	NEMO[c] (NF-κB essential modulator) deficiency or X-linked immunodeficiency with ectodermal dystrophy
IgA deficiency		IL-10R/IL-10 deficiency IL-2Rα (CD25) deficiency	
		LPS responsive beige-like anchor protein (LRBA) and cytotoxic T-lymphocyte-associated antigen 4 (CTLA4) deficiencies	

[a] WAS and interleukin (IL)-10R/IL-10 deficiency affect both innate and adaptive immune functions.
[b] T-cell dysfunction can also be an associated feature.
[c] NEMO also affects adaptive immune cell function.

regulates signaling through the pre-B-cell receptor (pre-BCR) and the BCR. BTK deficiency results in a block at the pro-B to pre-B-cell stage in B-cell differentiation in the bone marrow. Accordingly, patients with XLA have a severe reduction or absence of circulating B cells, associated with profound deficiency of all immunoglobulin isotypes. A similar phenotype can also be observed in patients with autosomal recessive forms of agammaglobulinemia, due to mutations of the mu (μ) heavy-chain gene; of the Igα, Igβ, and V pre-B components of the pre-BCR; or of the adaptor molecule B-cell linker protein, which is also involved in pre-BCR-mediated signaling.

The clinical phenotype of congenital agammaglobulinemia is mainly characterized by recurrent sinopulmonary infections that typically appear after 4 months of age, when levels of maternally derived IgG decrease, suggesting that among the immunoglobulin classes most associated with disease, the IgG class may be most relevant to the clinical manifestations of disease. The diagnosis is based on demonstration of pan-hypogammaglobulinemia and severe reduction (less than 0.1% of the normal level of immunoglobulins) in the number of circulating B lymphocytes. Treatment consists of regular administration of intravenous or subcutaneous immunoglobulins, possibly associated with antimicrobial prophylaxis. Administration of live viral vaccines should be avoided.

Gastrointestinal infections are relatively common in patients with XLA and other forms of congenital agammaglobulinemia, although their frequency is lower compared with other forms of antibody deficiency that are also characterized by concomitant defective T-cell function, such as common variable immunodeficiency and combined immunodeficiency. Common causes of gastrointestinal infection and the resulting chronic diarrhea in patients with agammaglobulinemia include *Giardia lamblia*, *Salmonella* spp., *Campylobacter* spp., and rotavirus. In addition, enteroviral infections (Coxsackie virus, echovirus) may disseminate to the central nervous system and cause life-threatening encephalitis. Intestinal biopsies show lack of plasma cells in the lamina propria and no germinal centers in gut-associated lymphoid tissue, consistent with the block in B-cell differentiation.

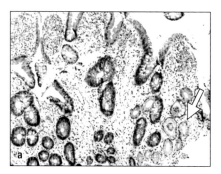
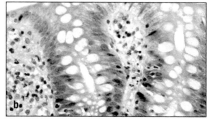

Figure 33.1 Intestinal inflammation in X-linked agammaglobulinemia (XLA). (**a**) Chronic changes and pyloric gland metaplasia (white arrow) in terminal ileum, ×100. (**b**) The absence of plasma cells and mild acute inflammation in colonic lamina propria, ×400. (Courtesy of Thomas Walker and Jeffrey Goldsmith.)

Small bowel Crohn's-like disease can occur in patients with XLA (Figure 33.1). In one recent survey, IBD-like disorders may occur in up to 10% of XLA patients. Nonetheless, since the majority of patients do not have an IBD-like disorder, primary humoral abnormalities alone are unlikely to explain IBD pathogenesis. Gastric adenocarcinoma and colorectal cancer may be more common in XLA than in the general population. A possible role for increased frequency of *Helicobacter pylori* infection has been hypothesized in the pathogenesis of the gastric tumors.

33.2 Common variable immunodeficiency (CVID) is the most common form of clinically significant primary immunodeficiency

CVID is defined by reduced levels of one or more immunoglobulin isotypes and impaired antibody production in response to immunization with antigens or to natural infections. It has been proposed to limit the diagnosis of CVID to individuals older than 4 years of age, with the idea that monogenic disorders of antibody production should be considered at a younger age. Patients with CVID are prone to recurrent respiratory infections, most often caused by *Haemophilus influenzae* and *Streptococcus pneumoniae*. In addition, autoimmune manifestations (cytopenias, inflammatory bowel disease), granulomas, lymphoid hyperplasia, and tumors (especially lymphoma) are also common.

The majority of patients with CVID have a normal number of B lymphocytes, but switched memory (CD27$^+$IgD$^-$) B lymphocytes are often significantly reduced, and there is a low rate of somatic hypermutation of immunoglobulin genes. Plasma cells are absent on histological examination of lymphoid tissue and bone marrow. Other significant laboratory abnormalities in patients with CVID include defects of T-cell activation and of cytokine production, increased cellular apoptosis, and impaired B-cell activation following ligation of endosomal toll-like receptors.

Although most cases of CVID represent sporadic forms of presentation, pedigrees consistent with autosomal dominant or autosomal recessive inheritance have also been described (Figure 33.2). Association with the human leukocyte antigen (HLA) region has been reported; however, the underlying gene defect has not been identified in most subjects with these disorders. Mutations of the transmembrane activator and calcium-modulator and cyclophilin ligand interactor (TACI) have been identified in approximately 10% of patients, most often in heterozygosity. TACI is expressed by B lymphocytes and interacts with the B-cell activating factor (BAFF) and with a proliferation inducing ligand (APRIL). In particular, APRIL and TACI interaction promotes B-cell activation and immunoglobulin class-switch recombination. Disruption of TACI in mice leads to lymphoid proliferation, resembling lymphoid hyperplasia that is often seen in humans with CVID. However, the significance of TACI mutations in CVID pathogenesis remains unclear; with few exceptions, it seems that they may represent susceptibility, rather than disease-causing mutations. A similar significance of CVID-predisposing factors has also been attributed to mutations of BAFF receptor (BAFF-R) and CD20, reported in a few patients with CVID. In contrast, a clear causal role has been demonstrated for mutations of ICOS, CD19, and CD81. ICOS is a costimulatory molecule, expressed by T cells, that interacts with ICOS-ligand (ICOS-L) expressed by B lymphocytes. CD19 and CD81

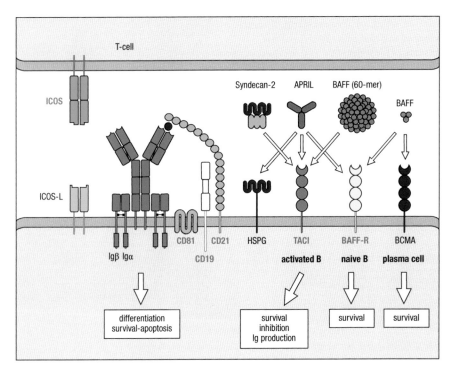

Figure 33.2 Genetic mutations associated with common variable immunodeficiency (CVID). Depicted are proteins involved in B-cell survival and differentiation in peripheral lymphoid organs. CD21, CD19, and CD81 are part of a macromolecular complex that associates with B-cell receptor (BCR) and contributes to intracellular signaling. TACI and BAFF-R are expressed by mature B lymphocytes and share binding to APRIL and BAFF. ICOS is a cell-surface molecule expressed by activated T cells and binds to ICOS ligand (ICOS-L) expressed by B lymphocytes. Molecules identified in blue have been associated with CVID. APRIL, a proliferation-inducing ligand; BAFF, B-cell-activating factor belonging to the tumor necrosis factor family; BAFF-R, BAFF receptor; BCMA, B-cell maturation antigen; HSPG, heparan sulfate proteoglycans; ICOS, inducible T-cell costimulator; Ig, immunoglobulin; TACI, transmembrane activator and calcium modulator cyclophilin ligand interactor.

form a cell-surface protein complex that participates in BCR-mediated signaling. Patients with CVID are highly prone to a variety of gastrointestinal manifestations (Table 33.2), including infections, autoimmunity, inflammation, and tumors. In addition, abnormalities of liver function have been reported.

Chronic diarrhea has been reported in 20%–60% of CVID patients. It has been suggested that diarrhea is more common in CVID patients with undetectable IgA, arguing for a protective role of mucosal IgA antibodies. Consistent with this observation, diarrhea does not usually improve upon immunoglobulin substitution therapy that does not replace IgA. The protozoan parasite *Giardia lamblia* is a frequent cause of diarrhea in CVID. Upon ingestion of the protozoan cysts, the trophozoites may diffusely colonize the mucosa of the small intestine, causing abdominal cramps, profuse diarrhea, and malabsorption. Identification of the pathogen is based on culture and examination of the stool but often requires duodenal aspiration and biopsies. Prolonged treatment with metronidazole is often required, but relapses

TABLE 33.2 GASTROINTESTINAL MANIFESTATIONS OF COMMON VARIABLE IMMUNODEFICIENCIES

Infections	Autoimmune complications	Malignancies
Giardia lamblia	Diarrhea/malabsorption	Gastric adenocarcinoma
Salmonella	Atrophic gastritis	Intestinal lymphoma
Campylobacter	Pernicious anemia	Colonic adenocarcinoma
Cryptosporidium	Small and/or large bowel inflammation	
Cytomegalovirus	Nodular lymphoid hyperplasia	
Hepatitis C virus	Hepatitis	
	Nodular regenerative hyperplasia	

are common. Other common pathogens seen in CVID include *Salmonella*, *Campylobacter*, cytomegalovirus, and *Cryptosporidium*.

Chronic gastritis is seen in one-third of patients and is associated with *H. pylori* infection that may also predispose to gastric adenocarcinoma, a malignancy with increased frequency in CVID patients. Atrophic gastritis with achlorhydria and pernicious anemia is also relatively common in CVID. Autoantibodies are not detected, but there is significant lymphocytic infiltration in the gastric biopsies, suggesting an immune-mediated pathogenesis.

Diarrhea and malabsorption may also reflect immune-mediated damage to the small intestine, with flattening of villi and crypt hyperplasia. This complication is seen in up to 50% of patients and may lead to significant weight loss, hypoproteinemia, and anemia. CD4 lymphopenia is also frequent. While these features are reminiscent of celiac disease, several pathologic elements are distinctive. In particular, plasma cells are absent in intestinal biopsies from CVID patients, whereas these are increased in patients with celiac disease. Furthermore, patients with CVID typically lack antibodies to gliadin and tissue transglutaminase (tTG), which are common biomarkers of celiac disease. Finally, introduction of a gluten-free diet is not beneficial in patients with CVID. Treatment of the diarrhea and malabsorption associated with CVID is often challenging, and is based on correction of metabolic disturbances, generous support of water-soluble vitamins, and the use of antisecretory agents. However, use of elemental diets and of parenteral nutrition may be necessary in severe cases. Judicious attempts with low-dose steroids are often considered.

IBD-like sequelae have been reported in up to 15% of CVID patients and include both Crohn's disease– and ulcerative colitis–like forms. Clinical manifestations include diarrhea, abdominal pain, and rectal bleeding. Duodenal and colonic biopsies demonstrate villous flattening and lymphocytic infiltration of the lamina propria, respectively (Figure 33.3). Apoptosis of epithelial cells is often prominent. In contrast to patients with Crohn's disease, there is often no increase in interleukin (IL)-17, IL-23, and tumor necrosis factor-α (TNF-α) production in the mucosa (but rather an increase in IL-12 and interferon-γ). Whether CVID and IBD have an overlapping genetic and environmental basis is unknown. Despite these distinct pathological features, treatment for the IBD-like disorder associated with CVID is similar to that used for Crohn's disease and ulcerative colitis and includes anti-inflammatory agents (e.g., 5-amino-salicylic acid compounds as well as topical and systemic steroids), immunomodulators (e.g., azathioprine, 6-mercaptopurine), and antibiotics (e.g., ciprofloxacin, metronidazole, and rifaximin). Beneficial effects have been reported with TN-α-blocking agents, such as infliximab, but judicious surveillance for mycobacterial, fungal, and other infections is required.

Intestinal nodular lymphoid hyperplasia (NLH) is seen in up to 10% of patients with CVID in association with villous flattening and diarrhea, which may be associated with malabsorption; however, when nodules are particularly prominent, they may also cause bowel obstruction. Histologically, large lymphoid follicles are found with germinal centers and an increase in B lymphocytes, but plasma cells are absent. It remains unclear whether NLH represents a lesion that precedes development of intestinal lymphoma. In any case, patients with CVID are at greater risk for non-Hodgkin's lymphoma of the intestine.

Finally, persistent elevation of liver enzymes can be seen in more than 50% of CVID patients. Most patients are found to have cholestasis without jaundice and, when liver biopsies have been available, have been found to

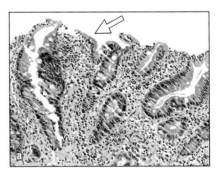

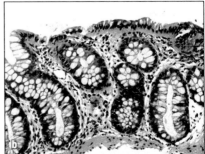

Figure 33.3 **Intestinal findings in common variable immunodeficiency (CVID).** (**a**) Villous atrophy and acute inflammation (white arrow) in the duodenum, ×100. (**b**) The absence of plasma cells in the colonic lamina propria. (Courtesy of Athos Bousvaros and Jeffrey Goldsmith.)

exhibit a nodular regenerative hyperplasia. Nodular regenerative hyperplasia is characterized by the presence of regenerating liver cell plates alternating with atrophic areas with no significant portal fibrosis, bridging fibrosis, or definite cirrhosis. Noncaseating granulomas are often an associated feature. Portal vein endotheliitis, intrasinusoidal inflammation with prominent T-cell infiltration, and portal hypertension in the absence of cirrhosis are additional common features. The presence of nodular regenerative hyperplasia has been correlated with other autoimmune phenomena, and it has been suggested that these liver modifications may represent the consequence of chronic gastrointestinal inflammation with excessive translocation of luminal antigens into the liver. Severe liver disease has been reported in the past in CVID patients who had acquired hepatitis C virus (HCV) infection through contaminated blood products (including immunoglobulin preparations). While this risk has been virtually eliminated with improvement in screening of blood products and the preparation of immunoglobulins, it appears that progression of liver disease is faster and more severe in patients with CVID who are infected with hepatotropic viruses.

33.3 IgA deficiency is the most common primary immunodeficiency

IgA deficiency is the most prevalent of the primary immunodeficiencies, affecting approximately 1 in 700 individuals of European ancestry and significantly fewer of those of Asian ancestry. Both partial and complete forms of IgA deficiency are known. Association with certain HLA alleles such as HLA-DR1 and a higher frequency within families with CVID and celiac disease have been reported, but the pathophysiology remains unknown. Mucosal IgA has recently been shown to target commensal microorganisms and may have a critical role in regulating microbial entry to mucosal compartments. Secretory IgA represents the most abundant mucosal immunoglobulin isotype. However, approximately two-thirds of IgA-deficient individuals are asymptomatic, while the remaining may suffer from recurrent infections, autoimmunity, or allergy. Respiratory and intestinal infections are more common when IgA deficiency is associated with deficiencies of some IgG subclasses. It is thought that IgM may compensate for the lack of IgA at the mucosal surface, because it can be efficiently transported by the polymeric immunoglobulin receptor binding to the J-chain of IgM.

Approximately one-third of adult individuals with IgA deficiency suffer from recurrent infections of the respiratory tract. Most of these infections are of viral origin and tend to be clinically mild. An increased incidence of pseudocroup (due to parainfluenza virus) has been observed in infants with IgA deficiency. Bacterial infections are less common and are predominantly due to *Haemophilus influenzae* and *Streptococcus pneumoniae*. Bronchiectasis and chronic lung damage have been rarely observed.

Gastrointestinal manifestations of IgA deficiency are similar to those of patients with CVID, but their frequency is lower. Lack of secretory IgA is often associated with giardiasis, resulting in diarrhea with steatorrhea and malabsorption. Patients with IgA deficiency have a 10- to 20-fold increased risk of celiac disease, with typical villous flattening and an excellent response to a gluten-free diet. The biochemical diagnosis may be challenging, because of the lack of anti-tTG IgA antibodies; however, anti-tTG IgG antibodies are elevated. Interestingly, both IgA deficiency and celiac disease are associated with the extended haplotype HLA-A1, Cw7, B8, DR3, DQ2.

NLH has been also reported in IgA deficiency, but it is far less common than in patients with CVID. An important distinction is that the NLH of patients with IgA deficiency is characterized by accumulation of IgM-expressing B cells, possibly a compensatory mechanism for the lack of IgA in response to chronic stimulation. Otherwise, the clinical symptoms are the same as reported for NLH in CVID, and include diarrhea, abdominal pain, and even obstruction. NLH is also uniquely sensitive to oral steroids. Whenever an underlying infection is identified, treatment should be targeted to the specific pathogen. Lymphoma and gastric adenocarcinoma have also been reported in patients with IgA deficiency. But IBD-like disease, pernicious anemia, and chronic gastritis are far less common in IgA deficiency than in CVID. Persistent elevations of liver enzymes and biliary cirrhosis have been described in IgA deficiency.

Combined immunodeficiencies

Combined immunodeficiencies are those disorders that are genetically determined and lead to loss of both B- and T-cell function. As such, afflicted individuals exhibit a much more profound immune defect along their mucosal surfaces and are thus susceptible to a much broader combination of pathogens and associated clinical problems.

33.4 Severe combined immunodeficiencies (SCIDs) are a heterogeneous group of genetically determined disorders

The diseases associated with SCID are characterized by profound defects in the development and function of T lymphocytes and are variably associated with impaired development of B and/or natural killer (NK) lymphocytes. SCID occurs in approximately 1 in 50,000–100,000 live births and is more common in males, reflecting the fact that X-linked SCID is the most common form of SCID in humans. SCIDs are usually classified according to the immunologic phenotype and are distinguished into (1) SCID with absence of T lymphocytes but presence of B lymphocytes ($T^- B^+$ SCID); and (2) SCID with absence of both T and B lymphocytes ($T^- B^-$ SCID). These distinct immunologic phenotypes correspond to an even greater degree of genetic heterogeneity (Table 33.3). In addition, atypical forms of the disease exist, in which development of T lymphocytes is severely compromised, but not completely abrogated. In these "leaky" variants, circulating T lymphocytes show an activated (CD45RO) phenotype, often display an oligoclonal T-cell receptor repertoire, and may infiltrate peripheral tissues, causing severe inflammatory manifestations. The prototype of such conditions is represented by Omenn syndrome. Patients with this variant present with significant skin rash and lymphadenopathy and are prone to autoimmune manifestations. Another condition in which circulating T lymphocytes may be detected in patients with SCID is represented by transplacental passage of maternally derived T cells, that occurs in as many as 50% of SCID infants and may cause graft-versus-host-like manifestations, including skin rash, severe diarrhea, elevation of liver enzymes, and bone marrow failure.

Irrespective of the nature of the genetic defect, patients with SCID present early in life with severe infections sustained by bacterial, viral, and fungal pathogens. Opportunistic respiratory infections (interstitial pneumonia due to *Pneumocystis jiroveci*, cytomegalovirus, adenovirus, and parainfluenza virus type 3) are particularly common. Infections of the gastrointestinal tract are a frequent cause of chronic diarrhea, leading to failure to thrive. Skin rashes may be present, especially in infants with maternal T-cell engraftment or with atypical variants of the disease. Additional clinical features may be present in specific forms of SCID. For example, bone abnormalities are often observed in patients with adenosine deaminase deficiency, whereas microcephaly and growth failure are common in infants with SCID due to defects in genes encoding for proteins (DNA ligase IV, Cernunnos) involved in DNA repair.

The diagnosis of SCID is based on typical clinical and laboratory features. Severe lymphopenia is the hallmark of the disease. The number of circulating T lymphocytes is severely decreased in 85%–90% of the patients, the exceptions

TABLE 33.3 GENETIC ABNORMALITIES ASSOCIATED WITH SEVERE COMBINED IMMUNODEFICIENCY

Cellular mechanism affected	Defective protein (gene)	Immunologic phenotype	Inheritance
Lymphoid cell survival	Adenosine deaminase (*ADA*)	T⁻ B⁻ NK⁻ SCID	AR
	Adenylate kinase (*AK2*)	T⁻ B^low NK⁻ SCID	AR
Cytokine-mediated signaling	Common γ chain (*IL2RG*)	T⁻ B⁺ NK⁻ SCID	XL
	Interleukin-7 receptor α chain (*IL7R*)	T⁻ B⁺ NK⁺ SCID	AR
	JAK3 (*JAK3*)	T⁻ B⁺ NK⁻ SCID	AR
V(D)J recombination	RAG proteins (*RAG1, RAG2*)	T⁻ B⁻ NK⁺ SCID	AR
	Artemis (*DCLRE1C*)	T⁻ B⁻ NK⁺ SCID	AR
	DNA-PKcs (*PRKDC*)	T⁻ B⁻ NK⁺ SCID	AR
	DNA ligase IV (*LIG4*)	T⁻ B⁻ NK⁺ SCID*	AR
	Cernunnos (*NHEJ1*)	T^low B^low NK⁺ SCID	AR
Signaling through pre-TCR and pre-BCR	CD3δ, ε, and ζ chains (*CD3D, CD3E, CD3Z*)	T⁻ B⁺ NK⁺ SCID	AR
	CD45 (*CD45*)	T⁻ B^{+/low} NK⁺ SCID	AR

Abbreviations: AR: autosomal recessive; BCR: B-cell receptor; DNA-PKcs: DNA protein kinase catalytic subunit; JAK: Janus-associated kinase; RAG: recombination-activating gene; TCR: T-cell receptor. XL: X-linked. *DNA ligase IV associates with Cernunnos.

being due to maternal T-cell engraftment or to atypical forms of SCID. When present, T cells are unable to proliferate *in vitro* in response to mitogens. Patients with SCID fail to produce specific antibodies upon immunization. However, total IgG serum levels may initially be within the normal range, reflecting transplacental passage of maternal IgG.

Unless immune reconstitution is provided by treatment, infants with SCID typically die at about 1–2 years of age, mostly due to infections. Recognition of SCID should prompt immediate and aggressive treatment of infections and use of broad antimicrobial prophylaxis, as well as immunoglobulin replacement therapy. Nutritional support is often necessary. The use of live virus immunizations should be strictly avoided, and all blood products should be irradiated and tested for cytomegalovirus before transfusion. Ultimately, treatment and immune reconstitution are based on hematopoietic cell transplantation (HCT). Survival is excellent (greater than 90%) when HCT is performed from HLA-identical donors; however, this option is available only to a minority (less than 15%) of patients. Similarly, good results are obtained when transplantation from HLA-mismatched related donors is performed early in life. Enzyme replacement therapy can be used in patients with adenosine deaminase deficiency, and gene therapy has shown promising results in patients with adenosine deaminase deficiency and with X-linked SCID, although leukemic proliferation due to insertional mutagenesis has been observed in several patients with X-linked SCID following gene transfer.

Persistent oral candidiasis and chronic diarrhea are common findings in patients with SCID. Gastrointestinal infections may be sustained by a variety of microorganisms, including viruses (enteroviruses, but also adenovirus and cytomegalovirus), bacteria (*Salmonella* spp., *Enterococcus*), candida, and protozoa (*G. lamblia, Cryptosporidium parvum*). In addition, rotavirus vaccine (that has recently become available) may cause severe diarrhea and failure to thrive in infants with SCID. Enteroviral infections in SCID infants may lead to further complications, including encephalitis.

In patients with Omenn syndrome, leaky SCID, or SCID with maternal T-cell engraftment, diarrhea may be particularly pronounced and lead to significant protein loss and generalized edema. Moreover, SCID patients can present with clinical and pathologic findings similar to IBD, including colonic inflammation and fistulae. Pathologic examination of small bowel biopsies of SCID patients frequently reveals prominent infiltrates with T lymphocytes and eosinophils, with villous flattening. Careful search for pathogens causing diarrhea, aggressive treatment of the infection (when possible), and nutritional support (often with use of parenteral nutrition) are the mainstay of treatment of diarrhea in infants with SCID. In addition, immunosuppression with steroids or cyclosporine A is often necessary to control the inflammatory reactions in the gut of infants with Omenn syndrome. The association of intestinal inflammation with Omenn syndrome implicates a critical role for T-cell regulation in mucosal homeostasis.

Elevation of liver enzymes is also common in infants with SCID and may reflect chronic infection in the intestine or in the liver. Inflammatory infiltrates of variable severity, with a predominance of mononuclear cells, may be observed in the lobules and in the periportal spaces in patients with Omenn syndrome or with SCID associated with maternal T-cell engraftment. Accumulation of toxic phosphorylated derivatives of adenosine and deoxyadenosine may cause increased apoptosis of hepatocytes in patients with adenosine deaminase deficiency.

33.5 CD40 ligand (CD40L) deficiency is a disorder with broad immunologic consequences due to the wide range of cell types that express its receptor, CD40

CD40 ligand (CD40L, CD154) is a cell-surface molecule predominantly expressed by activated CD4$^+$ T lymphocytes. Interaction of CD40L with its counter-receptor CD40 on the surface of B cells is essential for germinal center formation and class-switch recombination. Furthermore, CD40 is also expressed on dendritic cells, macrophages, and activated endothelial and epithelial cells. Interaction of CD40L-expressing CD4$^+$ T cells with these cell types promotes B- and T-cell priming triggering protective responses against intracellular pathogens. Mutations in the *CD40LG* gene, mapping at Xq26, result in X-linked hyper-IgM syndrome (also known as type 1 hyper-IgM syndrome), a combined immunodeficiency characterized by an increased occurrence of bacterial and opportunistic infections, neutropenia, a high incidence of liver and biliary tract disease, increased risk of malignancies, and a high mortality rate.

Patients with CD40L deficiency show undetectable or very low levels of serum IgG and IgA, and normal to increased levels of IgM. Chronic or intermittent neutropenia is common and is typically associated with a block in myeloid differentiation at the promyelocyte stage. Definitive diagnosis is based on the demonstration of defective expression of CD40L on the surface of T cells following *in vitro* activation and is eventually confirmed by mutation analysis.

Regular immunoglobulin replacement therapy, prophylactic trimethoprim-sulfamethoxazole, and hygiene measures represent the mainstay of therapy. Severe neutropenia may be treated with recombinant granulocyte colony-stimulating factor. In spite of these measures, the long-term prognosis is poor because of severe infections and liver disease. The only curative approach is represented by HCT, and better results are achieved when transplantation is performed prior to development of lung problems or of *Cryptosporidium parvum* infection.

Protracted or chronic diarrhea is common in patients with CD40L deficiency and is often due to *C. parvum* (leading to watery diarrhea), *G.*

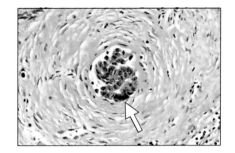

Figure 33.4 Sclerosing cholangitis in CD40L deficiency. Liver biopsy from a patient with CD40L deficiency and *Cryptosporidium parvum* infection, showing bile duct obliteration (white arrow) and periductular sclerosis; ×40. (Courtesy of Fabio Facchetti.)

lamblia, and *Salmonella* spp. infections. In a European series, the frequency of *Cryptosporidium* infection was as high as 70%, but lower figures have been reported in the United States. Both *Cryptosporidium* and cytomegalovirus infection may lead to sclerosing cholangitis in patients with CD40L deficiency (Figure 33.4). This severe complication is more common in European than in North American patients, presumably as a result of a different frequency of *C. parvum* infection in these populations. In any case, sclerosing cholangitis is a common cause of liver failure and death in patients with CD40L deficiency. The unique susceptibility to *Cryptosporidium*-related sclerosing cholangitis is also replicated in the mouse model of CD40L deficiency. It has been hypothesized that when infected with *C. parvum* or cytomegalovirus, bile duct cells may express CD40, and that CD40L-mediated signaling may induce apoptosis and block replication of the pathogen. Alternatively, because increased susceptibility to *C. parvum* infection is also seen in interferon-gamma (IFN-γ)-deficient mice, it is possible that CD40L deficiency may result in poor induction of IFN-γ and, hence, impaired clearance of the pathogen. Persistent infection of the bile ducts may eventually cause dysplasia and neoplastic transformation.

Sclerosing cholangitis is often associated with dilated bile ducts. Careful monitoring of infections and of the liver and biliary tract should be included in the management of patients with CD40L deficiency. Search for *C. parvum* should be restricted to detection of oocysts in the stool but should also be based on molecular detection by polymerase chain reaction. Approaches to cholangiography include endoscopic retrograde cholangiopancreatography (ERCP) and percutaneous transhepatic cholangiography (PTC). These also allow collection of bile fluid for microbiological and molecular analysis. Magnetic resonance cholangiopancreatography can also be used to visualize bile ducts, although it is not as sensitive as ERCP and PTC and may miss minor abnormalities of the third- and fourth-order intrahepatic ducts. It does not carry the risk of ascending infection that is potentially associated with ERCP and PTC. Finally, liver histology is not as sensitive as cholangiography in detecting sclerosing cholangitis.

Patients with CD40L deficiency are at increased risk of malignancies that often involve the gastrointestinal tract. Cholangiocarcinoma, hepatocellular carcinoma, peripheral neuroectodermal tumors, and lymphoma have been frequently reported. Epithelial tumors and dysplasia of the liver and bile ducts in patients with CD40L deficiency are often associated with sclerosing cholangitis, and with infections due to *Cryptosporidium*, cytomegalovirus, and hepatitis B or C viruses. Sclerosing cholangitis and tumors of the gastrointestinal tract are associated with poor outcomes in patients with CD40L deficiency. In the absence of immune reconstitution, attempts to treat *Cryptosporidium*-related sclerosing cholangitis with liver transplantation are usually not successful in patients with CD40L deficiency, due to reinfection and relapse of the liver disease. Successful outcome has been reported following combined liver transplantation and HCT in some patients, but even using this approach, disseminated *C. parvum* infection and death may occur before immune reconstitution is achieved.

A variety of therapeutic and preventive approaches to *Cryptosporidium* infection have been proposed in patients with CD40L deficiency, including use of paromomycin, azithromycin, and nitazoxanide, but none of them are supported by controlled studies. However, azithromycin and standard hygiene measures are frequently recommended. Ursodeoxycholic acid may promote bile flow and reduce the risk of ascending infections in the biliary tree.

Although protracted diarrhea in patients with CD40L deficiency is typically due to infections, inflammatory bowel disease and intestinal nodular lymphoid hyperplasia have been also reported. Finally, patients with CD40L deficiency frequently suffer from oral and perirectal ulcers. These complications are typically associated with the neutropenia that is frequently seen in this condition.

33.6 CD40 deficiency is a disorder that reproduces clinical phenotype of CD40L deficiency

CD40 deficiency is inherited as an autosomal recessive trait and has been reported in only few patients, most of whom were of Turkish or Middle-Eastern origin. The clinical and laboratory manifestations of the disease are indistinguishable from those observed in patients with CD40L deficiency, including a higher risk of sclerosing cholangitis, *Cryptosporidium* infection, and oral ulcers associated with chronic neutropenia. HCT may provide permanent cure.

33.7 Major histocompatibility complex class II deficiency leads to extensive abnormalities in humoral and cellular adaptive immunity

MHC-II deficiency is a rare combined immunodeficiency characterized by markedly reduced or absent expression of MHC-II antigens, severe reduction of the number of circulating CD4$^+$ T lymphocytes, impaired antibody production (frequently associated with hypogammaglobulinemia), and defective cell-mediated immune responses. Patients with MHC-II deficiency are highly prone to bacterial, viral, and opportunistic infections. The disease is more common in selected populations (Northern Africa, Arabic countries). It is inherited as an autosomal recessive trait and is due to mutations in any of four genes that encode for transcription factors that regulate MHC-II antigen expression (e.g., class II transcriptional activator). Treatment is based on HCT, but results are far less satisfactory than in other cases of combined immunodeficiency.

Intestinal infections due to *Salmonella* spp., *E. coli*, *Proteus*, enteroviruses, cytomegalovirus, adenovirus, *Giardia*, *Cryptosporidium*, and *Candida* occur in nearly all patients with MHC-II deficiency and are a frequent cause of chronic diarrhea, malabsorption, and failure to thrive. Histological examination shows various degrees of villous atrophy and increased intraepithelial lymphocytes, with a reduction of IgA plasmocytes.

Liver involvement is very common. Approximately one-third of patients develop biliary tract disease. Sclerosing cholangitis is frequently associated with *Cryptosporidium* or cytomegalovirus infection. Bacterial cholangitis has also been frequently described. The risk of biliary tract disease seems to increase with age. Viral infections (especially cytomegalovirus and adenovirus) are a cause of persistent elevation of liver enzymes. Drug-related toxicity and total parenteral nutrition may also lead to impaired liver function.

33.8 Deficiency of a nuclear protein, SP110, of unknown function leads to hepatic veno-occlusive disease with immunodeficiency

Veno-occlusive disease with immunodeficiency is an autosomal recessive, rare disorder characterized by profound immunodeficiency, liver fibrosis and veno-occlusive disease. Patients present with severe symptoms within the first year of life. The disease has been reported predominantly in patients of Lebanese origin and is due to mutations of the *SP110* gene that encodes for a nuclear protein of unclear function. Patients show profound hypogammaglobulinemia and a lack of germinal centers in the lymph nodes. T lymphocytes are present in normal numbers, but opportunistic infections (*P. jiroveci* pneumonia, cytomegalovirus infection, persistent candidiasis) are observed in most patients, indicative of a severe defect in cell-mediated immunity. Hepatomegaly and abnormalities of liver enzymes are often the first signs of severe liver disease that may progress to veno-occlusive disease and terminal liver failure. Liver biopsy shows sinusoidal congestion or obstruction and narrowing of terminal hepatic venules. Ultrasonography demonstrates hepatosplenomegaly, increased portal vein diameter, reduced hepatic vein diameter, ascites, and recanalization of the *ligamentum teres* due to shunting of blood flow from the liver. Doppler ultrasonography shows reduced portal venous flow and increased resistance in the hepatic artery. Prognosis is dismal, and most patients die in the first years of life.

Immunodeficiencies affecting regulatory factors and regulatory T cells

Regulatory T (T_{reg}) cells mediate suppression of immune responses in the periphery outside of the thymus. Generation of T_{reg} cells in the thymus is controlled by the transcription factor Foxp3. Foxp3-expressing T_{reg} cells can also be induced outside the thymus under the control of factors such as transforming growth factor-β (TGF-β) and retinoic acid ("induced T_{reg}"). T_{reg} cells, which are discussed in more detail in Chapter 9, secrete a variety of immunoregulatory factors and cytokines such as IL-10, IL-35, and TGF-β. Such responses are important in preventing autoimmunity and restricting excessive immune responses to pathogens, thus delimiting the extent of inflammation and consequent tissue injury.

33.9 Genetically mediated Foxp3 deficiency leads to the immune dysregulation, polyendocrinopathy, enteropathy, X-linked (IPEX) syndrome

Mutations of the *FOXP3* gene cause immune dysregulation, polyendocrinopathy, enteropathy, X-linked (IPEX) syndrome, which can present with a variety of presentations including: severe and early onset autoimmune enteropathy, insulin-dependent diabetes, and/or eczema. There is hyperactivation of T lymphocytes, associated with elevated IgE, eosinophilia, and autoantibodies. In typical cases, the disease evolves rapidly, unless treated by HCT.

Intractable diarrhea is a key element of IPEX and is present in virtually all cases with typical early presentation, often being severe enough to cause significant malabsorption and cachexia. Intestinal villous flattening and lymphoid infiltration can be seen in patients with IPEX (**Figure 33.5**). A recent

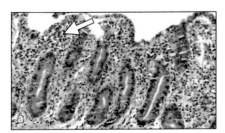
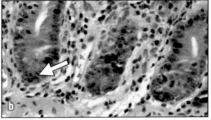

Figure 33.5 Autoimmune enteropathy in the duodenum of an immune dysregulation, polyendocrinopathy, and enteropathy (IPEX) patient. (a) Villous flattening and acute inflammation (white arrow) in the duodenum, ×100. (**b**) Lack of Paneth cells (white arrow) in the crypts, ×400. (Courtesy of Athos Bousvaros and Jeffrey Goldsmith.)

review of intestinal biopsies from a limited series of patients revealed that the majority exhibited abnormalities resembling acute graft-versus-host disease. Therapeutic strategies using elimination of presumed allergens from the diet were largely unsuccessful. Systemic autoimmunity with intestinal inflammation is also a hallmark of Foxp3-deficient mice.

33.10 Wiskott-Aldrich syndrome, due to deficiency in Wiskott-Aldrich syndrome protein, affects multiple lineages of hematopoietic cells and consequently diverse aspects of immune function

Wiskott-Aldrich syndrome (WAS) is an X-linked disorder characterized by eczema, congenital thrombocytopenia with small-sized platelets, and immune deficiency. The responsible gene, named *WAS*, maps at Xp11.2 and encodes for a protein involved in cytoskeleton reorganization in hematopoietic cells. Most patients with classical WAS have mutations that impair expression and/or function of Wiskott-Aldrich syndrome protein (WASp). However, some missense mutations are associated with a milder phenotype (isolated X-linked thrombocytopenia).

The immune deficiency of WAS leads to recurrent bacterial and viral infections, autoimmune manifestations, and increased occurrence of tumors (leukemia, lymphoma). Autoimmune or inflammatory sequelae are seen in 40%–70% of patients depending on the series, with a significant number of patients presenting with two or more autoimmune or inflammatory sequelae. The most common autoimmune disease or inflammatory manifestation includes hemolytic anemia, neutropenia, vasculitis, arthritis, IBD, and renal disease. Lymphopenia (particularly among $CD8^+$ T lymphocytes), impaired *in vitro* proliferation to immobilized anti-CD3, reduced serum IgM with increased levels of IgA and IgE, and inability to mount effective antibody responses, especially to T-independent antigens, are typical immunologic findings. The basis for the autoimmunity in WAS is still unclear but may represent aberrant effector T-cell, regulatory T-cell, NK cell, dendritic cell, macrophage, or other innate immune function.

Treatment of WAS is based on regular administration of intravenous immunoglobulins, antibiotic prophylaxis, topical steroids to control eczema, and use of vigorous immune suppression when autoimmune manifestations are present. Splenectomy may increase the platelet count in the majority of patients but carries significant risks of invasive infection by encapsulated pathogens. The only curative approach to WAS is represented by HCT, which gives optimal results when performed from HLA-identical family donors or early in life from matched unrelated donors. Gene therapy trials have been initiated with promising results.

WAS patients can present with life-threatening gastrointestinal bleeding associated with severe thrombocytopenia (over 25% of patients presenting with hematemesis or melena in one series). Chronic diarrhea is also a common feature of classical WAS, most often associated with infectious etiologies. Moreover, a significant fraction of patients (nearly 10% in one series) can present with an inflammatory disorder of the intestine resembling inflammatory bowel disease. These patients can present with inflammatory lesions anywhere in the gastrointestinal tract with disease resembling both Crohn's disease and ulcerative colitis. Rarely, patients can present with a clinical picture consistent with necrotizing enterocolitis. Similar to WAS patients, mice deficient in WASp can also develop severe colitis associated with helper-T (T_H) T_H2 cytokine skewing and defects in regulatory T cells.

33.11 Deficiency of IL-2Rα (CD25) chain has major effects on T_{reg} function

IL-2 signaling is essential to maintain peripheral immune homeostasis. Deficiency of the α chain of the IL-2 receptor (IL-2Rα, CD25) is inherited as an autosomal recessive trait and results in immune dysregulation and lymphoproliferation. Patients are at higher risk for early onset severe and recurrent viral and bacterial infections, oral thrush, and chronic diarrhea (that may lead to protein-losing enteropathy), associated with lymphadenopathy and hepatosplenomegaly. CD4[+] T-cell counts are low, and an *in vitro* proliferative response to mitogens is typically impaired. Treatment is based on HCT. Mice deficient in IL-2R (or IL-2) also develop severe intestinal inflammation that may result from defects in effector and/or regulatory T-cell homeostasis.

33.12 Genetically determined deficiency in the IL-10 receptor and IL-10 presents prominently with features associated with aberrant mucosal homeostasis

Rare cases of patients presenting with mutations in IL-10 or in the IL-10 receptor alpha or beta (IL-10Rα or IL-10Rβ) chains have been reported in individuals presenting with very early onset IBD. These patients have been found to also suffer from severe infections, folliculitis, and arthritis. Perianal abscesses, enteric fistula, and inflammation in the colon (**Figure 33.6**) and to a lesser extent in the small bowel have been described in such patients. Bone marrow transplant has been effective therapy in such patients. Patients with IL-10Rα or IL-10Rβ mutations have an increased risk of B-cell lymphomas in the absence of transplantation.

Genetic analyses have suggested an autosomal recessive inheritance pattern in the majority of patients. Mutations of IL-10Rα or IL-10Rβ abrogate interleukin-10-induced signaling, as shown by deficient STAT3 (signal transducer and activator of transcription 3) phosphorylation. Moreover, since IL-10-dependent inhibition of lipopolysaccharide-induced secretion of TNF-α and other pro-inflammatory cytokines is defective in monocytes from these patients, it has been proposed that a block in IL-10-dependent negative feedback is responsible for the inflammatory gastrointestinal disorder. IL-10 and IL-10R knockout mice also develop IBD. Numerous cell types express IL-10R, including innate immune cells, naive and regulatory T cells, and the IL-10Rβ chain can bind other cytokines, such as IL-22, IL-26, and IFN-λ; however, the specific cell type(s) that require IL-10 and IL-10R-dependent signaling and the role of these cytokines in maintaining mucosal homeostasis in these subjects remains unclear. Recent data suggest that patients with IL-10R-deficiency have enhanced IL-17A expression in mucosal compartments. Moreover, excessive IL-1β expression is associated with IL-10R-deficiency. In a small cohort of patients where bone marrow transplantation was not possible due to disease severity, IL-1 blockade was a successful bridge to transplantation.

Figure 33.6 Intestinal inflammation in IL-10RB deficiency. Lymphoplasmocytic infiltration of the colon; ×100. (Adapted from E.O. Glocker et al., *N. Engl. J. Med.* 2009, 361:2033–2045. With permission from Massachusetts Medical Society.)

33.13 LPS responsive beige-like anchor protein and cytotoxic T-lymphocyte-associated antigen 4 (CTLA4) deficiencies result in overlapping syndromes associated with autoimmunity

Mutations in LRBA and CTLA4 result in immune deficiencies with overlapping phenotypes since LRBA has been demonstrated to regulate CTLA4 by blocking endosomal trafficking of CTLA4 to lysosomes for degradation. CTLA4 is expressed on the surface of activated and regulatory T cells and is involved in blocking B7:CD28 interactions and delivering immunosuppressive signals in these cells. LRBA is a cytosolic protein expressed broadly and is involved in a variety of cellular processes including endosomal trafficking and cytoskeletal and transcriptional regulation. In addition to its role in regulating CTLA4 in T cells, LRBA regulates B-cell development and function. Patients with either CTLA4 or LRBA deficiencies present with a multitude of immunoregulatory dysfunctions including hypogammaglobulinemia, recurrent infections, organomegaly, enteropathy, neurologic involvement, and a variety of autoimmune diseases and malignancies. The most common autoimmune diseases include IBD, type I diabetes, polyarthritis, autoimmune thyroid disease, and Evan's syndrome. The most common malignancies include gastric cancer and lymphoma. Gastrointestinal involvement is common in both LRBA and CTLA4 deficiencies and can manifest as IBD-like inflammation, atrophic gastritis, celiac disease, acute pancreatitis, and pancreatic insufficiency. Currently, therapies for both LRBA and CTLA4 deficiencies focus on treatment of ongoing infections and treatment with either Abatacept (a CTLA4-Fc fusion protein) or mTOR inhibitors. In addition to therapy causing a reduction in autoimmune sequelae, responsive patients typically have a reduction in soluble CD25 and circulating follicular T-helper cells. HCT has been employed successfully for both CTLA4 and LRBA deficiency.

Immunodeficiencies affecting primarily innate immune cells

Both parenchymal cells, such as epithelial cells, and hematopoietic cells, such as polymorphonuclear cells (or phagocytes) that include neutrophils, eosinophils, and basophils, are essential components of innate immune function within mucosal tissues. Consequently, environmentally mediated (e.g., drug-induced neutropenia) or genetically determined (e.g., chronic granulomatous disease) dysfunction of these have important and deleterious consequences for both the maintenance of homeostasis and protection from pathogenic exposures.

33.14 Chronic granulomatous disease is an archetypal example of genetically specified disorder of phagocyte function

Phagocytes play a key role in the defense against bacteria and fungi; accordingly, patients with defects of phagocytic cell number and/or function exhibit significant problems at their mucosal surfaces. Chronic granulomatous disease (CGD) represents the prototype of defects associated with phagocyte function and is due to abnormalities in the NADPH oxidase complex. In 75% of the cases, CGD is inherited as an X-linked trait, due to mutations in the X-linked *CYBB* gene that encodes the gp91^{phox} component of NADPH oxidase. Autosomal recessive forms due to defects of the p22^{phox}, p47^{phox}, p67^{phox}, and p40^{phox} components are also known. The gp91^{phox} and p22^{phox} proteins are located in the phagosome membrane, where the cytosolic p47^{phox} and p67^{phox} isoforms are translocated to the membrane following phagocytic cell activation. Another subunit, p40^{phox}, plays an important role in phagocytosis-induced superoxide production. Assembly of the NADPH complex is regulated by two GTPases, Rac2 and Rap1, and defects of Rac2 have been identified in patients with CGD-like disease. Following phagocytosis, induction of the NADPH oxidase complex results in production of microbicidal compounds such as superoxide radicals and hydrogen peroxide and, even more importantly, activation of lytic enzymes (cathepsin G, elastase, myeloperoxidase) following potassium ion influx, ultimately permitting killing of bacteria and fungi. Patients with CGD suffer from recurrent and often severe infections of fungal or bacterial origin, leading to pneumonia, skin and perirectal abscesses (with frequent formation of fistulae), suppurative lymphadenitis, osteomyelitis, septicemia, and deep-seated abscesses in the liver and other organs. *Staphylococcus aureus, Serratia marcescens, Burkholderia cepacia*, and *Nocardia* spp. are responsible for most of the bacterial infections, whereas *Candida* spp. and *Aspergillus* spp. cause most of the fungal infections. Patients with CGD are also at higher risk for mycobacterial infections. The sustained inflammatory response observed in patients with CGD is also responsible for inflammatory manifestations (even in the absence of infection), presumably due to inappropriate responses to the commensal microbiota, that may cause obstruction in the gastrointestinal and/or urinary tract. Clinical symptoms tend to be more common, more severe, and occur earlier in life in patients with X-linked versus autosomal recessive forms of the disease. The diagnosis of CGD is based on clinical history and laboratory demonstration of impaired NADPH function, as shown by impaired production of dihydrorhodamine 123 (DHR123) by *in vitro* activated phagocytes in a flow-cytometry-based assay.

CGD carries a high mortality risk, especially due to infections. However, significant progress in the management of the disease now permits almost half of the patients to survive into the third decade. Treatment is based on prompt and aggressive investigation of any febrile episode in order to identify the responsible microorganism(s), and on continuous antimicrobial prophylaxis with trimethoprim-sulfamethoxazole (for bacterial infections) and itraconazole (for fungal infections). In addition, subcutaneous administration of IFN-γ has been shown to reduce occurrence of severe infections. However, the only definitive cure is represented by HCT, whose results are largely influenced by degree and duration of donor chimerism in the myeloid compartment, and by preexisting risk factors, such as lung aspergillosis. Results of HCT are better when the transplant is performed from an HLA-identical related donor. Use of a matched unrelated donor has been associated with increased risk of graft-versus-host disease, graft loss, and death. However, a successful outcome has been recently reported (including correction of inflammatory bowel disease) using reduced-intensity conditioning. Gene therapy has been attempted, but in the absence of conditioning, only modest and transient presence of gene-corrected phagocytes is achieved. Furthermore, myelodysplasia has been observed due to insertional mutagenesis.

Gastrointestinal manifestations are observed in approximately one-third of CGD patients. Common symptoms include abdominal pain, fever, diarrhea, vomiting, weight loss, and delayed growth. Virtually any segment of the gastrointestinal tract may be involved; liver disease is also common. Gastrointestinal pathologies include infection, chronic inflammation with strictures and fistulizing disease, ulcerations, granuloma formation, and dysmotility.

Stomatitis and oral candidiasis are frequent in CGD. Involvement of the esophagus is less common, but narrowing of the lumen, prestenotic dilatation, and dysmotility have been reported. Chronic gastritis and thickening of the stomach mucosa have been described more frequently and may lead to gastric outlet obstruction. Inflammation in the small or large bowel may cause pathologic features resembling IBD (**Figure 33.7**) and may also lead to obstruction. Histological examination of colonic biopsies reveals unique features compared with those typically seen in ulcerative colitis or Crohn's disease. In particular, there is a prominent presence of eosinophils

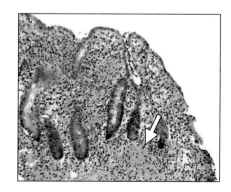

Figure 33.7 Active colitis and granulomas in chronic granulomatous disease (CGD). Acute inflammation and granuloma (white arrow) in colon, ×400. (Courtesy of John Watkins and Jeffrey Goldsmith.)

and pigmented macrophages in the basal mucosa or in the superficial submucosa, whereas neutrophils are very rare. Accumulation of eosinophils around the crypts is typically associated with damage to the fibroblastic pericryptal sheath and may ultimately cause reduced absorption of nutrients. There is increased expression of MHC-II antigens in epithelial and vascular endothelial cells and of intercellular adhesion molecule 1 expression in the lamina propria.

Thorough microbiological evaluation, imaging studies, and empiric aggressive use of antibiotics are needed in CGD patients who develop intestinal manifestations of possible infectious origin. However, in patients with colitis, frequently no pathogens are isolated and antimicrobial treatment alone is not sufficient. In these cases, immunomodulation with steroids and sulfasalazine is often required. Anti-TNF-α therapy has generally been avoided because of concern about disseminated infections. Cyclosporine and thalidomide have been also successfully used in some patients. Perirectal abscesses require surgical drainage. Emergency ileostomy may be required in patients with severe colitis refractory to immunosuppressive treatment.

Liver abscesses are also very common. Often multiple, they have a tendency to persist or relapse. *Staphylococcus aureus* is the most common organism involved. A high index of suspicion should be used, with a low threshold for the performance of radiologic imaging studies. Generous use of aggressive surgical debridement, associated with prolonged use of antibiotics, is important to facilitate definitive treatment of the abscesses.

Ultimately, successful correction of the disease, including gastrointestinal and hepatic complications, is based on HCT. However, controversies exist whether this should be reserved for patients with a more severe phenotype and lack of adequate response to medical management.

33.15 NEMO (NF-κB essential modulator) deficiency, or X-linked immunodeficiency with ectodermal dystrophy, is a disorder that typifies consequences of innate defects that affect epithelial barrier

The nuclear factor kappa B (NF-κB) is a multimeric transcription factor that regulates a variety of cellular processes, including innate and adaptive immune responses. Activation of NF-κB is regulated by a complex consisting of two catalytic protein subunits with kinase activity (IKKα and IKKβ) and a regulatory component, IKKγ (also known as NF-κB essential modulator, or NEMO), which function to phosphorylate and deactivate IκB, an inhibitor of NF-κB. NEMO is encoded by a gene (*IKBKG*) located on the X chromosome. In humans, null *IKBKG* gene mutations lead to incontinentia pigmenti in heterozygous females, and result in embryonic lethality in males. In contrast, hypomorphic mutations in males result in X-linked immunodeficiency with ectodermal dystrophy. Typical manifestations of the disease include scanty

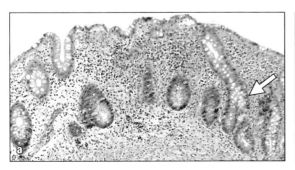
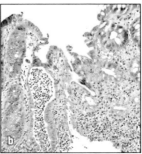

Figure 33.8 **Enterocolitis in nuclear factor κB essential modulator (NEMO) deficiency.** (**a**) Acute inflammation in colon with evidence of chronic changes as evidenced by branched crypt (white arrow), ×100. (**b**) Moderately active colitis. (Courtesy of Athos Bousvaros and Jeffrey Goldsmith.)

TABLE 33.4 MONOGENIC ETIOLOGIES OF VERY EARLY ONSET INFLAMMATORY BOWEL DISEASE

Group	Syndrome/disorder	Gene	Inheritance	Key laboratory findings	Other major clinical findings
IL-10 signaling defects	IL10R	IL10RA	AR	Defective STAT3 phosphorylation in response to IL10	Perianal fistula; Arthritis; Eczema; Folliculitis; Pyoderma; B-cell lymphoma
	IL10R	IL10RB	AR	Defective STAT3 phosphorylation in response to IL10	Perianal fistula; Arthritis; Eczema; Autoimmune hemolytic anemia; B-cell lymphoma
	IL10	IL10	AR	Decreased functional IL10	Perianal fistula
Immunoregulation	IPEX	FOXP3	XL	Elevated serum IgE; decrease in T_{reg} number; decreased Foxp3 expression	Atopic dermatitis; Arthritis; Multiple food allergies; Type 1 diabetes mellitus; Nephropathy, Hepatitis; Autoimmune hemolytic anemia; Hashimoto's thyroiditis; Neoplasia
	IPEX-like	IL2RA/ CD25	AR	Absent CD25 expression	Autoimmune endocrinopathy; Eczema; Short stature; Interstitial pneumonitis; Alopecia universalis; Bullous pemphigoid; CMV/EBV
	IPEX-like	STAT1 GOF	AD	Normal T_{reg} cell number and FOXP3 expression; STAT1 phosphorylation studies abnormal (increased and prolonged)	Autoimmune endocrinopathy; Chronic mucocutaneous candidiasis; Eczema, Recurrent sinopulmonary infections (RSV bronchiolitis common); Bronchiectasis
	MALT1 deficiency (IPEX-like)	MALT1	AR	High IgE; Low IgM; Normal IgG/IgA; Eosinophilia; Elevated T-cell number with decreased FOXP3 expression	Facial dysmorphism; Eczema-like dermatitis; Severe bacterial/viral/fungal infections
	STAT3	STAT3 GOF	AD	Hypogammaglobulinemia; Decreased class-switched memory B cells; Eosinophilia; Decreased NK and T_{reg} cells with increased CD4-CD8- T cells	Multisystem autoimmunity; Variable short stature; Lymphoproliferation; Delayed-onset mycobacterial disease
	JAK1	JAK1 GOF	AD	Increased CD56high/CD16; Low NK cells; Heightened STAT1/STAT3 phosphorylation	Membranous nephropathy; Chronic dermatitis; Short stature
T- and B-cell defects	LRBA deficiency	LRBA	AR	Hypogammaglobulinemia; Decreased class-switched memory B cells; Constrained T_{reg} population; Impaired vesicular trafficking of CTLA4; Lower CD25 levels on T_{reg} cells; Exaggerated T-follicular-helper cell response	Erythema Nodosum; Autoimmune Hemolytic anemia; Type 1 diabetes mellitus; Burkitt's lymphoma; Exocrine pancreatic insufficiency; Interstitial lung disease; Splenomegaly; Hepatitis; IgA deficiency; Uveitis
	CTLA4 deficiency	CTLA4	AD with incomplete penetrance	Hypogammaglobulinemia; Decreased class switched memory B cells	Type 1 diabetes mellitus; Autoimmune hypothyroidism; Demyelinating encephalopathy; Fibrosing interstitial lung disease; Chronic sinusitis; PSC; Hypogammaglobulinemia; Autoimmune cytopenias; Gastric adenocarcinoma
	IL21 deficiency (CVID-like)	IL21	AR	Decreased class switched B cells; Elevated IgE; Decreased IgG	Recurrent Cryptosporidial infections; Recurrent sinopulmonary infections; Chronic cholangitis; Aphthous stomatitis
	Wiskott-Aldrich syndrome	WAS	XL	Microthrombocytopenia; Variable lymphopenia; Absent class-switched memory B cells; High IgE; Low IgG, IgA, and IgM	Thrombocytopenia; Atopic dermatitis; Autoimmune hemolytic anemia; Arthritis; Bacterial/viral infections; Lymphoreticular malignancy
	Hyperimmunoglobulinemia	CD40LG	XL	Elevated/normal IgM; Neutropenia, absent class-switched memory B cells	Sinopulmonary infections; Cryptosporidium infections; Sclerosing cholangitis; Autoimmune hemolytic anemia; Aphthous stomatitis associated with neutropenia; Perianal ulcers/fistulae; Neurologic problems
		AICDA	AR	Elevated/Normal IgM; Low IgG and IgA	Sinopulmonary infections; Aphthous stomatitis associated with neutropenia; Perianal ulcers/fistulae; Autoimmune hemolytic anemia; Skin lesions
	Bruton's agammaglobulinemia	BTK	XL	Severe hypogammaglobulinemia; Absent B cells; decreased class-switched memory B cells	Recurrent infections; Lymphoreticular malignancies; Autoimmune hemolytic anemia
	Artemis-deficiency combined immunodeficiency	DCLRE1C/ ARTEMIS	AR	Profound lymphopenia; Variable changes to immunoglobulins; Variable specific antibody response	Autoimmune hemolytic anemia; Neutropenia; Thrombocytopenia; Recurrent sinopulmonary infections; Candidiasis; Hepatosplenomegaly
	PI3K Activation syndrome	PIK3R1	AR	Variable alterations in immunoglobulins; Decreased class-switched memory B cells	Recurrent respiratory infections; Lymphoproliferation; Systemic autoimmunity reminiscent of SLE; Erythema nodosum
	PI3K Activation syndrome	PIK3CD	AD	Variable alterations in immunoglobulins; Decreased class-switched memory B cells	Recurrent respiratory infections; Lymphoproliferation; Systemic autoimmunity reminiscent of SLE; Erythema nodosum
	SCID	ZAP70	AR	Absent CD8 cells; Defective CD4 T cells; Some have elevated IgE and eosinophilia	Profound immunodeficiency with recurrent infections; Other early onset autoimmune diseases: Nephrotic syndrome; Bullous pemphigoid; Rheumatoid arthritis
	Leaky SCID	RAG2	AR	Eosinophilia; T-cell lymphopenia; NK cells present	Profound immunodeficiency with recurrent infections; Erythroderma; Hepatosplenomegaly; Lymphadenopathy; Autoimmune hemolytic anemia
	SCID/Omenn	IL2RG	XL	Eosinophilia; T-cell lymphopenia; Decreased naive T cells	Profound immunodeficiency with recurrent infections; Candidal infections
	Leaky SCID	LIG4	AR	Variable lymphopenia	Variable presentation—from propensity to leukemia to Omenn syndrome; Aphthous stomatitis
	SCID	ADA	AR	Absent T, B, and NK cells	Profound immunodeficiency with recurrent infections; Erythroderma; Hepatosplenomegaly; Lymphadenopathy
	SCID	CD3-γ	AR	Defective TCR/CD3 expression on CD4/CD8 positive lymphocytes; Variable lymphopenia and hypogammaglobulinemia	Profound immunodeficiency with recurrent sinopulmonary infections; Multiple autoimmune diseases: Thyroiditis; AIHA; vitiligo, Nephrotic syndrome
	ICOS deficiency	ICOS	AR	Low to absent class-switched memory B cells; Loss of bone marrow plasma cells; Impaired vaccine response	Recurrent sinopulmonary infections; CMV infections; Arthritis; Psoriasis
	Hoyeraal-Hreidarsson syndrome	DKC1	XL	B- and T-cell lymphopenia (±NK cell reduction); Low IgA, IgG, IgM; Impaired proliferation to antigens	Skin pigmentation; Nail dystrophy; Hair thin, sparse, and discolored; Leukoplakia of tongue and gums; Aplastic anemia; Immunodeficiency; Motor delay; Cerebellar hypoplasia; Growth failure; Microcephaly
	Hoyeraal-Hreidarsson syndrome	RTEL1	AR	Extremely short telomers	IUGR; Low B, T, and NK cells; Hypogammaglobulinemia (IgA, IgM, IgG); Nail dysplasia; Cerebellar hypoplasia; Increased risk of leukemia
	Loeys-Dietz syndrome	TGFBR1	AD	Anemia; Hypergammaglobulinemia	Hypertelorism; Blue sclerae; Bifid uvula; Arachondactyly; Joint hyperplexity; Pes valgus; Dilation of aortic root; Fevers
	Loeys-Dietz-Syndrome	TGFBR2	AD	Anemia; Hypergammaglobulinemia	Hypertelorism; Proprtosis; Blue sclerae; Joint hyperlaxity; Mild dilation of ascending aorta
	Immunodeficiency, centromeric instability, facial anomalies (ICF) syndrome	ZBTB24	AR	Hypogammaglobulinemia; Decreased B cells and NK cells; Poor response to vaccines; Centromere instability; Facial anomalies	Hypertelorism; Epicanthal folds; Low-set ears; Developmental delay; Perianal fistula

(Continued)

TABLE 33.4 *(CONTINUED)* MONOGENIC ETIOLOGIES OF VERY EARLY ONSET INFLAMMATORY BOWEL DISEASE

	RIPK1 deficiency	RIPK1	AR	Defective differentiation of T and B cells; Lymphopenia; Increased inflammasome activity; Impaired response to TNFR1-mediated cell death	Hepatosplenomegaly; Maculopapular rash; Recurrent fevers; Neonatal infections
	Caspase-8 deficiency	CASP8	AR	Altered distribution of T cells; reduced T-cell proliferation and activation; Impaired B-cell maturation; Elevated IL1B with altered inflammasome activity	Susceptibility to viral and bacterial infections
Phagocyte defects	CGD	CYBB	XL	Decreased neutrophil oxidative burst study; Elevated IgG	Recurrent Infections for catalase positive organisms; Perianal fistula; Gastric outlet obstruction; Eczema; Chorioretinitis
		CYBA	AR		
		NCF1	AR		
		NCF2	AR		
		NCF4	AR		
	Glycogen Storage Disease Type 1 b	SLC37A4	AR	Neutropenia and/or neutrophil dysfunction with predisposition to infections; Hypoglycemia; Hyperuricemia; Hyperlipidemia	Hepatosplenomegaly; Nephrocalcinosis; Folliculitis
	Congenital neutropenia	G6PC3	AR	Severe neutropenia; T-cell lymphopenia; Elevated IgG	Cutaneous vascular malformation; Cardiac defects; Urogenital developmental defects; Bleeding tendency
	Leukocyte adhesion deficiency 1	ITGB2	AR	Leukocytosis, Absent CD11/CD18 complex expression	Neutrophilia; Recurrent bacterial infections; Delayed separation of umbilical cord; Poor wound healing; Folliculitis; Ulcers of skin; Periodontitis; Gingivitis
	Niemann-Pick type C disease	NPC1	AR	Abnormal bacterial handling	Hepatosplenomegaly; Cholestasis; Portal hypertension; Motor or cognitive delay; Seizures; White matter abnormalities; Perianal skin tags; Anal fissures; Arthritis
Hyperinflammatory and auto-inflammatory	X-linked lymphoproliferative syndrome 2 (XLP2)	XIAP	XL	Elevated IL-18; Decreased XIAP protein expression by flow; Little expression of IL-8 and MCP-1 in response to MDP stimulation	Perianal fistula; HLH; Splenomegaly; Cholangitis; Skin abscesses; Arthritis, EBV and CMV infections; Hypogammaglobinemia
	Hermansky-Pudlak syndrome	HPS1	AR	Decreased CD107a degranulation	Partial albinism; Bleeding tendency; Recurrent infections; Nystagmus; Pulmonary fibrosis; Bowel perforation
		HPS4	AR	Decreased CD107a degranulation; Absent platelet dense bodies	Partial albinism; Bleeding tendency; Recurrent infection; Pulmonary fibrosis; Variable hair and skin pigmentation
		HPS6	AR	Elevated IgD; Elevated urine mevalonate	Partial albinism; Bleeding tendency; Recurrent infection
	Mevalonate kinase deficiency	MVK	AR	Majority of cases with hyperimmunoglobulinemia D; Neutrophil-predominant leukocytosis during attacks with elevated ESR/CRP; Serum amyloid A when adult	Recurrent fevers; Rash; Macrophage activation syndrome; Recurrent febrile attacks; Polyarthritis; Edema; Urticarial rash; Hepatomegaly
	Phospholipase Cy2 defects	PLCG2	AD		Early onset recurrent blistering skin condition; Interstitial pneumonitis with respiratory bronchiolitis; Arthralgia, Eye inflammation; Cellulitis; Recurrent sinopulmonary infections
	Familial Mediterranean fever	MEFV	AR	Marked elevated ferritin and sIL-2R; Decreased CD107a degranulation	Recurrent fevers; Erythematous rash; Arthralgia; Pericardial effusion; Peritonitis
	Familial hemophagocytic lymphohistiocytosis type 5	STXBP2	AR	Elevated IL-18	HLH; Hypogammaglobulinemia; Sensorineural hearing loss
	TRIM22 defect	TRIM22	AR		Oral ulcerations, Fevers; Profound anemia; Hypogammaglobulinemia
	Prostaglandin transporter deficiency	SLCO2A1		Mild to severe T-cell immune deficiency; Hypogammaglobulinemia	Chronic nonspecific ulcers of the small intestine are distinct as multiple shallow, circular or oblique lesions, with sharp demarcation from surrounding normal mucosa; Can lead to primary heterotrophic osteoarthropathy
Epithelial barrier	TTC-7A deficiency	TTC7A	AR	Hypogammaglobulinemia; Decreased class switched memory B cells; Reduced NK-cell cytotoxic activity	Multiple intestinal atresia; Exfoliating and sloughing intestinal mucosa; Hypoplastic thymus; Severe combined immunodeficiency
	NF-κB essential modulator (NEMO) deficiency	IKBKG	XL	Hypogammagloulinemia; Defective antibody responses,; Impaired T-cell priming; Defective NK cell cytolytic function	Ectodermal dysplasia; Immunodeficiency
	Dystrophic epidermolysis bullosa	COL7A1	AR	Eosinophilia	Blistering disorder mainly of extremities—knees, elbows, feet, and hands
	Kindler syndrome	FERMT1 (AKA Kindlin-1)	AR	Moderately elevated IgE; Normal B- and T-cell subsets; T-cell infiltration in skin in absence of acute inflammatory response	Blistering skin; Photosensitivity; Progressive poikiloderma; Diffuse cutaneous atrophy
	ADAM-17 deficiency	ADAM17	AR	Hyponatremia; Hypokalemia; Hypomagnesemia; Hypocalcemia; Nutrient deficiencies	Neonatal onset inflammatory skin disease (facial scaling, pustules and erythema); Short and broken hair; Swollen distal phalanges; Abnormal nails with paronychia
	Familial diarrhea	GUCY2C	AD	High fecal sodium; Low urinary sodium	Strictures; Esophagitis
	Congenital Diarrhea	SLC9A3	AR	Elevated liver enzymes and cholestasis	
Other	Trichohepatoenteric syndrome	SKIV2L	AR	Low IgG levels or hypogammaglobulinemia; Decreased antigen-stimulated lymphocyte proliferation	Intrauterine growth restriction; Wooly, dry, easily removable hair that is poorly pigmented; Large forehead; Skin abnormalities; Mild mental retardation; Bronchiectasis; Cardiac abnormalities
	Trichohepatoenteric syndrome	TTC37	AR	Elevated liver enzymes; Low IgG, Poor vaccine response; Decreased antigen-stimulated lymphocyte proliferation	Facial dysmorphism, Hair anomalies; Low birth weight; Cardiac anomalies; Liver cirrhosis
		ARPC1B	AR	Microthrombocytopenia; Eosinophilia; Elevated IgA, IgG, TRECs, ANA and ANCA	Chronic infections; Eczema; Vasculitis
	CHAPLE syndrome	CD55	AR	Terminal complement activation in arterioles; Increased TNF; Reduced IL-10	Early onset protein losing enteropathy; Primary intestinal obstruction; Edema; Malabsorption; Recurrent infections; Angiopathic thromboemboli; Micronutrient deficiencies

Source: Expanded from Uhlig, H.H. et al. *Gastroenterology* 2014, 147:990–1007, and modified from Ouahed. J. et al., Inflamm Bowel Dis. 2019. pii: izz259. doi: 10.1093/ibd/izz259. [Epub ahead of print] PMID: 31833544.

Note: Crohn's-like disease refers to disease that has features of Crohn's including deep ulcerations, discontinuous inflammation, strictures, and fistulas.

Abbreviations: AD, autosomal dominant; AIHA, autoimmune hemolytic anemia; ANA, antinuclear antibodies; ANCA, antineutrophil cytoplasmic antibody; AR, autosomal recessive; CD, Crohn's disease; CMV, cytomegalovirus; CRP, C-reactive protein; CVID, common variable immune deficiency; EBV, Epstein-Barr virus; ESR, erythrocyte sedimentation rate; GOF, gain of function; HLH, hemphagocytic lymphohistiocytosis; Ig, immunoglobulin; IL, interleukin; IPEX, immune dysregulation, polyendocrinopathy enteropathy X linked; IUGR, intrauterine growth restriction; MDP, muramyl dipeptide; NK, natural killer; PSC, primary sclerosing cholangitis; R, receptor; sIL2R, soluble IL2 receptor; SLE, systemic lupus erythematosus; TNF, tumor necrosis factor; TREC, T-cell receptor excision circle; T_{reg}, T-regulatory cells; UC, ulcerative colitis; XL, linked.

hair, defective tooth formation with conical teeth, nail defects, hypohidrosis, increased susceptibility to recurrent infections (including mycobacterial disease), and aberrant inflammatory responses. Most patients show hypogammaglobulinemia, with reduced IgG levels and defective antibody responses. Impaired T-cell priming and defective cytolytic function of NK cells are also present. There is significant variability in the severity of the clinical phenotype. Treatment is based on immunoglobulin replacement therapy and prompt treatment of infections.

Gastrointestinal manifestations are common in patients with NEMO deficiency and may reflect either infections or excessive inflammatory responses. Since NEMO is required for functional responses through toll-like receptors, it is not unexpected that the functional response to infectious agents in the gastrointestinal tract is impaired. Accordingly, patients are more prone to bacterial and mycobacterial infections that may cause diarrhea, abdominal pain, fever, and malabsorption. However, NEMO is also critically involved in the generation and homeostasis of regulatory T cells and natural killer T (NKT) cells. Furthermore, NEMO functions in gut epithelial cells to control the integrity of the mucosal barrier and interaction of epithelial cells with the commensal microbiota. Consistent with this, NEMO deficiency also results in significant immune dysregulation and abnormal inflammatory responses. In the gastrointestinal tract, this aberrant pattern of response may lead to atypical enterocolitis (**Figure 33.8**), with abdominal pain and fever. Bloody stools and weight loss are common, and the erythrocyte sedimentation rate is markedly elevated. In such cases, no pathogens are cultured. Endoscopy reveals severe inflammation, especially in the colon, with exudates and mucosal bleeding. Histological examination shows superficial cryptitis, mucosal ulcerations, and edema associated with abundance of neutrophils within the lamina propria juxtaposed to the mucosal surface. Absence of granuloma and deep cryptitis, and paucity of lymphocytic infiltrates, distinguishes these lesions from those typically observed in IBD. Treatment is based on the use of anti-inflammatory agents.

33.16 Expanding causes of monogenic defects leading to very early onset inflammatory bowel disease

Over the past decade, an international effort focused on next-generation sequencing approaches has led to a marked increase in the identification of monogenic causes for inflammatory bowel diseases in infants and very young children (https:// veoibd.org). The majority of these monogenic defects result from a primary or secondary immune defect (with or without concomitant involvement of other affected cell populations). A summary of genes identified with mutations leading to very early onset inflammatory bowel disease (VEOIBD) is outlined in Table 33.4.

SUMMARY

The common presentation of mucosal and particularly gastrointestinal symptomatology in the setting of primary immunodeficiencies underscores the role of the immune system in regulating mucosal homeostasis. Moreover, genome-wide association studies evaluating patients with inflammatory bowel diseases have also identified polymorphisms in genes that affect either innate (e.g., *NOD2*) and/or adaptive immune function (e.g., *IL23R*, *IL10*). Primary humoral disorders (e.g., XLA, selective IgA deficiency) are most often associated with diarrhea and malabsorption resulting from infectious consequences and only rarely aggressive inflammatory pathologies. Immunodeficiencies that affect both B- and T-cell signaling (e.g., CVID, SCID) present with gastrointestinal infections, aggressive inflammatory responses, and malignancies, thus emphasizing the role of appropriate T-cell function in maintaining mucosal function. Aberrant immunoregulatory circuits also result in severe intestinal pathology as evidenced by several immunodeficiencies (e.g., IPEX, WAS, IL-10R deficiency). In these disorders, regulatory T-cell abnormalities, and/or myeloid cell dysfunction may play a significant role in the underlying pathology. Finally, immunodeficiencies that affect innate immune cells (e.g., CGD, NEMO deficiency) also present with both intestinal infections and aggressive intestinal inflammatory sequelae.

FURTHER READING

Alkhairy, O.K., Abolhassani, H., Rezaei, N. et al.: Spectrum of phenotypes associated with mutations in LRBA. *J. Clin. Immunol.* 2016, 36:33–45.

Allenspach, E., and Torgerson, T.R.: Autoimmunity and primary immunodeficiency disorders. *J. Clin. Immunol.* 2016, 36(Suppl 1):57–67.

Barmettler, S., Otani, I.M., Minhas, J. et al.: Gastrointestinal manifestations in x-linked agam-maglobulinemia. *J. Clin. Immunol.* 2017, 37:287–294.

Candotti, F.: Clinical manifestations and pathophysiological mechanisms of the Wiskott-Aldrich syndrome. *J. Clin. Immunol.* 2018, 38:13–27.

Cunningham-Rundles, C., and Bodian, C.: Common variable immuno-deficiency: Clinical and immunological features of 248 patients. *Clin. Immunol.* 1999, 92:34–48.

Dhillon, S.S., Fattouh, R., Elkadri, A. et al.: Variants in nicotinamide adenine dinucleotide phosphate oxidase complex components determine susceptibility to very early onset inflammatory bowel disease. *Gastroenterology* 2014, 147:680–689.

Hayward, A.R., Levy, J., Facchetti, F. et al.: Cholangiopathy and tumors of the pancreas, liver and biliary tree in boys with X-linked immunodeficiency with hyper-IgM. *J. Immunol.* 1997, 158:977–983.

Hou, T.Z., Verma, N., Wanders J. et al.: Identifying functional defects in patients with immune dysregulation due to LRBA and CTLA-4 mutations. *Blood* 2017, 129:1458–1468.

Huang, A., Abbasakoor, F., and Vaizey, C.J.: Gastrointestinal manifestations of chronic granulo-matous disease. *Colorectal Dis.* 2006, 8:637–644.

Khodadad, A., Aghamohammadi, A., Parvaneh, N. et al.: Gastrointestinal manifestations in patients with common variable immunodeficiency. *Dig. Dis. Sci.* 2007, 52:2977–2983.

Malamut, G., Ziol, M., Suarez, F. et al.: Nodular regenerative hyperplasia: The main liver disease in patients with primary hypogammaglobulinemia and hepatic abnormalities. *J. Hepatol.* 2008, 48:74–82.

Neven, B.L., Mamessier, E., Bruneau, J. et al.: A Mendelian predisposition to B-cell lymphoma caused by IL-10R deficiency. *Blood* 2013, 122:3713–3722.

Notarangelo, L.D.: Primary immunodeficiencies. *J. Allergy Clin. Immunol.* 2010, 125:S182–194.

Ouahed, J., Spencer, E., Kotlarz, D. et al.: Very Early Onset Inflammatory Bowel Disease: A Clinical Approach With a Focus on the Role of Genetics and Underlying Immune Deficiencies. *Inflamm Bowel Dis.* 2019. pii: izz259. doi: 10.1093/ibd/izz259. [Epub ahead of print] PMID: 31833544.

Picard, C., and Fischer, A.: Hematopoietic stem cell transplantation and other management strategies for MHC class II deficiency. *Immunol. Allergy Clin. North Am.* 2010, 30:173–178.

Schwab, C., Gabrysch, A., Olbrich, P. et al.: Phenotype, penetrance, and treatment of 133 cytotoxic T-lymphocyte antigen 4-insufficient subjects. *J. Allergy Clin. Immunol.* 2018, 142:1932–46.

Shouval, D.S., Biswas, A., Kang, Y.H. et al.: Interleukin 1β mediates intestinal inflammation in mice and patients with interleukin 10 receptor deficiency. *Gastroenterology* 2016, 151:1100–1104.

Shouval, D.S., Kowalik, M., and Snapper, S.B.: The treatment of inflammatory bowel disease in patients with selected primary immunodeficiencies. *J. Clin. Immunol.* 2018, 38:579–588.

Uhlig, H.H., Schwerd, T., Koletzko, S. et al.: COLORS in IBD Study Group and NEOPICS. The diagnostic approach to monogenic very early onset inflammatory bowel disease. *Gastroenterology* 2014, 147:990–1007.

Uzzan, M., Ko, H.M., Mehandru, S. et al.: Gastrointestinal disorders associated with common variable immune deficiency (CVID) and chronic granulomatous disease (CGD). *Curr. Gastroenterol. Rep.* 2016, 18:17.

Verbsky, J.W., and Chatila, T.A.: Immune dysregulation, polyendocrinopathy, enteropathy, X-linked (IPEX) and IPEX-related disorders: An evolving web of heritable autoimmune diseases. *Curr. Opin. Pediatr.* 2013, 25:708–714.

Westerberg, L.S., Klein, C., and Snapper, S.B.: Breakdown of T cell tolerance and autoimmunity in primary immunodeficiency: Lessons learned from monogenic disorders in mice and men. *Curr. Opin. Immunol.* 2008, 20:646–654.

Inflammatory bowel disease

34

GIOVANNI MONTELEONE, MARKUS F. NEURATH,
AND BRITTA SIEGMUND

Crohn's disease and ulcerative colitis are the major chronic inflammatory bowel diseases (IBDs) in humans. IBD affects about 3 million individuals in the United States and Europe with an increasing incidence worldwide. Both forms of IBD have a peak age of onset in males and females in the second to third decades of life. Crohn's disease usually involves the ileum and colon, but it can present anywhere in the alimentary tract, from the mouth to the anus. Inflammation is often transmural and discontinuous. Histologically, gut lesions are characterized by the accumulation of chronic inflammatory cells, and by deep-fissuring ulcers (Figure 34.1). The presence of noncaseating granuloma, although highly characteristic of Crohn's disease, is neither unique to Crohn's disease nor universally found. Dense lymphoid aggregates can be seen in the submucosa (see Figure 34.1). Fibrosis in the deeper layers of the gut wall contributes to the development of strictures, as a result of the overproduction of collagen by fibroblasts and smooth muscle cells. Ulcerative colitis, in contrast, typically involves the rectum, with the inflammation being confined to the mucosa or submucosal layers. Disease progresses proximally in a continuous fashion, resulting in variable degrees of involvement of the colon, with the most extreme form of total inflammation presenting as a pancolitis. Inflammation is characterized by a marked infiltration of the lamina propria with neutrophils, lymphocytes, plasma cells, and macrophages (see Figure 34.1). The neutrophils can invade the epithelium and cross into the lumen, particularly in the crypts, giving rise to cryptitis and crypt abscesses. Cryptitis is associated with a depletion of mucus from goblet cells and increased epithelial cell turnover. There is a marked increase in plasma cells in the lamina propria and an excess production of IgG, particularly IgG1 and IgG3; this contrasts markedly with Crohn's disease, in which there is also an increase in plasma cells, especially around ulcers, but in which the dominant isotype is IgG2.

Figure 34.1 **Histology of Crohn's disease and ulcerative colitis.** Both Crohn's disease and ulcerative colitis are associated with massive inflammatory lesions in the bowel. The Crohn's disease image (**a**) shows a dense mononuclear infiltrate into the full thickness of the gut wall. Ectopic lymphoid follicles are obvious in the deeper layers, and a deep ulcer penetrates from the lumen into the submucosa. The ulcerative colitis lesion (**b**) shows the characteristic neutrophilic mucosal lesions, although in this case disease is so fulminant that there is some submucosal involvement. There are very obvious crypt abscesses where neutrophils have crossed the glandular epithelium. There is also extensive epithelial cell damage. Hematoxylin and eosin stain. Original magnification ×40.

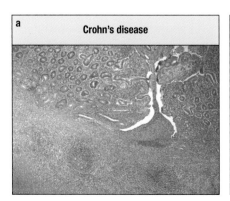

a   Crohn's disease

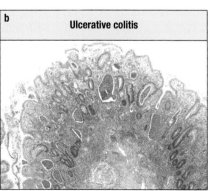

b   Ulcerative colitis

Patients with IBD suffer from frequent and chronically relapsing flares, with rectal bleeding, abdominal pain, and/or diarrhea. In both types of IBD, there can also be local (e.g., cancer) and extraintestinal (e.g., primary sclerosing cholangitis, spondylitis, or iritis) complications that influence the natural history of IBD patients. Some of these complications directly reflect the extent and severity of the inflammatory activity of these disorders (e.g., cancer), whereas others are comorbid conditions that follow their own course without regard necessarily to the intestinal inflammation (e.g., primary sclerosing cholangitis) and thus probably have a similar genetic and/or environmental foundation.

The etiology of both Crohn's disease and ulcerative colitis is unknown, but the rapid advancement of molecular techniques associated with genetics, microbiology, and immunology and the possibility of using a large number of animal models of intestinal inflammation have led to a better knowledge of factors that orchestrate the tissue-destructive inflammatory responses in these disorders. It is now well accepted that both forms of IBD develop in genetically predisposed individuals as a result of an exaggerated mucosal immune response directed against commensal microbiota that reside in the intestinal lumen. In this chapter, we focus on recent advances in our understanding of the involvement of genetic, bacterial, and immune factors in the pathogenesis of Crohn's disease and ulcerative colitis.

GENETIC BASIS OF INFLAMMATORY BOWEL DISEASE

Crohn's disease and ulcerative colitis have long been known to have a genetic basis; however, with the advent of population databases and modern molecular techniques for genetic analyses, the study of IBD has led the way in the discovery of gene polymorphisms in the development of complex disease.

34.1 Epidemiological observations support the idea of a genetic basis of IBD

In the 1980s, several epidemiological studies confirmed the early findings of familial aggregation in IBD, with a positive family history ranging from 5% to 20%. IBD is thus considered to be familial in roughly 10% of afflicted individuals and sporadic in the remaining 90%. The relative risk to siblings of affected individuals has been estimated to be 30- to 40-fold for Crohn's disease and 10- to 20-fold for ulcerative colitis. Studies of twins have provided additional evidence for a genetic contribution in IBD. The concordance rate is significantly greater in monozygotic than dizygotic twins for both Crohn's disease (20%–70% versus 0%–10%) and ulcerative colitis (10%–20% versus 0%–5%). In families containing multiple affected individuals with IBD, 75% are concordant for disease type (either Crohn's disease or ulcerative colitis), with the remaining 25% being "mixed" (both Crohn's disease and ulcerative colitis), which is consistent with an overlapping genetic and/or environmental basis as discussed further later. Overall, IBD is considered to be a polygenic disorder with complex inheritance except in a subset of individuals with early onset IBD (in the first 5 years of life), which may be due to monogenic (Mendelian) inheritance.

Genetic factors influence not only disease onset but also disease phenotype and course, particularly in Crohn's disease. A population-based study of Scandinavian cohorts of twins with IBD revealed a pronounced genetic impact on clinical characteristics of Crohn's disease in comparison with ulcerative colitis. Monozygotic twins with Crohn's disease seem to be concordant for age at onset of symptoms, age at diagnosis, extent of inflammation, and disease behavior both at diagnosis and at 10 years after diagnosis. In contrast, monozygotic twins with ulcerative colitis are concordant for age at diagnosis and symptomatic onset only. Genetic factors seem also to influence the disease location, given that the percentage of familial cases is lower in Crohn's disease confined to the colon than in ileal disease.

34.2 IBD susceptibility genes were identified before era of genome-wide association studies

To determine the IBD-associated susceptibility genes, investigators originally used two main types of genetic studies, namely, genetic linkage and candidate-gene association studies. Genetic linkage involves the study of the inheritance patterns of polymorphic genetic markers in families with multiple affected relatives. This approach led to the identification of at least nine disease loci designated IBD1–9 (Table 34.1); however, only five of these loci (IBD1, 2, and 4–6) showed significant linkage (LOD score >3.6 or $P < 2.2 \times 10^{-5}$), with little replication across the scans. Such linkage studies supported the polygenic basis of these disorders and predicted future observations—to be discussed later—showing that IBD probably results from the minor effects of a large number of genetic variants in a given individual. However, some genetic loci were unique and observed to contribute significant statistical risk for the development of IBD. One genetic locus that is worthy of being highlighted and was in fact the first genetic region to be identified was found within the pericentromeric region of chromosome 16. This locus was named IBD1 and was later shown to contain *NOD2* (nucleotide-binding oligomerization domain-containing 2), also known as caspase recruitment domain protein 15 (*CARD15*), the first susceptibility gene for Crohn's disease, but not ulcerative colitis.

DNA sequencing of *NOD2/CARD15* in Crohn's disease demonstrated that three main polymorphisms situated in or close to the leucine-rich repeat (LRR) ligand-binding domain of the molecule contribute to 80% of the risk associated with this important genetic risk factor. A gene dosing effect is seen in most studies, such that carriage of one or two copies of the risk allele(s) increases the risk of developing Crohn's disease two- to fourfold or 20- to 40-fold, respectively. Studies from France, Germany, the United Kingdom, and the United States have shown that up to 40% of patients with Crohn's disease (in contrast with 14% of controls) have one or more of these mutations. However, the allelic frequency is much lower in other European regions (e.g., Ireland, Scotland, Scandinavia, and Iceland), and NOD2 mutations are not found in patients with Crohn's disease in Asia, thus highlighting significant ethnic heterogeneity. Moreover, NOD2 mutations are associated with ileal Crohn's disease but not with colonic Crohn's disease or ulcerative colitis. Given that NOD2/CARD15 is an intracellular receptor for muramyl dipeptide (MDP) derived from peptidoglycan of gram-negative and gram-positive bacteria, and is a receptor for the glycolyl MDP of mycobacteria and single-stranded RNA of viruses and a member of the NOD-like pattern recognition receptors (NLR), these studies support a microbial basis for IBD, especially Crohn's disease.

Another instructive locus that has consistently been identified by linkage and association studies as important for both ulcerative colitis and Crohn's disease in many different populations is the major histocompatibility complex (MHC) region on chromosome 6p21.1–23 (IBD3 locus). The MHC locus is highly complex, with both human leukocyte antigen (HLA) and non-HLA alleles (e.g., tumor necrosis factor and complement genes), such that, despite extensive study, the specific genes and mechanism(s) involved in conferring genetic risk remain enigmatic. In ulcerative colitis, the most consistently replicated association is with the rare allele HLA-DRB1*0103. This

TABLE 34.1 CONTRIBUTION OF LINKAGE AND GENOME-WIDE ASSOCIATION STUDIES IN THE IDENTIFICATION OF CROHN'S DISEASE SUSCEPTIBILITY LOCI

Association mapping of linkage region		Genome-wide association studies			
Locus	Candidate gene(s)	Chromosomal region	Relevant genes	Chromosomal region	Relevant genes
IBD1	NOD2/CARD15	1p13	PTPN22	9p24	JAK2
IBD2		1p31	IL23R	9q32	TNFSF15
IBD3	HLA/TNF	1p36	PLA2G2E	9q33	TLR4
IBD4		1q23	ITLN1	9q34	CARD9
IBD5	T_H2 cytokines	1q32	IL10	10p11	CCNY
IBD6	STAT6	1q44	NLRP3	10q21	ZNF365
IBD7		2q37	ATG16L1	10q23	DLG5
IBD8		2p16	PUS10	10q24	NKX2-3
IBD9		2q11	IL18RAP	10q26	DMBT1
		3p12	CADM2	11p15	NELL1
		3p21	MST1/BSN	11q13	C11orf30
		4p13	PHOX2B	12q12	LRRK2/MUC19
		5p13.1	PTGER4	12q15	IFNG
		5q13	S100Z	15q13	HERC2
		5q31	SLC22A5	16q12	NOD2
		5q33	IRGM/IL12B	16q24	FAM92B
		6p21	BTNL2/HLA	17q21.2	OMDL3/STAT3
		6p22	CDKAL1	18p11	PTPN2
		6q23	TNFAIP3	20q13	TNFRSF6B
		6q27	CCR6	21q22	ISOSLG
		7q32	IRF5	22q12	NCF4/XBP1

variant is present in 0.2%–3.2% of the population, in 6%–10% of all ulcerative colitis cases, in 15.8% of severe and extensive ulcerative colitis cases, and in 14.1%–25% of severe ulcerative colitis cases requiring colectomy. HLA-DRB1*0103 is also associated with colonic Crohn's disease. The common HLA allele HLA-DRB1*0701 is associated with Crohn's disease, and ileal Crohn's disease in particular. The alleles HLA-B27, HLA-B35, and HLA-DRB1*0103 are also associated with extraintestinal manifestations of disease. Together, these studies predict the importance of both MHC class I- and MHC class II-related immune responses and consequently CD8 and CD4 T cells in the immunopathogenesis of these disorders.

34.3 Genome-wide association studies reveal large numbers of genes associated with IBD

The availability of the complete sequence of the human genome, the development of novel microarray technologies, and some limitations of linkage analysis (namely, difficulties encountered in trying to identify genes of weak effect and the necessity of testing a large set of affected siblings) have favored the advent of genome-wide association studies, which seek statistical evidence for association between variants distributed across the genome and disease susceptibility in a nonbiased manner. To provide adequate power, genome-wide association studies require large numbers of cases and controls, as well as the genotyping of several hundreds of thousands of markers to capture a majority of the common (more than 5% minor allele frequency) genomic variations, and replication in independent datasets on a different genotyping platform to confirm robust, true signals. So far, more than 100,000 single nucleotide polymorphisms (SNPs) have been used in the published genome-wide association studies. Overall these studies replicated the previously validated linkage and association studies (such as IBD1 and IBD3) and have identified new genetic risk factors. Up to now, over 200 risk-conferring loci have been identified for Crohn's disease and ulcerative colitis; some of these are described in Figure 34.2, which also highlights the complexity of these disorders. Genome-wide association studies and the associated deep DNA sequencing of candidate genes within identified loci and others have shown that (1) in the vast majority of identified examples, the causal genetic variants remain unknown because in only a minority of examples have nonsynonymous SNPs or other structural changes (e.g., variations in copy number) resulting in a change in the protein structure or promoter region been identified; (2) the genetic loci identified are also risk factors for many other immune-mediated disorders such as psoriasis, ankylosing spondylitis, celiac disease, and diabetes, demonstrating the similarities in disease pathogenesis with their attendant therapeutic implication; (3) the genetic loci so far identified by genome-wide association studies account for less than 20%–30% of the heritable risk, suggesting that the remainder of risk is conferred by rare genetic variants with potentially important biological effects that may account for the "missing heritability"; (4) most genetic variations identified are common to both forms of IBD, whereas some of them are unique for Crohn's disease or ulcerative colitis (see Figure 34.2); and (5) each genetic variant contributes a very modest risk such that IBD is the cumulative result of the combined risk conferred by the overall genetic composition of the host that elicits this risk through convergence on a variety of functional pathways described in more detail later (see Figure 34.2). Overall, these pathways seem to influence the genetically determined relationship between the commensal microbiota, the intestinal epithelial cells, and the immune system.

34.4 Environmental factors are clearly involved in IBD pathogenesis

Environmental factors, other than the commensal microbiota, are important in the pathogenesis of Crohn's disease and ulcerative colitis by modifying the genetic risk associated with disease. In particular, smoking is quite instructive in highlighting both the beneficial and the detrimental aspects of environmental factors. Smoking has been identified as a risk factor for an aggressive course of Crohn's disease, particularly after ileocecal resection, but seems to prevent the development of ulcerative colitis. Other studies indicate that environmental factors such as living in housing with hot water, fewer infections, and a low infant mortality rate coincide with a higher incidence of Crohn's disease; in contrast, early life residence on a farm may be protective for Crohn's disease and ulcerative colitis. These findings support the notion

IBD	Ulcerative colitis	Crohn's disease
Adaptive immunity	**Adaptive immunity**	**Adaptive immunity**
CARD9 MST1 PRDM1 REL SMAD3 IL1R2 ICOSLG RTEL1 PTPN2 YDJC TNFSF15 IL10	IFNγ/IL26 IL2/IL21 TNFRSF9 IL7R TNFRSF14 IRF5 LSP1 FCGR2A IL8RA / IL8RB	CCR6 PTPN22 IL27 IL2RA IL18RAP VAMP3 ITLN1 ERAP2 TNFS11 BACH2 CCL2/CCL7 TAGAP
HLA	**Gut epithelium**	**Innate immunity**
DRB03	HNF4A ECM1 LAMB1 GNA12 CDH1	LRRK2 IRGM NOD2 ATG16
T$_H$17	**Other**	**Other**
IL12B TYK2 STAT3 IL23R AK2	OTUD3/PLA2GE CAPN10 PIM3 DAP	SP140 THADA ZPF36L1 GCKR ZMIZ1 PRDX5 CPEB4 FADS1 DENNDIB MUC1/SCAMP3 5q31 (IBD5) DNNT3A
Others		
C11Orf30 CREM RTEL1 NKX2-3 ORMDL3 KIF21B ZNF365 CDKAL1 PTGER4		

Figure 34.2 The clustering of IBD susceptibility genes into groups according to disease and probable function. Genome-wide association studies have identified about 70 genes associated with IBD. Many of these genes are associated with innate immunity. Some are related to Crohn's disease, others to ulcerative colitis, and some are shared. It needs to be emphasized that, in the vast majority of cases, the causal variants are unknown and that the relative risk associated with these variants is extremely small, typically 1.2–1.3.

of the "hygiene hypothesis," although many other variables associated with a Westernized lifestyle may be more relevant and account for the increasing worldwide incidence of these disorders, especially Crohn's disease. It must be emphasized that whereas Crohn's disease is a disease of the developed world, and is increasing in the developing world as countries increase standards of living, ulcerative colitis shows a quite different distribution and is seen in developed as well as developing countries. In addition to these observations on the role of hygiene, further studies suggest that agents that disrupt mucosal barrier function and its relationship to the commensal microbiota such as nonsteroidal anti-inflammatory drugs (NSAIDs), antibiotics, and enteropathogenic exposures (viral, bacterial, and parasitic) that would be predicted to disrupt the normal composition of the commensal microbiota and function of epithelium and immune system may act as triggers of IBD in the genetically susceptible host. Finally, appendectomy before the age of 20 years is a protective factor for the future development of ulcerative colitis. Although the reasons for this finding are largely unknown, one can speculate that the appendix may shape the composition of the commensal microbiota in childhood and/or the education of T and B cells within organized structures of the gut-associated lymphoid tissue, thereby modulating susceptibility to ulcerative colitis. In any case, different patient subpopulations seem to have different genetic predispositions and exposures to environmental factors that modify this risk.

34.5 Bacteria and potentially viruses are major drivers of pathogenic inflammation in the gut

Studies in humans and animal models have identified important roles for the commensal bacteria in promoting the IBD-related immune response. Immunological tolerance exists toward resident intestinal flora in normal individuals but seems to be defective in IBD. Additionally, patients with IBD display aberrant T-cell activation, high levels of mucosal IgG antibodies, and cytokine responses to intestinal bacteria. Indirect evidence for a major role for the microflora also comes from clinical observations. For example, surgical reconstruction to divert the fecal stream is effective in ameliorating Crohn's disease, and antibiotics seem to provide some benefit in the treatment of Crohn's disease, especially disease in the distal colon. There is an increase in mucosa-associated bacteria in the neoterminal ileum after ileocecal resection for Crohn's disease, which may be associated with postoperative relapse.

Studies in murine models of IBD, such as interleukin (IL)-10-null mice, have provided additional insights into the influence of intestinal bacteria on the pathogenesis of these diseases. IBD is seen if the mice are maintained under conventional (specific pathogen-free) conditions. In these circumstances, disease is often ameliorated by treatment of the animals with antibiotics. In comparison, germ-free mice are resistant to experimentally induced colitis or manifest few signs of inflammation in models in which the inflammation develops as a result of targeted changes in the immune system. Mice deficient in genes associated with IBD risk in humans (e.g., *ATG16L1*) only manifest the disease-associated phenotype (namely, Paneth cell dysfunction) in the presence of specific viruses (such as norovirus), suggesting that infections may have a major role in the development of IBD in the susceptible host.

In normal subjects, the intestinal wall is covered with mucus, which prevents most bacteria from contacting the mucosal surface, but in patients with IBD, there is a dense coating of bacteria adjacent to the epithelial cell surface. Bacteria adhere to epithelial cells, enter crypts, and are occasionally found within cells. Of particular interest is an adherent and invasive *Escherichia coli* that seems particularly prominent in Crohn's disease. The use of next-generation DNA sequencing in the culture-independent rRNA sequence analysis of intestinal bacteria from patients with Crohn's disease or ulcerative colitis in comparison with non-IBD controls has shown that important differences can be detected between the microbiotas of patients with Crohn's disease or ulcerative colitis and those of non-IBD controls. Specifically, a subset of Crohn's disease and ulcerative colitis samples can be characterized by depletion of commensal bacteria, notably, members of the phyla Firmicutes and Bacteroidetes. It remains unknown whether such bacterial alterations contribute to the development of IBD or represent a consequence of the ongoing inflammation. This is underlined by the fact that data on fecal microbiota transfer (FMT) are indefinite at this point. In selected patients (more with ulcerative colitis), remission can be achieved by FMT. However, the factors identifying the "perfect" donor as well as the responsive recipients remain to be defined. Genetically determined inflammation can induce dysbiosis that is inflammatory in its own right as a result of the unregulated presence of certain bacteria such as Proteobacteria (e.g., *Serratia* and *Klebsiella* sp.) that function as pathobionts in animals that lack T-bet and an adaptive immune system (*Rag2*$^{-/-}$ mice). In contrast, inflammation may result in the reduction of other organisms such as *Faecalibacterium prausnitzii* in the gut of patients with Crohn's disease. This organism is interesting because it induces high levels of IL-10 in mononuclear cells and absence of this organism from the gut is associated with a higher risk of postoperative recurrence of ileal Crohn's disease. Along similar lines, administration of *Bacteroides fragilis* or its polysaccharide A prevents colitis in IL-10-deficient mice or SCID/Rag 2-deficient mice colonized with *Helicobacter hepaticus*, a commensal bacterium with pathogenic potential in the setting of immune deficiency. Taken together, these findings suggest that changes in the composition of commensal bacteria (e.g., increased Proteobacteria or decreased numbers of *F. prausnitzii*) may contribute to the amplification of the ongoing inflammation in IBD. It remains unknown whether IBD is the result of an inappropriate immune response to an individual pathobiont that is a commensal microorganism with pathogenic potential under specific circumstances (e.g., enteroadherent *E. coli*) or even whether it is due to an immune response to a true pathogen that has yet to be defined. However, it is interesting that the genetic basis for Crohn's disease shares many of the same genetic underpinnings as a true infectious disease (such as leprosy), blurring the differences between disease caused by a pathogen and that caused by either a pathobiont or a benign member of the commensal flora.

ASSOCIATION OF IBD WITH INNATE IMMUNITY ABNORMALITIES

Although the role of the vast majority of genes associated with IBD remains to be elucidated, an emerging theme is that many of the gene associations found since NOD2, itself a molecule that functions as a bacterial sensor, are also involved in the innate response to microbes.

34.6 Genes involved in innate immunity, endoplasmic reticulum stress, and autophagy are important in IBD

Autophagy is a cellular process that is classically responsible for the degradation of damaged organelles or long-lived proteins. It is activated by a variety of conditions associated with starvation and cellular stress such as that associated with the unfolded protein response as a consequence of stress caused by the accumulation of misfolded protein within the endoplasmic reticulum (ER) as well as the presence of cell-associated bacteria. Genes associated with bacterial sensing such as those encoding NOD2 (as discussed) and intelectin-1 have been associated with Crohn's disease, genes associated with autophagy such as autophagy-related gene 16 like-1 (*ATG16L1*) and that encoding immunity-related guanosine triphosphatase (*IRGM*) have been associated with Crohn's disease, and genes associated with ER stress such as those encoding X-box binding protein-1 (*XBP1*) and orsomucoid 1-like 3 (*ORMDL3*) have also been associated with both Crohn's disease and ulcerative colitis. Although the functional mechanisms for these susceptibilities are incompletely understood, as discussed later, they highlight the primary importance of innate immunity and the intestinal epithelium in the immunopathogenesis of IBD.

34.7 Autophagy-related proteins are linked to IBD

ATG16L1 is expressed in intestinal epithelial cells, T and B lymphocytes, and antigen-presenting cells. Very little is known about the function of the disease-associated *ATG16L1* gene product. However, it has recently been shown that macrophages isolated from mice deficient in *Atg16l1* produce elevated levels of IL-1 in response to lipopolysaccharide (LPS) and commensal bacteria such as *E. coli*, and mice develop more severe experimental colitis when the hematopoietic system is ATG16L1-deficient. Mice deficient for *Atg16l1*, and humans with a disease-associated risk variant, show decreased autophagy in the ileum and abnormalities in Paneth-cell granule structure with a pro-inflammatory phenotype including an increased expression of genes involved in peroxisome proliferator-activated receptor pathways and adipokines, such as leptin and adiponectin, which are known to be increased in IBD. Furthermore, NOD2 engagement by peptidoglycans induces autophagy and bacterial clearance. This process is dependent on the ability of NOD2 to recruit the autophagy protein ATG16L1 to the plasma membrane at the site of bacterial entry. *IRGM*, the human homolog of the mouse *Irgm/Lrg47*, encodes a GTP-binding protein that induces autophagy and is involved in the elimination of intracellular bacteria, including *Mycobacterium tuberculosis*. Taken together, these findings suggest that reduced function and/or activity of NOD2/ATG16L1/IRGM proteins could alter the innate immune response to luminal bacteria, thereby leading to persistence of intracellular bacteria and the development of pathogenic intestinal inflammation (Figure 34.3).

34.8 Impairment of intestinal epithelial cell function may also be a key factor for development and/or perpetuation of colitis

In sharp contrast to the prominent pro-inflammatory role of nuclear factor κ B (NF-κB) in myeloid cells, conditional ablation of IKKβ in epithelial cells results in increased inflammation in experimentally induced colitis; loss of IKKγ (NEMO [NF-κB essential modulator]) in epithelial cells results in spontaneous colitis. This is consistent with an overall protective function of NF-κB within the intestinal epithelium through either direct effects on epithelial cell survival or the ability to enhance the synthesis of chemokines, which may facilitate the mucosal recruitment of myeloid and T cells that in turn provide cytoprotective factors such as heat-shock proteins, IL-11, and IL-22 (Figure 34.4). Similarly, mice with an impaired ability to activate the inflammasome within epithelial cells, as a result of the lack of Nlrp3 (which in humans is a genetic risk factor for Crohn's disease), an adaptor protein, or caspase 1, are highly susceptible to experimentally induced colitis, which may be mediated by the ability of epithelial cells to secrete IL-18.

Studies in mice have shown that luminal bacteria stimulate epithelial cells to produce thymic stromal lymphopoietin (TSLP) and IL-25. These two proteins are directed to the lamina propria, where they exert anti-inflammatory effects as they render intestinal dendritic cells (DCs) and

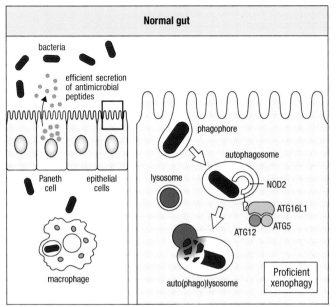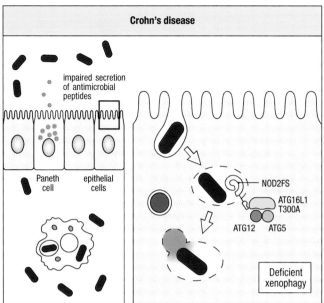

Figure 34.3 Autophagy and related genes seem to be important in Crohn's disease. A schematic view of the alterations that affect the NOD2–ATG16L1 protein interactions and xenophagy in Crohn's disease is shown. In the normal intestinal epithelium, Paneth cells secrete antimicrobial products into the intestinal lumen. In addition, resident macrophages mount a proficient xenophagic response, which is mediated by interaction between NOD2 and ATG16L1 and subsequent recruitment and activation of ATG5–ATG12 complexes. In Crohn's disease, Paneth cells have a reduced ability to produce and secrete antimicrobial peptides as a result of mutations of the genes encoding NOD2 and ATG16L1. The protein encoded by the T300A mutant of the *ATG16L1* gene retains the ability to recruit ATG5–ATG12 complexes, but it has less stability, which is responsible for a defective localization of the autophagic machinery to invading bacteria. Similarly, the xenophagic response to intracellular pathogens is highly impaired in macrophages displaying the NOD2 frameshift (FS) mutation, because NOD2FS prevents ATG16L1 from localizing at bacterial entry sites by retaining it in the cytoplasm.

macrophages unable to produce inflammatory cytokines such as IL-12 and IL-23 after bacterial stimulation. Production of both TSLP and IL-25 is diminished in IBD tissues, showing the relevance of this to human disease. Epithelial cells are also a major source of transforming growth factor (TGF)-β1, a powerful negative regulator of both macrophage and T-cell activation. In IBD, TGF-β1 is produced in excess, but its activity is markedly decreased as a result of high levels of Smad7, an intracellular inhibitor of TGF-β1 signaling. Epithelial cells contribute to the amplification of the ongoing mucosal inflammation by making chemokines, which attract more inflammatory cells into the inflamed tissue, and cytokines, which enhance T-cell growth and survival.

An excellent example of how primary, genetically encoded dysfunction of the epithelial cell can phenotypically reveal itself as IBD is through studies of the unfolded protein response in response to ER stress. Genetic defects in proteins involved in the unfolded protein response itself (e.g., *XBP1*), chaperones involved in protein folding in the ER (e.g., *AGR2*), or highly secretory proteins that make them susceptible to misfolding (e.g., *MUC2* and *HLAB27*) make the epithelium at risk for an unfolded protein

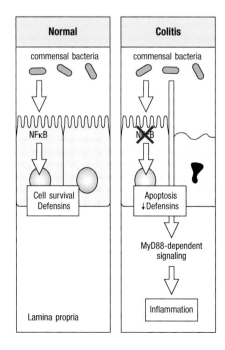

Figure 34.4 NF-κB activation in colonic epithelial cells maintains epithelial integrity. Conditional knockout of nuclear factor κB essential modulator (NEMO) in colonic epithelial cells results in a spontaneous colitis in mice. Mice lacking NEMO show increased epithelial apoptosis and reduced antibacterial defensin production, with penetration of commensal microbes into the lamina propria, triggering inflammation. Critically, in the absence of MyD88, the mice do not develop colitis, suggesting that toll-like receptor signaling is important in driving colitis.

response linked to the development of intestinal inflammation that may emanate directly from the epithelium. Mice with a gut epithelial deletion of *Xbp1*—a gene in the proximal effector pathway of the unfolded protein response and a genetic risk factor for Crohn's disease and ulcerative colitis—develop spontaneous enteritis. They also show increased susceptibility to experimental colitis and increased sensitivity of the epithelium to toll-like receptor (TLR) and cytokine signaling, a marked depletion of Paneth cells, and a reduction in the goblet cells. In mice that are genetically engineered to possess a mutant Muc2 protein that exhibits aberrant oligomerization, this is associated with the induction of ER stress in goblet cells and the development of ulcerative colitis-like colitis. Along similar lines, mice that carry a missense mutation in a membrane-bound protease (site 1 protease) that generates transcriptionally active ATF6, which is important for promoting the unfolded protein response, have enhanced susceptibility to experimental colitis.

34.9 Excessive innate immune response toward the microbiota causes chronic inflammation in the gut

The intestinal microbiota provides an abundant source of immunostimulatory organisms that can activate innate and adaptive immune responses. In IBD tissues, innate myeloid cells overexpress various membrane-bound and endosomal receptors for microbial components; these can signal to the cell, resulting in the production of enormous quantities of pro-inflammatory molecules. Thus, increased levels of pro-inflammatory cytokines, including IL-1β, IL-6, IL-18, and tumor necrosis factor (TNF-α), are detected in active IBD and correlate with the severity of inflammation. These cytokines are probably important in the IBD-associated chronic inflammation, given their ability to alter tight junctions and intestinal permeability, to stimulate the secretion of chemokines and the expression of adhesion molecules on endothelial vessels, and to induce stromal cells to secrete extracellular matrix-degrading enzymes, such as matrix metalloproteinase 1 (MMP1) and MMP3, which digest away the extracellular matrix, leading to ulceration.

The synthesis and secretion of pro-inflammatory cytokines are governed by germline-encoded receptors such as the TLRs and nucleotide-binding-domain and LRR-containing (NLR) proteins that can be activated by microbe-associated molecular patterns (MAMPs). For example, the production of active (mature) IL-1β and IL-18 occurs through a two-step process initiated by the transcriptional induction of a procytokine (e.g., a TLR stimulus) followed by caspase 1–mediated cleavage. In this process, NLRP3, also known as cryopyrin, associates with the NLR adaptor protein, apoptotic speck protein containing a CARD domain (ASC/PYCARD), to recruit procaspase 1. This complex is referred to as the inflammasome, and leads to the processing of procaspase 1 into active caspase 1. Caspase 1 is responsible for the subsequent cleavage of the IL-1β/IL-18 precursors into their functional forms. In addition to NLRP3, other NLRs, including NLRP1, NLRC4, and NAIP, also function in caspase-1 activation and IL-1β production through the formation of other inflammasomes in response to distinct sets of stimuli. Mice lacking NLRP3, ASC, or caspase 1 produce reduced levels of IL-1β and TNF-α and are protected from inflammation in acute, but not necessarily chronic, experimental colitis, suggesting that the NLRP inflammasome has a more critical role during the early phase of colitis.

TLR, NLR, and retinoic acid–inducible gene (RIG)-related molecules recognize "alarm" signals provided by MAMPs from bacteria and viruses as well as damage-associated molecular patterns (DAMPs) associated with cellular injury such as uric acid. Most TLRs use the adaptor protein MyD88

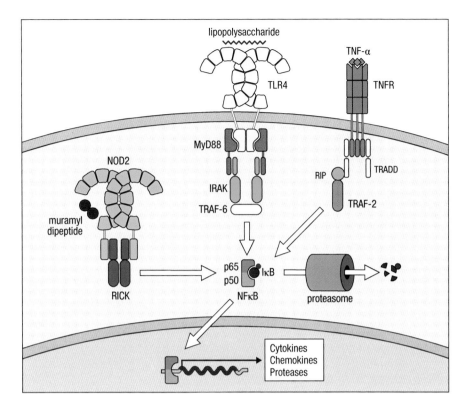

Figure 34.5 NF-κB activation in myeloid cells is strongly pro-inflammatory in the gut. One of the features of inflammatory bowel disease, especially Crohn's disease, is the influx of monocytes into diseased gut. In the inflamed tissues, monocytes potentially receive many pro-inflammatory signals, all of which can activate NF-κB. Three examples are shown here: lipopolysaccharide can activate NF-κB via TLR4; intracellular NOD2 can be activated by peptidoglycans; and cytokines such as TNF-α can signal via TNFR. In response to all these stimuli, monocytes produce extremely large amounts of TNF-α, IL-1β, IL-6, IL-8, and IL-12, as well as macrophage metalloelastase (MMP12), which serves to degrade the mucosa and maintain the continuing influx of white cells into the tissue.

to induce activation of the innate and adaptive immune systems. TLRs and MyD88 are critically involved in the pathogenesis of chronic intestinal inflammation on activation by the commensal microbiota. For instance, the absence of MyD88 prevents enterocolitis in IL-10–deficient mice in association with decreased activation of the T-cell–dependent immune responses directed against the commensal microbiota. However, TLR- and NLR-related pathways are also important in maintaining intestinal homeostasis and thus preventing intestinal inflammation. Examples include the suppression of the early onset of colitis in multidrug-resistance-α-deficient mice by TLR2 agonists, the enhanced inflammation observed in MyD88-deficient mice when experimentally challenged with a barrier disruptive agent (dextran sodium sulfate), and the inhibition of experimental colitis by activation of NOD2 signaling. Innate recognition of MAMPs is thus important both for the maintenance of homeostasis and for the induction of inflammation under the appropriate environmental and probably genetic contexts.

Given the importance of the transcription factor NF-κB in regulating TLR responses, it is important to consider the involvement of this transcription factor in the tissue-damaging inflammatory reaction associated with IBD. Increased NF-κB activity is seen in intestinal epithelial cells and lamina propria mononuclear cells in IBD and is associated with an enhanced production of inflammatory cytokines such as IL-1β, IL-6, and TNF-α (Figure 34.5). In contrast, deletion of *Ikkb* in macrophages and neutrophils, inhibition of NF-κB by antisense oligonucleotides directed against RelA/p65, and inhibition of IκB kinase by a small-molecule inhibitor (BMS-345541) can ameliorate disease in mouse models of intestinal inflammation. Although this highlights the important pro-inflammatory role of this transcription factor, NF-κB clearly has important homeostatic properties, especially within the intestinal epithelium, such that deletion of *Ikkg* (NEMO) specifically within the intestinal epithelium results in spontaneous colitis, which may account in part for the beneficial effects of TLR-related signaling in maintaining homeostasis.

34.10 Innate immunity dysfunction is linked to Crohn's disease and Crohn's-like disease

The pathophysiology of IBD, especially Crohn's disease, is currently undergoing a significant reassessment, because genetic studies have led to the identification of several susceptibility genes associated with innate immune dysfunction, suggesting that these may be central to disease pathogenesis. Another source of support for this idea comes from clinical observations showing that rare congenital disorders of innate immunity (and, in particular, phagocyte function) can associate with noninfectious bowel inflammation that shows similarities to Crohn's disease, and that some patients with various neutropenias (namely, leukocyte adhesion deficiency-1, glycogen storage disease type 1b, Hermansky-Pudlak syndrome, chronic granulomatous disease, and Chediak-Higashi syndrome) have been reported to develop chronic intestinal inflammation that looks similar to Crohn's disease. These results have been extrapolated to suggest that Crohn's disease is caused by a defect in macrophage function and that adaptive immune responses, as described later, are a secondary event. In reality, the situation is more complex in that there is unlikely to be a single unifying problem with macrophages in Crohn's disease; instead, a combination of subtle effects in innate immunity, differing between individuals, may allow the accumulation of sufficient microbial antigens in the gut wall to trigger T-cell activation. Therefore, although the rare phagocyte defects show that an inability to handle antigens from the gut microflora properly can result in disease that resembles Crohn's disease, it is unlikely that this is the complete story, and it is equally likely that some patients also have altered thresholds of T-cell activation.

34.11 NOD2/CARD15 is involved in the control of defensins in the gut

NOD2 is strongly expressed in Paneth cells, a subset of epithelial cells at the base of intestinal crypts of the small intestine, predominantly in the terminal ileum, and in the cytoplasm of monocytes and tissue macrophages. Mutations in NOD2 have been reported to associate with a diminished activation of NF-κB and decreased production of inflammatory cytokines and defensins in response to MDP. It is not clear how hypomorphic NOD2 function reconciles with intestinal inflammation and an absence of spontaneous intestinal inflammation in Nod2-deficient mice, although it is consistent with the absence of such inflammation in MyD88-deficient mice. The mechanisms by which hypomorphic NOD2 function predisposes to Crohn's disease are, however, likely to be multiple through effects on different cell types, which are, in turn, dependent on specific environmental factors such as enteropathogenic exposures. Such mechanisms potentially include decreased regulation of the inflammasome and consequently increased IL-1-β production in response to the commensal microbiota; diminished TLR-associated tolerance to microbial stimuli with increased production of IL-12 in response to TLR2 ligands; and lessened IL-10 production through a pathway involving NOD2 association with active, phosphorylated p38 mitogen-activated protein kinase, and the transcription factor heterogeneous nuclear ribonucleoprotein A1, which binds to and activates the IL-10 promoter. It has been suggested that mutations in NOD2 cause decreased α-defensin production by Paneth cells, leading to intestinal dysbiosis, decreased antimicrobial activity, and possibly decreased resistance to enteropathogens. As emphasized earlier, however, labeling Crohn's disease due to NOD2 mutations as being due to a defect in Paneth cells is likely to be an oversimplification of a complex disease. Other factors may also be involved, such as the failure of pro-inflammatory innate responses to be restrained by the zinc-finger protein A20, a ubiquitin-modifying enzyme also known as TNF-α-induced protein 3 (TNFAIP3), which is a negative regulator of signals downstream of TNF- and TLR-driven NF-κB activation and MDP-induced NOD2 signaling.

ADAPTIVE IMMUNITY IN PATHOGENESIS OF IBD

34.12 Genes associated with T-cell activation and differentiation are also involved in IBD

Gene variants related to specific pathways of inflammatory responses have been identified by genome-wide association studies in patients with IBD. One example is the signaling pathway associated with the IL-23 receptor, highlighting the importance of helper T (T_H) T_H17 cells, given the identification of polymorphisms in *IL23R* as well as IL-12p40 (*IL12B*), which forms a part of both IL-12 and IL-23, Janus-associated kinase 2 (*JAK2*), signal transducer of activation 3 (*STAT3*), and the *IL2/IL21* locus, which are associated with Crohn's disease and ulcerative colitis. Interestingly, both protective variants, such as the uncommon coding variant rs11209026 (Arg381Gln) of the IL-23R, and risk-associated variants have been identified, demonstrating that individual genes can contribute risk in quite opposite ways to others in an individual with these disorders (Figure 34.6).

34.13 Ongoing inflammation in IBD is driven by T cells

In IBD, antigen-presenting cells such as dendritic cells (myeloid, plasmacytoid, mature DCs) and macrophages seem to have a key role in shaping the mucosal T-cell responses that lead to polarized T-cell responses associated with chronic intestinal inflammation (Figure 34.7). The number of antigen-presenting cells is increased in IBD, and these cells display signs of local cell activation in the lamina propria. In addition to these classical cell subsets, cell populations with an intermediate phenotype have been described. A CD14$^+$ cell subset expressing both macrophage and dendritic cell markers has recently been identified in patients with Crohn's disease. This cell subset produces large quantities of pro-inflammatory cytokines such as IL-23, interferon-γ (IFN-γ), TNF-α, and IL-6 during active inflammation. In addition, antigen-presenting cells in IBD are known to produce cytokines such as IL-12 and Epstein-Barr virus-induced gene 3 (EBI3)-related cytokines such as IL-27 and IL-35 that control helper-T-cell polarization. IL-12 has been shown to have a key role in the induction of T_H1-cell differentiation by activating the signaling protein STAT4. The increased production of IL-12 p35/p40 in patients with Crohn's disease in comparison with controls and patients with ulcerative colitis is therefore likely to contribute to the different helper-T-cell polarization between Crohn's disease and ulcerative colitis. Interestingly, blockade of the p40 subunit of IL-12 is associated with marked therapeutic effects in various murine models of IBD. Because anti-p40 antibodies neutralize the IL-23 (p19/p40) in addition to IL-12, the anti-p40 effects might be due to neutralization of IL-12, IL-23, or both cytokines. In fact, several studies show that IL-23 has a prominent role in murine models of chronic intestinal inflammation: blockade of IL-23 and p19 deficiency protects from intestinal inflammation. The relevance of these findings to human IBD is underlined by the observation that lamina propria cells in IBD produce augmented amounts of IL-23 and that IL-23R mutations are frequently found in patients with IBD as described earlier in the chapter. The ultimate proof is the recent approval of an anti-p40 antibody for Crohn's disease patients.

The intestinal lamina propria contains many CD4 T lymphocytes that probably have a fundamental role in mucosal tolerance and host defense in the healthy intestine. Regulatory T cells in particular have been identified as a T-cell subset with anti-inflammatory properties in the gut. These cells

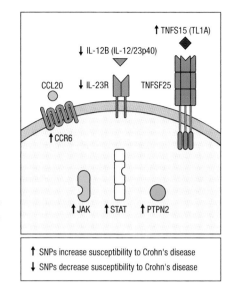

↑ SNPs increase susceptibility to Crohn's disease

↓ SNPs decrease susceptibility to Crohn's disease

Figure 34.6 Several of the loci identified in genome-wide association studies in Crohn's disease are involved in T-cell activation. Single nucleotide polymorphisms (SNPs) in *IL12B* (IL-12/23p40) and *IL23R* genes are associated with less risk of Crohn's disease, suggesting that the threshold of T_H1/T_H17 activation may be a determinant of disease. Similarly, TNFSF15 (TL1A) is a costimulatory molecule on antigen-presenting cells that, when it binds to its receptor on the T-cell surface (TNFSF25), increases interferon-γ production. *TL1A* SNPs are associated with increased risk of disease. *CCR6* SNPs are also associated with Crohn's disease; interestingly, this receptor binds CCL20, a chemokine made by epithelial cells to attract T cells into the gut. Finally, inside the cell, SNPs in the genes encoding the JAK/STAT pathways, which control cytokine signaling, and PTPN2, are associated with increased disease. The latter is interesting because the function of PTPN2 is to remove phosphates from signaling molecules and thus to dampen immune activation.

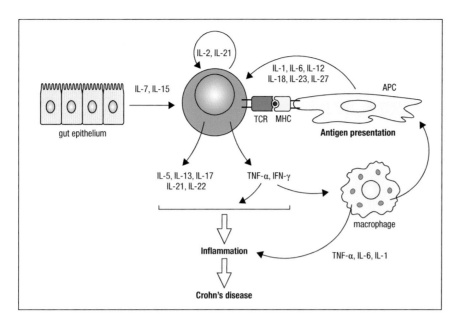

Figure 34.7 Chronic T-cell activation is a feature of inflammatory bowel disease (IBD), especially Crohn's disease. In the inflamed environment, multiple factors are probably responsible for keeping T cells alive and may vary from patient to patient. There is very little evidence for antigen-driven T-cell activation in human IBD because of the difficulty of identifying potential peptides from the myriad proteins made by gut bacteria. In contrast, there is a wealth of data to suggest that cytokines made by accessory cells are important in driving T-cell activation in IBD. APC, antigen-presenting cell.

express CD25 and the transcription factor Foxp3 and are known to counteract the function of mucosal effector T cells. The number of T cells with a normal regulatory function is increased in IBD. This is consistent with there being a marked expansion of effector T cells in IBD that cannot be adequately controlled by these regulatory T cells (see **Figure 34.7**). Effector T lymphocytes in IBD are activated in IBD via antigen-presenting cells and luminal antigens. They show elevated cell proliferation and the activation of T-cell receptor-inducible transcription factors such as NFATc2. Furthermore, specific signature cytokines are produced by these cells in active IBD that permit their classification into helper-T-cell subsets. For instance, lamina propria T cells in Crohn's disease produce larger amounts of T_H1 cytokines such as IFN-γ on stimulation and express the T_H1 transcription factor T-bet. In ulcerative colitis, however, T cells produce increased amounts of some T_H2 cytokines such as IL-5 and IL-13. In contrast, IL-4 production by these cells is lower than in controls, suggesting the presence of an atypical T_H2 cytokine profile. However, neutralization of IL-13 (tralokinumab) in patients with ulcerative colitis did not improve disease severity. In addition, a novel IL-9 producing T-cell subpopulation (T_H9) has been identified that is significantly increased in the mucosa of ulcerative colitis patients. In addition to prototypical T_H1, T_H2, or T_H9 cytokines, large amounts of disease-perpetuating cytokines such as TNF-α and IL-6 are produced by lamina propria T cells in IBD as well as in cells associated with the innate immune system.

A new subset of T cells, designated T_H17 cells, has recently been described; the development of these cells requires the presence of specific commensal microbes that are yet to be defined in humans. T_H17 cells are generated via naive cells that are exposed to TGF-β plus IL-6 or TGF-β plus IL-21. In addition, it is known that these cells require IL-23 for stabilization of their phenotype. Various studies have analyzed the presence of this T_H17 cell subset in IBD. In patients with IBD, T_H17-inducing cytokines such as IL-6 and IL-21, and T_H17-related cytokines such as IL-17A/F and IL-22, are present at increased levels in the inflamed mucosa. Furthermore, T_H17-related transcription factors such as IRF4, STAT3, and RORA/C are upregulated in CD4 T cells isolated from IBD mucosa. Taken together, these results indicate the presence of active T_H17 cells in patients with IBD. Inactivation of cytokines that induce or augment T_H17 responses such as IL-6 and IL-21 has been shown to suppress T-cell–dependent colitis. Together with the genetic information described showing an IL-23R/T_H17 genetic signature in IBD, T_H17 cells probably have an important role in the pathogenesis of chronic intestinal inflammation, but the precise

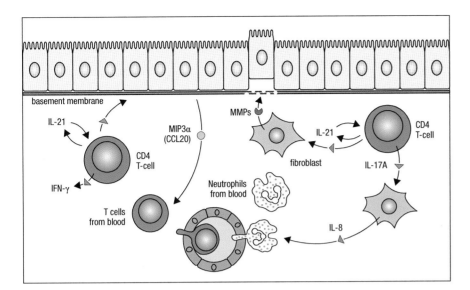

Figure 34.8 Newer cytokines such as IL-21 are strongly pro-inflammatory in the gut. IL-21 functions as an autocrine growth factor for both T_H1 and T_H17 cells but also induces epithelial cells to make CCL20 and fibroblasts to make matrix metalloproteinases (MMPs), which contribute to ulcer formation. IL-17A can also signal to many cell types and increase cytokine production. In the example shown here, IL-17A acts on fibroblasts to increase the production of neutrophil chemokines such as IL-8.

role remains to be defined. However, neutralization of IL-17A (secukinumab) did not show clinical efficacy in Crohn's disease. There is, however, no doubt that IL-21 is strongly pro-inflammatory in the gut (**Figure 34.8**), but whether IL-17F is pathogenic or protective in the gut remains to be determined.

One of the major advances in IBD has been the translation of anticytokine therapy into the clinic, spectacularly demonstrated by the dramatic effects of anti-TNF therapy in both Crohn's disease and ulcerative colitis. However, subsequent studies blocking IFN-γ or IL-17 A in Crohn's disease and similarly IL-13 in ulcerative colitis have not quite realized the same level of clinical efficacy, and many further studies are needed to identify which patients will respond best to individual therapies. In a similar fashion to the way in which the development of IBD in a single patient may be complex and even idiosyncratic and unique, it may well be that the disease-causing processes in individual patients may also be highly complex, and evolve as a result of therapy and disease duration, so that immune interventions will have to be tailored to the patient.

34.14 Cell migration to the site of inflammation

The expression of pro-inflammatory cytokines in the lamina propria of ulcerative colitis as well as Crohn's disease induces an upregulation of the adhesion molecules mucosal addressin cell adhesion molecule 1 (MAdCAM-1), intercellular adhesion molecule 1 (ICAM-1), and vascular cell adhesion molecule 1 (VCAM-1) on endothelial cells. During inflammatory bowel disease, immune cells expressing the integrins α4β7 and α4β1 are mostly relevant, entering the mucosa by binding to MAdCAM-1. While α4β1 is rather unspecific and allows for infiltration of lymphocytes in all organs via binding to VCAM-1, α4β7 mediates selectively gut-homing. This specific mechanism has been taken into the clinic, and the first therapeutic strategy blocking α4β7 (vedolizumab) has been approved for treatment of ulcerative colitis as well as Crohn's disease.

SUMMARY

Inflammatory bowel diseases are chronic conditions of unknown etiology. Recent data indicate that specific genetic factors predispose for disease development by modifying intestinal immune responses and barrier function. Changes in both innate and adaptive mucosal immune responses are associated with inflammatory bowel diseases and drive uncontrolled mucosal inflammation in the small and/or large bowel. In particular, activation of antigen-presenting cells and T cells in response to the commensal microflora is a key factor in disease perpetuation. Recent studies on mucosal immune responses have provided new insights into disease pathogenesis and will unequivocally result in new opportunities for optimized therapy.

FURTHER READING

Guiffrida, P., Biancheri, P., and Macdonald, T.T.: Proteases and small intestinal barrier function. *Curr. Op. Gastro.* 2014, 30:147–153.

Hosomi, S., Kaser, A., and Blumberg, R.S.: Role of endoplasmic reticulum stress and autophagy as interlinking pathways in the pathogenesis of inflammatory bowel disease. *Curr. Op. Gastro.* 2015, 31:81–88.

Kaser, A., Zeissig, S., and Blumberg, R.S.: Inflammatory bowel disease. *Annu. Rev. Immunol.* 2010, 28:573–621.

Lees, C.W., Barrett, J.C., Parkes, M. et al.: New IBD genetics: Common pathways with other diseases. *Gut* 2011, 60:1739–1753.

Macdonald, T.T., and Monteleone, G.: Immunity, inflammation, and allergy in the gut. *Science* 2005, 307:1920–1925.

McCarroll, S.A., Huett, A., Kuballa, P. et al.: Deletion polymorphism upstream of IRGM associated with altered IRGM expression and Crohn's disease. *Nat. Genet.* 2008, 40:1107–1112.

Monteleone, G., Pallone, F., and MacDonald, T.T.: Interleukin-21: A critical regulator of the balance between effector and regulatory T-cell responses. *Trends Immunol.* 2008, 29:290–294.

Ogura, Y., Bonen, D.K., Inohara, N. et al.: A frameshift mutation in NOD2 associated with susceptibility to Crohn's disease. *Nature* 2001, 411:603–606.

Rioux, J.D., Goyette, P., Vyse, T.J. et al.: Mapping of multiple susceptibility variants within the MHC region for 7 immune-mediated diseases. *Proc. Natl Acad. Sci. USA.* 2009, 106:18680–18685.

Sartor, R.B.: Genetics and environmental interactions shape the intestinal microbiome to promote inflammatory bowel disease versus mucosal homeostasis. *Gastroenterology* 2010, 139:1816–1819.

Sartor, R.B.: Key questions to guide a better understanding of host-commensal microbiota interactions in intestinal inflammation. *Mucosal Immunol.* 2011, 4:127–132.

Strober, W., Zhang, F., Kitani, A. et al.: Proinflammatory cytokines underlying the inflammation of Crohn's disease. *Curr. Opin. Gastroenterol.* 2010, 26:310–317.

Xavier, R.J., and Podolsky, D.K.: Unravelling the pathogenesis of inflammatory bowel disease. *Nature* 2007, 448:427–434.

Food allergies and eosinophilic gastrointestinal diseases

35

CATHRYN NAGLER AND GLENN T. FURUTA

Food-related diseases are clinical response(s) to a given food (or foods) that are governed by an aberrant immune response. Inappropriate responses to innocuous food antigens can be IgE mediated, non-IgE mediated, or a mixed variety. Food allergies are increasingly common, especially in children. Thirty-two million Americans are estimated to suffer from food allergies; disease prevalence is similar in other Westernized societies. A new group of diseases associated with aberrant responses to food, the eosinophilic gastrointestinal diseases (EGIDs), has also become increasingly recognized. EGIDs manifest as vague gastrointestinal complaints that occur in the setting of dense gastrointestinal mucosal eosinophilia. As for other allergic diseases, a genetic basis is involved in the susceptibility to allergic responses to food. However, the rapidly increasing prevalence of food allergies and EGIDs cannot be explained by genetic susceptibility alone, and environmental, particularly microbial, influences have been implicated. No treatment other than strict avoidance of potential food antigens is currently available for any of the food-related diseases, making potentially life-threatening anaphylactic responses to food particularly problematic in school and childcare settings.

BASIC PRINCIPLES OF IMMUNE RESPONSE TO FOODS

In the United States, the average adult consumes an estimated 0.85 tons of food per year. Thus, the volume and diversity of antigens to which we are exposed in food is enormous, yet most of us remain nonresponsive to this constant source of stimulation. Undoubtedly, the process of digestion eliminates the vast majority of antigens from the diet, but because the immune system is designed to respond to very small amounts of antigen, dose may not be a determinant of sensitivity.

35.1 Healthy individuals are tolerant to remarkable diversity of antigens present in food

Our immune system does not *ignore* dietary antigens—the induction of nonresponsiveness to food, referred to as oral tolerance in animal models, is an active immune-mediated process. The integrity of the mucosal barrier and the production of secretory IgA also serve to exclude most food antigens from gaining access to the antigen-presenting cells of the mucosal immune system, but this barrier is not impenetrable. Small amounts of food proteins are detectable in the bloodstream, and many healthy individuals have low levels of circulating food protein-specific IgG, IgA, and IgM in their serum.

Our understanding of the mechanism(s) by which nonresponsiveness to dietary antigens is induced (and why these mechanisms fail in food-allergic individuals) remains incomplete. The induction of a food antigen-specific immunoregulatory response is clearly one important component. Specialized populations of migratory CD103$^+$ dendritic cells capture antigen that crosses the epithelium covering the small intestinal lamina propria and carry it to the mesenteric lymph node. In this mucosal microenvironment, rich in retinoic acid and transforming growth factor (TGF)-β, the presentation of food antigens to naive T cells induces the upregulation of the gut homing receptors CCR9 and α4β7 and the differentiation of food antigen-specific regulatory T (T_{reg}) cells. These antigen-specific T_{reg} cells then traffic back to the small intestinal lamina propria to prevent sensitization following subsequent antigen exposure.

35.2 Aberrant immune responses to foods can be IgE- and non-IgE mediated

In some individuals, nonresponsiveness to particular food antigens is either not induced or abrogated. This failure to induce tolerance progresses to an aberrant or heightened immune response to particular food proteins. Food-related diseases can take a variety of forms and elicit responses that are predominantly allergic, inflammatory, or a mixture of the two (Table 35.1).

IgE-mediated (allergic) disease is characterized by an acute, systemic response to oral food challenges that can include gastrointestinal symptoms (nausea, vomiting, abdominal pain, and diarrhea), as well as the symptoms common to allergic responses at other sites (urticaria, pruritis, anaphylaxis). The mechanisms involved are straightforward, involving classical type I hypersensitivity, IgE, and mast cells. Interestingly, classical food allergies produce very little morphologic damage to the small bowel. The main finding is mucosal edema. However, local mediator release changes intestinal transit, increases epithelial permeability, and induces epithelial cells to become secretory instead of absorptive. The gut epithelium also expresses CD23, the

TABLE 35.1 TYPES OF IMMUNE RESPONSES TO FOOD

Type	Key features	Disorders	Symptoms
IgE mediated	Acute onset Systemic	Food allergy	Nausea Abdominal pain Vomiting Diarrhea Pruritis Urticaria Anaphylaxis
Non-IgE (cell mediated)	Delayed onset	Cow's milk allergy	Vomiting
	Confined to gastrointestinal tract	Food protein–induced enterocolitis	Diarrhea
		Celiac disease	
Mixed reaction (both IgE and cell mediated)	Chronic	Eosinophilic esophagitis Eosinophilic gastritis Eosinophilic gastroenteritis Eosinophilic colitis	Feeding difficulty Vomiting Abdominal pain Dysphagia Food impaction

low-affinity receptor for IgE, and compelling evidence indicates that allergens coupled to trace amounts of IgE can be transported across the epithelium into the lamina propria to propagate inflammation.

Non-IgE (cell-mediated) reactions are typically characterized by a delayed-onset inflammatory response that is confined to the gastrointestinal tract and includes disorders such as food protein–induced enterocolitis and celiac disease. This type of reaction may be T-cell mediated, as in celiac disease, which clearly involves a helper-T (T_H)–mediated response to gluten. However, the extent of the mucosal flattening in non-IgE-mediated food protein–induced enterocolitis is minor compared with celiac disease, with only some villous blunting. Late-onset responses are T_H2 mediated, involving eosinophils and cytokines such as interleukin (IL)-13 that increase intestinal permeability.

Mixed reactions to food allergens have both inflammatory and allergic components and are exemplified by the EGIDs, as discussed next.

35.3 Diagnosis of food allergies and eosinophilic gastrointestinal diseases are diagnosed by laboratory and clinical parameters

The diagnosis of food allergies relies on a combination of factors, including a complete medical history, radioallergosorbent testing (RAST) to detect circulating food allergen-specific IgE, skin prick testing with allergen, and, in the case of the EGIDs, endoscopy with mucosal biopsy. The diagnostic approach is dependent on the type of food-induced immune response and the symptoms at presentation. A detailed history is central to the diagnosis of these diseases. The patient must have a reproducible response to a specific food or group of foods. Historical elements that should be assessed include not only which food elicits the symptoms but also whether it is cooked or raw, the amount necessary to induce the response, the timing of symptoms after ingestion of the food, and the type, severity, and duration of intestinal and extraintestinal symptoms. The elimination of other diseases such as toxin-mediated reactions, inflammatory bowel diseases, and food intolerances is an important part of the diagnostic process. Food intolerances are non-immune-mediated physical responses to food products. A common food-intolerance condition is lactose intolerance, in which the patient has insufficient levels of lactase on the intestinal mucosa to digest lactose. This deficiency leads to maldigestion of the sugar with resultant gas production, diarrhea, and abdominal pain. Specific features important to the diagnosis of the different types of food-related diseases are discussed later.

While IgE-mediated food allergies can manifest as gastrointestinal symptoms, other allergic symptoms related to the skin (urticaria) and lungs (wheezing and cardiovascular collapse) usually dominate. Typically, symptoms occur within minutes following the ingestion of the food. RAST and skin prick testing with commercial food extracts are frequently positive for the suspected foods.

Non-IgE (cell-mediated) food-sensitive diseases include food protein–induced enterocolitis and celiac disease, which is induced by an inappropriate response to dietary gluten (see Chapter 31 for detailed description of celiac disease). Food protein–induced enterocolitis syndrome (FPIES) is an acute life-threatening disease; the diagnosis is based on history, physical, and biochemical analysis and does not require endoscopy with biopsy. FPIES can lead to profuse diarrhea, leading to dehydration and acidosis, that is typically caused by an immune-mediated response to milk protein. The diagnosis of celiac disease is based on the history and laboratory testing, including mucosal biopsy. Symptoms of celiac disease include diarrhea, abdominal

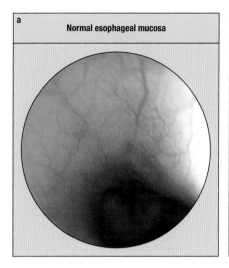

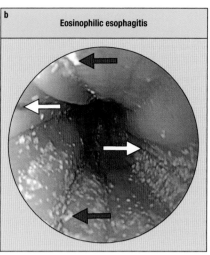

Figure 35.1 Endoscopic manifestations of eosinophilic esophagitis. (**a**) Normal esophageal mucosa. The mucosa is smooth and pink with visible vasculature underlying a translucent epithelial surface. (**b**) Mucosa from a patient with eosinophilic esophagitis. The mucosal surface is dull, without a visible vascular pattern, and contains small white patches of eosinophilic exudate (black arrows) and longitudinal furrows (white arrows).

pain, and slow growth. Serological testing for specific antibody titers (antitissue transglutaminase or endomysial antibodies) and an abnormal mucosal biopsy that shows villous blunting and lymphocytic inflammation are required to make a diagnosis of celiac disease.

Mixed food-sensitive enteropathies represent a broad range of diseases that include the EGIDs. EGIDs are characterized by various gastrointestinal symptoms that occur in the setting of mucosal eosinophilia. Involvement can occur at any location in the gastrointestinal tract and is designated accordingly as eosinophilic esophagitis, eosinophilic gastritis, eosinophilic gastroenteritis, or eosinophilic colitis. The best defined of these is eosinophilic esophagitis (EoE). EoE presents with feeding difficulties, vomiting, or abdominal pain in the young child and dysphagia or food impaction in the adolescent/adult. Endoscopic appearance of the mucosa in EoE includes mucosal edema, linear furrows, consecutive rings, exudates (Figure 35.1), and luminal narrowing due to stricture. Mucosal sampling of the esophagus reveals dense eosinophilia with over 15 eosinophils per high-power field, basal zone hyperplasia, and in some circumstances, microabscesses (Figure 35.2). Other diseases associated with esophageal eosinophilia, especially gastroesophageal reflux disease (GERD), must be ruled out before assigning a diagnosis of EoE.

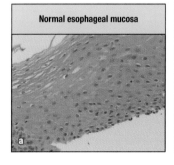

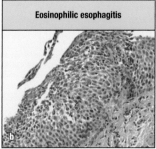

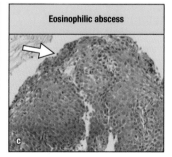

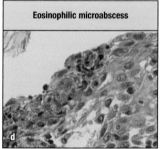

Figure 35.2 Histology of eosinophilic esophagitis. (**a**) Normal esophageal mucosa. The mucosa is composed of stratified squamous epithelia layered over a 2- to 3-cell-thick basal layer. A few scattered lymphocytes are present, and neutrophils or eosinophils are absent. (**b**) Eosinophilic esophagitis. The epithelium is hyperplastic with an expanded basal layer. Numerous eosinophils are present throughout the epithelial surface. (**c**) Eosinophilic abscess. In addition to a hyperplastic epithelium, an eosinophilic abscess (arrow) is present beneath a white exudate visualized endoscopically (see **Figure 35.1**). (**d**) Eosinophilic microabscess. A magnified view shows the luminal surface of the microabscess.

35.4 Induction of food-specific IgE can lead to mast cell degranulation and both local and systemic allergic symptoms

In individuals with a predisposition to allergic disease (atopy), exposure to food allergens induces an immune response biased to the production of the T_H2 cytokines IL-4 and IL-13, which support class switching to IgE.

Food antigen–specific IgE binds to cells bearing high-affinity Fcε receptors (FcεRI), including mast cells and basophils (**Figure 35.3**). A myeloid cell subset that bridges innate and adaptive immunity, mast cells are characterized by large numbers of preformed granules that contain pro-inflammatory, vasoactive, and neuroactive mediators. Mast cells typically reside at host-environment interfaces in vascularized connective tissue (notably skin) and at mucosal surfaces of the respiratory and gastrointestinal tracts. Binding of the ingested allergen to antigen-specific IgE cross-links the IgE bound to mast cell surface FcεRI, resulting in rapid mast cell degranulation. In the gastrointestinal tract, release of histamine by mucosal mast cells causes the contraction of smooth muscle and induces vomiting; transepithelial fluid loss into the gut lumen leads to diarrhea. Allergen rapidly absorbed from the gastrointestinal tract or introduced into the bloodstream also reaches the connective tissue mast cells associated with blood vessels; degranulation at these sites results in urticaria (hives). Widespread degranulation of connective tissue mast cells and the concomitant release of histamine can also result in systemic anaphylaxis and potentially fatal anaphylactic shock. The histamine-induced increase in vascular permeability results in a precipitous loss of blood pressure and airway constriction. Swelling of the epiglottis further impairs breathing and can lead to asphyxiation. Rapid injection of epinephrine typically prevents fatal anaphylactic shock by inhibiting smooth muscle contraction and vasodilation. Food-allergic individuals are advised to carry an auto-injectable epinephrine-containing EpiPen for protection against anaphylactic responses to accidental ingestion of food allergens.

Mast cells are also present in mucosal tissues of patients with EGIDs. Esophageal mast cells have not been well characterized, but their increased numbers and evidence of their degranulation are recognized as a feature of EoE. Speculation as to their role in EoE, as with other IgE-mediated diseases, includes the stimulation of smooth muscle contraction leading to esophageal

Figure 35.3 Cellular mechanisms of allergic responses to food. In healthy subjects, ingestion of innocuous antigens leads to systemic nonresponsiveness (oral tolerance). In allergic persons, oral tolerance is not induced. In the T_H2-biased mucosal microenvironment, antigen-specific T_H2 cells are generated when antigen fragments are presented to naive T cells by antigen-presenting cells, mainly dendritic cells. Once activated, T_H2 cells produce IL-4 and IL-13, which promote IgE production by B cells. IgE binds to the Fcε receptor on allergic effector cells (notably mast cells). Upon subsequent allergen ingestion, antigen presentation leads to rapid T-cell activation and secretion of T_H2 cytokines, triggering mediator release by eosinophils and basophils. At the same time, antigenic fragments may interact directly with receptor-bound IgE on mast cells, and aggregation of receptors triggers the release of preformed and newly formed chemical mediators (particularly histamine) responsible for clinical symptoms. IFN-γ, interferon-γ.

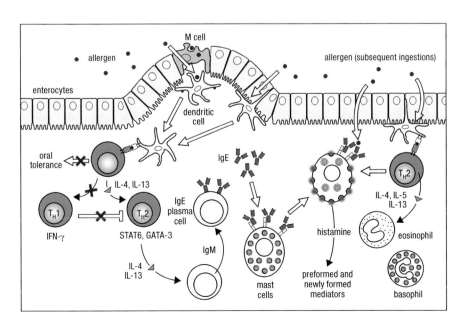

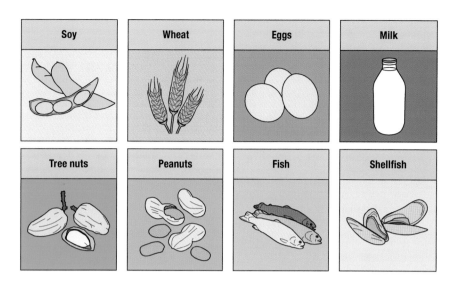

Figure 35.4 Most allergic responses are attributable to only eight foods. Allergic responses to some of these foods, including milk, eggs, wheat, and soy (blue shading), are transiently induced during childhood (typically ages 2–5 years) and outgrown with age. Other foods, including tree nuts, peanuts, fish, and shellfish (green shading), induce lifelong allergic disease with the risk of anaphylactic hyperreactivity.

dysmotility. Mast cells are also increased in the mucosa in other EGIDs of the distal intestinal tract. Murine models of intestinal eosinophilia support a role for mast cells in the generation of diarrhea and increase in intestinal permeability.

35.5 Eight types of food account for most allergic responses

Although any food can elicit an allergic response, 90% of aberrant reactions are attributable to only eight major types of food: milk, eggs, fish, shellfish (particularly crustaceans), peanuts, soybeans, tree nuts, and wheat.

Allergic responses to milk, soybeans, eggs, and wheat are typically transient and restricted to childhood, whereas allergic responses to peanuts, tree nuts, fish, and shellfish often result in lifelong anaphylactic hypersensitivity (Figure 35.4). Most of the immunodominant protein allergens for these foods have been identified and share some features in common; they are water-soluble, heat-stable glycoproteins generally resistant to digestion by gastric pepsin. Thus far, however, their biochemistry has not provided information to explain their potent allergenicity. Marked geographic variations in the prevalence of allergic responses to particular foods may relate to diet, ethnic background, or both. Some evidence suggests that regional methods of food preparation might also impact allergenicity.

35.6 Allergies initially manifest at peripheral sites such as skin, but progress to the airways

Food allergy is clinically linked to other allergic diseases, including asthma. Thus, atopic dermatitis, food allergy, and asthma may be viewed as cutaneous, gastrointestinal, and pulmonary manifestations of atopy as a systemic allergic disorder. All three conditions are characterized by eosinophilia and elevated levels of IgE. The progression of allergic manifestations from cutaneous and gastrointestinal sites to the respiratory tract has been called the "allergic march" and describes the natural history of atopy as a systemic disorder, whose manifestations peak sequentially at different sites (Figure 35.5). Pediatric patients who develop food allergy are at a higher risk of developing asthma over time.

Atopic dermatitis and food allergy predominate in early childhood and subside with age. In contrast, asthma, allergic rhinitis, and allergic responses to a subgroup of food allergens (notably peanuts, tree nuts, and shellfish)

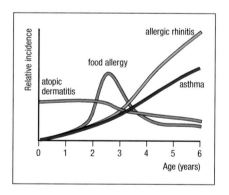

Figure 35.5 The allergic march. Atopic persons tend to present with site-specific manifestations of allergic disease that occur in an ordered sequence. Atopic dermatitis appears in infancy. Transient food allergy peaks in early childhood; a subset of food-allergic patients develops lifelong disease. Asthma and allergic rhinitis follow and continue to increase in prevalence in adulthood.

increase in prevalence with age. The mechanism(s) responsible for the spread of allergic disease from the gut to the lung is not well understood. Many children with food allergies have a family history of atopy, which is often not restricted to food. Some aspects of the relationship between food allergies and asthma are intriguing. In a study in London, half of the children who were intubated for life-threatening asthma also had food allergies, compared with only 10% of controls, and fatal food-induced anaphylaxis occurred almost exclusively in individuals with asthma. Peanuts or tree nuts accounted for 94% of these deaths.

35.7 Sensitization to food antigens may occur outside the gastrointestinal tract

Respiratory sensitization to inhaled antigens in plant pollens that cross-react with allergens in food can lead to oral allergy syndrome with symptoms similar to those induced by allergen ingestion. In both animal models and human subjects, increasing evidence links sensitization to food antigens via the skin to allergic responses to ingested food. Recent work has shown that epithelial cell–produced alarmins, including thymic stromal lymphopoietin (TSLP), IL-25, and IL-33, are present in high quantities in atopic skin lesions. Alarmins educate dendritic cells and type 2 innate lymphoid cells to promote T_H2-mediated immunity. Exposure via the skin through the use of oils or cream containing food allergens (particularly peanut oils) may lead to allergic sensitization as a result of the strong T_H2-polarizing environment induced by alarmin release.

35.8 Aberrant responses to food have a genetic component

The heritability of the predisposition to allergic disease suggests a genetic basis, but susceptibility genes for food allergy are as yet largely unidentified. Common (atopy) and unique (food allergy) genetic risk factors may contribute to allergic responses to food. Genome-wide association studies have identified intriguing candidates that may provide new insight into disease pathogenesis, particularly for EoE. The first genome-wide association study for EoE showed that eotaxin-3, a prominent eosinophil-recruiting chemokine, was the most upregulated gene in the esophageal epithelia of children with EoE. Further work has identified an epithelial cluster of EoE-associated genes, including filaggrin, an epidermal barrier protective protein. Loss-of-function variants in filaggrin are associated with atopic dermatitis in some patients. This finding is particularly relevant to EoE, since the esophagus (like the skin) is lined by squamous epithelium. If epithelial barrier function is impaired, as may occur with loss of filaggrin function, luminal allergens could enter the lamina propria and initiate an allergic response. Finally, variants in the epithelial cell alarmin TSLP have been implicated in susceptibility to EoE. The disease model that emerges involves a series of genes related to epithelial barrier defects, eosinophil chemotaxis, and predisposition to a T_H2 phenotype (**Figure 35.6**).

35.9 Role for microbiota in protection from food-allergic sensitization

The Centers for Disease Control and Prevention documented an 18% increase in the prevalence of food allergy reported for children under 18 years of age in the U.S. between 1997 and 2007; similar increases have been observed in other Western societies. One explanation attributes the increased prevalence of food allergy to improvements in hygiene. Reduced exposure to infectious diseases in childhood impairs the development of the immunoregulatory

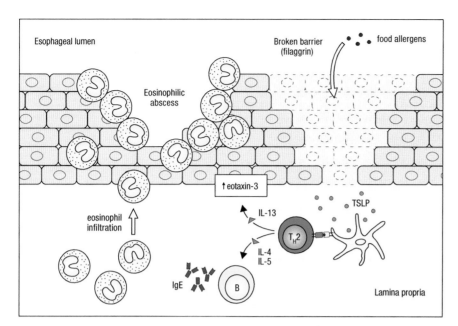

Figure 35.6 A model for the pathogenesis of EoE. Epithelial barrier dysfunction (possibly related to decreased expression of filaggrin) enhances allergen uptake. Antigen presentation by TSLP-primed dendritic cells induces an allergen-specific T_H2 response characterized by the production of IL-4, IL-13, and IL-5. IL-13 upregulates the expression of eotaxin-3 in the esophageal epithelium, which together with IL-5, mobilizes eosinophil recruitment. Eosinophil secretion of potent inflammatory mediators makes eosinophils key to both the initiation and perpetuation of EoE. IL-4 and IL-13 are switch factors for the production of allergen-specific IgE.

networks that protect against both autoimmune inflammatory diseases and allergic hyperreactivity to environmental antigens. Recent work suggests that non-infectious microbial exposures may play an equally important, or greater, role. Microbes (including bacteria, viruses, bacteriophage and fungi) colonize the skin and all mucosal surfaces. The composition of the bacterial microbiota has been the best characterized to date. Very large numbers of bacteria are present in the gut (approximately 10^{12} microbes/mL of luminal contents at its highest point in the colon) and are estimated to outnumber the cells of the body. The luminal bacterial microbiome exists in a dynamic symbiotic relationship with its host and has many influences on host physiology (see Chapter 19). The commensal microbiota also has a profound impact on the development of both systemic and mucosal immunity; germ-free mice, which lack a colonizing microbiome, exhibit profound defects. Initial colonization occurs during the birth process with founder bacteria derived from the mother's vagina and feces. Breast feeding provides dietary glycans that expand specific subpopulations of bacteria capable of metabolizing this food source. Nature's developmental strategy, honed by millions of years of co-evolution, is perturbed by Caesarean birth and formula feeding. Babies born by Caesarean section have founder bacteria derived from the skin of their mother or caregiver. Diet continues to shape the composition of the commensal microbiome throughout an individual's lifetime; changes are detectable on daily timescales. Commensal bacteria digest dietary fibers to produce short chain fatty acids (SCFA) critical to intestinal health. The microbial diversity and SCFA concentration are significantly elevated in the fecal microbiota of children living in rural Africa compared with that of children living in the developed world and consuming a Western-style high-fat, low-fiber diet of processed foods. Taken together, evidence suggests that twenty-first century lifestyle factors, including widespread use of antibiotics, reduced dietary fiber and increases in Caesarean birth and formula feeding, are changing the composition of the intestinal microbiota.

Recent work suggests that these changes are associated with the depletion of bacterial populations that protect against allergic and inflammatory disease. Murine model studies have demonstrated that mucosa-associated *Firmicutes* in the *Clostridia* class protect from allergic sensitization by inducing IL-22 dependent barrier protective responses in the intestine that limit systemic access of allergenic food proteins. Colonization with *Clostridia* has also been shown to promote the generation of intestinal Tregs. Bacteria-specific

Tregs may contribute to the prevention of allergic sensitization by inducing a strongly tolerogenic environment in the intestinal mucosa, promoting the development of food antigen-specific Tregs rather than T_H2 differentiation. Recent work analyzing the microbiota from infants with food allergy revealed key differences in composition, with allergic infants exhibiting an accelerated maturation of the intestinal microbial community. Healthy infants, by contrast, harbor intestinal bacteria that protect against food allergy.

35.10 Timing of the introduction of food antigens into the infant diet influences allergen sensitization

Pediatric guidelines had recommended that pregnant and lactating women with a family history of atopy avoid consumption of the major food allergens and delay introduction of these foods to their children. More recent work suggests that the reverse may be beneficial; early oral food exposure may be necessary to induce oral tolerance and avoid allergic responses to food. Delayed introduction of foods may increase the chances of allergen sensitization via the skin (particularly in individuals with defective barrier function) before oral tolerance has developed. Although trials of early food introduction have suggested that early food exposure limits allergic sensitization, more work is needed to determine the optimal dose and timing of food exposure.

The eosinophilic gastrointestinal diseases (EGIDs)

The advent of flexible endoscopy and biopsy in the 1960s heralded a new era in the study of mucosal immunology. With the ability to obtain mucosal samples from the esophagus, stomach, small intestine and colon, the morphological and immunological features of an array of mucosal diseases have been described and measured.

35.11 The definitions of mucosal eosinophilia are confusing

The identification of eosinophils in the mucosa requires several considerations. First, in healthy individuals, eosinophils are present at varying levels in all gastrointestinal organs, except the esophagus, where they are absent. Second, the "normal" number of eosinophils in the mucosa are still being defined and vary depending on the region of gastrointestinal tract. For example, cecal mucosa contains 40–50 eosinophils per high-power field, whereas the gastric mucosa contains less than 20. Third, the association between the presence of eosinophils with pathogenicity is under investigation. While current practice relies on the number of eosinophils per high-power field, other features that may help identify inflammatory patterns include the location of eosinophils within the mucosa and their state of degranulation, as well as epithelial disruption, goblet cell hyperplasia, and tissue remodeling. Radiographic features associated with mucosal eosinophilia are nonspecific and reflect evidence of either mucosal edema (flocculation) or remodeling (luminal narrowing). Pathological mucosal eosinophilia must be interpreted in the context of the clinical setting. The differential diagnosis for gastrointestinal mucosal eosinophilia includes gastroesophageal reflux disease, peptic disease, celiac disease, inflammatory bowel disease, hypersensitivity reactions, connective tissue diseases, immunodeficiency and the presence of bacterial, viral, parasitic or fungal infections (Table 35.2).

35.12 EoE is the most common form of EGID

Eosinophilic esophagitis is the most common manifestation of the EGIDs, occurring in 1–4 persons per 10,000 people in the U.S. More is known about

TABLE 35.2 ETIOLOGIES OF EOSINOPHILIC GASTROINTESTINAL INFLAMMATION
Gastroesophageal reflux disease
Peptic disease
Eosinophilic gastrointestinal diseases
Celiac disease
Inflammatory bowel diseases
Infections
Connective tissue diseases
Hypersensitivity reactions
Immunodeficiency
Cancer

the pathophysiology of EoE than other EGIDs. During the last decade, a series of basic and translational studies have identified the following features of EoE. The inflamed mucosa in EoE has an increased ratio of CD4$^+$ effector T cells and decreased numbers of Tregs. IL-5 is critical to the esophageal tissue remodeling associated with this eosinophilic inflammatory response. These findings provide support for the development of therapeutics that target IL-5 in the treatment of the disease. Initial studies have shown that anti-IL-5 antibody treatments are successful in significantly reducing mucosal eosinophilia but not symptoms. Murine model studies also support roles for TSLP, IL-13 and IL-15 in esophageal eosinophilia.

The downstream impact of chronic eosinophilic inflammation has focused on esophageal remodeling. IL-13 is a key mediator of epithelial barrier dysfunction. Additional studies suggest that the squamous epithelial cell barrier may be altered in patients with EoE, leading to the possibility of increased antigen sensitization or challenge. Clinical manifestations of esophageal fibrosis include esophageal strictures and food impactions that can be visualized radiographically or endoscopically.

35.13 Environmental allergens that contribute to EoE are dietary and airborne

Mechanistic and case studies support a role for environmental allergens in the etiology of EoE. Several reports note that some patients may develop both symptoms and esophageal inflammation during seasonal exposure to inhaled allergens. This finding may relate to sensitization to pollens and foods. However, more evidence supports a role for food allergy in the pathogenesis of EoE. Over two-thirds of patients with EoE have evidence of an allergic diathesis, including food allergic diseases. A number of studies have demonstrated that either the selected removal of suspected foods or the complete elimination of intact proteins and the use of an amino acid-based "elemental" dietary formula leads to clinical and pathological remission of EoE. These findings support the selective exclusion of the major dietary protein allergens as a successful treatment strategy (see **Figure 35.4**).

35.14 Eosinophils play a critical role in the pathogenesis of EoE

The exact impact of eosinophils in EoE remains uncertain but a number of factors support their potential role in promoting organ dysfunction. Eosinophil progenitors in the peripheral blood are increased in patients with active EoE and may serve as a surrogate biomarker. Eosinophils are biological "powerhouses" that contain large numbers of granules with preformed proteins and the capacity to synthesize and release newly generated mediators, including cytokines and leukotrienes. Highly charged eosinophil granule proteins such as major basic protein can decrease epithelial barrier function and stimulate smooth muscle contraction. By virtue of altering the epithelial barrier, eosinophils may facilitate transepithelial passage of luminal allergens to primed effector cells in the lamina propria and/or induce smooth muscle contraction that leads to dysphagia or food impaction.

35.15 Role of the esophageal microbiome in EoE

Recent studies have characterized the microbial contents of the esophagus in children and adults affected by EoE. These studies demonstrated that the esophageal microbiota in EoE is different compared with that of healthy control subjects. Samples from EoE patients were enriched with *Proteobacteria* and non-typeable *Haemophilus influenza* compared with control subjects. Dietary treatments did not lead to sustained changes in the microbiome. In support of the microbiome's role in EoE, a mouse study demonstrated a protective role for *Lactococcus lactis*. Future studies will need to dissect whether differences in bacterial populations initiate or perpetuate human esophageal inflammation.

35.16 Eosinophilic gastroenteritis is uncommon

Eosinophilic gastroenteritis is a rare disease that is characterized by a wide variety of vague intestinal symptoms and histological evidence of gastrointestinal eosinophilia. The diagnosis is based on the exclusion of other causes of intestinal eosinophilia. Since eosinophils reside in the normal gastrointestinal mucosa, the number of eosinophils that characterizes eosinophilic gastroenteritis has been difficult to determine. Eosinophilic gastroenteritis has three forms: mucosal, muscular, and serosal disease. Most patients with eosinophilic gastroenteritis have mucosal disease and symptoms related to epithelial dysfunction, including bleeding, pain and diarrhea. A smaller proportion of patients have muscular disease characterized by vomiting and bloating, suggestive of intestinal obstruction. An even smaller proportion have serosal disease that presents with abdominal bloating and ascites. Pathophysiologic mechanisms associated with eosinophilic gastroenteritis and the three phenotypes are not known. Treatment is based on either chronic topical or systemic steroid use or in some circumstances dietary elimination of suspected allergens.

35.17 Allergic reactions to food sometimes manifest only in colon

Allergic reactions in the colon to food, including cow's milk and soy proteins, can result in bleeding. The condition most commonly occurs in exclusively breastfed babies who become sensitized to dietary cow's milk present in their mother's milk. The main clinical features are diarrhea or rectal bleeding. The condition may affect any part of the large intestine, but the sigmoid colon and rectum are most commonly involved. Definitive diagnosis is achieved by elimination diet and challenge, but this is rarely implemented since most infants fail to thrive on a milk-free diet.

Colonoscopy reveals mucosal erythema, lymphoid hyperplasia, and in severe cases, ulceration. Colonic biopsy typically shows intense infiltration of eosinophils in the lamina propria and muscularis mucosa. Features of chronicity such as glandular distortion are not present. Peripheral eosinophilia is common, and the serum level of IgE may be increased.

The mechanism of allergic disease manifesting in the colon in some infants, especially breastfed babies who receive minute amounts of allergen that likely do not reach the distal colon intact to induce a local response, is not known. The immaturity of the intestinal mucosal barrier might allow allergens to enter the circulation to produce an IgE response, but genetic predisposition likely contributes to the condition, since all infants show some deficit in barrier function, and antigen penetrates into the blood in quite large amounts in all bottlefed babies. Very low doses of antigen might also predispose to an IgE-producing T_H2 response.

Changing to a hypoallergenic milk formula is often effective in producing a positive clinical response. For infants who are sensitized through breast milk, the mother can begin a milk-free diet. Allergic colitis of infancy is a self-limiting condition that presents with symptoms of blood-streaked stools during infancy. Babies otherwise appear well, have no anemia, and respond to eliminating cow's milk or soy protein from the formula or breast milk.

Treatment of food allergies

As mentioned earlier, food elimination diets are currently the most effective treatment of food allergy. Ongoing studies are examining whether oral or skin patch administration of small doses of food allergens can desensitize atopic individuals to protect against severe anaphylactic responses to accidental food exposure. The nature of the anaphylactic response makes this approach inherently risky and requires that allergen dosing be confined to a carefully controlled medical setting. The realistic goal of this therapy is not to make allergic individuals completely tolerant to food proteins but to use immunologic tolerance to raise the antigen threshold to a level beyond that induced by contaminants in poorly labeled foods. This is particularly relevant to peanut sensitization where raising the threshold for triggering anaphylaxis from trace amounts to gram amounts is potentially life-saving.

Omalizumab is a humanized anti-IgE monoclonal antibody licensed for use in moderate to severe allergic asthma. The antibody is beginning to be used in patients, including children, with IgE-mediated food allergies with beneficial effects. Omalizumab may be a useful adjunct to desensitization programs in the severely allergic child because it will help prevent anaphylaxis.

A better understanding of how bacteria-induced immunoregulatory signals impact susceptibility to food-allergic disease will inform strategies for the development of novel therapies. This new form of treatment could be administered as a preventative strategy early in life or as adjunctive to oral- or skin patch–based immunotherapies. Candidate microbiome modulating therapeutics include characterized bacterial consortia, bacterial metabolites such as short-chain fatty acids, and the dietary fibers fermented to produce these metabolites, all of which have the potential to reinforce epithelial barrier function.

SUMMARY

Healthy individuals develop tolerance to the many potential antigens present in food. Food-related diseases can present in a variety of forms. Allergic responses to food, in particular, appear to be increasing. Genome-wide association studies point to dysregulated epithelial barrier function and the predisposition to a T_H2-biased response to food antigens as key factors in disease pathogenesis. The elucidation of how environmental factors, particularly the commensal microbiome, contribute to food-related disease is an important area of study in order to devise new strategies to prevent and treat inappropriate responses to food.

FURTHER READING

Azouz, N.P., and Rothenberg, M.E.: Mechanisms of gastrointestinal allergic disorders. *J. Clin. Invest.* 2019, 130:1419–1430.

Branum, A.M., and Lukacs, S.L.: Food allergy among U.S children: Trends in prevalence and hospitalizations. *N C H S Data Brief* (10), 2008:1–8.

Burks, A.W., Sampson, H.A., Plaut, M., Lack, G., and Akdis, C.A. Treatment for food allergy. *J. Allergy Clin. Immunol.* 2018, 141:1–9.

De Filippo, C., Cavalieri, D., Di Paola, M. et al.: Impact of diet in shaping gut microbiota revealed by a comparative study in children from Europe and rural Africa. *Proc. Natl. Acad. Sci. USA.* 2010, 107:14691–14696.

Du Toit, G., Roberts, G., Sayre, P.H. et al.: LEAP Study Team: Randomized trial of peanut consumption in infants at risk for peanut allergy. *N. Engl. J. Med.* 2015, 372:803–813.

Feehley, T., Plunkett, C.H., Bao, R., Choi Hong, S.M., Culleen, E., Belda-Ferre, P., Campbell, E., Aitoro, R.,

Nocerino, R., Paparo, L., Andrade, J., Antonopoulos, D.A., Berni Canani, R., Nagler, C.R.: Healthy infants harbor intestinal bacteria that protect against food allergy. *Nat. Med.* 2019, 25:448–453.

Furuta, G.T., and Katzka, D.A.: Eosinophilic esophagitis. *N. Engl. J. Med.* 2015, 373:1640–1648.

Gupta, R.S., Warren, C.M., Smith, B.M., Jiang, J., Blumenstock, J.A., Davis, M.M., Schleimer, R.P. and K.C. Nadeau, 2019, Prevalence and severity of food allergies among U.S. adults. *JAMA Netw. Open.* 2: e185630

Iweala, O.I., and C.R. Nagler. The Microbiome and Food Allergy. *Annual Reviews of Immunology*, 2019, 37: 377–403

Pabst, O., and Mowat, A.M.: Oral tolerance to food protein. *Muc. Immunol.* 2012, 5:232–239.

Stefka, A.T., Feehley, T., Tripathi, P. et al.: Commensal bacteria protect against food allergen sensitization. *Proc. Natl. Acad. Sci. USA.* 2014, 111:13145–13150.

Index

Page numbers in *italics* and **bold** refer to figures and tables, respectively.